DIGITAL ELECTRONICS

(DIGITAL LOGIC DESIGN)

2nd Edition

DIGITAL ELECTRONICS

(DIGITAL LOGIC DESIGN)

Dr. Rishabh Anand

PostDoc (AI & ML), Ph.D. (Computer Science), MBA, M.Tech.(ECE), LMCEGR, ASR & IAENG

Global Service Delivery Manager -ITES

CPDT™,PMP®, PRINCE2®, DevOps-PM™, ITIL®, CSM© & Kanban-ASC™ Certified Professional

KHANNA BOOK PUBLISHING CO. (P) LTD.

Publisher of Engineering and Computer Books

4C/4344, Ansari Road, Darya Ganj, New Delhi-110002

Phone : 011-23244447-48 **Mobile:** +91-9910909320

E-mail : contact@khannabooks.com

Website : www.khannabooks.com

Dear Readers,

To prevent the piracy, this book is secured with HIGH SECURITY HOLOGRAM on the front title cover. In case you don't find the hologram on the front cover title, please write us to at contact@khannabooks.com or whatsapp us at +91-99109 09320 and avail special gift voucher for yourself.

Specimen of Hologram on front Cover title:

Moreover, there is a SPECIAL DISCOUNT COUPON for you with EVERY HOLOGRAM.

How to avail this SPECIAL DISCOUNT:

Step 1: Scratch the hologram
Step 2: Under the scratch area, your "coupon code" is available
Step 3: Logon to www.khannabooks.com
Step 4: Use your "coupon code" in the shopping cart and get your copy at a special discount
Step 5: Enjoy your reading!

Thanks,
Khanna Publishing House

Price: ₹ 795.00

ISBN: 978-93-82609-44-5

DIGITAL ELECTRONICS
(Digital Logic Design)
by Er. Rishabh Anand

Edition: 2026

To view complete list of books, please scan the QR Code:

Published by:
Khanna Book Publishing Co. (P) Ltd.
CIN: U22110DL1998PTC095547

Visit us at: www.khannabooks.com
Write us at: contact@khannabooks.com

Printed in Bharat.

Dedicated to
My Mother whose love and affection made
me capable of writing this book
And
Shri Guru Maharaj Ji of Shri Anandpur

✍ *Preface to the Second Edition*

The author feels great pleasure in presenting the third edition of the book with a marked improvement over the previous one. The book is thoroughly revised, rewritten and so designed as to meet the needs of A.M.I.E. Sec B. and Degree (both Electrical and Electronics based courses such as Electrical Engineering; Electrical and Electronics Engineering; Electronics; Electronics and Communication Engineering; Electronics and Instrumentation; Applied Electronics; Applied Electronics and Instrumentation; Electronics and Telecommunication Engineering; Computer Science; Computer Engineering; Computer Science and Engineering; Information Technology and Engineering) and Diploma students of Electrical Engineering and Electronics Engineering. The subject matter has been presented in an easily comprehensible manner and logical sequence using simple language and to provide a detailed exposition to the fundamentals. A large number of the old diagrams have been replaced by new ones and numerous new diagrams have added to make the subject more illustrative.

Every care has been taken to eliminate misprints, omissions and errors but it is too much to expect that no inaccuracy/error/misprint has crept in and the author would be grateful to the readers for bringing notice any of such errors/misprints/omissions they may come across while going through the book.

The author will like to thank the readers (both teachers and students) who sent their valuable suggestions that became the basis for the revision of the book.

The author shall feel satisfied if the book meets the needs of the students for whom it is meant. Suggestions and criticism for the further improvement of the book shall be welcomed.

Feb 3, 2021 **Dr. Rishabh Anand**

Preface to the First Edition

Today what we call "Digital Revolution" is governing human life. Desktop PC and Notebook usage for example is inevitable for all professionals as well as makes life easy for an ordinary person. These are examples of trivialusage of Digital Technology. There is also extensive usage of digital electronics which is naturally "hidden" or "embedded" and not being explicitly perceived by the user.

When we speak over a cell-phone, our voice is digitized and is transmitted via digital communication equipment. When we play a music CD, the music is recorded in digital form and processed by digital logic to recover media and recording errors and improve music quality. In modern cars an embedded microcontroller controls the fuel consumption, brake system, tire balance etc. We can give several more examples where digital electronics is implemented. Since digital electronics are incorporated in almost every technical application, it became compulsory for all non-electronic engineers to deal, at least to some extent, with some aspects of digital electronics.

One of the most important reasons for the unprecedented growth of digital electronics is the advent of integrated circuits (ICs). Developments in IC technology have made it possible to fabricate complex digital circuits, such as microprocessors, memory units, etc. on tiny chips of silicon.

The wonder chip—the microprocessor has been the most fantastic development of recent years. No other single development has affected our lives as much as the microprocessor in such a short time. Its ever-increasing applications have resulted in developments which were simply unheard of till a few years ago.

This has made it imperative for all those who will be required to design, develop, test, and maintain various electronic systems to learn the principles of modern digital devices and systems.

Availability of complex digital functions in ICs has created the need for a change in the teaching philosophy of digital electronics from the conventional style using discrete devices to a new style using modern digital ICs. For example, now it is no more important to minimize the number of gates for the design of a digital circuit, since a number of similar gates are available in a single IC chip; rather, it has become necessary to minimize the number of IC packages. Thus, the present-day designer of digital systems has to be thoroughly familiar with the principles of operation and flexibilities available in various ICs in order to optimize the design of systems from the point of view of cost, space, power requirement, speed of operation, etc.

With this in view, an attempt has been made to introduce the concepts of modern digital techniques and ICs (available) for the realization of various functions. This book is self-contained and is suitable for a course in digital systems for engineering and computer science programmers. Students of physics specializing in electronics will also find the book useful.

The book uses the plain, lucid and everyday language to explain the subject. The book has been systematically organized and the presentation has been kept at a level suitable for a student with a basic knowledge of circuit theory and electronics.

Chapter 1 is devoted to number systems. It introduces binary, octal, hexadecimal and BCD number systems. It explains how to convert numbers from decimal to binary, octal and hexadecimal and vice-versa.

Chapter 2 introduces concept of coding and explains various numeric and alpha numeric codes including various BCD codes, excess-3 code, biquinary code, unit distance codes, gray code and alphanumeric codes.

Chapter 3 is devoted to error detecting and correcting codes. It explains various error detecting and correcting codes such as a parity bits, linear block codes, hamming code, binary cyclic code, checksums and burst codes.

Chapter 4 deals with logical gates, Boolean algebra and Boolean theorems. It also explains how logical gates can be used to produce logical circuits and how these circuits can be described and analyzed using Boolean Algebra.

Chapter 5 explains various techniques for simplification of Boolean expressions such as Boolean theorems, Karnaugh maps, and Quine McCluskey tabulation method. It also explains Ex-OR and equivalence functions and NAND and NOR implementations of Boolean expressions.

Chapter 6 deals with the logic families. It introduces the TTL, CMOS, NMOS, Dynamic MOS, HMOS, and ECL families with their basic circuits, characteristics and interfacing.

Chapter 7 is devoted to the design of combinational logic circuits. It discusses various combinational circuits such as adders, subtractors, comparators and code converters.

Chapter 8 is devoted to combinational logic with MSI and LSI. It explains the application of typical TTL IC comparators like multiplexers(74150,74151,74157), encoder(74148), decoder(74138,7442), demultiplexers,and seven segment decoders. It also explains the design of arithmetic logic unit (ALU).

Chapter 9 deals with the sequential circuits. It introduces basic component of sequential circuits, flip-flops. It explains operation of all type of flip-flops with various triggering methods. It also explains design of registers and counters using flip-flops.

Chapter 10 explains analysis and design of clocked sequential circuits along with state reduction and state assignments.

Chapter 11 covers analog-to-digital (A/D) and digital-to-analog (D/A) converters form an important part of many digital systems.

Chapter 12 deals with semiconductor memories. It explains memory terminology, memory operation, CPU-Memory connections, ROM and its architecture, types of ROMs (PROM, EPROM, EEPROM), NVRAM, Flash Memory. It also explains architecture, operation, Read/Write Cycles of SRAM and DRAM.

Chapter 13 deals with programmable logic devices such as ROM, PLA and PAL. It explains their basic structure and implementation of combinational and sequential circuits using them.

Chapter 14 Experimental Work

Appendix-A Glossary of the important terms used in the book.

Appendix-B gives information about various TTL and CMOS ICs.

Appendix-C Multiple Choice Questions.

Just as the electronics industry continues to evolve, this book must also continue to evolve and improve if it is to remain useful.

The book explains the philosophy of the subject digital electronics. The book will be very useful not only to the students but also to the subject teachers. **The students have to omit nothing and possibly, have to cover nothing more.**

Er. Rishabh Anand

Acknowledgement

No one walks alone and when one is walking on the journey of life just where I start to thank those who joined me, walked besides me, and helped me along the way.

Over the years, those that I have met and worked with have continuously motivated me to write this book. So at last, here it is. So, perhaps this book and its pages will be seen as "thanks" to the tens of thousands of you who have helped bringing out this book in the form what is today.

There was, in and always will be the love of my students whom I taught, for what is written in this book has been the fruit of experience that I taught them. Finally, thanks to my mother, my wife and to rest of my family for their patience and support during the long hours of writing this book and above all there is the one almighty whose humble children we are. It is his blessings we cherish and pray for. It is the blessing I wish for you.

My special acknowledgment to all the authors whose books, journals and paper are referred while writing this book.

I deeply express my heartful thanks to the Management Team and Editorial Staff of Khanna Book Publishing Co., for publishing the book in such a beautiful get-up and well in time.

When you finish the journey through this book, having variety of experiences, ideas and thoughts. You are welcome to express your opinion at **rishabhanand009@gmail.com.**

Dr. Rishabh Anand

✍ *Message to My Students*

As you have signed up for this course and probably your teacher has adopted this book! You had no say about the choice, although you can form and express your opinion on the book after reading it. What I really meant is that although this is an easy book to study from (in my opinion), it is no-nonsense book. It does not have cartoons or nice-looking photographs to amuse you. From here on it is all business and hard work.

This book has many examples. Some of these are simple, for the purpose of illustrations of new ideas and subject matter. Some examples are made quite elaborate, in order to develop desired technical skill and confidence. Furthermore, the objective of this book is to present the complex subject in a clear and thorough manner.

I hope that you will enjoy this book, and it will be remain as major text book in your college career. My advice to you is not sell it back at the end of semester. If you do, not only will you be violating the "buy low and sell high "axiom of finance but definitely be missing it during your higher semester or project assignments.

When you finish the journey through this book, having variety of experiences, ideas and thoughts. You are welcome to express your opinion at **rishabh.anand009@yahoo.com**

Er. Rishabh Anand

Contents

1 Number System and Codes

1.1 INTRODUCTION

Electronic systems are classified in two categories generally referred to as analog and digital. There are in general two ways in which information is represented by using signals. In digital representation the information is denoted by a finite sequence of discrete value or digits. On the other hand, in an analog representation, a continuous value is used to denote the information. Digital concepts are more advantageous as compared to analog concepts.

In modern words of electronics, the term 'digital' is usually associated with a computer. Now a days, digital concepts are usually applied to various problems that could only by solved, by analog methods several years ago.

The term *digital* in digital circuits is derived from the way circuits perform operations by counting digits. A digital circuit operates with binary numbers, i.e., only in two states. The output of the circuits either low (0) or high (1) in a *positive logic system*. In general, 0 represents zero volts and 1 represents five volts. If the situation is reverse, it is known as a *negative logic system*.

***Note:** In this chapter, we shall consider only positive logic. Also, we shall assume that the logic 0 level corresponds to 0 volts and logic 1 level corresponds to + 5 V.*

In digital systems, as explained above, the data is usually in binary states (0 and 1) and is processed and stored electronically to prevent errors our to noise and interfering signals. At present, digital technology has progressed remarkably from vacum-tube circuits to integrated circuits, microprocessors and microcontrollers. Digital circuits find applications in computers, telephony, data processing, radar navigation, military systems, Medical instruments and consumer products. The general properties of number systems, methods of conversion from one to another, arithmetic operations, weighted codes, non-weighted codes, error detecting and correcting codes are discussed in this chapter.

1.2 NUMBER SYSTEM

A physical system whose behaviour is described by mathematical equations is simulated in a digital computer by means of "number systems." Number system is a language of digital system consisting of a set of symbols called digits with rules defined for their addition, multiplication and other mathematical operations.

The number systems are used quite frequently in the field of digital electronics and computers. However the type of number system used in computers could be different at different stages of the usage.

In general, in any number system, there is an ordered set of symbols known as digits. Basically there are two types of number systems.

1. Positional Number System 2. Non-Position Number Systems.

1.2.1 Decimal Numbers

Before considering any number system, let us consider familiar decimal number system system. In decimal number system we can express any decimal number in units, tens, hundreds, thousands and so on. When we write a decimal number say, 5678.9 we know it can be represented as

5000 + 600 + 70 + 8 + 0.9 = 5678.9

In power of 10, we can write as

$$\underbrace{5\times10^3}_{5}+\underbrace{6\times10^2}_{6}+\underbrace{7\times10^1}_{7}+\underbrace{8\times10^0}_{8}+\underbrace{9\times10^{-1}}_{.9}$$

This says that, the portion of a digit with reference to the decimal point determines its value/weight. The sum of all the digits multiplied by their weights gives the total number being represented. The leftmost digit, which has the greatest weight is called the most significant digit and the right most digit, which has the least weights is called the least significant digit.

Number system is a basic for counting various items. A number system defines a set of values used to represent quantity.

The decimal number system (0, 1, 2, ..., 9) is commonly used eventhough there are many other number systems like binary, octal, hexadecimal, etc. It is possible to express a number in any base or radix "X". In the binary system, the base is 2. In general, any number with radix X, having *m* digits to the left and *n* digits to the right of the decimal point, can be expressed as:

$$a_m(X)^{m-1} + a_{m-1}(X)^{m-2} + \ldots + a_2(X)^1 + a_1(X)^0. + b_1(X)^{-1} + b_2(X)^{-2} + \ldots + b_n(X)^{-n}$$

where a_m is the digit in *m*th position. The coefficient a_m is termed as the Most Significant Digit (MSD) and b_n is termed as the Least Significant Digit (LSD).

1.2.2 Binary Numbers

The Binary number system is simple because it consists of only two digits, i.e. 0 and 1. Just as the decimal system with its ten digits is a base-ten system, the binary system with its two digits is a base-two system. The position of 0 or 1 in a binary number indicates its "weight" within the number. In a binary number, the weight of each successively higher position to the left is an increasing *power* of two.

Do You Know?	The binary number system (0 and 1) is the basic system used in digital technology.

For example, in the decimal system.

$$(198)_{10} = 1 \times 10^2 + 9 \times 10^1 + 8 \times 10^0$$

Hundreds Tens Units

|-----------|-----------|

↑

Positional Weights

Similarly, binary numbers are also represented by positional weights.

For example,

$$\begin{aligned}(198)_{10} &= (11000110)_2\\ &= 1 \times 2^7 + 1 \times 2^6 + 0 \times 2^5 + 0 \times 2^4 + 0 \times 2^3 + 1 \times 2^1 + 0 \times 2^0\\ &= 128 + 64 + 0 + 0 + 0 + 4 + 2 + 0 = 198\end{aligned}$$

In the digital system, each of the binary digits is called a bit and a group of 4 and 8 bits are called a nibble and a byte respectively. The highest decimal number that can be represented by n-bits binary number is $2^n - 1$ (beginning with zero). Thus, with an 8-bit binary number, the maximum decimal number that can be represented is $2^8 - 1 = 255$.

1.2.3 Decimal–Binary Conversion

An easy method of converting a decimal number into a binary number is by dividing the decimal number by 2 progressively, until the quotient of zero is obtained. The binary number is obtained by taking the remainder after each division in the reverse order. This method is popularly known as the *double-dabble* method. The procedure for decimal to binary conversion is described in the following example.

Example 1.1. *Convert the decimal number 53.625 into an equivalent binary number.*

Solution:

Step 1: The integer is 53. The fraction is 0.625.

Step 2: Integer conversion

Division	Generated remainder
2) 53	
2) 26	→ 1
2) 13	→ 0
2) 6	→ 1
2) 3	→ 0
2) 1	→ 1
2) 0	→ 1 → MSB

Reading the remainders from bottom to top gives the binary equivalent. Thus, $(53)_{10} = (110101)_2$.

Step 3: Fractional conversion: If the decimal number is a fraction, its binary equivalent is obtained by multiplying the number continuously by 2, recording a *carry* in the integer position each time. The *carries* in the forward order give the required binary number.

Multiplication	Generated integer
$0.625 \times 2 = 1.25 \rightarrow$	$1 \rightarrow$ MSB
$0.250 \times 2 = 0.50 \rightarrow$	0
$0.500 \times 2 = 1.00 \rightarrow$	1
$0.000 \times 2 = 0.00 \rightarrow$	0

Example 1.2. *Convert the binary number $(101111.1101)_2$ into its decimal equivalent.*

Solution: A binary number can be converted into a decimal number by multiplying the binary numbers 1 or 0 by their weight and adding the products.

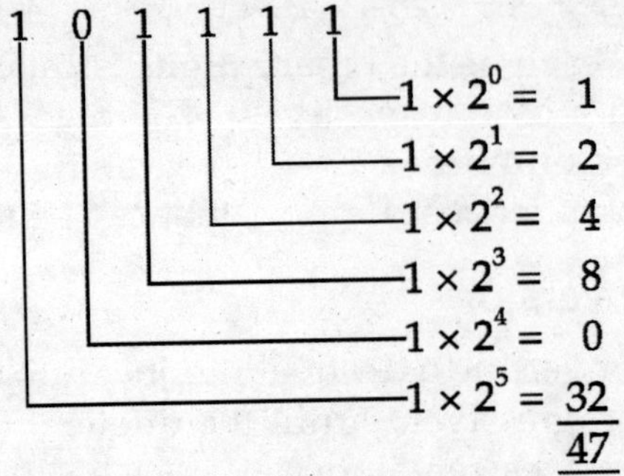

Therefore, $(101111)_2$ can be written as $(47)_{10}$.

Conversion of $(0.1101)_2$ is done in a similar manner.

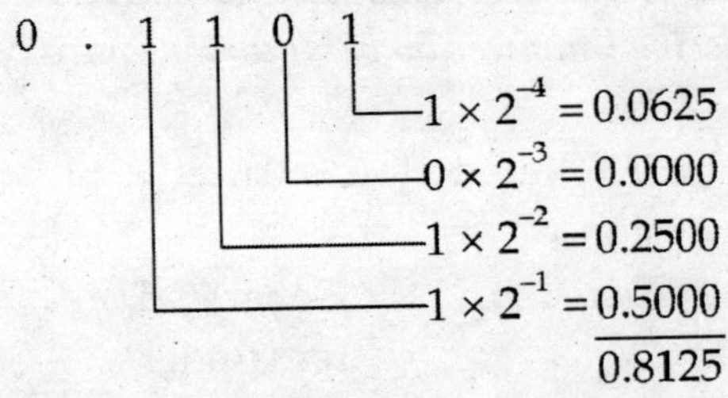

Thus, $(0.1101)_2$ is equal to $(0.8125)_{10}$.

Therefore, $(101111.1101)_2$ is equal to $(47.8125)_{10}$.

1.2.4 Octal Numbers

The Octal number system uses the digits 0, 1, 2, 3, 4, 5, 6 and 7. The *base* or *radix* of this system is eight. Each significant position in an octal number has a positional weight. The least significant position has a weight of 8^0, i.e. 1; the higher significant positions are given weights in the ascending powers of eight, i.e. 8^1, 8^2, 8^3, etc. respectively. The octal equivalent of a decimal number can be obtained by dividing a given decimal number by 8 repeatedly, until a equotient of 0 is obtained. The procedure is exactly the same as the *double-dabble* method explained earlier. The *decimal* to *octal* conversion method is explained in the following example.

Example 1.3. *Convert* $(555.456)_{10}$ *to an octal number.*

Solution: Integer conversion:

Division	Generated remainder
8) 444	
8) 55 →	4
8) 6 →	7
8) 0 →	6

Reading the remainders from bottom to top, the decimal number $(444)_{10}$ is equivalent to octal $(674)_8$.

Fractional conversion:

Multiplication	Generated integer
0.456 × 8 = 3.648 →	3
0.648 × 8 = 5.184 →	5
0.184 × 8 = 1.472 →	1
0.472 × 8 = 3.776 →	3
0.776 × 8 = 6.208 →	6

The process is terminated when significant digits are obtained.

Thus, the octal equivalent of $(444.456)_{10}$ is $(674.35136)_8$.

The conversion from an octal to decimal number can be done by multiplying each significant digit of the octal number by its respective weight and adding the products. The following example illustrates the conversion from octal to decimal.

Example 1.4. *Convert the octal numbers (a)* $(237)_8$ *and (b)* $(120)_8$ *to decimals.*

Solution: (*a*)

$$\begin{aligned}(237)_8 &= 2 \times 8^2 + 3 \times 8^1 + 7 \times 8^0 \\ &= 2 \times 64 + 3 \times 8 + 7 \times 1 \\ &= 128 + 24 + 7 \\ &= (159)_{10}\end{aligned}$$

(*b*)

$$\begin{aligned}(120)_8 &= 1 \times 8^2 + 2 \times 8^1 + 0 \times 8^0 \\ &= 1 \times 64 + 2 \times 8 + 0 \times 1 \\ &= 64 + 16 + 0 \\ &= (80)_{10}\end{aligned}$$

1.2.5 Octal–Binary Conversion

Conversion from octal to binary and vice versa can be easily carried out. For obtaining the binary equivalent of an octal number, each significant digit in the given number is replaced by its 3-bit binary equivalent. For example,

$$\begin{aligned}(376)_8 &= 3 \quad 7 \quad 6 \\ &= 011 \quad 111 \quad 110\end{aligned}$$

Thus, (376)8 = $(011111110)_2$. For converting a binary number to an octal, the reverse procedure is used, i.e. starting from the least significant bit, each group of 3 bits is replaced by its decimal equivalents. For example,

$$(10011010101)_2 = 010 \quad 011 \quad 010 \quad 101$$
$$= 2 \quad 3 \quad 2 \quad 1$$

Thus, $(10011010101)_2 = (2325)_8$

1.2.6 Hexadecimal Numbers

The Hexadecimal number system has a radix of 16 and uses 16 symbols, namely 0, 1, 2, 3, 4, 5, 6, 7, 8, 9, *A*, *B*, *C*, *D*, *E* and *F*. The symbols *A*, *B*, *C*, *D*, *E* and *F* represent the decimals 10, 11, 12, 13, 14 and 15 respectively. Each significant position in an hexadecimal number has a positional weight. The least significant position has a weight of 16^0, i.e. 1; the higher significant positions are given weights in the ascending powers of sixteen, i.e. 10^1, 16^2, 16^3, etc. respectively. The hexadecimal equivalent of a decimal number can be obtained by dividing the given decimal number by 16 repeatedly, until a quotient of 0 is obtained. The following example illustrates how the hexadecimal equivalent of a given decimal is obtained.

Example 1.5. *Convett (a) $(115)_{10}$ and (b) $(235)_{10}$ to hexadecimal numbers.*

Solution:

(*a*)

Division	Remainder
16) 115	–
) 7	3
16) 0	7

Reading the remainders from bottom to top, the decimal number $(115)_{10}$ is equivalent to the hexadecimal $(73)_{16}$. The hexadecimal $(73)_{16}$ can also be represented as 73 H.

(*b*)

Division	Remainder
16) 235	–
) 14	11 → B
16) 0	14 → E

Reading the remainders from bottom to top, the decimal number $(235)_{10}$ is equivalent to hexadecimal $(EB)_{16}$.

The conversion from an hexadecimal to a decimal number can be carried out by multiplying each significant digit of the hexadecimal by its respective weight and adding the products. This is illustrated in the following example.

Example 1.6. *Convert the following hexadecimal numbers into decimal numbers.*
(a) A3BH and (b) 2F3H

Solution: (*a*)

$$A3BH = (A3B)_{16} = A \times 16^2 + 3 \times 16^1 + B \times 16^0$$
$$= 10 \times 16^2 + 3 \times 16^1 + 11 \times 16^0$$
$$= 10 \times 256 + 3 \times 16 + 11 \times 1$$
$$= 2560 + 48 + 11$$
$$= (1619)_{10}$$

(*b*) $2F3H = (2F3)_{16} = 2 \times 16^2 + F \times 16^1 + 3 \times 16^0$

$= 2 \times 256 + 15 \times 16 + 3 \times 1$

$= 512 + 240 + 3$

$= (755)_{10}$

1.2.7 Hexadecimal–Binary Conversion

Conversion from hexadecimal to binary and vice versa can be easily carried out. For arriving at the binary equivalent of a hexadecimal number, each significant digit in the given number is replaced by its 4-bit binary equivalent.

For example,

$(2D55)_{16}$ = 2   *D*   %

= 0010   1101   0101

Thus, $(2D5)_{16} = (001011010101)_2$. The reverse procedure is used for converting a binary number to an hexadecimal, i.e. starting from the least significant bit, each group of 4 bits is replaced by its decimal equivalents.

For exaple,

$(111101101101)_2$ = 111   1011   0101

= 7   *B*   5

Thus,

$(11110110101)_2 = (7B5)_{16}$

Example 1.7. *Convert the following bases:*

(*a*) $(11011.011)_2 \rightarrow ()_{16}$ (*b*) $(2AC9)_{16} \rightarrow ()_7$

Solution: (*a*) Binary to Hexadecimal

$N = 11011.011$

3 zeros are added on LHS and 1 zero is added on RHS. Then a group of 4 bits are formed and then each group is converted into the corresponding hexadecimal digit.

0001 1011 . 0110

↓ ↓ ↓

1 B · 6

Therefore, $(11011.011)_2 = (1B.6)_{16}$

(*b*) Hexadecimal – Decimal – Base 7 Conversion

The given hexadecimal number is converted into decimal as follows:

N = | 2 | A | C | 9 |

$(2 \times 16^3) + (10 \times 16^2) + (12 \times 16^1) + (9 \times 16^0) = 10{,}953$

Therefore, $(2AC9)_{16} = (10{,}953)_{10}$

Then, the decimal number is converted into the number of base 7 as follows:

7	10,953		LSD
7	1,564	5	↑
7	223	3	
7	31	6	
7	4	4	
	0	4	LSD

Therefore, $(2AC9)_{16} = (44635)_7$

1.2.8 Hexadecimal–Octal Conversion

Conversion from hexadecimal to octal and vice versa is sometimes required. To convert a hexadecimal number to octal, the following steps can be applied.

(*i*) Convert the given hexadecimal number to its binary equivalent.
(*ii*) From groups of 3 bits, starting from the LSB (least significant digit).
(*iii*) Write the equivalent octal number for each group of 3 bits.

For example,

$$\begin{aligned}(47)_{16} &= (0100\quad 0111)_2\\ &= (01\ 000\ 111)_2\\ &= (107)_8\end{aligned}$$

Thus, 47 in hexadecimal is equivalent to 107 in the octal number system.

To convert an octal number to hexadecimal, the steps are as follows:

(*i*) Convert the given octal number to its binary equivalent.
(*ii*) From groups of 4 bits, starting from the LSB.
(*iii*) Write the equivalent hexadecimal number for each group of 4 bits.

For example,

$$\begin{aligned}(32)_8 &= (011\quad 010)_2\\ &= (01\quad 1010)_2\\ &= (1A)_{16}\end{aligned}$$

Thus 32 in octal is equivalent to 1 A in the hexadecimal number system.

Example 1.8. *Convert the hexadecimal* $(A6F.CD)_{16}$ *into octal number.*

Solution: Hexadecimal to Octal

$$N = (A6F \cdot CD)_{16}$$

The given hexadecimal number is converted into binary as follows:

A	6	F	·	C	D	Hex
1010	0110	1111	·	1100	1101	Binary

One 0 is added on LSB side and then groups of 3 bits are formed as follows:

101	001	101	111	·	110	011	011
5	1	5	7		6	3	2

Therefore, $(A6F \cdot CD)_{16} = (5157.632)_8$

Example 1.9. *Convert the number into desired base.*

(a) $(A6BF5)_{16} = ()_2$ (b) $(101.01)_2 = ()_{20}$

(c) $(7.FD6)_{16} = ()_8$ (d) $(345)_8 = ()_{10}$

(e) $(7864)_{10} = ()_{16}$

Solution: (*a*) Hexadecimal to Binary

Each digit is converted into a 4-bit binary as follows:

A	6	B	F	5
1010	0110	1011	1111	0101

Therefore, $(A6BF5)_{16} = (10100110101111110101)_2$

(*b*) Binary to Decimal

The given binary number is converted into decimla number as follows:

$$1\,0\,1 \cdot 0\,1$$

$$(1 \times 2^2) + 0 + (1 \times 2^0) \cdot 0 + (1 \times 2^{-2}) = (4 + 0 + 1) \cdot (0 + 0.25) = (5.25)_{10}$$

(*c*) Hexadecimal to Octal

The given Hexadecimal number is converted to binary number as follows:

$$(7.FD6)_{16} = (0111.111111010110)_2$$

Then the binary number is converted into Octal number as follows:

The binary bits are formed into groups of three and each group is converted into equivalent octal number.

000	111	·	111	111	010	110	Binary
0	7	.	7	7	2	6	Octal

Therefore, $(6 \cdot FD6)_{16} = (7.77266)_8$

(*d*) Octal to Decimal

The Given octal number is converted into decimal number as follows:

$$3\,4\,5$$

$$(3 \times 8^2) \times (4 \times 8^1) + (5 \times 8^0) = (3 \times 64) + (4 \times 8) + 5 = 229$$

Therefore $(345)_8 = (229)_{10}$

(*e*) Decimal to Hexadecimal

The given decimal number is converted into hexadecimal number as follows:

16	7864		
16	7864	8	→ 8 LSD
16	30	11	→ 8 ↑
	1	14	→ E MSD

Therefore, $(7864)_{10} = (EB8)_{16}$

1.3 FLOATING POINT REPRESENTATION OF NUMBERS

In the decimal system, very large and very small numbers are expressed in scientific notation as follows: 4.69×10^{23} and 1.601×10^{-19}. Binary numbers can also be expressed by the floating point representation. The floating point representation of a number consists of two parts: the first part represents a signed, fixed point number called the *mantissa* (m); the second part designates the position of the decimal (or binary) point and is called the *exponent* (e). The fixed point mantissa may be a fraction of an integer. The number of bits required to express the exponent and mantissa is determined by the accuracy desired from the computing system as well as its capability to handle such numbers. For example, the decimal number + 6132.789 is represented in floating point as follows:

sign		sign	
0	0.6132789	0	04
mantissa		exponent	

The mantissa has a 0 in the leftmost position to denote a plus. Here, the mantissa is considered to be a fixed point fraction. This representation is equivalent to the number expressed as a fraction 10 times by an exponent, that is $0.6132789 \times 10^{+04}$. Because of this analogy, the mantissa is sometimes called the *fraction part.*

Consider, for example, a computer that assumes integer representation for the mantissa and radix 8 for the numbers. The octal number + 36.754 = 36754×8^{-3} in its floating point representation will look like this.

sign		sign	
0	36754	0	03
mantissa		exponent	

When this number is represented in a register in its binary-coded form, the actual value of the register becomes 0 011 110 111 101 100 and 1 000 011.

Most computers and all elctronic calculators have a built-in capacity to perform floating-point arithmetic operations.

Example 1.10. *Determine the number of bits required to represent in floating point notation the exponent for decimal numbers in the range of* $10^{\pm 86}$.

Solution: Let n be the required number of bits to represent the number $10^{\pm 86}$.

$$2^n = 10^{86}$$

$$n \log 2 = 86$$

$$n = \frac{86}{\log 2} = \frac{86}{0.3010} = 285.7$$

$$\text{Therefore, } 10^{\pm 86} = 2^{\pm 285.7}$$

The exponent ± 285 can be represented by a 10-bit binary word. It has a range of exponent (+ 511 to – 512).

1.4 ARITHMETIC OPERATION

Arithmetic operation in a computer are done using binary numbers and not decimal number and these take place in its *arithmetic unit*. The electronic circuit of a binary adder with suitable shift register can perform all arithmetic operations.

1.4.1 Binary Arithmetic

The arithmetic rules for Addition, Subtraction, Multiplication and Division of binary numbers are given below:

	Addition	Subtraction	Multiplication	Division
(*i*)	$0 + 0 = 0$	$0 - 0 = 0$	$0 \times 0 = 0$	$0 \times 1 = 0$
(*ii*)	$0 + 1 = 1$	$1 - 0 = 1$	$0 \times 1 = 0$	$1 \times 1 = 1$
(*iii*)	$1 + 0 = 1$	$1 - 1 = 0$	$1 \times 0 = 0$	$0 \times 0 =$ not allowed
(*iv*)	$1 + 1 = 10$	$10 - 1 = 1$	$1 \times 1 = 1$	$1 \times 0 =$ not allowed

Binary addition: Two binary numbers can be added in the same way as two decimal numbers are added. The addition is carried out from the least significant bits and it proceeds to higher significant bits, adding the carry resulting from the previous addition each time. Consider the addition of the binary number 1010 and 1111.

MSB				LSB	Decimal
	1	1	1	1	15
	1	0	1	0	10
1	0	0	1	1	25

The addition carried out above can be explained as follow:

Step 1: The least significant bits are added, i.e. 0 + 1 = 1 with a carry 0.

Step 2: The carry in the previous step is added to the next higher significant bits, i.e. 1 + 1 + 0 = 0 with a carry 1.

Step 3: The carry in the above step is added to the next higher significant bits i.e. 0 + 1 + 1 = 0 with a carry 1.

Step 4: The preceding carry is added to the most significant bits, i.e. 1 + 1 + 1 + 1 = 1 with a carry 1.

Thus, the sum is 11001. The addition is also shown in the decimal number system in order to compare the results.

Binary Subtraction: Binary subtraction is also carried out in the same way as decimal numbers are subtracted. The subtraction is carried out from the least significant bits and proceeds to the higher significant bits. When 1 is subtracted from 0, a 1 is borrowed from the immediate higher significant bit. The following problem explains the steps involved. Suppose that 1001 is subtracted from 1101.

Case 1:

MSB			LSB	Decimal
1	1	0	1	13
1	0	0	1	9
0	1	0	0	4

The steps are described below.

Step 1: The LSB in the first column are 1 and 1. Hence the difference is 1 – 1 = 0.

Step 2: In the second column, the subtraction is performed as 0 – 0 = 0.

Step 3: In the third column, the difference is given by 1 – 0 = 1.

Step 4: In the fourth column (MSB), the difference is given by 1 – 1 = 0.

Thus, the difference between the two binary numbers is 0100.

Case 2:

	1			
	1	0		Decimal
1	0	0	1	9
0	1	1	1	7
0	0	1	0	2

The steps are described below.

Step 1: The least significant bits in the first column are 1 and 1. Hence, the difference is 1 – 1 = 0.

Step 2: In the second column, it is not possible to subtract the 1 from 0. So, a 1 has to be borrowed from the next MSB (3rd bit). But since the 3rd bit is also 0, borrowing has to be done from the MSB (4th bit). The borrowing of 1 from the 4th bit (with weight 8) results in 1 and 10 with weight 4 in the 3rd column and 0 in 4th column as shown above. Now, the subtraction is performed as 10 – 1 = 1.

Step 3: In the third column, the difference is given by 1 – 1 = 0.

Step 4: In the fourth column (MSB), the difference is given by 0 – 0 = 0.

Thus, the difference between the two binary numbers is 0010.

Binary multiplication: Binary multiplication is much simpler than decimal multiplication. The procedure is same as that of decimal multiplication. The binary multiplication procedure is as follows.

Step 1: The least significant bit of the multiplier is taken. If the multiplier bit is 1, the multiplicant is copied as such and, if the multiplier bit is 0, a 0 is placed in all the bit positions.

Step 2: The next higher significant bit of the multiplier is taken and the partial product is written with a shift to the left, as in step 1.

Step 3: Step 2 is repeated for all other higher significant bits and each time a left shift is given.

Step 4: When all the bits in the multiplier have been taken into account, the partial product terms are added, which give the actual product of the multiplier and the multiplicant.

The following examples illustrate the multiplication procedure.

Example 1.11. *Multiply the following binary numbers: (a) 1011 and 1101, (b) 100110 and 1001 and (c) 1.01 and 10.1.*

Solution: (*a*)

$$\begin{array}{cccccccc}
 & & & & 1 & 1 & 0 & 1 \\
 & & & \times & 1 & 1 & 1 & 1 \\
\hline
 & & & & 1 & 0 & 1 & 1 \\
 & & & 0 & 0 & 0 & 0 & \\
 & & 1 & 0 & 1 & 1 & & \\
 & 1 & 0 & 1 & 1 & & & \\
\hline
1 & 0 & 0 & 0 & 1 & 1 & 1 & 1 \\
\hline
\end{array}$$

(*b*)

$$\begin{array}{ccccccccc}
 & & & 1 & 0 & 0 & 1 & 1 & 0 \\
 & & & & \times & 1 & 0 & 0 & 1 \\
\hline
 & & & 1 & 0 & 0 & 1 & 1 & 0 \\
 & & 0 & 0 & 0 & 0 & 0 & 0 & \\
 & & 0 & 0 & 0 & 0 & 0 & 0 & \\
1 & 0 & 0 & 1 & 1 & 0 & & & \\
\hline
1 & 0 & 1 & 0 & 1 & 0 & 1 & 1 & 0 \\
\hline
\end{array}$$

(*a*)

$$\begin{array}{ccccc}
 & & 1. & 0 & 1 \\
 & \times & 1 & 0. & 1 \\
\hline
 & & 1 & 0 & 1 \\
 & 0 & 0 & 0 & \\
1 & 0 & 1 & & \\
\hline
1 & 1. & 0 & 0 & 1 \\
\hline
\end{array}$$

Binary division: Division in binary follows the same procedure as division in decimal. Division by 0 is meaningless. An example is given below.

Example 1.12. *Divise the following: (a) 11001 ÷ 101 (b) 11101 ÷ 1100*

Solution: (*a*) 11001 ÷ 101

$$\begin{array}{r}
101 \\
100\,\overline{)11001} \\
101 \\
\hline
0101 \\
101 \\
\hline
0000 \\
\hline
\end{array}$$

(*b*) 11101 ÷ 1100

```
           10·01101
     1100 )11101·0000
           1101
          ---------
            10100
             1100
          ---------
             10000
              1100
          ---------
             010000
                1100
          ---------
                0100
          ---------
```

1.5 1's AND 2's COMPLEMENTS

Subtraction of a number from another can be accomplished by adding the complement of the subtrahend to the minuend. The exact difference can be obtained with minor manipulations.

1.5.1 1's Complement Subtraction

Subtraction of binary numbers using the 1 's complement method allows subtraction only by addition. The 1 's complement of a binary number can be obtained by changing all 1s to 0s and all 0s to 1s. To subtract a smaller number from a larger number, the 1's complement method is as follows:

(*i*) Determine the 1 's complement of the smaller number.

(*ii*) Add this to the larger number.

(*iii*) Remove the carry and add it to the result. This carry is called end-around-carry.

Example 1.13. *Subtract $(1010)_2$ from $(1111)_2$ using the 1's complement method. Also subtract using direct method and compare.*

Solution:

Direct subtraction

```
    1 1 1 1
  - 1 0 1 0
  ---------
    0 1 0 1
  ---------
```

1's Complement method

```
                         1  1  1  1(+)
1's Complement →         0  1  0  1
                     ----------------
Carry          →     1   0  1  0  0
Add Carry      →                  1
                     ----------------
                         0  1  0  1
                     ----------------
```

Subtraction of a larger number from a smaller one by the 1's complement method involves the following steps:

(*i*) Determine the 1 's complement of the larger number,

(*ii*) Add this to the smaller number.

(*iii*) The answer is the 1's complement of the true result and is opposite in sign. There is no carry.

Example 1.14. *Subtract $(1010)_2$ from $(1000)_2$ using the 1 's complement method. Also subtract by direct method and compare.*

Solution:

Direct subtraction		1's Complement method
1 0 0 0		1 0 0 0(+)
– 1 0 1 0	1's Complement →	0 1 0 1
– 0 0 1 0		1 1 0 1

No carry is obtained. The answer is the 1 's complement of 1101 and is opposite in sign, i.e. – 0010.

The 1's complement method is particularly useful in arithmetic logic circuits because subtraction can be accomplished with the help of an adder.

1.5.2 2's Complement Subtraction

The 2's complement of a binary number can be obtained by adding 1 to its 1's complement. Subtraction of a smaller number from a larger one by the 2's complement method involves following steps:

(*i*) Determine the 2's complement of the smaller number.

(*ii*) Add this to the larger number.

(*iii*) Omit the carry (there is always a carry in this case).

Example 1.15. *Subtract $(1010)_2$ from $(1111)_2$ using the 2's complement method. Subtract by direct method also and compare.*

Solution:

Direct subtraction		2's Complement method
1 1 1 1		1 1 1 1(+)
– 1 0 1 0	1's Complement →	0 1 1 0
– 0 1 0 1		1 1 1 0 1

The carry is discarded. Thus, the answer is $(0101)_2$.

The 2's complement method for subtraction of a larger number from a smaller one is as follows:

(*i*) Determine the 2's complement of the larger number.

(*ii*) Add the 2's complement to the smaller number.

(*iii*) There is no carry. The result is in 2's complement form and is negative.

(*iv*) To get an answer in true form, take the 2's complement and change the sign.

Example 1.16. *Subtract $(1010)_2$ from $(1000)_2$ using 2's complement method. Subtract by direct method also and compare.*

Solution:

Direct subtraction		2's Complement method
1 0 0 0		1 0 0 0(+)
– 1 0 1 0	2's Complement →	0 1 1 0
0 0 1 0	No carry →	1 1 1 0

No carry is obtained. Thus, the difference is negative and the true answer is the 2's complement of $(1110)_2$, i.e. $(0010)_2$.

Though both 1's and 2's complement methods of subtraction seem complex compared to the direct method of subtraction, both have distinct advantages when applied using logic circuits, because they allow subtraction to be done using only addition. The 1's and 2's complements of a binary number can be easily arrived at using logic circuits; the advantage in 2's complement method is that the end-around-carry operation present in the 1's complement method is not involved here.

1.5.3 Signed Binary Number Representation

Binary numbers are represented with a separate sign bit along with the magnitude, as shown below. For example, in an 8-bit binary number, the MSB is the sign bit and the remaining 7 bits correspond to magnitude. The magnitude part contains true binary equivalent of the number for positive numbers, while 2's complement form of the number for negative numbers. For example, + 13, 0, – 46 are represented as follows:

	Sign	Magnitude
+ 13	0	000 1101
0	0	000 0000
– 46	1	010 1110

It is important to note that the number zero is assigned with the sign bit '0'. Therefore, the range of numbers that can be represented using 8-bit binary number is – 128 to + 127. In general, the range of numbers that can be represented by an n-bit number is (-2^{n-1}) to $(+2^{n-1})$.

1.5.4 Addition in the 2's Complement System

Addition can be explained with four possible cases: (*i*) when both the numbers are positive, (*ii*) when augend is a positive and addend is a negative number, (*iii*) when augend is a negative and addend is a positive number, or (*iv*) when both the numbers are negative.

Case 1: Two Positive Numbers

Consider the addition of + 29 and + 19:

+ 29	→	0	001	1101	(augend)
+ 19	→	0	001	0011	(addend)
		0	011	0011	(sum = 48)
		↑ Sign bit			

The sign bits of both augend and addend are zero and the sign bit of the sum is 0, indicating that when the sum is positive they have the same number of bits.

Case 2: Positive augend Number and Negative addend Number

Consider the addition of +39 and – 22. Remember that – 22 will be in its 2's complement form. Therefore, + 22 [00010110] must be converted to – 22 [11101010].

+ 39	→	0	010	0111	(augend)
– 22	→	1	110	1010	(addend)
	1	0	001	0001	(result = 17)
	↑	↑			
	Carry	Sign bit			

This carry is omitted and hence the result is 0001 0001.

In this case, the sign bit of addend is 1. Sign bits also participate in the process of addition; in fact, a carry is generated in the last position of addition. This carry is always omitted. Therefore, the final sum is 0001 0001, which is equivalent to + 17.

Case 3: Positive addend Number and Negative augend Number Consider the addition of – 47 and + 29.

– 47	→	1	101	0001	(augend)
+ 29	→	0	001	1101	(addend)
		1	110	1101	(result = – 18)
		↑ Sign bit			

The result has a sign bit of 1, indicating a negative number. It is in the 2's complement form. The last seven bits 110 1110 actually represent the 2's complement of the sum. The true magnitude of the sum can be found by taking the 2's complement of 110 1110; the result is 10010(+ 18). Thus, 1 110 1110 represents – 18.

Case 4: Two Negative Numbers Consider the addition of – 32 and – 44.

– 32	→	1	110	0000	(augend)
– 44	→	1	101	0100	(addend)
	1	1	011	0100	(result = – 76)
	↑	↑			
	Carry	Sign bit			

The carry is discarded and hence the result is 1011 0100

The true magnitude of the sum is the 2's complement of 011 0100, i.e. 1 1001100 (– 76). Thus, the 2's complement addition works in every case. This assumes that the decimal sum is within – 128 to + 127 range. Otherwise, we get an overflow.

1.5.5 Subtraction in the 2's Complement System

As in the case of addition, subtraction can also be carried out four possible cases. Subtraction by the 2's complement system involves addition.

Case 1: Both the Numbers are Positive

Consider the case where + 19 is to be subtracted from + 28.

+ 28	→ 0001	1100
+ 19	→ 0001	0011

To subtract + 19 from + 28, the computer will send the + 19 to a 2's compelling circuit to produce

– 19	→ 1110	1101

The system will then add + 28 and – 19 as follows:

+ 28	→	1110	1101
– 19	→	1110	1101
(Sum = 9)		10000	1001

↑ Omit this carry

Case 2: Positive Number and Smaller Negative Number

Consider that the minuend is + 39 and the subtrahend is – 21. In the 2's complement system, they appear as

+ 39	→	0010	0111
– 21	→	1110	1011

The computer sends – 21 to a 2's complement circuit to produce

+ 21 → 0001 0101

It then adds +39 and +21 as follows:

+ 39	→	0010	0111
+ 21	→	0001	0101
(Sum = 60)		0011	1100

Case 3: Positive Number and Larger Negative Number

Consider that the minuend is + 19 and the subtrahend is – 43. In the 2's complement system, they appear as

+ 19	→	0001	0011
– 43	→	1101	0101

The computer sends the 2's complement of – 43, i.e.

+ 43	→	0010	1011

It then adds +19 and +43 as shown below:

+ 19	→	0001	0011
+ 43	→	0010	1011
(Sum = 62) 0011	1110		

Case 4: Both the Numbers are Negative

Consider the subtraction of – 33 from – 57. In the 2's complement representation,

– 57	→	1100	0111
– 33	→	1101	1111

Taking the 2's complement of – 33,

+ 33	→	0010	0001

Then add + 33 to – 57. We have

– 57	→	1100	0111
– 33	→	0010	0001
(–24)		0011	1110

1.5.6 Arithmetic Overflow

When the number of bits in the sum exceeds the number of bits in each of the numbers added, *overflow* results. This appears in the ninth significant place, and is also called the excess-one. Overflow causes a sign change.

Assume that both the input numbers are in the range of – 128 to + 127. The problem arises only when the arithmetic circuit adds two positive numbers or two negative numbers. In such a case, it is possible for the sum to be outside the range of – 128 to + 127.

Case 1: Two Positive Numbers

Consider the addition of + 120 and + 65. As the decimal sum of + 120 and + 65 is + 185, an overflow occurs into the MSD position. This overflow forces the sign bit of the answer to change.

+ 120	→	0111	1000
+ 65	→	0100	0001
(185)		1911	1001

As the sign bit is 1, i.e. negative, the answer is not correct.

Case 2: Two Negative Numbers

Consider the addition of – 77 and – 122.

– 77	→	0111	1000		
+ (– 122)	→	+1000	0110		
– 199		10011	1001	→	00111001

The 8-bit answer is 0011 1001. Here the sign bit is positive. As the right answer has to contain a negative sign bit, the answer is not correct.

An overflow is a software problem and not a hardware problem. In digital computers, an overflow occurs when an operation results in a quantity beyond the capacity of the storage register. Therefore, a programmer must check the overflow after each addition or subtraction by looking for a change in the sign bit. Logic circuitry is used in each case to detect overflow.

1.5.7 Comparison Between 1's and 2's Complements

(*i*) The 1's complement can be easily obtained using an inverter. The 2's complement has to be arrived at by first obtaining the 1's complement and then adding one (1) to it.

(*ii*) The advantage in the 2's complement system is that only one arithmetic operation is required; the 1's complement requires two operations.

(*iii*) While the 1's complement is often used in logical manipulations for inversion operation, the 2's complement is used only for arithmetic applications.

1.6 9's COMPLEMENT

The 9's complement of a decimal number can be found by subtracting each digit in the number from 9. The 9's complement of decimal digits 0 to 9 is shown below:

Decimal digit	9's complement
0	9
1	8
2	7
3	6
4	5
5	4
6	3
7	2
8	1
9	0

Example 1.17. *Find the 9's complement of each of the following decimal numbers:*

(a) 19 (b) 146 (c) 469 and (d) 4397

Solution: Subtract each digit in the number from 9 to get the 9's complement.

(*a*)
$$\begin{array}{r} 99 \\ -19 \\ \hline 80 \\ \hline \end{array} \rightarrow \text{9's Complement of 19}$$

(*b*)
$$\begin{array}{r} 999 \\ -146 \\ \hline 853 \\ \hline \end{array} \rightarrow \text{9's Complement of 146}$$

(*c*)
$$\begin{array}{r} 999 \\ -469 \\ \hline 530 \\ \hline \end{array} \rightarrow \text{9's Complement of 469}$$

(*d*)
$$\begin{array}{r} 9999 \\ -4397 \\ \hline 5602 \\ \hline \end{array} \rightarrow \text{9's Complement of 4397}$$

1.6.1 9's Complement Subtraction

Subtraction of a smaller decimal number from a larger one in the 9's complement system is done by the addition of the 9's complement of the subtrahend to the minuend and then adding the carry to the result. Subtraction of a larger number from a smaller one does not produce a carry, and the result is a negative in the 9's complement form. This procedure has a distinct advantage in certain types of arithmetic logic.

Example 1.18. *Perform the following subtractions by using the 9's complement method:*
(a) 18–06 (b) 39–23 (c) 34–49 and (d) 49–84.

Solution:

	Regular subtraction	9's Complement subtraction	
(*a*)	18	18	
	– 06	+ 93	← 9's Complement of 6
	12	(1) 11	
		+ 1	Add carry to result
		12	
(*b*)	39	39	
	– 23	+ 76	← 9's Complement of 23
	16	(1) 15	
		+ 1	Add carry to result
		16	
(*c*)	34	34	
	– 49	+ 50	← 9's Complement of 49
	16	84	
		↓	
		– 15	← 9's Complement of 84
(*d*)	39	49	
	– 84	+ 15	← 9's Complement of 84
	– 35	64	
		↓	
		– 35	← 9's Complement of 64

1.7 10's COMPLEMENT

The 10's complement of a decimal number is equal to its 9's complement + 1.

Example 1.19. *Convert the following decimal numbers into its 10's complement form: (a) 9, (b) 46 and (c) 739.*

Solution: (*a*)

9	
9	
0	← 9's Complement of 9
+ 1	
1	← 10's Complement of 9

(*b*)

```
     99
   – 46
   ----
     53   ←  9's Complement of 46
    + 1
   ----
     54   ←  10's Complement of 46
```

(*c*)

```
    999
   –739
   ----
    260   ←  9's Complement of 739
    + 1
   ----
    261   ←  10's Complement of 739
```

1.7.1 10's Complement Subtraction

In the 10's complement method of subtraction, the minuend is added to the 10's complement of the subtrahend and the carry is dropped.

Example 1.20. *Subtract the following decimal numbers using the 10's complement method:*

(a) 9–4, (b) 24–09, (c) 69–32 and (d) 347–265.

Solution:

```
(a) Regular subtraction        10's Complement subtraction
          9                          9
        – 4                        + 6    ←  10's Complement of 4
        ---                      -----
          5                      (1) 5       Drop carry
(b)      24                         24
       – 09                       + 91    ←  10's Complement of 9
       ----                      ------
         15                     (1) 15       Drop carry
(c)      69                         69
       – 32                       + 68    ←  10's Complement of 32
       ----                      ------
         37                     (1) 37       Drop carry
(d)     347                        347
      – 265                      + 735    ←  10's Complement of 265
      -----                     ------
       – 82                     (1)082       Drop carry
```

1.8 BINARY CODED DECIMAL (BCD)

The Binary Coded Decimal (BCD) is a combination of four binary digits that represent decimal numbers. For example, the 8421 code is a type of binary coded decimal.. It has four bits and represents the decimal digits 0 to 9. The numbers 8421 indicate the binary weights of the four bits. The ease of conversion between the 8421 code numbers and the familiar decimal numbers is the main advantage of this code. To express any decimal number in BCD, each decimal digit should be replaced by the appropriate four-bit code. Table 1.1 gives the binary and BCD codes for the decimal numbers 0 to 15.

Table 1.1 Decimal numbers, binary equivalents and BCD

Decimal number	Binary number	Binary coded decimal (BCD)
0	0000	0000
1	0001	0001
2	0010	0010
3	0011	0011
4	0100	0100
5	0101	0101
6	0110	0110
7	0111	0111
8	1000	1000
9	1001	1001
10	1010	0001 0000
11	1011	0001 0001
12	1100	0001 0010
13	1101	0001 0011
14	1110	0001 0100
15	1111	0001 0101

1.8.1 BCD Addition

BCD is a numerical code. Many applications require arithmetic operations. Addition is the most important of these because the other three operations, namely subtraction, multiplication and division, can be done using addition. The rule for addition of two BCD numbers is given below.

(*i*) Add the two numbers using the rules for binary addition,

(*ii*) If a four-bit sum is equal to or less than 9, it is a valid BCD number,

(*iii*) If a four-bit sum is greater than 9, or if a carry-out of the group is generated, it is an invalid result. Add 6 (0110_2) to the four-bit sum in order to skip the six invalid states and return the code to BCD. If a carry results when 6 is added, add the carry to the next four-bit group.

Example 1.21. *Add the following BCD numbers: (a) 1001 and 0100 and (b) 00011001 and 00010100.*

Solution:

(*a*)

$$\begin{array}{rrl}
 & 1\;0\;0\;1 & \\
 & +\;0\;1\;0\;0 & \\ \hline
 & 1\;1\;0\;1 & \rightarrow \text{Invalid BCD number} \\
 & +\;0\;1\;1\;0 & \rightarrow \text{Add 6} \\ \hline
\underbrace{0\;0\;0\;1}_{1} & \underbrace{0\;0\;1\;1}_{3} & \rightarrow \text{Valid BCD number}
\end{array}
\qquad
\begin{array}{r}
9 \\ +\,4 \\ \hline 13_{10} \\ \hline
\end{array}$$

```
(b)    0 0 0 1      1 0 0 1
     + 0 0 0 1      0 1 0 0
     -------------------------
       0 0 1 0      1 1 0 1   → Right group is invalid     19
                  + 0 1 1 0   → Add 6                     + 14
     -------------------------                            ----
       0 0 1 1      0 0 1 1   → Valid BCD number          33₁₀
       \_____/      \_____/
          1            3
```

1.8.2 BCD Subtraction

Method I: Table 1.2 shows an algorithm for BCD subtraction. The 1′s complement of the BCD subtrahend is entered into adder 1, and the complement (true) of the result is transferred to adder 2, where either a 1010 or 0000 is added, depending on the sign of the total result. Examples of a positive and negative total result are given in Table 1.2. Arrows indicate EAC (end-around-carry) or carry to the next decade.

Table 1.2. Algorithm for BCD subtraction

Decade result	Sign of total result	
	(+) EAC = 1	**(–) EAC = 0**
	Transfer true results of adder	Transfer 1′s complement of result of adder 1
$C_n = 1$	0000 added in adder 2	1010 added in adder 2
$C_n = 0$	1010 added in adder 2	0000 added in adder 2

Total Result Positive]

```
                 10²       10¹   10⁰     EAC indicates a (+) total result
   835     |    1000      0011  0100 ─┐
 - 274     |    1101      1000  1011 ←┤  1's Complement of 0010 0111 0100
 + 561   1      0101      1011  0000  │
                 │           1     1 ←┘
                 ↓        1100  0001     Transfer true output of adder 1
                0100      1100  0001
                0000      1010  0000
                0101   1  0110  0001
                       ↑________________ Ignore this carry 1
                   5         6     1
```

```
           10^2  10^1    10^0
  429  0100  0010    1001    No EAC indicates (–) total result
– 476  1011  1000    1001 ←  1's Complement of 0100 0111 0110
  – 47  1111  1010    0010
         ↓     1↙      ↓
               1011
       0000  0100    1101    Transfer 1's Complement of adder 1 output
       0000  0000    1010
       0000  0100  1 0111
                   ↑_________ Ignore this carry
```

Method II: Another method in BCD subtraction is the addition of the 9's complement of the subtrahend to the minuend.

Example 1.22. *Subtract 748 from 983 using 9's complement method.*

Solution:

```
9's complement of 748 =     999
                          – 748
                            251
```

Direct method

```
  983          983
– 748        + 251   ← 9's Complement of 748
  235     1    234
         ↘       1   ← EAC
               235
```

Example 1.23. *Subtract the following using 9's complement method:*

(a) 649 – 387 and (b) 891 – 786

Solution:

```
  649          649
– 387        + 612   ← 9's Complement of 387
  262     1    261   ↑________ Ignore carry
         ↘       1   ← EAC
               262
  891          891
– 786        + 213   ← 9's Complement of 786
  105     1    104
         ↘       1   ← EAC
               105
```

Example 1.24. *Subtract the following using 10's complement method:*
(a) 786 – 427, (b) 473 – 438 and (c) 357 – 294

Solution:

```
  786           786
– 427         + 573   ← 10's Complement of 427
  359       1   359
            ↑
            └────────── Ignore carry

  473           473
– 438         + 562   ← 10's Complement of 438
   35       1   035
            ↑
            └────────── Ignore carry

  357           357
– 294         + 706   ← 10's Complement of 438
   63       1   063
            ↑
            └────────── Ignore carry
```

Example 1.25. *Carry out BCD subtraction for (68) – (61) using 10's complement method.*

Solution:

```
                                                  1
  68      68                                   0110  1000
– 61    + 39   ← 10's Complement of 61         0011  1001
  07  1   07                                   1010  0001
      ↑                                        0110  0110   Add 6
      └──────── Ignore carry                 1 0000  0111
                                             ↑
                                             └──────────── Ignore carry
```

1.9 CODES

Code is a symbolic representation of discrete information, which may be present in the form of numbers, letters or physical quantities. The symbols used are the binary digits 0 and 1 which are arranged according to the rules of codes. These codes are used to communicate information to a digital computer and to retrieve messages from it. A code is used to enable an operator to feed data into a computer directly, in: the form of decimal numbers, alphabets and special characters. The computer converts these data into binary codes and after computation, transforms the data into its original format (decimal numbers, alphabets and special characters).

When numbers, letters, or words are represented by a special group of symbols, this is called encoding, and the group of symbols is called a code. In Horse code, a; series of dots and dashes represent alphabet, numerals and special characters.

Codes are broadly classified into five groups, viz., (*i*) Weighted Binary Codes, (*ii*) Non-weighted Codes, (*iii*) Error-detecting Codes, (*iv*) Error-correcting Codes, and (*v*) Alphanumeric Codes.

1.9.1 Weighted Binary Codes

Weighted binary codes obey their positional weighting principles. Each position of a number represents a specific weight. In a weighted binary code, the bits are multiplied by the weights indicated; the sum of these weighted bits gives the equivalent decimal digit.

Straight binary coding is a method of representing a decimal number by its binary equivalent. The codes 8421,2421, 5421 and 5211 are weighted binary codes. Each decimal digit is represented by a four-bit binary word, the three digits for the left being weighted. Table 1.3 consists of a few weighted 4-bit binary codes with their decimal numbers and complements.

Table 1.3 Some weighted 4-bit binary codes

Decimal number	8421	5421	2421	9's complement of 2421 code
0	0000	0000	0000	1111
1	0001	0001	0001	1110
2	0010	0010	0010	1101
3	0011	0011	0011	1100
4	0100	0100	0100	1011
5	0101	1000	1011	0100
6	0110	1001	1100	0011
7	0111	1010	1101	0010
8	1000	1011	1110	0001
9	1001	1100	1111	0000

BCD (or) 8421 code: The Binary-coded Decimal (BCD) uses the binary number system to specify the decimal numbers 0 to 9. It has four bits. The weights are assigned according to the positions occupied by these digits. The weights of the first (right-most) position is 2^0(1), the second 2^1(2), the third 2^2(4), and the fourth 2^3(8). Reading from left to right, the weights are 8-4-2-1, and hence it is called 8421 code.

The binary equivalent of 7 is $[111]_2$, but the same number is represented in BCD in 4-bit form as $[0111]_2$. Also, the numbers from 0 to 9 are represented in the same way as in the binary system, but after 9 the representations in BCD are. different. For example, the decimal number 12 in the binary system is $[1100]_2$ but the same number is represented as [0001 0010] in BCD.

Example 1.26. *Give the BCD Code for the decimal number 874.*

Solution: Decimal number → 874

BCD code → 1000 0111 0100

Hence, $(874)_{10}$ = $(1000\ 0111\ 0100)_{BCD}$

Example 1.27. *Give the BCD code equivalent for the decimal number 96 × 42.*

Solution: Decimal number → 8 6 4 2

↑ ↑ ↑ ↑

BCDCode → 100101 1001000010

Hence, $[96.42]_{10}$ = $[10010110 \cdot 01000010]_{BCD}$.

2421 code: This is a weighted code; its weights are 2,4,2 and 1. A decimal number is represented in 4-bit form and the total weight of the four bits = 2 + 4 + 2 + 1 = 9. Hence the 2421 code represents decimal numbers from 0 to 9. Upto 4, the 2421 code is the same as that in BCD; however, it varies for digits from 5 to 9. This code is also a self-complementing code i.e. the 9's complement of a number 'N' is obtained by complementing the 0s and Is in the code word 'N'. For example, the 2421 code for 3 is 0011 and its natural complement 1100 gives 6 which is the 9's complement of 3. Table 1.3 gives the 2421 code of the decimal numbers and its complement. The bit combination 1101, when weighted by the reflective digits 2421, gives the decimal equivalent of 2 × 1 + 4 × 1 + 2 × 0 + 1 × 1 = 2 + 4 + 0 + 1 = 7.

Reflective codes: A code is said to be *reflective* when the code for 9 is the complement of the code for 0, 8 for 1, 7 for 2, 6 for 3, and 5 for 4. While the 2421, 5211 and Excess-3 codes are reflective codes, the 8421 code is not. While finding the 9's complement, such as in 9's complement subtraction, reflectivity is desirable in a code.

Sequential codes: A code can be said to be sequential when each succeeding code is one binary number greater than its preceding code. This greatly helps mathematical manipulation of data. While the 8421 and Excess-3 codes are sequential, the 2421 and 5421 codes are not.

1.9.2 Non-weighted Codes

Non-weighted codes are codes that are not positionally weighted. This means that each position within a binary number is not assigned a fixed value. Excess-3 codes and Gray codes are examples of non-weighted codes.

Excess-3 code: As the name indicates, the excess-3 represents a decimal number, in binary form, as a number greater than 3. An excess-3 code is obtained by adding 3 to a decimal number. For example, to encode the decimal number 6 into an excess-3 code, we must first add 3 in order to obtain 9. The 9 is then encoded in its equivalent 4-bit binary code 1001. The excess-3 code is a self-complementing code, and this helps in performing subtraction operations in digital computers, especially in the earlier models. The excess-3 code is also a reflective code.

Example 1.28. *Convert $[643]_{10}$ into its Excess-3 code.*

Solution:

Decimal number	6	4	3
Add 3 to each bit	+ 3	+ 3	+ 3
Sum →	9	7	6

Converting the above sum into its BCD code, we have

Sum $\rightarrow$ 9 7 6

$\downarrow$ $\downarrow$ $\downarrow$

BCD $\rightarrow$ 1001 0111 0110

Hence, the Excess-3 code for $[643]_{10}$ is 1001 0111 0110.

Table 1.4 lists the BCD, Excess-3 code and 9's'complement representations for decimal digits. Note that both codes use only 10 of the 16 possible 4-bit code groups. The excess-3 code, however, does not use the same code groups. Its invalid code groups are 0000, 0001, 0010, 1101, 1110, and 1111.

Table 1.4 Excess-3 codes

Decimal	8421(BCD) code	Excess-3 code	9's complement
0	0000	0011	1100
1	0001	0100	1011
2	0010	0101	1010
3	0011	0110	1001
4	0100	0111	1000
5	0101	1000	0111
6	0110	1001	0110
7	0111	1010	0101
8	1000	1011	0100
9	1001	1100	0011

Gray code: The *Gray code* belongs to a class of codes called minimum-change codes, in which only one bit in the code group changes when moving from one step to the next. The Gray code is a non-weighted code. Therefore, it is not suitable for arithmetic operations but finds applications in input/output devices and in some types of analog-to-digital converters. The Gray code is a reflective digital code which has a special property of containing two adjacent code numbers that differ by only one bit. Therefore, it is also called a unit-distance code.

Table 1.5 shows the Gray code representation for the decimal numbers 0 to 15, together with the straight binary code.

Table 1.5 Gray code

Decimal numbers	Binary code	Gray code
0	0000	0000
1	0001	0001
2	0010	0011
3	0011	0010
4	0100	0110
5	0101	0111
6	0110	0101

(Table 1.5 contd.)

Decimal numbers	Binary code	Gray code
7	0111	0100
8	1000	1100
9	1001	1101
10	1010	1111
11	1011	1110
12	1100	1010
13	1101	1011
14	1110	1001
15	1111	1000

Conversion of a binary number to gray code: A binary number can be converted to its Gray code when

(*i*) the first bit (MSB) of the Gray code is the same as the first bit of the binary number;

(*ii*) the second bit of the Gray code equals the exclusive-OR, of the first and second bits of the binary number, i.e. it will be 1 if these binary code bits are different and 0 if they are the same;

(*iii*) the third Gray code bit equals the exclusive-OR of the second and third bits of the binary number, and so on.

Example 1.29. *Convert $[10110]_2$ to Gray code.*

Solution:

Step 1: The first bit MSB of the Gray code is the same as the first bit of the binary number.

```
1  0  1  1  0    Binary
↓
1                Gray
```

Step 2: Add the first bit of the binary digit to the second bit of the binary. The addition of 1 and 0 is 1. The result is the second bit of the Gray code.

```
[1  ⊕  0]  1  1  0    Binary
       ↓
1      1              Gray
```

Step 3: Add the second bit 0 to the third bit 1 of the binary.

```
1  [0  ⊕  1]  1  0    Binary
          ↓
1   1     1           Gray
```

Step 4: Add the third bit 1 to the fourth bit 1 and omit the carry. The exclusive OR addition of 1 and 1 is 0. This is the fourth bit of the Gray code.

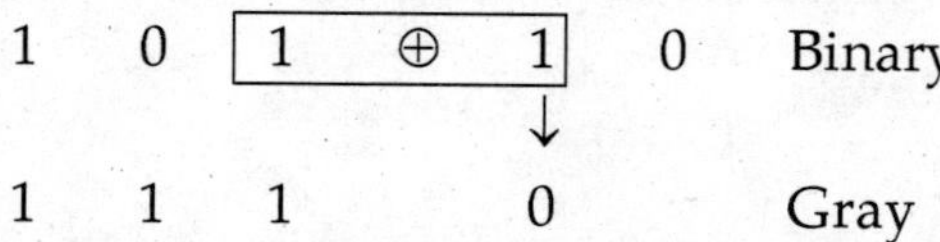

Step 5. Add the fourth bit 1 to the last bit of the binary.

1 0 1 [1 ⊕ 0] Binary

↓

1 1 1 0 1 Gray

As the conversion is complete, the Gray code of the binary 10110 is 11101.

Example 1.30. Convert the binary $[10101101]_2$ to its Gray code.

Solution:

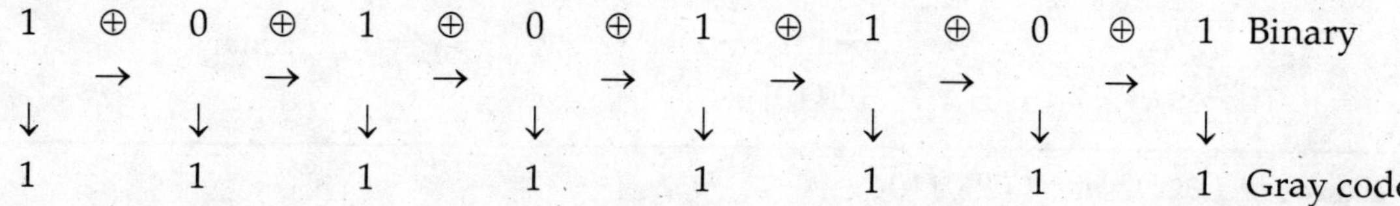

Conversion from gray code to binary: Conversion of a Gray code into its binary form involves the reverse of the procedure given above:

(*i*) The first binary bit (MSB) is the same as that of the first Gray code bit.

(*ii*) If the second Gray bit is 0, the second binary bit is the same as that of the first binary; if the second Gray bit is 1, the second binary bit is the inverse of its first binary bit.

(*iii*) Step 2 is repeated for each successive bit.

Example 1.31. *Convert the Gray code 110101 to binary form.*

Solution:

Step 1: Write the first binary bit 1 which is the MSB of the Gray code.

1 1 0 1 0 1 Gray

↓

1 Binary

Step 2: The second bit of the Gray code is 1 and therefore the second bit of the binary is 0, i.e. inverse of the first binary bit '1'.

1 1 0 1 0 1 Gray

↓

1 → 0 Binary

Step 3: The third bit of the Gray code is 0 and therefore the third bit of the binary is same as that of the second binary bit, i.e. 0.

1 1 0 1 0 1 Gray

↓

1 0 → 0 Binary

Step 4.

1	1			1	0	1 Gray
				↓		
1	0	0	→	1		Binary

Step 5.

1	1	0	1		0	1	Gray
					↓		
1	0	0	1	→	1		Binary

Step 6.

1	1	0	1	0		1	Gray
						↓	
1	0	0	1	1	→	0	Binary

Therefore, $[110101]_G = [100110]_2$.

Example 1.32. *Convert [100110]$_2$.*

Solution:

1		0		1		0		1		1		1	Gray
↓		↓		↓		↓		↓		↓		↓	
1	→	1	→	0	→	0	→	1	→	0	→	1	Binary

Therefore, $[1010111]_G = [1100101]_2$.

This process can be seen in another way. Each binary bit (except the first) can be obtained by taking the exclusive-OR of the corresponding Gray code bit and the previous binary bit. The reader should verify that this process give the proper result.

Example 1.33. *Convert [1011]$_G$ to binary.*

Solution:

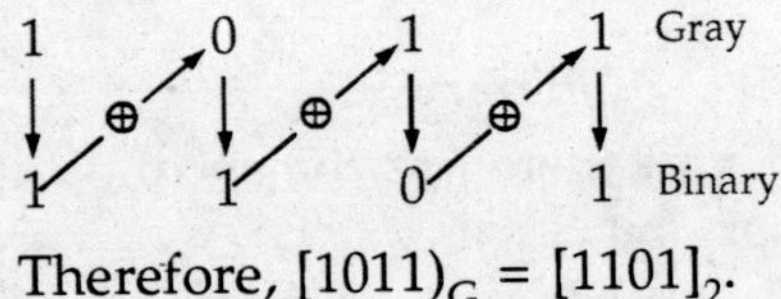

Therefore, $[1011)_G = [1101]_2$.

1.9.3 Error Detecting Codes

During the process of binary data transmission, errors may occur. In order to detect and correct such erros, two types of codes, namely (*i*) *error-detecting Codes* and (*ii*) *error-correcting codes,* may be used.

If a single error transforms a valid code word into an invalid one, it is said to be a single error-detecting code. The most simple and commonly used error detecting method is the *parity check,* in which an extra parity bit is included with the binary message, to make the total number of Is either odd or even, resulting in two methods, viz. (*i*) Even-parity method and (*ii*) Odd-parity method. In the even-parity method, the total number of 1s in the code group (including the parity bit) must be an even number. Similarly, In the odd-parity method, the total number of 1s (including the parity bit) must be an odd number. The parity bit can be placed at either end of the code word, such that the receiver should be able to understand the parity bit and the actual data. Table 1.6 shows a message of three bits and its corresponding even and odd parity bits.

Do You Know?	An error in a digital system is the corruption of data from its correct value to some other value.

Table 1.6 Parity-bit generation

Message	**Even-parity code**		**Odd-parity code**	
xyz	*xyz*	*p*	*xyz*	*p*
000	000	0	000	1
001	001	1	001	0
010	010	1	010	0
010	010	0	011	1
100	100	1	100	0
101	101	0	101	1
110	110	0	110	1
111	111	1	111	0

If a single error occurs, it transforms the valid code into an invalid one. This helps in the detection of single bit errors. Though the parity code is meant for single error detection, it can detect any odd number of errors. However, in both the cases, the original code word can not be found. If an even number of errors occur, then the parity check is satisfied, giving an erroneous result.

Check sums: The parity method can detect only a single error within a word and not double errors. Since the double error will not change the parity of the bits, the parity checker will not indicate any error. The check sum method is used to detect double errors and pinpoint erroneous bits. The working of this method is explained in the following lines.

Initially word A 10110111 is transmitted, next the word B 00100010 is transmitted. The binary digits in the two words are added and the sum obtained is retained in the transmitter. Then, a word C is transmitted and added to the previous sum and the new sum is retained. In the same manner, each word is added to the previous sum; after transmission of all the words, the final sum called the check sum is also transmitted. The same operation is done at the receiving end and the final sum obtained here is checked against the transmitted check sum. If the two sums are equal, then there is no error.

1.9.4 Error Correcting Codes

Hamming code: R. W. Hamming developed a system that provides a methodical way to add one or more parity bits to a data character in order to detect and correct errors. The

Hamming distance between two code words is defined as the number of bits changed from one code word to another.

Consider C_i and C_j to be any two code words in a particular block code. The Hamming distance d_{ij} between the two vectors C_i and C_j is defined by the number of components in which they differ. Assuming that dy is determined for each pair of code words, the minimum value of the dy can be called the Hamming distance, d_{min}. For linear block codes, minimum weight is equal to minimum distance.

For example,

$$C_i = 1\ 0\ 0\ 0\ 1\ 1\ 1$$
$$\downarrow \quad\quad \downarrow\ \downarrow$$
$$C_j = 0\ 0\ 0\ 1\ 0\ 1\ 1$$

Do You Know?	An error is caused by a physical failure. Failures can be either temporary or permanent.

Here, these code words differ in the leftmost bit position and in the fourth and fifth bit positions from the left. Accordingly, $d_{ij} = 3$.

From Hamming's analysis of code distances, the following important properties have been derived:

(*i*) A minimum distance of at least two is required for single error detection.

(*ii*) Since the number of errors, $E \le [(d_{min} - 1)/2]$, a minimum distance of three is required for single error correction.

(*iii*) Greater distances will provide detection and/or correction of more number of errors.

The 7-bit Hamming (7, 4) code word Aj $h_1\, h_2\, h_3\, h_4\, h_5\, h_6\, h_7$ associated with a 4-bit binary number $b_1\, b_2\, b_3\, b_0$ is:

$$h_1 = b_3 \oplus b_2 \oplus b_0 \qquad h_3 = b_3$$
$$h_2 = b_3 \oplus b_1 \oplus b_0 \qquad h_5 = b_2$$
$$h_4 = b_2 \oplus b_1 \oplus b_0 \qquad h_6 = b_1$$
$$h_7 = b_0$$

where $\oplus$ denotes the Exclusive-OR operation. Note that bits h_1, h_2 and h_4 are even parity bits for the bit fields $b_3\, b_2\, b_0$, $b_3\, b_1\, b_0$ and $b_2\, b_1\, b_0$, respectively. In general, the party bits (h_1, h_2, h_4, h_8, ...) are located in the positions corresponding to ascending powers of two (i.e., $2^0, 2^1, 2^2, 2^3$, ... = 1, 2, 4, 8, ...).

The h_1 parity bit has a 1 in the LSB of its binary representation. Therefore, it checks all bit positions, including it, that have 1's in the same location (i.e., LSB) in the binary representation (i.e., h_1, h_3, h_5 and h_7). The binary representation of h_2 has a 1 in the middle bit. Therefore, it checks all bit positions, including it, that have 1's in the same location (i.e., middle bit) in the binary representation (i.e., h_2, h_3, h_6 and h_7). The binary representation of h_4 has a 1 in the MSB. Therefore, it checks all bit positions, including it, that have 1's in the same location (i.e., MSB) in the binary representation (i.e., h_4, h_5, h_6 and h_7).

To decode a Hamming code, one must check for odd parity over the bit fields in which even parity was previously established. For example, a single bit error is indicated by a non-zero parity word $c_4\, c_2\, c_1$, where

$$c_1 = h_1 \oplus h_3 \oplus h_5 \oplus h_7$$
$$c_2 = h_1 \oplus h_3 \oplus h_6 \oplus h_7$$
$$c_4 = h_4 \oplus h_5 \oplus h_6 \oplus h_7$$

If $c_4\, c_2\, c_1 = 0\,0\,0$, then there is no error in the Hamming code. If it has a nonzero value, it indicates the bit position in error. For example, if $c_4\, c_2\, c_1 = 101$, then bit 5 is in error. To correct this error, bit 5 has to be complemented.

Example 1.34. *Encode data bits 0101 into a 7-bit even parity Hamming code.*

Solution: Given $b_3\, b_2\, b_1\, b_0 = 0101$.

Therefore,

$$h_1 = b_1 \oplus b_2 \oplus b_0 = 0 \oplus 1 \oplus 1 = 0 \quad h_3 = b_3 = 0$$
$$h_2 = b_3 \oplus b_1 \oplus b_0 = 0 \oplus 0 \oplus 1 = 1 \quad h_5 = b_2 = 0$$
$$h_4 = b_2 \oplus b_1 \oplus b_0 = 1 \oplus 0 \oplus 1 = 0 \quad h_6 = b_1 = 0$$
$$h_6 = b_1 = 0$$

h_1	h_2	h_3	h_4	h_4	h_6	h_7
0	1	0	0	1	0	1

Example 1.35. *A 7-bit Hamming code is received as 0101101. What is its correct code?*

Solution:

h_1	h_2	h_3	h_4	h_4	h_6	h_7
0	1	0	1	1	0	1

Now, to find the error,

$$c_1 = h_1 \oplus h_2 \oplus h_5 \oplus h_7 = 0 \oplus 0 \oplus 1 \oplus 1 = 0$$
$$c_2 = h_2 \oplus h_3 \oplus h_6 \oplus h_7 = 1 \oplus 0 \oplus 0 \oplus 1 = 0$$
$$c_4 = h_4 \oplus h_5 \oplus h_6 \oplus h_7 = 1 \oplus 1 \oplus 0 \oplus 1 = 1$$

Thus, $c_4 c_2 c_1 = 100$. Therefore, bit 4 is in error and the corrected code word can be obtained by complementing the fourth bit in the received code word as 0100 101.

1.9.5 Alphanumeric Codes

If a computer is to be useful, it must be capable of handling non-numerical information. Similarly, printers and other similar devices must be able to recognize codes that represent numbers (0 to 9), letters, and special symbols. The codes that represent numbers, alphabetic letters and special symbols are called alphanumeric codes.

A complete and adequate set of necessary characters includes (*i*) 26 lower case letters, (*ii*) 26 upper case letters, (*iii*) 10 numeric digits, and (*iv*) about 25 special characters, such as +, /, #, %, ', etc. These total up to about 87 symbols. The representation of 87 symbols with some type of binary code would require at least 7 bits. Each character is represented by a 7-bit code; usually an 8-bit is inserted for parity. With 7 bits, there are $2^7 = 128$ possible binary numbers; 87 of these arrangements of 0 and 1 bits serve as code groups representing 87 different characters. Hence, this code consists of 128 symbols. 95 (= 87 + 8) characters represent graphic symbols that include upper and lower case letters, numerals 0 to 9, punctuation marks and special symbols. 23 characters represent *format effectors* which are functional characters for controlling the layout of printing or display devices such as carriage return, line feed, horizontal tabulation and back space; the other 10 characters are used to direct the data communication flow and report its status.

ASCII code: A standardised code that has been widely accepted by the industry, the ASCII (pronounced "as-kee") code—*American Standards Code for Information Interchange,* used in most microcomputers by its manufacturers. The ASCII code represents a character with seven bits, which can be stored as one byte with one bit unused. The extra bit is often used to extend the ASCII code to represent an additional 128 characters. Table 1.7 partially lists the 7-bit ASCII code for each character and its octal and hexadecimal equivalents. The format of the ASCII code $X_2\,X_3\,X_2\,X_1\,X_0$, where each bit is a '0' or a '1'. For example, the letter A is coded as 100 0001.

EBCDIC code: Another alphanumeric code used in IBM equipment is the EBCDIC or *Extended Binary Coded Decimal Information Code.* It differs from ASCII only in its code grouping for the different alphanumeric characters. It uses eight bits for each character and a ninth bit for parity.

Hollerith code: The Hollerith code is used in punched cards. A punched card consists of 80 columns and 12 rows. Each column represents an alphanumeric character of 12 bits by punching holes in the appropriate rows. The presence of a hole represents a 1, its absence represents a 0. The 12 rows are marked starting from the top, as 12, 11, 0, 1, 2, ...9. The first three rows are called the zone punch and the last nine are called the numeric punch. Decimal digits are represented by a single hole in a numeric punch. The alphabets are represented by two holes in a column, one in the zone punch and the other in the numeric punch. Special characters are represented by one, two, or three holes in a column; while the zone is always used, the other two holes, if used, are in a numeric punch with the 8-th punch being commonly used. The 12-bit card code is inefficient with respect to the number of bits used. Most computers convert the input 12-bit card code into an internal 6-bit code to conserve bits of memory. The Hollerith code is BCD. Thus, the transmission from EBCDIC is simple. Since large computers use punched cards, the Hollerith code is used for their card readers, and punches and EBCDIC are used within the computer.

Table 1.7 Partial listing of ASCII code

Character	7-bit ASCII	Octal	Hexa
A	100 0001	101	41
B	100 0010	102	42
C	100 0011	103	43
D	100 0100	104	44
E	100 0101	105	45
F	100 0110	106	46
G	100 0111	107	47
H	100 1000	110	48
I	100 1001	111	49
J	100 1010	112	4A

Table 1.7 (contd.)

Character	7-bit ASCII	Octal	Hexa
K	100 1011	113	4B
L	100 1100	114	4C
M	100 1101	115	4D
N	100 1110	116	4E
O	100 1111	117	4F
P	101 0000	120	50
Q	101 0001	121	51
R	101 0010	122	52
S	101 0011	123	53
T	101 0100	124	54
U	1010101	125	55
V	101 0110	126	56
W	101 0111	127	57
X	101 1000	130	58
Y	101 1001	131	59
Z	101 1010	132	5A
0	101 0000	060	30
1	011 0001	061	31
2	011 0010	062	32
3	011 0011	063	33
4	011 0100	064	34
5	011 0101	065	35
6	0110110	066	36
7	011 0111	067	37
8	011 1000	070	38
9	011 1001	071	39
Blank	010 0000	040	20
.	0101110	056	2E
(	010 1000	050	28
)	010 1001	051	29
+	010 1011	053	2B
$	010 0100	044	24
*	010 1010	052	2A
–	010 1101	055	2D
/	010 1111	057	2F
,	010 1100	054	2C
=	011 1101	075	3D
!	010 0001	041	21
″	010 0010	042	22
%	010 0101	045	25
&	010 0110	046	26
	010 0111		
:	011 1010	072	3A
;	011 1011	073	3B
<	011 1100	074	3C
>	011 1110	076	3E
?	011 1111	077	3F

REVIEW QUESTIONS

1. What are the common features between the different number systems?
2. How is a number expressed in a general number system?
3. What is the base of the decimal number system?
4. How are binary digits used to express the integer and fractional parts of a number?
5. Explain how multiplication and addition can be performed in digital systems.
6. How are subtraction and division performed in digital systems?
7. List the salient features of the BCD, Excess-3 and Gray codes.
8. What is a BCD Code? What are its advantages and disadvantages?
9. What are meant by the 1's and 2's complements of a binary number?
10. Explain the rules for binary subtraction using the 1's and 2's complement methods.
11. Explain how BCD addition is carried out.
12. What is meant by overflow? Is it a software problem or a hardware problem?
13. Distinguish between 1's and 2's complements.
14. Distinguish between 9's and 10's complements.
15. How do you add two decimal numbers in the BCD form if the sum is greater than 9?
16. Define odd and even parity check codes.
17. Write a short note on "weighted and non-weighted codes".
18. What is a Gray code? Why is it important?
19. What is a Hamming code and how is it used?

PROBLEMS

1. Convert the following decimal numbers to their binary equivalents:
   (*a*) 37 (*b*) 14 (*c*) 167 (*d*) 72.45 (*e*) 0.4475 (*f*) 52 (*g*) 4097.188 (*h*) 2048.0625
   [**Ans.** (*a*) 100101 (*b*) 1110 (*c*) 10100111 (*d*) 1001000-01110 (*e*) 0.0111001 (*f*) 110100 (*g*) 1000000000001.001 (*h*) 100000000000.0001]
2. Convert the following binary numbers to their decimal equivalents:
   (*a*) 10110 (*b*) 10001101 (*c*) 10111.1011 (*d*) 0.011011 (*e*) 1101111.101
   [**Ans.** (*a*) 22 (*b*) 141 (*c*) 23.6875 (*d*) 0.421875 (*e*)111.625]
3. Convert the following numbers from octal to decimal.
   (*a*) 743_8 (*b*) 36.4_8 (*c*) 2376_8 (*d*) 64_8 (*e*) 557_8 (*f*) 1024_8 (*g*) 465_8 (*h*) 7765_8
   [**Ans.** (*a*) 483 (*b*) 30.5 (*c*) 1278 (*d*) 52 (*e*) 367 (*f*) 532 (*g*) 309 (*h*) 4085.
4. Convert the following binary digits to octal numbers:
   (*a*) 101101 (*b*) 101101110 (*c*) 10110111 (*d*) 110110.011 (*e*) 011.101101
   [**Ans.** (*a*) 55_8 (*b*) 556_8 (*c*) 267_8 (*d*) 66.3 (*e*) 3.55.]
5. Convert the following decimal numbers to octal numbers and then to binary:
   (*a*) 59 (*b*) .58 (*c*) 64.2 (*d*) 199.3
   [**Ans.** (*a*) 111011 (*b*) 0.100101000111 (*c*) 001000000.001100 (*d*) 011000111.010011001]

6. Convert each of the following octal numbers to binary:

(*a*) 15_8 (*b*) 24_8 (*c*) 167_8 (*d*) 234_8 (*e*) 173_8 (*f*) 157_8 (*g*) 4653_8 (*h*) 1723_8 (*i*) 2645_8

[**Ans.** (*a*) 0001 101 (*b*) 010 100 (*c*) 001.110 111 (*d*) 010 011 100] (*e*) 001 111 011 (*f*) 001 101 111 (*g*) 100 110 101 011 (*h*) 001 111 010011 (*i*) 010 110 100 101]

7. Convert the following hexadecimal numbers to binary:

(*a*) 49_{16} (*b*) 324_{16} (*c*) 649_{16} (*d*) ABC_{16} (*e*) $5C8_{16}$ (*f*) $FB17_{16}$ (*g*) $4A.67_{16}$ (*h*) $8109.4A_{16}$ (*i*) $EFF2.F_{16}$

[**Ans.** (*a*) 0100 1001 (*b*) 0011 0010 0100 (*c*) 0110 0100 1001 (*d*) 1010 1011 1100 (*e*) 0101 1100 1000 (*f*) 1111 1011 0001 0111 (*g*) 01001010-01100111 (*h*) 10000001 0000 1001-0100 1010 (*i*) 11101111111110010.1111]

8. Convert the following binary numbers to octal and then to hexadecimal:

(*a*) 101100110011 (*b*) 1011101.1011.

[**Ans.** (*a*) 5463, B33 (*b*) 135.54; 5D.B]

9. Convert the following hexadecimal numbers to their decimal equivalents:

(*a*) 49_{16} (*b*) 632_{16} (*c*) 54_{16} (*d*) 54_{16} (*e*) $BC2_{16}$ (*f*) FFF_{16} (*g*) 649_{16} (*h*) $54A_{16}$ (*i*) 1622_{16}.

[**Ans.** (*a*) 73_{10} (*b*) 1586_{10} (*c*) 84_{10} (*d*) 2736_{10} (*e*) 3010_{10} (*f*) 4095_{10} (*g*) 1609_{10} (*h*) 1354_{10} (*i*) 5666_{10}

10. Convert the following hexadecimal numbers to octal:

(*a*) $381B_{16}$ (*b*)$641A_{16}$ (*c*) $AB2_{16}$ (*d*) 1289_{16} (*e*) 2647_{16} (*f*) ABC_{16} (*g*) $1AF_{16}$ (*h*) $2AO_{16}$

[**Ans.** (*a*) 34033_8 (*b*) 62032_8 (*c*) 5262_8 (*d*) 11211_8 (*e*) 23107_8 (*f*)5274_8 (*g*) 657_8 (*h*)1240_8]

11. Convert the following octal numbers to hexadecimal:

(*a*) 137_8 (*b*) 4163_8 (*c*) 775_8 (*d*) 673_8 (*e*) 1275_8 (*f*) 3643_8 (*g*) 45558 (*h*) 4447_8

[**Ans.** (*a*) $5F_{16}$ (*b*) 873_{16} (*c*) $1FD_{16}$ (*d*) $1BB_{16}$ (*e*) $2BD_{16}$ (*f*)$7A3_{16}$ (*g*) $96D_{16}$ (*h*)927_{16}]

12. What are the decimal numbers represented by each BCD code?

(*a*) 101000111 (*b*) 1000001001 (*c*) 10011000110011 (*d*) 1000100100101000.00100111 (*e*) 100100010111.0011

[**Ans.** (*a*) 147_{10} (*b*) 209_{10} (*c*) 2633_{10} (*d*) 8928.27_{10} (*e*) 917.3_{10}]

13. Express the following decimal numbers in 2421 and 5421 codes:

(*a*) 169 (*b*) 264 (*c*) 6734 (*d*) 1993 (*e*) 9021

[**Ans.** (*a*) 0001 1100 1111, 0001 1001 1100 (*b*) 00101100 0100, 0010 1001 0100 (*c*) 1100 1101 0011 0100, 1001 1010 0011 0100 (*d*) 0001 1111 11110011, 0001 110011000011 (*e*) 1111 0000 0010 0001, 1100 0000 0010 0001

14. Express the following 2421 code numbers in decimal form:

(*a*) 1111 1011 0100 (*b*) 0010 1110 0001 (*c*) 1011 0100 1111 (*d*) 0010 1100 1110 (*e*) 1100 1111 0011 0010 (*f*) 1101 1100 1011

[**Ans.** (*a*) 954 (*b*) 281 (*c*) 549 (*d*) 268 (*e*) 6932 (*f*) 765]

15. Express the following decimal numbers in Excess-3 code form:

(*a*) 426 (*b*) 739 (*c*) 1234 (*d*) 5678 (*e*) 984 (*f*) 3421

[**Ans.** (*a*) 0111 0101 1001 (*b*) 1010 0110 1100 (*c*) 0100 0101 01100111 (*d*) 1000 1001 1010 1011 (*e*) 1100 1011 0111 (*f*) 0110 0111 0101 0100

16. Express the following Excess-3 codes as decimals:

(*a*) 0110 1011 1100 0111 (*b*) 0011 0101 1010 0100 (*c*) 0100 1000 1001 (*d*) 1001 0111 1100 (*e*) 1100 1010 0011

[**Ans.** (*a*) 3894 (*b*) 0271 (*c*) 156 (*d*) 649 (*e*) 970]

17. Convert the following binary numbers to Gray codes:
(*a*) 10110 (*b*) 1110111 (*c*) 101010001 (*d*) 101011101 (*e*) 110110011 (*f*) 10001110101 (*g*) 10101110001
[**Ans.** (*a*) 11101 (*b*) 1001100 (*c*) 11111101 (*d*) 111110011
(*e*) 101101010 (*f*) 11001001111 (*g*) 11111001001]

18. Express the following decimals in Gray code form:
(*a*) 4 (*b*) 7 (*c*) 82 (*d*) 324 (*e*) 15 (*f*) 457
[**Ans.** (*a*) 0110 (*b*) 0100 (*c*) 01111011 (*d*) 000111100110 (*e*) 1000(*f*)000100101101]

19. Convert the following Gray codes into binary digits:
(*a*) 10111 (*b*) 110011 (*c*) 1110011 (*d*) 1001101101 (*e*) 111011 (*f*) 101101 (*g*) 111011001 (*h*) 1010101101
[**Ans.** (*a*) 11010 (*b*) 100010 (*c*) 1011101 (*d*) 1110110110 (*e*) 101101
(*f*) 110110(*g*) 101101110 (*h*) 1100110110]

20. Encode the following binary digits into 7-bit even-parity Hamming codes:
(*a*) 1000 (*b*) 0101 (*c*) 1011
[**Ans.** (*a*) 1110000 (*b*) 0100101 (*c*) 0110011]

21. Add the following binary numbers using the binary addition method:
(*a*) 1010 + 1011 (*b*) 1011.0111 + 1101.101 (*c*) 1110.1011 + 1001.1110
[**Ans.** (*a*) 10101 (*b*) 11001.0001 (*c*) 11000.1001]

22. Subtract the following groups of binary numbers:
(*a*) 11110 – 11011 (*b*) 10110.1 – 1100.01 (*c*) 101.101 – 100.1
[**Ans.** (*a*) 0011 (*b*) 1010.01 (*c*) 1.001]

23. Multiply the following binary numbers:
(*a*) 1011 × 111 (*b*) 1101.1 × 1101 (*c*) 11111 × 101
[**Ans.** (*a*) 1001 101 (*b*) 10101111.1 (*c*) 10011011]

24. Perform the following divisions:
(*a*) 111111 ÷ 1001 (*b*)10111 ÷ 100 (*c*) 10110.1101 ÷ 11.1
[**Ans.** (*a*) 111 (*b*) 101.11 (*c*) 111.00011]

25. Convert the following numbers into floating point decimal notation:
(*a*) 81200 (*b*) 434.45 (*c*) 89.46 (*d*) 0.00379
[**Ans.** (*a*) 812×10^{2} (*b*) 43445×10^{-2} (*c*) 8946×10^{-2} (*d*) 379×10^{-5}]

26. Give the 1's complement of the following numbers:
(*a*) 1011011 (*b*) 1101101 (*c*) 1000111101 (*d*) 111001101 (*e*) 11001100 (*f*) 01010111
(*g*) 010011011 (*h*) 101011001
[**Ans.** (*a*) 0100100 (*b*) 0010010 (*c*) 0111000010 (*d*) 000110010
(*e*) 00110011 (*f*) 10101000 (*g*) 101100100 (*h*) 010100110]

27. Give the 2's complement of the following numbers:
(*a*) 1010111 (*b*) 11001101 [**Ans.** (*a*) 0101001 (*b*) 00110011]

28. Subtract the following binary numbers using the 1's complement method:
(*a*) 101 – 10 (*b*) 111 – 11 (*c*) 1010 – 111 (*d*) 1101 – 1001 (*e*) 110110 – 11001 (*f*) 1000010 – 100101
(*g*) 110111 – 11011 [**Ans.** (*a*) Oil (*b*) 100 (*c*) Oil (*d*) 100 (*e*) 11101 (*f*) 11101 (*g*) 11100]

29. Perform the following additions using the 2's complement method:
(*a*) 1101 + 1110 (*b*) + 39 and + 19 (*c*) + 40 and + 26 (*d*) + 63 and + 37
[**Ans.** (*a*) 0001 1011 (*b*) 0011 1010 (*c*) 0100 0010 (*d*) 0110 0100]

30. Add the following numbers using the 2's complement method:
(*a*) + 38 and – 22 (*b*) + 64 and – 29 (*c*) + 49 and – 37

[**Ans.** (*a*) 0001 0000 (*b*) 0010 0011 (*c*) 0000 1100]

31. Add the following numbers using the 2's complement method:
(*a*) – 48 and + 31 (*b*) – 64 and + 46 (*c*) – 99 and + 62

[**Ans.** (*a*) 1110 1111 (*b*) 1110 1110 (*c*) 1101 1010]

32. Add the following numbers using the 2's complement method:
(*a*) – 32 and – 16 (*b*) – 42 and – 31 (c) – 64 and – 13

[**Ans.** (*a*) 1101 0000 (*b*) 1011 0111 (*c*) 1011 0011]

33. Subtract the following numbers using the 2's complement method:
(*a*) + 39 – (+ 16) (*b*) + 49 – (+ 32) (*c*) + 62 – (+ 29)

[**Ans.** (*a*) 0001 0111 (*b*) 0001 0001 (*c*) 0010 0001

34. Give the 9's complement of the following decimal numbers:
(*a*) 49 (*b*) 245 (*c*) 7865

[**Ans.** (*a*) 50 (*b*) 754 (*c*) 2134]

35. Subtract the following decimal numbers using the 9's complement method:
(*a*) 19 – 12 (*b*) 39 – 15 (*c*) 349 – 436

[**Ans:** (*a*) 7 (*b*) 24 (*c*) – 87]

36. Give the 10's complement of the following decimal numbers:
(*a*)12 (*b*) 345 (*c*) 6789

[**Ans.** (*a*) 88 (*b*) 655 (*c*) 3211]

37. Convert the following decimal numbers to BCD:
(*a*) 26,0 (*b*) 37910 (*c*) 201910

[**Ans.** (*a*) 0010 0110 (*b*) 0011 0111 1001 (*c*) 0010 0000 0001 1001]

❑❑❑

2 Boolean Algebra and Minimization Techniques

2.1 INTRODUCTION

Binary logic deals with variables that take two discrete values—1 for TRUE and 0 for FALSE. A simple switching circuit containing active elements such as diode and the transistor can demonstrate the binary logic, which can either be ON (Switch closed) or OFF (Switch open). Electrical signals such as voltage or current exist throughout the digital system in 'either one of the two recognisable values', except during transition.

The Switching functions can be expressed with Boolean equations, Truth tables, Logic diagrams or Karnaugh maps. Truth tables offer an easy means of representing any switching function of n variables. The transformation of equations to truth tables or Karnaugh maps is little difficult. However, the reverse process of finding equations from tables and maps requires Minterms, Maxterms and Simplification methods.

Boolean algebra can be used to simplify the design of logic circuits. However, this method involves lengthy mathematical operations. An alternative method called the Karnaugh map can be used for the simplification of Boolean equations with up to four input variables. The use of Karnaugh map would become difficult if there are more than five input variables. Hence, it is better to employ the Quine-McCluskey method, which is a tabular method of minimization. These minimization techniques reduce the requirement of hardware.

2.2 DEVELOPMENT OF BOOLEAN ALGEBRA

Mathematician George Boole invented a new kind of algebra—the algebra of logic in the year 1854—popularly known as Boolean Algebra or Switching Algebra. He stated that symbols can be used to represent the structure of logical thought. Boolean algebra differs significantly from conventional algebra. This algebra deals with the rules by which the logical operations are carried out. Here, a digital circuit is represented by a set of input and output symbols and the circuit function expressed as a set of Boolean relationships between the symbols.

Do You Know?	In 1938 C.E. Shannon introduced a two-valued Boolean Algebra called Switching Algebra.

Boolean expressions are basically defined by stating that (*i*) a constant is a Boolean expression and (*ii*) a variable is a Boolean expression. For example, if A is a Boolean expression, so is $\overline{A}$. The combination of variables such as $\overline{A}B$ and $\overline{A}B + C$ are also Boolean expressions. However, $A - B$ is not a Boolean expression.

2.3 BOOLEAN LOGIC OPERATIONS

A Boolean function is an algebraic expression formed using binary constants, binary variables and basic logical operation symbols. Basic logical operations include the AND function (logical multiplication), the OR function (logical addition) and the NOT function (logical complementation). A Boolean function can be converted into, a logic diagram composed of the AND, OR and NOT (inverter) gates.

2.3.1 Logical AND Operation

The logical AND operation of two Boolean variables A and B, given as $Y = A \cdot B$, is represented by Table 2.1. The common symbol for this operation is the multiplication sign (·). The Table shows that the result of the AND operation on the variables A and B is logical 0 for all cases, except when both A and B are logical 1. Usually, the dot denoting the AND function is omitted and $A \cdot B$ is written as AB.

Table 2.1 Logical AND operation

Inputs	Output	
A	B	$Y = A \cdot B$
0	0	0
0	1	0
1	0	0
1	1	1

2.3.2 Logical OR Operation

The logical OR operation between two Boolean variables A and B, given as $Y = A + B$, is represented by Table 2.2. This table shows that the result of the OR operation on the variables A and B is logical 1 when A or B (or both) are logical 1. The common symbol used for this logical addition operation is the plus sign (+).

Table 2.2 Logical OR operation

Inputs	Output	
A	B	$Y = A + B$
0	0	0
0	1	1
1	0	1
1	1	1

2.3.3 Logical Complementation (Inversion)

The logical Inverse operation converts the logical 1 to the logical 0 and vice versa. This method is also called the NOT operation. The symbol used for this operation is a bar over the function or the variable. Several notations, such as adding an asterisk, a star, prime, etc. over the variable, are used to indicate the NOT operation. "NOT A" or the complement of A is represented by $\overline{A}$.

2.4 BASIC LAWS OF BOOLEAN ALGEBRA

Logical operations can be expressed and minimized mathematically using the rule: laws, and theorems of Boolean algebra. It is a convenient and systematic method of expressing and analyzing the operation of digital circuits and systems. Boolean al gebra uses binary arithmetic variables which have two distinct symbols 0 and 1. These are called levels or states of logic. For example, a binary 1 represents a Hig level and a binary 0 represents a Low level.

2.4.1 Boolean Addition

Addition by the Boolean method involves variables having values of either a binary 1 or a 0. The basic rules of Boolean addition are given below:

$$0 + 0 = 0$$
$$0 + 1 = 1$$
$$1 + 0 = 1$$
$$1 + 1 = 1$$

Boolean addition is same as the logical OR operation.

2.4.2 Boolean Multiplication

The basic rules of the Boolean multiplication method are as follows:

$$0.0 = 0$$
$$0.1 = 0$$
$$1.0 = 0$$
$$1.1 = 1$$

Boolean multiplication is same as the logical AND operation.

2.4.3 Properties of Boolean Algebra

Boolean Algebra is a mathematical system consisting of a set of two or more distinct elements, two binaiy operators denoted by the symbols (+) and (.) and one unary operator denoted by the symbol either bar (–) or prime ('). They satisfy the commutative, associative, distributive, absorption, consensus and idempotency properties of the Boolean Algebra.

Commutative property: Boolean addition is commutative, given by

$$A + B = B + A \qquad ...(1a)$$

According to this property, the order of the OR operation conducted on the variables makes no difference. Boolean algebra is also commutative over multiplication, given by

$$A \cdot B = B \cdot A \qquad (1b)$$

This means that the order of the AND operation conducted on the variables makes no difference.

Associative property: The associative property of addition is given by

$$A + (B + C) = (A + B) + C \qquad ...(2a)$$

The OR operation of several variables results in the same, regardless of the grouping of the variables. The associative law of multiplication is given by

$$A \cdot (B \cdot C) = (A \cdot B) \cdot C \qquad ...(2b)$$

According to this law, it makes no difference in what order the variables are grouped during the AND operation of several variables.

Distributive property: (*i*) The Boolean addition is distributive over Boolean multiplication, given by

$$A + BC = (A + B)(A + C) \quad \text{...(3a)}$$

This property states that the AND operation (multiplication) of several variables and then the OR operation (addition) of the result with a single variable is equivalent to the OR operation of the single variable with each of the several variables and then the AND operation of the sums.

Proof:

$$\begin{aligned} A + BC &= A \cdot 1 + BC && (\because\ A \cdot 1 = A) \\ &= A(1 + B) + BC && (\because\ 1 + B = 1) \\ &= A \cdot 1 + AB + BC && (\because\ A(B + C) = AB + BC) \\ &= A \cdot (1 + C) + AB + BC && (\because\ 1 + C = 1) \\ &= A \cdot 1 + AC + AB + BC && \\ &= A \cdot A + AC + AB + BC && (\because\ A \cdot A = A) \\ &= A(A + C) + B(A + C) && \\ &= (A + B)(A + C) && \end{aligned}$$

(*ii*) Boolean multiplication is also distributive over Boolean addition given by

$$A \cdot (B + C) = A \times B + A \cdot C \quad \text{...(3b)}$$

According to this property, the OR operation of several variables and then the AND operation of the result with a single variable is equivalent to the AND operation of the single variable with each of the several variables and then the OR operation of the products.

Absorption laws:

(*i*) $\boxed{A + AB = A}$ ...(4a)

Proof:

$$\begin{aligned} A + AB &= A \cdot 1 + AB \\ &= A(1 + B) \\ &= A \cdot 1 = A \end{aligned}$$

(*ii*) $\boxed{A \cdot (A + B) = A}$ ...(4b)

$$\begin{aligned} A(A + B) &= A \cdot A + A \cdot B \\ &= A + AB \\ &= A(1 + B) \\ &= A \cdot 1 \\ &= A \end{aligned}$$

(*iii*) $\boxed{A + \bar{A}B = A + B}$ ...(5a)

$$\begin{aligned} A + \bar{A}B &= (A + \bar{A})(A + B) && [\because\ A + BC = (A + B)(A + C)] \\ &= 1 \cdot (A + B) && (\because\ A + \bar{A} = 1) \\ &= A + B && \end{aligned}$$

(*iv*) $\boxed{A \cdot (\overline{A} + B) = AB}$ ...(5*b*)

Proof:

$$A \cdot (\overline{A} + B) = A \cdot \overline{A} + AB$$

$$= AB \quad (\because \; A\overline{A} = 0)$$

Consensus laws

(*i*) $\boxed{AB + \overline{A}C + BC = AB + \overline{A}C}$ ...(6*a*)

Proof:

$$AB + \overline{A}C + BC = AB + \overline{A}C + BC \cdot 1$$

$$= AB + \overline{A}C + BC(A + \overline{A}) \qquad (\because \; A + \overline{A} = 1)$$

$$= AB + \overline{A}C + ABC + \overline{A}BC$$

$$= AB(1 + C) + \overline{A}C(1 + B) \qquad (\because \; 1 + B = 1 = 1 + C)$$

$$= AB + \overline{A}C$$

(*ii*) $\boxed{(A + B)(\overline{A} + C)(B + C = (A + B)(\overline{A} + C)}$ ...(6*b*)

Proof:

$$(A + B)(\overline{A} + C)(B + C) = (A + B)(\overline{A} + C)(B + C + 0)$$

$$= (A + B)(\overline{A} + C)(B + C + A\overline{A})$$

$$[\because \; A + BC = (A + B)(A + C)]$$

$$= (A + B)(A + B + C)(\overline{A} + C)(\overline{A} + C + B)$$

$$= (A + B)(\overline{A} + C)$$

$$[\because \; A(A + B) = A]$$

The other basic laws (theorems) of Boolean algebra are given in Table 2.3. These theorems can be proved easily by adopting the truth table method or by using algebraic manipulation.

Table 2.3. Other laws of boolean algebra

No.	Boolean laws		
7	(*a*) $A + 0 = A$	(*b*) $A \cdot 1 = A$	—
8	(*a*) $A + 1 = 1$	(*b*) $A \cdot 0 = 0$	—
9	(*a*) $A + A = A$	(*b*) $A \cdot A = A$	Idempotency
10	(*a*) $A + \overline{A} = A$	(*b*) $A \cdot \overline{A} = 0$	Full set, null set
11		$\overline{\overline{A}} = A$	Double inversion or involution

2.4.4 Principle of Duality

From the above properties and laws of Boolean algebra, it is evident that they are grouped in pairs as (*a*) and (*b*). One expression can be obtained from the other in each pair by replacing every 0 with 1, every 1 with 0, every (+) with (.), and every (.) with (+). Any pair of expression satisfying this property is called dual expression. This characteristic of Boolean algebra is called the principle of duality.

Note: *It is possible to verify this theorem by constructing truth tables for both the sides of equations.*

An important property of Boolean Algebra is known as the duality principle. The duality principle has several applications in digital systems.

2.5 DEMORGAN'S THEOREMS

Two theorems that are an important part of Boolean algebra were proposed by DeMorgan. The first theorem states that the complement of a product is equal to the sum of the complements. That is, if the variables are *A* and *B*, then

$$\overline{AB} = \overline{A} + \overline{B}$$

The second theorem states that, the complement of a sum is equal to the product of the complements. In equation form, this can be written as

$$\overline{AB} = \overline{A} \cdot \overline{B}$$

The complement of a Boolean logic function or a logic expression may be expanded or simplified by following the steps of DeMorgan's theorem.

(*i*) Replace the symbol (+) with symbol (.), the symbol (.) with symbol (+) given in the expression,

(*ii*) Complement each of the terms or variables in the given expression.

Do You Know?	The postulates of a mathematical system form the basic assumptions from which it is possible to deduce the rules, theorems and properties of the system.

DeMorgan's theorems can be proved for any number of variables; proof of these two theorems for 2-input variables can be found in Table 2.4.

Table 2.4. Proof for DeMorgan's theorem by perfect induction method

1 A	2 B	3 $\overline{A}$	4 $\overline{B}$	5 $A+B$	6 $A \cdot B$	7 $\overline{A+B}$	8 $\overline{A} \cdot \overline{B}$	9 $\overline{A \cdot B}$	10 $\overline{A}+\overline{B}$
0	0	1	1	0	0	1	1	1	1
0	1	1	0	1	0	0	0	1	1
1	0	0	1	1	0	0	0	1	t
1	1	0	0	1	1	0	0	0	0

A study of Table 2.4 makes clear that columns 7 and 8 are equal. Therefore,

$$\overline{A+B} = \overline{A} \cdot \overline{B}$$

Similarly, columns 9 and 10 are equal. Therefore,

$$\overline{A \cdot B} = \overline{A} + \overline{B}$$

Also, De-Morgan's theorem can be proved by algebraic method as follows:

According to the first theorem, ($\bar{A}+\bar{B}$) is the complement of *AB*. From the Table 2.3, the Boolean Laws 10 (*a*) and 10 (*b*) are given as,

$$A+\bar{A} = 1 \quad \text{and} \quad A\bar{A} = 0$$

Substituting *AB* for *A* and ($\bar{A}+\bar{B}$) for $\bar{A}$ in the above expressions,

$$AB+\bar{A}+\bar{B} = 1 \quad \text{and} \quad AB(A+B) = 0$$

$$A+B+\bar{B} = 1 \quad \text{and} \quad A\,\bar{A}\bar{B}+A\,\bar{A}\,\bar{B} = 0$$

$$A+1 = 1 \quad \text{and} \quad 0+0 = 0$$

$$1 = 1 \quad \text{and} \quad 0 = 0$$

Thus De-Morgan's first theorem is proved algebraically.

Similarly, according to De-Morgan's second theorem, ($\bar{A}\cdot\bar{B}$) is the complement of ($A + B$). Substituting ($A + B$) for *A* and ($\bar{A}\cdot\bar{B}$) for $\bar{A}$ in 10 (*a*) and 10 (*b*).

$$A+B+\bar{A}\cdot\bar{B} = 1 \quad \text{and} \quad (A+B)\cdot(A\,.\,B) = 0$$

$$A+B+\bar{B} = 1 \quad \text{and} \quad A\bar{A}\,\bar{B}+B\bar{A}\,\bar{B} = 0$$

$$A+1 = 1 \quad \text{and} \quad 0+0 = 0$$

$$1 = 1 \quad \text{and} \quad 0 = 0$$

Thus De-Morgan's second theorem is proved algebraically.

Minimization (Simplification) of Boolean expressions using algebraic method: The switching or Boolean expressions can be simplified by applying properties, laws and theorems of Boolean Algebra. The simplification of different Boolean expressions are demonstrated in the following examples.

Example 2.1. *Prove that* $AB + BC + \bar{B}C = AB + C$.

Solution:

$$\begin{aligned} AB+BC+\bar{B}C &= AB+C(B+\bar{B}) \\ &= AB+C\,.\,1 \\ &= AB+C \end{aligned}$$

Example 2.2. *Simplify the expression* $\bar{A}\cdot B+A\cdot B+\bar{A}\cdot\bar{B}$.

Solution:

$$\begin{aligned} \bar{A}\cdot B+A\cdot B+\bar{A}\cdot\bar{B} &= (\bar{A}+A)\cdot B+\bar{A}\cdot\bar{B} \\ &= 1\cdot B+\bar{A}\cdot\bar{B} \\ &= B+\bar{A}\cdot\bar{B} \\ &= B+\bar{A} \quad (\because\ A+\bar{A}\times B = A+B) \end{aligned}$$

Example 2.3. *Simplify the given expression* $A+A\cdot\bar{B}+\bar{A}\cdot B$.

Solution:

$$\begin{aligned} A+A\cdot\bar{B}+\bar{A}\cdot B &= A(1+\bar{B})+\bar{A}\cdot B \\ &= A\times 1+\bar{A}\cdot B \\ &= A+B \end{aligned}$$

Example 2.4. *Complement the expression* $\overline{A}B+C\overline{D}$.

Solution:

$$\begin{aligned}\overline{\overline{A}B+C\overline{D}} &= \overline{(\overline{A}B)}\cdot\overline{(C\overline{D})}\\ &= (\overline{\overline{A}}+\overline{B})\cdot(\overline{C}+\overline{\overline{D}})\\ &= (A+\overline{B})\cdot(\overline{C}+D)\end{aligned}$$

Example 2.5. *Simplify the expression* $AB+\overline{AC}+A\overline{B}C(AB+C)$.

Solution:

$$\begin{aligned}AB+\overline{AC}+A\overline{B}C(AB+C) &= AB+\overline{AC}+A\overline{B}C\cdot AB+A\overline{B}C\cdot C\\ &= AB+\overline{AC}+A\overline{B}C \quad [\because C\cdot C=C \text{ and } \overline{B}\cdot B=0]\\ &= AB+\overline{A}+\overline{C}+A\overline{B}C\\ &= AB+\overline{A}+\overline{C}+\overline{B}C \quad [\because A+\overline{A}B=A+B]\\ &= \overline{A}+AB+\overline{C}+C\overline{B}\\ &= \overline{A}+B+\overline{C}+\overline{B}\\ &= \overline{A}+\overline{C}+1 \quad [\because B+\overline{B}=1]\\ &= \overline{A}+1\\ &= 1\end{aligned}$$

Example 2.6. *Simplify the expression* $Y=(\overline{A}+B)(A+B)$.

Solution:

$$\begin{aligned}Y &= (\overline{A}+B)(A+B)\\ &= B+A\overline{A} \quad [\because (A+B)(A+C)=A+BC]\\ &= B+0\\ &= B\end{aligned}$$

Example 2.7. *Simplify the expression* $\overline{\overline{A\overline{B}+ABC}+A(B+A\overline{B})}$.

Solution:

$$\begin{aligned}\overline{\overline{A\overline{B}+ABC}+A(B+A\overline{B})} &= \overline{\overline{A\overline{B}+ABC}+A(B+A)}\\ &= \overline{\overline{A(\overline{B}+C)}+AB+A\cdot A)}\\ &= \overline{\overline{A\overline{B}+AC}+AB+A}\\ &= \overline{\overline{A\overline{B}+AC}+A(B+1)}\\ &= \overline{\overline{A\overline{B}+AC}+A\cdot 1}\\ &= \overline{\overline{A\overline{B}+AC}+A}\\ &= \overline{\overline{(A\overline{B})}\cdot\overline{(AC)}+A}\end{aligned}$$

$$= \overline{(\overline{A}+B)\cdot(\overline{A}+\overline{C})+A}$$

$$= \overline{(\overline{A}+B\overline{C})+A} \qquad [\because \ (A+B)(A+C) = A + BC]$$

$$= \overline{\overline{A}+B\overline{C}+A}$$

$$= \overline{1+B\overline{C}}$$

$$= \overline{1} \qquad [\because \ 1 + X = 1]$$

$$= 0$$

Example 2.8. *Simplify* $Y = ABC + A\overline{B}C + AB\overline{C}$ *to* $Y = A(B + C)$.

Solution:

$$Y = ABC + A\overline{B}C + AB\overline{C}$$

$$= AC(B+\overline{B}) + AB\overline{C}$$

$$= AC\cdot 1 + AB\overline{C}$$

$$= A(C + B\overline{C})$$

$$= A(B + C)$$

Example 2.9. *Simplify the given Boolean expression* $\overline{A}\overline{B}\overline{C} + \overline{A}B\overline{C} + A\overline{B}\overline{C} + AB\overline{C}$.

Solution:

$$Y = \overline{A}\overline{B}\overline{C} + \overline{A}B\overline{C} + A\overline{B}\overline{C} + AB\overline{C}$$

$$= \overline{A}\overline{C}(B+\overline{B}) + A\overline{C}(B+\overline{B})$$

$$= \overline{A}\,\overline{C} + A\overline{C}$$

$$= \overline{C}(\overline{A}+A)$$

$$= \overline{C}\cdot 1$$

$$= \overline{C}$$

Example 2.10. *Simplify the expression* $Y = \overline{(AB+\overline{C})\,(\overline{A+B}+C)}$.

Solution:

$$Y = \overline{(AB+\overline{C})\,(\overline{A+B}+C)}$$

$$= \overline{(AB+\overline{C})\,(\overline{A}\cdot\overline{B}+C)}$$

$$= \overline{AB\cdot\overline{A}\overline{B} + ABC + \overline{A}\overline{B}\overline{C} + C\overline{C}}$$

$$= \overline{0 + ABC + \overline{A}\overline{B}\overline{C} + 0}$$

$$= \overline{ABC + \overline{A}\overline{B}\overline{C}}$$

$$= \overline{ABC} \cdot \overline{\bar{A}\bar{B}\bar{C}}$$

$$= (\bar{A} + \bar{B} + \bar{C}) \cdot (\bar{\bar{A}} + \bar{\bar{B}} + \bar{\bar{C}})$$

$$= (\bar{A} + \bar{B} + \bar{C}) \cdot (A + B + C)$$

Example 2.11. *Simplify the expression* $Y = \bar{A}C[\overline{\bar{A}BD}] + AB\bar{C}\bar{D} + A\bar{B}C.$

Solution:

$$Y = \bar{A}C[\overline{\bar{A}BD}] + AB\bar{C}\bar{D} + A\bar{B}C$$

$$= \bar{A}C[A + \bar{B} + \bar{D}] + \bar{A}B\bar{C}\bar{D} + A\bar{B}C$$

$$= \bar{A}CA + \bar{A}C\bar{B} + \bar{A}C\bar{D} + \bar{A}B\bar{C}\bar{D} + A\bar{B}C$$

$$= \bar{A}\bar{B}C + \bar{A}C\bar{D} + \bar{A}BC\bar{D} + A\bar{B}C \qquad [\because \quad A \times \bar{A} = 0]$$

$$= \bar{B}C(\bar{A} + A) + \bar{A}\bar{D}(C + B\bar{C})$$

$$= \bar{B}C + \bar{A}\bar{D}(B + C) \qquad [\because \quad A + \bar{A}B = A + B]$$

Example 2.12. *Prove the following Boolean expression* $(A + B)\,(\bar{A}\bar{C} + AC)\,(\overline{\bar{B} + AC}) = \bar{A}B.$

Solution:

$$(A + B)\,(\bar{A}\bar{C} + AC)\,(\overline{\bar{B} + AC}) = (A + B)\,(\bar{A}\bar{C} + C)\,(\bar{\bar{B}} \cdot \overline{AC})$$

$$= (A + B)\,(\bar{A}\bar{C} + C)\,(B \cdot \overline{AC})$$

$$= [A\bar{A}\bar{C} + AC + \bar{A}\bar{C}B + BC]\,[B(\bar{A} + \bar{C})]$$

$$= (AC + \bar{A}\bar{C}B + BC) \cdot (B\bar{A} + B\bar{C})$$

$$= AC \cdot B\bar{A} + AC \cdot B\bar{C} + \bar{A}\bar{C}B \cdot B\bar{A}$$
$$+ \bar{A}\bar{C}B \cdot B\bar{C} + BC \cdot B\bar{A} + BC \cdot B\bar{C}$$

$$= 0 + 0 + \bar{A}B\bar{C} + \bar{A}\bar{C}B + BC\bar{A} + 0$$

$$= \bar{A}B(\bar{C} + \bar{C} + C)$$

$$= \bar{A}B$$

Example 2.13. *Prove that* $\bar{A}\bar{B}\bar{C} + \bar{A}\bar{B}C + \bar{A}B\bar{C} + \bar{A}BC + A\bar{B}\bar{C} = \bar{A} + \overline{B + C}.$

Solution:

$$\bar{A}\bar{B}\bar{C} + \bar{A}\bar{B}C + \bar{A}B\bar{C} + \bar{A}BC = \bar{A}\bar{B}(\bar{C} + C) + \bar{A}B(\bar{C} + C) + A\bar{B}\bar{C}$$

$$= \bar{A}\bar{B} + \bar{A}B + A\bar{B}\bar{C}$$

$$= \bar{A}(\bar{B} + B) + A\bar{B}\bar{C}$$

$$= \bar{A} + A\bar{B}\bar{C}$$

$$= \bar{A} + \bar{B}\bar{C} \qquad [\because \quad A + \bar{A}B = A + B]$$

$$= A + \overline{B + C}$$

Example 2.14. *Find the complement of the expression* $Y = ABC + AB\bar{C} + \bar{A}\bar{B}C + \bar{A}BC.$

Solution:

$$\bar{Y} = \overline{AB\bar{C} + \bar{A}\bar{B}C + \bar{A}BC.}$$

$$= (\overline{ABC})(\overline{AB\bar{C}})(\overline{\bar{A}\bar{B}C})(\overline{\bar{A}BC})$$

$$= (\bar{A}+\bar{B}+\bar{C})(\bar{A}+\bar{B}+C)(A+B+\bar{C})(A+\bar{B}+\bar{C})$$

$$= (\bar{A}+\bar{B}+C\bar{C})(A+\bar{C}+B\bar{B}) \quad [\because \; (A+B)(A+C) = A+BC]$$

$$= (\bar{A}+\bar{B})(A+\bar{C})$$

Example 2.15. *Prove that* $BCD + A\bar{C}\bar{D} + ABD = BCD + A\bar{C}\bar{D} + AB\bar{C}.$

Solution:

$$BCD + A\bar{C}\bar{D} + ABD = BCD + A\bar{C}\bar{D} + (ABD).1$$

$$= BCD + A\bar{C}\bar{D} + (ABD)(C+\bar{C})$$

$$= BCD + A\bar{C}\bar{D} + ABCD + AB\bar{C}D$$

$$= BCD(1+A) + A\bar{C}(\bar{D}+DB)$$

$$= BCD + \overline{AC}(\bar{D}+B)$$

$$= BCD + A\bar{C}\bar{D} + AB\bar{C}$$

Hence, L.H.S. = R.H.S.

Example 2.16. *Simplify the given expression* $Y = \bar{A}B + ABD + A\bar{B}\,C\bar{D} + BC.$

Solution:

$$\bar{A}B + ABD + A\bar{B}C\bar{D} + BC = B(\bar{A}+AD) + C(B+\bar{B}A\bar{D})$$

$$= B(\bar{A}+D) + C(B+A\bar{D}) \qquad [\because \; A + \bar{A}B = A + B]$$

$$= \bar{A}B + BD + BC + AC\bar{D}$$

$$= \bar{A}B + BD + BC(A+A) + AC\bar{D} \qquad [\because \; A + \bar{A} = 1]$$

$$= \bar{A}B + BD + ABC + \bar{A}BC + AC\bar{D}$$

$$= \bar{A}B(1+C) + BD + ABC + AC\bar{D}$$

$$= \bar{A}B + BD + AC\bar{D} \qquad [\because \; AB + BC + \bar{A}C = AB + \bar{A}C]$$

[Here $A = D$; $B = B$; $C = AC$]

Example 2.17. *If* $\overline{A}B + C\overline{D} = 0$, *then prove that*

$$AB + \overline{C}(\overline{A} + \overline{D}) = AB + BD + \overline{B}\overline{D} + \overline{A}\overline{C}D.$$

Solution:

$$\begin{aligned}
\text{L.H.S.} &= AB + \overline{C}(\overline{A} + \overline{D}) + 0 \\
&= AB + \overline{C}(\overline{A} + \overline{D}) + \overline{A}B + C\overline{D} && (\text{given that } \overline{A}B + C\overline{D} = 0) \\
&= AB + \overline{A}\overline{C} + \overline{C}\overline{D} + \overline{A}B + C\overline{D} \\
&= B(A + \overline{A}) + \overline{D}(C + \overline{C}) + \overline{A}\overline{C} \\
&= B + \overline{D} + \overline{A}\overline{C}
\end{aligned}$$

$$\begin{aligned}
\text{R.H.S.} &= AB + BD + \overline{B}\overline{D} + \overline{A}\overline{C}D + 0 \\
&= AB + BD + \overline{B}\overline{D} + \overline{A}\overline{C}D + \overline{A}B + C\overline{D} && (\text{given that } A\overline{B} + C\overline{D} = 0) \\
&= B(A + \overline{A}) + BD + \overline{B}\overline{D} + \overline{A}\overline{C}D + C\overline{D} \\
&= B(1 + D) + \overline{B}\overline{D} + \overline{A}\overline{C}D + C\overline{D} \\
&= B + \overline{B}\overline{D} + \overline{A}\overline{C}D + C\overline{D} \\
&= B + \overline{D} + \overline{A}\overline{C}D + C\overline{D} && (\because\ A + \overline{A}B = A + B) \\
&= B + \overline{D}(1 + C) + \overline{A}\overline{C}D \\
&= B + \overline{D} + D\overline{A}\overline{C} \\
&= B + \overline{D} + \overline{A}\overline{C} && (\because\ A + \overline{A}B = A + B)
\end{aligned}$$

Hence, LHS. = R.H.S.

Example 2.18. *Simplify* $Y = A\overline{B} + (\overline{A} + B)C$

Solution:

$$\begin{aligned}
Y &= A\overline{B} + (\overline{A} + B)C \\
&= AB + (\overline{A\overline{\overline{B}}})C && (\because\ \overline{A} + B = \overline{A\overline{\overline{B}}}) \\
&= A\overline{B} + C && (\because\ X + \overline{X}Y = X + Y)
\end{aligned}$$

Example 2.19. *Simplify* $Y = A + \overline{A}B + \overline{A}\overline{B}C + \overline{A}\overline{B}\overline{C}D$

Solution:

$$\begin{aligned}
\overline{A}B + \overline{A}\overline{B}C + \overline{A}\overline{B}\overline{C}D &= A + B + \overline{A}\overline{B}C + \overline{A}\overline{B}\overline{C}D && [\because\ A + \overline{A}B = A + B] \\
&= A + B + \overline{B}C + \overline{A}\overline{B}\overline{C}D
\end{aligned}$$

$$= A + B + C + \overline{A}\overline{B}\overline{C}D$$

$$= A + B + C + \overline{B}\overline{C}D$$

$$= A + B + C + \overline{C}D$$

$$= A + B + C + D$$

Example 2.20. *If* $A\overline{B} + \overline{A}B = C$, *show that* $A\overline{C} + AC = B$

Solution:

$$A\overline{C} + AC = A(\overline{A\overline{B} + \overline{A}B}) + \overline{A}(A\overline{B} + \overline{A}B) \quad \text{(given that } C = A\overline{B} + \overline{A}B\text{)}$$

$$= A(\overline{A} + B)(A + \overline{B}) + \overline{A}A\overline{B} + \overline{AA}B$$

$$= (A\overline{A} + AB)(A + \overline{B}) + \overline{A}B$$

$$= AB + AB\overline{B} + \overline{A}B$$

$$= AB + \overline{A}B$$

$$= B(A + \overline{A})$$

$$= B$$

Example 2.21. *Using Boolean algebra, verify*

(i) $(A + B)(B + C)(C + A) = AB + BC + CA$

(ii) $(A + B)(\overline{A} + C) = AC + \overline{A}B + BC$

Solution:

(i) $(A + B)(B + C)(C + A) = AB + BC + CA$

$$\text{LHS} = (A + B)(B + C)(C + A) = (AB + AC + BB + BC)(C + A)$$

$$= (AB + AC + B + BC)(C + A)$$

$$= [B(A + 1) + AC + BC](C + A)$$

$$= [B + AC + BC](C + A)$$

$$= BC + ACC + BCC + AB + AAC + ABC$$

$$= BC + AC + BC + AB + AC + ABC$$

$$= BC + AB(1 + C) + AC$$

$$= BC + AB + AC$$

Hence Proved.

(ii) $(A + B)(\overline{A} + C) = AC + \overline{A}B + BC$

$$\text{LHS} = (A + B)(A + C) = A\overline{A} + AC + \overline{A}B + BC$$

$$= 0 + AC + \overline{A}B + BC$$

Hence Proved.

2.6 SUM OF PRODUCTS AND PRODUCT OF SUMS

Logical functions are generally expressed in terms of logical variables. Values taken on by the logical functions and logical variables are in the binary form. An arbitrary logic function can be expressed in the following forms:

(*i*) Sum of Products (SOP)

(*ii*) Product of Sums (POS)

Product term: The AND function is referred to as a *product*. In Boolean algebra, the word "product" loses its original meaning but serves to indicate an AND function. The logical product of several variables on which a function depends is considered to be a product term. The variables in a product term can appear either in complemented or uncomplemented form. $AB\overline{C}$, for example, is a product term.

Sum term: An OR function (+ sign) is generally used to refer a *sum*. The logical sum of several variables on which a function depends is considered to be a *sum term.* Variables in a sum term can appear either in complemented or uncomplemented form. $A + \overline{B} + C$, for example, is a sum term.

Sum of Products (SOP): The logical sum of two or more logical product terms, is called a Sum cf Products expression. It is basically an OR operation of AND operated variables such as:

(*i*) $Y = AB + BC + AC$

(*ii*) $Y = AB + \overline{A}C + BC$

Product of Sums (POS): A product of sums expression is a logical product of two or more logical sum terms. It is basically an AND operation of OR operated variables such as:

(*i*) $Y = (A + B)(B + C)(C + \overline{A})$

(*ii*) $Y = (A + B + C)(A + \overline{C})$

2.6.1 Minterm

A product term containing all the *K* variables of the function in either complemented or uncomplemented form is called a *Minterm.* A 2-variable function has four possible combinations, viz. $\overline{A}\overline{B}$, $\overline{A}B$, $A\overline{B}$, and AB. These product terms are called minterms or standard products or fundamental products. For a 3-binary input variable function, there are 8 minterms as shown in Table 2.5. Each minterm can be obtained by the AND operation of all the variables of the function. In the minterm, a variable appears either in uncomplemented form, if it possesses a value of 1 in the corresponding combination, or in complemented form, if it contains the value 0. The minterms of a 3-variable function can be represented by m_0, m_1, m_2, m_3, m_4, m_5, m_6 and m_7; the suffix indicates the decimal code corresponding to the minterm combination.

Table 2.5. The minterm table

A	B	C	Minterm
0	0	0	$\overline{A}\overline{B}\overline{C}$
0	0	1	$\overline{A}\overline{B}C$
0	1	0	$\overline{A}B\overline{C}$
0	1	1	$\overline{A}BC$
1	0	0	$A\overline{B}\overline{C}$
1	0	1	$A\overline{B}C$
1	1	0	$AB\overline{C}$
1	1	1	ABC

The main property of a minterm is that it possesses the value 1 for only one combination of K input variables; i.e., for a K variable function of the 2^K minterms, only one minterm will have the value 1, while the remaining 2^K-1 minterms will possess the value 0 for an arbitrary input combination. For example, as shown in Table 2.5, for input combination 010, i.e., for $A = 0$, $B = 1$ and $C = 0$, only the minterm $\overline{A}B\overline{C}$ will have the value 1, while the remaining seven minterms will have the value 0.

Canonical sum of product expression: It is defined as the logical sum of all the minterms derived from the rows of a truth table, for which the value of the function is 1. It is also called a minterm canonical form. The canonical sum of product expression can be given in a compact form by listing the decimal codes in correspondence with the minterm containing a function value of 1. For example, if the canonical sum of product form of a 3-variable logic function Y has three minterms $\overline{A}\overline{B}\overline{C}$, $A\overline{B}C$ and $AB\overline{C}$, this can be expressed as the sum of the decimal codes corresponding to these minterms as stated below:

$$\begin{aligned} Y &= \Sigma_m (0, 5, 6) \\ &= m_0 + m_5 + m_6 \\ &= \overline{A}\overline{B}\overline{C} + A\overline{B}C + AB\overline{C} \end{aligned}$$

where Σ_m (0,5,6) represents the summation of minterms corresponding to the decimal codes 0, 5 and 6.

Using the following procedure, the canonical sum of product form of a logic function can be obtained:

1. Examine each term in the given logic function. Retain it if it is a minterm; continue to examine the next term in the same manner.
2. Check for variables that are missing in each product which is not a minterm. Multiply the product by $(X + \overline{X})$, for each variable X that is missing.
3. Multiply all the products and omit the redundant terms.

The above procedures can be explained with the following examples.

Example 2.22. *Obtain the canonical sum of product form of the function*

$$Y(A, B) = A + B$$

Solution: The given function containing the two variables *A* and *B* has the variable *B* missing in the first term and the variable *A* missing in the second. Therefore, the first term has to be multiplied by $(B + \overline{B})$, the second term by $(A + \overline{A})$ as given below:

$$\begin{aligned} A + B &= A \cdot 1 + B - 1 \\ &= A \cdot (B + \overline{B}) + B \cdot (A + \overline{A}) \\ &= AB + A\overline{B} + BA + B\overline{A} \\ &= AB + A\overline{B} + \overline{A}B \quad (\because \quad AB + AB = AB) \\ Y(A, B) &= A + B = AB + \ A\overline{B} + \overline{A}B \end{aligned}$$

Example 2.23. *Obtain the canonical sum of product form of the function*

$$Y(A, B, C) = A + BC$$

Solution: Here, neither the first term nor the second term is a minterm. The variables *B* and *C* in the first term and the variable *A* in the second term are missing. Therefore, the first term has to be multiplied by $B + \overline{B}$ and $C + \overline{C}$, the second term by $(A + \overline{A})$ as shown below:

$$\begin{aligned} A + BC &= A(B + \overline{B})(C + \overline{C}) + BC(A + \overline{A}) \\ &= (AB + \overline{B})(C + \overline{C}) + BCA + BC\overline{A} \\ &= ABC + AB\overline{C} + A\overline{B}C + A\overline{B}\overline{C} + BCA + BC\overline{A} \\ &= ABC + AB\overline{C} + A\overline{B}C + A\overline{B}\overline{C} + \overline{A}BC \ (\because \quad ABC + ABC = ABC) \end{aligned}$$

Example 2.24. *Obtain the canonical sum of product form of the function*

$$Y = AB + ACD$$

Solution:

$$\begin{aligned} Y &= AB + ACD \\ &= AB(C + \overline{C})(D + \overline{D}) + ACD(B + \overline{B}) \\ &= (ABC + AB\overline{C}) + (D + \overline{D}) + ABCD + A\overline{B}CD \\ &= ABCD + ABC\overline{D} + AB\overline{C}D + AB\overline{C}\overline{D} + ABCD + A\overline{B}CD \\ &= ABCD + ABC\overline{D} + AB\overline{C}D + AB\overline{C}\overline{D} + A\overline{B}CD \ (\because \ ABCD + ABCD = ABCD) \end{aligned}$$

2.6.2 Maxterm

A sum term containing all the *K* variables of the function in either complemented or uncomplemented form is called a Maxterm. A 2-variable function has four possible combinations, viz. $A + B$, $A + \overline{B}$, $\overline{A} + B$ and $A + B$. These sum terms are called maxterms. So also, a 3-binary input variable function has 8 maxterms as shown in Table 2.6. Each maxterm can be obtained by the OR operation of all the variables of the function. In a maxterm, a variable appears either in uncomplemented form if it possesses the value 0 in the corresponding combination or in complemented form if it contains the value 1. The maxterms of a 3-variable function can be represented by M_0, M_1, M_2, M_3, M_4, M_5, M_6 and M_7; the suffix indicates the decimal code corresponding to the maxterm combination.

Table 2.6. The maxterra table

A	B	C	Maxterm
0	0	0	$A+B+C$
0	0	1	$A+B+\overline{C}$
0	1	0	$A+\overline{B}+C$
0	1	1	$A+\overline{B}+\overline{C}$
1	0	0	$\overline{A}+B+C$
1	0	1	$\overline{A}+B+\overline{C}$
1	1	0	$\overline{A}+\overline{B}+C$
1	1	1	$\overline{A}+\overline{B}+\overline{C}$

The most important property of a maxterm is that it possesses the value 0 for only one combination of K input variables; i.e., for a K variable function of the 2^K maxterms, only one maxterm will have the value 0, while all the remaining 2^{K-1} maxterms will have the value 1 for an arbitrary input combination. For example, for input combination 101, i.e., for $A = 1$, $B = 0$ and $C = 1$, only the maxterm $(\overline{A}+\overline{B}+\overline{C})$ will have the value 0, while the remaining seven maxterms will have the value 1. This can be studied in Table 2.6.

From Tables 2.5 and 2.6, it is found that each maxterm is the complement of the corresponding minterm. For example, if the maxterm is $(A + B + C)$, then its complement (i. e., $\overline{A+B+C})$ $\overline{A}\overline{B}\overline{C}$ is its corresponding minterm.

Canonical product of sum expression: This is defined as the logical product of all the maxterms derived from the rows of truth table, for which the value of function is 0. It is also known as the maxterm canonical form. The canonical product of sum expression can be given in a compact form by listing the decimal codes corresponding to the maxterms containing a function value of 0. For example, if the canonical product of sum form of a 3-variable logic function Y has four maxterms $(A + B + C)$, $(A + \overline{B} + C)$, $(\overline{A} + B + C)$ and $(\overline{A}+\overline{B}+\overline{C})$, then it can be expressed as the product of decimal codes as given below:

$$
\begin{aligned}
Y &= \Pi\,(0, 2, 4, 7) \\
&= M_0 \cdot M_2 \cdot M_4 \cdot M_L \\
&= (A + B + C)(A+\overline{B}+C)(\overline{A}+B+C)(\overline{A}+\overline{B}+\overline{C})
\end{aligned}
$$

The following procedure can be used to obtain the canonical product of the sum form of a logic function:

1. Examine each term in the given logic function. Retain it if it is a maxterm; continue to examine the next term in the same manner.
2. Check for variables that are missing in each sum, which is not a maxterm. Add $(X\bar{X})$ to the sum term, for each variable X that is missing.
3. Expand the expression using the distributive property and eliminate the redundant terms.

The above procedures can be explained with the following examples.

Example 2.25. *Obtain the canonical product of the sum expression of*

$$Y(ABC) = (A + \bar{B})(B + C)(A + \bar{C}).$$

Solution: In the given expression, the variable C is missing in the first term, the variable A in the second term and the variable B in the third term. Therefore, $C\bar{C}, A\bar{A}$ and $B\bar{B}$ have to be added with the first, second and third terms respectively as shown below:

$$Y(ABC) = (A + \bar{B})(B + C)(A + \bar{C})$$
$$= (A + \bar{B} + 0)(B + C + 0)(A + \bar{C} + 0)$$
$$= (A + \bar{B} + C\bar{C})(B + C + A\bar{A})(A + \bar{C} + B\bar{B})$$

Now, using the distributive property, each sum term can be expanded as

$$Y = (A + \bar{B} + C)(A + \bar{B} + \bar{C})(A + B + C)(\bar{A} + B + C)(A + B + \bar{C})(A + \bar{B} + \bar{C})$$
$$Y = (A + \bar{B} + C)(A + \bar{B} + \bar{C})(A + B + C)(\bar{A} + B + C)(A + B + \bar{C})$$
$$[\because (A + \bar{B} + \bar{C})(A + \bar{B} + \bar{C}) = (A + \bar{B} + \bar{C})]$$

This is called the maxterm canonical form or the canonical product of sum expression.

Example 2.26. *Express the function* $Y = A + \bar{B}C$ *in (a) canonical SOP and (b) canonical POS form.*

Solution: (*a*) Canonical sum of products form

$$Y = A + \bar{B}C$$
$$= A(B + \bar{B})(C + \bar{C}) + \bar{B}C(A + \bar{A})$$
$$= (AB + A\bar{B})(C + \bar{C}) + A\bar{B}C + \bar{A}\bar{B}C$$
$$= ABC + AB\bar{C} + A\bar{B}C + A\bar{B}\bar{C} + A\bar{B}C + \bar{A}\bar{B}C$$
$$= ABC + AB\bar{C} + A\bar{B}C + A\bar{B}\bar{C} + \bar{A}\bar{B}C \qquad [\because A\bar{B}C + A\bar{B}C + A\bar{B}C]$$
$$Y = m_7 + m_6 + m_5 + m_4 + m_1$$

Therefore, $Y = \Sigma(1, 4, 5, 6, 7)$

(*b*) Canonical product of sum form

$$Y = A + \bar{B}C$$

$$= (A+\overline{B})(A+C) \quad [\because \quad A + B.C = (A + B)(A + C)]$$

$$= (A+\overline{B}+C\overline{C})(A+C+B\overline{B})$$

$$= (A+\overline{B}+C)(A+\overline{B}+\overline{C})(A+B+C)(A+\overline{B}+C) \quad [\because \quad (A + B)(A + \overline{B}) = A]$$

$$Y = (A+\overline{B}+C)(A+\overline{B}+\overline{C})(A+B+C)$$

$$Y = M_2 M_3 M_0 \quad \text{or} \quad Y = M_0 M_2 M_3$$

Therefore, $Y = \Pi(0, 2, 3)$

Example 2.27. *Convert the following functions into canonical form*

(i) $AB + B\overline{C}D + \overline{A}D$

(ii) $(A+\overline{B})(\overline{C}+\overline{D})(\overline{B}+\overline{C})$

Solution: (i) $AB + B\overline{C}D + \overline{A}D$

$$= AB(C+\overline{C})(D+\overline{D}) + B\overline{C}D(A+\overline{A}) + \overline{A}D(B+\overline{B})(C+\overline{C})$$

$$= (ABC + AB\overline{C})(D+\overline{D}) + AB\overline{C}D + \overline{A}B\overline{C}D + (\overline{A}BD + \overline{A}\overline{B}D)(C+\overline{C})$$

$$= \overline{A}B\overline{C}D + \overline{A}BCD + AB\overline{C}D + \overline{A}B\overline{C}\overline{D} + AB\overline{C}D$$
$$+ \overline{A}B\overline{C}D + \overline{A}BCD + \overline{A}B\overline{C}D + \overline{A}\overline{B}CD + \overline{A}\overline{B}\overline{C}D$$

$$= ABCD + ABC\overline{D} + AB\overline{C}D + AB\overline{C}\overline{D} + \overline{A}B\overline{C}D$$
$$+ \overline{A}BCD + \overline{A}B\overline{C}D + \overline{A}\overline{B}CD + \overline{A}\overline{B}CD$$

(ii) $(A+\overline{B})(\overline{C}+\overline{D})(\overline{B}+\overline{C})$

$$= (A+\overline{B})+(C\overline{C})+(D\overline{D})$$
$$(\overline{C}+\overline{D})+(A\overline{A})+(B\overline{B})$$
$$(\overline{B}+\overline{C})+(A\overline{A})+(D\overline{D})$$

$$= (A+\overline{B}+C)(A+\overline{B}+\overline{C})+(D\overline{D})$$
$$(A+\overline{C}+\overline{D})(\overline{A}+\overline{C}+\overline{D})+(B\overline{B})$$
$$(A+\overline{B}+\overline{C})(\overline{A}+\overline{B}+\overline{C})+(D\overline{D})$$

$$= (A+\overline{B}+C+D)(A+\overline{B}+C+\overline{D})(A+\overline{B}+\overline{C}+D)(A+\overline{B}+\overline{C}+\overline{D})$$
$$(A+B+\overline{C}+\overline{D})(A+\overline{B}+\overline{C}+\overline{D})(\overline{A}+B+\overline{C}+\overline{D})(\overline{A}+\overline{B}+\overline{C}+\overline{D})$$

$$= (A+\overline{B}+\overline{C}+D)(A+\overline{B}+\overline{C}+\overline{D})(\overline{A}+\overline{B}+\overline{C}+D)(\overline{A}+\overline{B}+\overline{C}+\overline{D})$$
$$(A+\overline{B}+C+D)(A+\overline{B}+C+\overline{D})(A+\overline{B}+\overline{C}+D)(A+\overline{B}+\overline{C}+\overline{D})$$
$$(A+B+\overline{C}+\overline{D})(\overline{A}+B+\overline{C}+\overline{D})(\overline{A}+\overline{B}+\overline{C}+\overline{D})(\overline{A}+\overline{B}+\overline{C}+D)$$

2.6.3 Deriving Sum of Product (SOP) Expression from a Truth Table

The Sum of Product (SOP) expression for a Boolean function can be derived from its truth table by summing (OR operation) the product terms that correspond to the combinations containing a function value 1. In the product term, the input variable appears either in uncomplemented form if it possesses the value 1, or in complemented form if it contains the value 0.

Now, consider the truth table, shown in Table 2.7, for a 3-input function Y. Here, the Y value is 1 for the input combinations 010, 011, 101 and 111 and their corresponding product terms are $\bar{A}B\bar{C}$, $\bar{A}BC$, $A\bar{B}C$ and ABC respectively.

Table 2.7. Truth table

Inputs		Output		Product terms	Sum terms
A	B	C	Y		
0	0	0	0		$(A+B+C)$
0	0	1	0		$(A+B+\bar{C})$
0	1	0	1	$\bar{A}B\bar{C}$	
0	1	1	1	$\bar{A}BC$	
1	0	0	0		$(\bar{A}+B+C)$
1	0	1	1	$A\bar{B}C$	
1	1	0	0		$(\bar{A}+\bar{B}+C)$
1	1	1	1	ABC	

Now, the final SOP expression for the output Y is obtained by summing (OR operation of) the four product terms as follows:

$$Y = \bar{A}B\bar{C} + \bar{A}BC + A\bar{B}C + ABC$$

The procedure for obtaining the output expression in SOP form from a truth table can be summarised, in general, as follows:

1. Give a product term for each input combination in the table, containing an output value of 1.
2. Each product term contains its input variables in either complemented or uncomplemented form. If an input variable is 0, it appears in complemented form; if the input variable is 1, it appears in uncomplemented form.
3. All the product terms are OR operated together in order to produce the final SOP expression of the output.

2.6.4 Deriving Product of Sum (POS) Expression from a Truth Table

The Product of Sum (POS) expression for a Boolean (switching) function can also be obtained from a truth table by the AND operation of the sum terms corresponding to the combinations for which the function assumes the value 0. In the sum term, the input variable appears in an uncomplemented form if it has the value 0 in the corresponding combination and in the complemented form if it has the value 1.

Studying the truth table shown in Table 2.7, for a 3-input function Y, we find that the Y value is 0 for the input combinations 000, 001, 100 and 110 and that their corresponding sum terms are $(A + B + C)$, $(A + B + C)$, $(A + B + \bar{C})$, $(\bar{A} + B + C)$ and $(\bar{A} + \bar{B} + C)$ respectively.

Now the final POS expression for the output Y is obtained by the AND operation of the four sum terms as follows:

$$Y = (A + B + C)(A + B + \bar{C})(\bar{A} + B + C)(\bar{A} + \bar{B} + C)$$

The procedure for obtaining the output expression in POS form from a truth table can be summarised, in general, as follows:

1. Give a sum term for each input combination in the table, which has an output value of 0.
2. Each sum term contains all its input variables in complemented or uncomplemented form. If the input variable is 0, then it appears in an uncomplemented form; if the input variable is 1, it appears in the complemented form.
3. All the sum terms are AND operated together to obtain the final POS expression of the output.

The POS expression for a Boolean (switching) function can also be obtained from its SOP expression using $\overline{\overline{Y}} = Y$ given in the following example.

Consider a function,

$$Y = \bar{A}B\bar{C} + \bar{A}BC + A\bar{B}C + ABC$$

$$Y = \overline{\overline{\overline{\overline{Y}} = \bar{A}B\bar{C} + \bar{A}BC + A\bar{B}C + ABC}}$$

The complement $\bar{Y}$ can be obtained by the OR operation of the minterms which are not available in Y. Therefore,

$$\bar{Y} = \bar{A}\bar{B}\bar{C} + \bar{A}\bar{B}C + A\bar{B}\bar{C} + AB\bar{C}$$

$$Y = \overline{\bar{A}\bar{B}\bar{C} + \bar{A}\bar{B}C + A\bar{B}\bar{C} + AB\bar{C}}$$

$$= \overline{(\bar{A}\bar{B}\bar{C})} + \overline{(\bar{A}\bar{B}C)} + \overline{(A\bar{B}\bar{C})} + \overline{(AB\bar{C})}$$

$$= (A + B + C)(A + B + \bar{C})(\bar{A} + B + C)(\bar{A} + \bar{B} + C)$$

2.7 KARNAUGH MAP

The simplification of the switching functions using Boolean laws and theorems becomes complex with the increase in the number of variables and terms. The Karnaugh map technique provides a systematic method for simplifying and manipulating switching expressions. In this technique, the information contained in a truth table or available in the POS or SOP form is represented on the Karnaugh map (K-map). The K-map is actually a modified form of a truth table. Here, the combinations are conveniently arranged to aid the simplification process by applying the rule $Ax + Ax' = A$. In an n-variable K-map, there are 2^n cells. Each cell corresponds to one combination of n variables. Therefore, for each row of the truth table, i.e. for each minterm and for each maxterm, there is one specific cell in the K-map. The K-maps for 2, 3 and 4 variables are shown in Fig. 2.1. The decimal codes corresponding to the combination of variables are given inside the cells. The variables have been marke as A, B, C, and D, and the binary numbers formed by them are taken as AB, ABC, an $ABCD$ for 2, 3 and 4 variables respectively.

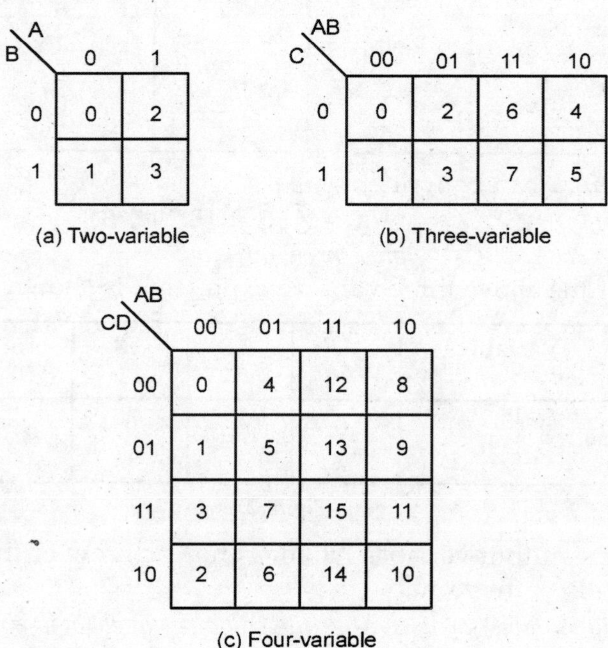

(a) Two-variable (b) Three-variable

(c) Four-variable

Fig. 2.1. Karnaugh maps.

The 3 and 4 variable K-maps show that the column and row headings, used in representing the cells, are cyclic or unit distance code which result in adjacent cells, differing in just one variable. This helps the grouping of the adjacent cells and in their simplification by the application of the rule $Ax + Ax' = A$. In addition, the let I and right-most cells of the 3-variable *K*-map are adjacent, for example, the cells 0 and 4 are adjacent, and the cells 1 and 5 are adjacent. This is because each pair differs in just a single variable. In the 4-variable *K*-map, the cells to the extreme left and right as well as those at the top and bottom-most position are adjacent.

A collection or group of 2^m cells, each adjacent to m cells, is called a group. This group can be expressed by a product containing n-m variables, where n is the number of variables in the K-map. For example, in a 4-variable K-map (i.e. $n = 4$), if a group of 4 (i.e., $2m = 4$; $m = 2$) 1s is formed, then this group can be expressed by $4 - 2 = 2$ variables. Similarly, if a group of eight Is are combined, then this group can be expressed by $4 - 3 = 1$ variable. This can be better understood by the examples given latter.

The entries in a truth table can be represented in a K-map given below. Consider the truth table shown in Table 2.8.

Table 2.8. Truth table of a digital system

Inputs			Output
A	*B*	*C*	*Y*
0	0	0	0
0	0	1	1
0	1	0	1
0	1	1	0
1	0	0	1
1	0	1	0
1	1	0	0
1	1	1	1

Here, the output *Y* can be written as

$$Y = \bar{A}\bar{B}C + \bar{A}B\bar{C} + A\bar{B}\bar{C} + ABC$$

$$Y(A, B, C) = m_1 + m_2 + m_4 + m_7$$

The *K*-map for the above three-variable expression is shown in Fig. 2.2.

Variables		$\bar{A}\bar{B}$ 00	$\bar{A}B$ 01	AB 11	$A\bar{B}$ 10
$\bar{C}$	0	0	1	0	1
C	1	1	0	1	0

Fig. 2.2.

The value of the output variable *Y* (0 or 1) for each row of the truth table is tered in the corresponding cells of the *K*-map.

Note: *K-map is ideally suitable for designing the combinational logic circuits using either a SOP method or a POS method.*

Simplification is based on the principle of combining the terms present in adjacent cells. The 1s in the adjacent cells can be grouped by drawing a loop around DSC cells following the given rules:

1. Construct the *K*-map and enter the 1s in those cells corresponding to the combinations for which function value is 1, then enter the 0s in the other cells.
2. Examine the map for 1 s that cannot be combined with any other 1 cells and form groups with such single 1.
3. Next, look for those 1s which are adjacent to only one other 1 and form groups containing only 2 cells and which are not part of any group of 4 or S cells. A group of 2 cells is called a pair.
4. Group the 1 s which results in groups of 4 cells but are not part of an 8-cell; group. A group of 4 cells is called a quad.
5. Group the 1s which results in groups of 8 cells. A group of 8 cells is called an octet.
6. Form more pairs, quads and octets to include these 1s that have not yet beer grouped, and use only a minimum number of groups. There can be overlap ping of groups if they include common 1s.
7. Omit any redundant group.
8. Form the logical sum of all the terms generated by each group.

When one or more than one variable appear in both complemented and uncomplemented form within a group, then that variable(s) is eliminated from the term corresponding to that group. Variables that are the same for all the cells of the group must appear in the term corresponding to that group.

A larger group of 1s eliminates more variables. To be precise, a group of two eliminates one variable; a group of four eliminates two variables; similarly a group of eight eliminates three variables.

Example 2.28. *Simplify the following expression using the Karnaugh map for the 4-variables A, B, C and D.*

$$Y = m_1 + m_3 + m_5 + m_7 + m_8 + m_9 + m_{12} + m_{13}$$

Solution: The *K*-map of the given equation is shown in Fig. E2.28. The expressior is minimized using the following steps:

Step 1: Construct the *K*-map and enter 1 in the cells corresponding to the minterms present in the expression and 0 in the other cells.

Step 2: There are no 1s which are not adjacent to other 1s.

Step 3: There are no pairs which are not part of any larger groups.

Step 4: There are 2 quads. Cells 1, 3, 5 and 7 are grouped to form one quad and the second quad is made up of cells 12, 13, 8 and 9. The combinations corresponding to the cells in the first quad are $\overline{A}\overline{B}\overline{C}D$, $\overline{A}\overline{B}CD$, $\overline{A}B\overline{C}D$ and $\overline{A}BCD$. In the above group of four combinations, the variables *AD* are common in all the cells while *B* and *C* appear both in complemented anc uncomplemented forms. From the preceding section, it is clear that only the variables that are the same in all the cells of the group must appear in the term corresponding to that group. Therefore, the minimized term for the first quad is $\overline{A}D$, and that of the second quad is $A\overline{C}$.

Step 5: There are no octets.

Step 6: All the 1s have already been grouped.

Step 7: The terms generated by the two groups are OR operated together to obtain the expression for *Y* as follows:

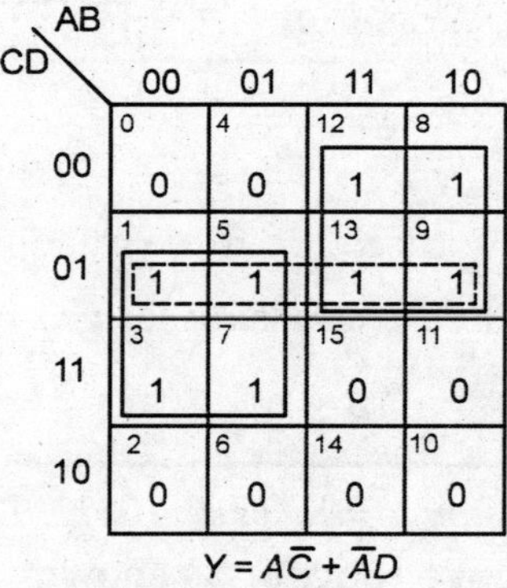

Fig. E2.28.

Note: *In the above K-map, if a third quad is formed as shown by the dotted lines, it results in a reductant expression because the 1s to be covered by the third quad are already covered by quads 1 and 2.*

Example 2.29. *Plot the logical expression* $ABCD + \overline{A}\overline{B}\overline{C}\overline{D} + A\overline{B}C + AB$ *on a 4-variable K-map; obtain the simplified expression from the map.*

Solution: To enter into a K-map, a logic expression must be either in the canonical SOP form or in the canonical POS form. The canonical SOP form of the given expression can be obtained as follows:

$$\begin{aligned} Y &= ABCD + A\overline{B}\overline{C}\overline{D} + A\overline{B}C + AB \\ &= ABCD + A\overline{B}\overline{C}\overline{D} + A\overline{B}C + (D+\overline{D}) + AB(C+\overline{C})(D+\overline{D}) \\ &= ABCD + A\overline{B}\overline{C}\overline{D} + A\overline{B}CD + A\overline{B}C\overline{D} + (ABC + AB\overline{C})(D+\overline{D}) \\ &= ABCD + A\overline{B}\overline{C}\overline{D} + A\overline{B}CD + A\overline{B}C\overline{D} + ABCD = ABC\overline{D} + AB\overline{C}D + AB\overline{C}\overline{D} \\ &= ABCD + A\overline{B}\overline{C}\overline{D} + A\overline{B}CD + A\overline{B}C\overline{D} + ABC\overline{D} + ABC\overline{D} + AB\overline{C}D + AB\overline{C}\overline{D} \\ &= m_{15} + m_8 + m_{11} + m_{10} + m_{14} + m_{13} + m_{12} \\ &= \Sigma_m(8, 10, 11, 12, 13, 14, 15) \end{aligned}$$

The *K*-map for the above expression is shown in Fig. E2.29.

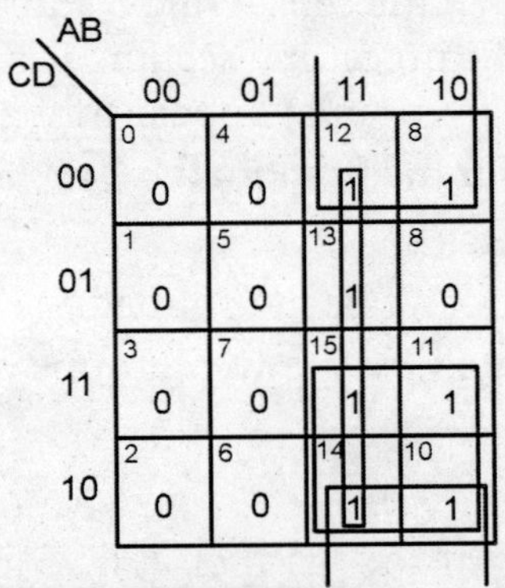

Fig. E2.29.

In the *K*-map in Fig. E2.29, there are three quads; the minimised terms for them are AB, AC and $A\overline{D}$ and the simplified expression is:

$$Y = AB + AC + A\overline{D}$$

Example 2.30. *Simplify the expression* $Y = S_m(7, 9, 10, 11, 12, 13, 14, 15)$, *using the K-map method.*

Solution: The *K*-map for the above function is shown in Fig. E2.30.

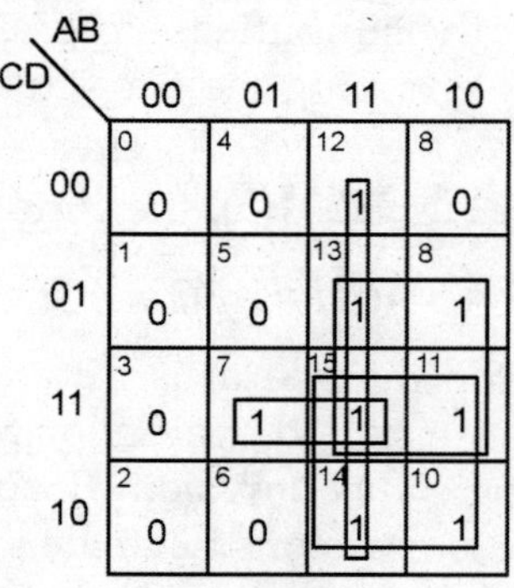

Fig. E2.30.

In the given *K*-map, there are three quads and one pair; the corresponding simplified terms are *AB*, *AD*, *AC* and *BCD*.

Now, the simplified expression is

$$Y = AB + AD + AC + BCD$$

Since the quads and pair formed in the above *K*-map overlap, the expression can be further simplified using the Boolean algebra as follows:

$$Y = AB + A + AC + BCD$$
$$= A(B + D + C) + BCD$$

Example 2.31. *Simplify the expression* $Y = m_1 + m_5 + m_{10} + m_{11} + m_{12} + m_{13} + m_{15}$, *using the K-map method.*

Solution: The *K*-map for the above expression is shown in Fig. E2.31 (*a*).

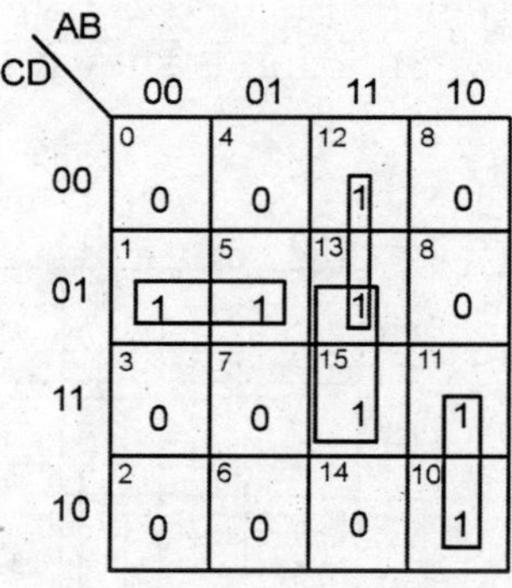

Fig. E2.31. (*a*)

As shown in Fig. E2.31 (*a*), the *K*-map contains four pairs but no quads or octets; the corresponding simplified expression is given by

$$Y = \bar{A}\bar{C}D + AB\bar{C} + ABD + A\bar{B}C \qquad ...(1)$$

It is important to note that the simplified expression obtained from the *K*-map is not unique. This can be explained by grouping the pairs in a different manner as shown in Fig. E2.31 (*b*).

From the *K*-map shown in Fig. E2.31(*b*), the simplified expression can be written as:

$$Y = \overline{A}\overline{C}D + AB\overline{C} + ACD + A\overline{B}C \qquad ...(2)$$

In equations (1) and (2), the third term is different, due to the different groupings done in Fig. E2.31(*b*). Though the simplified expression for any given function is not unique, both the above expressions are logically equivalent. Two expressions are said to be logically equivalent if and only if both the expressions have the same value for every combination of input variables.

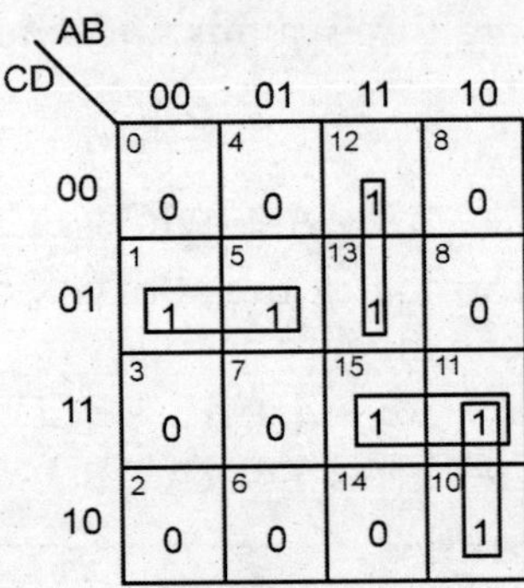

Fig. E2.31. (b)

Example 2.32. *Simplify the expression* $Y = S_m(3, 4, 5, 7, 9, 13, 14, 15)$, *using the K-map method.*

Solution: The *K*-map for the above function is shown in Fig. E2.32.

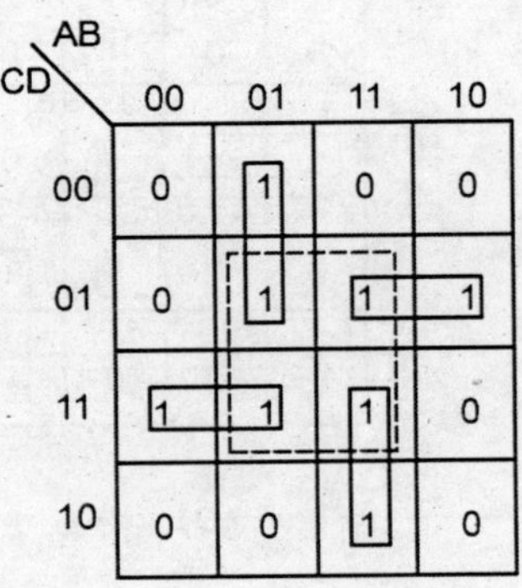

Fig. E2.32.

In the above *K*-map, the cells 5, 7, 13 and 15 can be grouped to form a quad as indicated by the dotted lines. In order to group the remaining 1 s, four pairs have to be formed as shown in Fig. E2.32. However, all the four Is covered by the quad are also covered by the pairs. So, the quad in the above K-map is redundant. Therefore, the simplified expression will be

$$Y = A\overline{C}D + ABC + A\overline{C}D + \overline{A}B\overline{C}$$

Example 2.33. *Simplify the expression Y = P (0, 1, 4, 5, 6, 8, 9, 12, 13, 14), using the K-map method.*

Solution: The given function is in the POS form. This can also be written as

$$Y = (A+B+C+D)(A+B+C+\overline{D})(A+\overline{B}+C+D)(A+\overline{B}+C+\overline{D})(A+\overline{B}+\overline{C}+D)$$
$$(\overline{A}+B+C+D)(\overline{A}+B+C+\overline{D})(\overline{A}+\overline{B}+C+D)(\overline{A}+\overline{B}+C+\overline{D})(\overline{A}+\overline{B}+\overline{C}+D)$$

To simplify a POS expression, for each maxterm in the expression, a 0 has to be entered in the corresponding cells and groups must be formed with 0 cells instead of 1 cells to get the minimal expression. The simplified term corresponding to each group can be obtained by the OR operation of the variables that are same for all cells of that group. Here, a variable corresponding to 0 has to be represented in an uncomplemented form and that corresponding to 1 in the complemented form.

The *K*-map for the function is shown in Fig. E2.33.

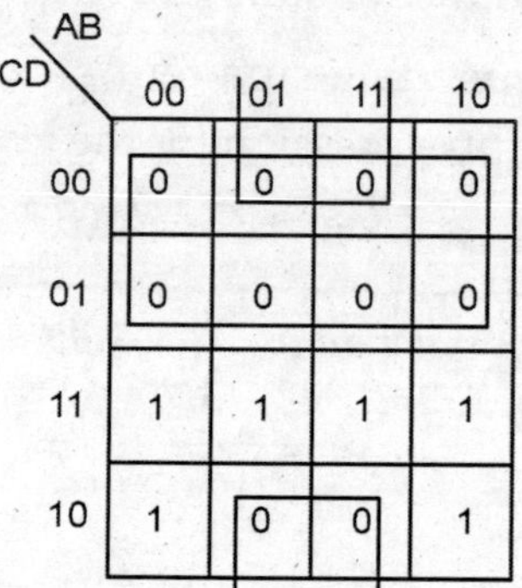

Fig. E2.33.

In the above *K*-map, one octet and one quad are produced by combining 0 cells. The simplified sum term corresponding to the octet is *C* whereas for the quad is ($\overline{B}$ + *D*). Hence, the simplified *POS* expression for the given function is

$$Y = C(\overline{B}+\overline{D})$$

Example 2.34. *Obtain (a) minimal sum of product and (b) minimal product of sum expressions for the function given below:*

$$F(A, B, C, D) = \Sigma_m(0, 1, 2, 5, 8, 9, 10)$$

Solution: Here, cells with 1 are grouped to obtain the minimal sum of product; cells with 0 are grouped to obtain minimal product of sum, as shown in Fig. E2.34.

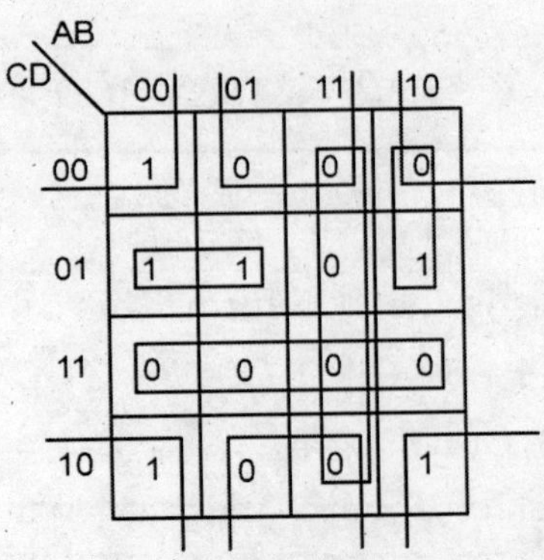

Fig. E2.34.

(*a*) To obtain minimal sum of products: A quad with four corner 1s and two pairs can be formed as shown in Fig. E2.34. hence, the minimal SOP expression is:

$$Y = \overline{B}\overline{D} + \overline{A}\overline{C}D + A\overline{B}\overline{C}$$

(*b*) To obtain minimal product of sum: Three quads can be formed as shown in Fig. E.2.34 with the corresponding sum terms $(\overline{A}+\overline{B}), (\overline{C}+\overline{D})$ and $(\overline{B}+D)$. Thus, the minimal product of sum expression for the given function is:

$$Y = (\overline{A}+\overline{B})(\overline{C}+\overline{D})(\overline{B}+D)$$

Example 2.35. *Minimize the given function using K-map and convert the minimized function into POS form f(A, B, C, D) = S(1, 3, 5, 7, 9, 10, 12, 13).*

Solution: $f(A, B, C, D) = (A+\overline{D})(\overline{A}+B+\overline{C}+D)(\overline{A}+C+\overline{D})(\overline{A}+\overline{B}+C)$

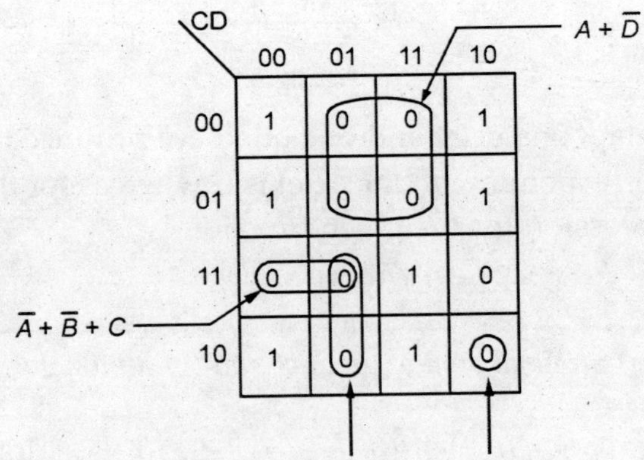

Fig. E2.35.

2.7.1 Five-variable K-map

A 5-variable *K*-map contains $32(2^5)$ cells which is used to simplify any 5-variable logic expression. A 5-variable *K*-map is shown in Fig. 2.3, with entries in each cell representing the decimal code of that cell. Three variables are used to mark the column headings and two variables are used to mark the row headings. The first four columns can be marked with headings in the same way as the 4-variable *K*-map, after which the remaining four columns can be marked with headings in the reverse order. In other words, the two least significant bits of headings in the last four columns are the mirror image of the corresponding bits in the first four columns. Add 0s to the first four column headings and Is to the remaining four columns. The simplification using the 5-variable *K*-map can be done in the same way as 4-variable *K*-map.

DE \ ABC	000	001	011	010	110	111	101	100
00	0	4	12	8	24	28	20	16
01	1	5	13	9	25	29	21	17
11	3	7	15	11	27	31	23	19
10	2	6	14	10	26	30	22	18

Fig. 2.3. 5-Variable K-map.

Example 2.36. *Simplify* $Y = S_m(3, 6, 7, 8, 10, 12, 14, 17, 19, 20, 21, 24, 25, 27, 28)$ *using the K-map method.*

Solution: In the 5-variable *K*-map shown in Fig.E2.36, there are three quads and three pairs; the simplified expression is given by

$$Y = BD\overline{E} + A\overline{C}E + \overline{A}B\overline{E} + A\overline{B}C\overline{D} + \overline{A}\overline{B}CD + \overline{A}\overline{B}DE$$

DE \ ABC	000	001	011	010	110	111	101	100
00	0: 0	4: 0	12: 1	8: 1	24: 1	28: 1	20: 1	16: 0
01	1: 0	5: 0	13: 0	9: 0	25: 1	29: 0	21: 1	17: 1
11	3: 1	7: 1	15: 0	11: 0	27: 1	31: 0	23: 0	19: 1
10	2: 0	6: 1	14: 1	10: 1	28: 0	30: 0	22: 0	18: 0

Fig. E2.36. 5-Variable K-map

Don't care combinations: In certain digital systems, some input combinations never occur during the process of a normal operation because those input conditions are guaranteed never to occur. Such input combinations are don't care combinations. We don't care what the function output is for such combinations. These combinations can be plotted on a map to provide further simplification of the function.

The functions considered so far in the various examples, for simplification using the *K*-map method, are completely specified, i.e., it assumes the value 1 for some input combinations and the value 0 for others. Also, there are functions which assume the value 1 for some combinations, the value 0 for some other, and either 0 or 1 for the remaining combinations. Such functions are called incompletely specified functions, and the combinations for which the value of the function is not specified are called don't care combinations. The don't care combinations are represented by *d* or *x* or ϕ.

When an incompletely specified function, i.e., a function with don't care combinations, is simplified to obtain minimal SOP expression, the value 1 can be assigned to selected don't care combinations. This is done in order to increase the number of 1s in the selected groups, wherever further simplification is possible. Also, a don't care combination need not be used in grouping if it does not cover a large number of 1s. In each case, the choice depends only on the simplification that has to be achieved. Similarly, when a function is simplified to obtain a minimal POS expression, the value 0 can be assigned to selected don't care combinations in order to increase the number of 0s in the selected groups which results in further simplification.

Example 2.37. *Simplify the Boolean function*

$$F(A, B, C, D) = \Sigma_m(1, 3, 7, 11, 15) + \Sigma_d(0, 2, 5)$$

Solution: The *K*-map for the given function is shown in Fig.E2.37 with entries d in cells corresponding to combinations 0, 2 and 5.

As discussed in the previous section, the 1s and d s are combined in order to enclose the maximum number of adjacent cells with 1. As shown in the K-map in Fig.E2,37, by combining the Is and d s, two quads can be obtained. The d in cell 5 is left free since it does not contribute in increasing the size of any group. Now, the simplified expression in SOP form is

$$Y = \overline{A}\overline{B} + CD$$

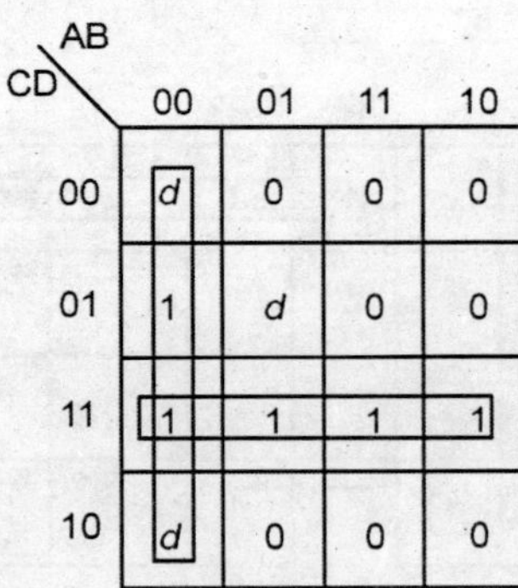

Fig. E2.37.

Example 2.38. *Using the K-map method, simplify the following Boolean function and obtain (i) minimal SOP and (ii) minimal POS expressions:*

$$Y = \Sigma_m(0, 2, 3, 6, 7) + \Sigma_d(8, 10, 11, 15)$$

Solution: The *K*-map for the above function is shown in Fig. E2.38.

Minimal SOP form: By combining the 1s and d s as shown in the *K*-map, there are two quads; the simplified SOP expression is given by

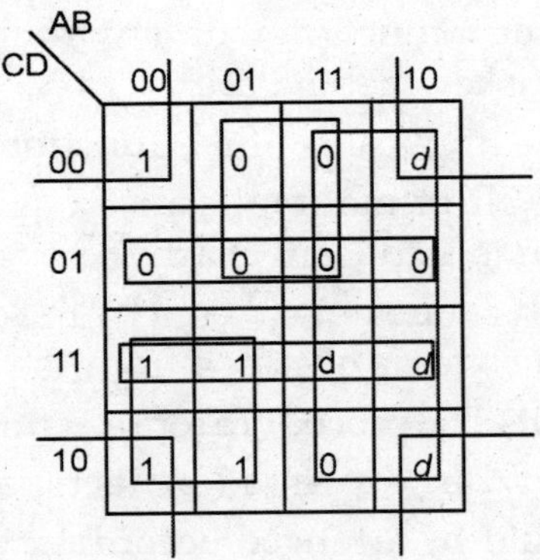

Fig. E.2.38.

Minimal POS form: One octet and two quads can be obtained by combining the 0s and d s as shown in the *K*-map; the simplified *POS* expression is given by

$$Y = \bar{A}(C+\bar{D})(\bar{B}+C)$$

Example 2.39. *Obtain the minimal SOP expression for the function*

$$Y = E_m\ (1, 5, 7, 13, 14, 15, 17, 18, 21, 22, 25, 29) + \Sigma_d(6, 9, 19, 23, 30)$$

Solution: The *K*-map for the given function is shown in Fig. E2.39.

DE \ ABC	000	001	011	010	110	111	101	100
00	(0) 0	(4) 0	(12) 0	(8) 0	(24) 0	(28) 0	(20) 0	(16) 0
01	(1) 1	(5) 1	(13) 1	(9) *d*	(25) 1	(29) 1	(21) 1	(17) 1
11	(3) 0	(7) 1	(15) 1	(11) 0	(27) 0	(31) 0	(23) *d*	(19) *d*
10	(2) 0	(6) *d*	(14) 1	(10) 0	(28) 0	(30) *d*	(22) 1	(18) 1

Fig. E2.39.

By combining the 1s and d s, one octet and two quads can be obtained as shown in Fig. E2.39. The simplified expression is

$$Y = \bar{D}E + \bar{A}CD + A\bar{B}D$$

2.8 QUINE-McCLUSKEY OR TABULAR METHOD OF MINIMIZATION OF LOGIC FUNCTIONS (i.e., TABULAR METHOD)

The *K*-map is a very effective tool for minimization of logic functions with 4 or less variables. For logic expressions with more than 4variables, the visualization of adjacent cells and the drawing of the *K*-map become more difficult. The Quine—Mc Cluskey method, also known as the tabular method, can be employed in such cases to minimize switching functions. This method employs a systematic, step-by-step procedure to produce a simplified standard form of expression for a function with any number of variables. The steps to be followed in the Quine—McCluskey method are:

Step 1: A set of all prime implicants of the function must be obtained.

Step 2: From the set of all prime implicants, a set of essential implicants must be determined by preparing a prime implicant chart.

Step 3: The minterms which are not covered by the essential implicants are taken into consideration and a minimum cover is obtained from the remaining prime implicants.

Selecting prime implicants The procedure for selecting prime implicants is given below:

(*i*) Each minterm should be expressed by its binary representation.

(*ii*) The minterms should be arranged according to increasing index (index can be defined as the number of 1s in a minterm). Separate each set of minterms possessing the same index by lines.

(*iii*) Each term of index '*n*' should be compared with each term of index ($n + 1$). For each pair of terms that can combine, the newly formed terms should be stated. If two minterms differ in only one variable, that variable should be removed and a dash placed at that position, thus a new term with one less literal is found. After all the pairs of terms with indices n and ($n + 1$) have been considered, a line should be drawn under the last term.

(*iv*) When the above process has been repeated for all the groups of terms, one stage of elimination will have been completed.

(*v*) The next stage of elimination or matching process should be repeated for the new terms. According to this stage, two terms can be combined only when they have dashes in the same positions.

(*vi*) The cycles have to be continued until no new list can be found (i.e. no further elimination of literals).

(*vii*) All terms which remain unchecked (do not match) during the process are considered to be prime implicants.

Do You Know?	Quine-McCluskey (QM) method is more effective since a digital computer can be used to simply the given Boolean switching function.

Prime implicant chart

(*i*) The prime implicants should be represented in rows and each minterm of the function in a column.

(*ii*) Crosses should be placed in each row to show the composition of minterms that makes the prime implicants.

(*iii*) A completed prime implicants table should be inspected for columns containing only a single cross. Prime implicants that cover minterms with a single cross in their column are called essential prime implicants.

Example 2.40. *Find the minimal sum of products for the Boolean expression, f = S(1, 2, 3, 7, 8, 9, 10, 11, 14, 15), using the Quine-McCluskey method.*

Solution: Firstly, these minterms are represented in the binary form as shown in Table E2.40(*a*). The above binary representation are grouped into a number of sections in terms of the number of 1s as shown in Table E2.40(*b*).

Table E2.40(*a*) Binary representation of minterms

Minterms	Variables			
	A	B	C	D
1	0	0	0	1
2	0	0	1	0
3	0	0	1	1
7	0	1	1	1
8	1	0	0	0
9	1	0	0	1
10	1	0	1	0
11	1	0	1	1
14	1	1	1	0
15	1	1	1	1

Table E2.40(*b*) Group of minterms for different number of 1s

Number of 1s	Minterms	Variables			
		A	B	C	D
1	1 ✓	0	0	0	1
	2 ✓	0	0	1	0
	8 ✓	1	0	0	0
2	3 ✓	0	0	1	1
	9 ✓	1	0	0	1
	10 ✓	1	0	1	0
3	10 ✓	0	1	1	1
	11 ✓	1	0	1	1
	11 ✓	1	1	1	0
	14 ✓	1	1	1	1
4	15 ✓	1	1	1	1

Any two numbers in these groups which differ from each other by only one variable can be chosen and combined, to get two-cell combinations, as shown in Table E2.40(*c*).

Table E2.40(*c*) 2-cell combinations

Combination	A	B	C	D
(1, 3) ✓	0	0	–	1
(1, 9) ✓	–	0	0	1
(2, 3) ✓	0	0	1	–
(2, 10) ✓	–	0	1	0
(8, 9) ✓	1	0	0	–
(8, 10) ✓	1	0	–	0
(3, 7) ✓	0	–	1	1
(3, 11) ✓	–	0	1	1
(9, 11) ✓	1	0	–	1
(10, 11) ✓	1	0	1	–
(10, 14) ✓	1	-	1	0
(7, 15) ✓	–	1	1	1
(11, 15) ✓	1	–	1	1
(14, 15) ✓	1	1	1	–

From the two-cell combinations, one variable and a dash in the same position can be combined to form 4-cell combinations as shown in Table E2.40(*d*).

Table E2.40(*d*) 4-cell combinations

Combination	A	B	C	D
(1, 3, 9, 11)	–	0	–	1
(2, 3, 10, 11)	–	0	1	–
(8, 9, 10, 11)	1	0	–	–
(3, 7, 11, 15)	–	–	1	
(10, 11, 14, 15)	1	–	1	–

Note that the cells (1, 3) and (9, 11) form the same 4-cell combination as the cells (1, 9) and (3, 11). The order in which the cells are placed in a combination does not have any effect. Thus, the (1, 3, 9, 11) combination may be given as (1, 9, 3, 11). Using Table E2.40(*d*), the prime implicants table can be plotted as shown in Table E2.40(*e*).

The columns having only one cross mark correspond to essential prime implicants. A tick mark is put against every column which has only one cross mark. A star mark is placed against every essential primary implicant. The sum of the prime implicants gives the function in its minimal SOP form, since all prime implicants are essential prime implicants.

Therefore, $$f = \overline{B} \cdot D + \overline{B} \cdot C + A \cdot \overline{B} + C \cdot D + A \cdot C$$

Table E2.40(*e*) Prime implicants table

Prime implicants	Minterms									
	1	2	3	7	8	9	10	11	14	15
(1, 3, 9, 11) *	×		×			×		×		
(2, 3, 10, 11) *		×	×				×	×		
(8, 9, 10, 11) *					×	×	×	×		
(3, 7, 11, 15) *			×	×				×		×
(10, 11, 14, 15) *							×	×	×	×
	✓	✓		✓	✓				✓	

Example 2.41. *Find the minimal sum of products for the Boolean expression, f(w, x, y, z) = S(1, 3, 4, 5, 9, 10, 11) + Sf(6, 8), using the Quine-McCluskey method.*

Solution: These minterms are firstly represented in binary form as shown in Table E2.41(*a*). The above binary representation is grouped into a number of sections in terms of a number of Is as shown in Table E2.41(*b*).

Table E2.41(*a*) Binary representation of minterms

Minterms	Variables			
	w	*x*	*y*	*z*
1	0	0	0	1
3	0	1	i	0
4	0	1	0	0
5	0	1	0	1
6	0	1	1	0
8	1	0	0	0
9	1	0	0	1
10	1	0	1	0
11	1	0	1	1

Table E2.41(*b*) Group of minterms for different number of Is

Number of 1s	Minterms	Variables			
		w	*x*	*y*	*z*
1	1 ✓	0	0	0	1
	4 ✓	0	1	0	0
	8 ✓	1	0	0	0
2	3 ✓	0	0	1	1
	5 ✓	0	1	0	1
	6 ✓	0	1	1	0
	9 ✓	1	0	0	1
	10 ✓	1	0	1	0
3	11 ✓	1	0	1	1

Any two numbers in these groups which differ from each other by only one variable can be chosen and combined to get two-cell combinations, as shown in Table E2.41(*c*).

Table E2.41(*c*) 2-cell combinations

Combination	*w*	*x*	*y*	*z*
(1, 3) ✓	0	0	–	1
(1, 5) ✓	0	–	0	1
(1, 9) ✓	–	0	0	1
(4, 5) ✓	0	1	0	–
(4, 6) ✓	0	1	–	0
(8, 9) ✓	1	0	0	–
(8, 10) ✓	1	0	–	0
(3, 11) ✓	–	0	1	1
(9, 11) ✓	1	0	–	1
(10, 11) ✓	1	0	1	–

From the two-cell combinations, one variable and a dash in the same position can be combined to form 4-cell combinations as shown in Table E2.41(*d*).

Table E2.41(*d*) 4-cell combinations

Combination	*w*	*x*	*y*	*z*
(1, 3, 9, 11)	–	0	–	1
(8, 9, 10, 11)	1	0	–	–

Note that the cells (1, 3) and (9, 11) form the same 4-cell combination as the cells (1, 9) and (3, 11). The order in which the cells are placed in a combination has no effect. Therefore, the (1, 3, 9, 11) combination may be written as (1, 9, 3, 11). Using Table E2.41(*d*), the prime implicants table can be plotted as shown in Table E2.41(*e*).

Don't care minterms cannot be listed as column headings in the chart because they do not have to be covered by a minimal expression.

Table E2.41(*e*) Prime implicants table

Prime	Minterms						
implicants	1	3	4	5	9	10	11
(1, 5)	X			×			
(4,5)			×	×			
(4, 6)			×				
(1, 3, 9, 11) *	×	×			×		×
(8, 9, 10, 11) *					×	×	×
		✓				✓	

The columns having only one cross mark correspond to the essential prime implicants. A tick mark is put against every column which has only one cross mark. A star mark is put against every essential primary implicant. The prime implicant which covers the minterm (1, 3, 9, 11) is the essential prime implicant. Therefore, in order to cover the remaining minterms, the reduced prime implicant chart is formed as shown in Table E2.41(*f*).

Table E2.41(*f*) Reduced prime implicants table

Prime implicants	Minterms						
	1	3	4	5	9	10	11
(1, 5)	X			X			
(4, 5)			X	X			
(4, 6)			X				

To cover the minterms (4, 5), the prime implicants (4, 5) can be selected, in addition to the essential prime implicants, for obtaining the minimal expression of the given function.

Therefore, $f(w, x, y, z) = \overline{x}y + \overline{w}x\overline{y} + w\overline{x}$

■ REVIEW QUESTIONS ■

1. State the methods used to simplify the Boolean equations.
2. State and explain the basic Boolean logic operations.
3. What are the applications of Boolean algebra?
4. Define the truth table.
5. How is the AND multiplication different from ordinary multiplication?
6. How does OR addition differ from the ordinary addition method?
7. What are the basic laws of Boolean algebra?
8. State and prove Absorption and Simplification theorems.
9. State and prove Associative and Distributive theorems.
10. What is meant by duality in Boolean algebra?
11. State DeMorgan's theorem.
12. State and explain the DeMorgan's theorems which convert a sum into a product form and vice versa.
13. Explain the terms: (*a*) prime implicant, (*b*) input variable, (*c*) minterm and (*d*) maxterm.
14. Prove DeMorgan's theorem for a 4-variable function.
15. Many cars produced in Japan have an interlock system that allows the engine to start only if both the front seat occupants have their seat-belts on. Construct a truth table to indicate whether the car may be started based upon whether a passenger is present and whether both the passenger and the driver have buckled their seat-belts.
16. Draw a truth function table for a person crossing a river based upon whether (*i*) the river is frozen over, (*ii*) the boat leaks, and (*iii*) the person can swim.

■ PROBLEMS ■

1. Indicate whether Y is a 0 or a 1 in the equation $Y = A\overline{B}\overline{C} + AB$, under the following conditions:
(*a*) A = 1, B = 0, C = l; (*b*) A = 0, B = 1, C = 1; (*c*) A = 0, B = 0, C = 0 **[Ans.** (*a*) 0, (*b*) 0, (*c*) 0]

2. Simplify the following expressions:

(*a*) $\mathrm{A\overline{B}\overline{C} + \overline{A}\overline{B}\overline{C} + \overline{A}B\overline{C} + \overline{A}\overline{B}C}$

(*b*) $\mathrm{ABC + \overline{A}BC + A\overline{B}C + AB\overline{C} + \overline{A}B\overline{C} + \overline{A}\overline{B}\overline{C}}$

(*c*) $\mathrm{A(A + B + C) \cdot (\overline{A} + B + C) \cdot (A + \overline{B} + C) \cdot (A + B + \overline{C})}$

(*d*) $\mathrm{(A + B + C) \times (A + \overline{B} + \overline{C}) \times (A + B + \overline{C}) \times (A + \overline{B} + C)}$

[Ans. (*a*) $\mathrm{\overline{B}\overline{C} + \overline{A}\overline{C} + \overline{A}\overline{B}}$, (*b*) $\mathrm{A + B + \overline{C}}$,
(*c*) A(B + C), (*d*) A is a minimal expression.

3. Draw a truth table for the equations given below:

(*a*) Y = AC + AB, (*b*) $\mathrm{Y = A(\overline{B} + \overline{C})}$

4. Draw a truth table for the equations given below:

(*a*) $\mathrm{Y = AB(B + C + \overline{D})}$ (*b*) $\mathrm{Y = (A + B + C)A\overline{B}}$

(*c*) $\mathrm{Y = ABC(C + \overline{D})}$ (*d*) Y = AB + BA + C(A + B)

5. Reduce the following Boolean expressions:

(*a*) ABC(ABC + 1) (*b*) $\mathrm{A + \overline{A} + B\ C}$

(*c*) AAB + ABB + BCC (*d*) AB + B + A + C

[Ans. (*a*) ABC, (*b*) 1, (*c*) AB + BC, (*d*) A + B + C]

6. Reduce the following Boolean expressions:

(*a*) $\mathrm{AB + BB + C + \overline{B}}$ (*b*) $\mathrm{A(\overline{A} + B)}$ (*c*) AB(B + C) (*d*) ABB(ABC + BC) (*e*) (AB + C)(AB + D)

[Ans. (*a*) 1, (*b*) AB, (*c*) AB, (*d*) ABC, (*e*) AB + CD]

7. Reduce the given Boolean expressions:

(*a*) $\mathrm{A\overline{B}C + \overline{A}\overline{B}C}$ (*b*) $\mathrm{A\overline{B}C + \overline{A}BC + ABC}$ (*c*) $\mathrm{(\overline{A}B)(AB) + AB}$ (*d*) (1 + B)(ABC)

(*e*) $\mathrm{(A\overline{B} + A\overline{C})(BC + B\overline{C})(ABC)}$

[Ans. (*a*) BC (*b*) AC + BC, (*c*) AB (*d*) ABC, (*e*) 0]

8. Reduce the given Boolean expressions:

(*a*). $\mathrm{A + \overline{B}C(A + \overline{B}C)}$ (*b*) $\mathrm{A(\overline{A}\overline{B}\overline{C} + A\overline{B}C)}$

(*c*) $\mathrm{\overline{ABC + \overline{A}B + BC}}$ (*d*) $\mathrm{A[B + C\overline{(AB + AC)}]}$

[Ans. (*a*) $\mathrm{A + \overline{B}C}$ (*b*) $\mathrm{A(\overline{B} + \overline{C})}$ (*c*) $\mathrm{\overline{B} + \overline{C}}$, (*d*) AB]

9. Derive the complements of the expressions given below:

(*a*) $(A + BC + AB)$ (*b*) $(A + B)(B + C)(A + C)$

(*c*) $AB + BC + CD$ (*d*) $\mathrm{AB(\overline{C}D + \overline{B}C)}$

(*e*) $\mathrm{AB(B + C)(\overline{C} + \overline{D})}$

[Ans. (*a*) $\mathrm{\overline{A}(\overline{B} + \overline{C})}$ (*b*) $\mathrm{\overline{A}\overline{B} + \overline{B}\overline{C} + \overline{A}\overline{C}}$ (*c*) $\mathrm{\overline{A}\overline{C} + \overline{B}\overline{C} + \overline{B}\overline{D}}$ (*d*) $\mathrm{\overline{ABD} + C}$ (*e*) $\mathrm{\overline{A} + \overline{B}\overline{C} + CD}$]

10. Simplify each of the following expressions:

(*a*) $Y = RST + RS + (\overline{T} + V)$ (*b*) $Y = (M + N)(\overline{M} + P)(\overline{N} + P)$

[**Ans.** (*a*) $RS + \overline{T} + V$ (*b*) $P(M + N)$

11. Using Boolean techniques, simplify the given expressions:

(*a*) $AB + A(B + C) + B(B + C)$ (*b*) $AB(C + B\overline{D}) + \overline{AB}$

[**Ans.** (*a*) $B + AC$, (*b*) $\overline{AB\overline{C}D}$]

12. Simplify the following expressions:

(*a*) $A + AB + A\overline{B}C$ (*b*) $(\overline{A} + B)C + ABC$

(*c*) $A\overline{B}C(BD + CDE) + A\overline{C}$ [**Ans.** (*a*) A, (*b*) $C(\overline{A} + B)$ (*c*) $A(\overline{C} + \overline{B}DE)$]

13. Apply DeMorgan's theorem to each of the given expressions:

(*a*) $\overline{A(B + C)}$ (*b*) $\overline{AB} + \overline{CD}$ (*c*) $\overline{AB + CD}$ (*d*) $(A + \overline{B})(\overline{C} + D)$

[**Ans.** (*a*) $\overline{A} + \overline{B} \cdot \overline{C}$ (*b*) $\overline{A} + \overline{B} + \overline{C} + \overline{D}$ (*c*) $(\overline{A} + \overline{B})(\overline{C} + \overline{D})$ (*d*) $\overline{A}B + C\overline{D}$]

14. Apply DeMorgan's theorem to the given expressions:

(*a*) $\overline{A\overline{B} + (C + D)}$ (*b*) $\overline{AB + (CD + EF)}$ (*c*) $\overline{(A + \overline{B} + C + D)} + \overline{ABCD}$

(*d*) $\overline{(\overline{A} + B + C + D) + A\overline{B}\overline{C}D}$ (*e*) $\overline{(\overline{AB}(CD + \overline{E}F)(\overline{AB} + \overline{CD})}$

[**Ans.** (*a*) $(\overline{A} + B)\overline{C}D$ (*b*) $(\overline{A} + B)(\overline{C} + \overline{D}) \cdot (\overline{E} + \overline{F})$

(*c*) $\overline{ABC\overline{D}}$ (*d*) $A\overline{BC}D$ (*e*) $AB + (E + \overline{F}) + (\overline{C} + \overline{D})$]

15. Simplify the given expressions using the Boolean Algebra method.

(*a*) $BD + B(D + E) + \overline{D}(D + F)$ (*b*) $\overline{A}\overline{B}C + \overline{(A + B + \overline{C})} + \overline{A}\overline{B}\overline{C}D$

(*c*) $(B + BC)(B + \overline{B}C)(B + D)$ (*d*) $ABCD + AB(\overline{CD}) + (\overline{AB})CD$

(*e*) AB AB + $\overline{C}(BC + AC)$

[**Ans.** (*a*) $BD + BE + \overline{D}E$ (*b*) $\overline{A}\overline{B}(C + D)$ (*c*) B

(*d*) $AB + CD$ (*e*) ABC

16. Give a Boolean expression for the following statements:

(*a*) *Y* is a 1 only if A is a 1 and B is a 1 or if A is a 0 and B is a 0.

(*b*) Y is a 1 only if A, B and C are all Is or i£only one of the variables is a 0.

[**Ans.** (*a*) $Y = AB + \overline{A}\overline{B}$ (*b*) $ABC + \overline{A}BC + A\overline{B}C + AB\overline{C}$]

17. Simplify the following logic expressions:

(*a*) $(\overline{A} + B + \overline{C})(\overline{A} + B + D + E)(C + D)$ (*b*) $F = \overline{A}\overline{B}C + BC + AC$

[**Ans.** (*a*) $(\overline{A} + B)(C + D) + \overline{C}D$ (*b*) C.]

18. Simplify the following expressions using the simplification theorem.

(*a*) $A + \overline{A}B + \overline{(A + B)}C + \overline{(A + B + C + D)}$ (*d*) $A\overline{B} + AC + BCD + \overline{D}$

[**Ans.** (*a*) $A+B + C + \overline{D}$ (*b*) $A\overline{B} + BC + \overline{D}$]

19. Using the absorption theorem, simplify the following expressions:

(*a*) $A + \overline{A}\overline{B} + BCD + B\overline{D}$ (*b*) $A\overline{B}C + (\overline{B} + \overline{C})(\overline{B} + \overline{D}) + \overline{A + C + D}$

[**Ans.** (*a*) $A + \overline{B} + \overline{D}$ (*b*) $\overline{B} + \overline{C}\overline{D}$]

20. Prove the following using Boolean theorems:

(*a*) $(A + C)(A + D)(B + C)(B + D) = AB + CD$

(*b*) $(\bar{A} + \bar{B} + \bar{D})(\bar{A} + B + \bar{D})(B + C + D)(A + \bar{C})(A + \bar{C} + D) = \overline{ACD} + AC\bar{D} + \overline{BCD}$

21. (*a*) Convert $f = ABCD + \bar{A}BC + \overline{BC}$ into a sum of minterms by algebraic method.

(*b*) Convert $f = AB + \bar{B}CD$ into a product of maxterms by algebraic method.

[**Ans.** (*a*) $f = A\bar{B}CD + A\bar{B}\bar{C}D + \bar{A}BC\bar{D} + \bar{A}\bar{B}\bar{C}D + A\bar{B}\bar{C}\bar{D} + \bar{A}\bar{B}C\bar{D}$

(*b*) $f = (A + B + C + D)(A + B + C + \bar{D})(A + B + \bar{C} + D)(A + \bar{B} + C + D)$

$(A + \bar{B} + C + \bar{D})(A + \bar{B} + \bar{C} + D)(A + \bar{B} + \bar{C} + \bar{D})(\bar{A} + B + C + D)$

$(\bar{A} + B + C + \bar{D})(\bar{A} + B + \bar{C} + D)$]

22. Obtain the canonical sum of products and product of sums of the following expression:

$f = x_1x_2x_3 + x_1x_3x_4 + x_1x_2x_4$

[**Ans.** $f = x_1x_2x_3x_4 + x_1x_2x_3\bar{x}_4 + x_1\bar{x}_2x_3x_4$]

[Canonical SOP]

$$f = (x_1 + x_2 + x_3 + x_4)(x_1 + x_2 + x_3 + \bar{x}_4)(x_1 + x_2 + \bar{x}_3 + x_4)(x_1 + x_2 + \bar{x}_3 + \bar{x}_4)$$
$$(x_1 + \bar{x}_2 + x_3 + x_4)(x_1 + \bar{x}_2 + x_3 + \bar{x}_4)(x_1 + \bar{x}_2 + \bar{x}_3 + x_4)(x_1 + \bar{x}_2 + \bar{x}_3 + \bar{x}_4)$$
$$(\bar{x}_1 + x_2 + x_3 + x_4)(\bar{x}_1 + x_2 + x_3 + x_4)(\bar{x}_1 + x_2 + \bar{x}_3 + x_4)(\bar{x}_1 + \bar{x}_2 + x_3 + \bar{x}_4)$$

[Canonical POS]

23. Using the K-map method, simplify the following function, obtain their (*i*) minimum sum of products, and (*ii*) minimum product of sums form.

$$f(w,x,y,z) = \Sigma(1, 3, 4, 5, 6, 7, 9, 12, 13)$$
$$f(w, x, y, z) = \Sigma(1, 5, 6, 7, 11, 12, 13, 15)$$

[**Ans.** (*a*) (*i*) $f = \bar{w}z + \bar{w}x + \bar{y}z + x\bar{y}$ (*ii*) $f = (\bar{w} + \bar{y})(x + z)$

(*b*) (*i*) $f = x(w \oplus y) + z(\overline{w \oplus y})$

(*ii*) $f = (w + y + z)(\bar{w} + x + y)(w + x + \bar{y})(\bar{w} + \bar{y} + z)$]

24. Using the Quine McCluskey method and K-map method, obtain the minimal sum of product expression of the following function.

$$Y = 1(0, 2, 3, 6, 7, 8, 10, 11, 12, 15)$$

[**Ans.** $Y = \bar{A}C + CD + \bar{B}\bar{D} + A\bar{C}\bar{D}$]

25. Determine the don't care condition in the following Boolean expression $BE + \bar{B}D\bar{E}$ which js a simplified version of the expression $\bar{A}BE + BCDE + B\bar{C}\,\bar{D}E + \bar{A}\bar{B}D\bar{E} + \bar{B}\bar{C}D\bar{E}$.

[**Ans.** don't care combinations:

$A\bar{B}CD\bar{E} + AB\bar{C}DE + ABC\bar{D}E$]

26. Using the Quine McCluskey method and *K*-map method, simplify the following functions into minimal sum of products:

$$f(u, w, x, y, z) = 1(0, 2, 5, 7, 9, 11, 13, 15, 16, 18, 21, 23, 25, 27, 29, 31)$$

[**Ans.** $f = wz + xz + \overline{wxz}$]

27. A corporation having 100 shares entitles the owner of each share to cast one vote at the shareholders' meeting. Assume that *A* has 40 shares, *B* has 30 shares, *C* has 20 shares and *D* has 10 shares. A two-third majority is required to pass a resolution in a shareholders' meeting. Each of these four men has a switch which he closes to vote YES and opens to vote NO for his percentage of shares. When the resolution is passed the output, LED must be ON. Derive a truth table for the output function and give the sum of product equation for it.

[**Ans.** $f = AB\bar{C}\bar{D} + A\bar{B}CD + ABC\bar{D} + AB\bar{C}D + ABCD = AB + ACD$]

28. (*a*) Express the following function as a product of maxterms $f = Z(1, 3, 5, 7)$

(*b*) Express the complement of the function as a sum of minterms.

(*c*) Express the complement of the function as a product of maxterms.

[**Ans.** (*a*) $(A + B + C)(\bar{A} + B + C)(\bar{A} + \bar{B} + C)$

(*b*) $\bar{A}\bar{B}\bar{C} + \bar{A}B\bar{C} + A\bar{B}\bar{C} + AB\bar{C}$

(*c*) $(A + B + \bar{C})(A + \bar{B} + \bar{C})(\bar{A} + B + \bar{C})(\bar{A} + \bar{B} + \bar{C})$]

29. Prepare Karnaugh maps for the following functions:

(*a*) $f = ABC + \bar{A}BC + \bar{B}\bar{C}$ (*b*) $f = A + B + \bar{C}$ (*c*) $f = AB + \bar{B}CD$

$f(u, w, x, y, z) = \Sigma(0, 2, 5, 7, 9, 11, 13, 15, 16, 18, 21, 23, 25, 27, 29, 31)$

[**Ans.** $f = wz + xz + \overline{wxz}$]

❑❑❑

3 Logic Gates

3.1 INTRODUCTION

Boolean algebra is used in describing and simplifying logic circuits. Simplification of Boolean logic expressions is very important because it reduces the hardware required to design a specific system. The Boolean expression corresponding to a given gate network can be derived by systematically progressing from the input to the output of gates. The gating or logic network can be formed by interconnecting the OR, AND and NOT gates.

Combinational circuit: A combinational circuit consists of input variables, logic gates and output variables. The design of combinational circuits starts from the verbal outline of the problem or from a set of Boolean functions, and ends in a logic circuit diagram. The steps involved in the design of combinational circuits are as follows:

(*i*) State the problem in words.

(*ii*) Find the number of input and output variables.

(*iii*) Assign letter symbols to the input and output variables.

(*iv*) Obtain the truth table using the word statement.

(*v*) Obtain Boolean expressions for each output from the truth table.

(*vi*) Simplify the Boolean expressions to minimise the number of variables by using laws of Boolean algebra or Karnaugh map method or McCluskey's method.

(*vii*) Draw the logic circuit diagram corresponding to the simplified Boolean expression.

3.2 POSITIVE AND NEGATIVE LOGIC DESIGNATION

Logics 1 and 0 are generally represented by voltage levels. In a Positive logic system, the most positive voltage level (HIGH) represents the logical 1 state, and the most negative voltage level (LOW) represents the logical 0 state. In a Negative logic system, the most positive (HIGH) voltage level represents logical 0 state and the most negative (LOW) voltage level represents logical 1 state. For example, if the voltage levels are – 0.1 V and – 5 V, then in a positive logic system the – 5 V level represents a 0 state and the – 0.1 V level represents a 1 state; in a negative logic system, – 0.1 V level represents a 0 state and – 5 V represents a 1 state. Conversely, if the voltage levels are 0. 1 V and 5 V, then in a positive logic system the

5 V level represents a 1 state and the 0. 5 V represents a 0 state; in a negative logic system, the 0. 1 V represents a 1 state and the 5 V level represents a 0 state.

The effect of changing from one logic designation to the other is equivalent to complementing the logic function. The simple method of converting the logic designation (i.e. from positive to negative logic or vice versa) is that all 0s are replaced with 1s and all 1s with 0s in the Truth Table. The resulting logic function is determined accordingly. For example, when 0s and 1s are interchanged in the truth table, the positive logic AND gate becomes negative logic OR gate, and positive logic NAND gate becomes negative logic NOR gate. As there is no real advantage to either designation, the choice of positive or negative logic is made by the individual logic designer.

3.3 LOGIC GATES

A logic gate is an electronic circuit which makes logical decisions. To arrive at these decisions, the most common logic gates used are OR, AND, NOT, NAND and NOR gates. The NAND and NOR gates are called as the Universal gates. The exclusive-OR gate is another logic gate which can be constructed using basic gates such as AND, OR and NOT gates.

Do You Know?	The logic circuit is constructed by using logic gates. Minimum number of gates must be used to reduce the hardware and this can be achieved by simplification of Boolean functions. The minimization of Boolean functions for logic gates is known as gate level minimization.

Logic gates have two or more inputs and only one output except for the NOT gate, which has only one input. The output signal appears only for certain combinations of the input signals. The manipulation of binary information is done by the gates. The logic gates are the building blocks of hardware which are available in the form of various IC families. Each gate has a distinct logic symbol and its operation can be described by means of an algebraic function. The relationship between input and output variables of each gate can be represented in a tabular form called a truth table.

3.3.1 OR Gate

The OR gate performs logical addition, commonly known as OR function. The OR gate has two or more inputs and only one output. The operation of OR gate is such that a HIGH(1) on the output is produced when any of the inputs is HIGH(1). The output is LOW(0) only when all the inputs are LOW(0).

If A and B are the input variables of an OR gate and Y is its output, then

$$Y = A + B$$

Similarly, for more than two variables, the OR function can be expressed as

$$Y = A + B + C + D + \ldots$$

An OR gate using diodes is shown in Fig. 3.1(*a*) in which *A* and *B* represent the inputs and Y the output. The resistance R_L is the load resistance.

If $A = 0$ and $B = 0$, both the diodes will not conduct, and hence the output $Y = 0$.

If $A = 1$ and $B = 0$, diode D_1 conducts, then $V_0 = 5$ V and so $Y = 1$.

If $A = 0$ and $B = 1$, diode D_2 conducts and hence $Y = 1$.

If A = 1 and $B = 1$, both the diodes conduct and hence $Y = 1$.

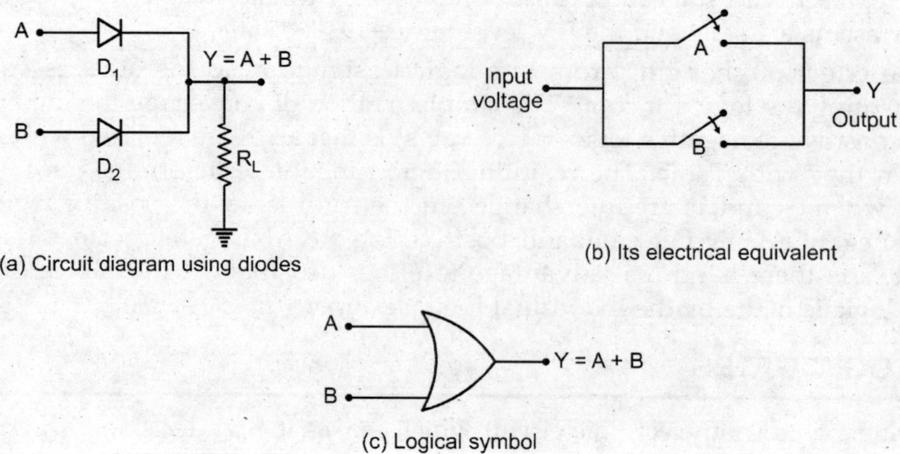

(a) Circuit diagram using diodes

(b) Its electrical equivalent

(c) Logical symbol

Fig. 3.1. 2-input OR gate.

The electrical equivalent circuit of an OR gate is shown in Fig. 3.1(*b*) where switches *A* and *B* are connected in parallel. If either *A* or *B* is closed or if both are closed, then the output is high. The logic symbol for a 2-input OR gate is shown in Fig. 3.1(*c*). The logical operation of the two-input OR gate is described in the truth table shown in Table 3.1.

Table 3.1. Truth table of OR gate

Input		Output
A	B	$Y = A + B$
0	0	0
0	1	1
1	0	1
1	1	1

The same idea can be extended to an OR gate with more than two inputs. Fig. 3.2 shows a 3-input OR gate. Table 3.2 gives the truth stable of a 3-input OR gate.

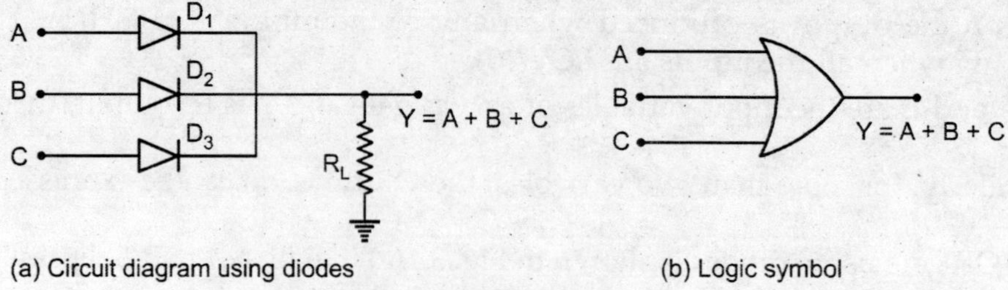

(a) Circuit diagram using diodes

(b) Logic symbol

Fig. 3.2. 3-input OR gate.

Table 3.2. Truth table of 3-input OR gate

Inputs			Output
A	B	C	$Y = A + B + C$
0	0	0	0
0	0	1	1
0	1	0	1
0	1	1	1
1	0	0	1
1	0	1	1
1	1	0	1
1	1	1	1

In general, if n is the number of input variables, then there will be 2^n possible combinations, since each variable can take on either of two values.

3.3.2 AND Gate

The AND gate performs logical multiplication, commonly known as AND function. The AND gate has two or more inputs and a single output. The output of an AND gate is HIGH only when all the inputs are HIGH. Even if any one of the inputs is LOW, the output will be LOW.

If A and B are the input variables of an AND gate and Y is its output, then

$$Y = A \cdot B$$

where the dot (.) denotes the AND operation. Moreover, one typically deletes the dot and writes as $Y = AB$.

A 2-input AND gate using diodes is shown in Fig. 3.3(*a*) in which A and B represent the inputs and Y the output.

If $A = 0$ and $B = 0$, both the diodes conduct as they are forward biased, and hence the output is $Y = 0$.

If $A = 0$ and $B = 1$, the diode D_1 conducts and D_2 does not conduct, and hence the output is $Y = 0$.

If $A = 1$ and $B = 0$, the diode D_1, does not conduct and D_2 conducts, and hence the output is Y = 0.

If $A = 1$ and $B = 1$, both the diodes do not conduct as they are reverse biased, and hence the output is $Y = 1$.

The electrical equivalent circuit of an AND gate is shown in Fig. 3.3(*b*) where two switches A and B are connected in series. If both A and B are closed, then the output is high. Logic symbol of the 2-input AND gate is shown in Fig. 3.3(*c*). The logical operation of the two input AND gate and the three-input AND gate are described in the truth tables shown in Tables 3.3. and 3.4.

Table 3.3. Truth table of a 2-input AND gate

Input		Output
A	*B*	$Y = A \cdot B$
0	0	0
0	1	0
1	0	0
1	1	1

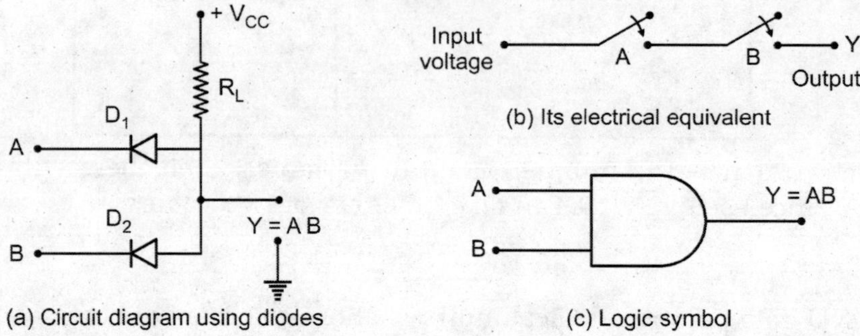

Fig. 3.3. 2-input AND gate.

From Table 3.4, it is seen that the AND gate has a HIGH output only when A, B and C are HIGH. When there are more inputs, all inputs must be HIGH for a HIGH output. For this reason, the AND gate is also called an ALL Gate.

Table 3.4. Truth table of a 3-input AND gate

Inputs			Output
A	*B*	*C*	$Y = A \cdot B \cdot C$
0	0	0	0
0	0	1	0
0	1	0	0
0	1	1	0
1	0	0	0
1	0	1	0
1	1	0	0
1	1	1	1

3.3.3 NOT Gate (Inverter)

The NOT gate performs the basic logical function called inversion or complementation. The purpose of this gate is to convert one logic level into the opposite logic level. It has one input and one output. When a HIGH level is applied to an inverter, a LOW level appears at its output and vice versa.

A NOT gate using a transistor is shown in Fig. 3.4(*a*) in which A represents the input and *Y* represents the output, i.e. $Y = \overline{A}$. When the input is HIGH, the transistor is in the ON state and the output ($V_C = V_{CE(sat)}$) is LOW. If the input is LOW, the transistor is in the OFF state and the output $V_C = V_{CC}$ is HIGH. The symbol for the inverter is shown in Fig. 3.4(*b*). The truth table of a NOT gate is given in Table 3.5.

Table 3.5. Truth table of an inverter

Input A	Output $Y = \overline{A}$
0	1
1	0

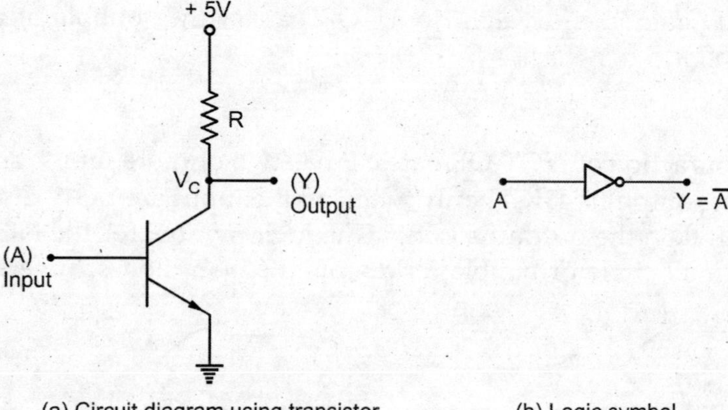

(a) Circuit diagram using transistor (b) Logic symbol

Fig. 3.4. NOT gate.

3.3.4 NAND Gate

NAND is a contraction of the NOT–AND gates. It has two or more inputs and only one output, i.e. $Y = \overline{A \cdot B}$. When all the inputs are HIGH, the output is LOW. If any one or both the inputs are LOW, then the output is HIGH. the logic symbol for the NAND gate is shown in Fig. 3.5(*a*). The small circle or bubble represents the operation of inversion.

(a) Logic symbol of NAND gate

(b) NAND gate ≡ Bubbled OR gate

Fig. 3.5. NAND gate.

The truth table for the NAND gate is shown in Table 3.6.

Table 3.6. Truth table of a 2-input NAND gate.

Input		Output
A	B	$Y = \overline{A \cdot B}$
0	0	1
0	1	1
1	0	1
1	1	0

The NAND gate is equivalent to an OR gate with a bubble at its inputs which is shown in Fig. 3.5(*b*).

3.3.5 NOR Gate

NOR is contraction of NOT-OR gates. It has two or more inputs and only one output, i.e. $Y = \overline{A+B}$. The output is HIGH only when all the input are LOW. If any one or both the inputs are HIGH, then the output is LOW. The logic symbol for the NOR gate is shown in Fig. 3.6(*a*). The small circle or bubble represents the operation of inversion.

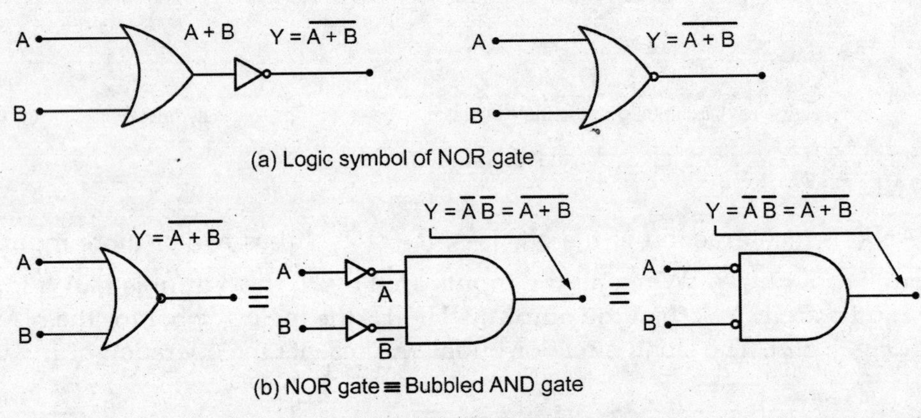

Fig. 3.6. NOR gate.

The truth table of a two input NOR gate is shown in Table 3.7.

Table 3.7. Truth table of a 2-input NAND gate.

Input		Output
A	B	$Y = \overline{A + B}$
0	0	1
0	1	0
1	0	0
1	1	0

The NOR gate is equivalent to an AND gate with a bubble at its input. This is shown in Fig. 3.6(*b*).

3.3.6 Universal Gates/Universal Building Blocks

NAND and NOR gates are called Universal gates or universal building blocks because both can be used to implement any gate like AND, OR and NOT gates or any combination of these basic gates. Fig. 3.7 shows how a NAN gate can be used to realise various logic gates while Fig. 3.8 shows how a NOR gate can be used for the same.

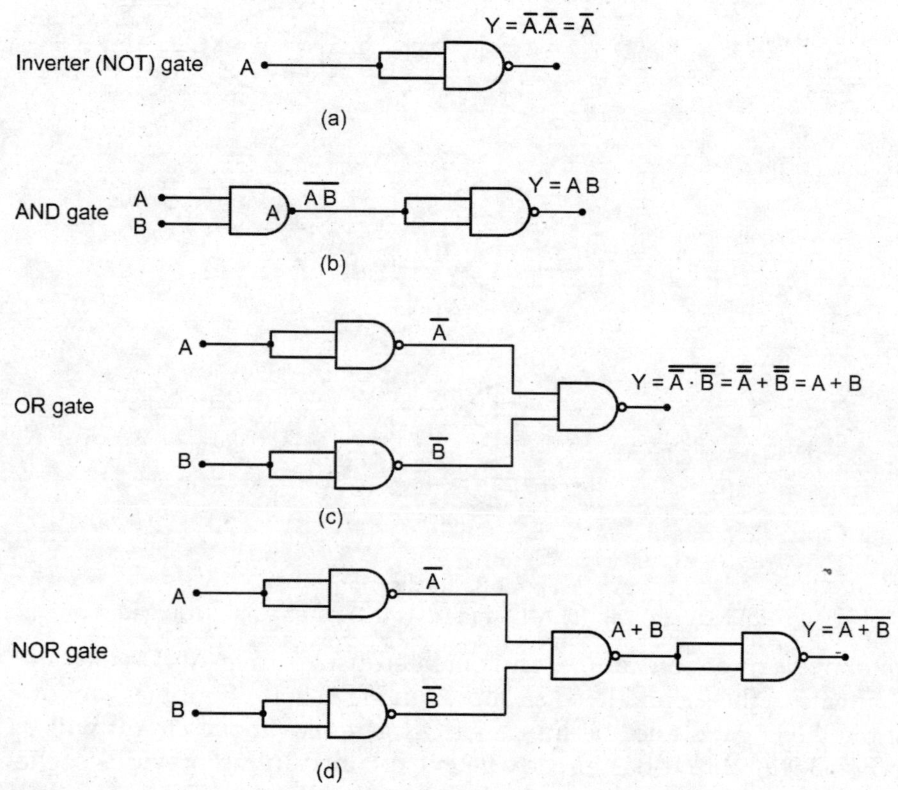

Fig. 3.7. Realisation of (*a*) NOT, (*b*) AND, (*c*) OR and (*d*) NOR gates using NAND gates.

Realisation of logic function using NAND gates: Any logic function can be implemented using NAND gates. To achieve this, first the logic function has to be written in SOP (Sum of Products) form. Then, this can be easily realised using NAND gates. In other words, a logic gate circuit with AND gates in the first level and OR gates in the second level can be converted into a NAND-NAND gate circuit. To understand this concept, consider the following SOP expression.

$$Y = ABC + BCD + ACD$$

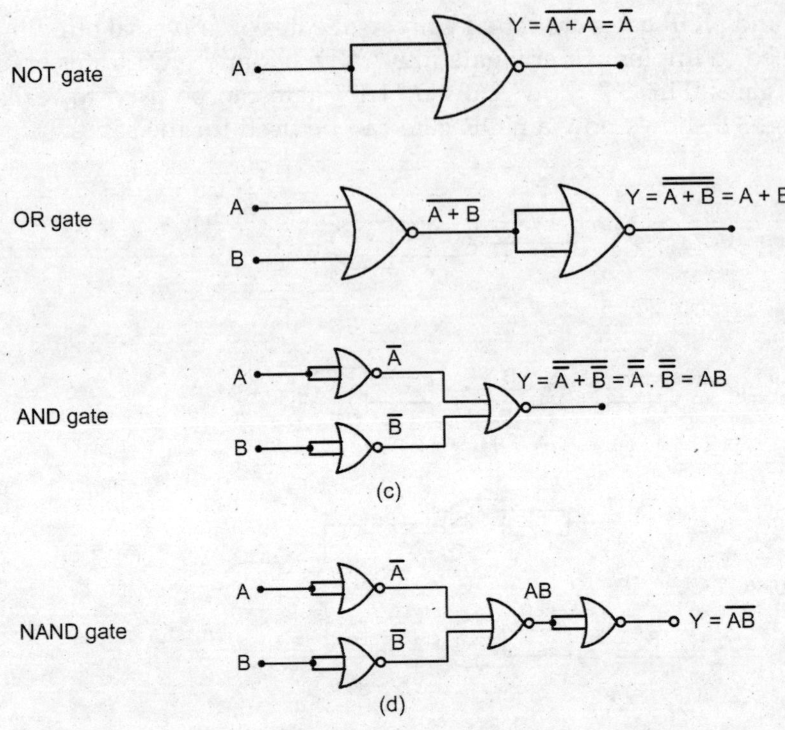

Fig. 3.8. Realisation of
(*a*) NOT, (*b*) OR, (*c*) AND and (*d*) NAND gates using NOR gates.

The above expression can be implemented using three AND gates in the first level and one OR gate in the second level as shown in Fig. 3.9(*a*). If bubbles are introduced at the output of the AND gates and the inputs of OR gate, the above circuit will be modified as shown in Fig. 3.9(*b*). But, it has already been explained in the previous section that an OR gate with bubbles at its inputs in equivalent to a NAND gate. Therefore, the OR gate with bubbled inputs in Fig. 3.9(*b*) can be replaced by a NAND gate as shown in Fig. 3.9(*c*). Now, the above SOP expression is implemented using only NAND gates.

Realisation of logic functions using NOR gates: Any logic function can also be implemented using NOR gates. To achieve this, first the logic function has two be written in POS (product of Sums) form. Then, this can be easily realised using only NOR gates. In other words, a logic gate circuit with OR gates in the first level and AND gates in the second level can be converted into a NOR-NOR gate circuit. To understand this concept, consider the following POS expression.

$$Y = (A + B + C)(B + C + D)(A + B + D)$$

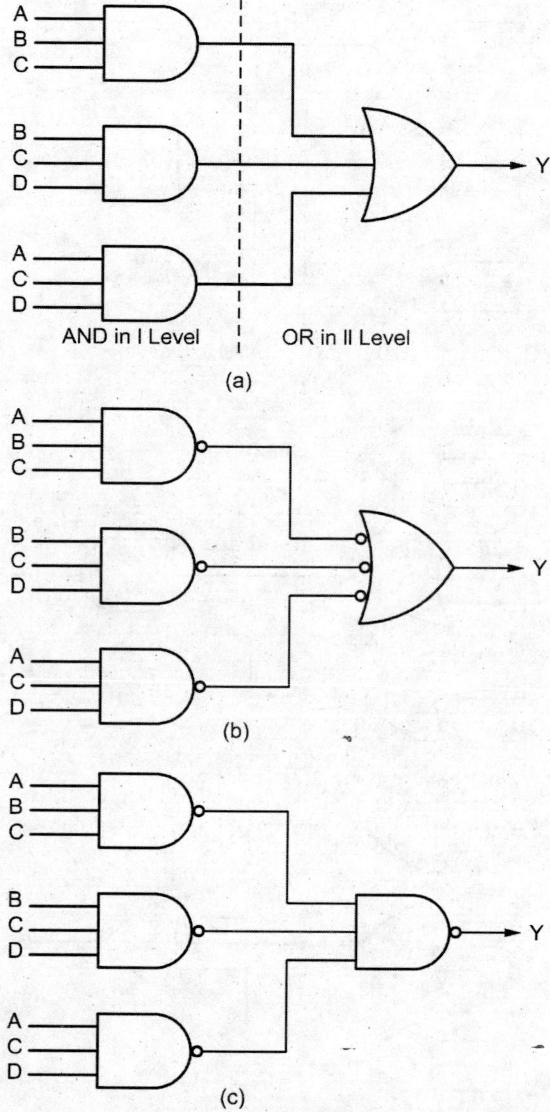

Fig. 3.9. Realisation of a logic expression (SOP form) using NAND gates.

The POS expression can be implemented using three OR gates in the first level and one AND gate in the second level as shown in Fig. 3.10(*a*). If bubbles are introduced at the output of the OR gates and the inputs of the AND gate, the above circuit will be modified as shown in Fig. 3.10(*b*). But, an AND gate with bubble at its inputs is equivalent to a NOR gate. Therefore, the AND gate with bubbled input in Fig. 3.10(*b*) can be replaced by a NOR gate as shown in Fig. 3.10(*c*). Now, the above POS expression is implemented using only NOR gates.

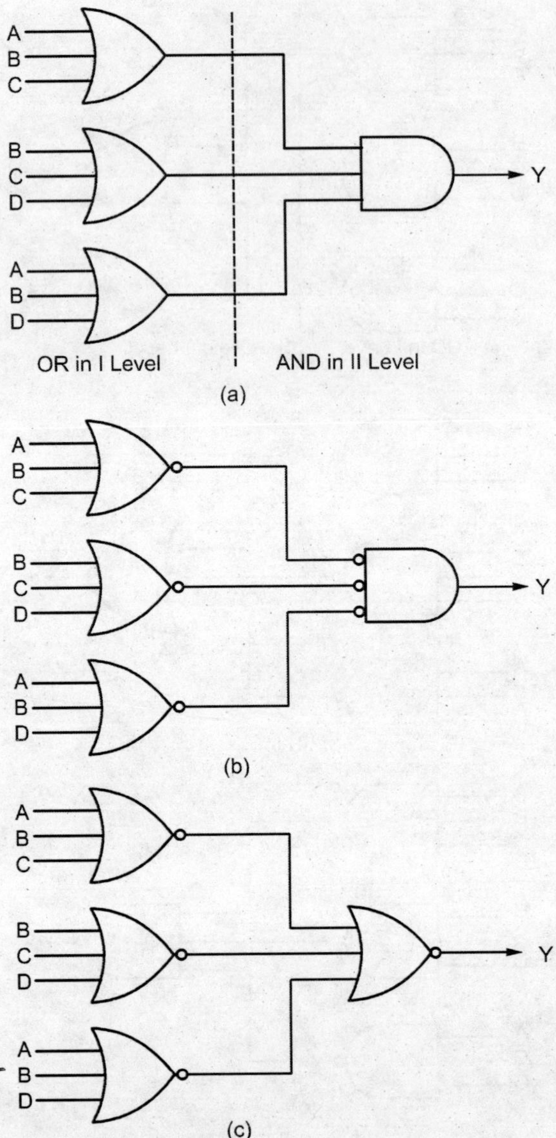

Fig. 3.10. Realisation of a logic expression (POS form) using NOR gates.

3.3.7 Exclusive-OR (Ex-OR) Gate

An Exclusive-OR gate is a gate with two or more inputs and one output. The output of a two-input Ex-OR gate assumes a HIGH state if one and only one input assumes a HIGH state. This is equivalent to saying that the output is HIGH if either input A or input B is HIGH exclusively, and low when both are 1 or 0 simultaneously.

The logic symbol for the Ex-OR gate is shown in Fig. 3.11(*a*) and the truth table for the Ex-OR operation is given in Table 3.8.

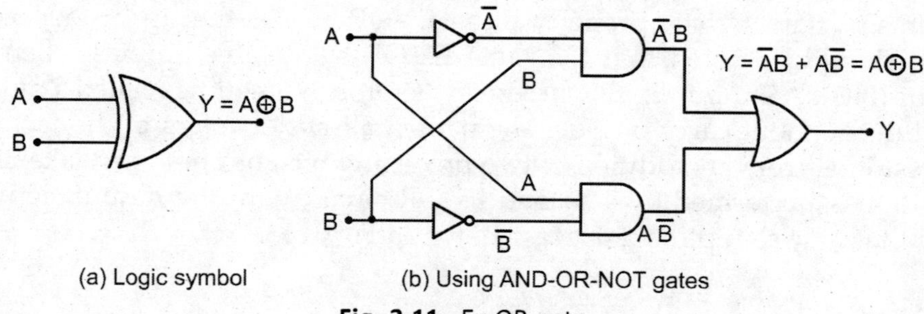

(a) Logic symbol (b) Using AND-OR-NOT gates

Fig. 3.11. Ex-OR gate.

Table 3.8. Truth table of a 2-input Ex-OR gate

Input		Output
A	*B*	$Y = A \oplus B$
0	0	0
0	1	1
1	0	1
1	1	0

The truth table of the Ex-OR gate shows that the output is HIGH when any one, but not all, of the inputs is at 1. This exclusive feature eliminates a similarity to the OR gate. The Ex-OR gate responds with a HIGH output only when an odd number of inputs is HIGH. When there is an even number of HIGH inputs, such as two or four, the output will always be LOW. From the truth table of a 2-input Ex-OR gate, the Ex-OR function can be written as $Y = \bar{A}B + A\bar{B} = A \oplus B$.

The above expression can be read as Y equals A Ex-OR B. Using the above expression, a 2-input Ex-OR gate can be implemented using basic gates like AND, OR and NOT gates as shown in Fig. 3.11(*b*).

The 2-input Ex-OR gate can also be implemented using NAND gates as shown in Fig. 3.12.

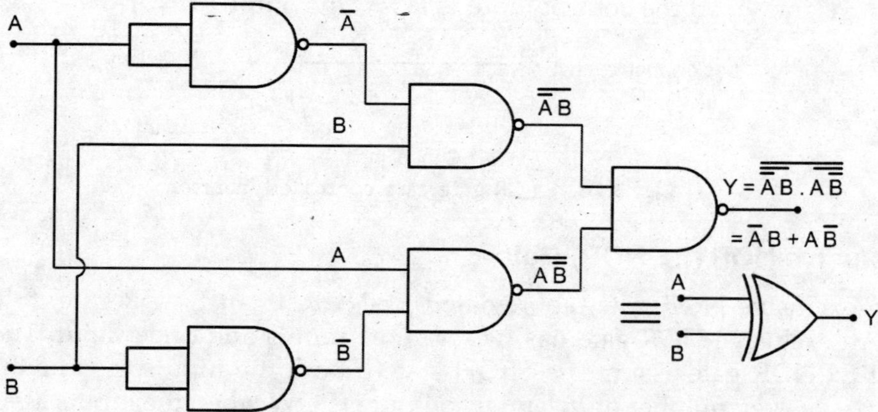

Fig. 3.12. Ex-OR gates using NAND gates.

The main characteristic property of an Ex-OR gate is that it can perform modulo-2 addition. It should be noted that the same Ex-OR truth table applies when adding two binary digits (bits). A 2-input Ex-OR circuit is, therefore, sometimes called a modulo-2 adder or a half-adder without carry output. The name half-adder refers to the fact that possible carry-bit, resulting from an addition of two preceding bits, has not been taken into account. A full addition is performed by a second Ex-OR circuit with the output signal of the first circuit and the carry as input signals, as shown in Fig. 3.13.

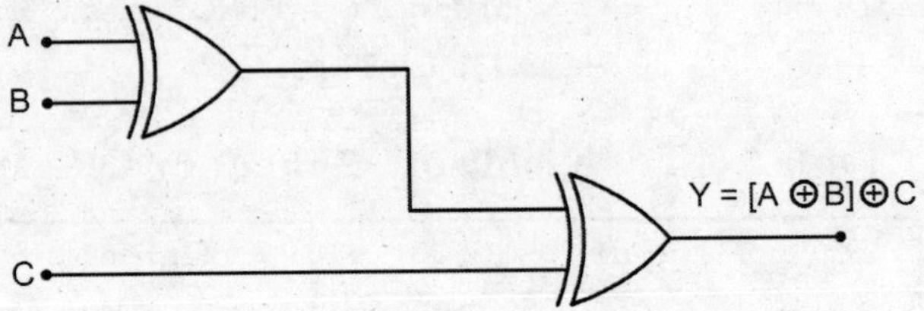

Fig. 3.13. Cascading of two Ex-OR circuits.

The configuration of Fig. 3.13 is a cascading of two Ex-OR circuits resulting in an Ex-OR operation of three variables A, B, and C. Consequently, the sum output of a full adder for two bits is an Ex-OR operation of the 2 bits to be added and the carry of the preceding adding stage. The logic expression of the Ex-OR operation of three variables A, B, and C is given by

$$A \oplus B \oplus C = (A\overline{B} + \overline{A}B)\overline{C} + \overline{(A\overline{B} + \overline{A}B)}C$$

$$= (A\overline{B} + \overline{A}B)\overline{C} + (\overline{A}\overline{B} + AB)C$$

$$A \oplus B \oplus C = A\overline{B}\overline{C} + \overline{A}B\overline{C} + \overline{A}\overline{B}C + ABC$$

In general, Ex-OR operation of n variables results in a logical 1 output if an odd number of the input variables are 1s. An Ex-OR operation of n variables can be obtained by cascading 2-input Ex-OR gates.

Another important property of an Ex-OR gate is that it can be used as a controlled inverter, i.e., by using an Ex-OR gate, a logic variable can be complemented or allowed to pass through it unchanged. This is done by using one Ex-OR input as a control input and the other as the logic variable input as shown in Fig. 3.14. When the control input is HIGH, the output $Y = \overline{A}$ and when the control input is LOW, the output $Y = A$.

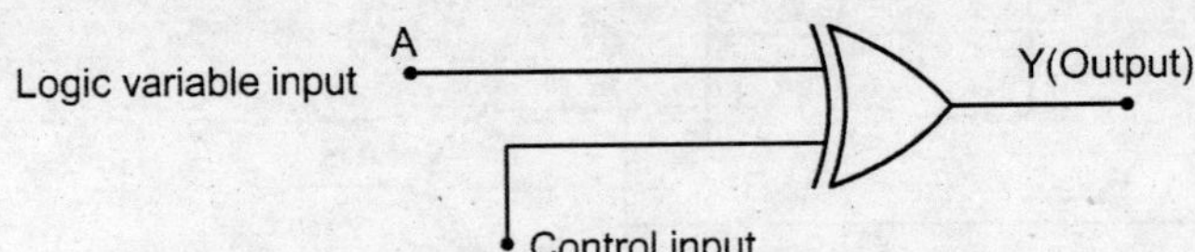

Fig. 3.14. Ex-OR gate as a controlled inverter.

3.3.8 Exclusive-NOR (Ex-NOR) Gate

The exclusive-NOR gate, abbreviated Ex-NOR, is an Ex-OR gate, followed by an inverter. An exclusive-NOR gate has two or more inputs and one output. The output of a two-input Ex-NOR gate assumes a HIGH state if both the inputs assume the same logic state or have an even number of 1s, and its output is LOW when the inputs assume different logic states or have an odd number of 1s. The logic symbol of Ex-NOR gate is shown in Fig. 3.15 and its truth table is given in Table 3.9. From the truth table, it is clear that the Ex-NOR output is the complement of the Ex-OR gate. The Boolean expression for the Ex-NOR gate is

$$Y = \overline{A \oplus B}$$

Read the above expression as "Y equals A Ex-NOR B." According to DeMorgan's theorem,

$$\begin{aligned} \overline{A \oplus B} &= \overline{\overline{A}B} \cdot \overline{A\overline{B}} \\ &= \overline{\overline{A}B} \cdot \overline{A\overline{B}} \\ &= (A + \overline{B})(\overline{A} + B) \\ &= AB + \overline{A}\overline{B} \end{aligned}$$

Table 3.9. Truth table of 2-input Ex-NOR gate

Input		Output
A	B	$Y = \overline{A \oplus B}$
0	0	1
0	1	1
1	0	0
1	1	1

Fig. 3.15. Logic symbol of 2-input Ex-NOR gate.

An important property of the Ex-NOR gate is that it can be used for bit comparison. The output of an Ex-NOR gate is 1 if both the inputs are similar, i.e., both are 0 or 1; otherwise, its output is 0. Hence, it can be used as a one-bit comparator. It is also called a coincidence circuit.

Another property of the Ex-NOR gate is that it can be used as an even-parity checker. The output of the Ex-NOR gate is 1 if the number of 1 s in its inputs is even; if the number of 1s is odd, the output is 0. Hence, it can be used as an even/odd parity checker. Hence, the 2-input Ex-NOR gate is immensely useful for bit comparison and parity checking.

Example 3.1. *Realise the logic expression* $y = \overline{BC} + \overline{AC} + \overline{AB}$ *using basic gates.*

Solution: In the given expression, there are 3 product terms each with two variables which can be implemented using three 2-input AND gates, and the product terms can be OR operated together using a 3-input OR gate. The complemented form of individual variable can be obtained by 3 NOT gates. Thus, the circuit for the given expression is realised as shown in Fig. E3.1.

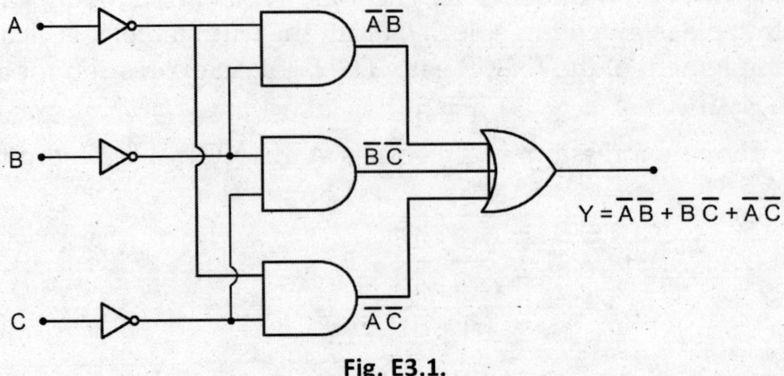

Fig. E3.1.

Example 3.2. *Realise the logic expression $Y = (A + B)\ (\overline{A} + C)\ (B + D)$ using basic gates.*

Solution: In the given expression, there are 3 sum terms which can be implemented using three 2-input OR gates and their outputs are AND operated together by a 3-input AND gate. A NOT gate can be used to obtain the inverse of A. Now, the realised circuit is shown in Fig. E3.2.

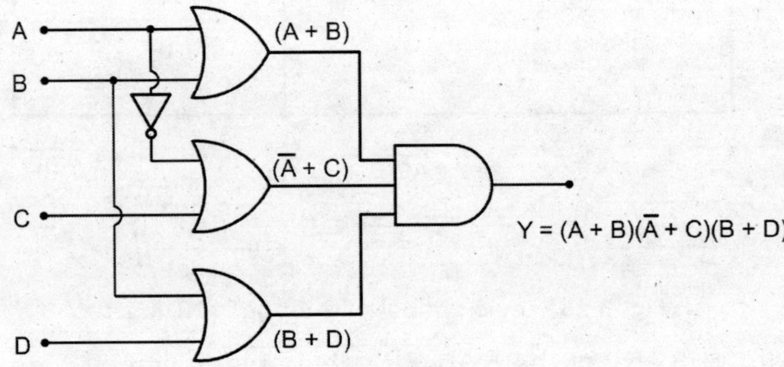

Fig. E3.2.

Example 3.3. *Implement $Y = \overline{AB} + A + \overline{(B + C)}$ using NAND gates only.*

Solution: The implementation of the given function is shown in Fig. E3.3.

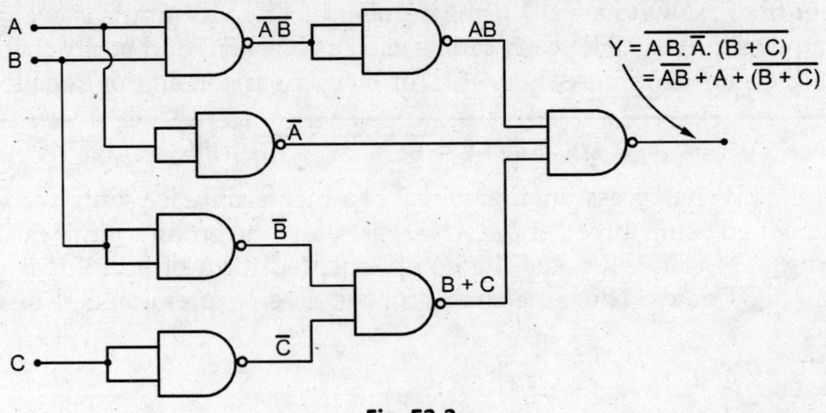

Fig. E3.3.

Example 3.4. *Simplify the logic circuit shown in Fig. E3.4(a).*

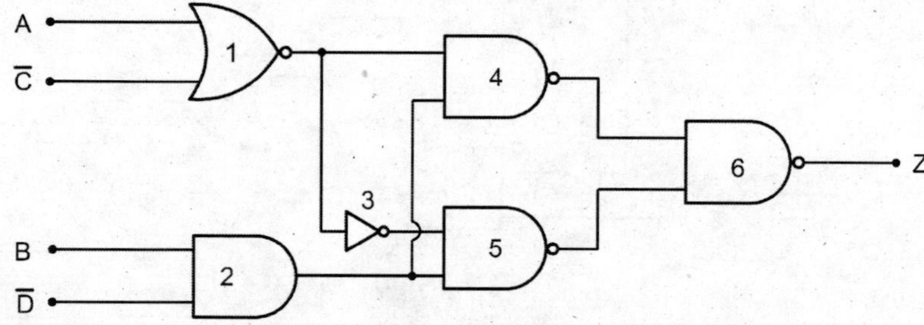

Fig. E3.4 (*a*)

Solution: From the given logic circuit, the expression for Z can be written as

$$Z = \overline{\overline{\overline{(A+\overline{C})}\cdot B\overline{D}}\cdot\overline{(A+\overline{C})\cdot B\overline{D}}}$$

$$= \overline{[\overline{\overline{(A+\overline{C})}}+\overline{B\overline{D}}]\cdot[\overline{(A+\overline{C})}+\overline{B\overline{D}}]}$$

$$= \overline{[(A+\overline{C})+\overline{B\overline{D}}]\cdot[\overline{(A+\overline{C})}+\overline{B\overline{D}}]}$$

$$= \overline{\overline{B\overline{D}}+(A+\overline{C})\overline{(A+\overline{C})}} \qquad [\because \quad (A+B)(A+C) = A+BC]$$

$$= \overline{\overline{B\overline{D}}} \qquad [\because \quad A\overline{A} = 0]$$

$$= B\overline{D}$$

Therefore, the above logic circuit can be simplified as shown in Fig. E3.4(*b*).

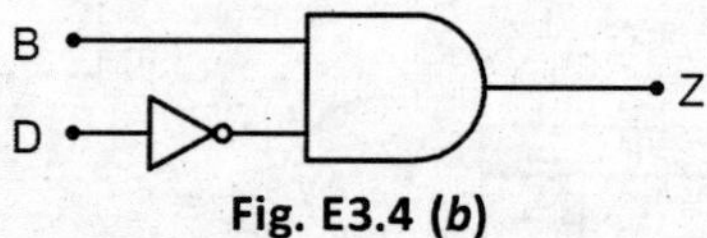

Fig. E3.4 (*b*)

Instead of using Boolean algebra, the logic circuit can be simplified directly as shown below. In the given logic circuit shown in Fig. E3.4(*a*), the NAND gate (6) can be replaced by an OR gate with a bubble at its inputs as shown in Fig.E.3.4(*c*).

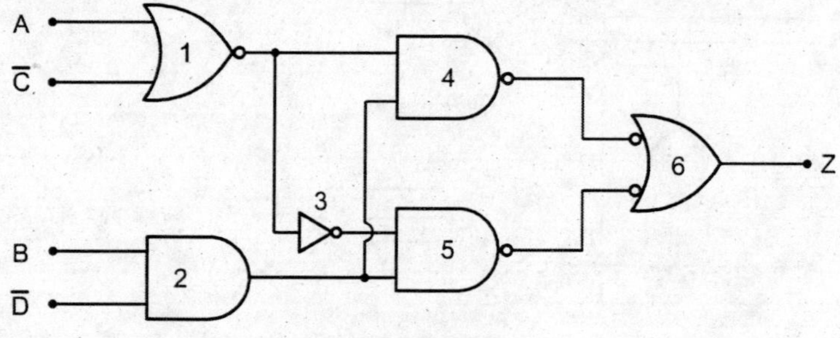

Fig. E3.4 (*c*)

Now, using $\overline{\overline{A}} = A$, the bubble at the outputs of gates 4 and 5 get cancelled with the bubble at the inputs of gate 6 as shown in Fig. E3.4(*d*).

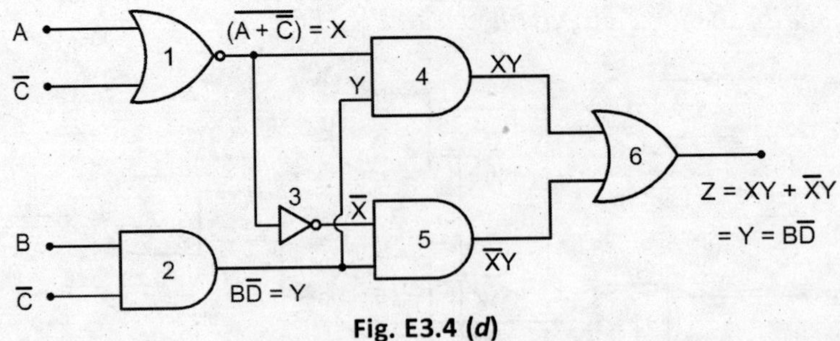

Fig. E3.4 (*d*)

In the above figure, if we assume the output of gate 1$\overline{(A+\overline{C})} = X$ and the output of gate 2 $B\overline{D} = Y$, then the output of gate 4 is XY and the output of gate 5 is $\overline{X}Y$. If XY and $\overline{X}Y$ are OR operated in gate 6, then $XY + \overline{X}Y = Y(X+\overline{X}) = Y = B\overline{D}$. Therefore, the above circuit can be simplified as shown in Fig. E3.4(*e*).

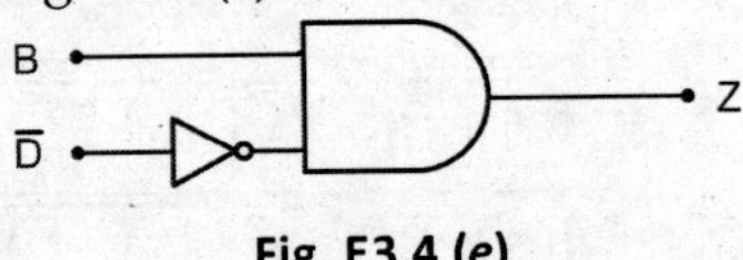

Fig. E3.4 (*e*)

Example 3.5. *Realise* (*a*) $Y = A + BC\overline{D}$ *using NAND gates and*

(*b*) $Y = (A+C)(A+\overline{D})(A+B+\overline{C})$ *using NOR gates*

Solution: (*a*) Using NAND gates

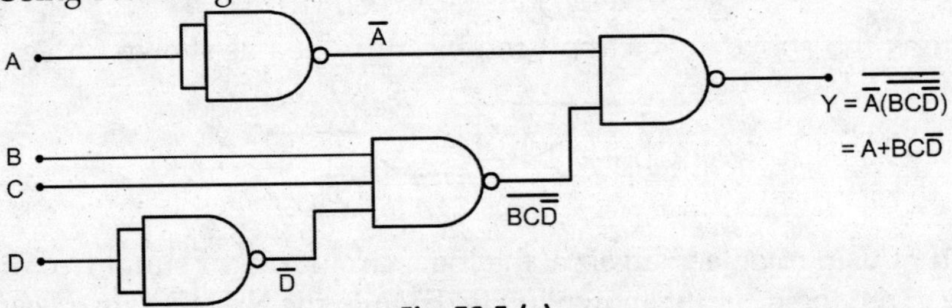

Fig. E3.5 (*a*)

(*b*) Using NOR gates

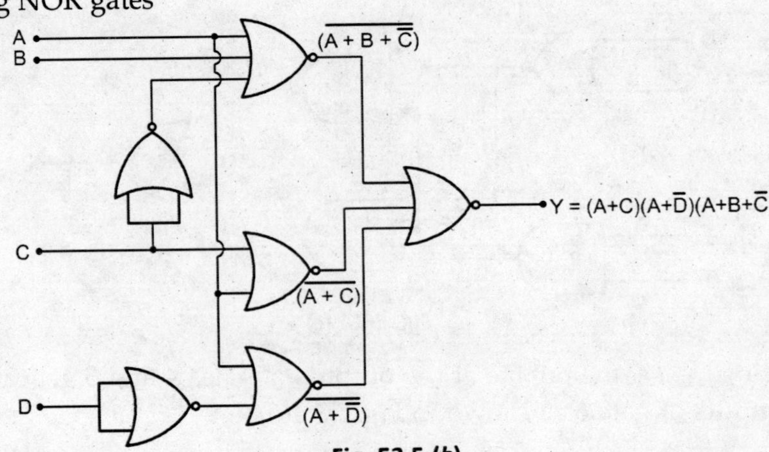

Fig. E3.5 (*b*)

Example 3.6. *Minimize the given Boolean function using K-map.*

$f(A, B, C, D) = \Sigma m(3, 4, 5, 7, 9, 13, 14, 15) + \Sigma_d(0, 2, 8)$

Implement the minimized function only using NAND gates.

Solution:

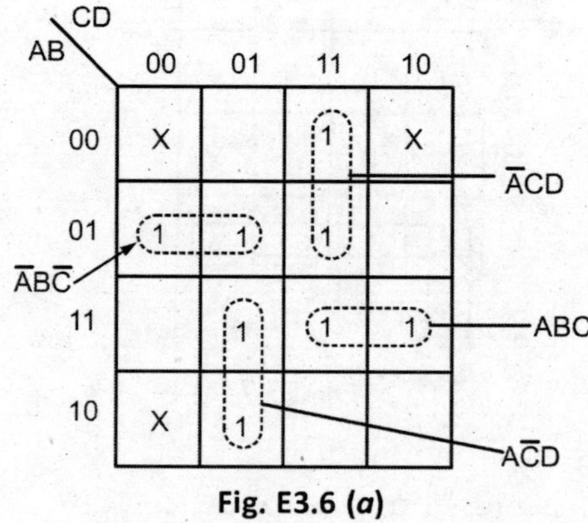

Fig. E3.6 (*a*)

Therefore, $F(A, B, C, D) = \bar{A}B\bar{C} + \bar{A}CD + A\bar{C}D + ABC$

The sum of product function can be implemented directly by NAND gates.

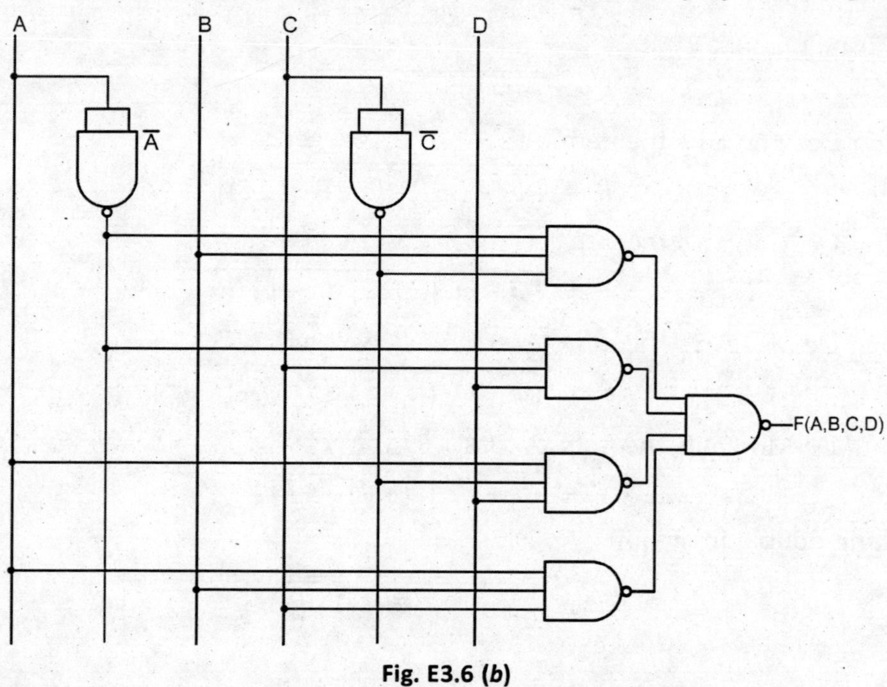

Fig. E3.6 (*b*)

Example 3.7. *Minimize the given Boolean function using K-map and implement the simplified function using NOR gates only.*

$$F(w, x, y, z) = \Sigma m(0, 1, 2, 9, 11, 15) + \Sigma_d(0, 10, 14)$$

Solution: K map and simplification

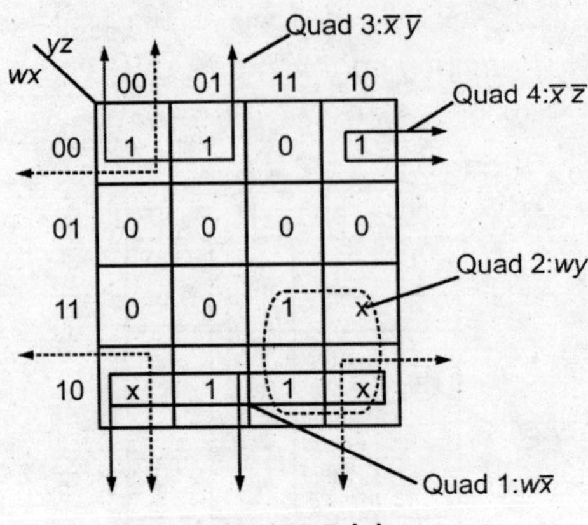

Fig. E3.7 (*a*)

There are four quads as shown in Fig. E3.7(*a*).

Hence the simplified expression is as follows:

$$F = w\overline{x} + wy + \overline{x}\,\overline{y} + \overline{x}\,\overline{z}$$

Conversion into NOR:

$$F = w\overline{x} + wy + \overline{x}\,\overline{y} + \overline{x}\,\overline{z}$$

Take double inversion to get

$$F = \overline{\overline{w\overline{x} + wy + \overline{x}\,\overline{y} + x\,\overline{z}}}$$

Using De-Morgan's theorem $\overline{A+B} = A \cdot \overline{B}$, we get

$$F = \overline{\overline{(w\overline{x})} \cdot \overline{(wy)} \cdot \overline{((\overline{x}\,\overline{y}))} \cdot \overline{((\overline{x}\,\overline{z}))}}$$

Using De-Morgan's theorem $\overline{A \cdot B} = \overline{A} + \overline{B}$, we get

$$F = \overline{(\overline{w} + x) \cdot (\overline{w} + \overline{y}) \cdot (x + y) \cdot (x + z)}$$

Using De-Morgan's theorem $\overline{A \cdot B} = \overline{A} + \overline{B}$, we get

$$F = \overline{\overline{(\overline{w} + x)} \cdot \overline{(\overline{w} + \overline{x})} \cdot (x + y) \cdot (x + z)}$$

Using De-Morgan's theorem $\overline{A \cdot B}\ \overline{A} + \overline{B}$, we get

$$F = \overline{(\overline{w} + x)} \cdot \overline{(\overline{w} + \overline{y})} \cdot \overline{(x + y)} \cdot \overline{(x + z)}$$

Taking double inversion, we get

$$F = \overline{\overline{(\overline{w} + x) + \overline{(\overline{w} + \overline{x})} + \left(\overline{x + y + \overline{(x + z)}}\right)}}$$

Implementation: The above expression is implemented as shown in Fig. E3.7(*b*).

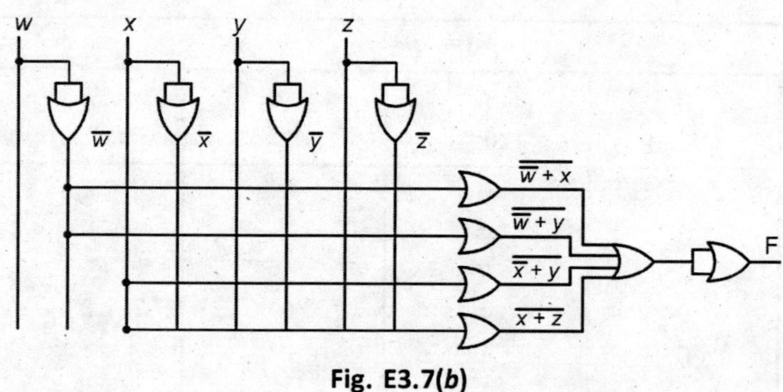

Fig. E3.7(*b*)

Example 3.8. *A company has five directors, namely A, B, C, D and E, and their corresponding percentage of shares in the company are 30, 25, 20, 15 and 10 respectively. The directors are eligible to vote according to their percentage of shares (e.g. A can cast 30 votes; B can cast 25 votes and so on) in the board of directors meeting and two-third majority is required to pass any resolution. Design a combinational circuit to indicate whether a resolution is passed or not.*

Solution: From the word description of the problem, the truth table for the output Y, i.e. whether a resolution is passed or not, can be written as follows.

Table E3.8. Truth table

Inputs					Output
A (30%)	*B* (25%)	*C* (20%)	*D* (15%)	*E* (10%)	*Y*
0	0	0	0	0	0
0	0	0	0	1	0
0	0	0	1	0	0
0	0	0	1	1	0
0	0	1	0	0	0
0	0	1	0	1	0
0	0	1	1	0	0
0	0	1	1	1	0
0	1	0	0	0	0
0	1	0	0	1	0
0	1	0	1	0	0
0	1	0	1	1	0
0	1	1	0	0	0
0	1	1	0	1	0
0	1	1	1	0	0
0	1	1	1	1	1
1	0	0	0	0	0

Table E3.8 (contd.)

Inputs					Output
A (30%)	*B* (25%)	*C* (20%)	*D* (15%)	*E* (10%)	*Y*
1	0	0	0	1	0
1	0	0	1	0	0
1	0	0	1	1	0
1	0	1	0	0	0
1	0	1	0	1	0
1	0	1	1	0	0
1	0	1	1	1	1
1	1	0	0	0	0
1	1	0	0	1	0
1	1	0	1	0	1
1	1	0	1	1	1
1	1	1	0	0	1
1	1	1	0	1	1
1	1	1	1	0	1
1	1	1	1	1	1

From the above truth table, the expression for *Y* can be written as

$$Y = \Sigma_m(15, 23, 26, 27, 28, 29, 30, 31)$$

This five variable functions can be simplified using Karnaugh map method as shown in Fig. E3.8(*a*).

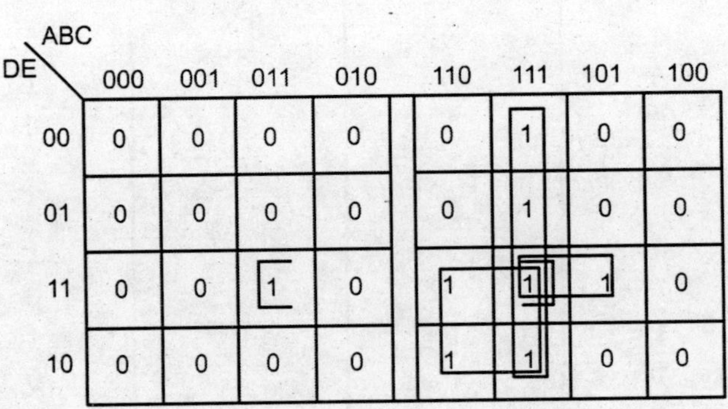

Fig. E3.8. (*a*) Karnaugh simplification.

From the above *K*-map, the simplified expression for Y is given by

$$= ABC + ABD + ACDE + BCDE$$

Now, the above expression can be implemented as shown in Fig. E3.8(*b*).

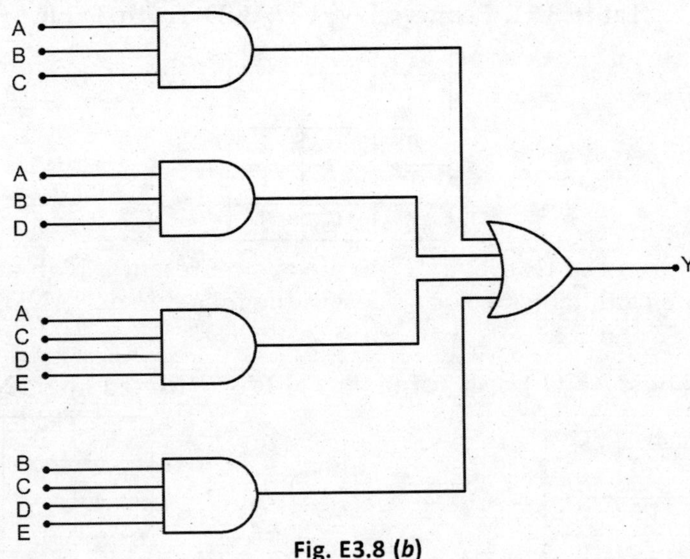

Fig. E3.8 (*b*)

3.4 MIXED LOGIC

In a positive logic, a '1' (i.e., TRUE) is assigned to + 5 V and a '0' (i.e., FALSE) is assigned to 0 V and in a negative logic, a ' 1' is assigned to 0 V and a '0' is assigned to +5 V. But, in mixed logic, the assignment of logical values to voltage values is not fixed, and is left to the discretion of designers. The notation of mixed logic provides a simplified mechanism for the analysis and design of digital circuits. Correct use of mixed logic notation provides logic expressions and logic diagrams that are analogue to each other. Additionally, a mixed logic diagram provides clear documentation as to the operation of a circuit. The positive logic interpretation of the "NAND" gate is that the output will be a "low" logic level for two high inputs. That is, if A is high "AND" B is high, the output will be low; if A is low "OR" B is low, the output will be high. By viewing both the "AND" and "OR" functionality associated with each logic gate, it becomes easy to design and analyze digital circuits.

3.4.1 Basic Mixed Logic Operators

In positive logic, the basic building blocks are "AND", "NAND", "NOR", "OR" and the "Inverter." The "Exclusive "OR" can be obtained from a combination of "AND", "OR" and "Inverter". The truth tables associated with the "AND" and "NAND" positive logic operations are given in Tables 3.10 and 3.11 to show the difference in the positive logic and mixed logic "one" and "zero" notation.

Table 3.10. Positive Logic AND Truth Table

A	B	C
0	0	0
0	1	0
1	0	0
1	1	1

Table 3.11. Positive Logic NAND Truth Table

A	B	C
0	0	1
0	1	1
1	0	1
1	1	0

Mixed logic does not restrict a "1" to always represent a high and a "0" to always represent a low. The truth tables associated with the mixed logic "AND" and "NAND" are shown in Tables 3.12 and 3.13.

Table 3.12 Mixed Logic AND Truth Table

A	B	C
0	0	0
0	1	0
1	0	0
1	1	1

Table 3.13 Mixed Logic NAND Truth Table

A	B	C
0	0	0
0	1	0
1	0	0
1	1	1

The generality of the mixed logic approach allows all "AND" operations, regardless of input and output, assertion levels to be represented by a single truth table. The same generalization applies to the "OR" function. The general notational technique used in mixed logic makes it possible to easily visualize "AND" and "OR" operations.

3.4.2 Mixed Logic Symbols/Alternate GATE Symbols

The correct use of mixed logic requires the placement of the "AND" or "OR" symbol in the logic diagram anytime an "AND" or "OR" operator is called for in the Boolean Expression. The voltage requirements are met by the assignment of assertion levels.

Alternate gate symbols have been developed to clearly denote "AND" and "OR" operations in a logic diagram. The basic positive logic symbols for the AND, NAND, OR, NOR and NOT gates and their corresponding alternate gate symbols are shown in Fig.3.16.

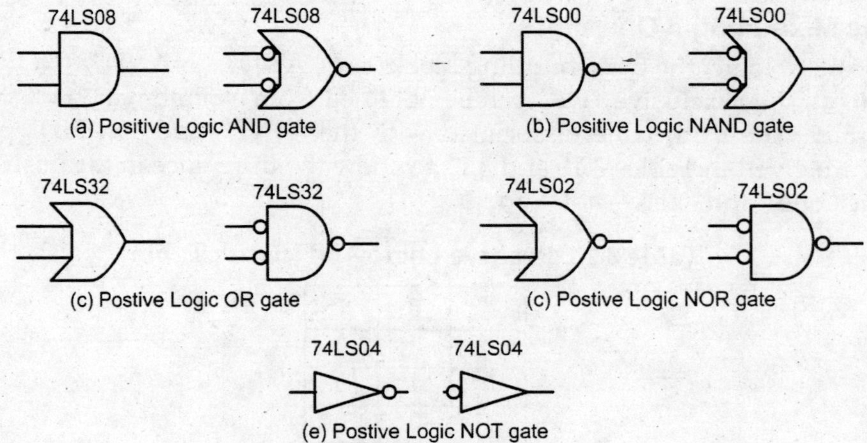

Fig. 3.16. Mixed Logic Symbol Pairs.

The alternate logic symbol is obtained by changing the operation, "AND" to "OR" or "OR" to "AND" and complementing assertion levels. The circle, or "bubble" on the output of the "AND" gate shown in Fig.3.16(*b*) means that the output will be low when asserted. The absence of bubbles on the inputs means that both of the gate inputs require a logic high to be asserted. The mixed logic interpretation for the logical "AND" circuit shown in Fig.3.16(*b*) is given by the expression $F = (A.B)_{-L}$. This expression is much clearer than the positive logic expression $F = \overline{A \cdot B}$ since $\overline{A \cdot B} = \overline{A} + \overline{B}$.

By conversion of "AND" and "OR" symbols, it is easy to realise any logic circuits and the operation of the circuit is always clearly communicated by the logic diagram. Two circuits using mixed logic notations that realise the expression $(A \cdot C + B)_H$ are shown in Fig.3.17. Here, the mixed logic circuit produces a logic high output signal when inputs *A* "AND" *C* are low "OR" *B* is high.

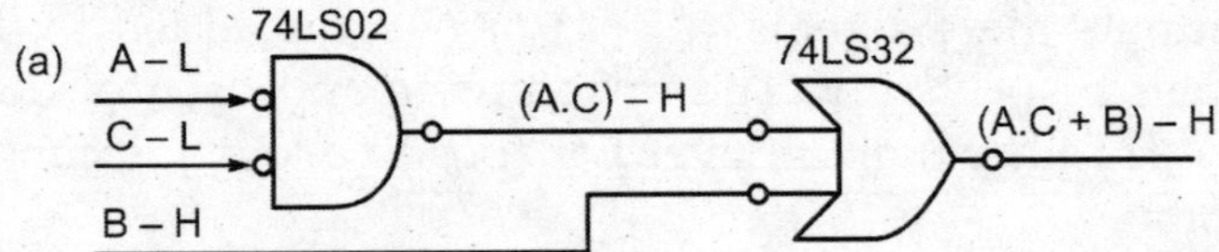

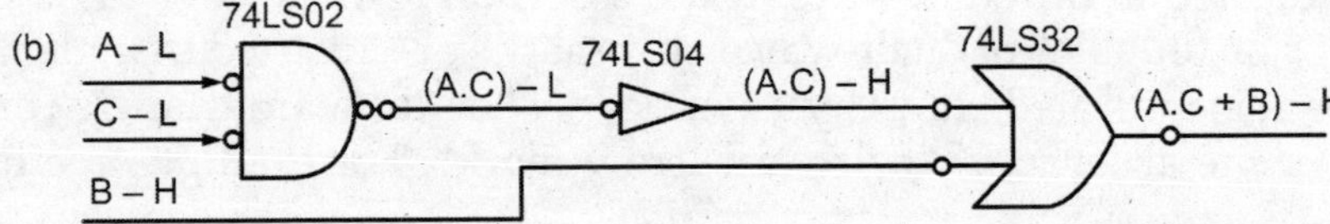

Fig. 3.17. Two circuits that realize the expression $(A \cdot C + B) - H$.

The form of the "OR" operation depends on the choice made for the "AND" realisation. Regardless of the final realisation, the desired logic operation is clearly denoted by the logic diagram. Here, a positive logic "AND" gate is not required to implement the circuit.

3.4.3 Assertion Levels and Polarity Indication

The mixed logic design approach allows the freedom to produce circuits that generate and respond to either high or low logic levels.

In terms of the logic diagram, a bubble on the output means that a "low" will be obtained when the gate conditions are met. Here, the logic expression at the gate output should contain – *L*, i.e., $F = (A.B)_{-L}$ at the end of the expression. Likewise, in a logic "high" output, the output expression should be labeled with a – *H* i.e., $F = (A.B)_{-H}$. These notations are shown in Fig.3.17.

In mixed logic, the requirements of input and output voltage levels are expressed in terms of their assertion levels. For example, a positive logic NAND gate is expressed as being an "AND" function with an "asserted low" output and "asserted high" inputs. Assertion means "the affirmative position of an action related to a Boolean variable." This concept is demonstrated with the circuits shown in Fig.3.18.

Let us assume that the switches, shown in Fig.3.18, are spring loaded so that in order" for contact to be made, they must be pressed.

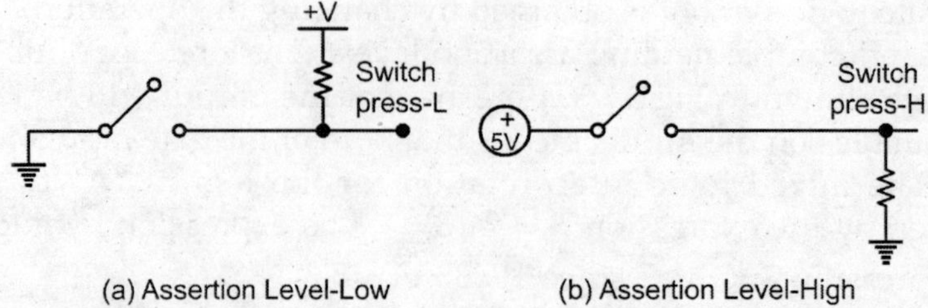

Fig.3.18. Correct mixed logic assertion level labeling.

In both cases, the output signal of both circuits is named "switch pressed". If the circuits are examined, the signal generated by one circuit will be the opposite logic level of the signal generated by the other circuit. In one case when the switch is pressed, the output will go to a logic "low" and in the other case the output of switch circuit will be a logic "high". To clearly differentiate these signals, a (– *L*) or (– *H*) is added to denote asserted low or asserted high signals respectively. The mixed logic approach of denoting voltage levels using either – *L* or – *H* rather than an over bar (positive logic) is consistent with the expected Boolean expression.

In general, the assertion levels i.e., the polarity at an input to a gate or output of a gate can be indicated in different ways. It the assertion level is "low", it is indicated either by a symbol "*L*" or "bubble" or "half-way arrow mark" or "inverted triangle". If the assertion level is "high", it is indicated cither by a symbol "*H*" or "upward triangle" or the absence of the "bubble" or the absence of "half-way arrow mark". For example, the logic expression $Y = \overline{\bar{A}B}$ is represented as shown in Fig. 3.19.

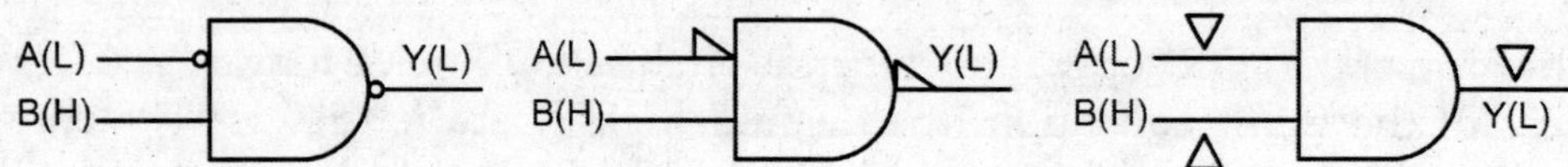

Fig. 3.19. Representation of $Y = \overline{\bar{A}B}$.

Easy interpretation of a circuits operation is possible by maintaining the consistent use of assertion levels. An example is provided using the circuit shown in Fig.3.20.

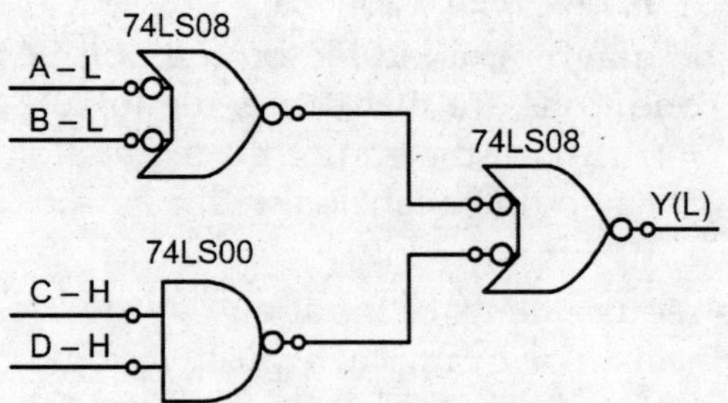

Fig. 3.20. Gate circuit demonstrating an asserted low output signal.

The output expression of the circuit is *Y* $[(A + B)\ 4\ (C.\ D)]_{-L}$. Here, the output will be a logic "low" when the input conditions are met. The input conditions are that either *A* "OR" *B* must be asserted "OR" *C* and *D* must be asserted in order that the output will be asserted low. Inspection of the input assertion levels indicates that in order for *A* to be asserted it must be low, likewise with *B*. The assertion of the *C* and *D* signals require logic highs. In contrast, a positive logic approach to the analysis of this circuit would begin by "barring" of all the variables which assert low as shown in Fig. 3.21.

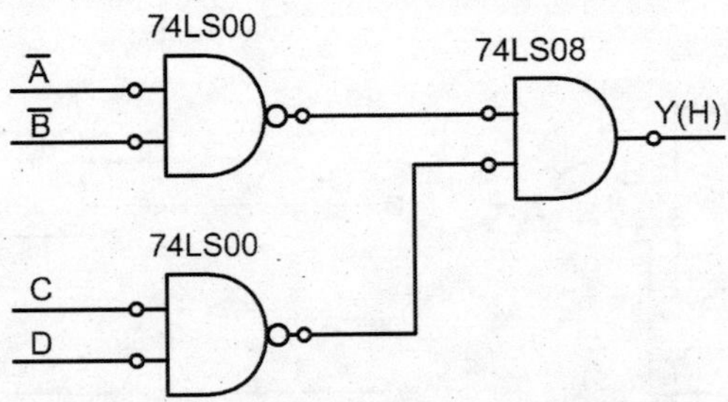

Fig. 3.21. Gate circuit demonstrating an asserted high output signal.

The "barring" of the terms in the positive logic approach obscures the functionality of the circuit. The diagram shown in Fig.3.21 implies that no "OR" operations occur in the circuit. Positive logic can obscure circuit understanding.

Mixed logic is a useful tool for logic design and analysis, and circuit documentation. The mixed logic designer provides design documentation with greater clarity, accuracy and comprehension of the digital logic design to the end users.

3.5 MULTILEVEL GATING NETWORKS

The maximum number of gates cascaded in series between an input and output is called level of gates. For example, a Sum of Product (SOP) expression can be implemented using a two level gate network i.e., AND gates in the first level and a OR gate in the second level as shown in Fig.3.9(*a*) in section 3.3.6. Similarly, a Product of Sum (POS) expression can be implemented using a two level gate network i.e., OR gates in the first level and AND gate in the second level as shown in Fig.3.10(*a*) in section 3.3.7. It is important to note that the inverter gates are not considered to decide the level of the gate network.

Do You Know?	NAND and NOR gates are easier to fabricate with electric components and are the basic gates used in all IC digital logic families.

The number of levels can be increased by factoring the Sum of Products (SOP) expression for AND-OR network or by multiplying out some terms in the Product of Sum (POS) expression for OR-AND network. If a switching expression is implemented using gates in more than two levels, then it is called Multilevel gate network. Generally, the propagation delay through a multi-level gate network is proportional to the number of levels in it. However, sometimes by increasing the number of levels of logic, the number of gates and number of gate inputs are decreased.

3.5.1 Implementation of Multilevel Gate Network

The implementation of the multilevel gate network can be explained with the following two cases.

Case 1: Consider the switching function, $Y = B\overline{C} + \overline{A}B + D$. This expression can be implemented using a two level AND-OR gate network as shown in Fig.3.22(*a*). It requires two 2-input AND gates and one 3-input OR gate with a total number of five literals.

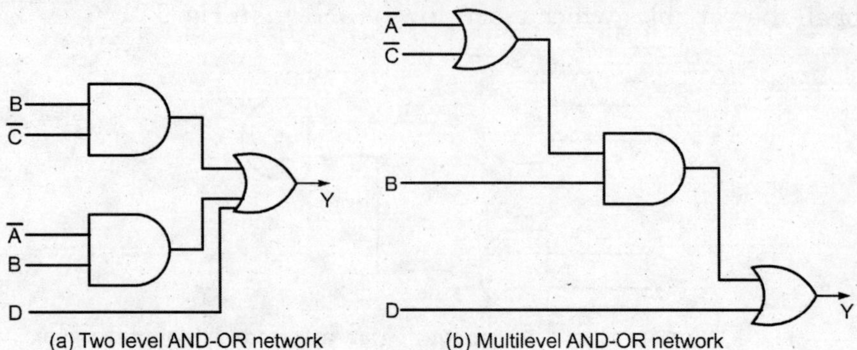

Fig. 3.22

If the above switching expression is factored into a different form as $Y = B(\overline{A} + \overline{C}) + D$, then it can be implemented using a three level gate network as shown in Fig.3.22(*b*). Now, this implementation requires two 2-input OR gates and one 2-input AND gate with a total number of four literals. Thus, it reduces the number of gate inputs by one.

Case 2: Next consider the following function

$$f(A, B, C, D, E, F, G) = ADF + AEF + BDF + BEF + CDF + CEF + G$$

It is already in its minimal sum of products form. Its implementation as a two-level network of AND and OR gates requires six 3-input AND gates and one 7-input OR gate, a total of seven gates and 19 literals as shown in Fig. 3.23(*a*).

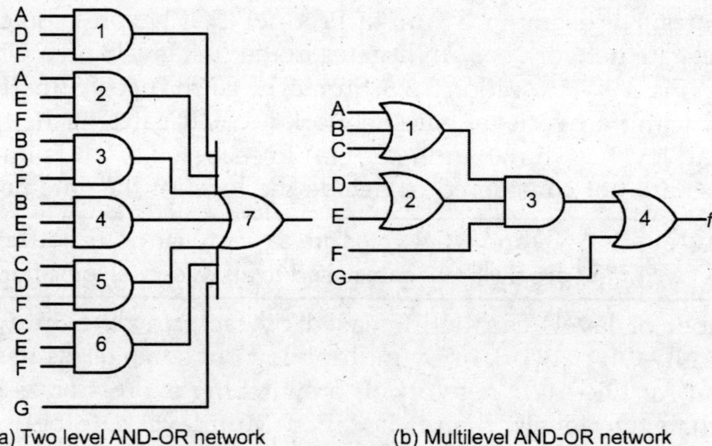

Fig. 3.23

The above two-level expression can be replaced with a so-called factored form by factoring out common literals from the product terms whenever possible. By recursively factoring out the common literals, the expression can be written as:

$$= (AD + AE + BD + BE + CD + CE)\,F + G$$
$$= [(A + B + C)\,D + (A + B + C)\,E]\,F + G$$
$$= (A + B + C)(D + E)\,F + G$$

Now, the function requires one 3-input OR gate, two 2-input OR gates, and a; input AND Rate, a total of four gates and nine literals. The implementation of tli factored form is shown in Fig. 3.23(*b*). This implementation significantly reduces the number of wires and gates needed to implement the function, but it probably has worse delay because of the increased levels of logic. Compared to two level implemer tations, the multi-level implementations require less number of gates, thereby increas ing the propagation delay.

3.5.2 Conversion to NAND-NAND and NOR-NOR Gate Networks

From De-Morgan's theorem, it is known that

$$\overline{A+B} = \overline{AB} \text{ and } \overline{AB} = \overline{A} + \overline{B}$$

These expressions can also be written as

$$A+B = \overline{\overline{A}\overline{B}} \text{ and } AB = \overline{\overline{A}+\overline{B}}$$

It means that an OR gate is equivalent to a NAND gate with bubbles at its input and an AND gate is equivalent to a NOR gate with bubbles at its inputs. Also, a NANI gate is equivalent to an OR gate with bubbles at its inputs and a NOR gate is equiva lent to an AND gate with bubbles at its inputs, as discussed in sections 3.3.4 and 3.3.5 The above facts can be summarized in Fig.3.24(*a*) and (*b*).

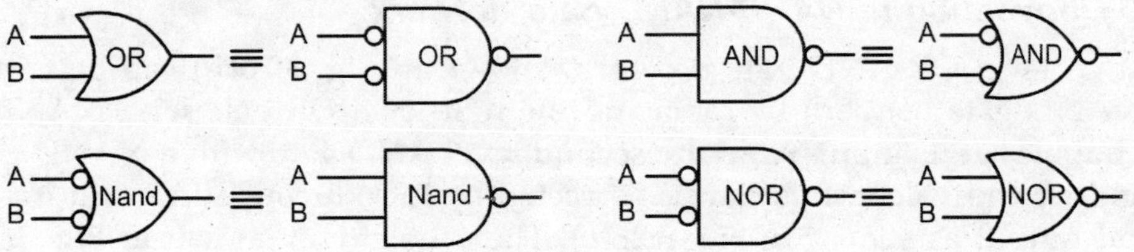

A B OR ≡ A B OR A B AND ≡ A B AND

A B Nand ≡ A B Nand A B NOR ≡ A B NOR

Fig. 3.24

The schematic symbols on either side of the Fig.3.24(*a*) and (*b*) can be freely exchanged without changing the truth table or logical value of the function.

AND-OR conversion to NAND-NAND gate networks

A two level AND-OR gate network can be easily converted to a NAND-NAND gate network as discussed in section 3.3.6.

AND-OR conversion to NOR-NOR gate networks

A two level AND-OR gate network, shown in Fig. 3.25(*a*) can be converted into NOR NOR gate network by replacing the first level AND gate with NOR gates (i.e., AND wit! bubbles at its inputs) and the second level OR gate with a NOR gate. But, this is no logically equivalent. This can be corrected by introducing additional inverters at tht inputs and the output, as shown in Fig. 3.25(*b*). Then the circuit shown ir Fig. 3.25(*b*) can be modified as shown in Fig. 3.25(*c*).

OR-AND conversion to NOR-NOR gate networks

A two level OR-AND gate network can be easily converted to a NOR-NOR gate network as discussed in section 3.3.6.

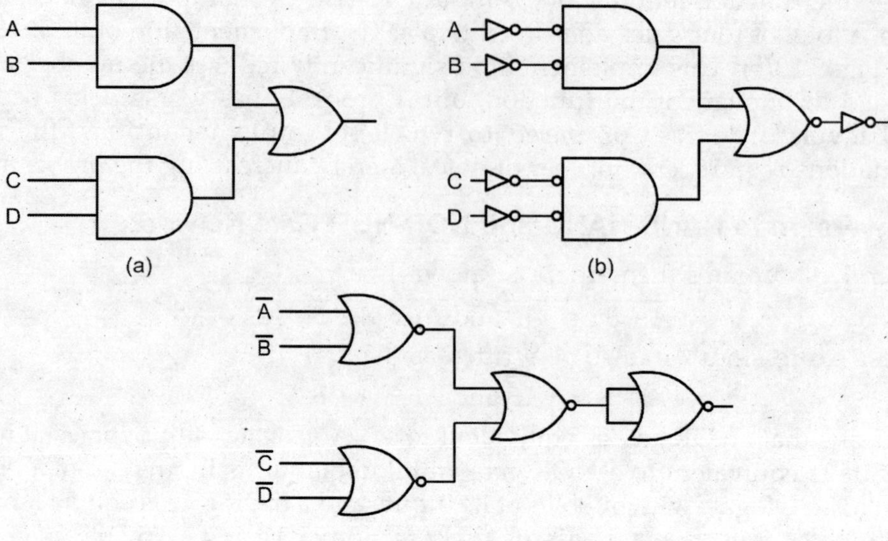

Fig. 3.25

OR-AND Conversion to NAND-NAND gate network

A two level OR-AND gate network, shown in Fig. 3.26(***a***) can be converted into NAND-NAND gate network by replacing the first level OR gate with NAND gates (i.e., OR with bubbles at its inputs) and the second level AND gate with a NAND gate. But this is not logically equivalent. this can be corrected by introducing additional inverters at the inputs and output as shown in Fig. 3.26(*b*). Then the circuit shown in Fig. 3.26(*b*) can be modified as shown in Fig. 3.26(*c*).

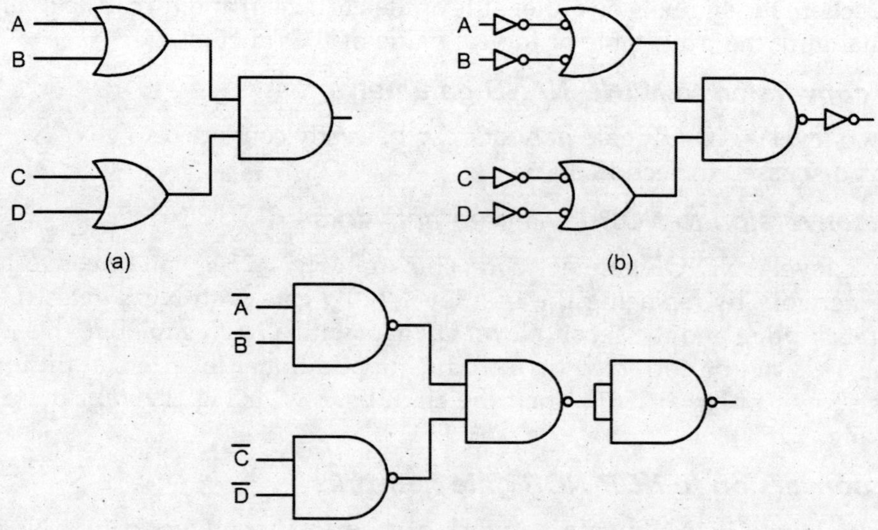

Fig. 3.26

Multilevel AND-OR conversion to NAND-NAND gate networks

A multilevel AND-OR gate network can be easily converted into a gate network with NAND-NAND gates. Whenever a gate has a complemented switching variable at its input, it is logically equivalent to a gate with the un-complemented switching variable and a bubble at the input of that gate to which the complemented switching variable is connected. Also, two bubbles can be introduced at both ends of a line connecting two gates because their logical effect is nil. Using the above facts, the circuit shown in Fig. 3.27(*a*) can be modified as shown in Fig. 3.27(*b*).

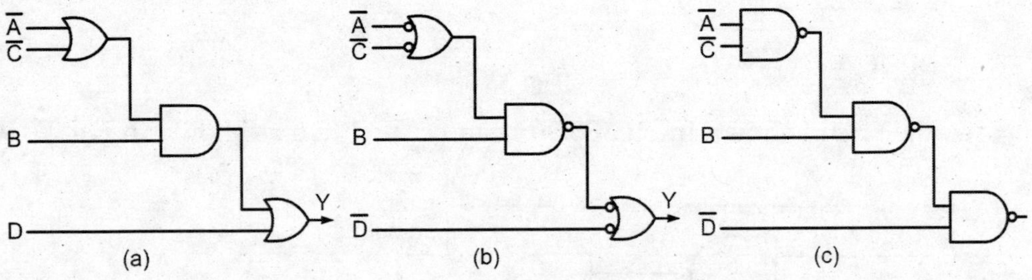

Fig. 3.27

It is known that an OR gate with bubbles at its inputs is equivalent to NAND gate. Now, the circuit can be drawn as shown in Fig. 3 .27(*c*).

Example 3.9. *Realise the following function as (i) multilevel NAND-NAND gate network and (ii) multilevel NOR-NOR network.*

$$f = (A + CD)\ A\overline{C}$$

Solution: The given expression can be implemented as a four level AND-OR gate network as shown in Fig. E3.9(*a*).

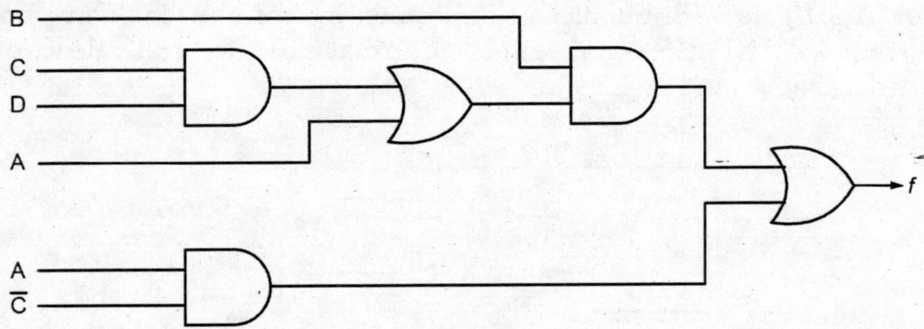

Fig. E3.9(*a*)

(*i*) Multilevel NAND-NAND implementation: For NAND-NAND implementation, (*i*) each AND gate is replaced by a NAND gate followed by an inverter and (*ii*) each OR gate is replaced by OR gate with bubbles at its inputs. In addition, the corresponding input variables to bubbled OR gate is complemented so that the logical equivalence of the function is maintained as shown in Fig. E3.9(*b*).

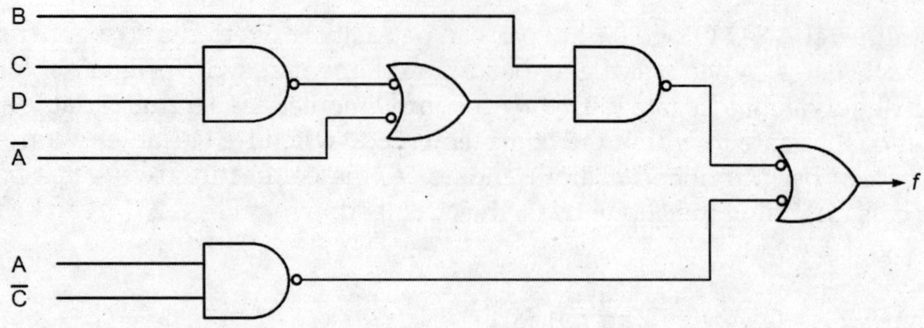

Fig. E3.9(*b*)

Now, the circuit shown in Fig. E3.9(*b*) can be modified as shown in Fig. E3.9(*c*).

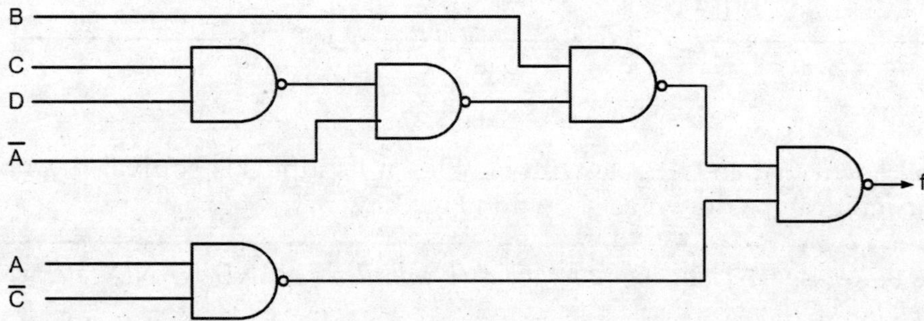

Fig. E3.9(*c*)

(*ii*) Multilevel NOR-NOR implementation: For NOR-NOR implementation, (*i*) each OR gate is replaced by a NOR gate followed by an inverter and (*ii*) each AND gate is replaced by AND gate with bubbles at its inputs. In addition, the corresponding input variables to bubbled AND gate is complemented so that the logical equivalence of the function is maintained as shown in Fig. E3.9(*d*).

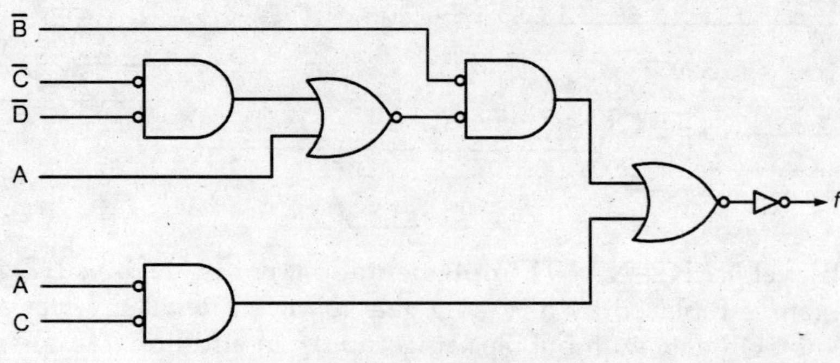

Fig. E3.9(*d*)

Now, the circuit shown in Fig. E3.9(*d*) can be modified as shown in Fig. E3.9(*e*).

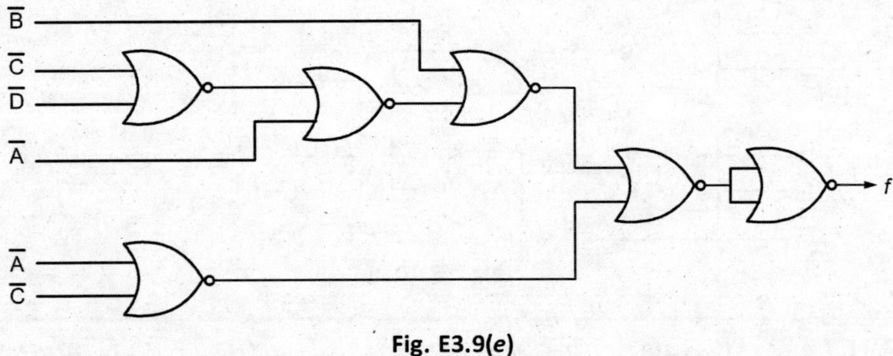

Fig. E3.9(*e*)

Example 3.10. *Realize the following switching function using a multilevel gate network. Also, obtain the logically equivalent multilevel NAND-NAND gate circuit.*

$$Y = A + (B + \overline{C})(\overline{D}E + F)$$

Solution: The given expression can be implemented as a four level gate network using basic AND and OR gates as shown in Fig. E3.10(*a*).

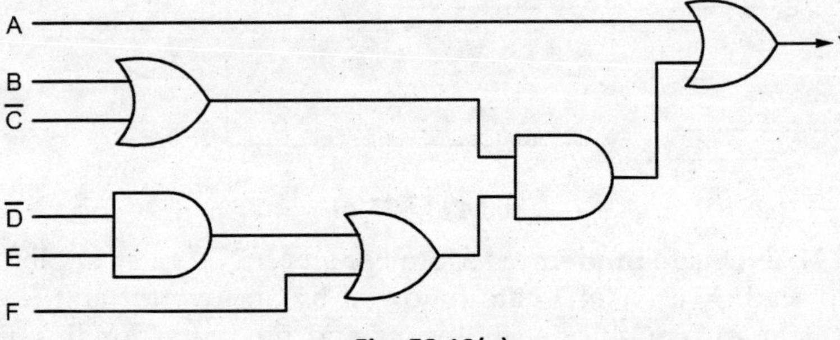

Fig. E3.10(*a*)

The four level gate network in Fig. E3.10(*a*) can be modified as shown in Fig. E3.10(*b*), so that it can be realised using universal gates.

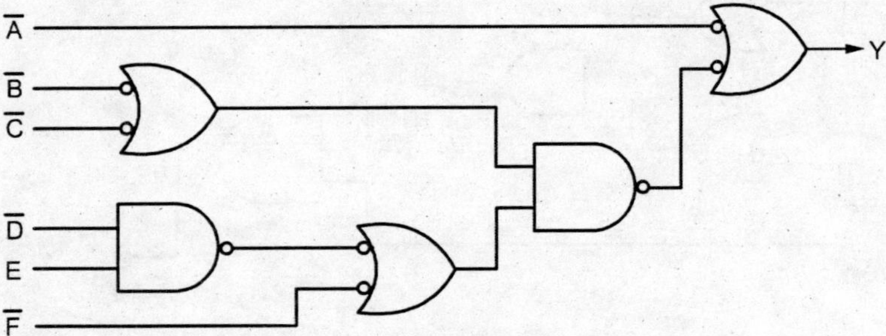

Fig. E3.10(*b*)

Now, the above circuit can be implemented using NAND gates as shown in Fig. E3.10(*c*).

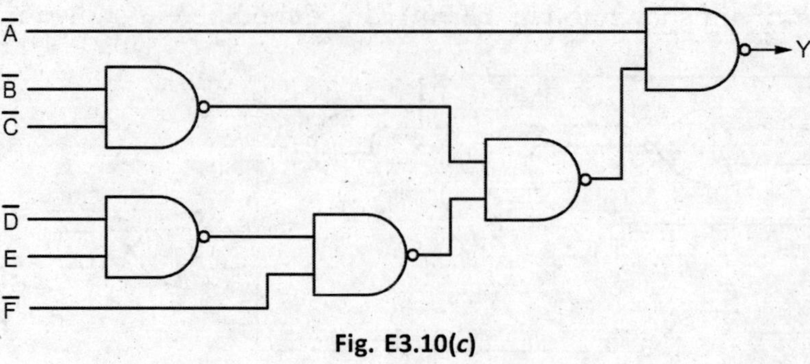

Fig. E3.10(*c*)

Example 3.11. *Convert the gate network, shown in Fig. E3.11(a) (i) all NAND gates network (ii) all NOR gates network by adding bubbles and inverters wherever necessary.*

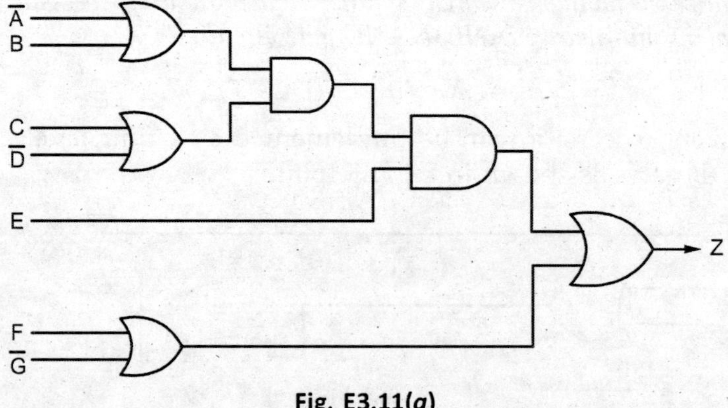

Fig. E3.11(*a*)

Solution: (*i*) All NAND gate implementation: For all NAND gates implementation, (*i*) each AND gate is replaced by a NAND gate followed by an inverter and (*ii*) each OR gate is replaced by OR gate with bubbles at its inputs and the corresponding input variables to bubbled OR gate is complemented. Using this procedure, the multilevel gate network, shown in Fig. E3.11(*a*) is modified as shown in Fig. E3.11(*b*).

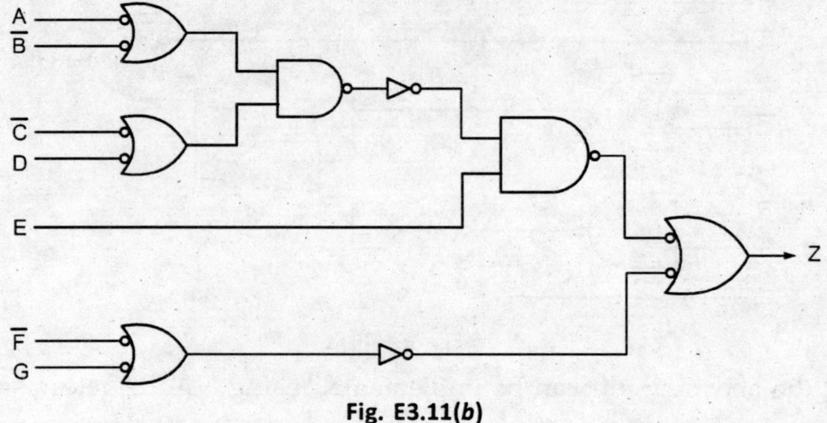

Fig. E3.11(*b*)

It is known that a bubbled OR gate is equivalent to a NAND gate. Also, a NOT gate can be implemented using a NAND gate. Therefore, the circuit shown in Fig. E3.11(*b*) can be modified as shown in Fig. E3.1 l(*c*).

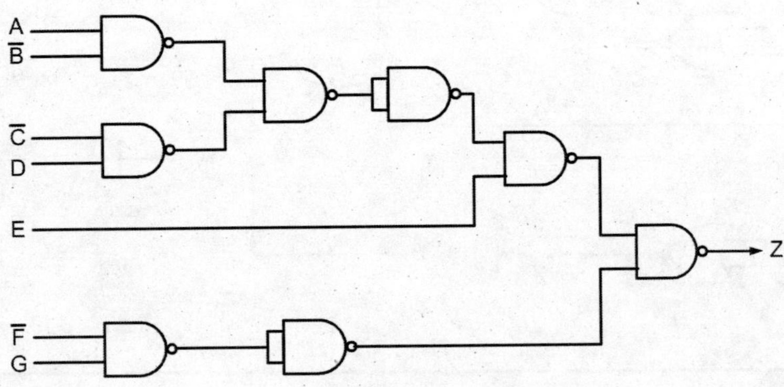

Fig. E3.11(*c*)

***(ii)* All NOR gates implementation:** For all NOR gates implementation, (*i*) each OR gate is replaced by a NOR gate followed by an inverter and (*ii*) each AND gate is replaced by AND gate with bubbles at its inputs and the corresponding input variables to bubbled AND gate is complemented. Using this procedure, the multilevel gate network shown in Fig. E3.11(*a*) can be modified as shown in Fig. E3.11(*d*).

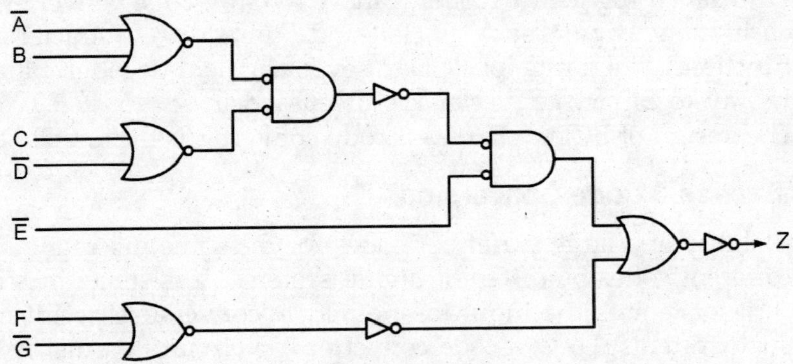

Fig. E3.11(*d*)

It is known that a bubbled AND gate is equivalent to a NOR gate and a NOT gate can be implemented using NOR gate. Therefore, the circuit shown in Fig. E3.11 (*d*) can be modified as shown in Fig. E3.11(*e*).

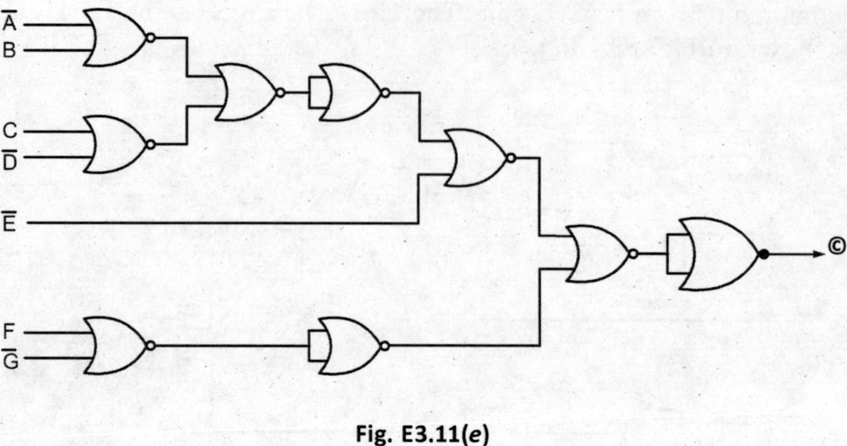

Fig. E3.11(*e*)

3.6 MULTIPLE OUTPUT GATE NETWORKS

A switching function with more than one output is called Multioutput function and the corresponding logic circuit is called Multiple output gate network. A multi-output function can be completely described or specified by a truth table in which for all possible combinations of inputs, the multi-outputs are specified. Code converters such as BCD to Excess-3 code converter, BCD to seven segment decoder, Binary to Gray code converter, Gray to Binary converter and arithmetic circuits such as Half and Full adder/subtracter are examples for multi-output functions. The design of BCD to seven segment decoder, Binary to Gray code converter and Gray to Binary converter are discussed in section 6.5.9, 6.10.2 and 6.10.3 respectively. The design of BCD to Excess-3 code converter is discussed below.

3.6.1 BCD to Excess-3 Code Conversion

The availability of a large variety of codes for the same discrete information results in the use of different codes by different digital systems. It is sometimes necessary to use the output of one system as the input to another. A conversion circuit must be inserted between two such systems. Thus, a code converter is a circuit that makes the two systems compatible eventhough each uses a different binary code.

The following example illustrates the conversion of BCD-to-excess-3 code. Table 3.14 shows the input in BCD and the output in Excess-3. The four input combinations from 10 to 15 never occur and therefore they are don't care combinations.

Table 3.14. BCD-to-excess-3 code conversion

Decimal number	Inputs				Outputs			
	A	*B*	*C*	*D*	*W*	*X*	*Y*	*Z*
0	0	0	0	0	0	0	1	1
1	0	0	0	1	0	1	0	0
2	0	0	1	0	0	1	0	1
3	0	0	1	1	0	1	1	0
4	0	1	0	0	0	1	1	1
5	0	1	0	1	1	0	0	0
6	0	1	1	0	1	0	0	1
7	0	1	1	1	1	0	1	0
8	1	0	0	0	1	0	1	1
9	1	0	0	1	1	1	0	0
10	1	0	1	0	d	d	d	d
11	1	0	1	1	d	d	d	d
12	1	1	0	0	d	d	d	d
13	1	1	0	1	d	d	d	d
14	1	1	1	0	d	d	d	d
15	1	1	1	1	d	d	d	d

From the truth table, one can write the logic expressions for Excess-3 code outputs *W*, *X*, *Y* and *Z* as follows.

$$W = \Sigma_m(5, 6, 7, 8, 9) + \Sigma_d(10, 11, 12, 13, 14, 15)$$
$$X = \Sigma_m(1, 2, 3, 4, 9) + \Sigma_d(10, 11, 12, 13, 14, 15)$$
$$Y = \Sigma_m(0, 3, 4, 7, 8) + \Sigma_d(10, 11, 12, 13, 14, 15)$$
$$Z = \Sigma_m(0, 2, 4, 6, 8) + \Sigma_d(10, 11, 12, 13, 14, 15)$$

where Σ_d represents the summation of don't care combinations.

Now, the above expressions for Excess-3 code outputs can be simplified using K-map method.

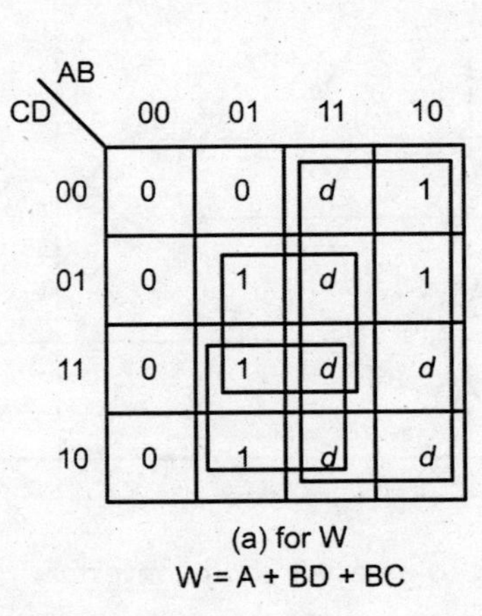

(a) for W

$W = A + BD + BC$

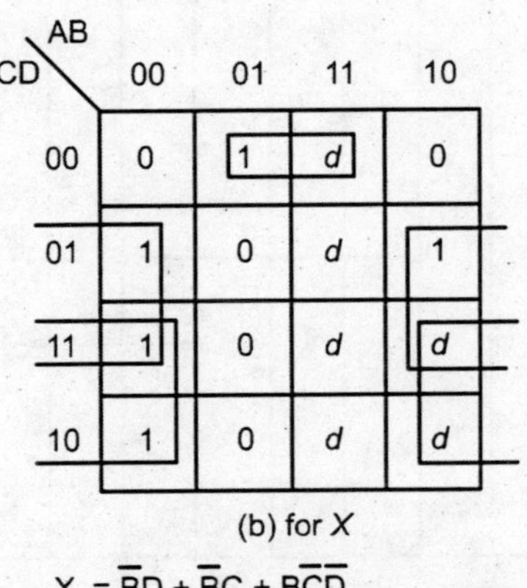

(b) for *X*

$X = \overline{B}D + \overline{B}C + B\overline{C}\,\overline{D}$

$= \overline{B}(C + D) + B(\overline{C.D}) = \overline{[B \oplus (\overline{CD})]}$

$[\because A\overline{B} + \overline{A}B = A \oplus B]$

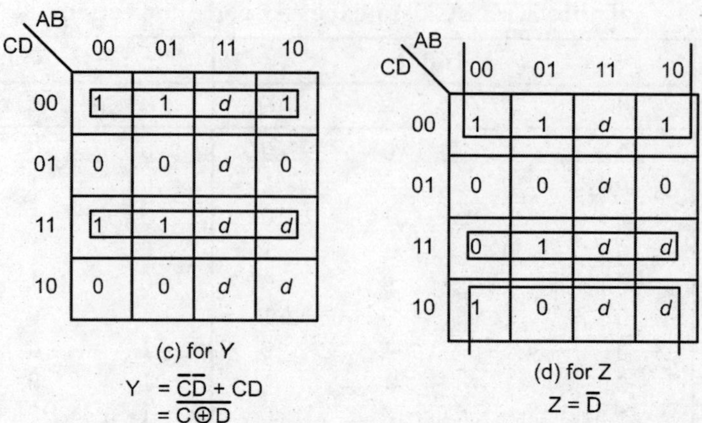

Fig. 3.28. K-map simplification for BCD-to-excess-3 code conversion.

Using the above simplified expressions, the circuit diagram for BCD-to-excess-3 code converter can be drawn as shown in Fig. 3.29.

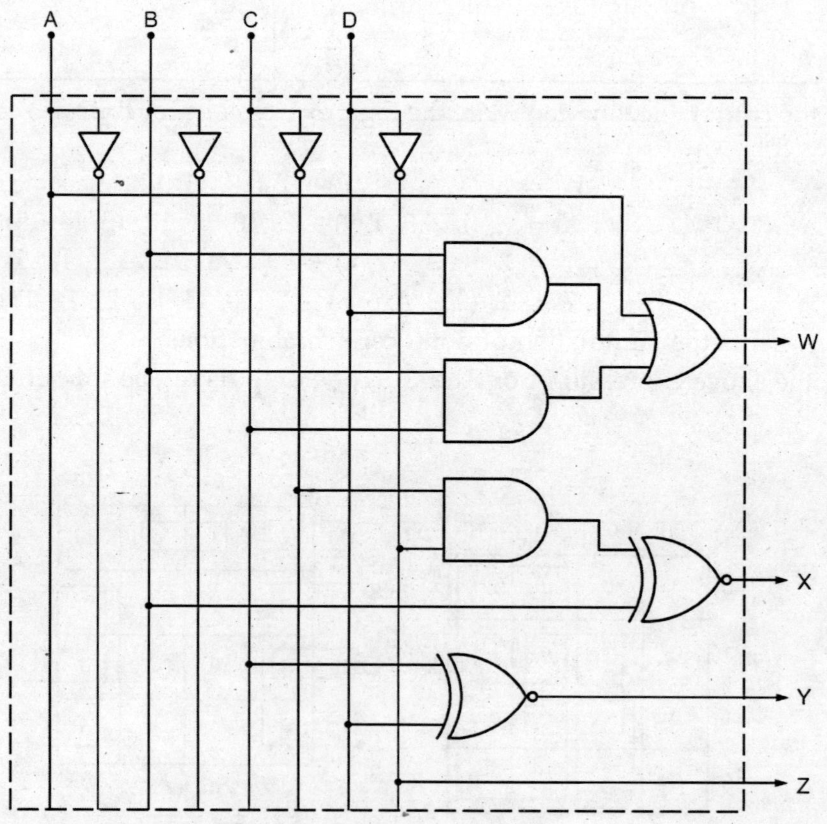

Fig. 3.29. Circuit diagram for BCD-to-excess-3 code converter.

■ REVIEW QUESTIONS ■

1. Write the truth table and logic symbol of a three-input OR gate.
2. What is the only condition under which an OR gate output will be 0?
3. Under what conditions will the output of an AND gate be 0?
4. Write the expression for a 4-input AND gate. Construct the complete truth table showing the output for all possible cases.
5. Is there such a thing as a 3-input NOT gate?
6. Differentiate positive and negative logic system.
7. Define the NAND and NOR gates through their truth tables.
8. Explain the function of an OR gate using diodes.
9. Explain the function of an AND gate using diodes.
10. What is a logic gate? Explain logic designation.
11. Draw the logic diagram of an Ex-OR gate and discuss its operation.
12. What is an Ex-NOR gate? Write its truth table.
13. Using only NAND gates, design a circuit to provide an output of logic 1 when only one of three inputs is logic 1.
14. Explain the term "universal gate."
15. Name the gates that are used as universal gates. Explain.
16. Explain how the basic gates can be realised using NAND gates.
17. Explain how the basic gates can be realised using NOR gates.
18. Give the action of the 2-input Ex-OR gate and construct it using NAND gates.
19. Write a short note on logic gates.
20. Explain the transistor inverter circuit and verify its truth table.
21. Verify that the following operations are commutative and associative (*a*) AND, (*b*) OR, (*c*) Ex-OR.
22. Verify that the following operations are commutative but not associative (*a*) NAND, (*b*) NOR.
23. Realise the logic expression $Y = A \oplus B \oplus C$ using Ex-OR gates.
24. Realise the logic expression $Y = A \oplus B \oplus C \oplus D$ using Ex-OR gates.
25. State and explain DeMorgan's theorems which transform a sum into a product form and vice versa. Draw the logic equivalent circuits representing the theorems using basic gates.
26. Obtain the minimal sum of products expression for the following function and implement the same using universal gates.
$$f(A, B, C, D) = \Sigma(0, 2, 3, 5, 7, 8, 13) + \Sigma_d(1, 6, 12)$$
27. Obtain the minimal sum of products expression for the following function and implement the same using only NAND gates.
$$f(A, B, C,D) = \Sigma(1, 4, 7, 8, 9, 11) + \Sigma(0, 3, 5)$$
28. $f(A, B, C, D) = \Sigma(0, 2, 3, 5, 6, 7, 8, 9)$ with (10, 11, 12, 13, 14, 15) as don't cares. Realise the minimised function using only NOR gates.
29. Draw a logic circuit for the following function using NOR gates:
$(A + B)(B + C)(A + C)$

30. A combinational switching network has four inputs (A, B, C and D) and one output Z. The output is to be '0' if the input combination is a valid Excess-3 coded decimal digit. If any other combination of inputs is given, the output is to be 1 . Design the network using basic gates.
31. Seven switches operate a lamp in the following way: If switches 1, 3, 5 and 7 are closed and switch 2 is opened, or if switches 2, 4 and 6 are closed and switch 3 is opened, or if all seven switches are closed, the lamp will light. Use NOT, AND and OR gates to show how the switches must be connected.
32. What is mixed logic?
33. Compare positive, negative and mixed logics.
34. Give the mixed logic symbol pairs of AND, OR and NOT gates.
35. Draw the alternate gate symbols for basic and universal gates.
36. Explain the concept of mixed logic with necessary diagrams.
37. Explain assertion levels.
38. How polarities can be indicated in different ways?
39. What is level of a gate network?
40. What are the advantages and disadvantages of multilevel gate network?
41. Write down the procedure to convert a given AND-OR gate network to all NAND gates network.
42. How will you convert a given multilevel AND-OR gate network to all NOR gates network?
43. Simplify the following function and implement it as (*i*) Two level AND-OR gate network (*ii*) Two level OR- AND gate network, (*iii*) Multilevel NAND-NAND gate network and (*iv*) multilevel NOR-NOR gate network.

$$f = \Sigma(0, 2, 3, 6, 8, 10, 11, 14, 15)$$

44. Realise the following function using (*i*) multilevel NAND-NAND network and (*ii*) Multilevel NOR-NOR network.

$$Y = \bar{A}B + B(C + D) + E\bar{F}(\bar{B} + \bar{D})$$

45. Differentiate multilevel and multi-output gate networks.

Logic Families

4.1 INTRODUCTION

Logic gates and memory devices are fabricated as *integrated circuits* (ICs) because the components used such as resistors, diodes, bipolar junction transistors and the insulated gate or metal-oxide semiconductor field-effect transistors are the integral parts of the chip. The various components are interconnected within the chip to form an electronic circuit during assembly. The ICs result in an increase in reliability and a reduction in weight and size.

Small Scale Integration (SSI) refers to ICs with fewer than 10 gates on the same chip. Medium Scale Integration (MSI) includes 12 to 100 gates per chip. Large Scale Integration (LSI) refers to more than 100 upto 5000 gates per chip. Very Large Scale Integration (VLSI) devices contain several thousand gates per chip.

Do You Know?	When we use a bipolar device, like transistor in the IC fabrication technology, it is known as biopolar technology. In unipolar technology we make use of unipolar devices e.g., MOSFETs.

Integrated circuits are classified into two general categories: (*i*) *Linear* and (*ii*) *Digital.* Linear integrated circuits operate with continuous signals and are used to construct electronic circuits such as amplifiers, voltage comparators, etc. Digital integrated circuits operate with binary signals and are invariably constructed with integrated circuits.

4.2 DIGITAL INTEGRATED CIRCUITS

The various logic families can be placed into two broad categories according to the IC fabrication process: (*i*) Bipolar and (*ii*) Metal-oxide semiconductor (MOS).

ICs come in the following types of packages:

(*i*) Dual-in-Line Package (DIP)
(*ii*) Leadless Chip Carrier (LCC)
(*iii*) Plastic Leaded Chip Carrier (PLCC)
(*iv*) Plastic Quad Flat Pack (PQFP) and
(*v*) Pin Grid Array (PGA)

4.2.1 Bipolar Logic Families

The important elements of a bipolar IC are resistors, transistors and diodes (varactor diodes used as capacitors). Based on the two main operations of bipolar ICs, i.e., saturated and non-saturated, bipolar families are classified into

(*i*) saturated logic and (*ii*) non-saturated logic.

The following are the saturated bipolar logic families:

1. Resistor-Transistor Logic (RTL)
2. Direct-coupled Transistor Logic (DCTL)
3. Diode-Transistor Logic (DTL)
4. High Threshold Logic (HTL)
5. Transistor-Transistor Logic (TTL)
6. Integrated-injection Logic (I^2L)

The following are the non-saturated logic families:

1. Schottky TTL
2. Emitter-coupled Logic (ECL)

4.2.2 MOS Families

The MOS families include:

1. PMOS p-channel MOSFETs
2. NMOS H-channel MOSFETs
3. CMOS Complementary MOSFETs

4.3 CHARACTERISTICS OF DIGITAL ICs

Some of the important parameters or properties of various logic families are listed as follows:

1. Speed of operation (Propagation delays)
2. Power dissipation
3. Fan-in
4. Fan-out
5. Noise immunity
6. Operating temperature
7. Power supply requirements

The comparison of performance of digital ICs may be made with reference to the above properties.

4.3.1 Speed of Operation

The speed of operation of an IC is expressed in terms of propagation delay. *Propagation delay* is defined as the time taken for the output of a gate to change after the inputs have changed.

A logic signal always experiences a delay in going through a circuit. The two propagation delay times shown in Fig. 4.1 are defined as follows:

t_{PHL}: It is the propagation delay time in going from logical LOW (0 state) to logical HIGH (1 state).

t_{PHL}: It is the propagation delay time in going from logical HIGH (1 state) to logical LOW (0 state).

It is evident that t_{PHL} is the delay in the *output* response as it goes from LOW state to a HIGH state, and vice versa for t_{PHL}. The delay times are measured between the 50% voltage levels of the input and output waveforms. In general, the two delays t_{PHL} and t_{PLH} are not necessarily equal and will vary depending on load conditions. The values of propagation times are a measure of the relative speed of logic circuits. The average of the above two propagation delays $(t_{PLH} + t_{PHL})/2$ is called the *average propagation delay* and is used to rate the circuit. It is a function of the switching time of the individual transistors or MOSFETs in the circuit.

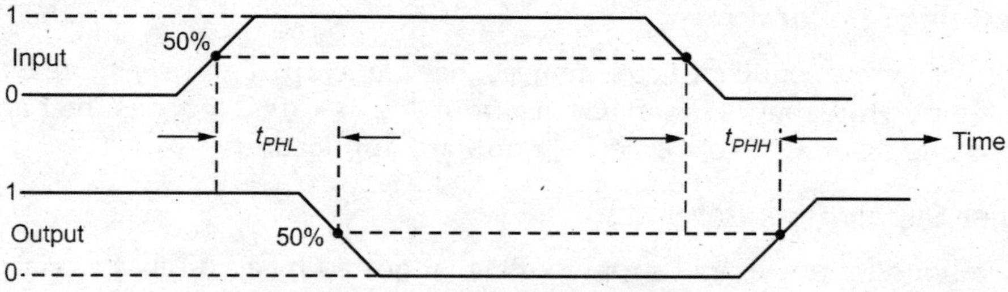

Fig. 4.1. Propagation delays.

4.3.2 Power Dissipation

Power dissipation is a measure of the power consumed by the logic gate when fully driven by all its inputs, and it is expressed in milliwatts or nanowatts. The d.c. or average power dissipation is the product of d.c. supply voltage and the mean current taken from that supply.

4.3.3 Fan-in

The fan-in of a gate is the number of inputs connected to the gate without any degradation in the voltage levels. For example, an eight-input gate requires one Unit Load (UL) per input. Its fan-in is 8. This parameter determines the functional capabilities of a logic circuit.

4.3.4 Fan-out

Fan-out is the maximum number of similar logic gates that a gate can drive without any degradation in voltage levels. Very often a gate will drive several other gates. Each driven gate requires a certain current which must be supplied by the driving gate. The driving gate must be capable of supplying this current while maintaining the required voltage level. In part, this is a function of the output impedance of the driving gate and the input impedance of the driven gates. Usually, in a given logic family, gates drive others of the same type. If their output impedance is low while their input impedance is high, then one gate can often drive many others.

4.3.5 Noise Immunity or Noise Margin

The term noise denotes an unwanted signal voltage, e.g., hum, transients and glitches. Noise can sometimes cause the input voltage of a logic gate to drop below V_{IH} (min) or rise above V_{IL} (max), which leads to unreliable operation. *Noise immunity* is the maximum noise voltage that may appear at the input of a logic gate without changing the logical state of its output. A quantitative measure of noise immunity is called *noise margin.*

The difference between the operating input-logic voltage level and the threshold voltage is called the noise margin of the circuit. The manufacturer usually quotes the noise margin, which refers to the amplitude of the noise voltage that may cause the logic level to change. In the worst case, a TTL gate functions properly as long as the noise margin is kept less than 0.4 V.

4.3.6 Operating Temperature

All IC gates are semiconductor devices that are temperature-sensitive by nature. The operating temperature ranges for an IC vary from 0°C to + 70°C for consumer and industrial applications and from 55°C to + 125°C for military applications.

4.3.7 Power Supply Requirements

The amount of power and supply voltage required by an IC are the main parameters to be taken into consideration while choosing a proper power supply.

4.3.8 Current and Voltage Parameters

The following currents and voltages are very important in designing digital systems. The values given below are for TTL gates only.

High-level input voltage (V_{IH}) [$V_{in\ (1)}$]: It is the minimum voltage level required for a logical 1 at an input. Its minimum value is 2 V.

Low-level input voltage (V_{IL}) [$V_{in\ (0)}$]: It is the maximum input voltage required for a logical 0 (LOW) at an input. Its maximum value is 0.8 V.

High-level output voltage (V_{IH}) [$V_{out\ (0)}$]: It is the minimum voltage required for a logical 1 state at the output. Its minimum value is 2.4 V.

Low-level output voltage (V_{OL}) [$V_{out\ (0)}$]: It is the maximum voltage available at the circuit's output corresponding to the logical 0 state. Its maximum value is 0.4 V.

High-level input current (V_{IH}) [$V_{in\ (1)}$]: The current that flows through an input when a specified high-level voltage is applied to that input.

Low-level input current (V_{IL}) [$V_{in\ (0)}$]: The current that flows through an input when a specified low-level voltage is applied to that input.

High-level output current (V_{OH}) [$V_{out\ (1)}$]: The current that flows from an output in the logical 1 state under specified load conditions.

Low-level output current (V_{OL}) [$V_{out\ (0)}$]: The current that flows from an output in the logical 0 state under specified load conditions.

4.4 CURRENT-SOURCING AND CURRENT-SINKING LOGIC

Logic families can be categorised depending upon the flow of current from the output of one logic circuit to the input of another. Current-sourcing and current-sinking logic gates are illustrated in Fig. 4.2(*a*) and (*b*) respectively.

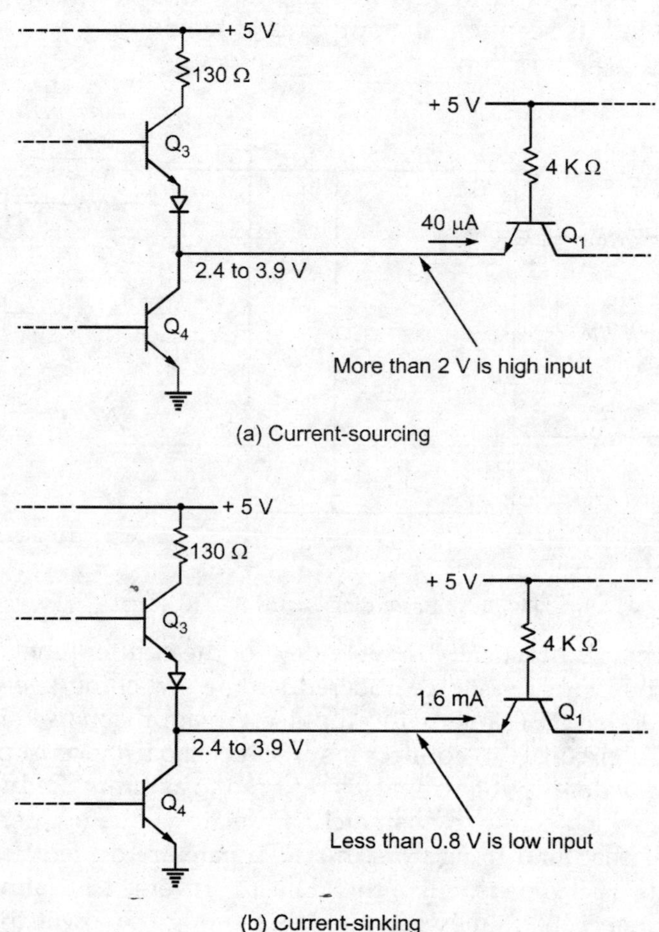

Fig. 4.2

When a standard TTL gate output is HIGH as shown in Fig. 4.2(*a*), a reverse emitter current of 40 mA flows from transistor Q_3 of driver gate to the emitter of transistor Q_1 of load gate, and hence Q_3 acts as a *current source.*

When a standard TTL gate output is LOW as shown in Fig. 4.2(*b*), an emitter current of 1.6 mA flows from the emitter of transistor Q_1 of load gate to the collector of transistor Q_4 of driver gate. As Q_4 is saturated, current flows through it to the ground, and hence Q_4 acts as a *current sink*.

4.5 RESISTOR-TRANSISTOR LOGIC (RTL)

The basic diagram of an RTL NOR gate consisting of resistors and transistors is shown in Fig. 4.3. When the inputs *A*, *B* and *C* are at 0 V (or logic 0), the transistors are turned OFF. Hence the output goes to + V_{CC}, i.e., logic 1. If either one or all input terminals are at + V_{CC} (or logic 1), one transistor or all would be fully turned ON, thereby reducing the output voltage to almost 0 V. It is seen that the output is at logic 1 only when all the inputs are at logic 0, i.e. the NOR logic function.

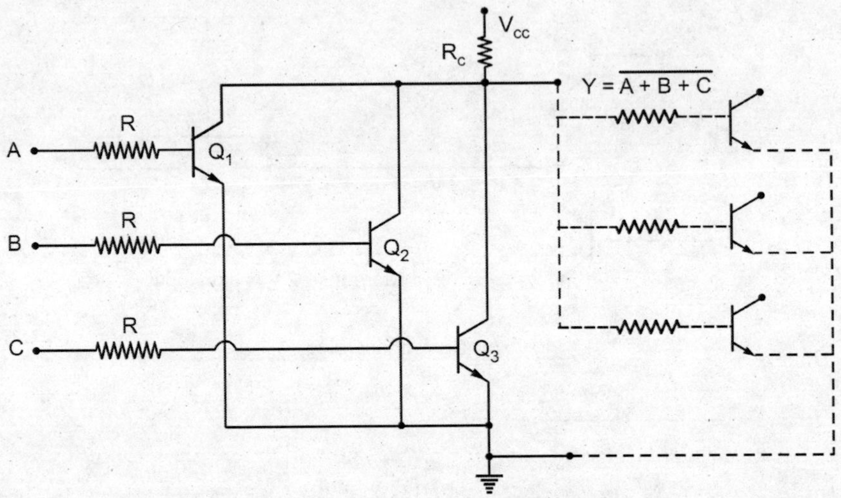

Fig. 4.3. Basic diagram of RTL NOR gate.

The base current is practically independent of the emitter junction characteristic. The resistors increase the input resistance and reduce the switching speed of the circuit. This degrades the rise and fall times of any input pulse. An approach used in practice to increase the speed of an RTL circuit is to connect a speed-up capacitance in parallel with the base resistance. The number of input terminals is referred to as *fan-in*. Reducing current-hogging by load transistors, which is due to mismatch of junction voltages, permits a larger fan-out. Another problem is that load transistors in a RTL gate are driven heavily into saturation, resulting in long turn-off delays. Also, the collector reverse saturation current of a driver transistor at high temperatures may become large enough to lower the already low output voltage. An Integrated Injection Logic (I^2L) circuit, which is a modified version of the basic RTL circuit, alleviates all these problems.

Do You Know?	Although RTL has become obsolete now, because of its simplicity and for historical reasons, it is proper to devote some attention to it and develop concepts useful in connection with all types of gates.

The following are the characteristics of the RTL family.

(*a*) Speed of operation is low, i.e., the propagation delay is of the order of 500 ns; it cannot operate at speeds above 4 MHz.

(*b*) Fan-out is 4 or 5 with a switching delay of 50 ns, and fan-in is 4.

(*c*) Poor noise immunity.

(*d*) High average power dissipation. Elimination of base resistors in RTL will reduce the power dissipation which results in Direct-coupled Transistor Logic (DCTL).

(*e*) The noise margin from zero to the threshold voltage is about 0.5 V, and from one to the threshold voltage is only 0.2 V.

(*f*) Sensitive to temperature.

The RTL family of ICs includes NOR gates with two, three or four inputs, flip-flops and four-bit shift registers.

4.6 RESISTOR-CAPACITOR-TRANSISTOR LOGIC (RCTL)

The RCTL circuit employs a capacitor in parallel with an input resistor to increase the speed and to improve noise immunity. The basic circuit of a RCTL NOR gate is shown in Fig. 4.4. During the transient, the capacitor bypasses the resistor, with the result that the base currents grow and the input capacitance discharges more quickly. The use of the capacitor also allows higher values of resistance, making possible lower power dissipation per gate. In compression with the RTL family, the RCTL circuits have low propagation delay time t_{pd}, ranging from 10 to 15 ns, although having the same values of fan-out and fan-in. In RCTL circuit, manufacturing of PN junction capacitor is difficult and also occupies larger area. The RCTL circuit is not ideal for fabrication because it includes a high proportion of resistors and capacitors.

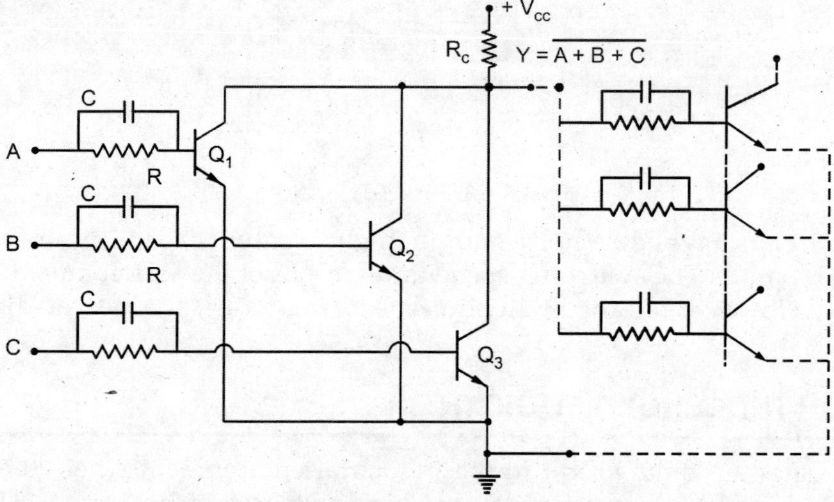

Fig. 4.4. Basic circuit of RCTL NOR gate.

4.7 DIODE-TRANSISTOR LOGIC (DTL)

The DTL family eliminates the problem of decreasing output voltage with increasing load. The basic circuit of a DTL NAND gate is shown in Fig. 4.5(*a*). Two diodes in this circuit, D_A and D_B, perform the logic AND operation followed by a transistor inverter which results in a NAND gate. When both the inputs are at logic HIGH level, the diodes D_A and D_B are reverse-biased. Diodes D_1 and D_2 and transistor are switched ON and hence the output is LOW. The additional diode D_2 increases the noise margin. If any of the inputs drops to ground potential (logic 0), the corresponding input diode will conduct, and current will flow through the diode and R_D, causing a voltage drop at the input of the diode D_1. The base voltage becomes low and the transistor remains cut off and hence the output is HIGH.

The DTL family has the following characteristics.

Propagation delay: The turn-off delay is considerably larger than the turn-on delay, often by a factor of 2 or 3. The propagation delay of DTL is 25 ns.

Fan-out: A fan-out as high as 8 is possible with the DTL family because of the high input impedance of the subsequent gates in the logic 1 state.

Fan-in: It has a fan-in of 8.

Noise immunity: The noise margin is high due to the additional diode (D_2) connected in series with Z_1. The noise margin of the DTL NAND gate circuit shown in Fig. 4.5(*a*) is 0.8 V when the output is low and 3.4 V when the output is high.

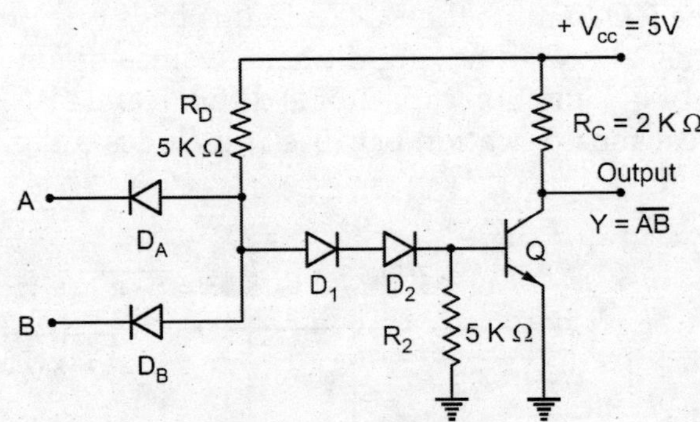

Fig. 4.5 (*a*) Basic DTL NAND gate.

DTL circuits have better noise margin, higher fan-out capability and faster response than the RTL family. The switching speed and fan-out of the DTL family are improved in TTL family. However, as the switching speed increases, the power dissipation also increases.

4.8 HIGH THRESHOLD LOGIC (HTL)

HTL gates are quite useful in the industrial environment where the noise level is usually high due to the presence of motors, high voltage switches, etc. A HTL NAND gate can be derived from an ordinary DTL NAND gate by replacing diode D_2 by a 6.9 V Zener diode and using a higher supply voltage (+ 15 V instead of 5 V) as shown in Fig. 4.5(*b*). The resistor values are also increased so that the same currents are obtained in both ordinary DTL and HTL NAND gates. Because of the use of 6.9 V Zener diode, the noise margin of this circuit is increased to 7 V.

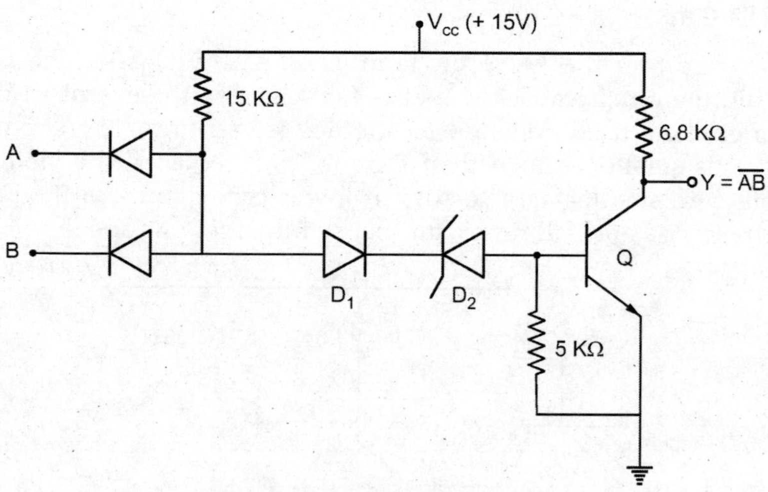

Fig. 4.5 (*b*) Basic HTL NAND gate.

4.9 TRANSISTOR-TRANSISTOR LOGIC (TTL OR T²L)

The most commonly used saturating logic family called the Transistor-Transistor Logic, (TTL or T^2-L), has the fastest switching speed when compared to other logic-families that utilize saturated transistors. The series 54/74 TTL family has grown and evolved into five major divisions:

(*i*) standard (SN 54/74)
(*ii*) high-speed (SN54H/74H)
(*iii*) low-power (SN54L/74L)
(*iv*) schottky-diode-clamped (SN54S/74S)
(*v*) low power schottky. (SN54LS/74LS)

Although the high-speed and low-power series were designed for specific applications, all four families are compatible and are capable of interfacing directly with one another.

Do You Know?	Transistor-transistor logic, TTL, is named for its dependence on transistors alone to perform basic logic opereations. The first version, which is now known as standard TTL, was developed in 1965 and is rarely used in today's systems.

They have the following typical characteristics in common:

(*i*) Supply voltage : 5.0 V
(*ii*) Logical 0 output voltage : 0 V to 0.4 V
(*iii*) Logical 1 output voltage : 2.4 V to 5 V
(*iv*) Logical 0 input voltage : 0 V to 0.8 V
(*v*) Logical 1 input voltage : 2 V to 5 V
(*vi*) Noise immunity : 0.4 V.

4.9.1 TTL NAND Gate

The basic circuit for the TTL logic family is the NAND gate. The TTL circuit uses a special single multi-emitter transistor that is fabricated with several emitters at its input. The number of emitters used depends on the desired fan-in of the circuit. Since a multi-emitter transistor is smaller in area than the diodes it replaces, the yield from a wafer is increased. Moreover, smaller area results in lower capacitance to the substrate, thereby reducing the circuit rise and fall times and hence increasing its speed.

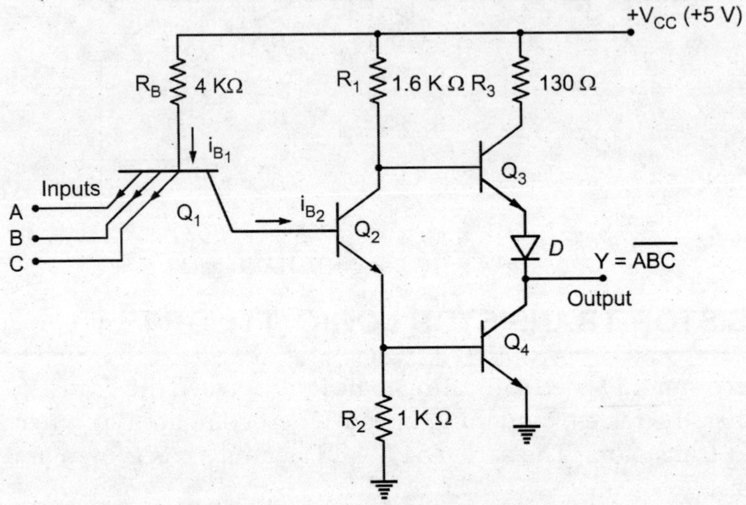

Fig. 4.6. 3-input TTL NAND gate.

Circuit operation: The basic circuit of the TTL NAND gate is shown in Fig. 4.6. The output is taken from the collector of transistor Q_4. Each emitter of Q_1 acts like a diode. Therefore, transistor Q_1, and the 4 kΩ resistor act like a 3-input AND gate and the rest of the circuit inverts the signal. Hence, the overall circuit acts like a 3-input NAND gate.

When either or all inputs (*A*, *B* and *C*) are at 0 V (logic 0), the corresponding emitter-base junction of Q_1 becomes forward-biased. The value of *RB* is selected so as to ensure that Q_1 is turned ON. However, the value of current iB_2 flowing through the base of Q_2 reduces the potential at the base of Q_2 and hence transistors Q_2 and Q_4 are cutoff so that the output voltage is at V_{CC} (logic 1).

If all the inputs are high (logic 1), the emitter-base junction of Q_1 is reverse-biased so that it has no base current. Hence, Q_1 is OFF. However, its collector-base junction is forward-biased supplying base current iB_2 to Q_2. The current i_{B2}, will be sufficiently large to saturate Q_2. As a result, transistor Q_2 is turned ON and the drop across R_2 is sufficient to forward bias the base-emitter junction of Q_4, thereby turning Q_4 ON. Hence, the output at its collector is low (logic 0). The function of diode *D* is to prevent both Q_3 and Q_4 from being ON simultaneously.

In the absence of diode *D*, the transistor Q_3 will conduct slightly when the output is LOW. In order to prevent this, the diode is connected between the emitter of Q_3 and the collector of Q_4. The voltage drop across the diode keeps the base-emitter junction of Q_3 reverse-biased. In this way, transistor Q_4 only conducts when the output is LOW, which confirms the conditions for NAND operation.

As TTL input circuits require higher drive currents than DTL, they are designed to have high power output stages. The open collector gates are used in three major applications: driving a lamp and relay, performing a wired logic and for the construction of common bus system.

4.9.2 Other TTL Series

The TTL gate has a number of circuits in its series. They have been developed to provide a wider choice of speed and power-dissipation characteristics. There are five series in the TTL family which are listed in Table 4.1 together with their respective propagation delay and power dissipation characteristics.

Table 4.1. Characteristics of 5 TTL versions

Version	Abbreviation delay (ns)	Propagation dissipation (mW)	Power clock rate (MHz)	Maximum	Fan-out
Standard	TTL	10	10	35	10
Low power	LTTL	33	1	3	10
High speed	HTTL	6	22	50	10
Schottky	STTL	3	19	125	10
Low power Schottky	LSTTL	9.5	2	45	10

Standard series 54/7400 TTL: The Standard TTL gate was the first version of the TTL family. The basic standard gate circuit shown in Fig. 4.6 features a multiple-emitter input and an active pull-up output configuration. The use of multiple emitters is a major contribution to derive fast-switching speeds of TTL. Low output impedance is attained with totem-pole output stage consisting of transistors Q_3 and Q_4, which also results in improved noise immunity and faster switching.

Standard series 54/74 line includes shirt registers, counters, decoders, memories, data selectors and arithmetic elements in addition to SSI devices.

Low power TTL, 54/74LOO series: Low power TTL circuits designated as the 74LOO series have essentially the same basic circuit as the standard 7400 series except that all the resistor values are increased.

Since an increase in resistance results in the reduction of power dissipation, the power requirements of low power gates are less than one-tenth of those of standard ICs. Series 54L/74L devices have a power dissipation of only 1 mW per gate and an average propagation delay of 33 ns. Low frequency, battery-operated circuits, such as calculators, are well suited for this version of the TTL series.

High-speed TTL, 54H/74HOO series: The basic circuitry for this series is essentially the same as the standard 7400 series except that smaller resistor values are used and the emitter-follower transistor Q_3 is replaced by a Darlington pair. Hence, the output section consists of a Darlington transistor pair Q_3 and Q_4 as shown in Fig. 4.7. This arrangement provides slightly higher speed (6 ns per gate) than the standard gate. The smaller resistances, however, increase power dissipation to about 22 mW.

Series 54H/74H master-slave flip-flops and edge-triggered flip-flops are capable of operating with clock input frequencies as high as 50 MHz.

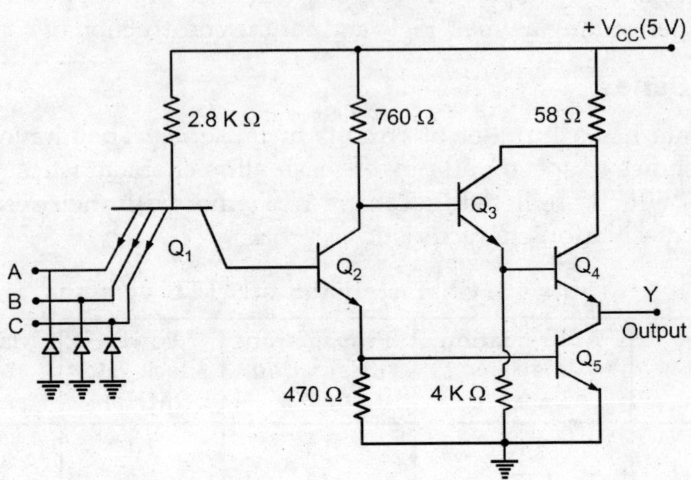

Fig. 4.7. High speed TTL gates. SN 54H/74H.

Schottky TTL, 54S/74SOO series: TTL 54S/74S series have the *highest* speed among TTL gates. This is achieved by using a Schottky-barrier diode (SBD) as a clamp from base to collector of each circuit transistor as shown in Fig. 4.8(*a*). The symbol for a transistor with a Schottky-barrier diode clamp is shown in Fig. 4.8(*b*). The characteristics of the SBD that make it useful are its low forward voltage drop (0.25 V) and its fast switching speed. When used as a clamp, the SBD diverts most of the excess base current and prevents the transistor from reaching saturation. This reduces the average propagation delay to 3 ns for a typical NAND gate.

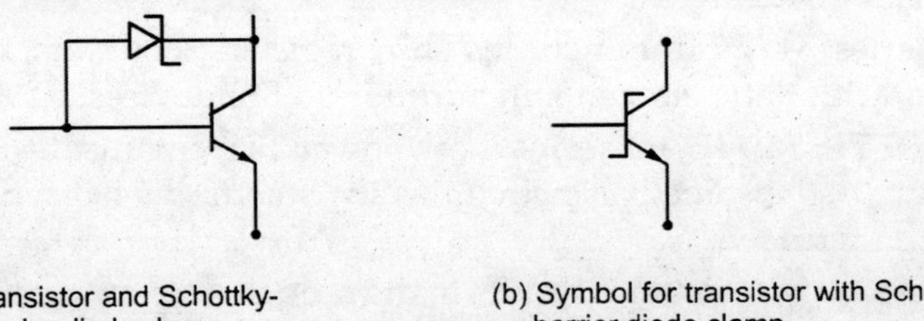

(a) Transistor and Schottky-barrier diode clamp

(b) Symbol for transistor with Schottky-barrier diode clamp

Fig. 4.8

Schottky TTL devices are very fast, and are capable of operating at 100 MHz. The 74S00 series has a propagation delay of 3 ns and an average power dissipation of about 20 mW per gate.

Low power Schottky TTL, 74LS00 series: By increasing internal resistance as well as using Schottky diodes, manufacturers have come up with a compromise between low power and high speed: low power Schottky TTL. It compares favourably with the standard TTL in speed and requires considerably less power.

These devices are numbered 74LS00, 74LS01, 74LS02, etc. A low power Schottky gate has a power dissipation of around 2 mW and a propagation delay of 9.5 ns.

4.9.3 TTL Circuit Output Connections

A number of output connections are provided using TTL logic gates. Each of (he five TTL versions comes in one of three output circuit configurations commonly referred to as:

1. Totem-pole output
2. Open-collector output
3. Tri-state output.

Totem-pole output: The circuit of Fig. 4.6 shows a TTL NAND gate with totem-pole output. The totem-pole output is the standard output of a TTL. gate and is specifically designed to reduce the propagation delay in the circuit and to provide sufficient output power for a high fan-out. The output in circuit of Fig. 4.6 is obtained as a high voltage level when Q_3 is ON or a low voltage level when Q_4 is ON. The circuit is designed in such a way that both Q_3 and Q_4 can never be ON at the same time. When either Q_4 or Q_3 conducts, the output impedance is low and hence the totem-pole output of a standard TTL circuit cannot be connected to any other output without causing a serious loading problem. This loading problem can be eliminated by a modified TTL circuit, called a Tri-state TTL circuit.

The totem-pole output: configuration has the advantage of low output impedance in both logical states. When Q_3 is conducting, the output impedance is around 70 ohms; when Q_4 is saturated, the output impedance is only 12 ohms. Because of such low output impedance, any stray output capacitance is rapidly charged and discharged, thereby changing the output voltage quickly from one state to the other. This lower output impedance is also responsible for the capability of the gate to drive high capacitive loads, for low-noise susceptibility and high-speed performance characteristics of Series 54/74. Totem-pole outputs cannot be connected together to form an AND function as in open-collector outputs.

Open-collector output: The open-collector TTL gate needs an external resistor that must be connected between the collector of a pull-down transistor and the supply voltage for proper operation. The TTL NAND gate with open-collector output is obtained by removing the following components: transistor Q_3 diode D and resistor R_3 of Fig. 4.6. The resulting open collector TTL NAND gate is shown in Fig. 4.9(*a*). As the collector of Q_4 is open, this open collector gate will not work properly unless an external pull-up resistor is connected as shown in Fig. 4.9(*b*). The output is taken at the collector of transistor Q_4. A high voltage level will appear at the output in the HIGH state.

Open-collector gates provide the versatility of wire AND operation or OR operation with a large number of logic variables. For instance, Fig. 4.10 shows three TTL devices connected to the common pull-up resistor. This has the advantage of combining the output of the three devices without using a finai OR gate (or AND gate). This combining is done by a direct connection of the three outputs to the lower end of the common pull-up resistor. This is very useful when many devices are wire-ORed together. For instance, in some systems, the outputs of 16 open-collector devices are connected to a pull-up resistor.

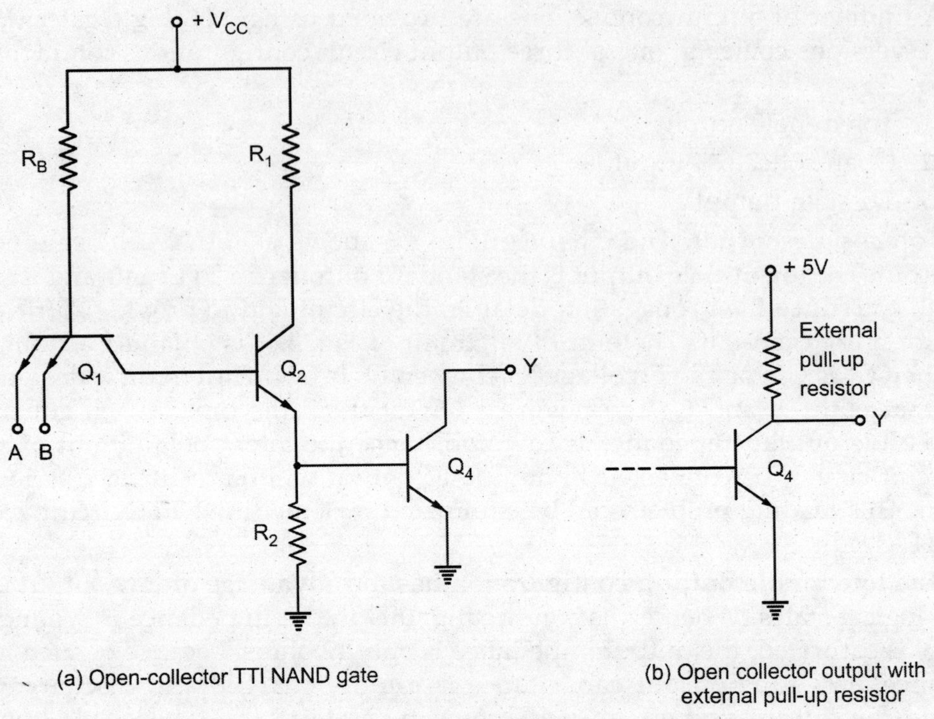

(a) Open-collector TTI NAND gate

(b) Open collector output with external pull-up resistor

Fig. 4.9.

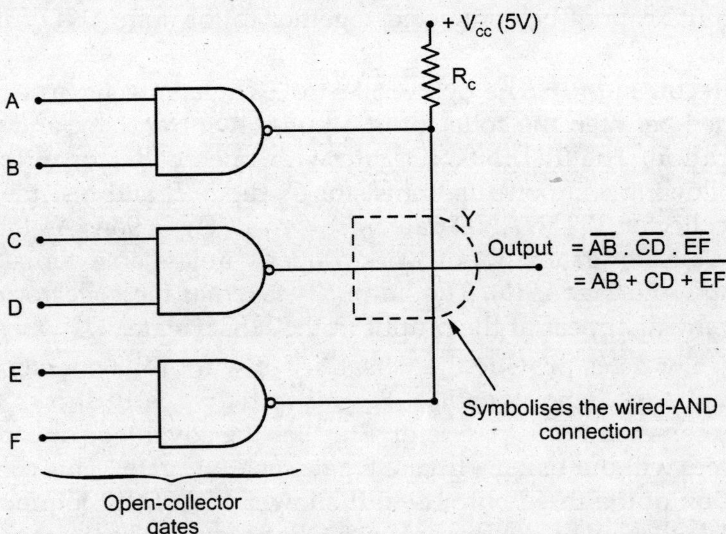

Fig. 4.10. Wired-AND operation using open-collector gates.

The main disadvantage of open-collector gates is that switching time dealy is increased because the pull-up resistance is few KΩ, which results in a relatively long time constant when it is multiplied by the stray output capacitance. The slow switching speed of open-collector TTL devices becomes worse when the output goes from low to high. When the output transistor Q_4 in the circuit of Fig. 4.9 goes into cutoff, then any capacitance across the output has to charge through the pull-up resistor. This charging produces a relatively slow exponential rise between the low and high state. Also, the circuit is more sensitive to noise at the output.

Tri-state output: A very popular output connection that incorporates the benefits of totem-pole and open-collector in the single circuit is the tri-state output connection.

The tri-state (three state) output exhibits three possible output-state conditions as shown in Table 4.2. Two of these states are the conventional logic 0 and logic 1. The third state is a high impedance (open circuit) state. This means, for all practical purposes, the circuit behaves as if the output is disabled. As a consequence, the output cannot affect or be affected by any external signal at its input terminals. The third state is controlled by a separate control input as shown in Fig. 4.11(*b*). A control (Select or Inhibit) input terminal is provided to allow the output to be switched into (or out of) its high impedance condition. When the control input C is 1, the gate behaves like a normal NAND gate providing states of 0 and 1. When the control input C is 0, the output is disabled irrespective Of the values of the normal inputs. Tri-state gates are also available with the control input having a complementary effect, i.e. disabling the gate when the control input is 1 and enabling it when it is 0. The high-impedance state of a tri-state gate allows the possibility of making a direct wire connection from many outputs to a common bus line, in which only one output line will be enabled while all other outputs are disabled by their respective control inputs.

Table 4.2. Truth table for tri-state TTL NAND gate

A	B	C	Q_1	Q_2	Q_3	Q_4	Q_5	Y
0	0	1	ON	OFF	OFF	ON	ON	1
0	1	1	ON	OFF	OFF	ON	ON	1
1	0	1	ON	OFF	OFF	ON	ON	1
1	1	1	OFF	ON	ON	OFF	OFF	0
X	X	0	ON	OFF	OFF	OFF	OFF	Open-circuit

Operation of tri-state TTL NAND gate Table 4.2 shows the truth table summarizing the operation of the tri-state NAND circuit of Fig. 4.11(*a*). When the control input C is HIGH (1) and any input A or B is LOW, Q_1 is ON and both Q_2 and Q_3 are OFF. Hence, Q_4 and Q_5 will be turned ON and the output will be at the HIGH level (nearly + 3.6 V). When the control input C is HIGH and both inputs A and B are HIGH, transistor Q_1 becomes OFF, and thereby drives both the transistors, Q_2 and Q_3. ON. Hence, Q_4 and Q_5 are OFF and the output is LOW(0). Thus, when the control input C is HIGH, the circuit operates like a totem-pole output circuit. When the control input is in LOW state, then diode D_1 conducts and therefore the voltage at the base of transistor Q_4 is 0.7 V which is not enough to make both the transistors, Q_4 and Q_5, to switch to the ON state. Also, since Q_1 is conducting, the transistor Q_2 is in a cutoff state and therefore Q_3 is also OFF. So, neither the output transistor

Q_5 nor Q_3 is ON and the output is open circuited or in HIGH impedance state. Therefore, it is concluded that there are three states of the output—LOW, HIGH and Open circuit as determined by the inputs.

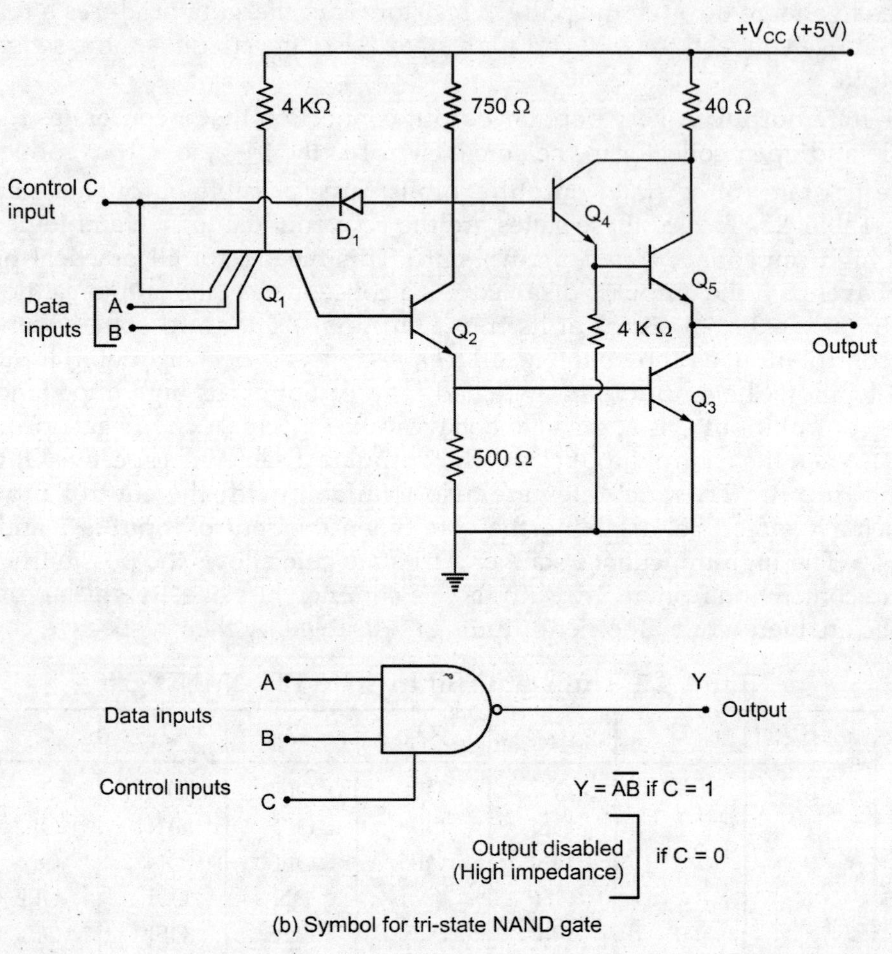

(b) Symbol for tri-state NAND gate

Fig. 4.11

4.9.4 TTL Parameters

Series 54/7400 devices work reliably over a temperature range of 0 to 70°C and over a supply range of 4.75 to 5.25 V.

Floating inputs When a TTL input voltage is HIGH (ideally + 5 V) as shown in Fig. 4.12(*a*), the emitter current is zero. When a TTL input is unconnected (floating) as shown in Fig. 4.12(*b*), there is no flow of emitter current because of the open circuit. Hence, a floating TTL input is equivalent to a high input.

Also, when an input terminal is left open, it acts like a small antenna and picks up stray electromagnetic noise voltages leading to malfunctioning or erroneous operation of the gate. Therefore, it is a must to connect the unused TTL inputs either to the ground or to the V_{CC}, depending upon the gate. For example, in AND and NAND gates, the unused input must be connected to V_{CC}, while in OR and NOR gates the unused inputs should be connected to ground.

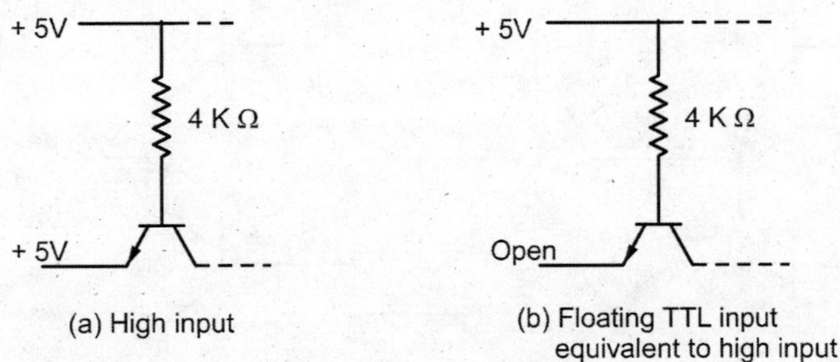

Fig. 4.12

Current sourcing and current sinking: When the output of a gate is HIGH, thereby providing current to the input of the gate being driven, the output is said to act as a *current source.* For a TTL circuit, the maximum current drawn by an input from a high output is 40 μA.

When the output of a TTL gate goes LOW, it must be capable of sinking current drawn from gate inputs which are driven LOW. The driver then operates as a *current sink.* In a standard TTL gate, when one of its inputs is low, a current of 1.6 mA flows out of the device. Thus,

$$I_{IL}(\text{max}) = -1.6 \text{ mA and } I_{IH}(\text{max}) = +40\ \mu\text{A}.$$

Here, the negative sign indicates that the current flows out of the device. [Refer to Fig. 4. 12(*a*) and (*b*).]

Standard loading A TTL device can source (high output) or sink current (low output). Specification sheets of standard TTL devices show that any 54/7400 series can sink upto 16 mA and is denoted by I_{OL} (max) = 16 mA and can source up to 400 μA, denoted by I_{OH}, (max) = – 400 μA. The negative sign indicates that the conventional current is out of the device. Since the maximum output currents, i.e. I_{OL} (max) and I_{OH} (max) , are 10 times larger than the input currents, i.e. I_{IL} (max) and I_{IH}(max), we can connect upto 10 TTL emitters to any TTL output.

Fan-out: The maximum number of TTL loads that can be driven by a TTL driver is called fan-out. As discussed in the previous section, 10 TTL inputs can be connected to the output of a standard TTL. Thus, the fan-out of standard TTL is 10. When the totem-pole output of a TTL gate goes HIGH, it reverse-biases another gate input with the resulting current (40 mA, maximum) as shown in Fig. 4.13(*a*). The TTL output going low must sink a current from the gate being driven, as shown in Fig. 4.13 (*b*). The current from one standard TTL load is 1.6 mA, while from an LS circuit the load current is only 0.36 mA. Using the standard unit as reference, one unit load is then the same as a current of 1.6 mA into a low output. Since a standard output drive is capable of sinking the current of 16 mA, it can drive upto 10 loads.

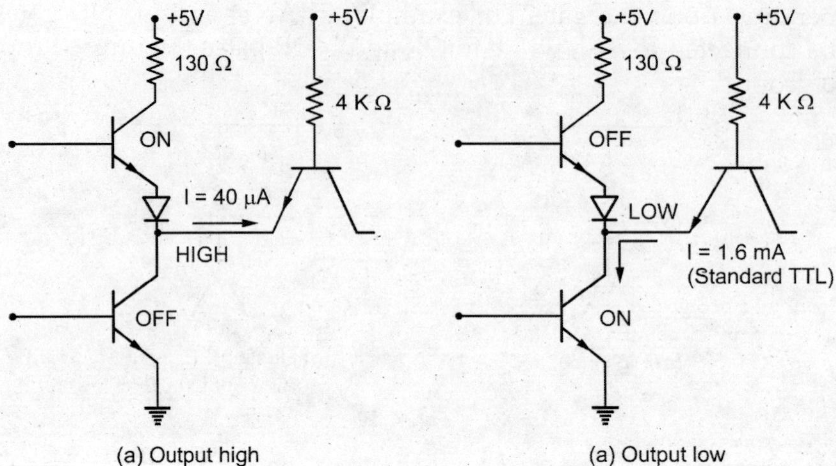

Fig. 4.13. Fan-out operation.

For LOW power TTL,

I_{IL} (max) = – 0.18 mA; I_{IH} (max) = 10 μA

I_{OL} (max) = 3.6 mA; I_{OH} (max) = – 200 μA

Considering HIGH output state: $\dfrac{I_{OH}\text{ max}}{I_{IH}\text{ max}} = \dfrac{200\ \mu\text{A}}{10\ \mu\text{A}} = 20$

Considering LOW output state: $\dfrac{I_{OL}\text{ max}}{I_{IL}\text{ max}} = \dfrac{3.6\text{ mA}}{0.18\text{ mA}} = 20$

Therefore, twenty LOW power TTL gate inputs can be connected to the output of another LOW power TTL gate.

For LOW power Schottky TTL,

$$I_{IL}(\text{max}) = -0.36\text{mA};\ I_{IH}(\text{max}) = 20\ \mu\text{A}$$
$$I_{OL}(\text{max}) = 8\text{ mA};\ I_{OH}(\text{max}) = -400\ \mu\text{A}$$

Therefore, for Lowpower Schottky TTL,

Fan-out = $\dfrac{8\text{ mA}}{0.36\text{ mA}} = 22$ (or) $\dfrac{400\ \mu\text{A}}{20\ \mu\text{A}} = 20$, whichever is less.

Also, a particular type of TTL gate can be connected with other types of TTL. For example, if a standard TTL is connected with a HTTL, the fan-out is 8; if it is connected with a LTTL, the fan-out is 40; if it is connected to a STTL, the fan-out is 8 and with a LS series, the fan-out is 20. The above data is summarised in Table. 4.3.

Table 4.3. Fan-outs

TTL	TTL load				
Driver	**74**	**74 H**	**74 L**	**74 S**	**74 LS**
74	10	8	40	8	20
74 H	12	10	50	10	25
74 L	2	1	20	1	10
74 S	12	10	100	10	50
74 LS	5	4	40	4	20

Switching speed: The TTL circuit has the fastest switching speed of any saturated logic. Two switching parameters are tested on TI 54/74 TTL gates: propagation delay time t_{PHL} from a logical 1 to a logical 0 level at the output and propagation delay time t_{PLH} from a logical 0 to a logical 1 level at the output. These switching tests are performed at the following nominal conditions: $V_{CC} = 5$ V, $T_A = 25$ °C and $N = 10$. Acceptable devices have $t_{PHL} \leq 15$ ns and $t_{PHL} \leq 22$ns. Note that t_{PHL} decreases with increasing temperature, and t_{PHL} is independent of temperature. The propagation delay time of a standard TTL gate is approximately 10ns.

Supply current characteristics: Power supply current requirements for all series 54/74 circuits are specified as maximum current drains with maximum permissible power-supply voltage, V_{CC}. Maximum I_{CCL} [$I_{CC}(0)$] per gate is specified as 5.5 mA and maximum I_{CCH}[$I_{CC}(1)$] per gate is specified as 2.0 mA. At the nominal supply voltage of 5 V, typical I_{CCL} per gate is 3mA and typical I_{CCH} is 1 mA. Thus, I_{CCL} is about 3 times larger than I_{ccH}.

Worst case input and output voltages: Theoretically, logic LOW state is 0 V and a logic HIGH state is 5 V. But, practically for TTL gates, there is a window or a range of low voltages which is still recognised as LOW state and a range of high voltages which is still recognised as HIGH state. Also, this range of LOW and HIGH state voltages is different at inputs and outputs of a TTL gate.

For a TTL gate, the worst case input voltages are:

V_{IL}, max = 0.8 V (It means a voltage from 0 to 0.8 V without changing the output is recognized as LOW state.)

V_{IH}, min = 2 V (It means a voltage from 5 V down to 2 V without changing the output is recognized as HIGH state.)

A LOW voltage greater than 0.8 V and a HIGH voltage lower than 2 V lead to unpredictable input state. Similarly, the worst case output voltages are:

V_{OL}, max = 0.4 V (It means a LOW state output having any value from 0 to 0.4 V.)

F_{OH}, min = 2.4 V (It means a HIGH state output having any value from 5 to 2.4 V.)

Thus, as far as TTL output is concerned, a LOW voltage greater than 0.4 V and a HIGH voltage less than 2.4 V leads to unpredictable output state.

Noise immunity: It is the maximum induced noise voltage a TTL device can withstand without a false change in the output state. The rating of the circuit depends upon the smallest noise voltage that will perturb it. TTL gate has less noise immunity. From the above section, the worst case LOW voltages are:

$$V_{OL}(\text{max}) = 0.4 \text{ V}$$
$$V_{IL}(\text{max}) = 0.8 \text{ V}$$

and the worst case HIGH voltage are:

$$V_{OH}(\text{min}) = 2.4 \text{ V}$$
$$V_{IH}(\text{min}) = 2 \text{ V}$$

One can observe that there is a difference of 0.4 V in both cases. This difference between maximum input LOW voltage and maximum output LOW voltage is called noise *immunity*. The smallest magnitude of noise voltage that would perturb the input signal is 0.4 V.

Power dissipation: A standard TTL gate is operated with a power supply of 5 volts, which draws an average supply current of 2mA, resulting in a power dissipation of 2 mA × 5 V = 10 mW.

Loading rules A single TTL output in the LOW state connected to several TTL inputs is shown in Fig. 4.14(*a*). Transistor Q_4 is ON and is acting as a current sink for all the currents (I_{IL},) coming back from each input. Although Q_4 is saturated, its ON state resistance is some value other than zero, so the current I_{OL} produces an output voltage drop V_{QL}. The value of V_{OL} must not exceed 0.4 V for TTL, and this limits the value of I_{OL} and thus the number of loads that can be driven.

Fig. 4.14(*b*) shows the HIGH-state situation. The transistor Q_3 is acting as an emitter-follower and is sourcing current to each TTL input. If too many loads are driven, however, the total output current I_{OH} can become too large, causing larger drops across R_2, Q_3 and D, thereby lowering V_{QH} below the minimum allowable voltage 2.4 V.

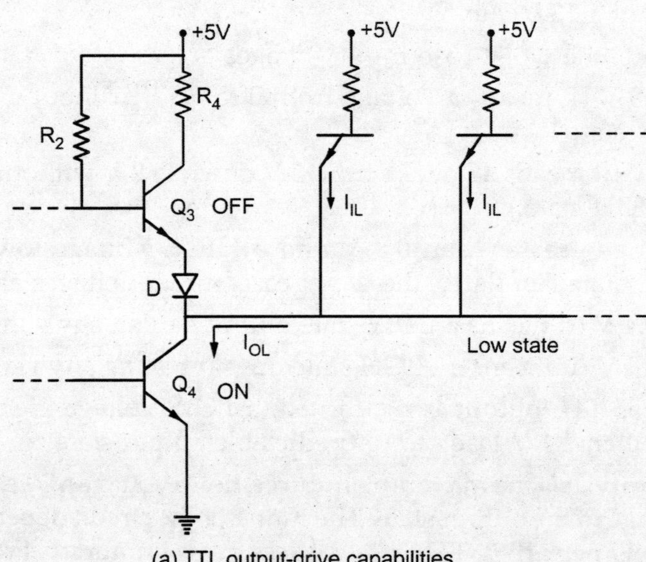

(a) TTL output-drive capabilities

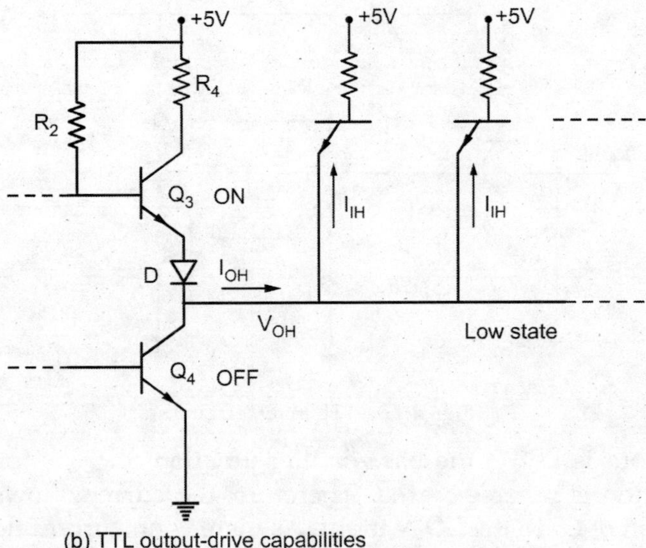

(b) TTL output-drive capabilities

Fig. 4.14

Protective (clamping) diodes: The input signals for the TTL circuit are always positive. If negative signals were to be inadvertently applied, excess input current could result which might damage the circuit. In a general circuit, the load inductance coupled with stray capacitance can result in damped sinusoidal transients.

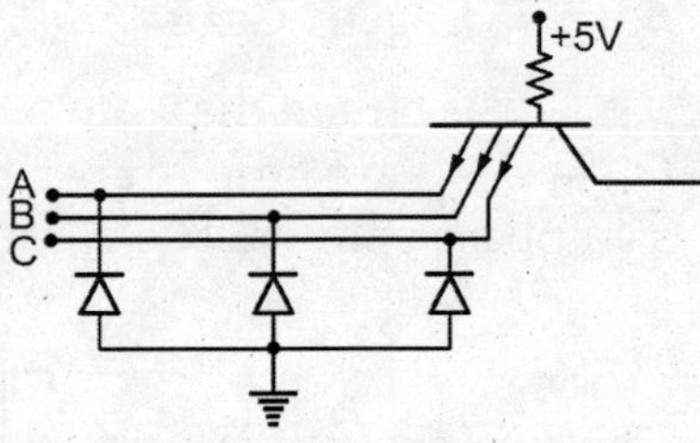

Fig. 4.15. A protected TTL.

Such signals might result in values V_A, V_B or V_C which are negative for short periods of time. To prevent these transient negative voltages (swings) from becoming substantial, diodes are usually connected from each input to ground as shown in Fig. 4.15. These diodes do not affect positive signals. The protective diodes limit negative excursions to about – 0.7 V.

4.9.5 TTL Inverter

The circuit diagram of a TTL inverter is shown in Fig. 4.16. The transistor Q_1 is the input coupling transistor and Z_1 is the protective diode. Transistor Q_2 is called a phase splitter, and the combination of transistors Q_3 and Q_4 forms the totem-pole output circuit.

Operation: When a HIGH (+ 5V) voltage is applied at the input (A), the base-emitter junction of Q_1 becomes reverse-biased and the base-collector junction is reverse-biased. The current then flows through R_1 and base-collector junction of Q_1 into the base of Q_2. Therefore, the transistor Q_2 drives into saturation and a voltage drop across R_3 turns ON the transistor Q_4. So, the output is nearer to LOW potential.

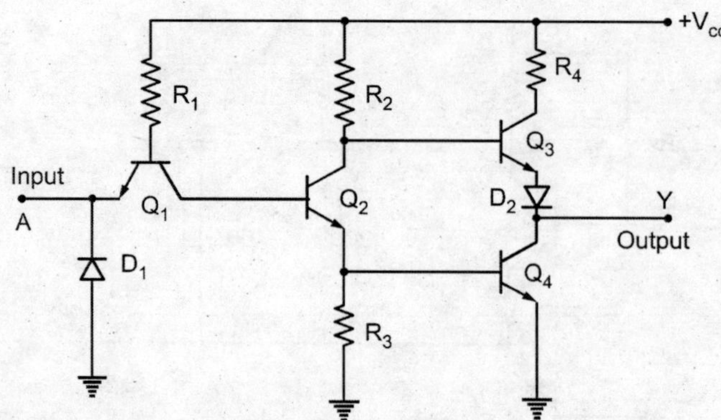

Fig. 4.16. TTL inverter circuit.

When the input is LOW, the base-emitter junction of Q_1 is forward-biased and the base-collector junction is reverse-biased. Therefore, the current flows through R_1 and the base-emitter junction of Q_1 to the LOW input. As there is no current flowing into the base of Q_2, it is OFF. The collector of Q_2 is HIGH thus turning Q_3 ON. A saturated Q_3 provides low impedance path from V_{CC} to the output. Hence, the output is HIGH.

4.9.6 TTL NOR Gate

The circuit diagram of a two-input TTL NOR gate is shown in Fig. 4.17. In this circuit, Q_1, and Q_2, are input transistors, and transistors Q_3 and Q_4 that are connected in parallel act as a phase splitter. The combination of Q_5 and Q_6 forms a totem-pole output in the circuit..

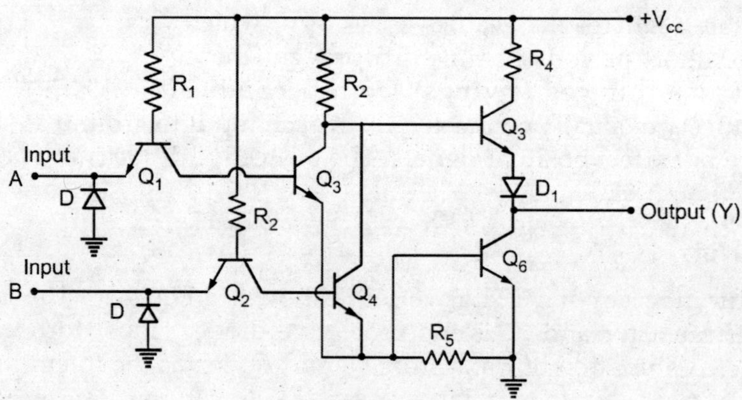

Fig. 4.17. TTL NOR gate circuit.

Operation: When the inputs are LOW, the base-emitter junctions of Q_1 and Q_2 are forward-biased and pull current away from transistors Q_3 and Q_4, keeping them OFF. As a result, Q_5 is ON and Q_6 is OFF, producing a HIGH output.

When input *A* is LOW and input *B* is HIGH, Q_3 is OFF and Q_4 is ON. The transistor Q_4 turns ON Q_6 and turns OFF Q_5, producing a LOW output.

When input *A* is HIGH and input *B* is LOW, Q_3 is ON and Q_4 is OFF. The transistor Q_3 turns ON Q_6 and turns OFF Q_5, producing a LOW output.

When both inputs *A* and *B* are HIGH, transistors Q_3 and Q_4 are ON. This has the same effect as either one being ON, turning Q_6 ON and Q_5 OFF. The result is a LOW output. Thus, this circuit functions as a NOR gate.

4.9.7 TTL AND and OR Gate

The circuits of a TTL two-input AND gate and a two-input OR gate are shown in Fig. 4.18(*a*) and (*b*) respectively. In each diagram, the arrangement of transistors provides an inversion. That is, a NAND becomes an AND and a NOR gate becomes an OR.

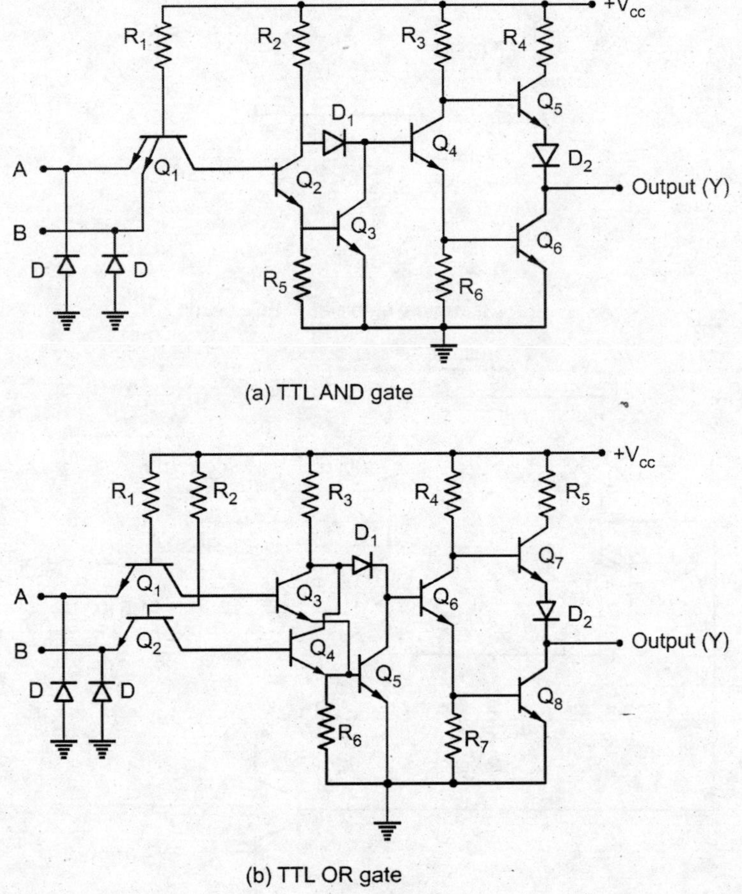

(a) TTL AND gate

(b) TTL OR gate

Fig. 4.18

4.10 EMITTER-COUPLED LOGIC (ECL)—NON-SATURATING LOGIC

Emitter-coupled Logic (ECL) is a Current-Mode Logic (CML) or non-saturated digital logic family, which eliminates the turn-off delay of saturated transistors by operating in the active mode. At present, the ECL family has the fastest switching speed among the commercially available digital ICs. The propagation delay time of a typical ECL gate is 1 ns. Also, it requires a relatively large silicon area and dissipates high power.

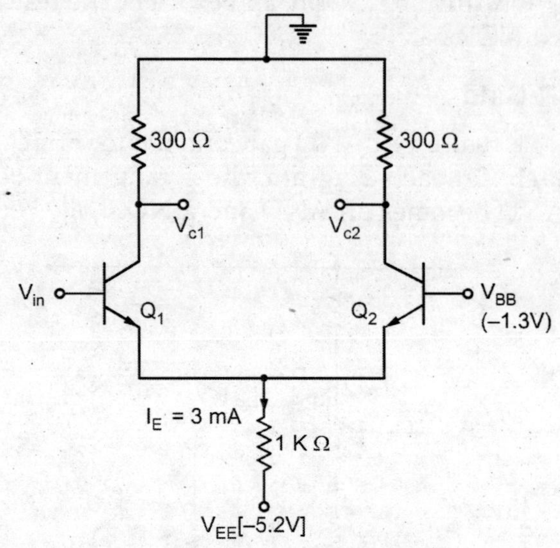

(a) Differential amplifier of ECL circuit

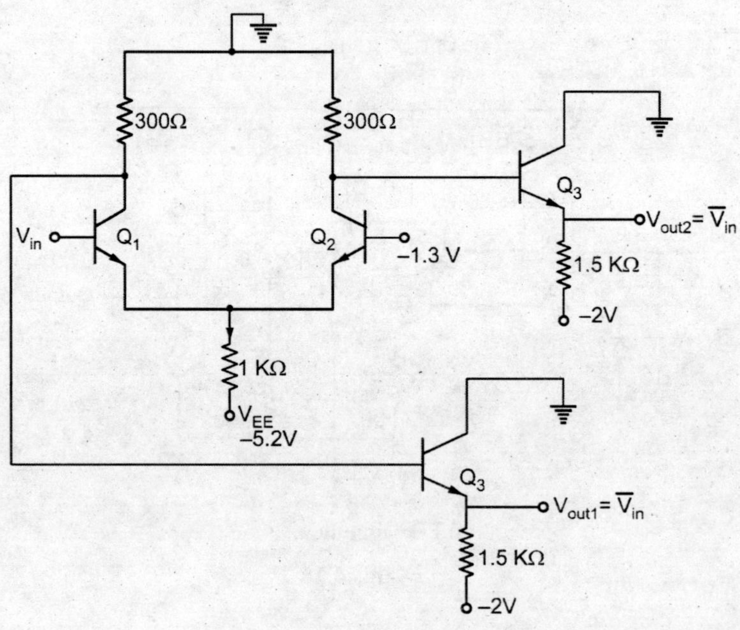

(b) Basic ECL circuit

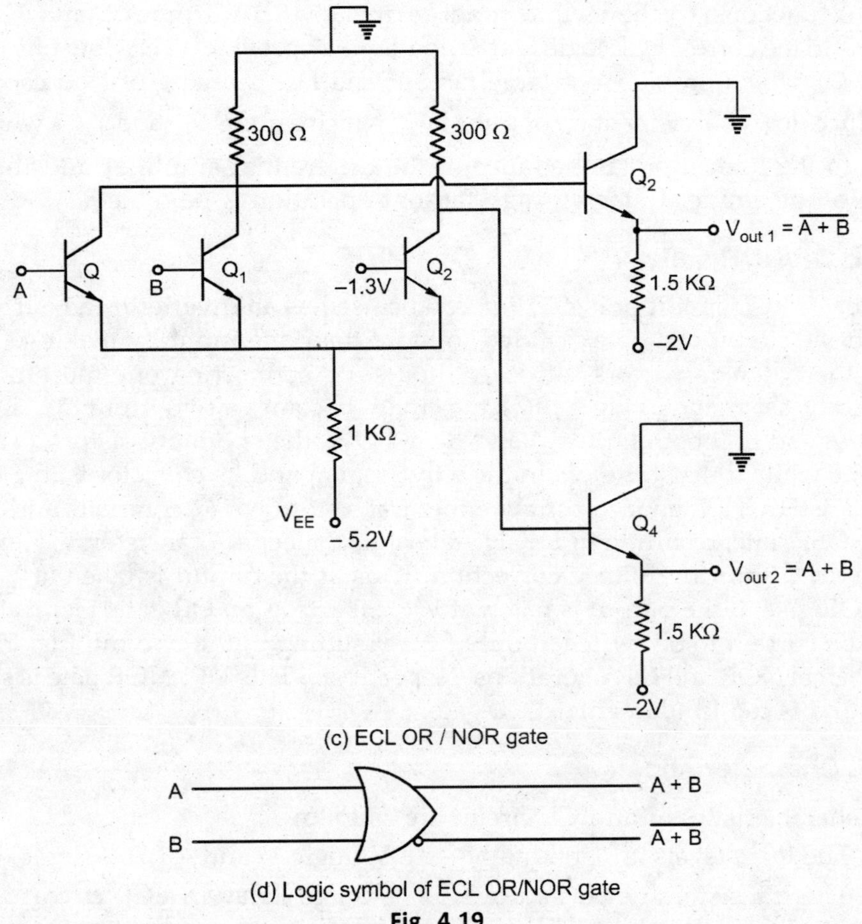

(c) ECL OR / NOR gate

(d) Logic symbol of ECL OR/NOR gate

Fig. 4.19

The basic circuit of emitter-coupled logic is a differential amplifier as shown in Fig.4.19(*a*). As a constant current is drawn by the differential amplifier even during transition from one state to another, the power supply line can be free of noise and spikes. Because of the active mode of operation of the differential amplifier, there is no storage delay in switching between ON and OFF states of the transistors in the differential pair.

Do You Know?	Unlike TTL and CMOS families, ECL does not produce a large voltage swing between the LOW and HIGH levels.

The V_{EE} supply produces a fixed current I_E which remains around 3mA during normal operation. This current is allowed to flow through Q_1 or Q_2, depending on the voltage level at V_{in}. In other words, this current switches between the collectors of Q_1 and Q_2 as V_{in} switches between its two logic levels of – 1.7 V (logical 0 for ECL) and – 0.8 V (logical 1 for ECL). V_{c1} and F_{c2} are the complements of each other, and the output voltage levels are not the same as the input logic levels.

The output voltage levels are made equal to the input logic level by connecting V_{c1} and V_{c2} to emitter-follower stages (Q_3 and Q_4), as shown in Fig.4.19 (*b*). The emitter followers perform two functions: (*i*) they subtract approximately 0.8 V from V_{C1} and V_{C2} to shift the output levels to the correct ECL logic levels, and (*ii*) they provide a very low output impedance (typically 7 Ω), which provides for large fan-out and fast charging of load capacitance. This circuit produces low complementary ouputs: V_{out1}, which equals, $\overline{V_{in}}$ and Y_{out2}, which equals V_{in}.

Due to the high input impedance of the differential amplifier and the low output impedance of the emitter follower, high fan-out operation is possible.

4.10.1 ECL OR/NOR Gate

The basic ECL circuit of Fig.4.19(*b*) can be used as an inverter if the output is taken at V_{outl}. This basic circuit can be expanded to more than one input by making. Transistor Q_1 parallel to the other transistors for other inputs. By connecting one more transistor *Q* in parallel with Q_1 as shown in Fig.4.19(*c*), the circuit becomes a two-input ECL OR/NOR gate with inputs *A* and *B*. If both inputs *A* and *B* are LOW, then both transistors, *Q* and Q_1, are in the OFF state while transistor Q_2 is in the ictive region and its collector is in a LOW state. If either *A* or *B* is HIGH, then iccordingly either transistor *Q* or Q_1 conducts and the transistor Q_2 is in the OFF state, resulting in HIGH state at its collector. Transistors Q_3 and Q_4 provide the iccessary d.c. shift for voltage correction. Thus, if the ouptut is taken at V_{out1} the ;ircuit acts as a NOR gate; if the output is taken at V_{out2}, it acts as an OR gate. Here, either Q_1 or Q_2 can cause the current to be switched out of Q_2, resulting in the two mtputs, V_{out1} and V_{out2} being the logical NOR and OR operations, respectively. This DR/NOR gate is symbolized in Fig.4.19(*d*) and is the fundamental ECL gate.

4.10.2 ECL Characteristics

The characteristics of an ECL circuit are as follows:

(*i*) The logic levels are nominally – 0.8 V (logic 1) and – 1.70V (logic 0).

(*ii*) The transistors never saturate, i.e. storage delay in ECL circuit is eliminated, and hence switching speed is very high. Typical propagation delay time is 1 ns, which makes ECL faster than advanced Schottky TTL (74AS series).

(*iii*) Because of the low noise margin, 250 milli-volt, ECL circuits are not reliable in heavy industrial environments.

(*iv*) An ECL logic block usually produces an output and its complement. This eliminates the need for inverters.

(*v*) Fan-outs are typically around 25, owing to the low-impedance emitter-follower ouputs. Such a small fan-out is a limitation compared with the saturating logic families or the MOS logic families.

(*vi*) Typical power dissipation for a basic ECL gate is 40 mW, somewhat higher than the 74AS series. This is true because all the transistors are in the active mode.

(*vii*) The total current flow in an ECL circuit remains relatively constant regardless of its logic state. This helps to maintain an unvarying current drain on the circuit power supply even during switching transistions. Thus, no noise spikes will be internally generated like those produced by TTL totem-pole circuits.

The ECL family is not as widely used as the TTL and MOS families except in very high frequency applications, where its speed is superior. Its relatively low noise margins and high power drain are disadvantages when compared with other logic families. Another drawback is its negative supply voltage and logic levels, which are not compatible with the other logic families; this makes ECL difficult to use in conjunction with TTL and MOS circuits.

4.11 INTEGRATED-INJECTION LOGIC (I²L)

Integrated-Injection Logic, IIL or I^2L, is a new LSI technique also called Merged Transistor Logic (MTL) that uses both *n-p-n* and *p-n-p* bipolar junction transistors to form a large number of IC gates on a chip. It reduces the number of metal inter-connections. When operated at low speeds, I^2L dissipates less power (5 mW) than any logic family including CMOS. At high speeds (5 ns), it only dissipates 5 mW per gate. Because of its high speed and less power dissipation, it is used in large computers.

Operation: The basic I^2L inverter circuit is shown in Fig. 4.20(*a*). Due to the absence of resistors, the *PL* inverter occupies much smaller area than a I^2L, inverter. The base of Q_1 and emitter of Q_2 are internally merged; the collector of Q_1 and the base of Q_2 are merged, and hence only four separate regions are required to form the two transistors. Thus, the entire I^2L gate takes only the space of a single TTL multiple-emitter transistor.

The *p-n-p* transistor Q_1 acts as a current source and active pull-up, and the multiple-collector *n-p-n* transistor Q_2 operates as an inverter. Most of the current leaving from the emitter of Q_1 is injected directly into the base of Q_2, and hence the emitter of Q_1, is known as the *injector*, and the integrated gate structure is called the Integrated Injection Logic.

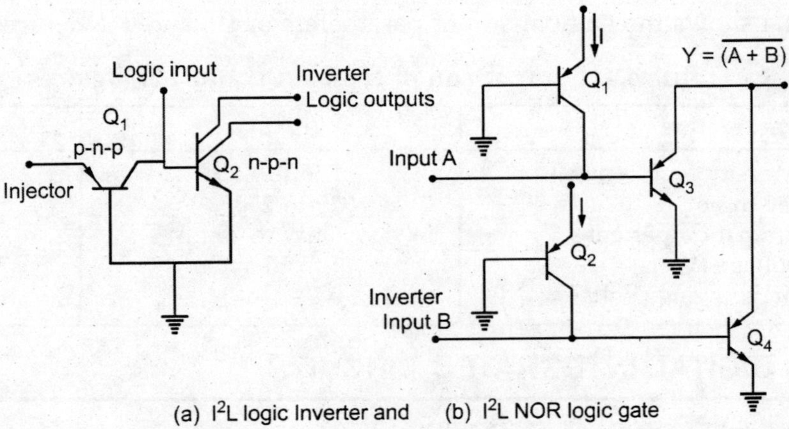

(a) I²L logic Inverter and (b) I²L NOR logic gate

Fig. 4.20

The base of Q_1 and emitter of Q_2 are internally merged and the collector of Q_1 and base of Q_2 are merged, the entire I^2L gate takes only the space of a single TTL multiple-emitter transistor.

A low input voltage of 0.1 V, the saturation voltage of the preceding multi-collector transistor, pulls injector current out of the input of the *n-p-n* transistor Q_2. Then, this injector current flows through the driving transistor Q_1. With no base current, the inverter transistor Q_2 turns OFF and its output goes HIGH. A high input voltage of 0.8 V or more to the logic input terminal allows the injector current to hold the inverter transistor ON, and hence its output goes LOW.

Fig. 4.20(*b*) shows the circuit of I^2L NOR gate wkh two inputs *A* and *B* and output *Y*. If either or both of the inputs are HIGH, one or both *n-p-n* transistors (Q_3 and Q_4) are ON, and hence the output goes LOW. If both the inputs are LOW, then both Q_3 and Q_4 are turned OFF, and hence the output is HIGH. Thus, the circuit acts as a NOR gate.

The P^2L circuit resembles the DCTL circuit. The P^2L gate with three Schottky diodes in the base region that realise the AND function and with three collector leads forming a branched output is shown in Fig. 4.21.

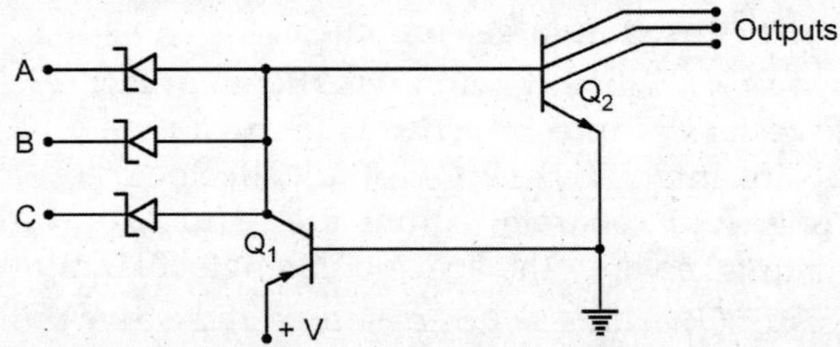

Fig. 4.21. Schottky I^2 L gate.

In summary, the P^2L family is one of the more promising of the bipolar families. It has already found application in video games, watches, television tuning and control, and memory and microprocessor chips.

Table 4.4 shows the comparison of parameters of I^2L and T^2L devices.

Table 4.4. Comparison of typical PL and T2L devices

Parameters	I^2L	T^2L
Packing density (gates/mm)	1500	150
Gate delay (nsec)	25 to 250	10
Power dissipation per gate	5 nW to 75 mW	15 nW
Supply voltage (Volts)	1 to 15	3 to 7
Logic voltage swing (Volts)	0.6	5

4.12 MOS DIGITAL INTEGRATED CIRCUITS

MOS (Metal-Oxide Semiconductor) technology derives its name from the basic MOS structure of a metal electrode on an oxide insulator over a semiconductor substrate. The most common applications of MOS devices are digital, such as logic gates and registers, or memory arrays. MOS ICs can accommodate a much larger number of circuit elements on a single chip than bipolar ICs. The principal disadvantage of MOS IC is its relatively low operating speed when compared to bipolar IC families.

4.12.1 MOSFET

In MOSFET, the channel can be of *p* or *n* type, depending on whether the majority carriers are either holes or electrons. The mode of operation can be *enhancement* or *depletion,* depending on the state of the channel region at zero gate voltage. If the channel is initially doped lightly with *p*-type impurity, a conducting channel exists at zero gate voltage and the device is said to operate in the *depletion mode.* In this mode, current flows unless the channel is depleted by an applied gate field. If the region beneath the gate is left initially uncharged, a channel must be induced by the gate field before current can flow. Thus, the channel current is enhanced by the gate voltage and such a device is said to operate in the *enhancement mode.*

The schematic symbols for the *n*-channel and *p*-channel enhancement MOSFETs are shown in Fig. 4.22. The direction of the arrow indicates the type of channel *p* or *n*. The symbols show a broken line between the *source, substrate* and *drain* to indicate that there is *normally* no conducting channel among these electrodes. The symbol also shows a separation between the gate and the other terminals to indicate the very high resistance (> 10,000 MΩ) between the gate and channel.

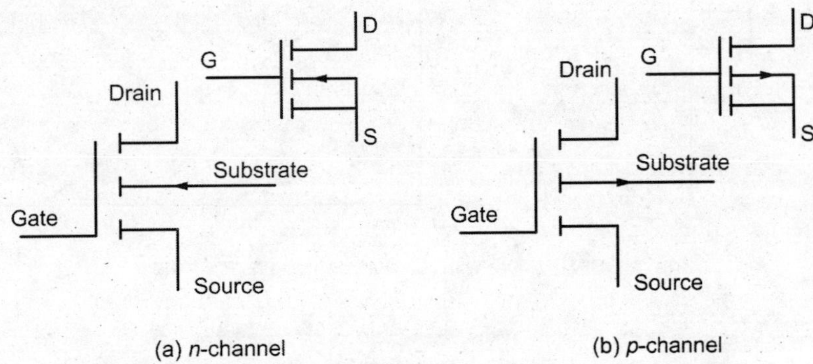

Fig. 4.22. Schematic symbols for the enhancement MOSFpTs.

PMOS: It uses only *p*-channel enhancement MOSFET. A resistor at the output of a PMOS circuit could be used to drop the high level voltage to one suitable for CMOS circuit. *Holes* are the current carriers for PMOS.

NMOS: It uses only *n*-channel enhancement MOSFET. NMOS has a greater packing density than PMOS. *Free electrons* are the current carriers in NMOS. These circuits require voltage typically ranging from 5 V to 12 V.

CMOS: It uses both *p*-and *n*-channel devices. It has the greatest complexity and lowest packing density among the MOS families. It possesses the important advantages of much lower power dissipation, very high input impedance and high noise immunity. The CMOS logic gates are used in battery-operated portable equipment. Its main disadvantage is its low speed due to high input impedance.

PMOS and NMOS digital ICs have a greater packing density and are therefore more economical than CMOS. NMOS is also about twice as fast as PMOS. PMOS and NMOS find their widest applications in LSI (microprocessors, memories, ROMs, etc.) while CMOS is widely used in MSI applications.

4.12.2 MOSFET Logic Circuits

For an *n*-channel MOS, supply voltage is of about + 5 V. The two voltage levels are a function of the threshold voltage V_T. The low level ranges anywhere from zero to V_T, and the high level ranges from V_T to V_{DD}. The *n*-channel gates usually employs positive logic.

Fig. 4.23 shows a MOSFET driver with a passive load resistor (R_D). MOSFET acts as a switch depending on V_{in}, either low or high. It switches between saturation and cutoff. When V. is low, MOSFET is cutoff and hence V_{out} is high. When V_{in} is high. When V_{in} is high, MOSFET is saturated and hence V_{out} is low.

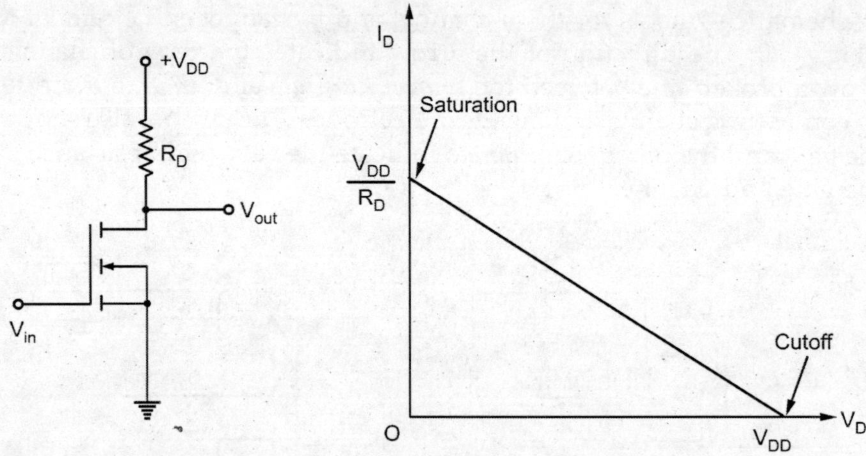

(a) MOSFET driver with passive load and its load line

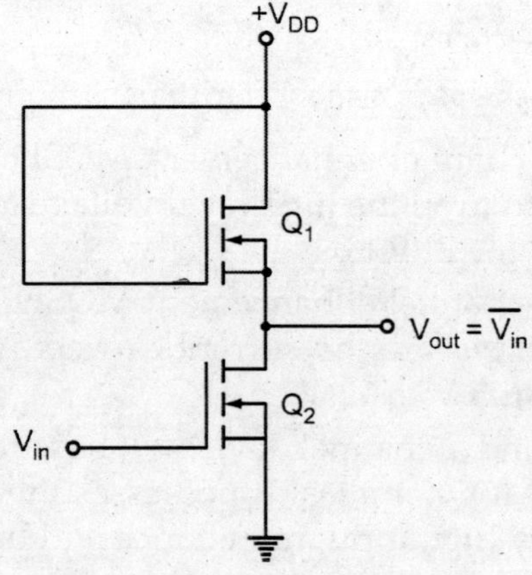

(a) MOSFET driver with active load

Fig. 4.23

In MOS technology, MOSFET fabrication is easier than resistors. For this reason, the resistor R_D in Fig. 4.23(*a*) is replaced by an *n*-channel MOSFET as shown in Fig. 4.23(*b*).

The gate of the upper MOSFET is connected to the drain; this MOSFET always conducts and behaves like the resistor, R_D. The upper MOSFET has a resistance ten times greater than that of the lower MOSFET. When the input voltage is low (below V_T), Q_2 turns OFF. Since Q_1 is always ON, the output voltage is at about V_{DD}. When the input voltage is high (above V_T), Q_2 turns ON. Current flows from V_{QD} through the active load resistor, Q_1, and into Q_2. Thus, the circuit behaves as an inverter. The geometry of the two MOS devices must be such that the resistance of Q_2 when conducting is much less than the resistance of Q_1, to maintain the output Fat a voltage below V_T.

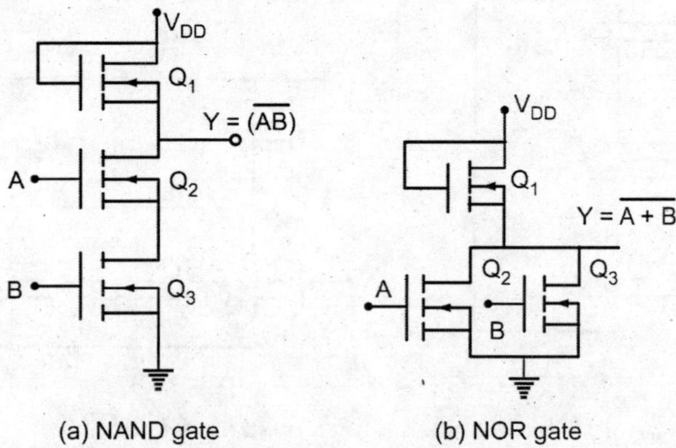

Fig. 4.24. *n*-channel MOS logic circuits.

The circuit of NMOS NAND gate that uses three transistors in series is shown in Fig. 4.24(*a*). Inputs *A* and *B* must be high for all transistors to conduct and cause the output to go LOW. If either of the inputs is LOW, the corresponding transistor is turned OFF and the output is HIGH. Again, the series resistance formed by the two active MOS devices must be much less than the resistance of the load of MOSFET.

The NOR gate that uses two NMOS transistors in parallel is shown in Fig. 4.24(*b*). If either of the inputs is HIGH, the corresponding MOS transistor conducts and the output is LOW. If the inputs are LOW, both MOSFETs are OFF, and hence the output is HIGH.

Fig. 4.25(*a*) illustrates a 3-input NMOS AND gate with an additional output inverter. The transistors Q_1 and Q_2 serve as loads. If any one input is at 0 V, then the corresponding input transistor is in OFF state, resulting in conduction of Q_6. This leads to 0 V at the output. If all inputs are at a high voltage, the MOSFETs Q_3, Q_4, and Q_5 will be ON and they offer a low resistance compared to the resistance of the load transistor Q_1, then the transistor Q_6 goes OFF. The output is now at a high level. Hence, the circuit acts as an AND gate.

A 3-input *n*-channel MOS OR gate with an output inverter is shown in Fig. 4.25(*b*). Transistors Q_1, Q_4, Q_5 and Q_6 form a NOR gate, and transistors Q_2 and Q_3 form an inverter. Transistors Q_1 and Q_2 act as loads. If the voltage is equal to V_{DD} at one of the inputs, the corresponding transistor is ON and allows the current to flow. The voltage at point *K* then drops close to zero, the transistor Q_3 switches OFF, and the output voltage approaches the level V_{DD}. If all the inputs are LOW (at 0 V), the transistors Q_4

Q_5 and Q_6 turn OFF, and the voltage at point K comes close to V_{DD}. The transistor Q_3 then switches ON and the output drops to zero. Hence, this circuii performs the OR function.

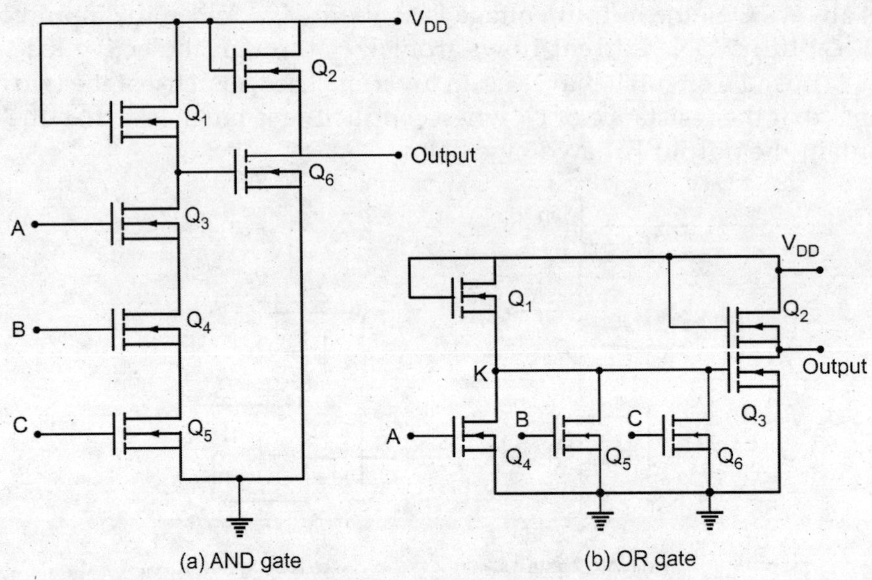

Fig. 4.25. *n*-channel MOS logic circuits.

4.12.3 Characteristics of MOS Logic

MOS logic families are slower in operating speed, require much less power, have a better noise margin and a higher fan-out.

Operating speed: A typical NMOS NAND gate has a propagation delay of 50 ns. The combination of large R_{out} and large C_{load} serves to increase switching time.

Noise margin: Typically NMOS noise margins are around 1V.

Fan-out: The fan-out capabilities of MOS logic would be virtually unlimited be cause of the extremely high input resistance at each MOSFET input. MOS logic car easily operate at a fan-out value of 50.

Power drain: MOS logic circuits draw small amount of power because of the relatively large resistance being used.

Process complexity: MOS logic is the simplest logic family when it comes to fabrication since it uses only one basic element, an NMOS (or PMOS) transistor. It does not require resistors, diodes, etc. This characteristic together with its lower power dissipation (P_D) makes it ideally suited for LSI, and this is where MOS logic has made its greatest impact in the digital field.

4.13 COMPLEMENTARY MOS LOGIC

Complementary Symmetry Metal-oxide Semiconductors, COSMOS or CMOS, are logic gates made using both PMOS and NMOS transistors. The basic gates employ both *p*- and *n*-channels enhancement mode complementary symmetry MOSFETs. The power consumption of CMOS under static conditions is extremely low. CMOS logic circuits excel PMOS and NMOS logic circuits in a number of features like extremely small d.c. power dissipation, enhanced noise immunity, high fan-out capability and ease of interfacing (compatibility with other logic circuits). CMOS circuits are used both in logic circuits and memory devices. The source terminal of the *p*-channel device is at V_{DD}, and the source terminal of the *n*-channel device is at ground. The systems employing CMOS transistors require only one power supply source of a wide range of voltages, from + 3 to +15 V. The CMOS fabrication process is simpler than TTL and has a greater packing density, therefore, permitting more circuitry in a given area and reducing the cost per function.

4.13.1 CMOS Inverter

Just as in the case of ordinary MOS gate, the inverter is basic to the CMOS gate. A basic inverter connection is shown in Fig. 4.26. The driver is transistor Q_2 which is the *n*-channel, and Q_1 (the *p*-channel device) acts as the load. Notice that drains are connected together to provide the output and that the source and substrate are connected together. The source of *p*-channel device is connected to + V_{DD} and the source of *n*-channel is connected to ground. The gates of the two devices are connected together as a common input. It is important to note that the output voltage is equal to the supply voltage and the current flows through the circuit only during switching of the input voltage from one level to the other.

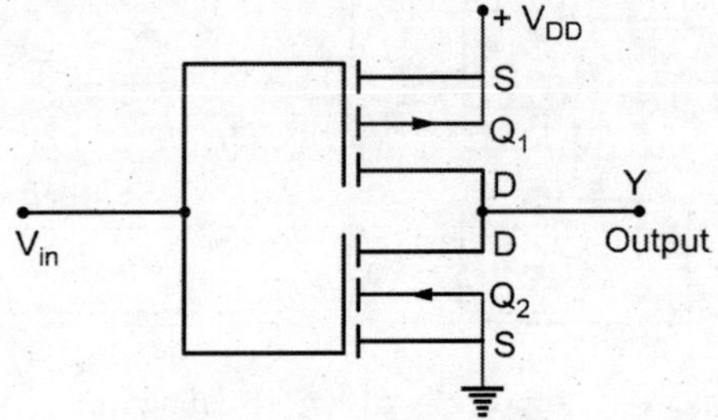

Fig. 4.26. CMOS inverter.

Operation When V_{in} is LOW, Q_2 is OFF but Q_2, is ON. This means the output voltage is HIGH. On the other hand, when V_{in} is HIGH, Q_2, is ON and Q_1 is OFF. In this case, the output voltage is LOW. Since the output voltage is always opposite in phase to the input voltage, the circuit acts as an inverter.

Do You Know?	CMOS circuits contain both NMOS and PMOS devices to speed the switching of capacitive loads. It consumes lower power and an be operated at high voltages, resulting in improved noise immunity.

The operation of CMOS transistors can be described by the following rules:

n-channel MOSFETs are turned ON by a *positive* gate voltage.

p-channel MOSFETs are turned ON by a *negative* gate voltage.

The CMOS inverter can be modified to build other CMOS logic circuits. A CMOS circuit is ideal in a number of ways. First, it needs extremely low power to operate the circuit. Since either one of the MOS devices is OFF when the input is in LOW or HIGH state, only the leakage current in the order of nanoamperes flows through the circuit and the power dissipation of the CMOS devices is typically in the range of nanowatts. This low power consumption is the reason for the popularity of CMOS devices in pocket calculators, digital wristwatches, and portable microcomputers.

Fan-out for CMOS circuits is ideally infinite since no loading occurs when it is connected to the gate of enhancement MOSFET. Practical values of fan-out greater than 50 are typical.

Propagation delay of a CMOS gate is typically about 25 to 100 ns, depending on the particular device. It increases with greater load capacitance.

4.13.2 CMOS NAND Gate (74COO)

A two-input NAND gate which consists of two *p*-type units in parallel and two *n*-type units in series is shown in Fig. 4.27. Notice that Q_1 and Q_2 form one complementary connection; Q_3 and Q_4 form another.

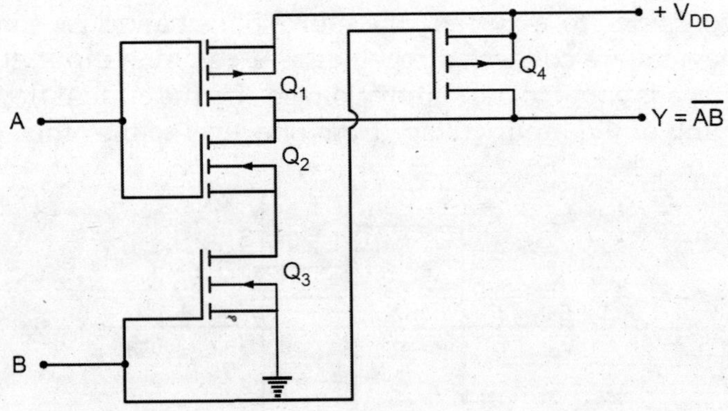

Fig. 4.27. CMOS NAND gate.

If both inputs are HIGH, both *p*-channel transistors turn OFF and both *n*-channel transistors turn ON. The output has a low impedance to ground and produces a LOW state. If any input is LOW, the associated *n*-channel transistor is turned OFF and the associated *p*-channel transistor is turned ON. The output is coupled to V_{DD} and goes to the HIGH state. This functions as a-logic NAND gate. The 74C00 is a quad 2-input NAND gate.

To produce the positive AND function, the output of the CMOS NAND gate can be connected to a CMOS inverter.

4.13.3 CMOS NOR Gate

A two-input CMOS NOR gate using a pair of PMOS transistors (Q_1 and Q_2) and NMOS transistors (Q_3 and Q_4) is shown in Fig. 4.28. Of the two inputs, *A* and *B* either of the inputs can turn ON the PMOS or NMOS device connected to it.

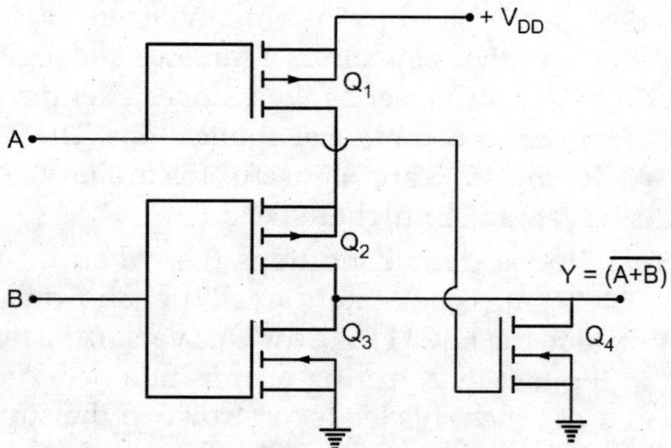

Fig. 4.28. CMOS NOR gate.

When both inputs are LOW, both PMOS devices are driven ON and both NMOS devices OFF. The output is coupled to KDD and goes to the HIGH state. If any input is HIGH, the associated *p*-channel transistor is turned OFF and the associated *n*-channel transistor turns ON. This connects the output to the ground causing a LOW output. Thus this circuit functions as a NOR gate. A CMOS OR gate can be formed by combining the output of the CMOS NOR gate with a CMOS inverter.

4.13.4 CMOS Series

There are several series of the CMOS digital logic family. The original design of CMOS ICs is recognised from the 4000 number designation. The 74C series are pin-and-function compatible with TTL devices having the same number. The performance characteristics of the 74C series are about the same as the 4000 series. CMOS IC type 74C04 has six inverters with the same pin configuration as TTL type 7404. The 74HC series operates at higher speeds than the 74C series. The 74HCT series is electrically compatible with TTL ICs. This means that the circuit in this series can be connected to inputs and outputs of TTL ICs without the need of additional interfacing circuits. The commercially available CMOS series are listed in Table 4.5.

Table 4.5. Various series of the CMOS logic family

CMOS series	Prefix	Example
Original CMOS	40	4009
Pin compatible with TTL	74 C	74 C04
High-speed and pin compatible with TTL	74 HC	74 HC04
High-speed and electrically compatible with TTL	74 HCT	74 HCT04

4.14 CHARACTERISTICS OF CMOS

The first CMOS logic series was produced by RCA and is known as the 4000 series, which was later developed by other manufacturers. At present, several manufacturers have developed a CMOS series which is pin-for-pin compatible with TTL. This is the 74C00 series and it contains devices that have the same pin assignments and logic operations as their TTL counterparts. Any device in the 74C00 series works over a temperature range of – 40°C to + 85°C, which is sufficient for most commercial applications. The 54C00 series works over a temperature range of – 55°C to 125°C and is useful for military applications. The 74HC00 series of devices has the advantage of higher speed.

Power dissipation: Under static conditions (i.e. when the output is constant), the power consumed by a CMOS gate is extremely small (in nanowatts). When a CMOS output changes from the LOW state to the HIGH state (or vice versa), the average power dissipation increases. This is due to the fact that, during a transition between states, both MOSFETs conduct for a small period of time. This leads to a spike in the supply current.

Therefore, during the transition, the drain current becomes appreciable. Moreover, any stray capacitance across the output has to be charged before the output voltage can change. This capacitive charging draws additional current from the supply, thereby increasing the instantaneous power dissipation.

The average power dissipation of a CMOS device whose output is continuously changing is called the *active power dissipation*. This power dissipation per gate increases with frequency and supply voltage. The power consumption of a CMOS gate is around lOmW in the MHz region. Thus, CMOS loses its advantages at higher frequencies.

Propagation delay time: The propagation delay of a standard CMOS gate ranges from 25 to 150 ns, with the exact value depending on the power supply voltage and other factors. A CMOS NAND gate typically has a propagation delay time of about 25 ns when $V_{DD} = 10$ V, and 50ns when $V_{DD} = 5$ V.

Voltage levels: CMOS can be operated over a supply voltage range of 3 to 15 V. A supply voltage of 9 to 12 V can be used to obtain the overall best performance of a CMOS gate in respect of high speed and noise immunity. When CMOS is being used with TTL, the V_{DD} supply voltage is made 5 V so that the voltage levels of the two families are the same.

Noise margin: In CMOS series, the noise margin is typically about 45% of the supply voltage V_{DD}. They have the same noise margin in both HIGH and LOW states. A V_{DD} of 5 V guarantees a 2.25 V noise margin.

Floating inputs: A floating TTL input is equivalent to a high input. If a CMOS input is floated, a possible noise problem is set up and there is excessive power dissipation. Therefore, it is necessary to connect all the input pins of the CMOS devices to some voltage level, preferably to ground or V_{DD}.

Sourcing and sinking: When a standard CMOS driver output is LOW, the current from the CMOS load to the driver is only 1 μA. This indicates that the CMOS driver has to sink only 1μA. Similarly, when the CMOS driver output is HIGH, the driver is sourcing 1 uA to an input of the load gate. The worst case input currents for CMOS devices are:

$$I_{IL(max)} = -1\ \mu A;\ I_{IH(max)} = 1\ \mu A$$
$$I_{OL(max)} = 10 \mu A;\ I_{OH(max)} = -10\ \mu A$$

Fan-out: The fan-out of CMOS gates depends on the type of load being connected. If a standard CMOS drives another standard CMOS, the fan-out can be calculated from the input and output currents of the standard CMOS gate given above.

$$\text{Considering low output state: } \frac{I_{OL,\,\max}}{I_{IL,\,\max}} = \frac{10\ \mu\text{A}}{1\ \mu\text{A}} = 10$$

$$\text{Considering high output state: } \frac{I_{OH,\,\max}}{I_{IH,\,\max}} = \frac{10\ \mu\text{A}}{1\ \mu\text{A}} = 10$$

Therefore, 10 standard CMOS gates can be connected to the output of another standard CMOS gate. Thus, the fan-out of standard CMOS gate is 10.

4.15 BICMOS LOGIC CIRCUITS

BiCMOS logic circuits employ the recent development of digital technology in the silicon fabrication process that combines the speed and driving capability of bipolar junction transistors with the density and low power dissipation of CMOS devices. BiCMOS technology has been used to develop low voltage analog circuits, VLSI circuits and ASICS (Application Specific Integrated Circuits). BiCMOS circuits have little degradation in density. Because of the low output impedance and increased charging and discharging currents of BJTs, the propagation delay of the BiCMOS gate does not increase much as in the CMOS gates. Also, it has good compatibility of the technology with ECL and TTL levels with ready interfacing and little loss of switching speed. With high current capability and faster response, BiCMOS circuits can be used in place of CMOS buffers. They are widely used in ASICS and high density gate arrays. In terms of cost, power and density, BiCMOS technology can be compared with ECL.

Basic BiCMOS inverter: The circuit of a basic BiCMOS inverter is shown in Fig. 4.29. The complementary pair of PMOSFET (Q_1) and NMOSFET (Q_2) form the inverter, and the pair of matching NPN transistors T_1 and T_2 form the active pull-up/pull-down driver stage. The circuit operation is analysed for a capacitive load as explained below.

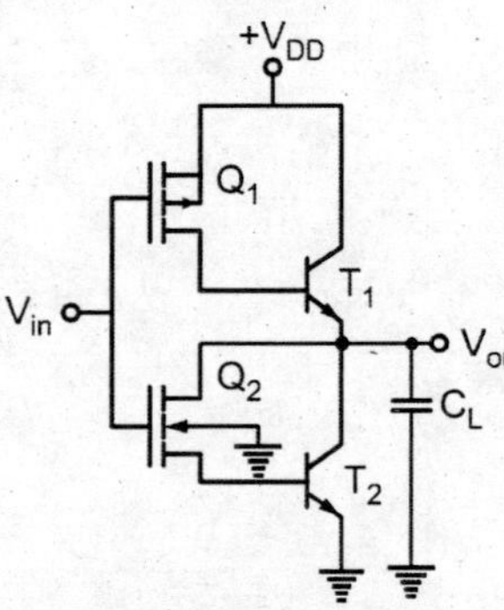

+V_{DD} Q_1 T_1 V_{in} V_{out} Q_2 C_L T_2

Fig. 4.29 Basic BiCMOS inverter

When the logic input is LOW, the PMOS (Q_1) is turned ON and NMOS (Q_2) is in cut-off. As no channel is formed in Q_2, the base current for T_2 is zero and hence T_2 is OFF. At the same time, a low resistance conducting channel is formed in O_1 which provides base curccent for T_1. Thus, T_1 operates in the active mode and it supplies (β + 1) times the base current at the emitter to charge the load capacitance. Hence the output voltage is less than the supply voltage V_{DD} by the base emitter junction voltage V_{BE} of T_1. Therefore, $V_{OH} = V_{DD} - V_{BE}$. Thus, for a LOW input, the output V_0 is in HIGH state.

When the logic input is HIGH, Q_2 is turned ON while Q_1 is turned OFF. Hence, a low resistance conducting channel is formed in Q_2 that supplies base current to switch ON the transistor T_2. T_2 causes a large discharge current from the capacitor. Thus, the high-to-low transition occurs rapidly. When the output voltage reaches the base-emitter cut-in voltage of T_2, i.e. $V_{OL} \approx V_{BE}$, further discharge takes place. Thus, for a HIGH input, the output V_{out} is in LOW state and acts as an inverter.

BiCMOS NAND and NOR gates: The transfer characteristics and switching speed of the basic inverter can be improved by providing paths for discharging the excess carriers from the bases of T_1 and T_2 with additional NMOS devices, resulting in each BIT OFF while the other is ON. The circuit of the conventional BiCMOS inverter is shown in Fig. 4.30. This circuit can be modified to implement BiCMOS NAND and NOR gates as shown in Fig. 4.31 (*a*) and (*b*) respectively.

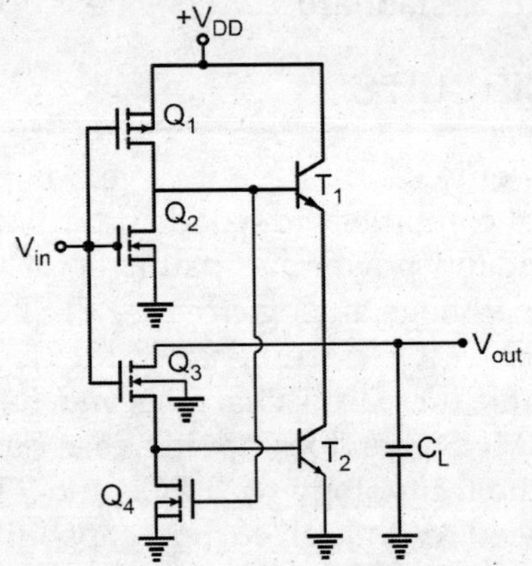

Fig. 4.30. Conventional BiCMOS inverter.

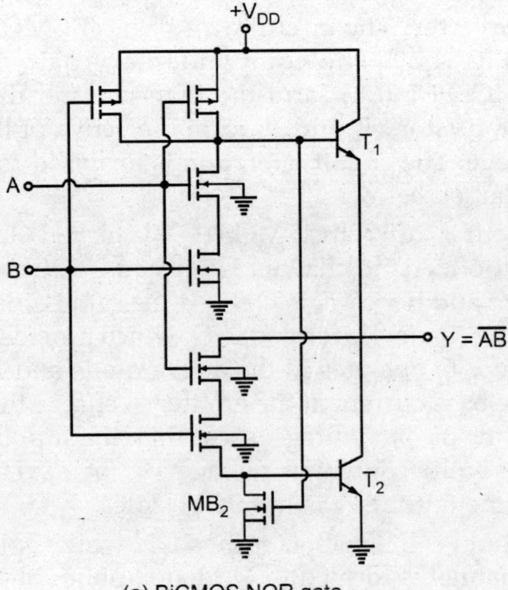

(a) BiCMOS NOR gate

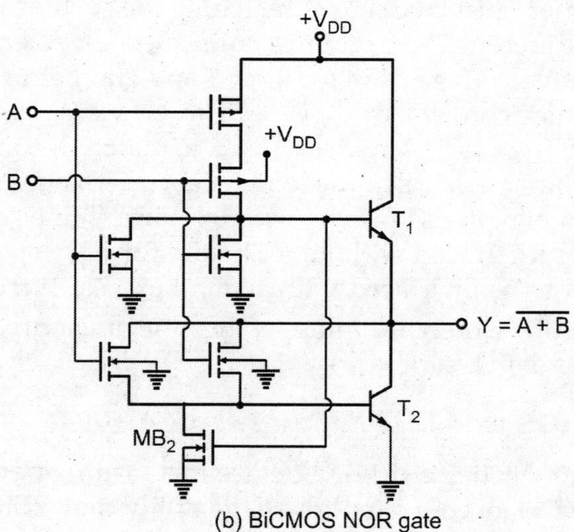

(b) BiCMOS NOR gate

Fig. 4.31

The circuit diagram of a BiCMOS NAND gate shown in Fig. 4.31 (*a*), the base of the bipolar pull-up transistor T_1 is being driven by two parallel-connected PMOS transistors. Therefore, the pull-up device is turned ON when either one or both of the inputs are logic-low. On the other hand, the bipolar pull-down transistor T_2 is driven by two series-connected NMOS transistors between the output node and the base. Therefore, the pull-down device can be turned on only if both of the inputs are logic-high. Hence, for the removal of the base charges of T_1 during turn-OFF, two series-connected NMOS transistors are used, whereas only one NMOS transistor is utilized for removing the base charge of T_2.

The circuit diagram of a BiCMOS NOR gate shown in Fig. 4.31(*b*), the base of the bipolar pull-up transistor T_1 is being driven by two series-connected PMOS transistors. Therefore, the pull-up device can be turned ON only if both of the inputs are logic-low. The base of the bipolar pull-down transistor T_2 is driven by two parallel-connected NMOS transistors. Therefore, the pull-down device can be turned on if either one or both of the inputs are logic-high. Also, the base charge of the pull-up device is removed by two minimum-size NMOS transistors connected in parallel between the base node and the ground. Here, only one NMOS transistor, MB_2, is being used for removing the base charge of T_2, when both inputs are logic-low.

4.16 COMPATIBILITY OR INTERFACING

The output(s) of a circuit or a system should match the input(s) of another circuit or system that has different electrical characteristics. This is referred to as *compatibility*. Interfacing between different logic families is important for compatibility. An interface circuit is one that is connected between the driver and the load; its function is to take the driver output signal and condition it so that it is compatible with the requirements of the load.

The circuit designer must take care in matching the current and voltage characteristics of the two circuits of different logic families connected together. For example, when one circuit using only negative voltage (– V) for logic 1 and another using positive voltage (+ V) for logic 1 are to be inter-connected, they need interface circuit. An interface or buffer circuit between two different logic families is used to match the output characteristics of the driver with the input characteristics of the load. Also, interfacing may be needed between a low speed family and a high speed family. Examples are (*i*) the serial output of a TTL shift register may need to be interfaced with an ECL gate for high speed processing and (ii) an ECL counter may need to be interfaced with a standard TTL circuit that drives a display.

Table 4.6 contains the worst case values of the input and output parameters for standard devices in the CMOS and TTL series.

4.16.1 Interfacing CMOS with TTL

The method of connecting a driving device to a loading device is known as interface. We know that TTL devices need a supply voltage of 5 V and CMOS devices need a supply voltage that ranges from 3 to 15 V. As the supply requirements are different, several interfacing schemes may be used.

Table 4.6. Worst case values of the input and output parameters for standard devices in the CMOS and TTL series

	CMOS			TTL			
Parameter	**4000 B**	**74 HC**	**74 HCT**	**74**	**74 LS**	**74 AS**	**74 ALS**
V_{IH}(min)	3.5 V	3.5 V	2.0 V	2.0 V	2.0 V	2.0 V	2.0 V
$_{IL}$(max)	1.5 V	1.0 V	0.8 V	0.8 V	0.8 V	0.8 V	0.8 V
V_{OH}(min)	4.95 V+	4.9 V+	4.9 V+	2.4 V	2.7 V	2.7 V	2.7 V
V_{OL}(max)	0.05 V+	0.1 V+	0.1 V+	0.4 V	0.5 V	0.5 V	0.4 V
I_{IH}(min)	1 mA	1 mA	1 mA	40 mA	20 mA	200 mA	20 mA
I_{IL}(max)	1 mA	1 mA	1 mA	1.6 mA	0.4 mA	2 mA	100 mA
I_{OH}	0.4 mA	4 mA	4 mA	0.4 mA	0.4 mA	2 mA	100 mA
I_{OL}	0.4 mA	4 mA	4 mA	16 mA	8 mA	20 mA	8 mA

Supply voltages = 5 V.
\+ CMOS driving only CMOS inputs.

4.16.2 TTL Driving CMOS

Supply voltage at 5 V One approach to TTL/CMOS interfacing is to use + 5 V as the supply voltage for both the TTL driver and the CMOS load. When a TTL gate output drives a CMOS input, there is no problem in the LOW state since $V_{OL(max)}$ = 0.4 V for the TTL output and the CMOS input will accept voltage anywhere upto 1.5 V for a LOW state. Thus, the CMOS load always interprets the TTL low-state drive as a LOW state. In the HIGH state, a problem can occur because the minimum TTL output voltage is 2.4 V [i.e. $V_{OH,min}$ = 2.4 V]. This 2.4 V results in indeterminate action at the CMOS input because it requires a minimum of 3.5V for a HIGH state [i.e. $F_{IH,\,min}$ = 3.5 V]. Therefore, it is recommended that an external pull-up resistor R_p can be used as shown in Fig. 4.32. The effect of R_p is to raise V_{QH} from the TTL circuit to approximately + 5 V. For instance, when the TTL output is LOW, the lower end of R_p = 3.3 kΩ is grounded. Therefore, the TTL driver sinks a current of roughly 1.5 mA.

$$I = \frac{5\text{ V}}{3.3\text{ k}\Omega} = 1.52\text{ mA}$$

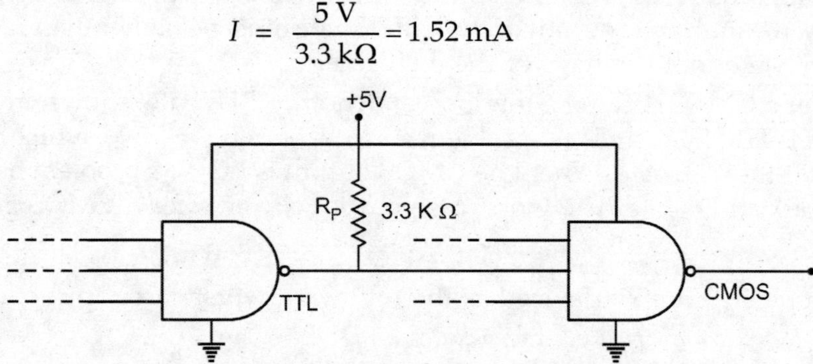

Fig. 4.32. TTL driver and CMOS load.

The gate capacitance of the CMOS load has to be charged through the pull-up resistor, R_p . This slows down the switching speed. In order to increase the speed of action, it is important to decrease the value of the resistance. The minimum resistance is then determined by the maximum sink current of the TTL device, i.e., $I_{0L,\text{ max}}$ = 16 mA and the maximum supply voltage (V_{CC}) is 5.25 V. Then,

$$R_{\text{min}} = \frac{5.25\text{ V}}{16\text{ mA}} = 328\ \Omega \simeq 330\ \Omega$$

Different supply voltages: The performance of CMOS gates deteriorate at lower voltages. This is due to the increase in propagation delay time and decrease in noise immunity. Therefore, a supply voltage that ranges from 9 to 12 V can be applied to CMOS devices for better performance. One way of using higher supply voltage is with an open-collector TTL driver, as shown in Fig. 4.33. The open-collector is connected to a supply voltage of + 12 V through a pull-up resistance of 6.8 kΩ.

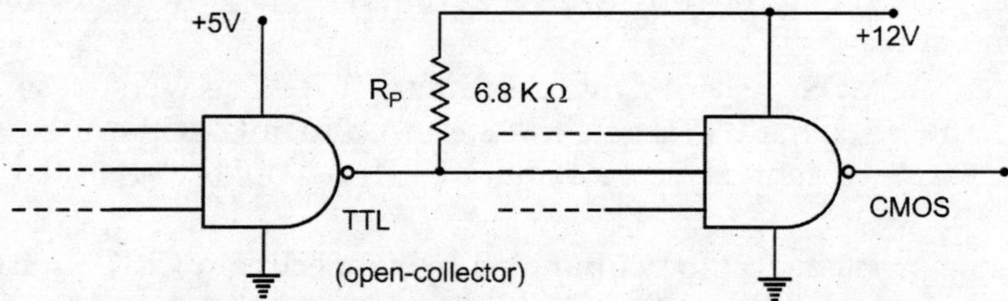

Fig. 4.33. Open-collector TTL driver allows higher CMOS supply voltage.

When the TTL output is LOW, the lower end of R_p is approximately at 0 V. The TTL device then has to sink a current approximately equal to 1.76 mA.

$$I_{\text{sink}} = \frac{12\text{ V}}{6.8\text{ K}\Omega} = 1.76\text{ mA}$$

When the TTL output is HIGH, the open-collector output increases passively to + 12 V. In conclusion, TTL outputs are compatible with CMOS input states.

4.16.3 CMOS Driving TTL

To interface a CMOS gate with a TTL gate, the low state output voltage of the CMOS gate must be less than 0.8 V, the maximum allowable low state input voltage of the TTL gate. Similarly, the high state output of the CMOS gate must be higher than 2 V, the minimum allowable high state input voltage of the TTL gate.

Consider a CMOS driving a low power Schottky TTL gate, shown in Fig. 4.34. When CMOS output is HIGH, there is no problem since $V_{OH} = V_{DD} = +5$ V, which is an acceptable HIGH input for the TTL gate. When the CMOS output is LOW, a problem occurs because of I_{IL} of the Schottky TTL gate. The worst case output currents for a CMOS gate are:

$$I_{OL,\ max} = 360 \text{ mA};\ I_{OH,\ max} = -360\ \mu\text{A}$$

A low power Schottky device has the following worst-case input currents:

$$I_{IL,\ max} = -360\ \mu\text{A};\quad I_{IH,\ max} = 20\ \mu\text{A}$$

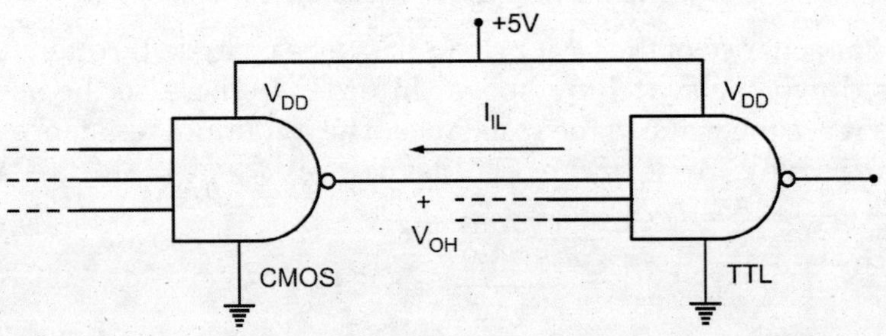

Fig. 4.34. CMOS driver and low power schottky TTL load.

From the above output and input currents of CMOS and Schottky TTL gates respectively, it is clear that a CMOS driver can sink 360 μA current which is exactly the low state input current for the low power Schottky TTL gate. But, the CMOS driver can source 360 μA current which is much higher than the high state input current for the low power Schottky TTL. Hence, only one low power Schottky TTL can be connected to a CMOS driver (i.e. fan-out = 1).

However, a CMOS can not drive a standard TTL. The worst-case low state input current (i.e., I_{IL}) of a standard TTL gate is – 1.6 mA. But, a CMOS driver with low output can sink only 360 μA current. Hence, a standard TTL can not be connected to the output of a CMOS gate.

The above problems can be eliminated by connecting a CMOS buffer, a chip with larger output currents, directly to the CMOS driver output. For example, if a CMOS buffer IC 74C902 is connected at the output of the CMOS driver, then two standard TTL gates can be connected. This is due to the following worst-case output currents of CMOS buffer:

$$I_{OL,\ max} = 3.6 \text{ mA},\quad I_{OH',\ max} = -800\ \mu\text{A}$$

Since $I_{IL,\ max} = -1.6$ mA for a standard TTL gate, two standard TTL gates can be connected.

4.17 COMPARISON OF LOGIC GATES

Each logic family projects different characteristics in terms of static and dynamic performance, size and cost. Table 4.7. provides a brief comparison of typical performance characteristics of the more commonly used families. Note that some of the values differ because the table contains only average values.

Table 4.7. Comparison of logic families

Characteristics	RTL	DTL	TTL (Active pull-up)	STTL (High speed)	ECL	I^2L	MOS	CMOS
Basic gate(s)	NOR	NAND (NOR)	NAND	NAND	NOR-OR	NAND	NAND-NOR	NAND-NOR
Logic swing with 5 V power supply(V)	2.5	4.7	3.8	3.8	3.6	3.6	3.8	5
Fan-out	5	8	12	12	16	12	12	12
Power dissipation (mW)	20	9	10	2	25	0.1 to 100	0.1	0.002
Propagation delay (ns)	12	12	10	3	2	0.7 to 20	1	1
Noise immunity (V)	0.3	0.3	0.4	0.5	0.25	0.4	2.5	2.5

Example 4.1. *For the DTL NAND gate shown in Fig. E4.1, $V_{BE(sat)} = 0.8$ V, $V_g = 0.5$ V, $V_{CE} = 0.2$ V, the drop across the conducting diode is 0.7 V and V_g (diode) = 0.6 V. The inputs of this switch are obtained from the outputs of similar gates.*

(a) *Verify that the circuit functions as a positive NAND and calculate $h_{FE\ (min)}$.*

(b) *Will the circuit operate properly if D_2 is not used?*

(c) *Calculate the noise margin if all the inputs are HIGH.*

(d) *Calculate the noise margin if at least one input is LOW.*
Assume for the moment that Q is not loaded by a following stage.

(e) *Calculate fan-out.*

(f) *Obtain average power.*

Solution: (*a*) The logic levels are $V_{CE(sat)} = 0.2$ V for the logic 0 state and $V_{CC} = 5$ V for the logic 1 state. If at least one input is in the 0 state, then the corresponding diode *D* conducts and $V_P = 0.2 + 0.7 = 0.9$ V. For the diodes D_1 and D_2, to conduct, a voltage of (2) (0.7) = 1.4 V is required. Hence, these diodes are cut-off and $V_{BE} = 0$. Therefore, *Q* is OFF and the output voltage is 5 V, i.e. *Y* = 1. This confirms the first three rows of the NAND truth table.

If all inputs are HIGH, then all input diodes are OFF. Hence, both D_1 and D_2, conduct, and *Q* is turned ON. The voltage at *P* is

$$V_P = V_{D1} + V_{D2} + V_{BE(sat)}$$
$$V_P = 0.7 + 0.7 + 0.8 = 2.2 \text{ V}$$

Then, the voltage across each input diode is 5V – 2.2 V = 2.8 V in the opposite direction, thus confirming that the input diode *D* is OFF.

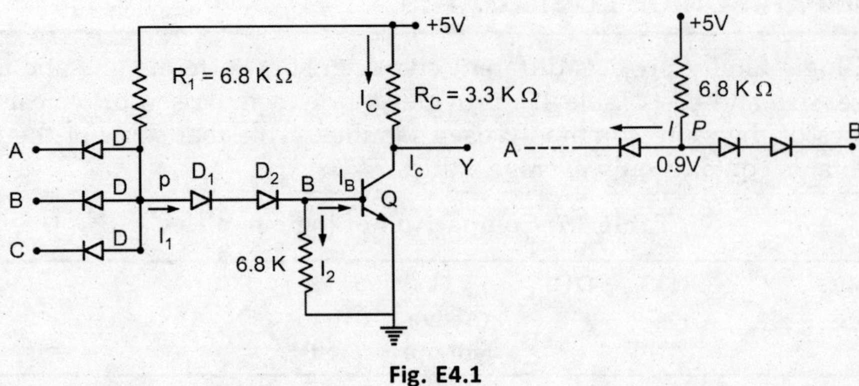

Fig. E4.1

To find $h_{FE(min)}$,

$$I_1 = \frac{V_{CC} - V_P}{R_1} = \frac{5 - 2.2}{6.8 \times 10^3} = 0.412 \text{ mA}$$

$$I_2 = \frac{V_{BE(sat)}}{R_2} = \frac{0.8}{6.8 \times 10^3} = 0.112 \text{ mA}$$

$$I_B = I_1 - I_2 = 0.412 - 0.112 = 0.3 \text{ mA}$$

$$I_C = \frac{V_{CC} - V_{CE(sat)}}{R_C} = \frac{5 - 0.2}{3.3 \times 10^3} = 1.45 \text{ mA}$$

$$\therefore \quad h_{FE(min)} = \frac{I_C}{I_B} = \frac{1.45 \text{ mA}}{0.3 \text{ mA}} = 4.83$$

If $h_{FE} > h_{FE}$(min), then the transistor Q is in saturation and, therefore, $Y = 0$. Thus, the last line of the NAND truth table is verified.

(*b*) If one input is at LOW, then $V_p = 0.2 + 0.7 = 0.9$ V. Hence if one diode is connected in between P and B, then $V_{BE} = 0.9 - 0.6 = 0.3$ V, where 0.6 V is the diode cut-in voltage. Since the cut-in voltage of a transistor $V_\gamma = 0.5$ V, then theoretically Q is OFF. But, sometimes, a small spike of noise may turn Q ON.

(*c*) If all inputs are HIGH, then the output is LOW. As $V_p = 2.2$ V, each input diode is reverse-biased by 2.8 V. A diode starts conducting when it is forward-biased by 0.6 V. Hence, a negative noise spike in excess of 2.8 + 0.6 = 3.4 V must be present at the input before the circuit malfunctions. Therefore, the noise margin is 3.4 V, if all the inputs are HIGH.

(*d*) If one input is LOW, then the output is HIGH. From part (*a*), $V_P = 0.9$ V and Q is OFF. For the transistor Q to enter just into its active region, the voltage at V_P should be 1.7 V, i.e. $V_P = V_\gamma$ (diode 1) + V_γ (diode 2) + $V_\gamma(Q) = 0.6 + 0.6 + 0.5 = 1.7$ V. Hence, the noise marein 's 1,7 – 0.9 = 0.8 V.

It is important to note that, if only one diode is used, the noise margin is reduced by 0.6 V to 0.2 V (i.e. 0.8 V – 0.6 V). This confirms the value obtained in part (*b*).

(*e*) Initially, we assumed that the NAND gate is unloaded, and we found in part (*a*) that the unloaded collector current is 1.45 mA. If we assume that the NAND gate drives *N* similar gates, then the fan-out is *N*. Now, the output transistor (*Q*) acts as a sink for the current it drives from the input of succeeding gates. The input current *I*, from a single following stage is given by

$$I = \frac{V_{CC} - V_P}{R} = \frac{5-0.9}{6.8\times 10^3} = 0.60 \text{ mA}$$

This is called *standard load.* Now, the total collector current of *Q* is

$$I_C = 0.6 \text{ mA} \times 7 \text{ V} + 1.45 \text{ mA}$$

where 1.45 mA is the unloaded collector current.

If we assume *hpg* (min) = 30, then

$$I_C = h_{FE} \times I_B = 30 \times 0.3 \text{ mA} = 9 \text{ mA}$$

Now, $$N = \frac{9 \text{ mA} - 1.45 \text{ mA}}{0.6 \text{ mA}} = 12.58 \approx 12$$

Therefore, the fan-out N = 12. The fan-out can be further increased by replacing diode D_1 by a transistor.

(*f*) When the output is LOW, the power *P*(0) is given by

$P(0) = V_{CC}(I_1 + I_C) = 5$ (0.412 mA + 1.45 mA) = 5.31 lmW

When the output is HIGH, at least one of the input diodes conducts. Therefore, I_1, = 0.6 mA and I_C = 0. Hence *P*(1) = 0.6 mA × 5 V = 3 mW.

Therefore, average power, $P_m = \frac{P(0)+P(1)}{2} = \frac{9.31+3}{2}$ = 6.155 mW.

Example 4.2. *Refer to the data sheet for the 7400 Quad two-input NAND gate IC. Determine the typical average power dissipation and average propagation delay of a simple NAND gate.*

Solution: We know that this IC contains four NAND gates. Under the "electrical characteristics", the typical values for I_{CC} are 4 mA and 1 2 mA in the HIGH and LOW state respectively. The average supply current of 8 mA is assumed to be divided equally among the four gates. Thus, the average power dissipation of a single gate is 2 mA × 5 V = 10 mW.

The typical values for t_{PLH} and t_{PHL} are llns and 7 ns. Hence, an average delay time = 9 ns.

Example 4.3. *Determine the maximum output currents for the 7400 NAND gate in both states.*

Solution: The 7400 N AND output has a fan-out of 10UL. In the HIGH state, 1UL is 40μA, so this gate output can supply 10 × 40 μA = 0.4 mA, while still maintaining $V_{OH} \geq 2.4$ V.

In the LOW state, 1UL = 1.6 mA so this gate output can sink 10 × 1.6 mA = 16 mA while maintaining $V_{OL} \leq 0.4$ V.

Example 4.4. *(a) Determine the fan-out of a 74S00 NAND gate, (b) How many 74S00 inputs can a 74S00 output drive? (c) How many 74S00 inputs can a 7400 output drive?*

Solution: (*a*) The data sheet for the 74S00 IC does not specify the fan-out explicitly. However, the output current values can be obtained under the test conditions for V_{OH} and V_{OL}. They are given as:

$$I_{OH,max} = -1\ \mu A;\ I_{OL',max} = 20\ mA$$

$$I_{OH,max} = 50\ \mu A;\ I_{IL,max} = -2\ mA$$

From the above, input and output currents for Schottky TTL, the fan-out for high and low state outputs can be calculate follows:

$$\text{Fan- out (HIGH state)} = \frac{I_{OH,max}}{I_{IH,max}} = \frac{1\ mA}{50\ \mu A} = 20\ UL$$

$$\text{Fan- out (LOW state)} = \frac{I_{OL,max}}{I_{IL,max}} = \frac{20\ mA}{2\ mA} = 10\ UL$$

Thus, the fan-out of a 74S00 NAND gate is 10.

(*b*) From the above calculation, it is clear that a 74S00 output can drive ten 74S00 inputs.

(*c*) For a standard 7400 gate, I_{OL},max = 16 mA , $I_{OH,max} = -400\ \mu A$; for a 74S00 gate, $I_{IL',max} = -2mA$, $I_{IH,max} = 50\ \mu A$.

$$\text{Fan-out (LOW state)} = \frac{I_{OL,max}}{I_{IL,max}} = \frac{16\ mA}{2\ mA} = 8$$

$$\text{Fan-out (HIGH state)} = \frac{I_{OH,max}}{I_{IH,max}} = \frac{400\ \mu A}{50\ \mu A} = 8$$

Therefore, a 7400 output can drive 8 numbers of 74S00 gate inputs.

Example 4.5. *What voltage levels should be used to measure the rise time and fall time for a signal going from 0.2 V to 3.9 V?*

Solution: The overall voltage is 3.9 V – 0.2 V = 3.7 V

The 10% voltage point is at 0.2 V + 10%(3.9 V) = 0.59 V

The 90% voltage point is at 0.2 V + 90% (3.9 V) = 3.71 V.

■ REVIEW QUESTION ■

1. Explain what is meant by logic family?
2. Explain the parameters used to characterise logic families.
3. Describe one major difference between a bipolar integrated circuit and a MOS integrated circuit.
4. Explain the difference between current-sinking and current-sourcing logic circuits.
5. Give two advantages and one disadvantage of the totem-pole output arrangement.

6. Write notes on (*i*) noise immunity (*ii*) fan-out (*iii*) fan-in (*iv*) transition times and (*v*) propagation delays.
7. Write a brief note on characteristics of digital ICs.
8. What is a logic gate load?
9. Draw the circuit diagram of negative logic AND gate with 3 inputs.
10. Draw and explain the circuit diagram of a 3-input I^2L NOR gate.
11. Which type of logic gate is the fastest? Explain.
12. Explain why an open TTL input acts as a HIGH.
13. What is the chief advantage of ECL over other IC technologies? What is the second advantage?
14. What are the two advantages of using the wired-AND connection?
15. Why ECL logic is faster than TTL?
16. What is meant by multiple-emitter transistor?
17. Write a brief note on MOS digital ICs.
18. Explain briefly the characteristics of MOS logic, and write a note on CMOS logic.
19. Discuss why wired logic should not be used with active pull-up outputs.
20. Why is the CMOS switching speed greater than PMOS/NMOS?
21. Which *bipolar* family is best suited for LSI?(a) Show the circuit of a 4-input NAND gate using CMOS transistors, (*b*) Repeat for a 4-input NOR gate.
22. Write a note on interfacing CMOS with TTL.
23. Write a brief note on interfacing TTL with CMOS.
24. What is a tri-state gate?
25. What is the function of diode in the path of totem-pole output stage in a standard TTL gate?
26. Explain, with the aid of a circuit diagram, the operation of a TTL 3-input NAND gate.
27. Explain the purpose of the totem-pole output stage used in a TTL gate.
28. What are the advantages of BiCMOS logic circuits?
29. Draw the basic BiCMOS inverter and explain its operation.
30. Describe the operation of BiCMOS NAND and NOR gates.
31. What is the necessity of interfacing in logic circuits?
32. Construct a circuit with TTL as a driver and CMOS as a load.
33. Construct a circuit with CMOS as a driver and TTL as a load.
34. Compare the characteristics of different logic families.

■ PROBLEMS ■

1. In the basic TTL NAND gate of Fig.4.6, the function of *D* is to ensure that Q_3 will not turn ON when both the inputs *A* and *B* are HIGH. Assume that Q_2 saturates for this condition and determine the approximate voltage at the base of Q_3. Then show that this amount of voltage is not enough to forward-bias Q_3.
2. What is the fan-out of a standard TTL device driving 74LS device? **[Ans.** 20]
3. What is the output of AND-OR-NOT schematic diagram given in Fig. P4.3 for the following inputs?

   (*a*) $ABCD$ = 0000 (*b*) $ABCD$ = 0101.

   (c) $ABCD$ = 1100 (*d*) $ABCD$ = 1111. **[Ans.** (*a*) 1, (*b*) 1, (*c*) 0, (*d*) 0]

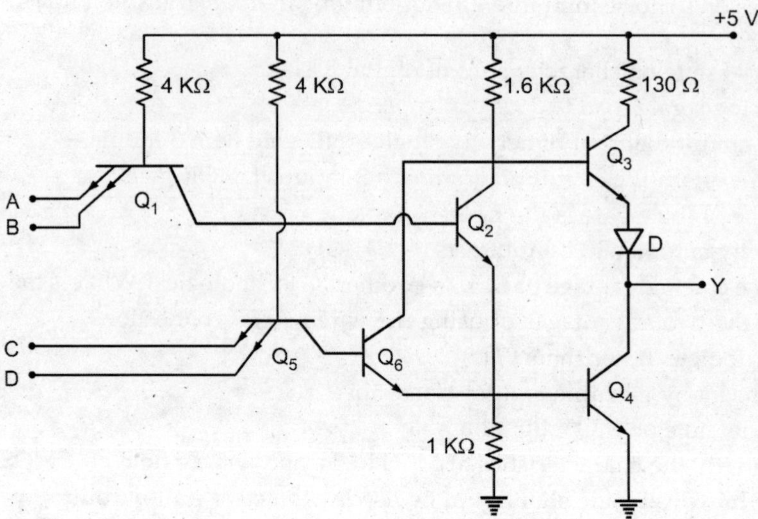

Fig. P4.3

4. Show that the output transistor of the DTL gate of Fig. 4.5 (*a*) goes into saturation when all inputs are HIGH. Assume that $h_{FE} = 20$.
5. The 7409 TTL IC is a quad two-input AND with open collector outputs. Show how 7409s can be used to implement the operation $Y = A.B.C.D.E.F.G.H.I.J.K.M.$
6. What is the current drain through the pull-up resistors when both switches are closed in Fig. P4.6? What is the time constant for each input when the switches are open?

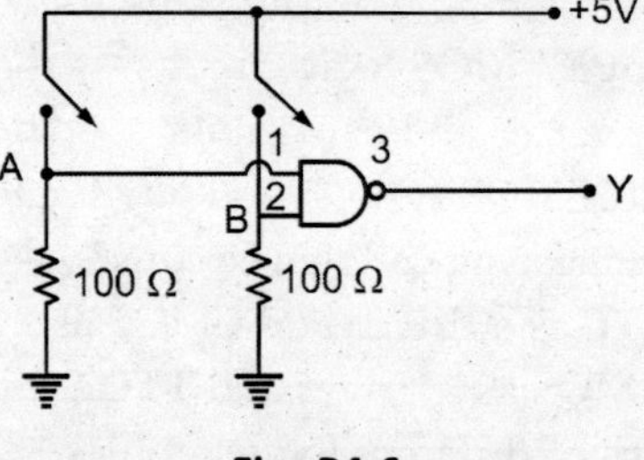

Fig. P4.6

7. In Fig. P4.7, the TTL output voltage is 0.4 V, and the LED voltage is 2 V. What is the sink current when the LED is ON? **[Ans.** 14.4 mA]
8. What is the LED current in Fig. P4.8 if the LED voltage drop is 2 V and the TTL output is HIGH? If the TTL output is 0.4 V, what is the LED output? **[Ans.** 2.27 mA]

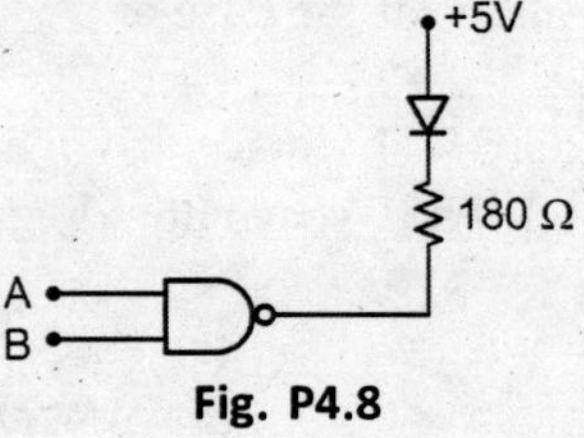

Fig. P4.8

9. Calculate the noise margin of the ECL gate.
10. Using the NOR outputs of the ECL gates, show that, when connected together to an external resistor and negative supply voltage, the wired connection produces an OR function.
11. The MOS transistor is bilateral, i.e., current may flow from source to drain or from drain to source. Using this property, derive a circuit that implements the Boolean function

$$Y = \overline{(AB + CD + AED + CEB)}$$

Use six MOS transistors.
12. (*a*) Show the circuit of a four-input NAND gate using CMOS transistors,

(*b*) Repeat for a four-input NOR gate.
13. Construct an Exclusive-NOR circuit using MOS transistor and explain its operation.

5 Arithmetic Circuits

5.1 INTRODUCTION

Digital computers and calculators consist of arithmetic and logic circuits, which contain logic gates and flip-flops that add, subtract, multiply and divide binary numbers. The basic building blocks of the arithmetic unit in a digital computer are *adders*. These circuits perform operations at speeds less than 1   s.

A digital system consists of two types of circuits, namely

(*i*) Combinational logic circuit,

(*ii*) Sequential logic circuit.

In a combinational circuit, the output at any time depends only on the input values at that time. In a sequential circuit, the output at any time depends on the present input values as well as the past output values. The basic building blocks of an arithmetic unit such as half-adder and full-adder, are combinational circuits.

5.2 PROCEDURE FOR THE DESIGN OF COMBINATIONAL CIRCUITS

Any combinational circuit can be designed by following the design procedure given below:

1. From the word description of the problem, identify the inputs and outputs and draw a block diagram.
2. Draw a truth table such that it completely describes the operation of the circuit for different combinations of inputs.
3. Write down the switching expression(s) for the output(s).
4. Simplify the switching expression using either algebraic or *K*-map method.
5. Implement the simplified expression using logic gates.

5.3 HALF-ADDER

The simplest combinational circuit which performs the arithmetic addition of two binary digits is called a *half-adder*. As shown in Fig. 5.1 (*a*), the half-adder has two inputs and two outputs. The two inputs are the two 1-bit numbers *A* and *B*, and the two outputs are the sum (*S*) of *A* and *B* and the carry bit denoted by *C*. From the truth table of the half-adder shown in Table 5.1, one can understand that the Sum output is 1 when either of the inputs (*A* or *B*) is 1, and the Carry output is 1 when both the inputs (*A* and *B*) are 1.

Table 5.1. Truth table of half-adder

Inputs		**Outputs**	
Augend	**Addend**	**Sum**	**Carry**
A	B	S	C
0	0	0	0
0	1	1	0
1	0	1	0
1	1	0	1

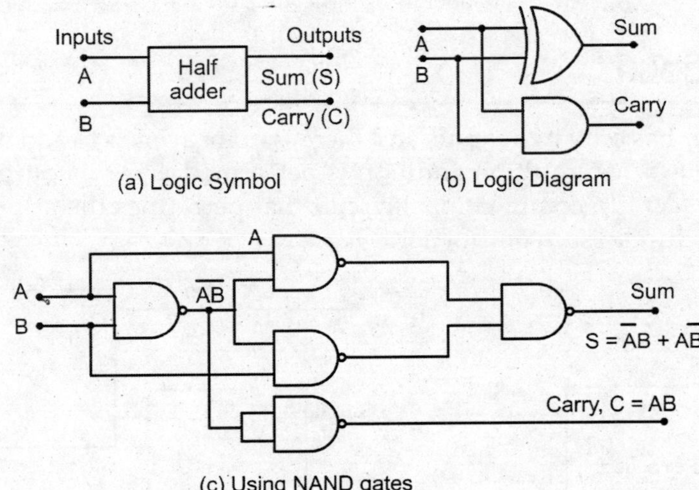

(a) Logic Symbol (b) Logic Diagram

(c) Using NAND gates

Fig. 5.1. Half-adder.

From Table 5. 1, the logic expression for the Sum output can be written as a Sum of Product expression by summing up the input combinations for which the sum is equal to 1.

In the truth table, the sum output is 1 when $AB = 01$ and $AB = 10$. Therefore, the expression for sum is

$$S = \overline{A}B + A\overline{B}$$

Now, this expression can be simplified as

$$S = A \quad B$$

Similarly, the logic expression for Carry output can be expressed as a Sum of Product expression by summing up the input combinations for which the carry is equal to 1. In the truth table, the carry is 1 when $AB = 11$. Therefore,

$$C = AB$$

This expression for C cannot be simplified. The sum output corresponds to a logic Ex-OR function while the carry output corresponds to an AND function. So, the half-adder circuit can be implemented using Ex-OR and AND gates as shown in Fig.5.1(*b*). Fig. 5.1(*c*) gives the realisation of the half-adder using minimum number of NAND gates. The implementation of the half-adder circuit using basic gates AND, OR and NOT is shown in Fig. 5.2.

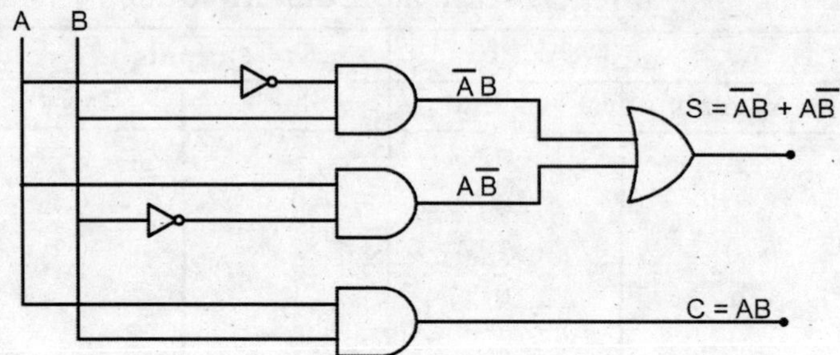

Fig. 5.2. Half-adder using basic AND, OR and NOT gates.

5.4 FULL-ADDER

A half-adeer has only two inputs and there is no provision to add a carry coming from the lower order bits when multibit addition is performed. For this purpose, a full-adder is designed. A *full-adder* is a combinational circuit that performes the arithmetic sum of three input bits and produces a sum output and a carry.

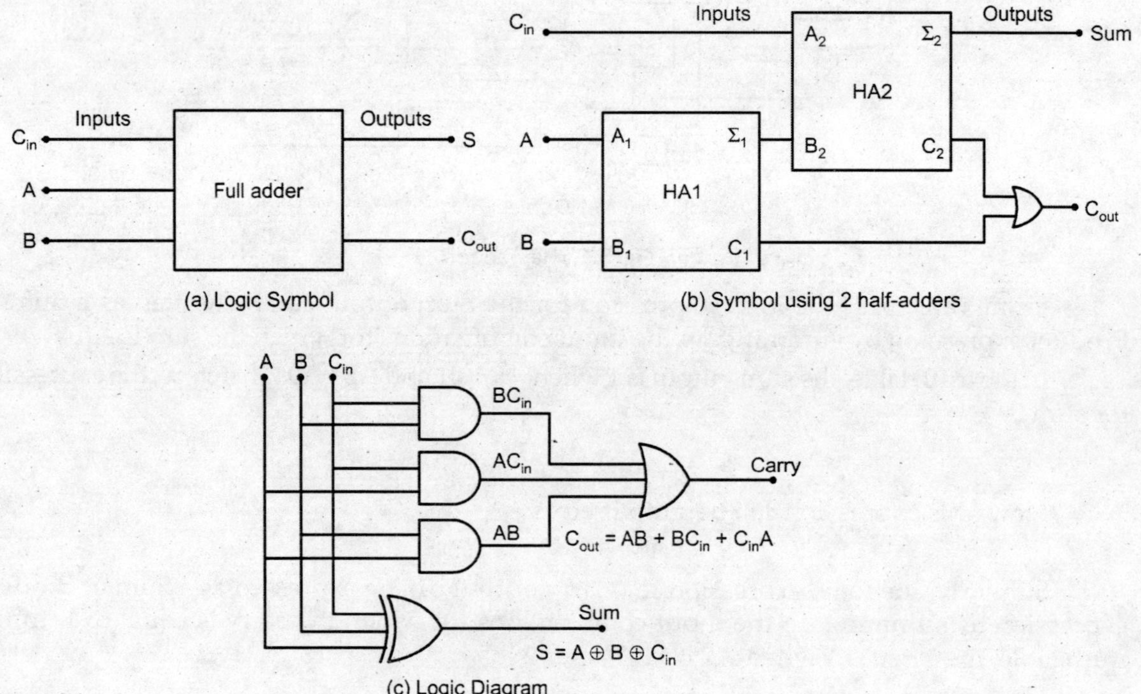

Fig. 5.3. Full-adder.

The logic symbol of the full adder is shown in Fig.5.3(*a*). It consists of three inputs and two outputs. The two input variables denoted by *A* (Augend bit) and *B* (Addend bit) represent the two significant bits to be added. The third input, C_{in}, represents the carry from the previous lower significant position. The outputs are designated by the symbols *S* (for sum) and C_{out} (for carry). The truth table for the full-adder circuit is shown in Table 5.2. The binary variable *S* gives the value of the LSB of the sum, and the binary variable C_{out}, gives the output carry. A full-adder can be formed using two half-adder circuits and an OR gate as shown in Fig. 5.3 (*b*).

Table 5.2. Truth table of full-adder

Inputs			Outputs	
Augend bit *A*	**Addend bit** *B*	**Carry input** C_{in}	**Sum** *S*	**Carry output** C_{out}
0	0	0	0	0
0	0	1	1	0
0	1	0	1	0
0	1	1	0	1
1	0	0	1	0
1	0	1	0	1
1	1	0	0	1
1	1	1	1	1

As shown in Table 5.2, there are eight possible input combinations for the three inputs and for each case the *S* and C_{out} values are listed. From the truth table, the logic expression for *S* can be written by summing up the input combinations for which the sum output is 1 as:

$$S = \overline{A}\overline{B}C_{in} + \overline{A}B\overline{C}_{in} + A\overline{B}\overline{C}_{in} + ABC_{in}$$

Do You Know?	In digital systems, many devices must often share the same data path. This data path is often called a data bus.

Simplifying the above expression, we get

$$S = \overline{A}(\overline{B}C_{in} + B\overline{C}_{in}) + A(\overline{B}\overline{C}_{in} + BC_{in})$$

$$= \overline{A}(B \oplus C_{\overline{in}}) + A(\overline{B \oplus C_{in}})$$

Let $B \quad C_{in} = X$

Now, $S = \overline{A}X + A\overline{X} + A \oplus X$

Replacing *X* by $B \quad C_{in}$ the above expression, we have

$$S = A \quad B \quad C_{in}$$

Similarly, the logic expression for C_{out} can be written by summing up the input combinations for which C_{out} is 1, as given below:

$$C_{out} = \overline{A}BC_{in} + AB\overline{C}_{in} + AB\overline{C}_{in} + ABC_{in}$$

$$= BC_{in}(A + \overline{A}) + A\overline{B}C_{in} + AB\overline{C}_{in}$$

$$= BC_{in} + A\overline{B}C_{in} + AB\overline{C}_{in}$$

Now, the ABC_{in} term is added twice for simplification.

$$
\begin{aligned}
C_{out} &= BC_{in} + A\overline{B}C_{in} + AB\overline{C}_{in} + ABC_{in} + ABC_{in} \\
&= BC_{in} + AC_{in}(B+\overline{B}) + AB(C_{in} + \overline{C}_{in}) \\
&= BC_{in} + AC_{in} + AB
\end{aligned}
$$

From the simplified expressions of S and C_{out}, the full-adder circuit can be implemented using one 3-input Ex-OR gate, three 2-input AND gates and one 3-input OR gate as shown in Fig. 5.3(*c*).

5.5 K-MAP SIMPLIFICATION

K-map method can also be used for simplifying the logic expressions for S and C_{out}. The *K*-maps for the outputs S and C_{out} are given in Fig. 5.4.

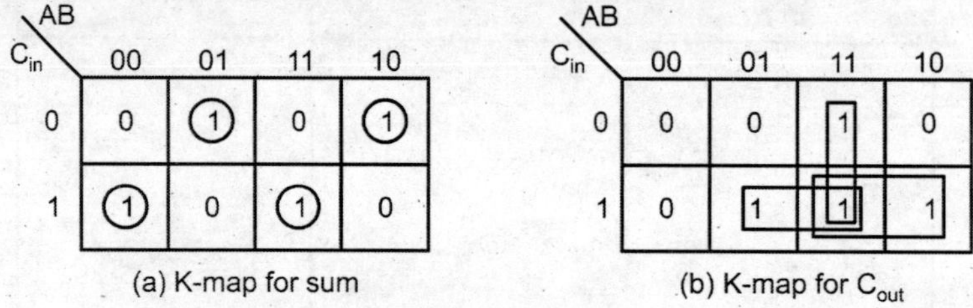

Fig. 5.4 Karnaugh maps

From the *K*-maps, the simplified expressions for S and C_{out} can be written as follows:

$$
\begin{aligned}
S &= \overline{A}\overline{B}C_{in} + \overline{A}B\overline{C}_{in} + A\overline{B}\overline{C}_{in} + ABC_{in} \\
C_{out} &= AB + BC_{in} + C_{in}A
\end{aligned}
$$

Using the above expressions, the full-adder can be implemented using basic gates as shown in Fig. 5.5.

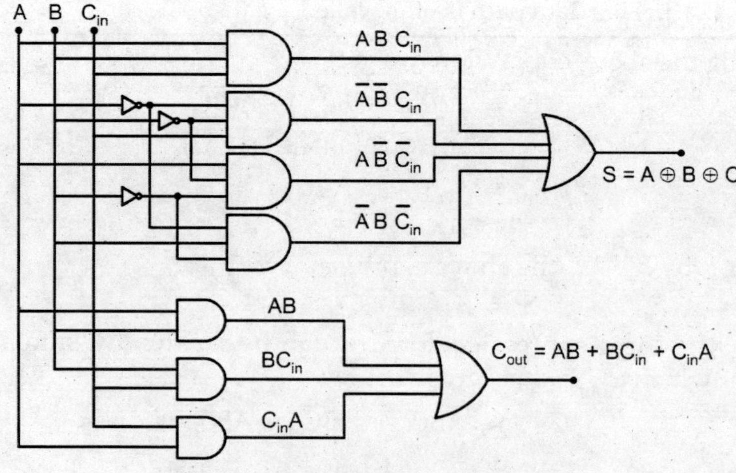

Fig. 5.5 Full-adder using basic gates

5.6 HALF-SUBTRACTOR

The half-subtractor is a combinational circuit which is used to perform subtraction of two bits. It has two inputs, X (minuend) and Y (subtrahend) and two outputs D (difference) and B_{out} (borrow out). The logic symbol for a half-subtractor is shown in Fig. 5.6(*a*). The truth table for half-subtractor is shown in Table 5.3. From the truth table, it is clear that the difference output is 0 if $X = Y$ and 1 if $X \quad Y$; the borrow output B_{out} is 1 whenever $X \quad Y$. If X is less than Y, then subtraction is done by borrowing 1 from the next higher order bit.

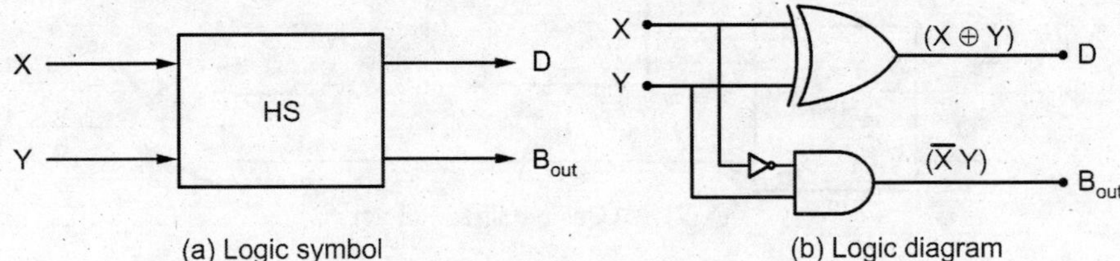

(a) Logic symbol (b) Logic diagram

Fig. 5.6 Half-subtractor

Table 5.3 Truth table of half-subtractor

Inputs		**Outputs**	
Minuend	**Subtrahend**	**Difference**	**Borrow**
X	Y	D	B_{out}
0	0	0	0
0	1	1	1
1	0	1	0
1	1	0	0

From Table 5.3, as discussed earlier, the Boolean expressions for difference (D) and Borrow out (B_{out}) can be written as follows:

$$D = \overline{X}Y + X\overline{Y} = X \oplus Y$$

$$B_{out} = \overline{X}Y$$

From the above equations, the half-subtractor can be implemented using an Ex-OR gate, a NOT gate and an AND gate as shown in Fig. 5.6(*b*).

5.7 FULL-SUBTRACTOR

A full-subtracter is a combinational circuit that performs subtraction involving three bits, namely minuend bit, subtrahend bit and the borrow from the previous stage. The logic symbol for full-subtractor is shown in Fig. 5.7(*a*).

It has three inputs, X (minuend), Y (subtrahend) and B_{in} (borrow from previous stage), and two outputs D (difference) and 6_{out} (borrow out). The truth table for the full-subtractor is given in Table 5.4. The full-subtractor can be implemented using two half-subtractors and an OR gate as shown in Fig. 5.7(*b*).

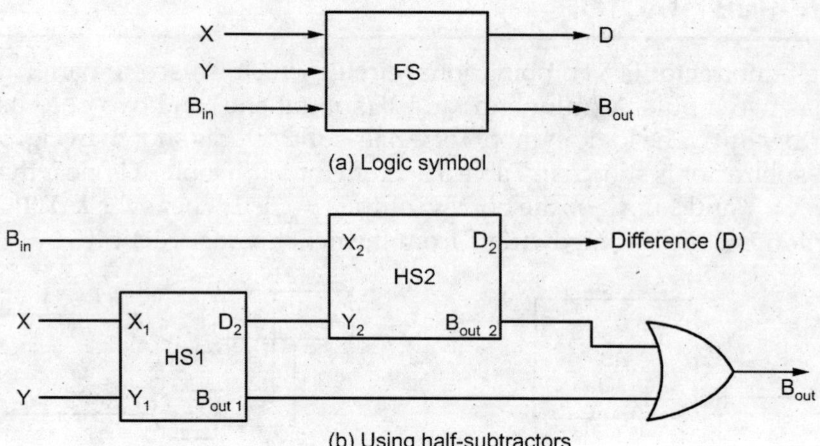

Fig. 5.7 Full-subtractor

Table 5.4 Truth table of full-subtractor

Inputs			Outputs	
Minuend bit ***X***	**Subtrahend bit** ***Y***	**Borrow in** B_{in}	**Difference** ***D***	**Borrow out** B_{out}
0	0	0	0	0
0	0	1	1	1
0	1	0	1	1
0	1	1	0	1
1	0	0	1	0
1	0	1	0	0
1	1	0	0	0
1	1	1	1	1

From Table 5.4, the Sum of Product expression for the difference (*D*) output can be written as:

$$D = \bar{X}\bar{Y}B_{in} + \bar{X}Y\overline{B_{in}} + X\overline{Y\,B_{in}} + XY\,B_{in}$$

Simplifying the above expression,

$$D = (\bar{X}\bar{Y} + XY)B_{in} + (\bar{X}\bar{Y} + X\bar{Y})\,\overline{B_{in}}$$

$$= (\overline{X \oplus Y})\,B_{in} + (X \oplus Y)\,\overline{B_{in}}$$

$$D = X \quad Y \quad B_{in}$$

Similarly, the sum of product expression for B_{out} can be written from the truth table as:

$$B_{out} = \bar{X}\bar{Y}\,B_{in} + \bar{X}\bar{Y}B_{in} + \bar{X}Y\,\overline{B_{in}} + XY\,B_{in}$$

The equation for B_{out} can be simplified using Karnaugh map as shown in Fig. 5.8.

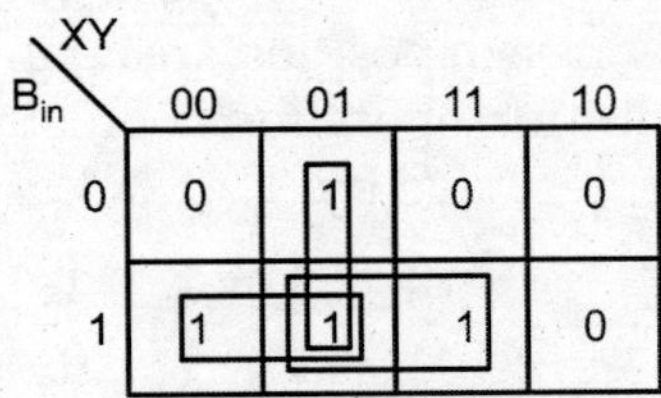

Fig. 5.8 *K*-map for *B*

Now, $$B_{out} = \bar{X}Y + \bar{X}B_{in} + YB_{in}$$

Using the above simplified expressions, the full-subtractor can be realised as shown in Fig. 5.9.

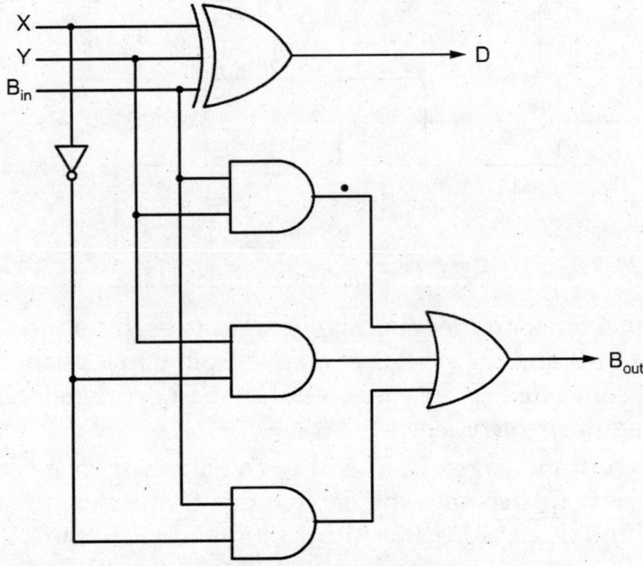

Fig. 5.9 Realisation of full-subtractor

One can notice that the equation for *D* is the same as the sum output for a full-adder, and the borrow output B_{out} resembles the carry output for full-adder except that one of the inputs is complemented. From these similarities, it is possible to convert a full-adder into a full-subtractor by merely complementing that input prior to its application to the input of gates which form the borrow output.

5.8 PARALLEL BINARY ADDER

In most logic circuits, addition of more than 1-bit is carried out. For example, modern computers and calculators use numbers ranging from 8 to 64-bits. The addition of multibit numbers can be accomplished using several full-adders. The 4-bit adder using full-adder circuits is capable of adding two 4-bit numbers resulting in a 4-bit sum and a carry output as shown in Fig. 5.10. Since all the bits of the augend and addend are fed into the adder circuits simultaneously and the additions in each position are taking place at the same time, this circuit is known as *parallel adder*. The addition operation is illustrated in the following example: Let the 4-bit words to be added be represented by $A_3A_2A_1A_0 = 1111$ and $B_3B_2B_1B_0 = 0011$.

Significant place		4	3	2	1
Input carry		1	1	1	0
Augend word *A*:		1	1	1	1
Addendword *B*:		0	0	1	1
	1	0	0	1	0	Sum

Outputcarry

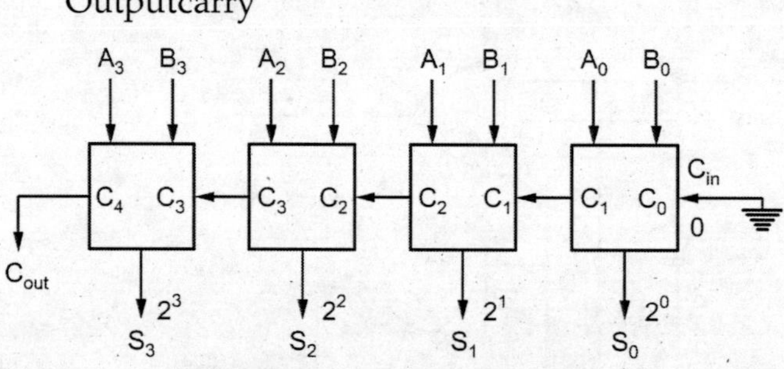

Fig. 5.10 4-bit binary parallel adder

In a 4-bit parallel binary adder circuit, the input to each full-adder will be A_i, B_i and C_i, and the outputs will be S_i and C_{i+1}, where *i*. varies from 0 to 3. Also, the carry output of the lower order stage is connected to the carry input of the next higher order stage. Hence, this type of adder is called *ripple-carry adder*.

In the least significant stage, A_0, B_0 and C_0 (which is 0) are added resulting in Sum S_0 and Carry C_1 This carry C_1 becomes the carry input to the second stage. Similarly, in the second stage, A_1, B_1 and C_1 are added resulting in S_1 and C_2; in the third stage, A_2, B_2 and C_2 are added resulting in S_2 and C_3; in the fourth stage, A_1, B_1 and C_3 are added resulting in S_3 and C_4 which is the output carry. Thus, the circuit results in a sum $(S_3S_2S_1S_0)$ and a carry output (C_{out}).

Though the parallel binary adder is said to generate its output immediately after the inputs are applied, its speed of operation is limited by the carry propagation delay through all stages. In each full-adder, the carry input has to be generated from the previous full-adder which has an inherent propagation delay. The propagation delay (t_p) of a full-adder is the time difference between the instant at which the inputs (A_i, B_i and C_i) are applied and the instant at which its outputs (S_i and C_{i+1}) are generated. Therefore, in a 4-bit binary adder, the output in LSB stage is generated only after t_p seconds. Similarly, the output in the second stage will be generated only after *t* seconds from the time the outputs of the first stage are generated, i.e., after $2t_p$ seconds from the time the inputs are applied; the third stage will generate outputs only after $3t_p$ seconds and the fourth stage will generate outputs only after $4t_p$ seconds. Thus, in a 4-bit binary adder, where each full adder has a propagation delay of 50 ns, the output in the fourth stage will be generated only after $4t_p = 4 \times 50$ ns = 200 ns. The magnitude of such delay is prohibitive for high-speed computers. However, there are several methods to reduce this delay.

One of the methods of speeding up this process is look-ahead carry addition which eliminates the ripple-carry delay. This method is based on the carry generate and the carry propagate functions of the full-adder.

This scheme utilises logic gates to look at the lower order bits of the augend and addend if a higher-order carry is to be generated. This requires extra circuitry for getting high speed adders. This is not a significant consideration with the present day availability of integrated circuits.

5.8.1 IC 7483—4-bit Parallel Binary Adder

The IC 7483 is a commonly available TTL 4-bit parallel binary adder chip. It contains four interconnected full-adders and a look-ahead carry circuitry for its operation. The logic symbol of IC 7483 is shown in Fig. 5.11(*a*). It has two 4-bit inputs $X_3X_2X_1X_0$ and $Y_3Y_2Y_1Y_0$, and, a carry input C_{in} in the LSB stage. The outputs are a 4-bit sum $S_3S_2S_1S_0$ and a carry output C_{out} from the most significant bit stage.

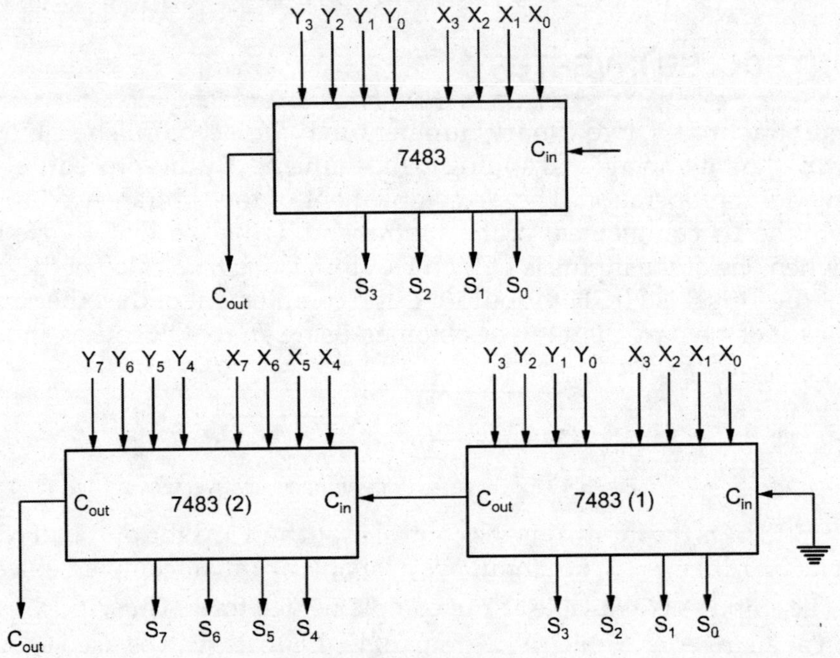

Fig. 5.11

Two or more parallel adder blocks can be connected in cascade to perform the addition operation on larger binary numbers. Fig. 5.11(b) shows the cascading connection of two 7483 adders. The four least significant bits of the numbers are added in the first adder. The carry output of this adder is given as the carry input to the second adder, which adds four most significant bits of the numbers. The output carry of the second adder is the final carry output.

5.8.2 4-bit Parallel Binary Subtracter

Just as a parallel binary adder can be implemented by cascading several full-adders, a parallel binary subtracter can also be implemented by cascading several full-subtracters. A 4-bit parallel binary subtracter that subtracts a 4-bit number $Y_3Y_2Y_1Y_0$ from another 4-bit number $X_3X_2X_1X_0$ is shown in Fig. 5.12. It has 4 difference outputs ($D_3D_2D_1D_0$) and a borrow output (B_{out}). Note that the B_{in} of the LSB full-subtracter is connected to 0 and B_{out} of *i*th full-subtracter is connected to B_{in} of $(i + 1)$th full-subtracter.

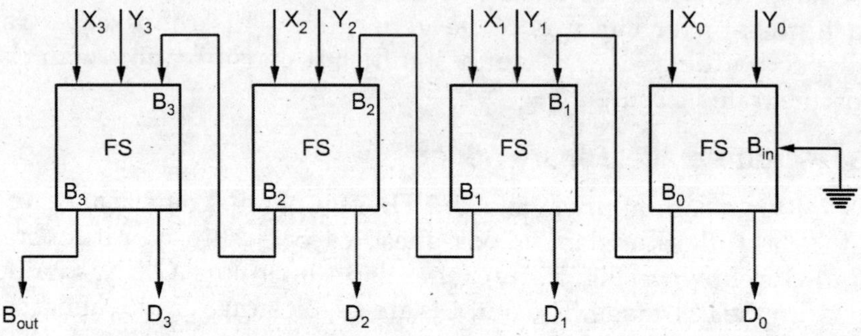

Fig. 5.12 4-bit parallel binary subtractor

5.9 CONTROLLED INVERTER

The subtraction of two binary numbers may be accomplished by taking the 2's complement of the subtrahend and adding to the minuend. By this procedure, the subtraction becomes an addition operation. The 2's complement of the subtrahend can be obtained by adding a 1 to the 1's complement of the subtrahend. From the Ex-OR gate truth table, we know that when one of the inputs is LOW the output is the true value of the other input and when one of the inputs is HIGH the output is the complement of the other input. Therefore, the complement of a binary digit can be obtained using an Ex-OR gate as shown in Fig. 5.13.

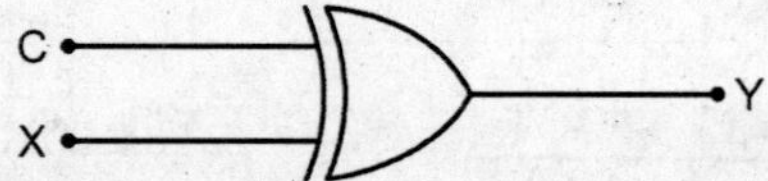

Fig. 5.13 Ex-OR gate functioning as an inverter

In Fig. 5.13, X is the input, C is the control input and Y is the output. From the figure, it is clear that if $C = 0$, then $Y = X$, i.e., input X is available at Y in uncomplemented form; if $C = 1$, then $Y = \overline{X}$, i.e., input X is available at Y in complemented form. Thus, if $C = 1$, the Ex-OR gate can function as an inverter. Similarly, a group of Ex-OR gates can be used to invert a group of bits. Fig. 5.14 shows the complementing process of a 4-bit binary number ($Y_3Y_2Y_1Y_0$) using a controlled inverter.

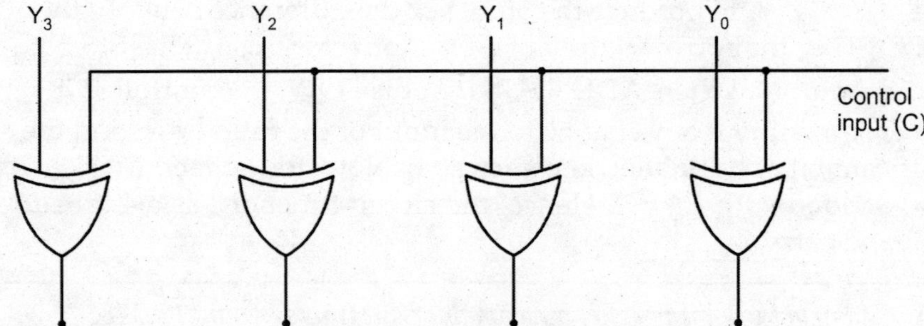

Fig. 5.14 Controlled inverter

When the control input is low, the output will be the input, i.e. $Y_3Y_2Y_1Y_0$. When the control input is high, the output will be the complement of the input $\overline{Y}_3\overline{Y}_2\overline{Y}_1\overline{Y}_0$.

5.10 4-BIT PARALLEL ADDER/SUBTRACTOR

The 4-bit parallel binary adder/subtractor circuit shown in Fig. 5.15 performs the operations of both addition and subtraction. It has two 4-bit inputs $X_1X_2X_3X_0$ and $Y_3Y_2Y_1Y_0$. The $\overline{\text{ADD}}$ /SUB control line, connected with C_{in} of the least significant bit of the full-adder, is used to perform the operations of addition and subtraction. The Ex-OR gates are used as controlled inverters.

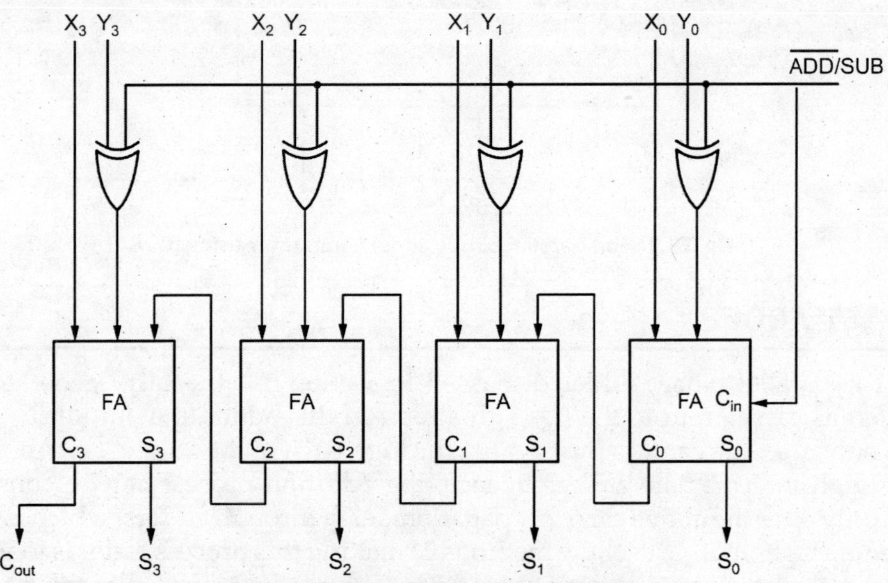

Fig. 5.15 4-bit parallel adder/subtracter

To perform subtraction, the $\overline{\text{ADD}}$ /SUB control input is kept high. Now, the controlled inverter produces the 1's complement of the addend $(\overline{Y}_3\overline{Y}_2\overline{Y}_1\overline{Y}_0)$. Since 1 is given to C_{in} of the least significant bit of the adder, it is added to the complemented addend producing 2's complement of the addend before addition. Now, the data $X_3X_2X_1X_0$ will be added to the 2's complement of $Y_3Y_2Y_1Y_0$ to produce the Sum, i.e., the difference between the addend and the augend, and C_{out}, i.e., the borrow output of 4-bit subtractor. Also, it has $S_3S_2S_1S_0$ as sum output and C_{out} as carry output. When $\overline{ADD}$ /SUB input is LOW, the controlled inverter allows the addend ($Y_3Y_2Y_1Y_0$) without any change to the input of the full-adder, and the carry input C_{in} of least significant bit of full-adder, becomes zero. Now, the augend ($X_3X_2X_1X_0$) and addend ($Y_3Y_2Y_1Y_0$) are added with C_{in} = 0. Hence, the circuit functions as a 4-bit adder resulting in sum $S_3S_2S_1S_0$ and carry C_{out}.

Example 5.1. *Construct a 4-bit parallel binary adder/subtractor using IC 7483.*

Solution: IC7483 is a 4-bit parallel binary adder. A 4-bit parallel binary adder/subtractor can be implemented using IC7483, controlled inverter IC7486 and a control line $\overline{ADD}$ /SUB as shown in Fig. E5.1.

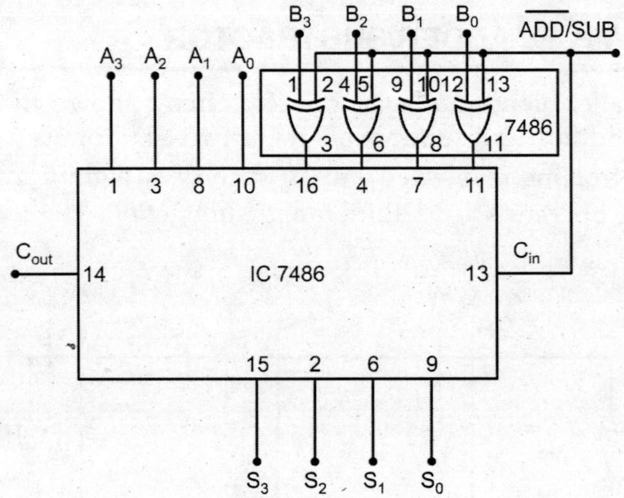

Fig. E5.1 4-bit parallel binary adder/subtractor using IC 7843

5.11 FAST ADDER

In the parallel binary adder discussed in section 5.7, the carry generated by the *i*th adder is fed as carry input to the (*i* + 1)th adder. In this adder, the output (C_{out} $S_3S_2S_1S_0$) is available only after the carry is propagated through each of the adders i.e., from LSB to MSB adder through intermediate adders. Hence, the addition process can be considered to be complete only after the above carry propagation delay through adders, which is proportional to number of stages in it. One of the methods of making this process faster is look ahead carry addition, which eliminates the ripple carry delay. This method is based on the carry generating and the carry propagating functions of the full adder.

5.11.1 4-bit Carry Look Ahead Adder

The carry look ahead adder is based on the principle of looking at the lower order bits of the augend and addend if a high order carry is generated. This adder reduces the carry delay by reducing the number of gates through which a carry signal must propagate. To explain its operation, consider the truth table of full adder, shown in Table 5.5.

Table 5.5 Truth table of Full adder, emphasizing the conditions at which carry generation occurs

Row	A	B	C_{in}	S	C_{out}	
0	0	0	0	0	0	No carry generation i.e. $C_{out} = 0$
1	0	0	0	1	0	
2	0	1	1	1	0	Carry propagation i.e. $C_{out} = C_{in}$
3	0	1	1	0	1	
4	1	0	0	1	0	
5	1	0	0	0	1	
6	1	1	1	0	1	Carry generation i.e. $C_{out} = 1$
7	1	1	1	1	1	

In rows 0 and 1, the carry output (C_{out}) is always zero' and independent of carry input (C_{in}), while in rows 6 and 7, the C_{out} is always one' and independent of C_{in}. These are known as carry generate combinations. In rows 2, 3, 4 and 5 the carry output is equal to the carry input i.e. $C_{out} = 1$ only when $C_{in} = 1$. These are carry propagate combinations. Let G_i represents the unity carry (i.e. $C_{out} = 1$) generate condition and P_i represents the carry propagate condition of the *i*th stage of a parallel adder.

From the Truth Table 5.5 G_i is obtained by summing up the combinations corresponding to 6th and 7th rows as follows:

$$G_i = A_iB_iC_{in} + A_iB_i\overline{C_{in}}$$

$$= A_iB_i(C_{in} + \overline{C_{in}})$$

$$G_i = A_iB_i$$

Similarly the carry propagation condition (P_i) occurs when either $A_i = 1$ and $B_i = 0$ or viceversa as shown in Truth Table 5.5. Now P_i is given by

$$P_i = A_i\overline{B_i} + \overline{A_i}B_i = A_i \oplus B_i$$

Consider the addition of two 4-bit binary numbers $A(A_3A_2A_1A_0)$ and $B(B_3B_2B_1B_0)$. The unity carry output of the *i*th stage can be expressed in terms of G_i, P_i and C_{i-1} which is the unity carry output of the $(i-1)$th stage as follows:

$$C_i(C_{out}) = G_i + P_iC_i - 1$$

where C_{i-1} for LSB stage is C_{in} which is assumed to be zero. In a 4-bit binary adder, four stages of addition are required to add A_0B_0, A_1B_1, A_2B_2, and A_3B_3. Therefore, for $i = 0, 1, 2$ and 3, the C_i's are given by

$$C_0 = G_0 + P_0C_{in} \text{ ....where } G_0 = A_0B_0;\ P_0 = A_0 \oplus B_0 \text{ and } C_{in} = 0$$

$$C_1 = G_1 + P_1C_0$$

$$= G_1 + P_1(G_0 + P_0C_{in})$$

$$= G_1 + P_1G_0 + P_1P_0C_{in} \text{ ....where } G_1 = A_1B_1 \text{ and } P_1 = A_1 \oplus B_1$$

$$C_2 = G_2 + P_2C_1$$

$$= G_2 + P_2(G_1 + P_1G_0 + P_1P_0C_{in})$$

$$= G_2 + P_2G_1 + P_2P_1G_0 + P_2P_1P_0C_{in} \text{...where } G_2 = A_2B_2 \text{ and } P_2 = A_2 \oplus B_2$$

$$C_3 = G_3 + P_3C_2$$

$$= G_3 + P_3(G_2 + P_2G_1 + P_2P_1P_0C_{in})$$

$$= G_3 + P_3G_2 + P_3P_2P_1G_0 + P_3P_2P_1P_0C_{in} \text{...where } G_3 = A_3B_3 \text{ and } P_3 = A_3 \oplus B_3$$

The sum of A and B is given by

$$S = C_3S_3S_2S_1S_0$$

where

$$S_i = A_i \oplus B_iC_{i-1} \text{ for } i = 0, 1, 2, 3$$

i.e.

$$S_0 = A_0 \oplus B_0 \oplus C_{in}$$

$$S_1 = A_1 \oplus B_1 \oplus C_0$$

$$S_2 = A_2 \oplus B_2 \oplus C_1$$

$$S_3 = A_3 \oplus B_3 \oplus C_2$$

Using the above equation, a 4-bit carry look ahead adder can be realized as shown in Fig. 5.16. From the diagram, one can easily understand that the addition of two 4-bit numbers can be done by a carry look ahead adder in a four gate propagation time. Also, it is important to note that the addition of n-bit binary numbers takes the same four gate propagation delay.

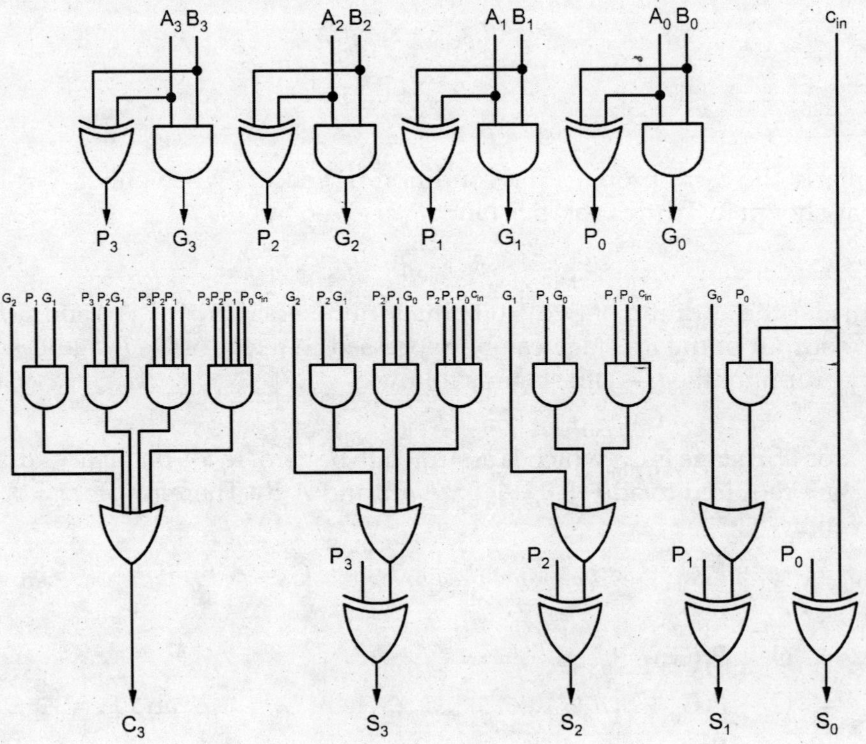

Fig. 5.16 4-bit serial adder

5.12 SERIAL ADDER

Though the parallel adder performs the addition of two binary numbers at a relatively faster rate, the disadvantage of the parallel addition is that it requires a large amount of logic circuitry. This increases in direct proportion with the number of bits in the numbers being added. On the other hand, in *serial addition*, the addition operation is carried out bit-by-bit. Therefore, the serial adder requires simpler circuitry than a parallel adder but results in a low speed of operation.

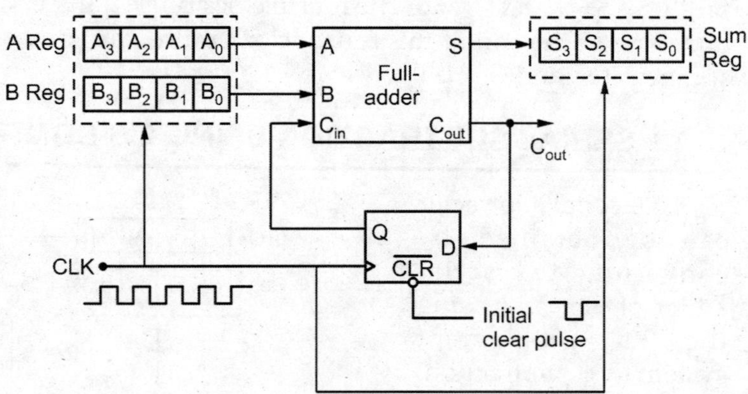

Fig. 5.17 4-bit serial subtractor using full-adder

The diagram of a 4-bit serial adder is shown in Fig. 5.16. The two shift registers *A* and *B* are used to store the numbers to be added serially. A single full-adder is used to add one pair of bits at a time along with the carry. The *D* flip-flop, i.e. carry flip-flop, is used to store the carry output of the full-adder so that it can be added to the next significant position of the numbers in the registers. The contents of the shift registers shift from left to right and their outputs starting from A_0 and B_0 are fed into a single full-adder along with the output of the carry flip-flop upon application of each clock pulse.

The sum output of the full-adder is fed to the Most Significant Bit (S_1) of the sum register. For each succeeding clock pulse, the contents of both the shift registers are shifted once to the right, and new sum bit and new carry bit are transferred to sum register and carry flip-flop respectively. This process continues until all the pairs of bits are added.

The following example explains the operation of a serial adder. Let the augend ($A_3A_2A_1A_0$) be 0111 and the addend ($B_3B_2B_1B_0$) be 0010, stored in shift registers *A* and *B* respectively. Also, the carry flip-flop has been initially cleared to the 0 state, so that $C_{in} = 0$.

First clock pulse: Before the first clock pulse occurs, as the inputs to the Full-adder are $A_0 = 1$, $B_0 = 0$ and $C_{in} = 0$. The full-adder outputs will be $S = 1$ and $C_{out} = 0$. When the first clock pulse occurs, the value in the *A* and *B* registers shift from left to right by one bit. In addition, the sum (*S*) is transferred to S_3 of sum register, and the C_{out} is transferred to the carry flip-flop, whose output becomes 0, which is the cany input of the full-adder.

Second clock pulse: Now, $A_0 = 1$, $B_0 = 1$ and $C_{in} = 0$, at the full-adder input. Therefore, S = 0 and $C_{out} = 1$. When the second clock pulse occurs, *A*, *B* and sum registers again shift right; $S = 0$ is transferred to S_3, and $C_{out} = 1$ is transferred to the carry flip-flop.

Third clock pulse: The values of A_0 and B_0 are now 1 and 0 respectively, and C_{in} is 1. These values produce $S = 0$ and $C_{out} = 1$ at the full-adder outputs. On the occurrence of the third clock pulse, the *A*, *B* and sum registers shift right, sum = 0 goes to S_3, and $C_{out} = 1$ goes to the carry flip-flop.

Fourth clock pulse: Both A_0 and B_0 are now zero, and $C_{in} = 1$. Therefore, the full-adder produces $S = 1$ and $C_{in} = 0$. The fourth clock pulse transfers $S = 1$ to S_3 and initiates all the other transfers. At the end of this fourth clock pulse, the sum value (1001) will be available in the sum register, and C_{out} is 0.

5.13 SERIAL SUBTRACTION USING 2'S COMPLEMENT

A serial subtracter can be obtained by converting the serial adder of Fig. 5.17 using the 2's complement system. For subtraction, the subtrahend is stored in the *B* register and must be 2's-complemented before it is added to the minuend stored in the *A* register. One simple way to accomplish this is to complement the *B* register and to make the initial $C_{in} = 1$ instead of 0 prior to the first clock pulse. This can be easily accomplished by feeding the inverted output $\overline{B_0}$ into the full-adder instead of B_0 and initially setting the carry flip-flop to 1 instead of clearing it. The remaining circuitry is the same as serial adder.

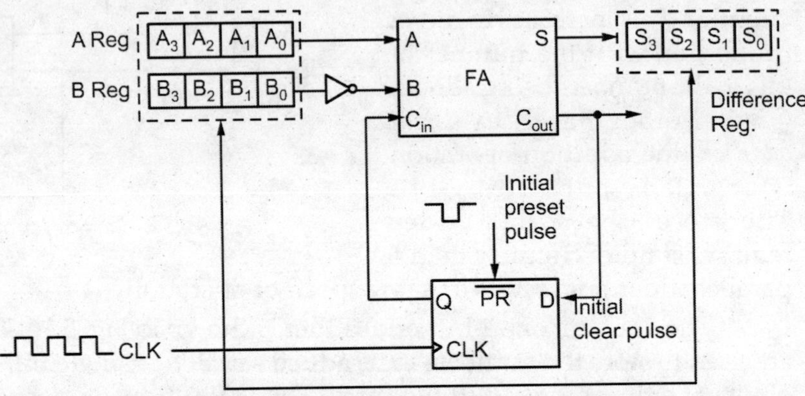

Fig. 5.18 4-bit serial subtractor using full-adder

The circuit for 4-bit serial subtractor using full-adder is shown in Fig. 5.18.

5.14 4-BIT SERIAL ADDER/SUBTRACTOR

A 4-bit serial adder/subtractor can be constructed in a manner similar to its parallel counterpart as shown in Fig. 5.19.

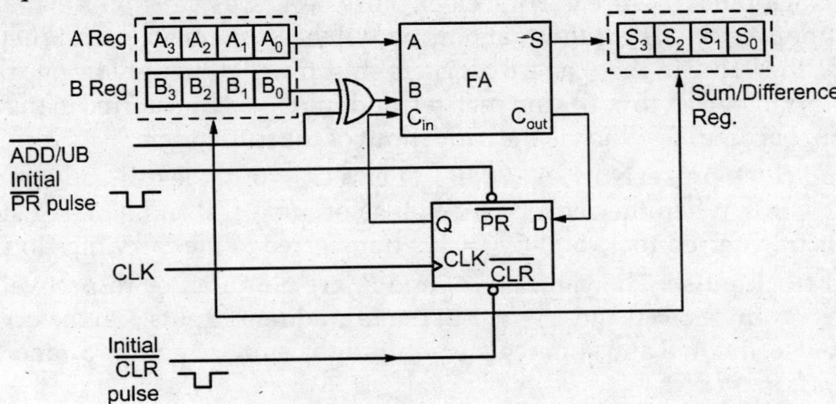

Fig. 5.19 4-bit serial adder/subtractor

In the circuit shown in Fig. 5.19, when $\overline{\text{ADD}}/\text{SUB} = 0$, the uncomplemented $B_3B_2B_1B_0$ will be applied to the full-adder. The carry flip-flop is initially cleared by applying a low pulse at $\overline{\text{CLR}}$ input and the circuit thus functions as a 4-bit serial adder. When $\overline{\text{ADD}}/\text{SUB} = 1$, the complemented output of *B* register $(\overline{B}_3\overline{B}_2\overline{B}_1\overline{B}_0)$ will be applied to the full-adder. The carry flip-flop is initially set to 1 so as to get the 2's complement of the subtrahend by applying a low pulse at $\overline{PR}$ input and thus the circuit functions as a 4-bit serial subtractor. The comparison of serial adder with parallel adder is shown in Table 5.5.

Table 5.5 Serial adder Vs parallel adder

Serial adder	**Parallel adder**
Serial adder is less fast	Parallel adder is generally faster
It requires fewer components	It requires more components compared to serial adder
Addition is performed bit by bit starting from the LSB	All the bits are added simultaneously

5.15 BCD ADDER

A BCD adder is a circuit that adds two BCD digits in parallel and produces a sum digit which is also in BCD. A BCD adder must include the correction logic in its internal construction. A block diagram for the BCD adder is shown in Fig. 5.20. This adder has two 4-bit BCD inputs $X_3X_2X_1X_0, Y_3Y_2Y_1Y_0$ and a carry input (C_{in}). It also has a 4-bit sum output ($_3$ $_2$ $_1$ $_0$) and a carry output (C_{out}). Here, the sum output is also in BCD form.

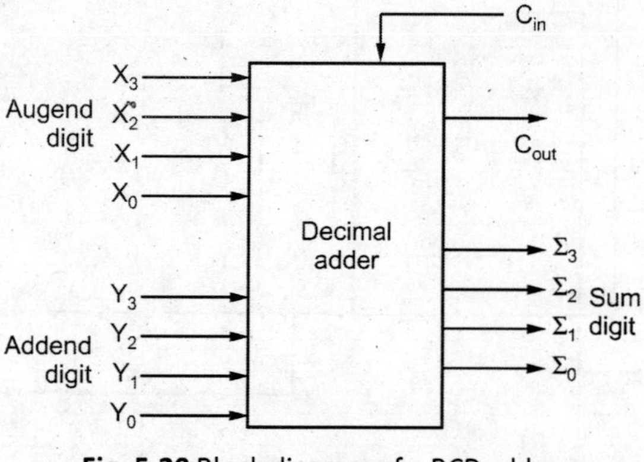

Fig. 5.20 Block diagram of a BCD adder

A BCD adder circuit must be able to do the following:

1. Add two 4-bit BCD numbers using straight binary addition.
2. If the four-bit sum is equal to or less than 9, the sum is in proper BCD form and no correction is needed.
3. If the four-bit sum is greater than 9 or if a carry is generated from the sum, the sum is not in the BCD form. Then, the digit 6 (0110) should be added to the sum to produce the BCD results. The carry may be produced due to this addition and it is added to the next decimal position.

Table 5.6 shows the results of BCD addition with corrections indicated.

Table 5.6 Results of BCD addition with corrections indicated

Decimal digit	Uncorrected BCD sum					Corrected BCD sum					
	C_3	S_3	S_2	S_1	S_0	C_{out}	S_3	S_2	S_1	S_0	
0		0	0	0	0		0	0	0	0	
1		0	0	0	1		0	0	0	1	
2		0	0	1	0		0	0	1	0	
3		0	0	1	1		0	0	1	1	No
4		0	1	0	0		0	1	0	0	correction
5		0	1	0	1		0	1	0	1	required
6		0	1	1	0		0	1	1	0	
7		0	1	1	1		0	1	1	1	
8		1	0	0	0		1	0	0	0	
9		1	0	0	1		1	0	0	1	
10		1	0	1	0	1	0	0	0	0	
11		1	0	1	1	1	0	0	0	1	
12		1	1	0	0	1	0	0	1	0	
13		1	1	0	1	1	0	0	1	1	
14		1	1	1	0	1	0	1	0	0	Correction
15		1	1	1	1	1	0	1	0	1	required
16	1	0	0	0	0	1	0	1	1	0	
17	1	0	0	0	1	1	0	1	1	1	
18	1	0	0	1	0	1	1	0	0	0	
19	1	0	0	1	1	1	1	0	0	1	

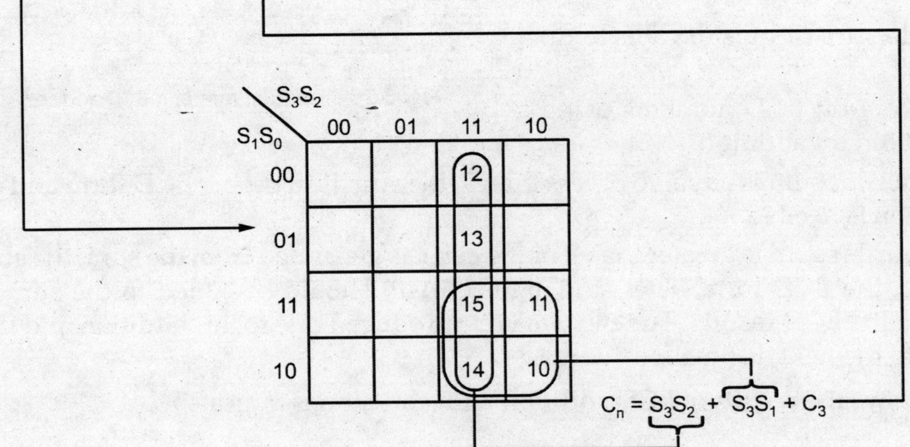

Fig. 5.21 *K*-map simplification

From the Table 5.6, it is clear that correction is required. When the sum output $(S_3S_2S_1S_0)$ is greater than 9, i.e., when $C_3 = 1$ (OR) $S_2 = 1$ (AND) [$S_2 = 1$ (OR) $S_1 = 1$], add 0110 to get the BCD result. Note that when $C_3 = 1$, the result is 16 and above; when $5, = 1$ (AND) $S_2 = 1$, the result is 12 and above; when $S_1 = 1$ (AND) $S_1 = 1$, the result is 10 and above; when $S_3 = 1$ (AND) [$S_2 = 1$ (OR) $S_1 = 1$], the result is 14 and above. Therefore, the condition for correction can be written as an expression as follows:

Alternatively, the above condition for correction can also be obtained by *K*-map method as shown in Fig. 5.21.

As discussed above, a BCD adder must be capable of adding two 4-bit BCD numbers, and the result has to be corrected to BCD, if the above condition is satisfied, by adding 0110. The circuit diagram for BCD adder using full-adders is shown in Fig. 5.22.

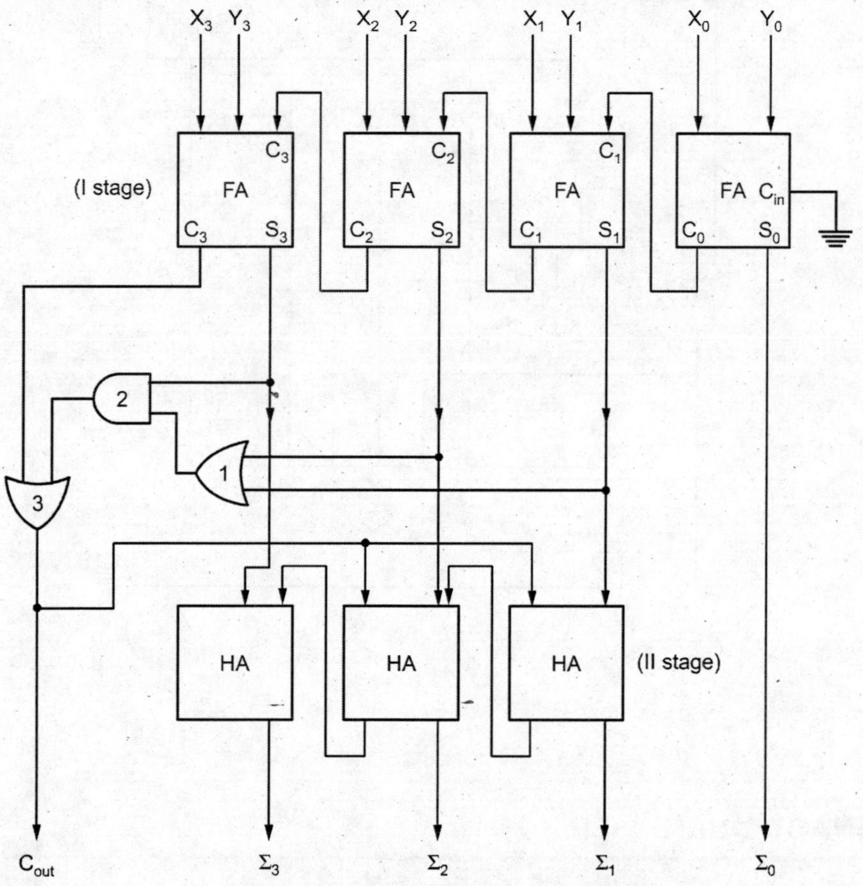

Fig. 5.22 BCD adder using full-adders

The operation of BCD adder shown in Fig. 5.22 is explained as follows: the first stage of full-adders adds two 4-bit BCD numbers, and its sum $(S_3S_2S_1S_0)$ and carry (C_3) are checked to ascertain whether the result exceeds 9 by AND-OR gate combinations. If the output of OR gate (3) is equal to 1, then correction is required and this is accomplished by adding 0110 in the second stage of adders as shown in Fig. 5.22. Now, the BCD result is $_3\ _2\ _1\ _0$ with carry output (C_{out}).

The BCD adder can also be implemented using two 7483 ICs as shown in Fig. 5.23. Here, $X_3X_2X_1X_0$ and $Y_3Y_2Y_1Y_0$ are the BCD inputs. The outputs of adder 1 ($S_3S_2S_1S_0$ and C_{out}) are checked to ascertain whether the output is greater than 9 by AND-OR gate combinations. If correction is required, then a 0110 is added with the output of adder 1. Now, the adder 2 output forms the BCD result ($S_3S_2S_1S_0$) with carry output (C_{out}).

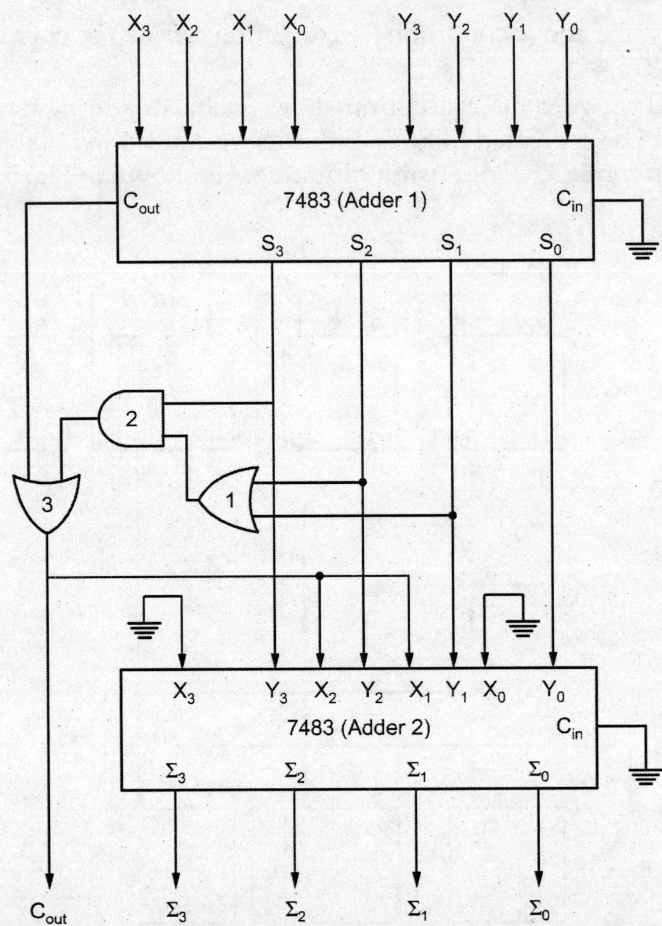

Fig. 5.23 BCD adder using 7483 ICs

5.16 BINARY MULTIPLIER

Multiplication operation can be carried out by (*i*) Multipliers using partial product addition and shifting and (*ii*) Parallel multipliers.

5.16.1 Multiplier using Shift Method

To understand the multiplication process using shift method, consider the multiplication of two 4-bit binary numbers 1010 and 1011, as an example.

			1	0	1	0	Multiplicand
		×	1	0	1	1	Multiplier
			1	0	1	0	Partial product 1
		1	0	1	0		Partial product 2
	0	0	0	0			Partial product 3
1	0	1	0				Partial product 4
1	1	0	1	1	1	0	

From the above multiplication process, one can easily understand that if the multiplier bit is 1, then the multiplicand is simply copied as a partial product; if the multiplier bit is 0, then the partial product is 0. Whenever a partial product is obtained, it is shifted one bit to the left of the previous partial product. This process is continued until all the multiplier bits are checked, and then the partial products are added. This multiplication process, i.e. multiplication by partial product addition and shifting, can be implemented using the block diagram shown in Fig. 5.24.

In the diagram shown in Fig. 5.24, the 4-bit multiplier is stored in register Y ($Y_3Y_2Y_1Y_0$); the 4-bit multiplicand is stored in register M ($M_3M_2M_1M_0$), and the X register ($X_4X_3X_2X_1X_0$) is initially cleared to 00000. Here, to perform multiplication, the least significant bit of the multiplier bit (Y_0) is checked whether it is 0 or 1. If $Y_0 = 1$, the number in the multiplicand register (M) is added with the least significant 4-bits of X register ($X_3X_2X_1X_0$; X_4 is to store carry in addition process) and the combined X and Y register is shifted to the right by 1-bit. If $Y_0 = 0$, the combined X and Y register is shifted to the right by 1-bit without performing any addition. This process has to be repeated four times to perform 4-bit multiplication. Now, the multiplication result ($R_7R_6R_5R_4R_3R_2R_1R_0$) will be available inland/registers ($X_3X_2X_1X_0Y_3Y_2Y_1Y_0$).

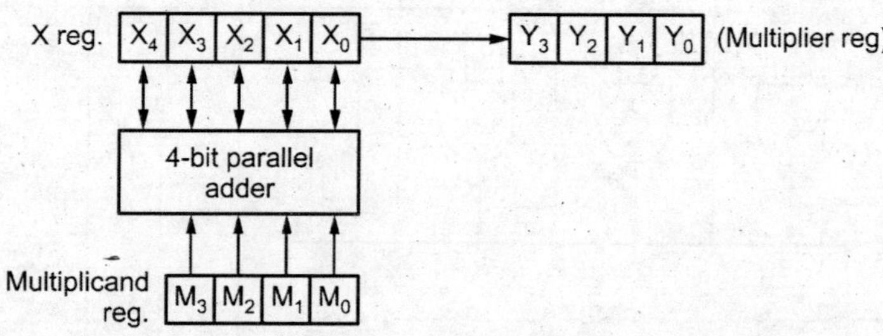

Fig. 5.24 4-bit binary multiplier using shift method

5.16.2 Parallel Multiplier

The 4-bit multiplier using shift method requires 4 cycles of addition and shifting operations, but it requires only a single 4-bit parallel adder. The speed of multiplication process can be increased considerably in parallel multiplier at the extra cost of increased hardware. The circuit diagram for a 4-bit parallel multiplier is shown in Fig. 5.25.

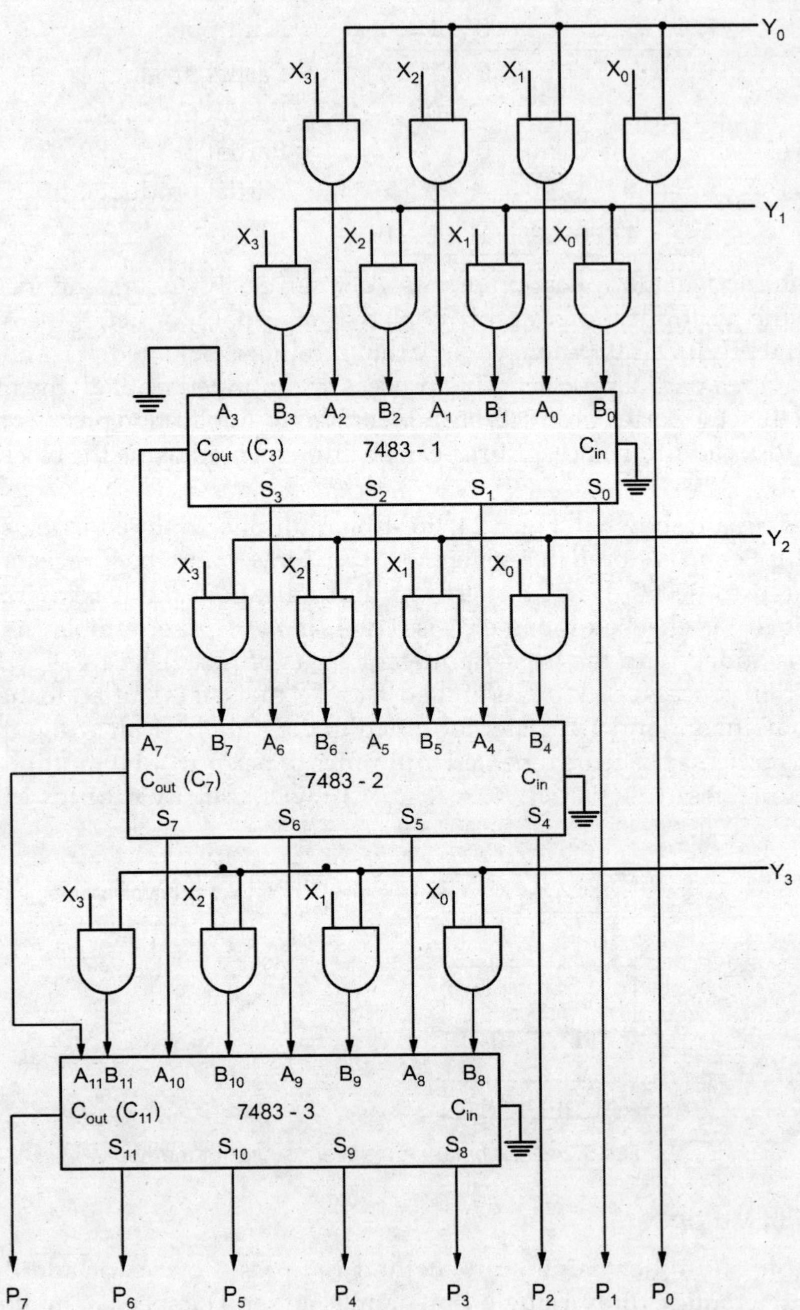

Fig. 5.25 4-bit parallel multiplier

It requires three 4-bit parallel binary adders and 16 numbers of 2-input AND gates. Here, each group of 4 AND gates is used to obtain partial products while 4-bit parallel adders are used to add the partial products. Since the generation of partial products and their additions are performed in parallel in the group of AND gates and 4-bit adders respectively, the multiplication result ($P_7P_6P_5P_4P_3P_2P_1P_0$) will be available at the output immediately after the propagation delay in the multiplier circuit.

The operation of the parallel multiplier can be understood in a better manner from the symbolic form of binary multiplication process shown in Fig. 5.26.

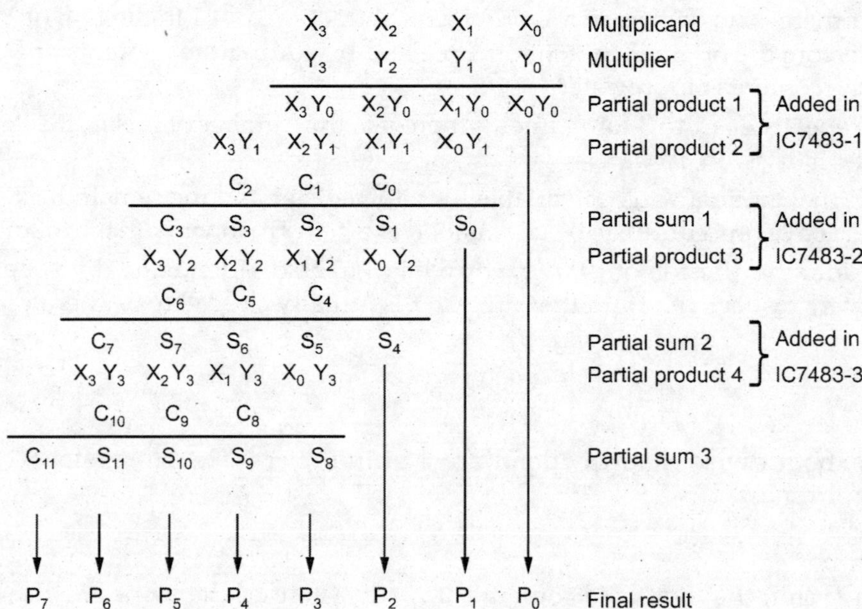

Fig. 5.26 Symbolic form of binary multiplication process

5.17 BINARY DIVIDER

Division is the most difficult and time consuming operation for a general purpose computer. Ablock diagram of a binary divider unit using restoring technique for division is shown in Fig. 5.27.

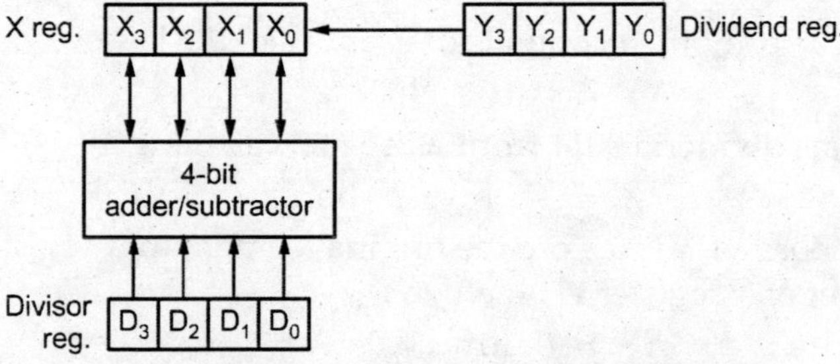

Fig. 5.27 Block diagram of a 4-bit binary divider

Here, the dividend is stored in the dividend register $Y(Y_3Y_2Y_1Y_0)$, the divisor is stored in the divisor register $D(D_3D_2D_1D_0)$ and initially the X register ($X_3X_2X_1X_0$) is cleared.

The procedure for division operation using the circuit shown in Fig. 5.27 is explained as follows:

1. Shift the combined of *X* and *Y* registers to the left by one bit.
2. Perform trial subtraction by subtracting the content of *D* register from the content of *X* register.
3. If there is no borrow in the previous subtraction, put 1 in the LSB of *Y* register, else restore the original content of *X* register by adding the contents of *D* register with the contents of *X* register.
4. Repeat steps 1 to 3 for *n* times, where *n* is the number of bits in the dividend. For a 4-bit division, $n = 4$.

Now, the quotient will be available in the *Y* register and the remainder will be in the *X* register. The above procedure can be understood in a better manner with the following example.

Consider the division of 1011 (11) by 0011 (3). The dividend and divisor are stored in Y and *D* registers respectively, and the *X* register is initially cleared to 0. Therefore,

$$Y_3Y_2Y_1Y_0 = 1011$$
$$X_3X_2X_1X_0 = 0000$$
$$D_3D_2D_1D_0 = 0011$$

Here, both divisor and dividend are 4-bit numbers. Therefore, steps 1-3 have to be repeated four times.

I cycle

Step 1: Shift the combined contents of *X* and *Y* to the left by one bit. Therefore,

$$XY = 0001 \quad 0110$$

Step 2: Subtract dividend 0011 from register *X* resulting in

$$X = 1110 \quad \text{with borrow} = 1$$

Step 3: Since borrow = 1, restore the original content in *X* register by adding dividend 0011 with the content of *X* register 1110. Now,

$$XY = 0001 \quad 0110$$

II cycle

Step 1: Shift the combined contents of *X* and *Y* to the left by one bit. Therefore,

$$XY = 0010 \quad 1100$$

Step 2: Subtract dividend 0011 from *X* register resulting in

$$X = 1111 \quad \text{with borrow} = 1$$

Step 3: Since borrow = 1, restore the original content in *X* register by adding dividend 0011 with the content of *X* register 1111. Now

$$XY = 0010 \quad 1100$$

III cycle

Step 1: Shift the combined contents of *X* and *Y* to the left by one bit. Therefore,

$$XY = 0101 \quad 1000$$

Step 2: Subtract dividend 0011 from register X which results in

X = 0010 with borrow = 0

Step 3: Since borrow = 0, put 1 in the LSB of Fregister (Y_0). Therefore,

XY = 0010 1001

IV cycle

Step 1: Shift the combined contents of X and Y to the left by one bit. Therefore,

XY = 0101 0010

Step 2: Subtract dividend 0011 from Jf register resulting in

X = 0010 with borrow = 0

Step 3: Since borrow = 0, put 1 in the LSB of Fregister (Y_0). Therefore,

XY = 0010 0011

Now, the quotient 0011(3) is available in the Fregister and the remainder 0010(2) is available in the X register.

■ REVIEW QUESTIONS ■

1. What is the need of arithmetic circuits?
2. What is a half-adder? Write its truth table.
3. Design a half-adder using only NOR gates.
4. Describe the working of a half-adder.
5. What is a full-adder?
6. Draw the full-adder circuit using NAND gates only. Explain the functioning of the circuit and show that the output is that of a full-adder.
7. Design a full-adder circuit using only NOR gates. What relation has it to the half-adder circuit?
8. Design a full-subtractor using only NAND gates.
9. Design a full-subtractor using only basic gates.
10. Design a half-subtractor using only NAND gates.
11. Design a half-subtractor using only basic gates.
12. Design a half-subtractor using only NOR gates.
13. Design a full-subtractor using only NOR gates.
14. What is the difference between a full-adder and a full-subtractor?
15. Design the logic diagram of a circuit for addition/subtraction. Use a control variable w and a circuit that functions as a full-adder when $w = 0$, as a full-subtractor when $w = 1$.
16. Show how a full-adder can be converted to a full-subtractor with the addition of an inverter circuit.
17. What are the advantages of complement arithmetic?
18. Design a parallel binary multiplier that multiplies a 4-bit number $B = B_3B_2B_1B_0$ by a 3-bit number $A = A_2A_1A_0$ to form the product $C = C_6C_5C_4C_3C_2C_1C_0$.

    [Hint: This can be done with 12 gates and two 4-bit parallel adders. The AND gates are used to form the products of pairs of bits. The partial products formed by the AND gates are added with the parallel adder.]

❑❑❑

6 Combinational Circuits

6.1 INTRODUCTION

Combinational logic circuits are circuits in which the output at any time depends upon the combination of the input signals present at that instant only, and does not depend upon any past conditions. A combinational circuit designed with individual gates can be implemented with SSI circuits that contain several independent gates. The number of gates in an SSI circuit is limited by the number of pins in the IC package. There are many combinational circuits that are employed in the design of digital systems and are available in the form of MSI components. The following are the most important combinational circuit-type MSI components that are readily available in IC packages: (*i*) adders, (*ii*) subtracters, (*iii*) comparators, (*iv*) decoders and encoders (devices that change data from one type of code to another), (*v*) multiplexers (selecting one of several groups of data), (*vi*) demultiplexers (distributing data to one among several destinations), (*vii*) parity generators/checkers and (*viii*) code converters. In Chapter 5, arithmetic circuits like adders and subtracters were discussed. In this chapter, the other types of combinational circuits are discussed in detail.

Do You Know?	Each of the basic gates (AND, OR, NAND, NOR) can be used to enable or disable the passage of an input signal to its output.

6.2 MULTIPLEXERS (DATA SELECTORS)

The term 'multiplex' means *"many into one."* Multiplexing is the process of transmitting a large number of information over a single line. A digital multiplexer (MUX) is a combinational circuit that selects one digital information from several sources and transmits the selected information on a single output line. A multiplexer is also called a *data selector* since it selects one of many inputs and steers the information to the output.

The multiplexer has several data-input lines and a single output line. The selection of a particular input line is controlled by a set of selection lines. The block diagram of a multiplexer with *n* input lines, *m* select signals and one output line is shown in Fig. 6.1. The selection lines decide the number of input lines of a particular multiplexer. If the number of *n* input lines is equal to 2^m, then m select lines are required to select one of the n input lines. For example, to select 1 out of 4 input lines, two select lines are required; to select 1 of 8 input lines, three select lines are required and so on.

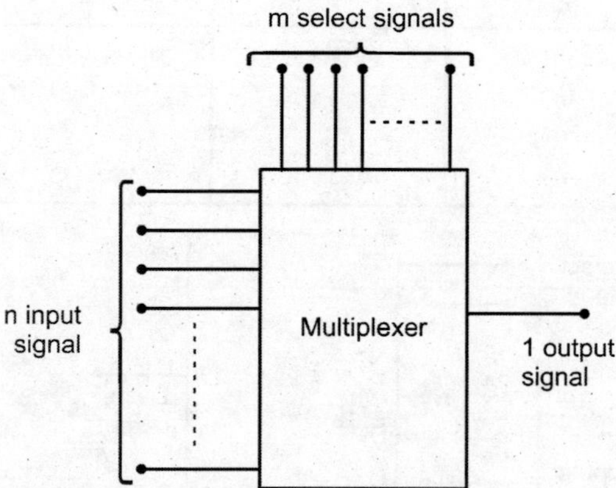

Fig. 6.1 Block diagram of multiplexer

The multiplexer acts like a digitally controlled multiposition switch where binary code applied to the select inputs, controls the data input that will be switched on to the output.

6.2.1 Basic Four-input Multiplexer

The logic symbol of a 4-to-1 multiplexer is shown in Fig 6.2(*a*). It has four data input lines ($D_0 - D_3$), a single output line (Y) and two select lines (S_0 and S_1) to select one of the four input lines. The truth table for a 4-to-1 multiplexer is shown in Table. 6.1.

From the truth table of Table 6.1 a logical expression for the output in terms of the data input and the select inputs can be derived as follows:

The data output Y = data input D_0, if and only if $S_1 = 0$ and $S_0 = 0$,

Therefore, $Y = D_0\overline{S}_1\overline{S}_0 = D_0\overline{0}.\overline{0} = D_0 1.1 = D_0$

The data output $Y = D_1$, if and only if $S_1 = 0$ and $S_0 = 1$.

Therefore, $Y = D_1\overline{S}_1 S_0 = D_1$, when $S_1S_0 = 01$

Similarly, $Y = D_2\overline{S}_1 S_0 = D_2$, when $S_1S_0 = 10$ and

$Y = D_3S_1S_0 = D_3$, when $S_1S_0 = 11$

If the above terms are ORed, then the final expression for the data output is given by

$$Y = D_0\overline{S}_1\overline{S}_0 + D_1\overline{S}_1 S_0 + D_2 S_1\overline{S}_0 + D_3 S_1 S_0$$

Using the above expression, the 4-to-1 multiplexer can be implemented using two NOT gates, four 3-input AND gates and one 4-input OR gate as shown in Fig. 6.2(*b*). Here, each of the four data input lines is applied to any one input of an AND gate and the AND gate outputs are connected with the inputs of OR gate to generate the output Y.

Table 6.1 Truth table of 4-to-1 multiplier

Date select inputs		Output
S_1	S_0	Y
0	0	D_0
0	1	D_1
1	0	D_2
1	1	D_3

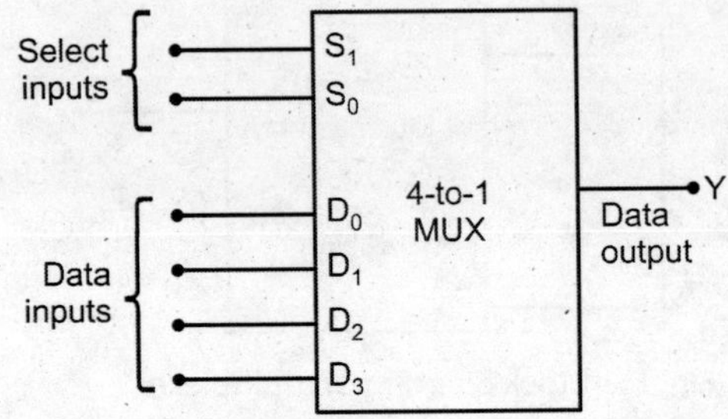

(a) Logic symbol

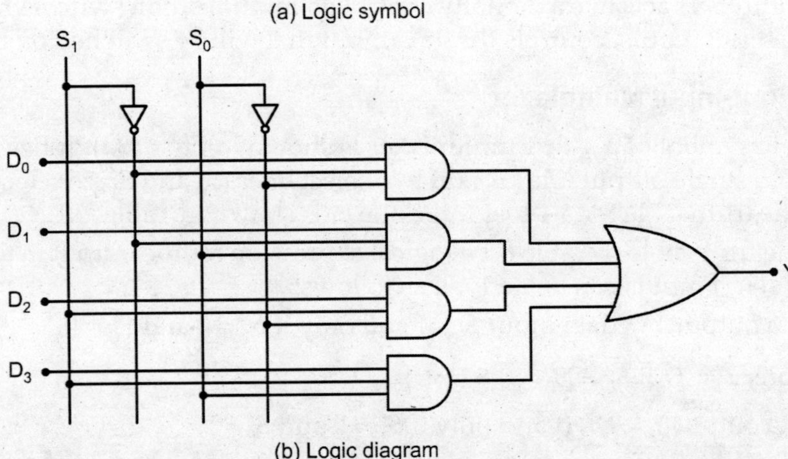

(b) Logic diagram

Fig. 6.2 4-to-1 multiplexer

To demonstrate the operation of the circuit, consider the case when $S_1S_0 = 00$. If $S_1S_0 = 00$ is applied to the select lines, the AND gate associated with D_0 will have two of its inputs equal to 1 and the third input connected to D_0. The other three AND gates have 0 in atleast one of their inputs, which makes their output equal to 0. Hence, the OR output (Y) is equal to the value of D_0. Thus, it provides a path from the selected input (i.e. D_0) and the data on the input D_0 appears on the data-output line (Y). If $S_1S_0 = 01$ (binary 1) is applied to the select lines, the data on the input D_1 appears on the data output line. If $S_1S_0 = 10$ (binary 2) is applied, the data on the input line D_2 appears on the output line (Y). Similarly, if $S_1S_0 = 11$ is applied, the data on D_3 is switched to the output line (Y).

Do You Know?	A decoder is a device whose output is activated only when a unique binary combination (code) is presented on its inputs.

The AND gates and the inverters resemble a decoder circuit, and indeed they decode the input select lines. In general, a 2^n-to-1 multiplexer is constructed from an n-to-2^n decoder by adding to it 2^n input lines, one to each AND gate. The outputs of the AND gates are applied to a single OR gate to provide a single output. The size of the multiplexer is specified by the number 2^n of input lines and the single output line. Multiplexer ICs have an enable input to control the operation of the unit. The enable input (also called *strobe*) can be used to cascade two or more multiplexer ICs to construct a multiplexer with larger number of inputs.

6.2.2 IC 74151 — 8-to-1 Multiplexer

IC 74151 is an 8-to-1 multiplexer with eight data inputs ($D_0 - D_7$), three select input lines ($S_2 - S_0$) and a single output (Y). It also has an enable input $\bar{E}$ and provides both normal and inverted outputs (i.e. Y and $\bar{Y}$). Since $2^3 = 8$, three bits are required to select any one of the eight data bits. When $\bar{E} = 0$, the select inputs $S_2S_1S_0$ will select one of the data input to pass through the output Y. When $\bar{E} = 1$, the multiplexer is disabled. The logic symbol of IC 74151 is shown in Fig. 6.3(a) and its logic diagram is shown in Fig. 6.3(b). The operation of this IC is summarised in truth Table 6.2.

Table 6.2 Truth table of IC 74151 — 8-to-1 multiplexer

Inputs				Outputs	
$\bar{E}$	S_2	S_1	S_0	Y	$\bar{Y}$
1	X	X	X	1	0
0	0	0	0	D_0	$\bar{D}_0$
0	0	0	1	D_1	$\bar{D}_1$
0	0	1	0	2	$\bar{D}_2$
0	0	1	1	D_3	$\bar{D}_3$
0	1	0	0	4	$\bar{D}_4$
0	1	0	1	5	$\bar{D}_5$
0	1	1	0	6	$\bar{D}_6$
0	1	1	1	7	$\bar{D}_7$

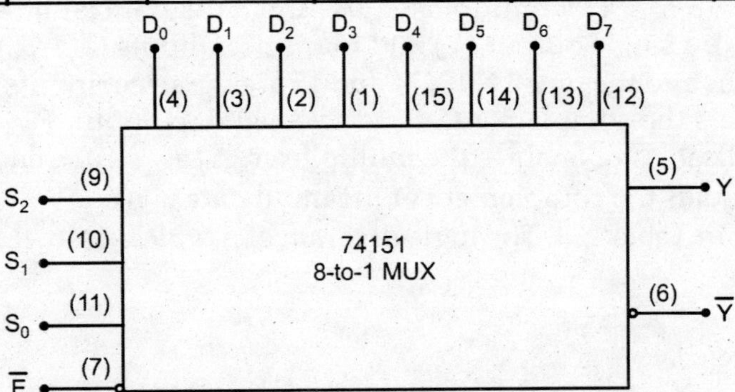

Fig. 6.3(*a*) Logic symbol of IC 74150 — 8-to-1 multiplexer

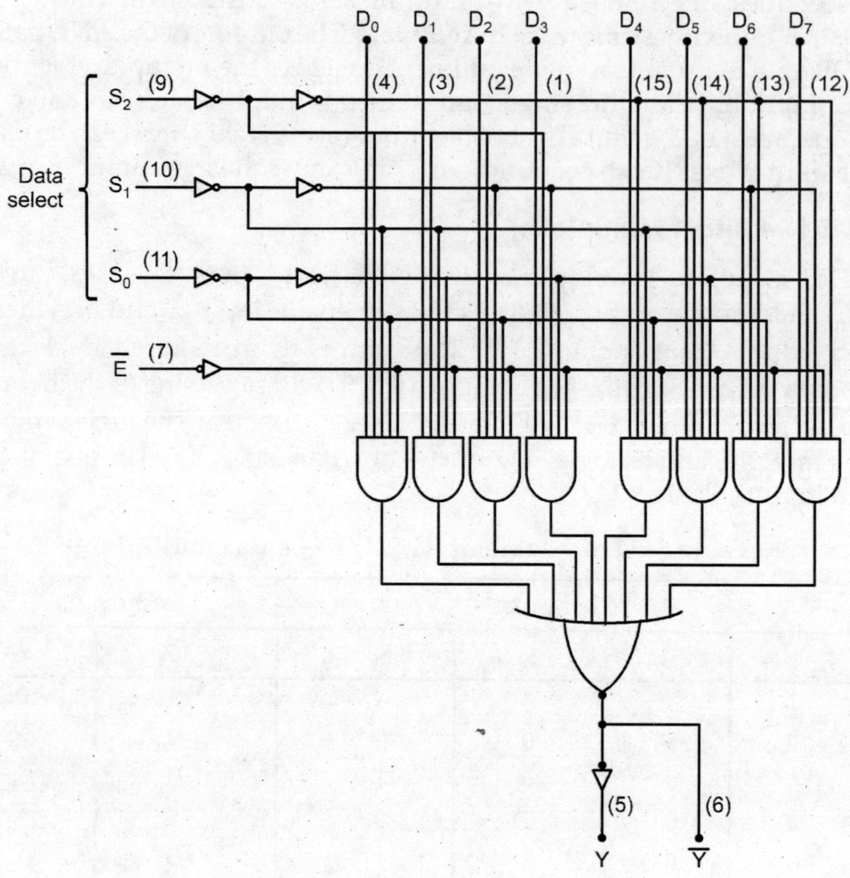

Fig. 6.3(*b*) Logic diagram of IC 74151 — 8-to-1 multiplexer

6.2.3 IC 74150 — 16-to-1 Multiplexer

IC 74150 is a 16-to-1 TTL multiplexer. Its pinout diagram is shown in Fig. 6.4(*a*). It has 16 inputs ($D_0 - D_{15}$), a single output (Y) and four select inputs ($S_1 - S_0$). Pins 1 to 8 and 16 to 23 are the input pins and the pins 11, 13, 14 and 15 are the select inputs $S_3S_2S_1S_0$. Pin 10 is the output and it equals the complement of the selected data input. Pin 9 is for the *strobe*, an input signal that disables or enables the multiplexer. A low strobe enables the multiplexer, so that output Y equals the complement of the input data bit (i.e. $Y = \bar{D}_n$). The truth table of IC 74150 is shown in Table. 6.3. The logic diagram of a typical 16-to-1 multiplexer is shown in Fig. 6.4(*b*).

Table 6.3 Truth table of IC 74150 — 16-to-1 multiplexer

Strobe	Select inputs				Output
$\overline{S}$	S_3	S_2	S_1	S_0	Y
0	0	0	0	0	$\overline{D}_0$
0	0	0	0	1	$\overline{D}_1$
0	0	0	1	0	$\overline{D}_2$
0	0	0	1	1	$\overline{D}_3$
0	0	1	0	0	$\overline{D}_4$
0	0	1	0	1	$\overline{D}_5$
0	0	1	1	0	$\overline{D}_6$
0	0	1	1	1	$\overline{D}_7$
0	1	0	0	0	$\overline{D}_8$
0	1	0	0	1	$\overline{D}_9$
0	1	0	1	0	$\overline{D}_{10}$
0	1	0	1	1	$\overline{D}_{11}$
0	1	1	0	0	$\overline{D}_{12}$
0	1	1	0	1	$\overline{D}_{13}$
0	1	1	1	0	$\overline{D}_{14}$
0	1	1	1	1	$\overline{D}_{15}$
1	X	X	X	X	1

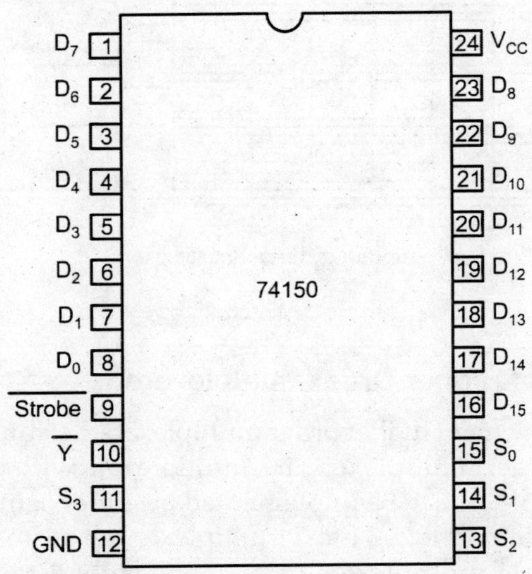

(a) Pinout diagram of IC 74150 — 16-to-1 multiplexer

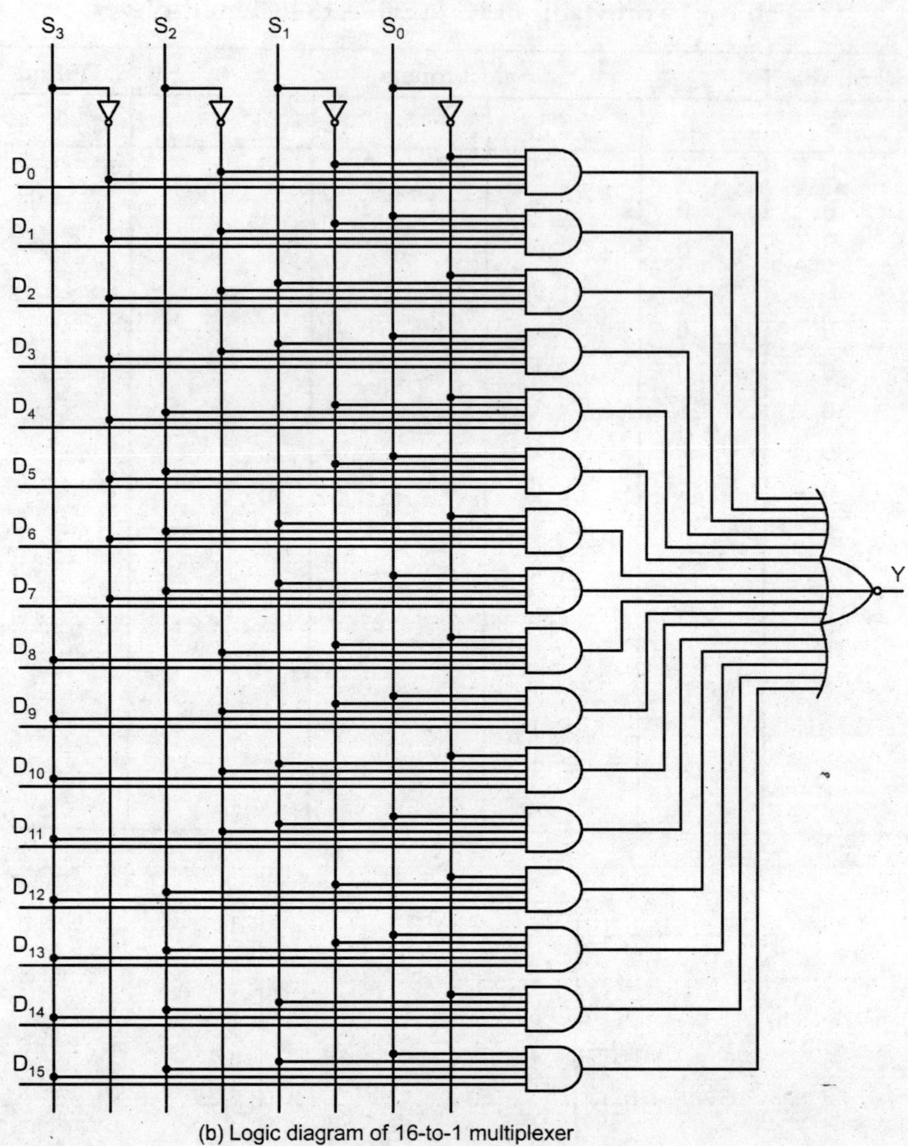

(b) Logic diagram of 16-to-1 multiplexer

Fig. 6.4

6.2.4 Implementation of Higher Order Multiplexers

It is possible to implement higher order multiplexers, i.e. multiplexers with more number of inputs, using lower order multiplexers, i.e. multiplexers with lesser number of inputs. For example, a 16-to-1 multiplexer can be implemented by either using two 8-to-1 multiplexers or four 4-to-1 multiplexers. Similarly, a 32-to-1 multiplexer can be implemented using two 16-to-1 multiplexers or four 8-to-1 multiplexers or by using eight 4-to-1 multiplexers. This can be explained with the following example.

Example 6.1. *Implement a 16-to-1 multiplexer using two 8-to-1 multiplexer ICs (74151).*

Solution. A 16-to-1 multiplexer can be implemented using two IC 74151—8-to-1 multiplexers and one 2-input OR gate as shown in Fig. E6.1.

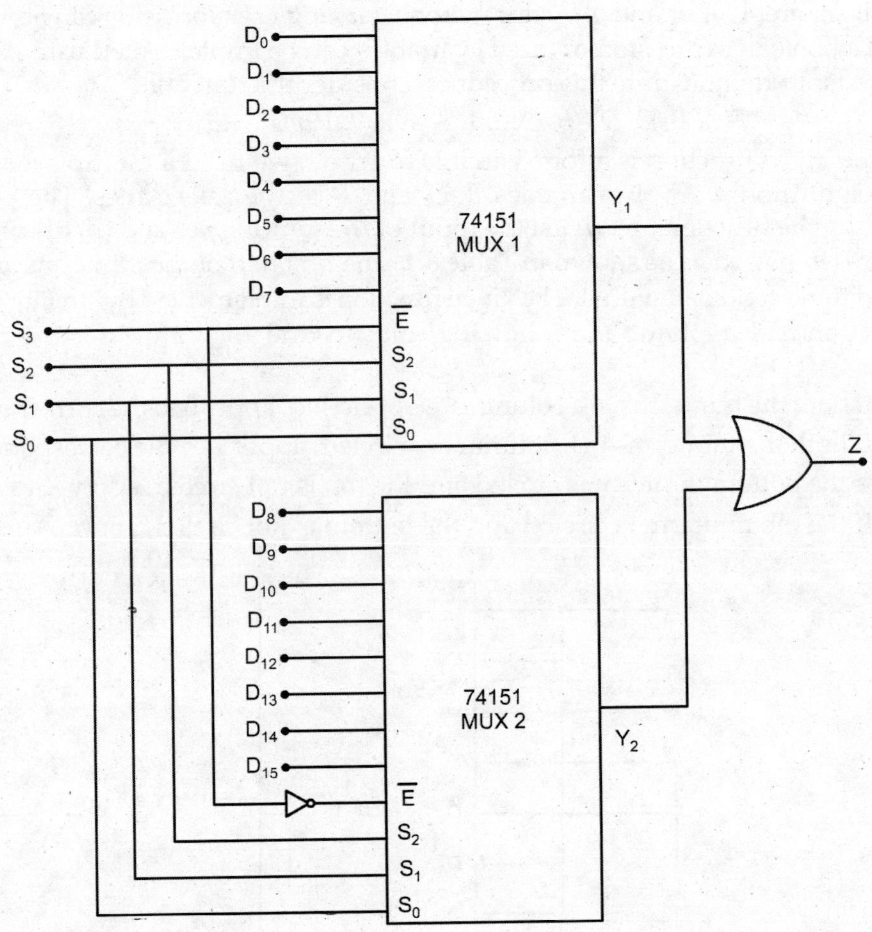

Fig. E6.1 A 16-to-1 multiplexer using two IC 74151 — 8-to-1 multiplexer

Here, to select one of the 16 inputs, four select lines ($S_3S_2S_1S_0$) are required. Among the four select lines, the least significant three select lines ($S_2S_1S_0$) are connected with three select inputs of both the multiplexer ICs. The most significant select line S_3 is connected directly to the $\overline{E}$ input of MUX1 while the same is connected through an inverter to the $\overline{E}$ input of MUX2. Therefore, when $S_1 = 0$, MUX1 is selected and the inputs (D_0 to D_7) are multiplexed to the output Z and MUX2 is disabled. When $S_3 = 1$, the MUX1 is disabled while MUX2 enabled and the inputs (D_0 to D_7) are multiplexed to the output Z. Also, note that the outputs of MUX1 and MUX2 (i.e. Y_1 and Y_2) are ORed using an OR gate to generate output Z.

6.2.5 Implementation of Boolean Expression using Multiplexers

Any boolean or logical expression can be easily implemented using a multiplexer. If a boolean expression has $(n + 1)$ variables, then n of these variables can be connected to the select lines of the multiplexer. The remaining single variable along with constants 1 and 0 is used as the input of the multiplexer. For example, if A is the single variable, then the inputs of the multiplexer are A, $\overline{A}$, 1 and 0. By this method, any logical expression can be implemented. In general, a boolean expression of $(n + 1)$ variables can be implemented using a multiplexer with 2^n inputs. To demonstrate this procedure, consider the function.

$$F(A, B, C, D) = (0, 1, 3, 4, 8, 9, 15)$$

As the given function is a four-variable function, we need a multiplexer with 3 select lines and eight inputs. Apply variables B, C, and D to the select lines. The procedure for implementing the function are: (*i*) list the input of the multiplexer and (*ii*) list under them all the minterms in two rows as shown in Table 6.4. The first half of the minterms are associated with $\overline{A}$ and the second half with A. The given function is implemented by circling the minterms of the function and applying the following rules to find the values for the inputs of the multiplexer.

1. If both the minterms in a column are not circled, apply 0 to the corresponding input.
2. If both the minterms in a column are circled, apply 1 to the corresponding input.
3. If the bottom minterm is circled and the top is not circled, apply A to the input.
4. If the top minterm is circled and the bottom is not circled, apply $\overline{A}$ to the input.

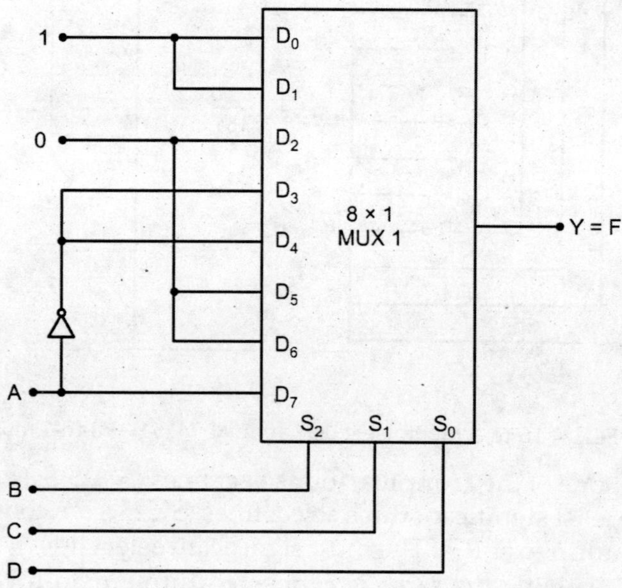

Fig. 6.5 Implementation of F(A, B, C, D) = (0, 1, 3, 4, 8, 9, 15) using 8-to-1 MUX

Table 6.4 Procedure for implementation of the function

	D_0	D_1	D_2	D_3	D_4	D_5	D_6	D_7
$\bar{A}$	(0)	(1)	2	(3)	(4)	5	6	7
A	(8)	(9)	10	11	12	13	14	(15)
	1	1	0	$\bar{A}$	$\bar{A}$	0	0	A

Now, using the procedure and the table, the given function can be implemented using an 8-to-1 multiplexer as shown in Fig. 6.5.

It is not necessary to choose the most significant variable as an input to the multiplexer. One can choose any one of the variables as an input and accordingly the multiplexer implementation table has to be modified.

Example 6.2. *Implement the following function using a multiplexer:*

$$F(A, B, C) = Z(1, 3, 5, 6)$$

Solution: The given function has three variables. Hence, it can be implemented using a multiplexer with two select inputs and four data inputs. The implementation table of the given function is shown in Table E6.2.

Table E6.2

	D_0	D_1	D_2	D_3
$\bar{A}$	0	(1)	2	(3)
A	4	(5)	(6)	7
	0	1	A	$\bar{A}$

Now, the given three variable functions can be implemented using 4-to-1 multiplexer as shown in Fig. E6.2.

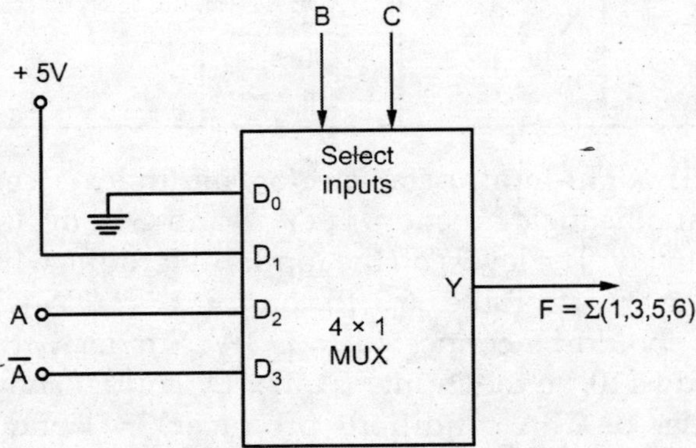

Fig. E6.2 Implementation of F(A, B, C) =   (1, 3, 5, 6) using 4-to-1 MUX

6.3 APPLICATIONS OF MULTIPLEXER

Multiplexer circuits find numerous applications in digital systems. Some of the fields where multiplexing finds immense use are data selection, data routing, operation sequencing, parallel-to-serial conversion, waveform generation and logic function generation.

Data routing: Multiplexers can be used to route data from one of several sources to one destination.

Logic function generator: It can also be used to implement logic functions in sum-of-products form directly from a truth table without the need for simplification. The logic variables are used as the select inputs and each data input is connected permanently HIGH or LOW. This is explained with an example 6.3.

Control sequencer: A multiplexer can also be used as a part of control sequencer.

Parallel-to-serial converter: Digital systems that process data in parallel form take very less time. In order to transmit the information over long distances, the parallel arrangement is undesirable as it requires a large number of transmission lines. Therefore, data in parallel form is converted to serial form using multiplexers.

Example 6.3. *Generate the logic function given in Table E6.3 using IC 74151 — 8-to-1 MUX.*

Table E6.3

Inputs			Outputs
C	B	A	Y
0	0	0	0
0	0	1	1
0	1	0	1
0	1	1	0
1	0	0	0
1	0	1	1
1	1	0	1
1	1	1	1

Solution: The IC 74151, eight-input data selector/multiplexer can be used to implement any specified three-variable logic function. Let A, B and C be the input variables connected to S_0, S_1 and S_2 respectively. The levels on the inputs determine which data input appears at the output Y. According to truth Table E6.3, the output Y is LOW when $CBA = 000$. Thus, the multiplexer input D_0 should be connected to LOW. Similarly, the output Y is LOW for $CBA = 011$, 100, 101 and 110; so the inputs D_3, D_4, D_5 and D_6 should also be connected to LOW. The other values of CBA conditions produce $Y = 1$ and therefore the inputs of multiplexer D_1, D_2 and D_7 should be connected to HIGH. Now, the three variable function, given in Table E6.3, can be implemented using IC 74151— 8-to-1 multiplexer as shown in Fig. E6.3.

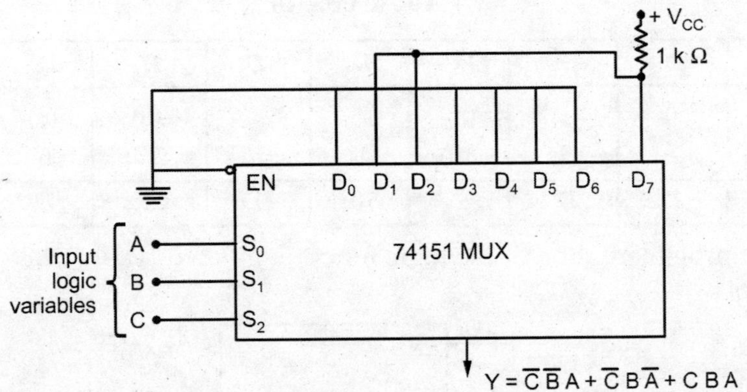

Fig. E6.3 3-variable logic function generator using IC 74151 — 8-to-1 MUX

Example 6.4. *Implementation the logic function in Table E6.4(a) using a 74151 A eight-input data selector/MUX.*

Table E6.4(*a*)

Inputs				Outputs
D	*C*	*B*	*A*	*Y*
0	0	0	0	0
0	0	0	1	1
0	0	1	0	1
0	0	1	1	0
0	1	0	0	0
0	1	0	1	1
0	1	1	0	1
0	1	1	1	1
1	0	0	0	1
1	0	0	1	0
1	0	1	0	1
1	0	1	1	0
1	1	0	0	1
1	1	0	1	1
1	1	1	0	0
1	1	1	1	1

Solution: Let *CBA* be the data-select inputs. In the first row of Table E6.4(*a*), *CBA* = 000 and therefore *Y*= *D* = 0. In the ninth row of the table, *CBA* = 000, and *Y*= *D* = 1. Thus, *D* is connected to the 0 input. In the second and tenth rows, *CBA* = 001 and *Y* = *D* = 1 and *Y* = *D* = 0 respectively. Thus, *D is* inverted and connected to the 1 input. The implementation table to decide the inputs of the multiplexer is shown in Table E6.4(*b*).

Table E6.4(*b*)

	D_0	D_1	D_2	D_3	D_4	D_5	D_6	D_7
$\bar{D}$	0	(1)	(2)	3	4	(5)	(6)	(7)
D	(8)	9	(10)	11	(12)	(13)	14	(15)
	D	$\bar{D}$	1	0	D	1	$\bar{D}$	1

Now, the implementation of the logic function given in Table E6.4(*a*) using IC 74151 is shown in Fig. E6.4.

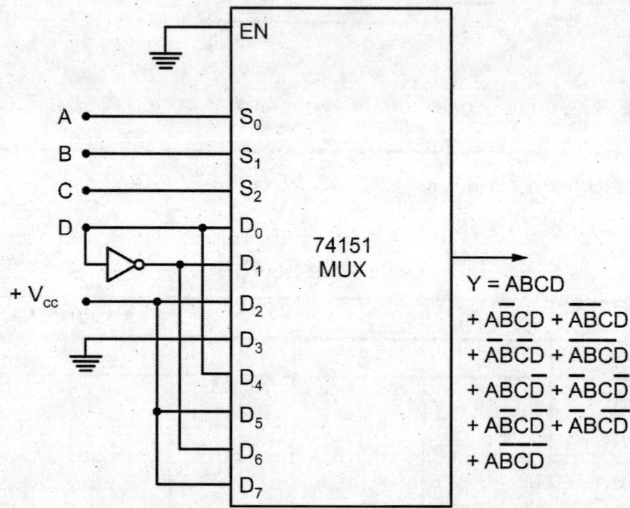

Fig. E6.4 4-variable logic function generator using IC 74151 — 8-to-1 MUX

Some of the multiplexer ICs are given in Table 6.5.

Table 6.5 Multiplexer ICS

No.	IC number	Inputs
1.	SN 74157	Quad 2-input
2.	SN 74153	Dual 4-input
3.	SN 74151	Eight-input
4.	SN 74152	Eight-input
5.	SN 74150	16-input

6.4 DEMULTIPLEXER (DATA DISTRIBUTIONS)

The word "demultiplexer" *means one into many*. Demultiplexing is the process of taking information from one input and transmitting the same over one of several outputs.

Do You Know?	Demultiplexers are used at the other end of the data path to separate the signals that are sharing a data path and distribute them to their respective destinations.

A demultiplexer is a logic circuit that receives information on a single input and transmits the same information over one of several (2^n) output lines.

The block diagram of a demultiplexer which is opposite to a multiplexer in its operation is shown in Fig. 6.6. The circuit has one input signal, m select signals and n output signals. The select inputs determine to which output the data input will be connected. As the serial data is changed to parallel data, i.e the input caused to appear on one of the n output lines, the demultiplexer is also called a *distributor* or a *serial-to-parallel converter*.

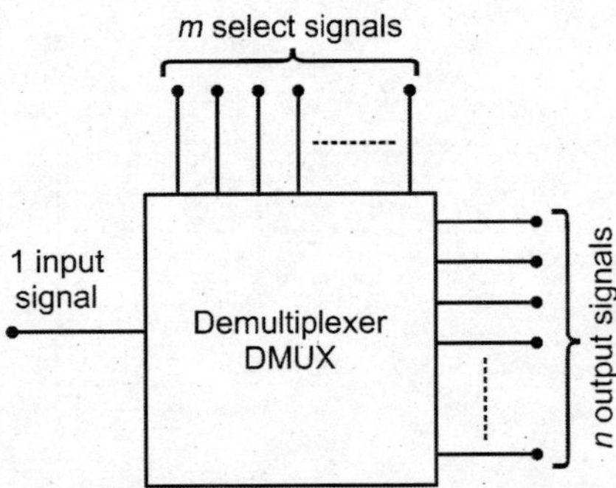

Fig. 6.6 Block diagram of demultiplexer

6.4.1 1-to-4 Demultiplexer

A 1-to-4 demultiplexer has a single input (D), four outputs (Y_0 to Y_3) and two select inputs (S and S_0). The truth table of the 1-to-4 demultiplexer is shown in Table 6.6.

Table 6.6 Truth table of 1-to-4 demultiplexer

Data Input	Select Inputs		Outputs			
D_0	S_1	S_0	Y_3	Y_2	Y_1	Y_0
D	0	0	0	0	0	D
D	0	1	0	0	D	0
D	1	0	D	D	0	0
D	1	D	0	0	0	0

From the truth table, it is clear that the data input is connected to output Y_0 when $S_1 = 0$ and $S_0 = 0$ and the data input is connected to output Y_1 when $S_1 = 0$ and $S_0 = 1$. Similarly, the data input is connected to outputs Y_2 and Y_3 when $S_1 = 1$ and $S_0 = 0$ and when $S_1 = 1$ and $S_0 = 0$, respectively. Also, from the truth table, the expressions for outputs can be written as follows:

$$Y_2 = \overline{S}_1\overline{S}_0D$$
$$Y_1 = \overline{S}_1S_0D$$
$$Y_2 = S_1\overline{S}_0D$$
$$Y = S_1S_0D$$

Now, using the above expressions, a 1-to-4 demultiplexer can be implemented using four 3-input AND gates and two NOT gates as shown in Fig. 6.7. Here, the input data line is connected to all the AND gates. The two select lines S_1S_0 enable only one gate at a time and the data that appears on the input line passes through the selected gate to the associated output line.

6.4.2 1-to-8 Demultiplexer

A 1-to-8 demultiplexer has a single input (D), eight outputs (Y_0 to Y_7) and three select inputs (S_2, S_1 and S_0). It distributes one input line to eight output lines based on the select inputs. The truth table of 1-to-8 demultiplexer is shown in Table 6.7.

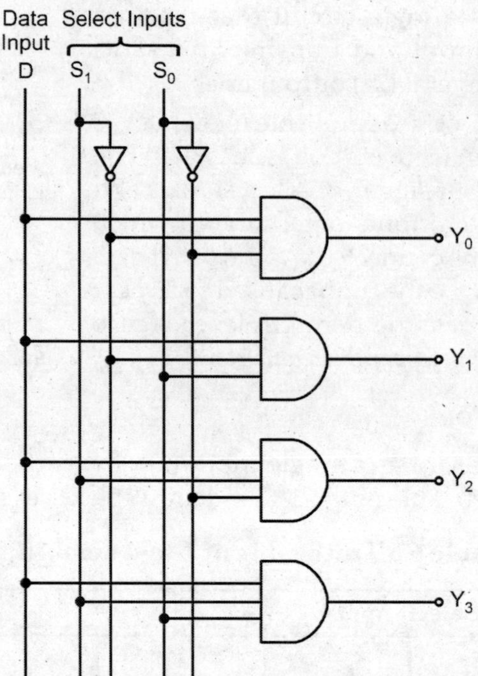

Fig. 6.7 Logic diagram of 1-to-4 demultiplexer

Table 6.7 Truth table of 1-to-8 demultiplexer

Data input	Select inputs			Outputs							
D	S_2	S_1	S_0	Y_7	Y_6	Y_5	Y_4	Y_3	Y_2	Y_1	Y_0
D	0	0	0	0	0	0	0	0	0	0	D
D	0	0	1	0	0	0	0	0	0	D	0
D	0	1	0	0	0	0	0	0	D	0	0
D	0	1	1	0	0	0	0	D	0	0	0
D	1	0	0	0	0	0	D	0	0	0	0
D	1	0	1	0	0	D	0	0	0	0	0
D	1	1	0	0	D	0	0	0	0	0	0
D	1	1	1	D	0	0	0	0	0	0	0

From the above truth table, it is clear that the data input is connected with one of the eight outputs based on the select inputs. Now, from this truth table, the expressions for eight outputs can be written as follows:

$$Y_0 = \overline{S}_2\overline{S}_1\overline{S}_0 D; \quad Y_1 = \overline{S}_2\overline{S}_1 S_0 D; \quad Y_3 = \overline{S}_2 S_1\overline{S}_0 D; \quad Y_3 = \overline{S}_2 S_1 S_0 D;$$

$$Y_4 = S_2\overline{S}_1\overline{S}_0 D; \quad Y_5 = S_2\overline{S}_1 S_0 D; \quad Y_6 = S_2 S_1\overline{S}_0 D; \quad Y_3 = S_2 S_1 S_0 D$$

Now, using the above expressions, the logic diagram of a 1-to-8 demultiplexer can be drawn as shown in Fig. 6.8. Here, the single data input line D is connected to all the eight

AND gates, but only one of the eight AND gates will be enabled by the select input lines. For example, if $S_2S_1S_0 = 000$, then only AND gate-0 will be enabled and thereby the data input D will appear at Y_0. Similarly, for different combinations of select inputs, input D will appear at the respective output.

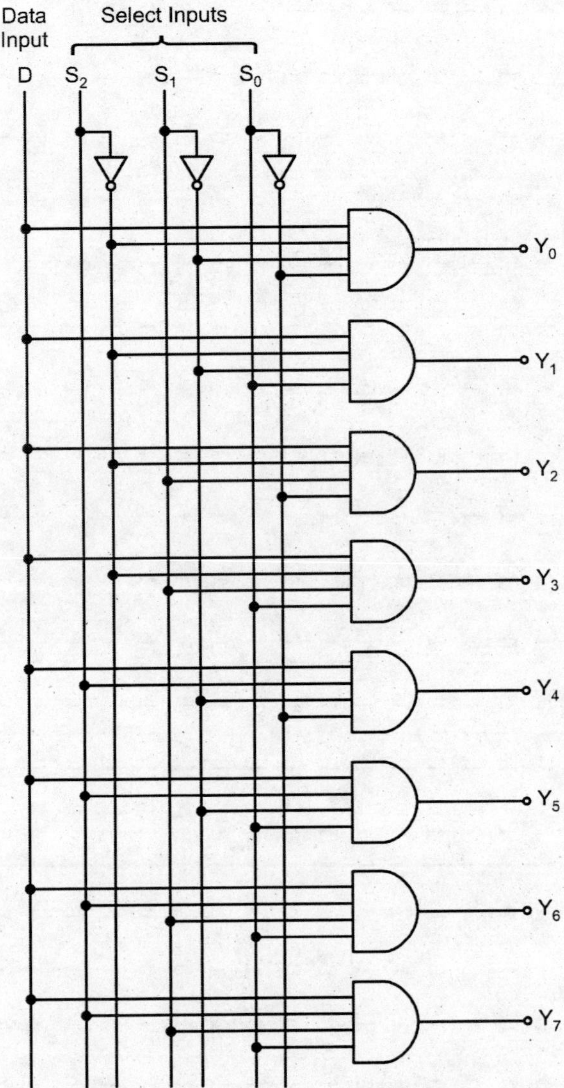

Fig. 6.8 Logic diagram of 1-to-8 demultiplexer

6.4.3 IC 74154 —1-to-16 Demultiplexer

IC 74154 is a 1-to-16 demultiplexer with single input (D), sixteen active low outputs (Y_0 to Y_{15}), four select inputs (S_3 to S_0) and a strobe input. Its pinout diagram is shown in Fig. 6.9. The input strobe must be LOW to activate the IC 74154. The logic symbol of IC 74154 is shown

Table 6.8 Truth table of IC 74154 — 1-to-16 Demus

Strobe	Data	S_3	S_2	S_1	S_0	Y_0	Y_1	Y_2	Y_3	Y_4	Y_5	Y_6	Y_7	Y_8	Y_9	Y_{10}	Y_{11}	Y_{12}	Y_{13}	Y_{14}	Y_{15}
0	0	0	0	0	0	0	1	1	1	1	1	1	1	1	1	1	1	1	1	1	1
0	0	0	0	0	1	1	0	1	1	1	1	1	1	1	1	1	1	1	1	1	1
0	0	0	0	1	0	1	1	0	1	1	1	1	1	1	1	1	1	1	1	1	1
0	0	0	0	1	1	1	1	1	0	1	1	1	1	1	1	1	1	1	1	1	1
0	0	0	1	0	0	1	1	1	1	0	1	1	1	1	1	1	1	1	1	1	1
0	0	0	1	0	1	1	1	1	1	1	0	1	1	1	1	1	1	1	1	1	1
0	0	0	1	1	0	1	1	1	1	1	1	0	1	1	1	1	1	1	1	1	1
0	0	0	1	1	1	1	1	1	1	1	1	1	0	1	1	1	1	1	1	1	1
0	0	1	0	0	0	1	1	1	1	1	1	1	1	0	1	1	1	1	1	1	1
0	0	1	0	0	1	1	1	1	1	1	1	1	1	1	0	1	1	1	1	1	1
0	0	1	0	1	0	1	1	1	1	1	1	1	1	1	1	0	1	1	1	1	1
0	0	1	0	1	1	1	1	1	1	1	1	1	1	1	1	1	0	1	1	1	1
0	0	1	1	0	0	1	1	1	1	1	1	1	1	1	1	1	1	0	1	1	1
0	0	1	1	0	1	1	1	1	1	1	1	1	1	1	1	1	1	1	0	1	1
0	1	1	0	1	1	1	1	1	1	1	1	1	1	1	1	1	1	1	1	0	1
0	0	1	1	1	1	1	1	1	1	1	1	1	1	1	1	1	1	1	1	1	0
0	1	X	X	X	X	1	1	1	1	1	1	1	1	1	1	1	1	1	1	1	1
1	0	X	X	X	X	1	1	1	1	1	1	1	1	1	1	1	1	1	1	1	1
1	1	X	X	X	X	1	1	1	1	1	1	1	1	1	1	1	1	1	1	1	1

in Fig. 6.10. The data bit *D* is automatically steered to the output line whose subscript is equivalent, of S_1, S_2, S_1 and S_0. The truth table of a 74154 is shown in Table 6.8. When the data input is high, all the output lines are high. Similarly, when the strobe is high, all the output lines are high.

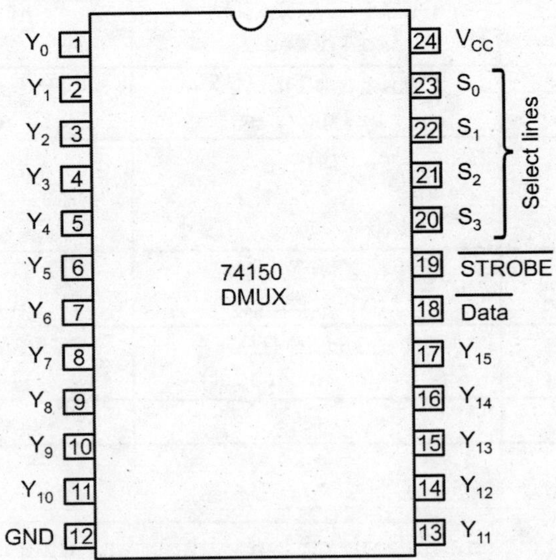

Fig. 6.9 Pinout diagram of IC 74154—1-to-16 DEMUX

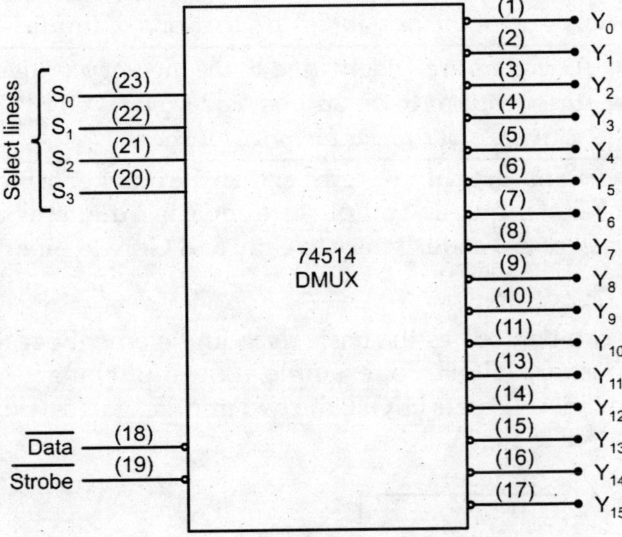

Fig. 6.10 Logic symbol of IC 74154—1-to-16 DEMUS

A list of some important Demultiplexer ICs is given in Table 6.9.

Table 6.9 Demultiplexer ICs

IC no.	Description	Output (Y)
74139	Dual 1 to 4 DEMUX (2 to 4 decoder)	Inverted input
74155	Dual 1 to 4 DEMUX (2 to 4 decoder)	1*Y* – inverted input 2*Y* – same as input
74156	-do-	Open-collector 1*Y* – inverted input 2*Y* – same as input
74138	1 to 8 DEMUX (3 to 8 decoder)	Inverted input
74154	1 to 16 DEMUX (4 to 16 line decoder)	Inverted input
74159	-do-	Open-collector

6.5 DECODERS

A decoder is similar to demultiplexer but without any data input. Most digital systems require the decoding of data. Decoding is necessary in applications such as data demultiplexing, digital display, digital-to-analog converters and memory addressing. A decoder is a logic circuit that converts an n-bit binary input code (data) into 2^n output lines, such that each output line will be activated for only one of the possible combinations of inputs.

Do You Know?	Decoders are widely used in the memory system of a computer where they respond to the address code generated by the central processor to activate a particular memory location.

In a decoder, the number of output is greater than the number of inputs. Here, it is important to note that if the number of inputs and outputs are equal in a digital system then it can be called *converters*, e.g. BCD to Excess-3 code, Binary to Gray and Gray to Binary code converters.

6.5.1 Basic Binary Decoder

An AND gate can be used as the basic decoding element because its output is HIGH only when all the inputs are HIGH. For example, if the input binary number is 1001, then, to make all the inputs to the AND gate HIGH, the two middle bits (0s) must be inverted by using two NOT gates as shown in Fig. 6.11.

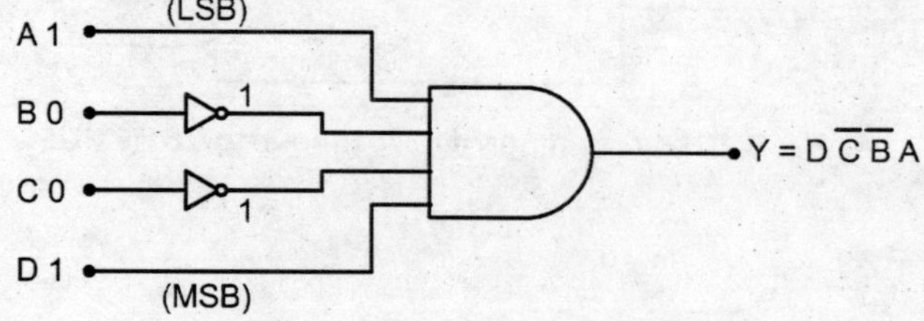

Fig. 6.11 Decoding logic for input 1001 with an active HIGH output

If a NAND gate is used in place of the AND gate, then a LOW output is generated to indicate the presence of the proper binary code.

6.5.2 3-to-8 Decoder

A 3-to-8 decoder has three inputs (A, B, C) and eight outputs (D_0 to D_7). Based on the 3 inputs, one of the eight outputs is selected. The truth table for 3-to-8 decoder is shown in Table. 6.10.

Table 6.10 Truth table of 3-to-8 decoder

Inputs			Outputs							
A	B	C	D_0	D_1	D_2	D_3	D_4	D_5	D_6	D_7
0	0	0	1	0	0	0	0	0	0	0
0	0	1	0	1	0	0	0	0	0	0
0	1	0	0	0	1	0	0	0	0	0
0	1	1	0	0	0	1	0	0	0	0
1	0	0	0	0	0	0	1	0	0	0
1	0	1	0	0	0	0	0	1	0	0
1	1	0	0	0	0	0	0	0	1	0
1	1	1	0	0	0	0	0	0	0	1

From the above truth table, it is clear that only one of eight outputs (D_0 to D_1) is selected based on the three select inputs. Also, from the truth table, the logic expressions for the outputs can be written as follows:

$$D_0 = \bar{A}\bar{B}\bar{C}; \quad D_1 = \bar{A}\bar{B}C; \quad D_2 = \bar{A}B\bar{C}; \quad D_3 = \bar{A}BC$$

$$D_4 = A\bar{B}\bar{C}; \quad D_5 = A\bar{B}C; \quad D_6 = AB\bar{C}; \quad D_7 = ABC$$

Using the above expressions, the circuit of a 3-to-8 decoder can be implemented using three NOT gates and eight 3-input AND gates as shown in Fig. 6.12. The three inputs, A, B and C are decoded into eight outputs, each output representing one of the minterms of the 3-input variables. The three inverters provide the complement of the inputs and each one of the eight AND gates generates one of the minterms. This decoder can be used for decoding any 3-bit code to provide eight outputs, corresponding to eight different combinations of the input code.

This is also called a *1-of-8 decoder*, since only one of eight output lines is HIGH for a particular input combination. For example, when $ABC = 010$, only the AND gate-2 has HIGH at all its inputs, and therefore D_2 = HIGH. Similarly, if $ABC = 110$, the AND gate 6 has all its inputs in HIGH state and thereby D_6 = HIGH. It is also called a *binary-to-octal decoder* since the inputs represent three-bit binary numbers and the outputs represent the eight digits in the octal number system.

Enable inputs: Some decoders have one or more enable inputs which are used to control the operation of the decoder. With the enable line held HIGH, the decoder functions normally and the input code, A, B and C will determine which output is HIGH. Hence, the decoder is enabled only if the enable line is HIGH.

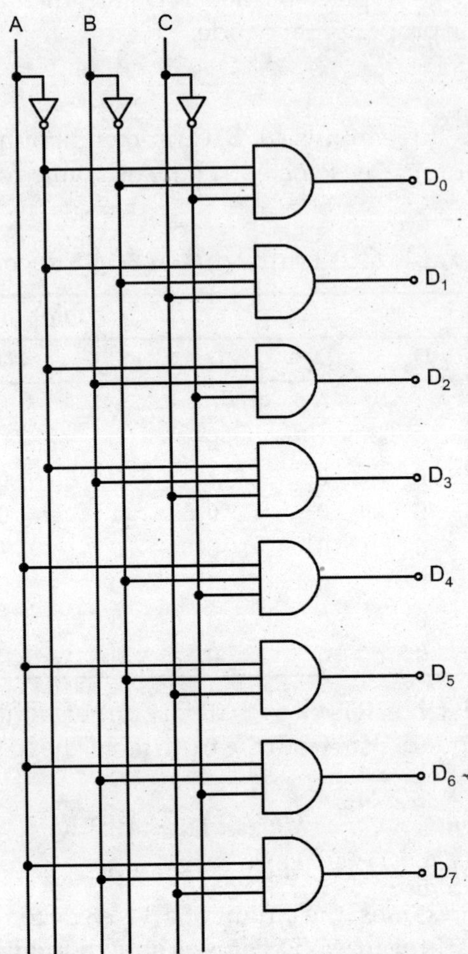

Fig. 6.12 Logic diagram of 3-to-8 decoder

6.5.3 4-to-16 Decoder

A 4-to-16 decoder is one with four inputs and sixteen outputs. The truth table for a 4-to-16 decoder is shown in Table 6.11. From the truth table, it is clear that, for a particular input combination, only one output is in LOW state while the remaining outputs are in HIGH state. As discussed in the previous section, the 4-to-16 decoder can be implemented using four NOT gates and 16 decoding NAND gates as shown in Fig. 6.13. Here, sixteen decoding gates ($2^4 = 16$) are required to decode all possible combinations of four bits. Since it has four inputs and 16 outputs, it is commonly called a *4-to-16 decoder*. It is important to note that NAND gates are used in the implementation to have active LOW outputs.

The 4-to-16 decoder is also called a *1-of-16 decoder*, since only one of 16 outputs is selected based on a particular input combination. It is also called *a binary-to-hexadecimal decoder* since the inputs represents four-bit binary number and the outputs represent the sixteen digits in the hexadecimal number system.

Table 6.11 Truth table of 4-to-16 decoder

Decimal digit	Binary inputs				Logic function	Outputs															
	D	C	B	A		0	1	2	3	4	5	6	7	8	9	10	11	12	13	14	15
0	0	0	0	0	$\bar{D}\bar{C}\bar{B}\bar{A}$	0	1	1	1	1	1	1	1	1	1	1	1	1	1	1	1
1	0	0	0	1	$\bar{D}\bar{C}\bar{B}A$	1	0	1	1	1	1	1	1	1	1	1	1	1	1	1	1
2	0	0	1	0	$\bar{D}\bar{C}B\bar{A}$	1	1	0	1	1	1	1	1	1	1	1	1	1	1	1	1
3	0	0	1	1	$\bar{D}\bar{C}BA$	1	1	1	0	1	1	1	1	1	1	1	1	1	1	1	1
4	0	1	0	0	$\bar{D}C\bar{B}\bar{A}$	1	1	1	1	0	1	1	1	1	1	1	1	1	1	1	1
5	0	1	0	1	$\bar{D}C\bar{B}A$	1	1	1	1	1	0	1	1	1	1	1	1	1	1	1	1
6	0	1	1	0	$\bar{D}CB\bar{A}$	1	1	1	1	1	1	0	1	1	1	1	1	1	1	1	1
7	0	1	1	1	$\bar{D}CBA$	1	1	1	1	1	1	1	0	1	1	1	1	1	1	1	1
8	1	0	0	0	$D\bar{C}\bar{B}\bar{A}$	1	1	1	1	1	1	1	1	0	1	1	1	1	1	1	1
9	1	0	0	1	$D\bar{C}\bar{B}A$	1	1	1	1	1	1	1	1	1	0	1	1	1	1	1	1
10	1	0	1	0	$D\bar{C}B\bar{A}$	1	1	1	1	1	1	1	1	1	1	0	1	1	1	1	1
11	1	0	1	1	$D\bar{C}BA$	1	1	1	1	1	1	1	1	1	1	1	0	1	1	1	1
12	1	1	0	0	$DC\bar{B}\bar{A}$	1	1	1	1	1	1	1	1	1	1	1	1	0	1	1	1
13	1	1	0	1	$DC\bar{B}A$	1	1	1	1	1	1	1	1	1	1	1	1	1	0	1	1
14	1	1	1	0	$DCB\bar{A}$	1	1	1	1	1	1	1	1	1	1	1	1	1	1	0	1
15	1	1	1	1	$DCBA$	1	1	1	1	1	1	1	1	1	1	1	1	1	1	1	0

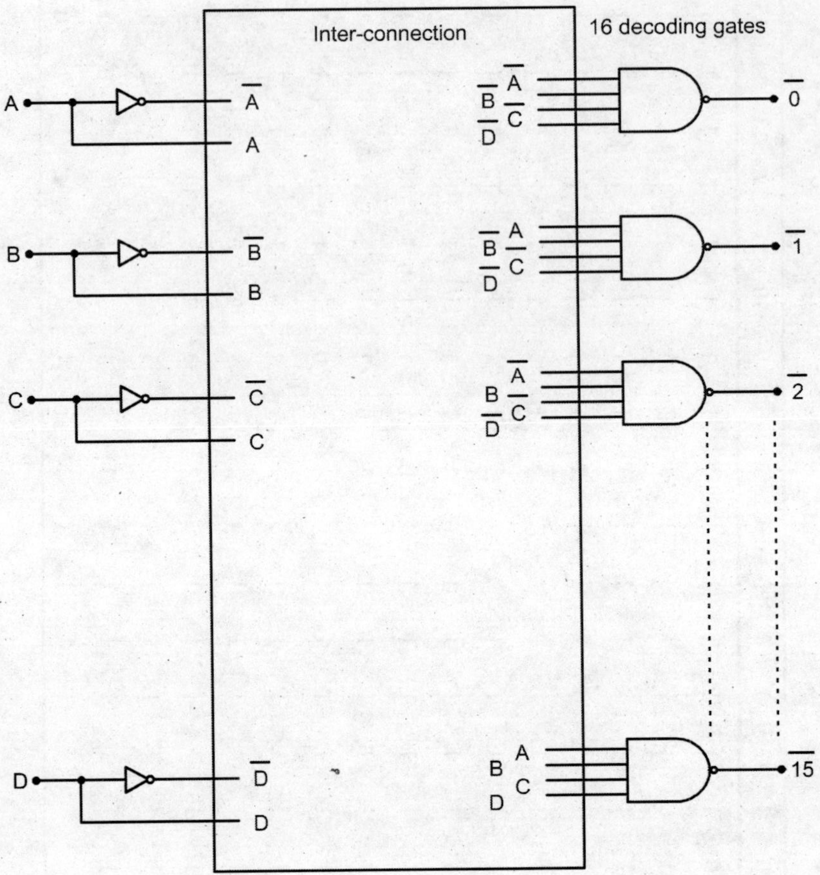

Fig. 6.13 Logic diagram of 4-to-16 decoder

6.5.4 IC 74139—Dual 2-to-4 Decoder

IC 74139 consists of two individual 2-to-4 decoders/demultiplexers in a single package. The logic symbol of IC 74139 is shown in Fig. 6.14. Each decoder has two inputs, four active LOW outputs and one active LOW enable input. This active LOW enable input can be used as the data input in demultiplication applications.

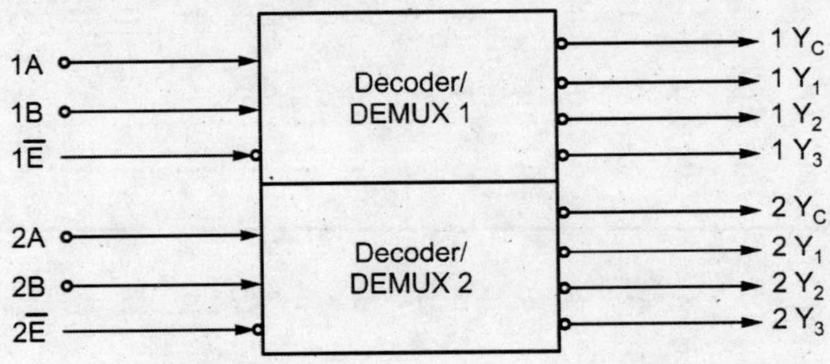

Fig. 6.14 Logic symbol of IC 74139—dual 2-to-4 decoder

6.5.5 IC 74154—4-to-16 Decoder

IC 74154 is called a 4-to-16 decoder/demultiplexer because it is capable of decoding as well as demultiplexing. In order to use this IC 74154 as a decoder, the inputs Data and Strobe must be grounded as shown in Fig. 6.15.

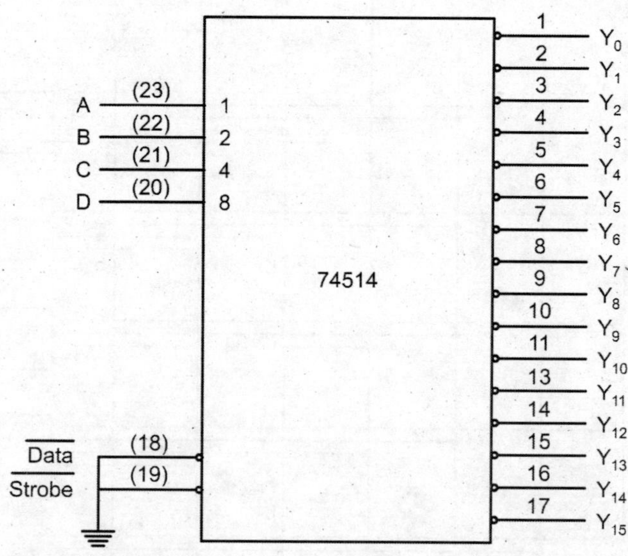

Fig. 6.15 Logic symbol of IC 74154 used as decoder

6.5.6 BCD-to-Decimal Decoder

A decoder that takes a 4-bit BCD as the input code and produces 10 outputs corresponding to the decimal digits is called a BCD-to-decimal decoder. The logic diagram of a BCD to decimal decoder using AND gates is shown in Fig. 6.16. Here, each output goes HIGH when its corresponding BCD code is applied at its inputs. For example, the output Y_5 will go HIGH only when 0101 (BCD for 5) occurs at inputs *ABCD*. This type of decoder is also called a *1-of-10 decoder* because only one of the 10 output lines is HIGH. Here, the subscript of the HIGH output always equals the decimal equivalent of the input BCD digit.

6.5.7 IC 7445—BCD-to-Decimal Decoder

The TTL IC 7445 is a BCD-to-decimal decoder/driver. The term *driver* is added to its description because this IC has open collector outputs that can operate at higher current, and voltage limits than a normal TTL output. The outputs of 7445 can sink upto 80 mA in the LOW state and can be raised up to 30 V in the HIGH state. This makes it possible for them to drive loads such as indicator LEDs or lamps, relays or DC motors. This IC is functionally equivalent to 1-of-10 decoder, shown in Fig. 6.16, except that the outputs are in the active LOW state. For a valid BCD code, the corresponding output will be in the LOW state while all other output lines are in the HIGH state. It is important to note that an invalid BCD input (1010 to 1111) forces all output lines into the HIGH state. The pinout diagram of IC 7445 is shown in Fig. 6.17.

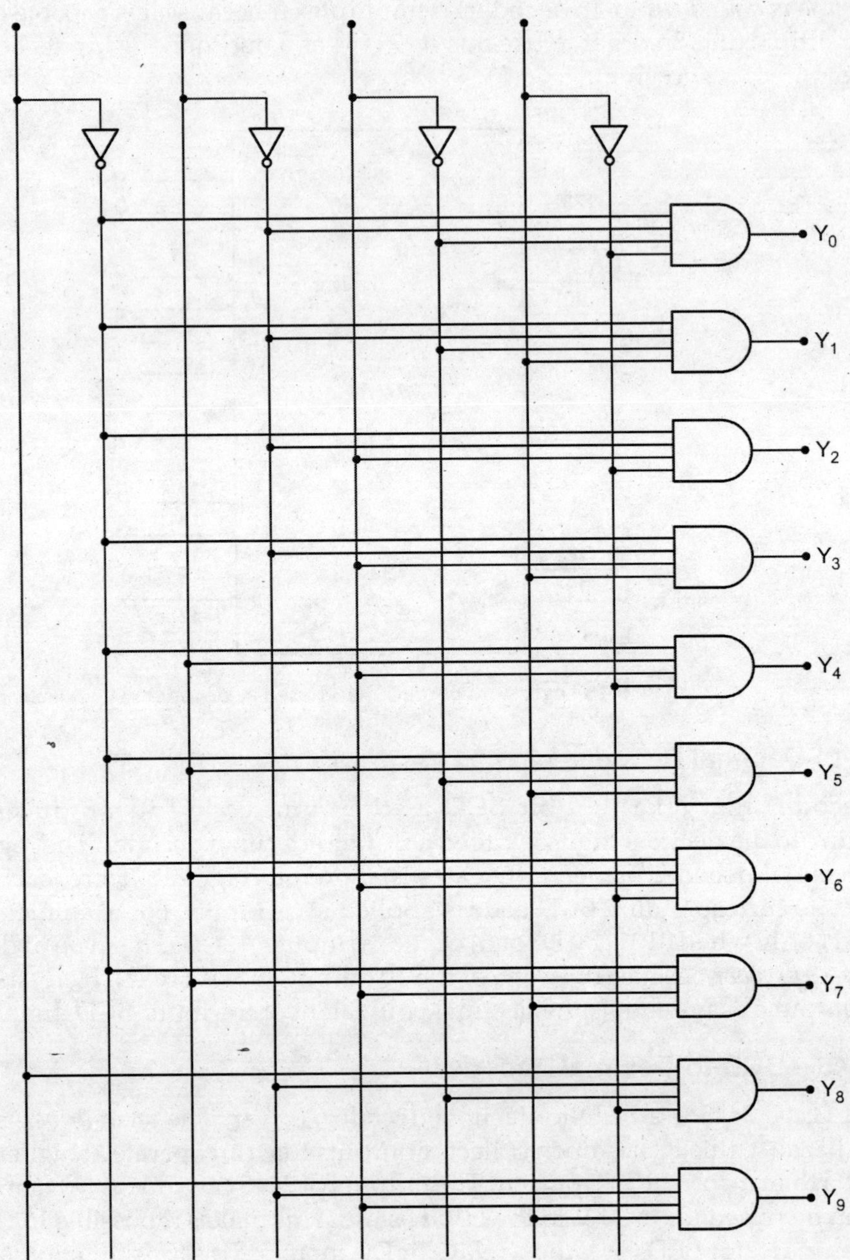

Fig. 6.16 Logic diagram of BCD-to-decimal decoder

A list of BCD-to-decimal decoder ICs is given in Table 6.12.

Table 6.12 BCD-to-decimal decoder/driver ICs

IC no.	Output circuits	Applications
7441	Open-collector	Nixie tube driver
7442	Totem-pole	LED driver
7445	Open-collector	Indicator/relay driver
74141	-do-	Nixie tube driver
74145	-do-	Indicator/relay driver
74445	-do-	Indicator/relay driver

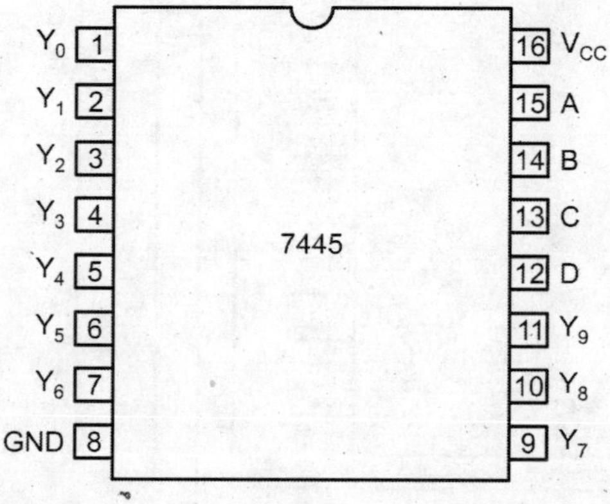

Fig. 6.17 Pinout diagram of IC 7445 — BCD-to-decimal decoder

6.5.8 Implementation of Higher Order Decoders using Lower Order Decoders

As in the case of multiplexer, it is possible to implement a decoder having more number of outputs using decoders with lesser number of outputs. For example, a 1-of-16 decoder can be implemented using two 1-of-8 or four 1-of-4 decoders. Similarly, a 1-of-32 decoder can be implemented using two 1-of-16 decoders or four 1-of-8 decoders or eight 1-of-4 decoders. This can be explained with the following example.

Example 6.5 *Construct 1-of-32 or 5-to-32 decoder using two 1-of-16 decoder ICs.*

Solution: IC 74154 is a 4-to-16 or 1-of-16 decoder with four inputs and 16 outputs. By connecting two 74154 ICs as shown in Fig. E6.5, it is possible to realise 1-of-32 or 5-to-32 decoder. To select 1-of-32 outputs, 5 select lines are required. Of these, four select lines (D, C, B, A) are connected to both the decoders to select 1-of-16 outputs in the individual decoders. The most significant select line E is connected directly to the Strobe and data inputs of IC 74154-1, while the same is connected through an inverter to the Strobe and data inputs of IC 74154-2, Now, when $E = 0$, the IC 74154-1 is selected and one of its outputs ($Y_0 - Y_{15}$) is in LOW state and IC 74154-2 is disabled. When $E = 1$, IC 74154-1 is disabled while IC 74154-2 is enabled and one of its outputs (Y_{16} to Y_{31}) is selected, i.e. in LOW state.

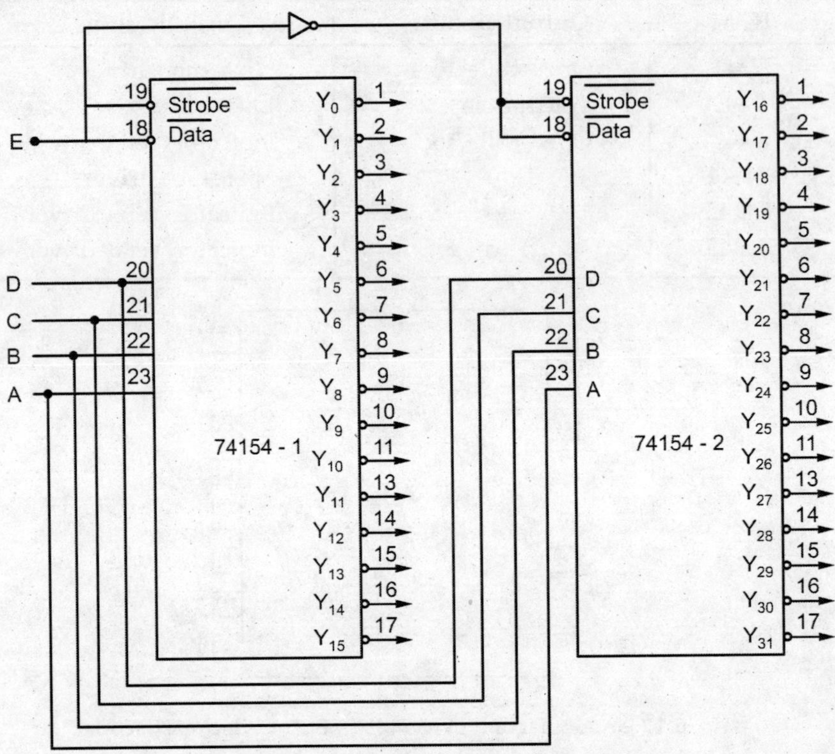

Fig. E6.5 1-of-32 decoder using two 74154 ICs

Here, all the output lines are HIGH, except the decoded output line. Hence, a bubble is included on each output line. Similarly, bubbles are added to the Strobe and Data inputs of each 74154 to indicate active LOW inputs.

6.5.9 BCD-to-Seven-segment Decoder/Driver

A seven-segment display is normally used for displaying any one of the decimal digits, 0 through 9. A BCD-to-seven segment decoder accepts a decimal digit in BCD and generates the corresponding seven-segment code.

Figure 6.18 shows a seven-segment display composed of seven elements or segments. Each segment is made up of a material that emits light when current is passed through it. Most commonly used displays are LEDs and incandescent filaments. Note, that letters *a*, *b*, *c*, *d*, *e*, *f*, and *g* run clockwise from the top of each segment For instance, to display a 1, the segments *b* and *c* have to be illuminated; to display a 0, the segments *a*, *b*, *c*, *d*, *e* and *f* have to be illuminated by properly forward biasing the LEDs in the selected segments.

(a) Seven-segment display (b) Display of decimal digits in a seven segment display

Fig. 6.18

Design of BCD-to-seven-segment decoder: A BCD-to-seven-segment decoder can be designed using logic gates. A block diagram of BCD-to-seven-segment decoder with four BCD inputs (*A, B, C* and *D*) and seven outputs (*a, b, c, d, e, f* and *g*), corresponding to seven segments of a display, is shown in Fig. 6.19. The truth table of the BCD-to-7-segment decoder is shown in Table. 6.13.

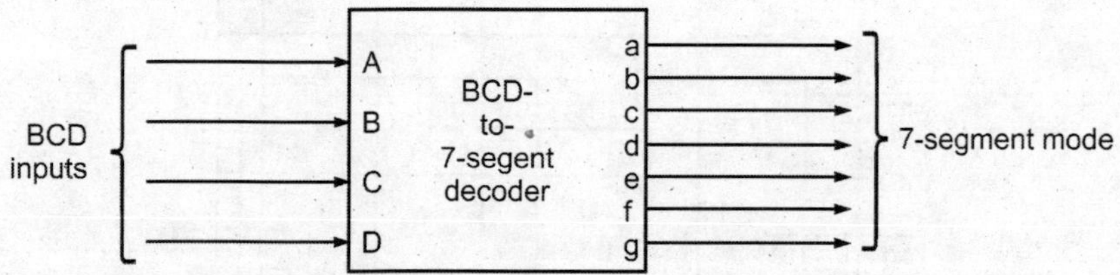

Fig. 6.19 Block diagram of BCD-to-7-segment decoder

Table 6.13 Truth table of BCD-to-7-segment decoder

BCD inputs				Seven segment outputs						
A	*B*	*C*	*D*	*a*	*b*	*c*	*d*	*e*	*f*	*g*
0	0	0	0	1	1	1	1	1	1	0
0	0	0	1	0	1	1	0	0	0	0
0	0	1	0	1	1	0	1	1	0	1
0	0	1	1	1	1	1	1	0	0	1
0	1	0	0	0	1	1	0	0	1	1
0	1	0	1	1	0	1	1	0	1	1
0	1	1	0	1	0	1	1	1	1	1
0	1	1	1	1	1	1	0	0	0	0
1	0	0	0	1	1	1	1	1	1	1
1	0	0	1	1	1	1	1	0	1	1

Since only BCD inputs are valid combinations, the other input combination of four variables corresponding to 10,11,12,13, 14 and 15 can be termed as don't care combinations to aid the simplification of logic expressions. Now, the logic expressions corresponding to seven-segment can be written from the truth table shown in Table 6.13 as follows:

$$a = {}_m(0, 2, 3, 5, 6, 7, 8, 9) + {}_d(10, 11, 12, 13, 14, 15)$$
$$b = {}_m(0, 1, 2, 3, 4, 7, 8, 9) + {}_d(10, 11, 12, 13, 14, 15)$$
$$c = {}_m(0, 1, 3, 4, 5, 6, 7, 8, 9) + {}_d(10, 11, 12, 13, 14, 15)$$
$$d = {}_m(0, 2, 3, 5, 6, 8, 9) + {}_d(10, 11, 12, 13, 14, 15)$$
$$e = {}_m(0, 2, 6, 8) + {}_d(10, 11, 12, 13, 14, 15)$$
$$f = {}_m(0, 4, 5, 6, 7, 8, 9) + {}_d(10, 11, 12, 13, 14, 15)$$
$$g = {}_m(2, 3, 4, 5, 6, 8, 9) + {}_d(10, 11, 12, 13, 14, 15)$$

The above expressions can be simplified using *K*-map method as shown in Fig. 6.20.

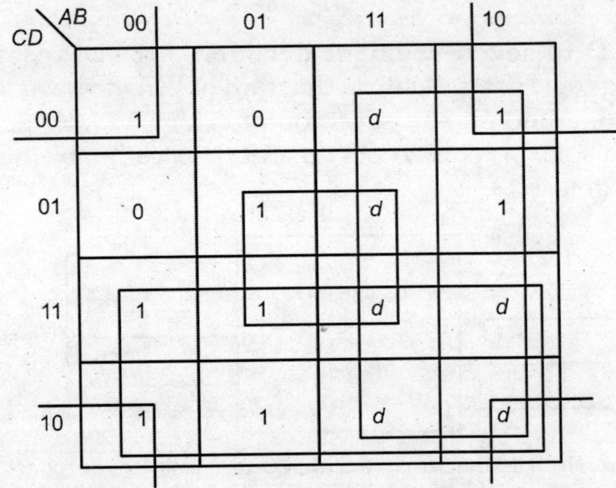

(*a*) *K*-map for 7-segment output '*a*'

From the above *K*-map, $a = A + C + BD + \overline{B}\overline{D} = A + C + \overline{B \oplus D}$

CD \ AB	00	01	11	10
00	1	1	1	1
01	1	0	d	1
11	1	1	d	d
10	1	0	d	d

(*b*) *K*-map for 7-segment output '*b*'

From the above *K*-map, $b = \overline{B} + CD + \overline{C}\overline{D} + \overline{B} + \overline{C \oplus D}$

CD \ AB	00	01	11	10
00	1	1	*d*	1
01	1	1	*d*	1
11	1	1	*d*	*d*
10	0	1	*d*	*d*

(*c*) *K*-map for 7-segment output '*c*'

From the above *K*-map, $c = B + \overline{C} + D$

CD \ AB	00	01	11	10
00	1	0	*d*	1
01	0	1	*d*	1
11	1	0	*d*	*d*
10	1	1	*d*	*d*

(*d*) *K*-map for 7-segment output '*d*'

From the above *K*-map,

$$d = A + \overline{B}\overline{D} + C\overline{D} + B\overline{C}D + \overline{C \oplus D} = A + C\overline{D} + B(C + \overline{D}) + B\overline{C}D$$

$$= A + C\overline{D} + B \oplus (C + \overline{D})$$

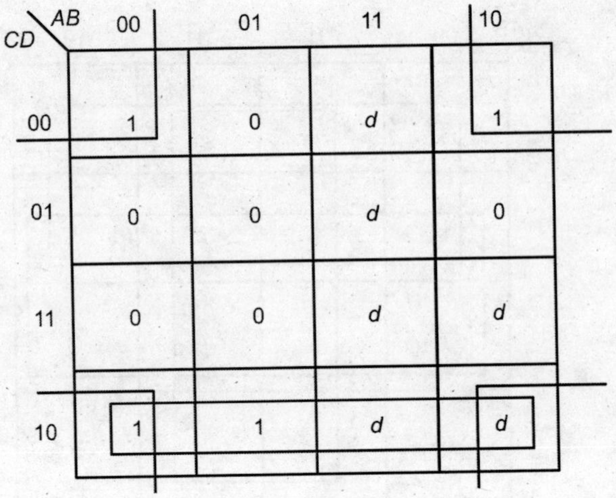

CD \ AB	00	01	11	10
00	1	0	d	1
01	0	0	d	0
11	0	0	d	d
10	1	1	d	d

(*e*) K-map for 7-segment output '*e*'

From the above K-map, $e = \overline{B}\overline{D} + C\overline{D} = \overline{D}(\overline{B} + C)$

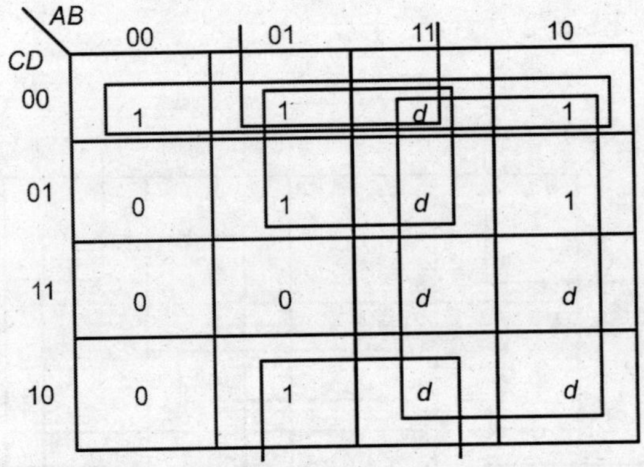

CD \ AB	00	01	11	10
00	1	1	d	1
01	0	1	d	1
11	0	0	d	d
10	0	1	d	d

(*f*) K-map for 7-segment output '*f*'

From the above K-map, $f = A + \overline{C}\overline{D} + B\overline{C} + B\overline{D} = A + \overline{C}\overline{D} + B(\overline{C} + \overline{D})$

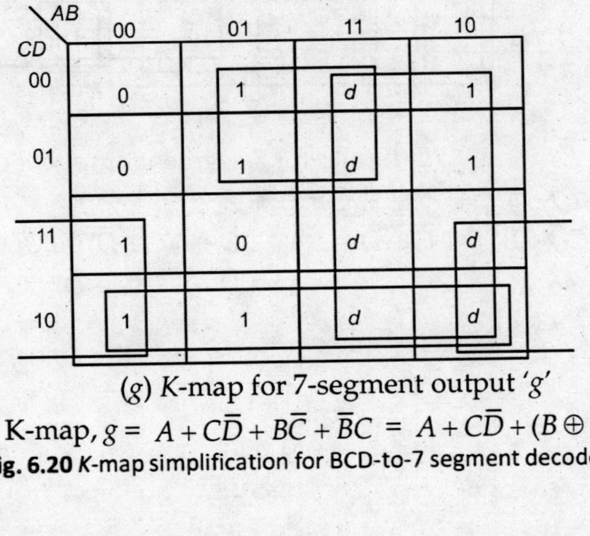

CD \ AB	00	01	11	10
00	0	1	d	1
01	0	1	d	1
11	1	0	d	d
10	1	1	d	d

(*g*) K-map for 7-segment output '*g*'

From the above K-map, $g = A + C\overline{D} + B\overline{C} + \overline{B}C = A + C\overline{D} + (B \oplus C)$

Fig. 6.20 *K*-map simplification for BCD-to-7 segment decoder

Now, using the above simplified expressions for seven-segment outputs, the BCD-to-seven-segment decoder can be implemented using logic gates as shown in Fig. 6.21.

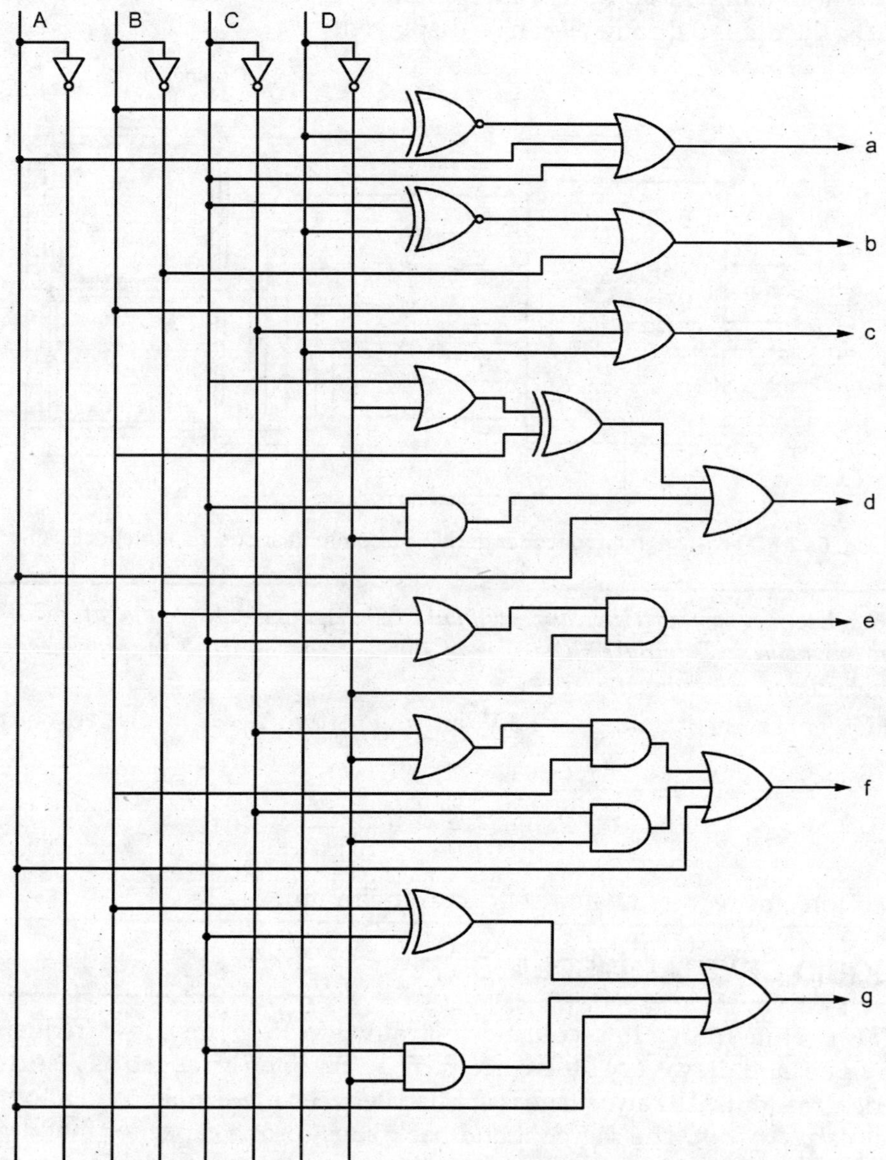

Fig. 6.21 Logic diagram of BCD-to-7-segment decoder

IC 7446/7447 and IC 744817449 — BCD-to-seven-segment decoders ICs: 7446 and 7447 are BCD to seven segment decoders with active LOW open collector outputs designed for driving common anode seven-segment displays. ICs 7448 and 7449 are with active HIGH outputs for driving common cathode seven segment displays.

Fig. 6.22 shows a BCD to seven segment decoder/driver (TTL 7446 or 7447) used to drive a common anode 7-segment LED display. Each segment consists of one LED and the anodes of all LEDs are connected to + V_{CC} (5 V). The cathodes of the LEDs are connected through current-limiting resistors to the appropriate outputs of the decoder/driver. By forward-biasing LEDs, the digits 0 through 9 can be displayed.

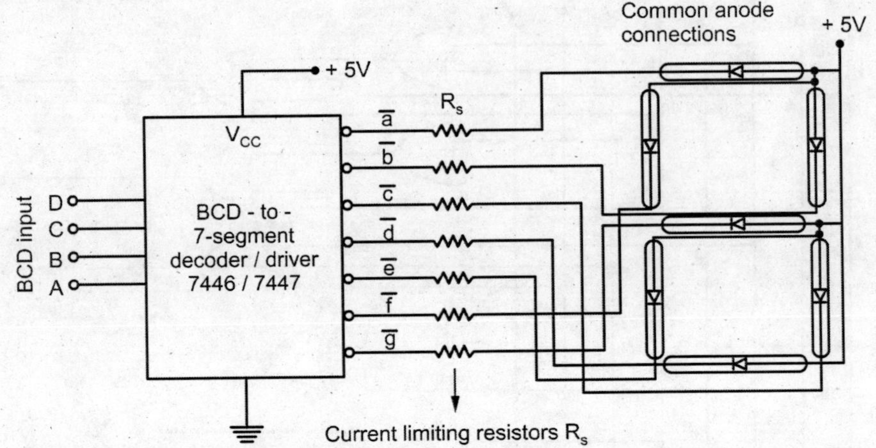

Fig. 6.22 BCD-to-7 segment decoder deriving a common anode 7-segment LED display

Example 6.6. *Each segment of a typical 7-segment LED display is rated to operate at 10 mA and 2.8 V for normal brightness. Calculate the value of the current limiting resistor needed to produce approximately 10 mA per segment.*

Solution: Given, segment voltage = 2.8 V, current = 10 mA, V_{CC} = applied voltage = 5 V. By Ohm's law,

$$R_3 = \frac{(5-2.8)\text{ V}}{10\text{ mA}} = \frac{2.2\text{ V}}{10\text{ mA}} = 220\ \Omega$$

Therefore, the required value of the current limiting is 220 .

6.6 LIQUID CRYSTAL DISPLAYS

LCDs operate from a low-voltage (typically 3 to 15 V rms), low frequency (30 Hz to 60 Hz) ac signal and draw very little current. They are often arranged as 7-segment displays for numerical readouts. The ac voltage is applied between the segment and a common element called the *backplane* (bp). The segment and backplane form a capacitor that draws very low current as long as the frequency of the ac signal is kept low. An LCD does not emit light energy like LED, and so it requires an external source of light.

Each segment in the display is driven by an exclusive-OR gate with one input connected to the output of a seven-segment decoder and the other input connected to the signal source. A 40 Hz square wave is applied to the backplane and also to the input of a CMOS 4070 exclusive-OR gate. Fig. 6.23 shows the basic operation of an LCD segment.

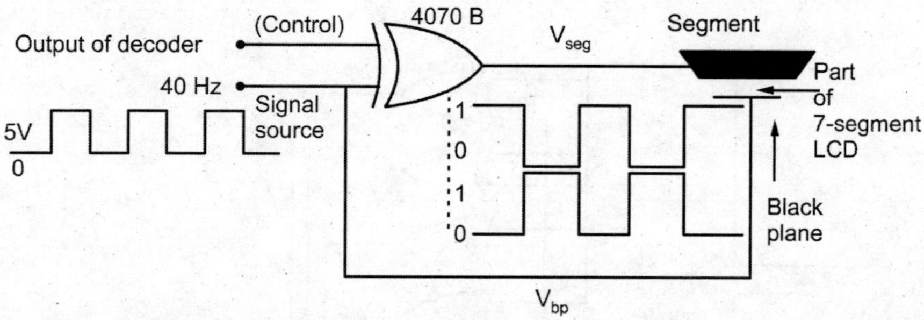

Fig. 6.23 Basic operation of a LCD segment

When the control input of the EX-OR gate is LOW, its output will be exactly the same as the 40 Hz square wave, so that the signals applied to the segment and backplane are equal. Since there is no difference in voltage, the segment will be OFF. When the control input is HIGH, the EX-OR output is a square wave that is 180° out-of-phase with the signal applied to the backplane. As a result, the segment voltage will be alternatively at +5 V and –5 V relative to the backplane.

The same logic can be applied to a complete 7-segment LCD display as shown in Fig. 6.24. In this case, the CMOS 4511B—BCD-to-7-segment decoder/driver supplies the control signals to each of the seven EX-OR gates for the seven segments. The IC 4511 has active-HIGH outputs since a HIGH is required to turn ON the segment.

Applications of decoders

(1) Decoders are used in counter systems.
(2) They are used in analog-to-digital converters.
(3) Decoder outputs can be used to drive a display system.

6.7 ENCODERS

An encoder is a digital circuit that performs the inverse operation of a decoder. Hence, the opposite of the decoding process is called *encoding*. An encoder is a combinational logic circuit that converts an active input signal into a coded output signal.

Do You Know?	An encoder is a device that generates a unique binary code in respond to the activation of individual input.

It has n input lines, only one of which is active at any time and m output lines. It encodes one of the active inputs to a coded binary output with m bits. In an encoder, the number of outputs is less than the number of inputs. The block diagram of an encoder is shown in Fig. 6.25.

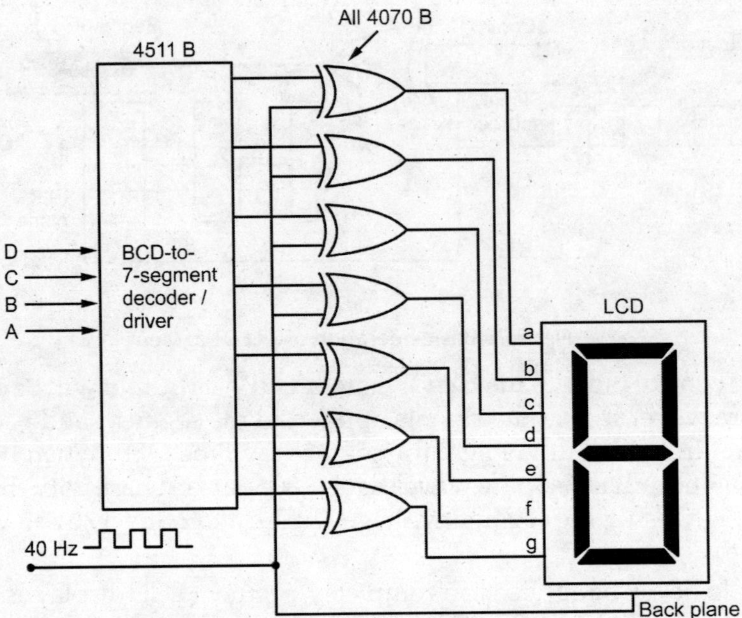

Fig. 6.24 BCD-to-7-segment decoder driving a LCD

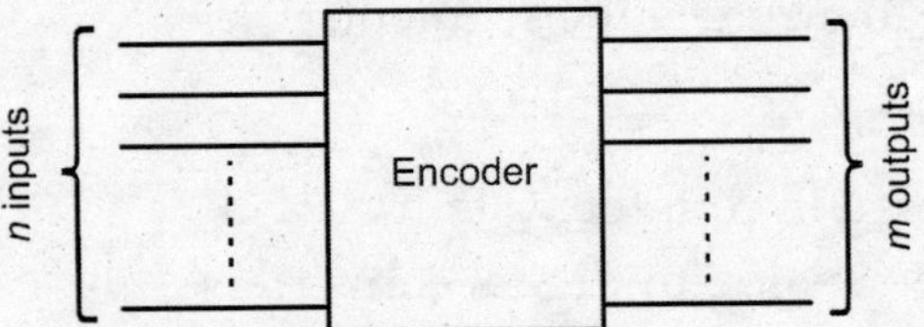

Fig. 6.25 Block diagram of an encoder

6.7.1 Octal-to-Binary Encoder

It is well-known that a binary-to-octal decoder (3-to-8 decoder) accepts a 3-bit input code and activates one of eight output lines corresponding to that code. An octal-to-binary encoder performs the opposite function; it accepts eight inputs and produces a 3-bit output code corresponding to the activated input. The truth table for the octal-to-binary encoder is shown in Table 6.14.

Table 6.14 Truth table of octal-to-binary encoder

Inputs								Outputs		
D_0	D_1	D_2	D_3	D_4	D_5	D_6	D_7	Y_2	Y_1	Y_0
1	0	0	0	0	0	0	0	0	0	0
0	1	0	0	0	0	0	0	0	0	1
0	0	1	0	0	0	0	0	0	1	0
0	0	0	1	0	0	0	0	0	1	1
0	0	0	0	1	0	0	0	1	0	0
0	0	0	0	0	1	0	0	1	0	1
0	0	0	0	0	0	1	0	1	1	0
0	0	0	0	0	0	0	1	1	1	1

The truth table 6.14 shows that Y_0 (LSB of output code) must be 1 whenever the input D_1 OR D_3 OR D_5 OR D_7 is HIGH. Thus,

$$Y_0 = D_1 + D_2 + D_3 + D_4$$

Similarly,

$$Y_1 = D_2 + D_3 + D_6 + D_7$$

$$Y_2 = D_4 + D_5 + D_6 + D_7$$

Using the above expressions, the octal-to-binary encoder can be implemented using three 4-input OR gates as shown in Fig. 6.26. The circuit is designed in such a way that, when D_0 is HIGH, the binary code 000 is generated; when D_1 is HIGH, the binary code 001 is generated, and so on.

The design is made simple by the fact that only eight out of the total of 2^8 possible input conditions are used.

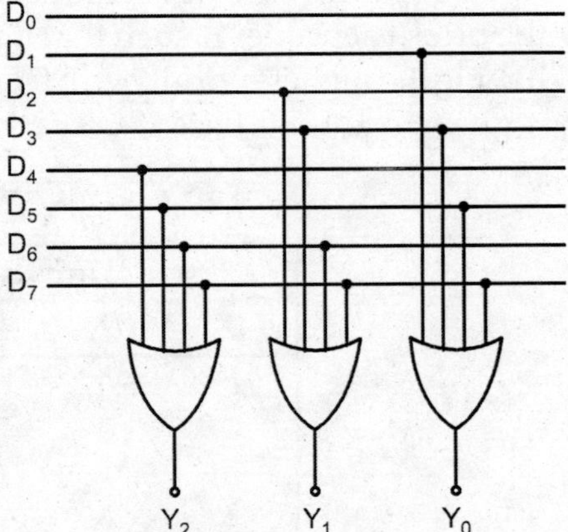

Fig. 6.26 Octal-to-binary encoder

6.7.2 Decimal-to-BCD Encoder

A decimal-to-BCD encoder is one with ten inputs corresponding to ten decimal digits (0 to 9) and four outputs (*A*, *B*, *C*, *D*) representing the BCD value of input decimal digit. The truth table for a decimal-to-BCD encoder is shown in Table. 6.15.

Table 6. 15 Truth table of decimal-to-BCD encoder

Decimal Inputs										BCD Outputs			
0	1	2	3	4	5	6	7	8	9	A	B	C	D
1	0	0	0	0	0	0	0	0	0	0	0	0	0
0	1	0	0	0	0	0	0	0	0	0	0	0	1
0	0	1	0	0	0	0	0	0	0	0	0	1	0
0	0	0	1	0	0	0	0	0	0	0	0	1	1
0	0	0	0	1	0	0	0	0	0	0	1	0	0
0	0	0	0	0	1	0	0	0	0	0	1	0	1
0	0	0	0	0	0	1	0	0	0	0	1	1	0
0	0	0	0	0	0	0	1	0	0	0	1	1	0
0	0	0	0	0	0	0	0	1	0	1	0	1	0
0	0	0	0	0	0	0	0	0	1	1	0	1	1

From the above truth table, it is clear that the output^ is HIGH whenever the input 8 OR 9 is HIGH. Therefore,

$$A = 8 + 9$$

The output *B* is HIGH whenever the input 4 OR 5 OR 6 OR 7 is HIGH. Therefore,

$$B = 4 + 5 + 6 + 7$$

Similarly, $C = 2 + 3 + 6 + 7$ and $D = 1 + 3 + 5 + 7 + 9$

Now, the above expressions for BCD outputs can be implemented as shown in Fig. 6.27 using four OR gates. Here, the inputs are ten push button type switches, each representing one of the decimal digit.

6.7.3 Priority Encoder

A priority encoder is an encoder that includes the priority function. The operation of the priority encoder is such that if two or more inputs are equal to 1 at the same time, the input having the highest priority will take precedence. The truth table of a four-input priority encoder is given in Table 6.16.

Table 6.16 Truth table of a four-input priority encoder

Inputs				Outputs		
D_0	D_1	D_2	D_3	Y_2	Y_1	V
0	0	0	0	X	X	0
1	0	0	0	0	0	1
X	1	0	0	0	1	1
X	X	1	0	1	0	1
X	X	X	1	1	1	1

The Xs are don't care conditions that designate the fact that the binary values they represent may be equal to 0 or 1. Input D_3 has the highest priority; so regardless of the values of the other inputs, when this input is 1, the output Y_1, Y_2 is 11 (i.e. 3). D_2 has the next priority level. The output is 10 if $D_2 = 1$ and $D_3 = 0$, irrespective of the values of the other two lower-priority inputs. The output for D_1 is generated only if higher priority inputs are 0, and so on down the priority level. A *valid* output indicator, V, is set to 1 only when one or more of the inputs are equal to 1. If all the inputs are 0, V is equal to 0, and the other two outputs of the circuit are not used. IC 74147 and IC 74148 are some of the priority encoder ICs.

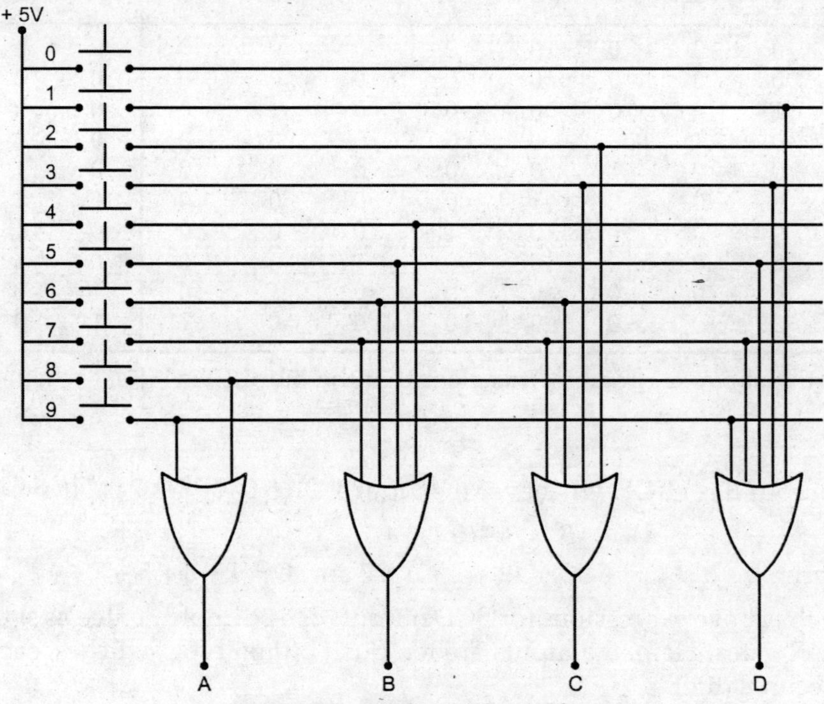

Fig. 6.27 Decimal-to-BCD encoder

6.7.4 IC 74148—8-to-3 Priority Encoder

IC 74148 encodes 8 data lines to 3-bit binary (octal). It is also a priority encoder because it gives priority to the highest order input. Here, both the data inputs and outputs are active LOW. In addition, an enable input (EI) and an enable output (EO) are provided to cascade 74148 ICs. This allows octal expansion without the need for an external circuitry. The enable input must be asserted to LOW state to enable the chip while the enable output goes to LOW state only when all its inputs are inactive (i.e. in HIGH state). Also, there is one more output, namely GS, which goes LOW whenever one of its inputs is active (i.e. in LOW state). Its operation is explained in the truth table shown in Table 6.17. Its logic symbol is shown in Fig. 6.28.

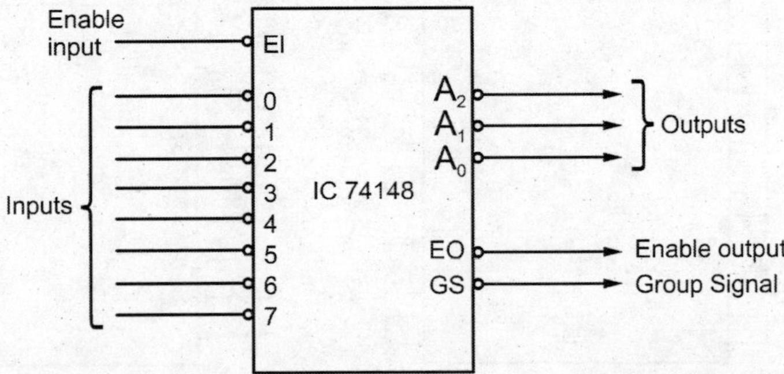

Fig. 6.28 Logic symbol of IC 74148—8-to-3 priority encoder

Table 6.17 Truth table of IC 74148—8-to-3 priority encoder

Inputs									Outputs				
EI	0	1	2	3	4	5	6	7	A_2	A_1	A_0	G_s	*EO*
1	X	X	X	X	X	X	X	X	1	1	1	1	1
0	1	1	1	1	1	1	1	1	1	1	1	1	0
0	X	X	X	X	X	X	X	0	0	0	0	0	1
0	X	X	X	X	X	X	0	1	0	0	1	0	1
0	X	X	X	X	X	0	1	1	0	1	0	0	1
0	X	X	X	X	0	1	1	1	0	1	1	0	1
0	X	X	X	0	1	1	1	1	1	0	0	0	1
0	X	X	0	1	1	1	1	1	1	0	1	0	1
0	X	0	1	1	1	1	1	1	1	1	0	0	1
0	0	1	1	1	1	1	1	1	1	1	1	0	1

6.7.5 IC 74147—Decimal-to-BCD Priority Encoder

The pinout diagram of IC 74147, a decimal-to-BCD priority encoder is shown in Fig. 6.29(*a*) and its logic symbol is shown in Fig. 6.29(*b*). Its truth table is given in Table 6.18. It has 9 active LOW inputs representing the decimal digits, 1 through 9 and produces the inverted BCD code corresponding to the highest order activated input.

When all the inputs ($X_1 - X_9$) are HIGH, all the outputs are HIGH (i.e. 1111) which is the inverse of 0000, the BCD code for 0. When X_9 is LOW, the *ABCD* output is 0110, which is the inverse of 1001, the BCD code for 9; when X_8 is LOW, the *ABCD* output is 0111, the inverse of 1000, the BCD code for 8. Hence, the outputs of 74147 will normally be HIGH when none of the inputs is activated. This corresponds to the decimal 0 input condition. Since there is no X_0 input, the encoder assumes the decimal 0 input state when all the inputs are HIGH.

The 74147 is called a priority encoder because it gives priority to the highest-order input. For example, at a particular instant, if both the inputs X_3 and X_5 are activated, then the highest priority of these two inputs (i.e. X_5) is encoded as 1010 which is the inverse of 0101.

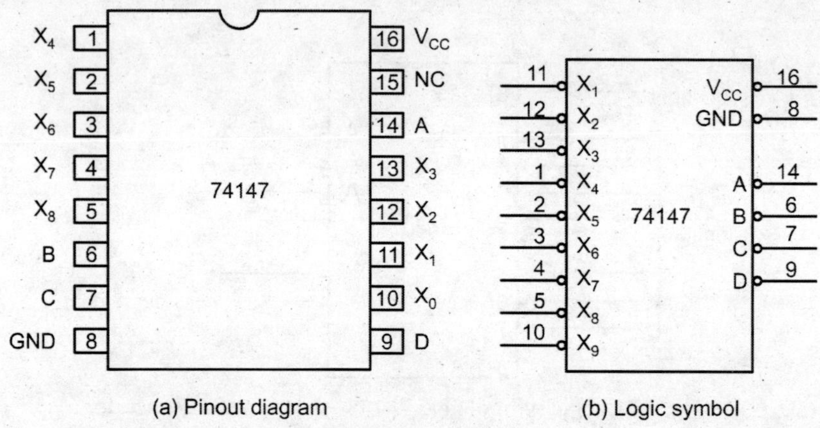

Fig. 6.29 IC 74147

Table 6.18 Truth table of IC 74147—decimal-to-BCD priority encoder

Decimal Inputs									BCD Outputs			
X_1	X_2	X_3	X_4	X_5	X_6	X_7	X_8	X_9	*A*	*B*	*C*	*D*
1	1	1	1	1	1	1	1	1	1	1	1	1
X	X	X	X	X	X	X	X	0	0	1	1	0
X	X	X	X	X	X	X	0	1	0	1	1	1
X	X	X	X	X	X	0	1	1	1	0	0	0
X	X	X	X	X	0	1	1	1	1	0	0	1
X	X	X	X	0	1	1	1	1	1	0	1	0
X	X	X	0	1	1	1	1	1	1	0	1	1
X	X	0	1	1	1	1	1	1	1	1	0	0
X	0	1	1	1	1	1	1	1	1	1	0	1
0	1	1	1	1	1	1	1	1	1	1	1	0

Example 6.7. *Determine the states of the outputs in Fig. 6.29(b) when X_5, X_7 and X_3 are LOW and all other inputs are HIGH.*

Solution: The truth table 6.18 shows that, when X_7 is LOW, the levels at X_5 and Z_3 do not matter. Thus, the outputs will be 1000 which is the inverse of 0111, the BCD code for 7.

Example 6.8. *Design a hexadecimal to binary priority encoder using 74148 encoders and a 74157 multiplexer.*

Solution: A hexadecimal to binary priority encoder can be implemented using two 74148 ICs and one 74157 multiplexer IC as shown in Fig. E6.8. In a hexadecimal number system, there are 16 symbols ranging from 0 to *F*. Therefore, two 74148 encoders are required. Hexadecimal inputs 0 through 7 are applied to IC 74148-1 and inputs 8 through *F* are applied to IC 74148-2. Whenever one of the inputs of IC 74148-2 is active (LOW), IC 74148-1 must be disabled. If all the inputs of IC 74148-2 are HIGH, then IC 74148-1 is enabled. This is achieved by connecting the EO line of IC 74148-2 to the EI line of IC 74148-1. In order to get 4-bit binary outputs, a quad 2-to-1 multiplexer is required.

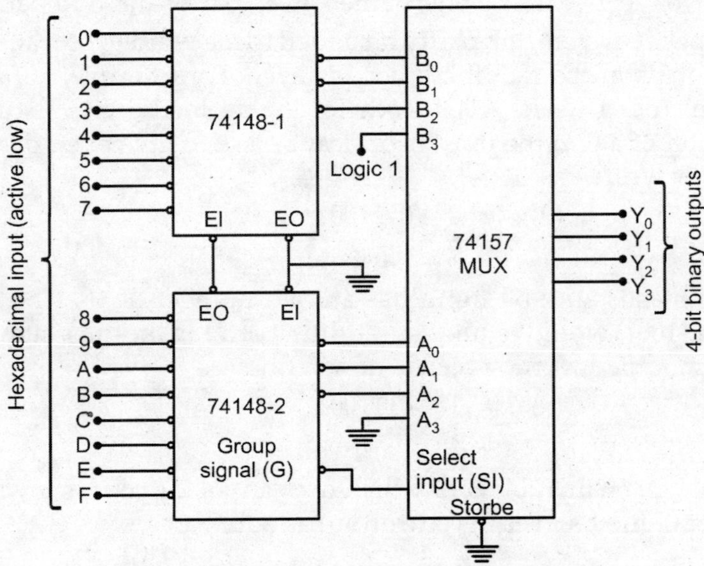

Fig. E6.8 Hexadecimal-to-binary priority encoder

IC 74157 is a quad 2-to-1 multiplexer. The inputs to the four multiplexers are A_0B_0, A_1B_1, A_2B_2 and A_3B_3 respectively. All the four multiplexers have a common select input (SI). When SI = LOW, the *A* input in each multiplexer will be connected to the respective output; when SI = HIGH, the *B* input in each multiplexer will be connected to the respective output. Here, the outputs of the priority encoder IC 74148-1 are connected to $B_2B_1B_0$ inputs of the multiplexers, while the outputs of IC 74148-2 are connected to $A_2A_1A_0$ inputs of the multiplexers. The B_3 line is connected to the HIGH state while the A_3 line is connected to the LOW state.

The Group Signal (GS) output of IC 741-48-2 goes LOW whenever one of its inputs is active. Therefore, GS of IC 74148-2 is connected to the select input (SI) of 74157, which selects *A* inputs if it is LOW; otherwise *B* inputs are selected.

Now, if input '*A*' is in active LOW state and higher order inputs *B* through *F* are in HIGH state, then IC 74148-2 is selected and *A* is encoded as 101 at its outputs which are connected with $A_2A_1A_0$ inputs of MUX IC 74157. Moreover, the EO of IC 74148-2 is in HIGH state and, therefore, IC 74148-1 is disabled. Also, the GS output of IC 74148-2 is in a LOW state which connects $A_3A_2A_1A_0$ of MUX inputs to its outputs $Y_3Y_2Y_1Y_0$ (i.e. $Y_3Y_2Y_1Y_0$ = 0101, which is the inverse of 1010 = *A*). Thus, the outputs are binary outputs and they are active LOW.

6.8 PARITY GENERATORS/CHECKERS

When digital data is transmitted from one location to another, it is necessary to know at the receiving end whether the received data is free of error. A simple form of error detection is achieved by adding an extra bit to the transmitted word. This additional bit is known as the parity bit and it decides whether the data transmitted is error free or not.

Do You Know?	The circuit that generates the parity bit in the transmitters is called a parity generator and the circuit that checks the parity in the receiver is called a parity checker.

There are two types of parity bits, namely even parity and odd parity.

In an even parity system, the parity bit added to the word to be transmitted is chosen so that the number of 1's in the modified word is even. For example, 101011 has even parity because it contains four 1's. The ASCII code for the decimal digit 9 is 0111001. This would require the addition of a 0 in the most significant place to give even parity in the modified word, which is now written as

$$(9)_{10} = \underset{\uparrow \text{Parity bit}}{0} \quad 0111001$$

Odd parity means an n-bit input has an odd number of 1's. The ASCII code for the decimal digit 9 (0111001) would require the addition of a 1 in the most significant place to give odd parity in the modified word, which is now written as

$$(9)_{10} = \underset{\uparrow \text{Parity bit}}{1} \quad 0111001$$

The example given in Table 6.19 which uses an 8 4 2 1 code, shows a parity bit which makes the number of the 1's in each code group as odd.

Table 6.19 Odd-parity

Decimal	BCD	Odd-parity bit
0	0000	1
1	0001	0
2	0010	0
3	0011	1
4	0100	0
5	0101	1
6	0110	1
7	0111	0
8	1000	0
9	1001	1

6.8.1 Design of Parity Checkers

The circuit for generating parity bits and checking the parity of a given word can be designed using gates. The technique of parity checking is the most popular method of detecting errors in stored code groups, especially for storage devices such as magnetic tapes, paper tapes and even core and drum systems.

Exclusive-OR gates are ideal for checking the parity of a binary number because they produce an output when the input has an odd number of 1's. Therefore, an even-parity input to an exclusive-OR gate produces a low output, while an odd-parity input produces a high output.

A two-input EX-OR gate produces 1 at its output only when the number of 1's in the inputs is odd. For even number of 1's in the inputs, a 0 output is generated. Similarly, to check the parity of given *n*-bit number, $(n-1)$ two-input EX-OR gates can be cascaded. For example, to detect the parity of a 4-bit number (*WXYZ*), three 2-input EX-OR gates can be cascaded as shown in Fig. 6.30(*a*). Here, the output is 1 when the number of 1's in the inputs *W*, *X*, *Y* and *Z* is odd and 0 when the number of 1's in the inputs is even.

A 16-input exclusive-OR gate is shown in Fig. 6.30(*b*). This exclusive-OR gate produces an output 1 because the input has odd parity (an odd number of 1's). If the 16-bit input changes to another value, the output becomes 0 for even-parity numbers and 1 for odd-parity numbers.

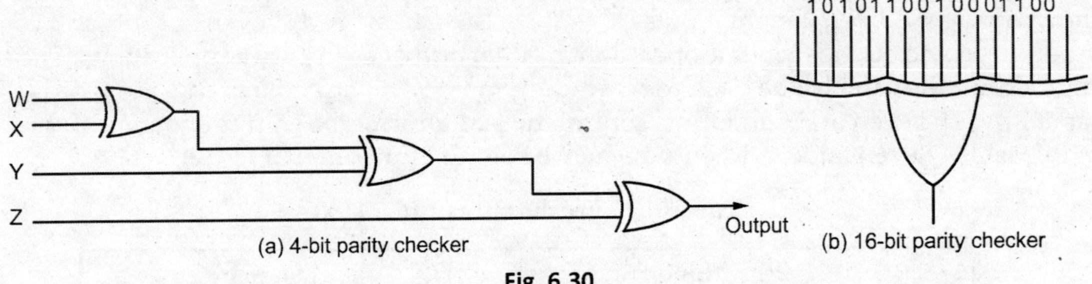

Fig. 6.30

6.9 PARITY GENERATION

A binary number may represent an instruction that tells the computer to add, subtract, and so on. The binary number may also represent data to be processed, like a number, letter, etc., in a computer. In either case, an extra bit is added to the original binary number to produce a binary number with even or odd parity. Such an extra parity bit can be easily generated using an EX-NOR gate.

6.9.1 Odd-parity Generator

Consider the odd-parity generator using an EX-OR gate and a NOT gate, shown in Fig. 6.31. Let the 8-bit number $X_7 X_6 X_5 X_4 X_3 X_2 X_1 X_0$ be equal to 0100 0001. This number has even parity. It means that, when it is applied to an exclusive-OR gate, it will produce an output of 0. Because of the inverter, $X_8 = 1$ and the final 9-bit output is 1 0100 0001. Now, it has odd parity. Thus, an EX-NOR can be used to generate parity bits.

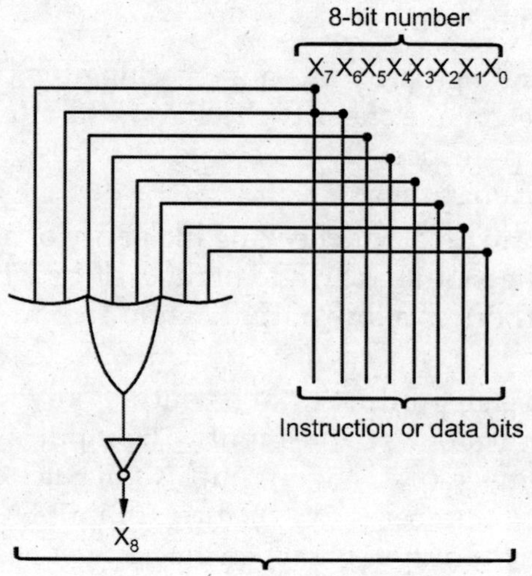

Fig. 6.31 Odd-parity generation

IC 74180—9-bit odd/even-parity generator/checker: IC 74180 is a TTL 9-bit odd/even-parity generator/checker circuit, available as a MSI chip. The pinout diagram of IC 74180 is shown in Fig. 6.32. It has eight inputs ($X_0 - X_7$). Also, it has odd, even outputs and control inputs even and odd to facilitate operation in either odd or even parity applications. The even and odd inputs can also be used for cascading more 74180 ICs to increase the word length capability. The even and odd inputs control the operation of the chip as shown in Table 6.20. In this table, even and odd outputs may be either LOW or HIGH.

Table 6.20 Truth table of IC 74180

Inputs			Outputs	
Number of 1s in X_7 to X_0	**Even**	**Odd**	**Even**	**Odd**
Even	1	0	1	0
Odd	1	0	0	1
Even	0	1	0	1
Odd	0	1	1	0
X	1	1	0	0
X	0	0	1	1

Fig. 6.33 gives the block diagram of IC 74180 in which there are eight parity inputs, X_0 through X_7, and two cascading inputs. There are two outputs even and odd.

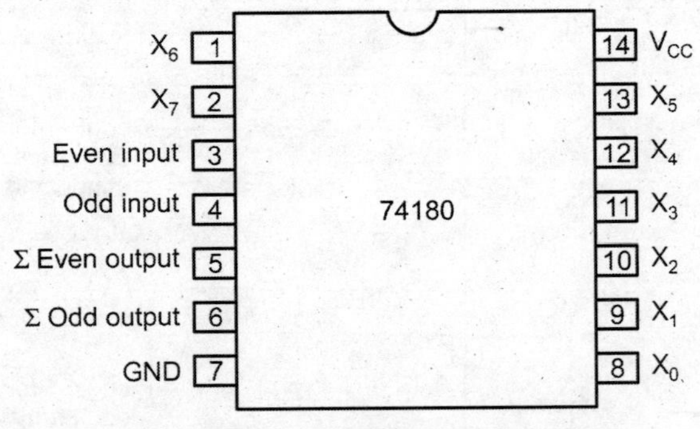

Fig. 6.32 Pinout diagram of IC 74180

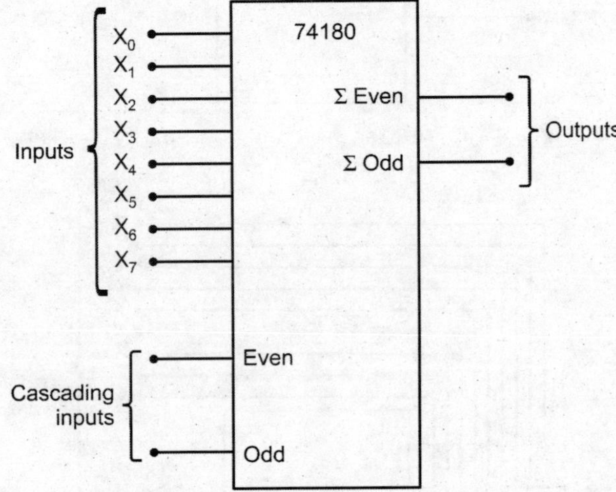

Fig. 6.33 Block diagram of IC 74180

IC 74180 as 9-bit odd/even-parity checker: IC 74180 can be used as a 9-bit odd/even parity checker by connecting the odd input to the HIGH state and the even input to the LOW state as shown in Fig. 6.34. If the parity of 9 input bits i.e. eight data input bits and one odd input bit is odd, then a logic 1 is generated in odd output and a 0 in even output. If there are even number of 1s in the 9 bits, then a 1 is generated in even output and a 0 in odd output. Thus, IC 74180 can be used as 9-bit odd/even-parity checker.

IC 74180 as 9-bit odd/even-parity generator: IC 74180 can also be used to generate odd/even-parity bit for a given 9-bit number as shown in Fig. 6.35. If an 8-bit input number is applied at (X_7 -X_0) and the 9th bit X_8 and its complement X_8 are applied at even and odd inputs respectively, then for the given 9-bit number ($X_7 - X_0$), IC 74180 will generate an even or odd-parity bit in even or odd output. As shown in Table 6.20, if the total number of 1s in the 9-input bits is odd (i.e. $X_7 - X_0$ is even and even input = 1 OR $X_7 - X_0$ is odd and even input = 0) then a 1 is generated as even-parity bit at even output. This results in even parity in 10 bits (i.e. $X_7 - X_0$ and even output). Similarly, if the total number of 1's in the 9-input bits is even (i.e. $X_7 - X_0$ is odd and even input = 1 OR $X_7 - X_0$ is even and even input = 0), then a 0 is generated as even parity bit at even output. This also results in even parity in 10 bits (i.e. $X_8 - X_0$ and even output).

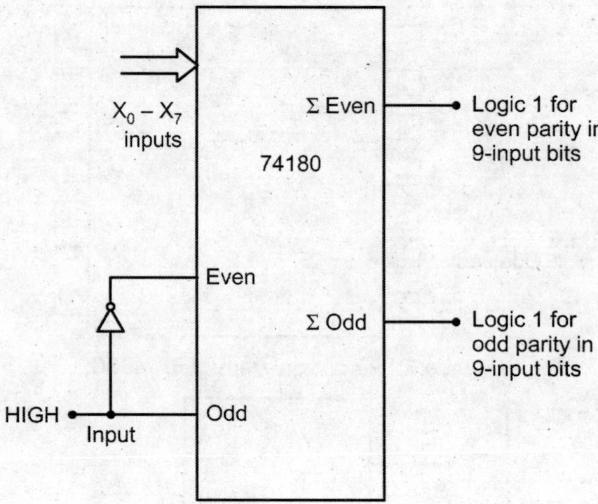

Fig. 6.34 IC 74180 as 9-bit odd/even-parity checker

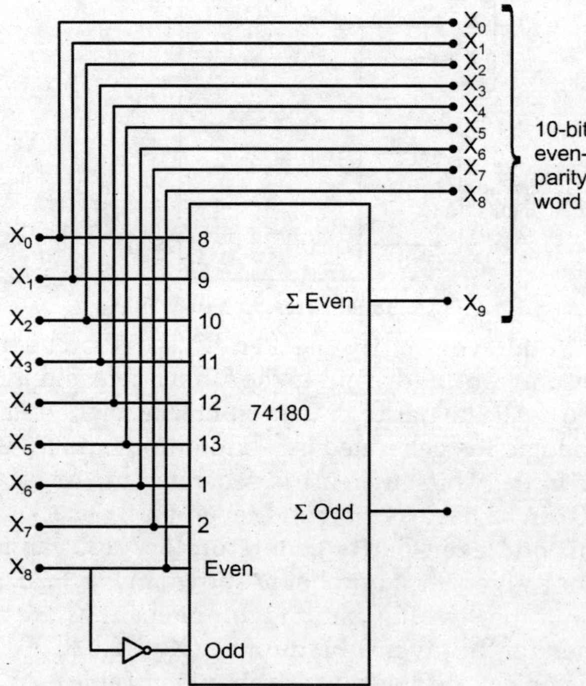

Fig. 6.35 IC 74180 as 9-bit odd/even-parity generator

Similarly, if the output at odd is taken, then it can be used as an odd-parity generator. This case results in odd parity in 10 bits (i.e. $X_8 - X_0$ and odd output). Its function as a parity generator is given in Table 6.21.

Table 6.21 Truth table of 74180 as a 9-bit parity generator

Parity of inputs X_0 through X_7	Cascading inputs		Parity of X_0 through X_7 and Even	Parity of X_0 through X_7 and Odd
	Even	Odd		
Odd	1	0	Odd	Even
Even	1	0	Odd	Even
Odd	0	1	Even	Odd
Even	0	1	Even	Odd

Cascading of 74180 ICs: 74180 ICs can be cascaded to increase the word length capability of odd/even parity checker. Two 74180 ICs can be cascaded to check odd/even parity in the given 16-bit number ($X_{15} - X_0$) by connecting even and odd outputs of IC 74180-1 with even and odd inputs of IC 74180-2 respectively as shown in Fig. 6.36. Also, logics 1 and 0 should be applied at even and odd inputs of IC 74180-1 respectively. Now, if the total number of 1's in the 16-bit number is even, then a 1 is generated at even output of IC 74180-2. If the number of 1's is odd in the given 16-bit number, then a 1 is generated at odd output.

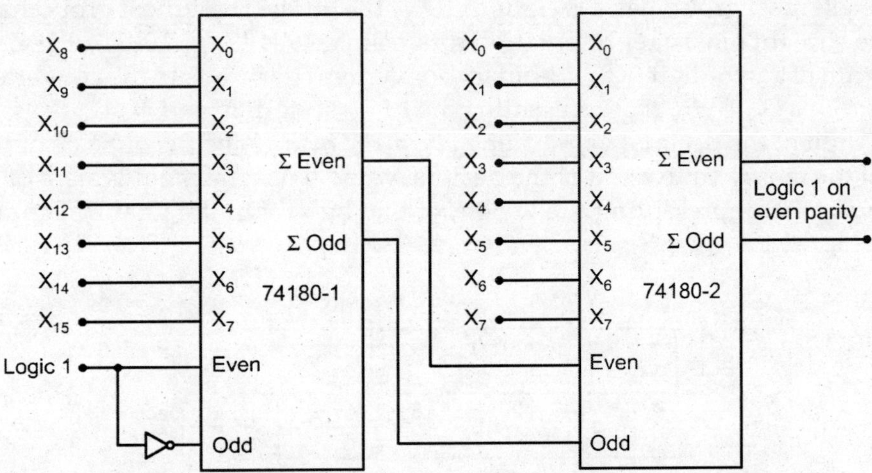

Fig. 6.36 Cascading of 74180 ICs as 16-bit odd/even-parity checker

It is important to note that the unused parity inputs must be tied to logic 0 level and the cascading inputs must not be equal.

6.10 CODE CONVERTERS

A code converter is a logic circuit that changes data presented in one type of binary code to another type of binary code. The BCD to 7-segment decoder, discussed in the section 6.5.9, is also a code converter. The following are some of most commonly used code converters:

(*i*) BCD-to-binary

(*ii*) Binary-to-BCD

(*iii*) Binary-to-Gray code
(*iv*) Gray-code-to-binary
(*v*) ASCII-to-EBCDIC
(*vi*) EBCDIC-to-ASCII.

6.10.1 BCD-to-Binary Converters

The steps involved in the BCD-to-binary conversion process are as follows:

1. The value of each bit in the BCD number is represented by a binary equivalent or weight.
2. All the binary weights of the bits that are 1's in the BCD are added.
3. The result of this addition is the binary equivalent of the BCD number. Two-digit decimal values ranging from 00 to 99 can be represented in BCD by two 4-bit code groups. For example, 19_{10} is represented as

The left-most four-bit group represents 10 and right-most four-bit group represents 9. That is, the left most group has a weight of $10^1 = 10$ and the right most group has a weight of $10^0 = 1$. The straight binary representation for decimal 19 is $19_{10} = 10011_2$. Fig. 6.37 shows the block diagram of a two-digit BCD-to-binary converter. The inputs to the converter are the two 4-bit code groups D_0, C_0, B_0, A_0, representing the 10^0 or units digit and D_1, C_1, B_1, A_1 representing 10^1 or tens digit of the decimal value. Let $b_6, b_5, b_4, b_3, b_2, b_1, b_0$ be the outputs of the converter, the 7 bits of the binary equivalent of the decimal value. The binary equivalent of each BCD bit is a binary number representing the weight of that bit within the total BCD number. This is given in Table 6.22.

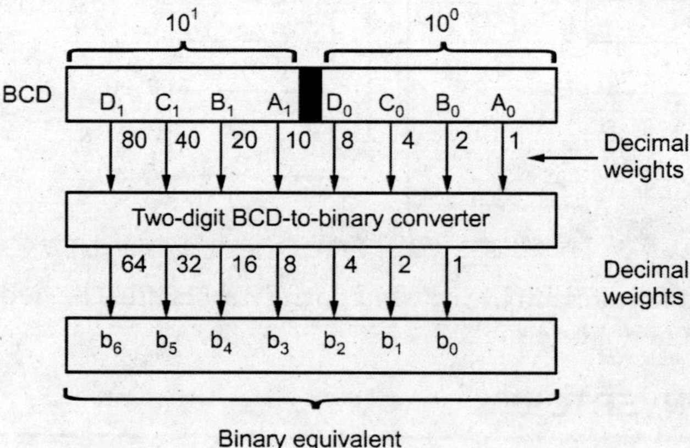

Fig. 6.37 Block diagram of a 2-digit BCD-to-binary converter

Table 6.22 Binary representation of BCD bit weights

BCD bit	BCD weight or decimal weight	Binary equivalent or weight						
		b_6 (=64)	b_5 (=32)	b_4 (=16)	b_3 (=8)	b_2 (=4)	b_1 (=2)	b_0 (=1)
A_0	1	0	0	0	0	0	0	1
B_0	2	0	0	0	0	0	1	0
C_0	4	0	0	0	0	1	0	0
D_0	8	0	0	0	1	0	0	0
A_1	10	0	0	0	1	0	1	0
B_1	20	0	0	1	0	1	0	0
C_1	40	0	1	0	1	0	0	0
D_1	80	1	0	1	0	0	0	0

The bits in a *BCD* representation have decimal weights which are 8, 4, 2, 1 within each code group but which differ by a factor of 10 from one code group (decimal digit) to the next. Using the weights, the BCD to binary conversion can be done easily by computing the binary sum of the binary equivalents or weights of all those bits in the BCD representation that are 1's. Example 6.9 illustrates this.

Example 6.9. *Convert the BCD number 00101001 (decimal 29) and 01100111 (decimal 67) to binary.*

Solution:

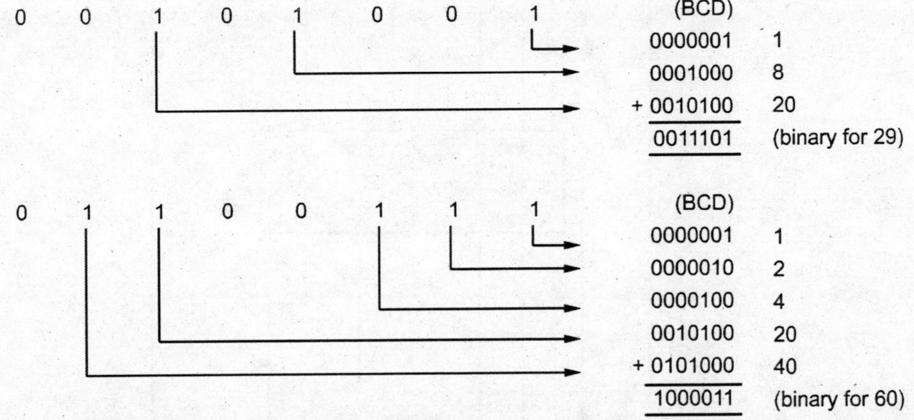

From truth table 6.22, it is clear that A_0 contributes to b_0; B_0 and A_1 contribute to b_1; C_0 and B_1 contribute to D_2; D_0, A_1 and C_1 contribute to by B_1 and D_1 contribute to b_4; C_1 contributes to b_5 and D_1 contributes to b_6. This is represented as shown below:

D_1	C_1	B_1	D_0	C_0	B_0	A_0
		D_1	A_1	B_1	A_1	
			C_1			
b_6	b_5	b_4	b_3	b_2	b_1	b_0

From the above representation, it is clear that A_0 directly contributes to b_0. To generate the remaining b_1 to b_6, addition has to be performed in two stages using two 4-bit parallel binary adders (IC 7483s) as shown in Fig. 6.38(a) and (b).

		B_1	D_0	C_0	B_0	A_0
		0	A_1	B_1	A_1	↓
0	C_{out}	S_3	S_2	S_1	S_0	↓
D_1	C_1	D_1	C_1	↓	↓	↓
Σ_3	Σ_2	Σ_1	Σ_0	↓	↓	↓
↓	↓	↓	↓			
b_6	b_5	b_4	b_3	b_2	b_1	b_0

Fig. 6.38(*a*)

The $B_1D_0B_0$ and 0 A_1 B_1 A_1 of BCD inputs are added in IC 7483-1 and their outputs S_1 and S_0 directly give b_2 and b_1 respectively. Now, the remaining outputs, 0 C_{out} S_3S_2 of adder 1 and D_1 C_1 D_1 C_1 of BCD inputs, are added in IC 7483-2 and their outputs $_3$, $_2$, $_1$ and $_0$ correspond to b_6, b_5, b_4 and b_3 of binary outputs respectively.

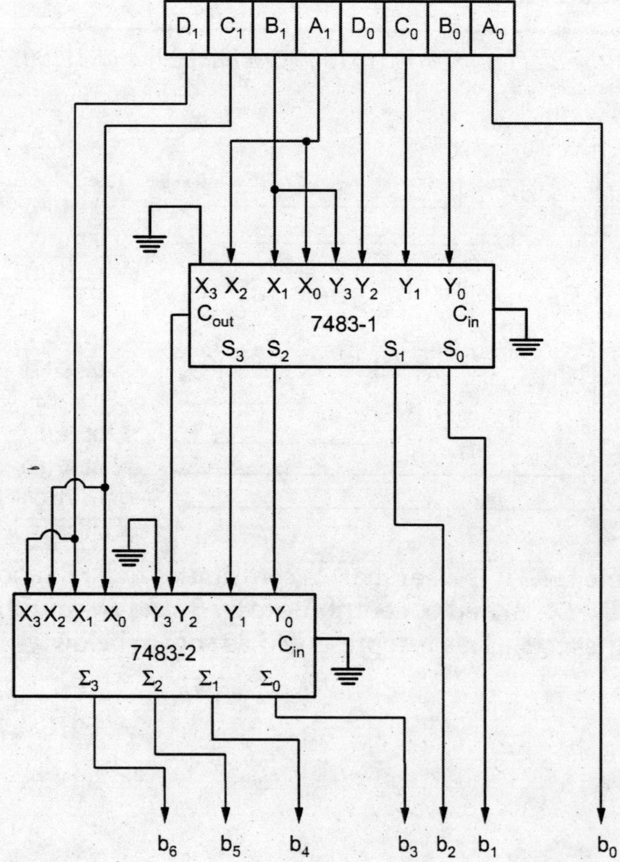

Fig. 6.38(*b*) BCD-to-binary converter using IC 7483—4-bit parallel adders

6.10.2 Binary-to-Gray Code Converters

The block diagram of a 4-bit binary-to-gray code converter is shown in Fig. 6.39. It has four inputs ($B_3B_2B_1B_0$) representing 4-bit binary numbers and four outputs ($G_3G_2G_1G_0$) representing 4-bit gray code. The truth table for the binary-to-gray code converter is shown in Table 6.23.

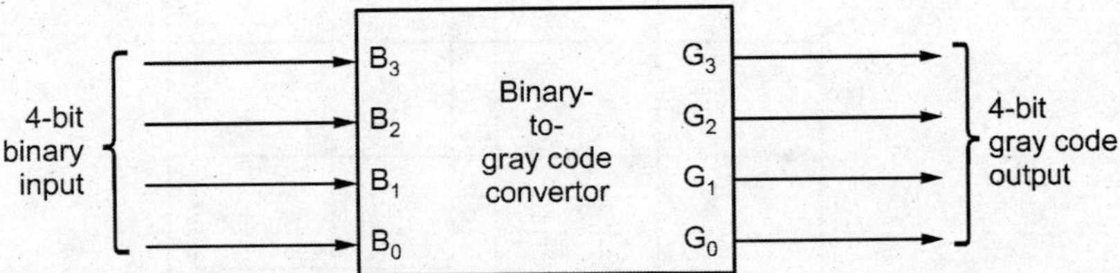

Fig. 6.39 Block diagram of 4-bit binary-to-gray code converter

Table 6.23 Truth table of binary-to-gray code converters

Binary inputs				Gray code outputs			
B_3	B_2	B_1	B_0	G_3	G_2	G_1	G_0
0	0	0	0	0	0	0	0
0	0	0	1	0	0	0	1
0	0	1	0	0	0	1	1
0	0	1	1	0	0	1	0
0	1	0	0	0	1	1	0
0	1	0	1	0	1	1	1
0	1	1	0	0	1	0	1
0	1	1	1	0	1	0	0
1	0	0	0	1	1	0	0
1	0	0	1	1	1	0	1
1	0	1	0	1	1	1	1
1	0	1	1	1	1	1	0
1	1	0	0	1	0	1	0
1	1	0	1	1	0	1	1
1	1	1	0	1	0	0	1
1	1	1	1	1	0	0	0

From the truth table shown in Table 6.23, the logic expressions for the gray code outputs can be written as:

$$G_3 = {}_m(8, 9, 10, 11, 12, 13, 14, 15)$$
$$G_2 = {}_m(4, 5, 6, 7, 8, 9, 10, 11)$$
$$G_1 = {}_m(2, 3, 4, 5, 10, 11, 12, 13)$$
$$G_0 = {}_m(1, 2, 5, 6, 9, 10, 13, 14)$$

The above expressions can be simplified using *K*-map method as shown in Fig. 6.40.

B_1B_0 \ B_3B_2	00	01	11	10
00	0	0	1	1
01	0	0	1	1
11	0	0	1	1
10	0	0	1	1

(a) K-map for G_3

From the above *K*-map, $G_3 = B_3$

B_1B_0 \ B_3B_2	00	01	11	10
00	0	1	0	1
01	0	1	0	1
11	0	1	0	1
10	0	1	0	1

(b) K-map for G_2

From the above *K*-map, $G_2 = \overline{B}_3B_2 + B_3\overline{B}_2 = B_3 \oplus B_2$

B_1B_0 \ B_3B_2	00	01	11	10
00	0	1	1	0
01	0	1	1	0
11	1	0	0	1
10	1	0	0	1

(c) K-map for G_1

From the above *K*-map, $G_1 = \overline{B}_2B_1 + B_2\overline{B}_1 = B_2 \oplus B_1$

From the above *K*-map, $G_0 = \overline{B}_1 B_0 + B_1 \overline{B}_0 = B_1 \oplus B_0$

B_1B_0 \ B_3B_2	00	01	11	10
00	0	0	0	0
01	1	1	1	1
11	0	0	0	0
10	1	1	1	1

(d) K-map for G_0

Fig. 6.40 *K*-map simplification for binary-to-gray code converter

Now, the above expressions can be implemented using EX-OR gates as shown in Fig. 6.41.

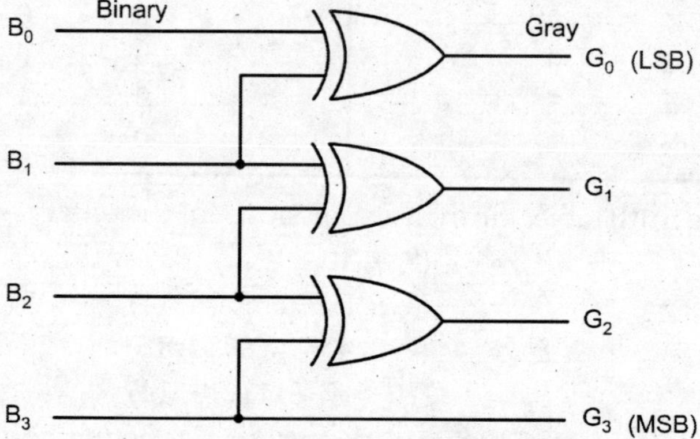

Fig. 6.41 Logic diagram of 4-bit binary-to-gray code converter

The block diagram a 4-bit gray-code-to binary converter is shown in Fig. 6.42. It has four inputs ($G_3G_2G_1G_0$) representing 4-bit gray code and four outputs ($B_3B_2B_1B_0$) representing 4-bit binary numbers. The truth table for the gray code-to-binary converter is shown in Table 6.24.

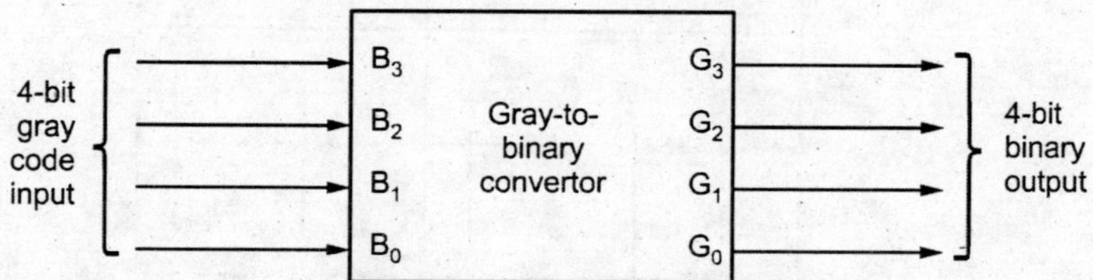

Fig. 6.42 Block diagram of 4-bit gray code-to-binary converter

Table 6.24 Truth table of gray code-to-binary converter

Gray code outputs				Binary outputs			
G_3	G_2	G_1	G_0	B_3	B_2	B_1	B_0
0	0	0	0	0	0	0	0
0	0	0	1	0	0	0	1
0	0	1	0	0	0	1	1
0	0	1	1	0	0	1	0
0	1	0	0	0	1	1	1
0	1	0	1	0	1	1	0
0	1	1	0	0	1	0	0
0	1	1	1	0	1	0	1
1	0	0	0	1	1	1	1
1	0	0	1	1	1	1	0
1	0	1	0	1	1	0	0
1	0	1	1	1	1	0	1
1	1	0	0	1	0	0	0
1	1	0	1	1	0	0	1
1	1	1	0	1	0	1	1
1	1	1	1	1	0	1	0

From the above truth table, the logic expressions for the binary outputs can be written as:

$$B_3 = {}_m(8, 9, 10, 11, 12, 13, 14, 15)$$
$$B_2 = {}_m(4, 5, 6, 7, 8, 9, 10, 11)$$
$$B_1 = {}_m(2, 3, 4, 5, 8, 9, 14, 15)$$
$$B_0 = (1, 2, 4, 7, 8, 11, 13, 14)$$

The above expressions can be simplified using *K*-map method as shown in Fig. 6.43.

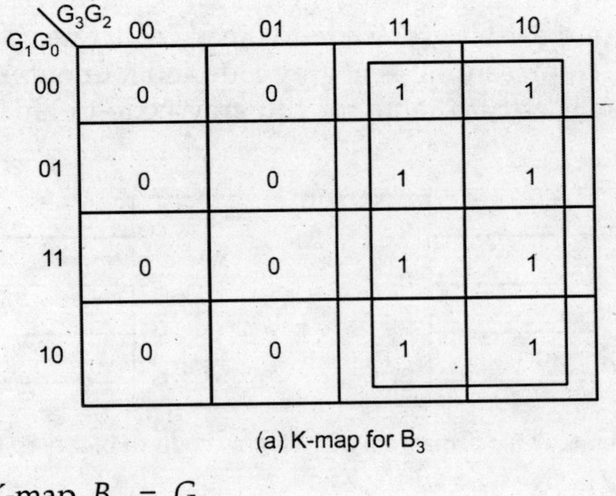

G_1G_0 \ G_3G_2	00	01	11	10
00	0	0	1	1
01	0	0	1	1
11	0	0	1	1
10	0	0	1	1

(a) K-map for B_3

From the above *K*-map, $B_3 = G_3$

G_1G_0 \ G_3G_2	00	01	11	10
00	0	1	0	1
01	0	1	0	1
11	0	1	0	1
10	0	1	0	1

(b) K-map for B_2

From the above *K*-map, $B_2 = \overline{G}_3G_2 + G_3\overline{G}_2 = G_3 \oplus G_2 = G_3 \oplus G_2$

G_1G_0 \ G_3G_2	00	01	11	10
00	0	1	0	1
01	0	1	0	1
11	1	0	1	0
10	1	0	1	0

(c) K-map for B_1

From the above *K*-map,

$$\begin{aligned} B_2 &= \overline{G}_3\overline{G}_2G_1 + \overline{G}_3G_2\overline{G}_1 + G_3G_2G_1 + G_3\overline{G}_2\overline{G}_1 \\ &= \overline{G}_3(\overline{G}_2G_1 + G_2\overline{G}_1) + G_3(G_2G_1 + \overline{G}_2\overline{G}_1) \\ &= \overline{G}_3(G_2 \oplus G_1) + \overline{(G_2 \oplus G_1)} \\ &= G_3 \oplus G_2 \oplus G_1 \\ B_1 &= B_2 \oplus G_1 \end{aligned}$$

G_1G_0 \ G_3G_2	00	01	11	10
00	0	1	0	1
01	1	0	1	0
11	0	1	0	1
10	1	0	1	0

(d) K-map for B_0

From the above *K*-map,

$$
\begin{aligned}
B_0 &= \bar{G}_3\bar{G}_2\bar{G}_1G_0 + \bar{G}_3\bar{G}_2G_1\bar{G}_0 + G_3G_2\bar{G}_1G_0 + G_3G_2G_1\bar{G}_0 \\
&\quad + \bar{G}_3G_2\bar{G}_1\bar{G}_0 + G_3\bar{G}_2\bar{G}_1\bar{G}_0 + \bar{G}_3G_2G_1G_0 + G_3\bar{G}_2G_1G_0 \\
&= \bar{G}_3\bar{G}_2(\bar{G}_1\bar{G}_0 + G_1\bar{G}_0) + G_3G_2(\bar{G}_1G_0 + G_1\bar{G}_0) + \bar{G}_1\bar{G}_0(\bar{G}_3G_2 + G_3\bar{G}_2) \\
&= + G_1G_0(\bar{G}_3G_2 + G_3\bar{G}_2) \\
&= \bar{G}_3\bar{G}_2(\bar{G}_0 \oplus G_1) + G_3G_2(G_0 \oplus G_1) + \bar{G}_1\bar{G}_0(G_2 \oplus G_3) + G_1G_0(G_2 \oplus G_3) \\
&= (G_0 \oplus G_1)(\bar{G}_3\bar{G}_2 + G_3G_2) + (G_2 \oplus G_3)(\bar{G}_1\bar{G}_0 + G_1G_0) \\
&= G_0 \oplus G_1 \oplus G_2 \oplus G_3 \\
B_0 &= G_0 \oplus B_1
\end{aligned}
$$

Fig. 6.43 *K*-map simplification for 4-bit gray code-to-binary converter

Now, the above expressions can be implemented using EX-OR gates as shown in Fig. 6.44.

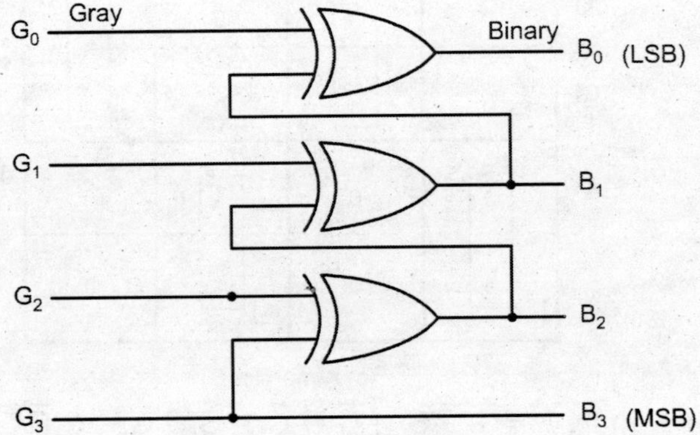

Fig. 6.44 Logic diagram of 4-bit gray code-to-binary converter

6.11 MAGNITUDE COMPARATOR

A Magnitude Comparator is a combinational circuit that compares the magnitude of two numbers (*A* and *B*) and generates one of the following outputs: $A = B$, $A < B$ and $A > B$. The block diagram of a single-bit magnitude comparator is shown in Fig. 6.45.

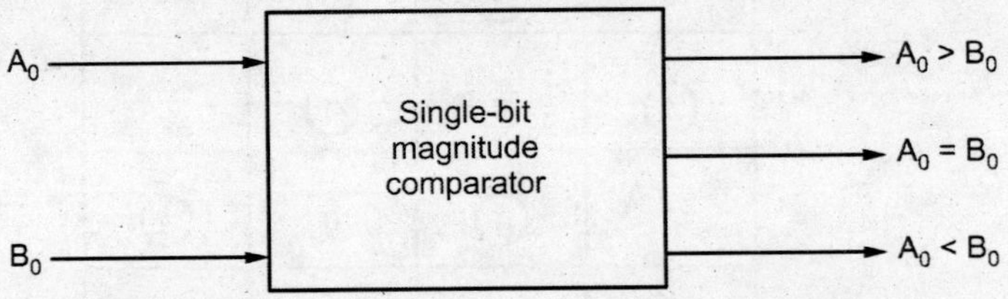

Fig. 6.45 Block diagram of single-bit magnitude comparator

To implement the magnitude comparator, the EX-NOR gates and AND gates are used. The property of the EX-NOR gate can be used to find whether the two binary digits are equal or not, and the AND gates are used to find whether a binary digit is less than or greater than another bit. Fig. 6.46(*a*) shows an EX-NOR gate with two inputs A_0 and B_0. If $A_0 = B_0$, then the output of EX-NOR gate will be 1. If A_0 B_0, the output will be 0.

Fig. 6.46(*b*) and (*c*) show two AND gates, one with A_0 and $\overline{B}_0$ as inputs and another with $\overline{A}_0$ and B_0 as their inputs. The AND gate output shown in Fig. 6.46(*b*) is 1 if $A_0 > B_0$ (i.e. $A_0 = 1$ and $B_0 = 0$) and 0 if $A_0 < B_0$ (i.e. $A_0 = 0$ and $B_0 = 1$). Similarly, the AND gate output shown in Fig. 6.46(*c*) is 1 if $A_0 < B_0$ (i.e. $A_0 = 0$ and $B_0 = 1$) and 0 if $A_0 > B_0$ (i.e. $A_0 = 1$ and $B_0 = 0$).

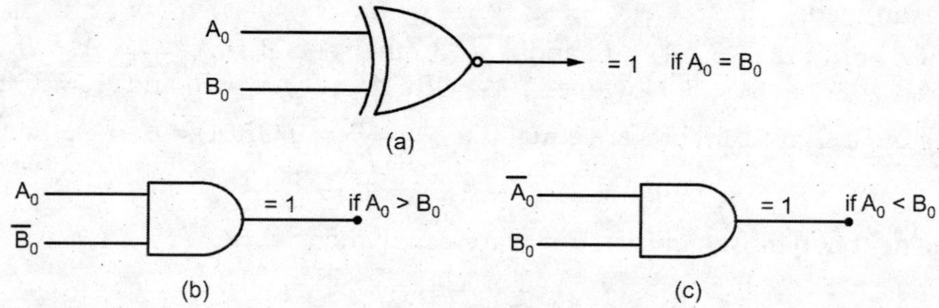

Fig. 6.46

If the above EX-NOR gate and two AND gates are combined as shown in Fig. 6.47, the circuit will function as a single bit magnitude comparator.

The same principle can be extended to an *n*-bit magnitude comparator. The design of a 4-bit magnitude comparator is discussed in the next section.

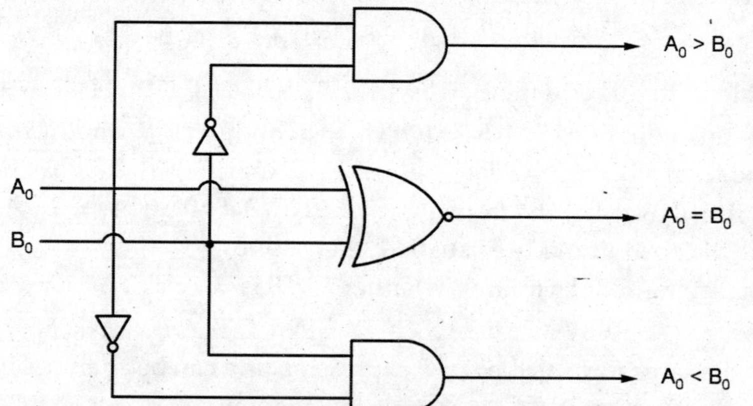

Fig. 6.47 Single-bit magnitude comparator

6.11.1 4-bit Magnitude Comparator

A 4-bit magnitude comparator compares two 4-bit numbers *A* and *B* and gives one of the following outputs: $A = B$, $A < B$ and $A > B$. Let $A = A_3A_2A_1A_0$ and $B = B_3B_2B_1B_0$ be the two 4-bit numbers to be compared. The steps involved in comparing two such numbers can be used as the basis for a hardware implementation. The steps involved in comparing two 4-bit numbers are:

(*a*) Examine the two most significant bits (A_3 and B_3). If $A_3 > B_3$, then $A > B$; if $A_3 < B_3$, then $A < B$. If $A_3 = B_3$, no decision can be made regarding the relative magnitudes of the two numbers and the next pair of bits (A_2 and B_2) must be examined.

(*b*) If $A_3 = B_3$ and $A_2 > B_2$, then A > B; if $A_3 = B_3$ and $A_2 < B_2$, then $A < B$. However, if $A_1 = B_3$ and $A_2 = B_2$, no conclusion can be drawn regarding the relative magnitudes of the two numbers and the next pair of bits (A_1 and B_1) must be examined.

(*c*) If $A_3 = B_3$, $A_2 = B_2$ and $A_1 > B_1$; then $A > B$; if $A_3 = B_3$, $A_2 = B_2$ and $A_1 < B_1$, then $A < B$. However, if $A_3 = B_2$, $A_2 = B_2$ and $A_1 = B_1$, no conclusion can yet be drawn regarding the relative magnitudes of the two numbers and the LSBs (A_0 and B_0) must be examined.

(*d*) If $A_3 = B_3$, $A_2 = B_2$, $A_1 = B_1$ and $A_0 > B_0$, then $A > B$; if $A_3 = B_3$, $A_2 = B_2$, $A_1 = B_1$ and $A_0 < B_0$, then $A < B$. However, if $A_3 = B_3$, $A_2 = B_2$, $A_1 = B_1$ and $A_0 = B_0$, then $A = B$.

If the most significant bits are equal (i.e., $A_3 = B_3 = 0$ OR $A_3 = B_3 = 1$), then

$$E_3 = \bar{A}_3\bar{B}_3 + A_3B_3 = \overline{A_3 \oplus B_3}$$

If the next two most significant bits are equal, then

$$E_2 = \bar{A}_2\bar{B}_2 + A_2B_2 = \overline{A_2 \oplus B_2}$$

If the next two most significant bits are equal, then $E_1 = \bar{A}_1\bar{B}_1 + A_1B_1 = \overline{A_1 \oplus B_1}$

If the two least significant bits are equal, then $E_0 = \bar{A}_0\bar{B}_0 + A_0B_0 = \overline{A_0 \oplus B_0}$.

Hence, if $A = B$, then

$$E = E_3E_2E_1E_0 = (\overline{A_3 \oplus B_3}).(\overline{A_2 \oplus B_2}).(\overline{A_1 \oplus B_1}).(\overline{A_0 \oplus B_0}) = 1$$

The expression for determining whether $A > B$ is

$$A < B = A_3\bar{B}_3 + E_3A_2\bar{B}_2 + E_3E_2A_1\bar{B}_1 + E_3E_2E_1A_0\bar{B}_0$$

The first term in this equation $A_3\bar{B}_3 = 1$ if $A_3 > B_3$ and, if that is the case, then $A > B$. The second term in this equation $E_3A_2\bar{B}_2 = 1$ if $A_3 = B_3$ and, $A_2 > B_2$ and, if that is the case, then $A > B$. The third term $E_3E_2A_1\bar{B} = 1$ if $A_1 = B_3$, $A_2 = B_2$ and $A_1 > B_1$; if these three conditions are satisfied, then $A > B$. Finally, the fourth term $E_3E_2E_1A_0\bar{B} = 1$ if $A_3 = B_3$, $A_2 = B_2$, $A_1 = B_1$ and $A_0 = B_0$; if these four conditions are satisfied, then $A > B$.

The expression for determining whether $A < B$ is

$$A < B = \bar{A}_3B_3 + E_3\bar{A}_2B_2 + E_3E_2\bar{A}_1B_1 + E_3E_2E_1\bar{A}_0B_0$$

This has the same form as the $A > B$ expression and can be analysed in a similar manner.

The implementation of a 4-bit magnitude comparator using EX-NOR and AND gates using the above expression is shown in Fig. 6.48.

6.11.2 IC 7485-4-bit Magnitude Comparator

IC 7485 is a 4-bit magnitude comparator with two 4-bit inputs $A_3A_2A_1A_0$ and $B_3B_2B_1B_0$ and three outputs, viz. $A = B$, $A < B$ and $A > B$. In addition, it has three cascading inputs which allow several comparators to be cascaded. By cascading several such comparators, any number of bits can be compared. The logic symbol for the 4-bit magnitude comparator is shown in Fig. 6.49. Its truth table is given in Table. 6.25.

Table 6.25 Truth table of IC 7485 — 4-bit magnitude comparator

Comparing inputs				Cascading inputs			Outputs		
A_3, B_3	A_2, B_2	A_1, B_1	A_0, B_0	$A > B$	$A < B$	$A = B$	$A > B$	$A < B$	$A = B$
$A_3 > B_3$	X	X	X	X	X	X	1	0	0
$A_3 < B_3$	X	X	X	X	X	X	0	1	0
$A_3 = B_3$	$A_2 > B_2$	X	X	X	X	X	1	0	0
$A_3 = B_3$	$A_2 < B_2$	X	X	X	X	X	0	1	0
$A_3 = B_3$	$A_2 = B_2$	$A_1 > B_1$	X	X	X	X	1	0	0
$A_3 = B_3$	$A_2 = B_2$	$A_1 < B_1$	X	X	X	X	0	1	0
$A_3 = B_3$	$A_2 = B_2$	$A_1 = B_1$	$A_0 > B_0$	X	X	X	1	0	0
$A_3 = B_3$	$A_2 = B_2$	$A_1 = B_1$	$A_0 < B_0$	X	X	X	0	1	0
$A_3 = B_3$	$A_2 = B_2$	$A_1 = B_1$	$A_0 = B_0$	1	0	0	1	0	0
$A_3 = B_3$	$A_2 = B_2$	$A_1 = B_1$	$A_0 = B_0$	0	1	0	0	1	0
$A_3 = B_3$	$A_2 = B_2$	$A_1 = B_1$	$A_0 = B_0$	0	0	1	0	0	1

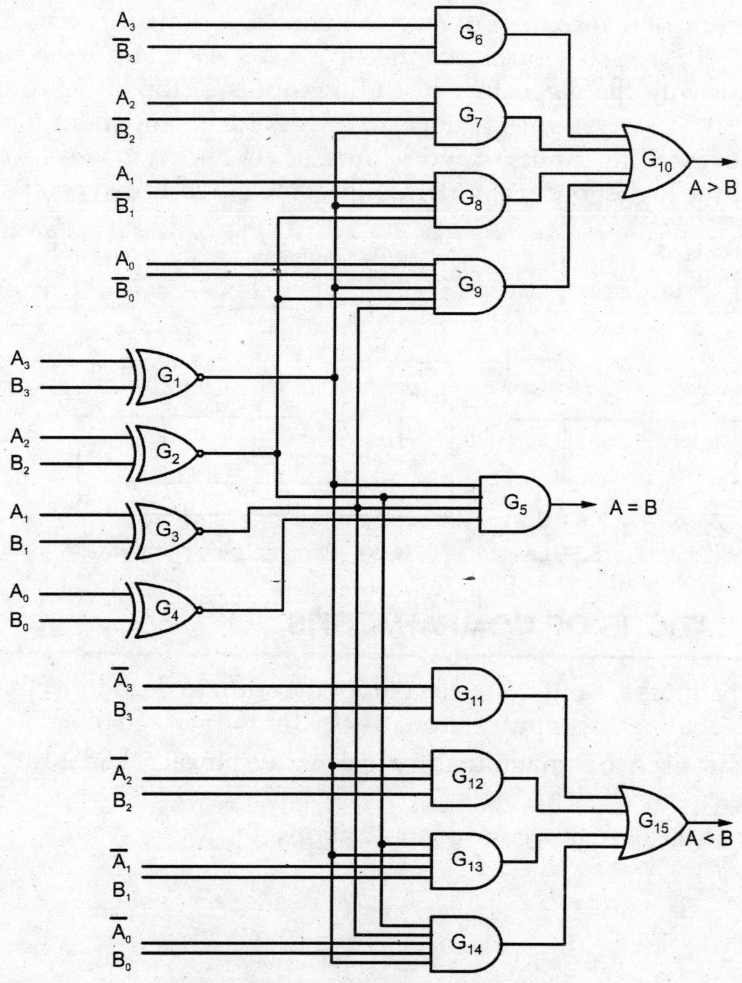

Fig. 6.48 4-bit magnitude comparator

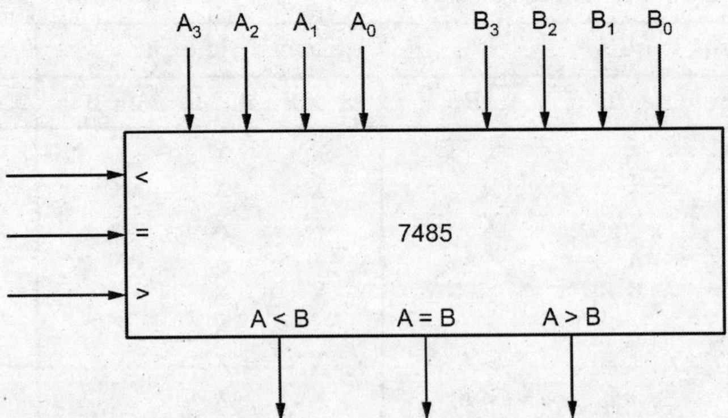

Fig. 6.49 Logic symbol of IC 7485—4-bit magnitude comparator

6.11.3 Cascading of IC 7485s

The cascading of two comparators to compare two 8-bit numbers ($A_7A_6A_5A_4A_3A_2A_1A_0$ and $B_7B_6B_5B_4B_3B_2B_1B_0$) can be done by connecting $A < B$, $A = B$ and $A > B$ outputs of the lower order comparator with the respective cascading inputs of the higher order comparator as shown in Fig. 6.50. The cascading input (=) of lower order comparator must be connected to HIGH, while the cascading inputs (< and >) must be connected to LOW. The $A < B$, $A = B$ and $A > B$ outputs of the higher order comparator become the cascaded comparator outputs.

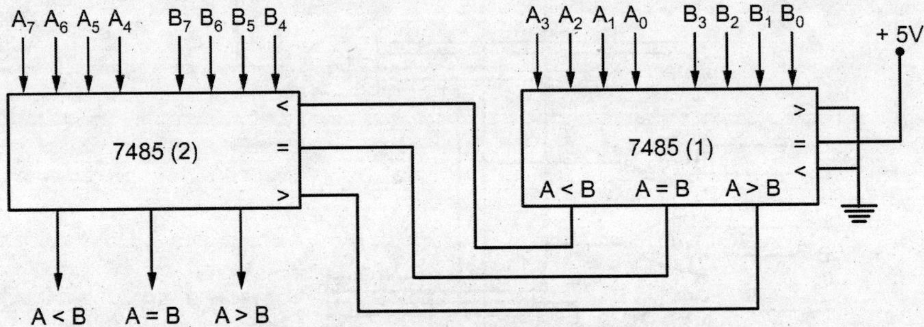

Fig. 6.50 Cascading of two 4-bit magnitude comparators

6.12 APPLICATIONS OF COMPARATORS

(1) Comparators are often used as part of the address decoding circuitry in computers to select a specific input/output device for the storage of data.

(2) They are used to actuate circuitry to drive the physical variable toward the reference value.

(3) They are also used in control applications.

■ REVIEW QUESTIONS ■

1. What is a multiplexer? Explain: (*a*) Two-channel multiplexer and (*b*) Four-channel multiplexer.
2. What is the function of a multiplexer's select inputs?
3. What are some of the major applications of multiplexers?
4. Identify each MSI device (*a*) 74157 (*b*) 74151 and (*c*) 74150.
5. Generally, how can an MSI decoder be used as a demultiplexer?
6. Design a 32-to-1 multiplexer using 8-to-1 multiplexer ICs.
7. Draw a multiplexer using only NAND gates which selects from four inputs A_0 to A_1 using two select inputs S_0 and S_1.
8. What is a demultiplexer? Explain the difference between a DEMUX and MUX.
9. Briefly describe the purpose of each of the following ICs: (*a*) 74157 (*b*) 7449 and (*c*) 74139.
10. What are nibble multiplexers?
11. Explain (*a*) 1-to-8 demultiplexer (*b*) 1-to-16 demultiplexer.
12. Design a 1-of-40 decoder using BCD to decimal decoders.
13. Describe the applications of multiplexers.
14. Explain how would you construct a logic function generator using multiplexers.
15. What is meant by a decoder? Explain it with a block diagram.
16. Explain how decoders are designed.
17. Give the logic diagram and logic symbol of IC 74154—4-to-16 decoder and briefly explain its functions.
18. Explain the BCD-to-decimal decoder.
19. Explain briefly the BCD-to-seven segment decoder.
20. Explain the basic operation of an LCD that is driven by a decoder.
21. Design the logic circuitry for the '*a*' output of a BCD-to-7 segment decoder with active LOW outputs.
22. Can more than one decoder output be activated at one time? Justify your answer.
23. What is the function of a decoder's enable input(s)?
24. How does an encoder differ from a decoder?
25. How does a priority encoder differ from an ordinary encoder?
26. How do you realise a parity-bit checker?
27. Draw the logic diagram of a 2-to-4 decoder with an ENABLE input using (*a*) NAND gates (*b*) NOR gates. Show that the realisation using NAND gates is more convenient to distinguish the selected output with a value of 0.
28. Design a 6-bit odd/even-parity checker using IC 74180.
29. What is a parity-bit? Design a parity generator circuit using IC 74180 to add odd parity bit to a 7-bit word.
30. Draw the logic diagram of IC 74180 parity generator/checker and explain its operation with the help of a truth table.
31. Define a code converter logic circuit.
32. How do you convert binary numbers to corresponding gray codes using a converter?
33. How do you convert gray code numbers to corresponding binary numbers using a converter?
34. What is meant by a magnitude comparator?
35. What is the purpose of cascading inputs of 7485?

■ PROBLEMS ■

1. A certain multiplexer can switch one of 32 data inputs to its output. How many different inputs does this MUX have?

   Ans: 32 data inputs and 5 select inputs.

2. An 8-to-1 MUX has inputs A, B and C connected to the selection inputs S_2, S_1 and S_0 respectively. The data inputs, D_0 through D_7, are as follows: $D_1 = D_2 = D_1 = 0$; $D_3 = D_5 = 1$; $D_0 = D_4 = D$; $D_6 = \bar{D}$. Determine the Boolean expression that the MUX implements.

   Ans: $F(A, B, C, D) = S\,(1, 6, 7, 9, 10, 11, 12)$

3. Implement the following Boolean function with a 4-to-1 MUX and external gates. Connect inputs A and B to the selection lines. The input requirements for the four data lines will be a function of variables C and D. These values are obtained by expressing F as a function of C and D for each of the four cases when AB – 00, 01, 10 and 11. These functions may have to be implemented with external gates.

$$F(A, B, C, D) = (1, 3, 4, 11, 12, 13, 15)$$

   Ans: When $AB = 00$, $F = D$; when $AB = 01$, $F = \overline{(C + D)}$ [use a NOR gate]; when $AB = 10$, $F = CD$ (use an AND gate); when $AB = 11$, $F = 1$

4. Implement the logic function specified in Table P6.4 using IC 74151 data selector.

Table P6.4

Inputs				Outputs
D	*C*	*B*	*A*	(*Y*)
0	0	0	0	0
0	0	0	1	0
0	0	1	0	1
0	0	1	1	1
0	1	0	0	0
0	1	0	1	0
0	1	1	0	1
0	1	1	1	1
1	0	0	0	1
1	0	0	1	0
1	0	1	0	1
1	0	1	1	1
1	1	0	0	0
1	1	0	1	1
1	1	1	0	0
1	1	1	1	1

Ans: $Y = \bar{D}\bar{C}B\bar{A} + \bar{D}\bar{C}BA + \bar{D}CB\bar{A} + \bar{D}CBA + D\bar{C}\bar{B}\bar{A}$
$+ D\bar{C}B\bar{A} + D\bar{C}BA + DC\bar{B}A + DCBA$

5. Show how a 16-input MUX such as IC 74150 is used to generate the function

$$Y = \overline{A}\overline{B}\overline{C}D + BCD + A\overline{B}\overline{C} + AB\overline{C}D$$

Ans: Connect D_1, D_7, D_8, D_9, D_{13}, D_{15} to V_{CC}.
All other inputs must be grounded.

6. The circuit of Fig. P6.6 shows how an 8-bit MUX can be used to generate a four-variable logic function eventhough the MUX has only 3 SELECT inputs. Three of the logic variables *A, B* and *C* are connected to the SELECT inputs.

The fourth variable *D* and its inverse $\overline{D}$ are connected to selected data inputs of the MUX as required by the desired logic function. The other MUX data inputs are tied to a LOW or a HIGH as required by the function,

(*a*) Setup a truth table showing the output *Y* for the 16 possible combinations of input variables.
(*b*) Write the sum-of-products expression for *Y* and simplify it to verify that

$$Y = \overline{C}B\overline{A} + D\overline{C}\overline{B}\overline{A} + \overline{D}C\overline{B}\overline{A}$$

Ans: *Y*= HIGH for *DCBA* = 0010, 0100, 1000, 1010

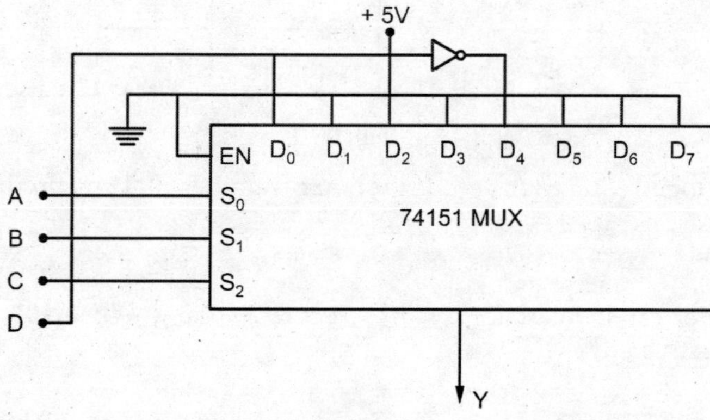

Fig. P6.6

7. If the data-select inputs to the multiplexer in Fig. P6.7(*a*) are sequenced as shown by the waveforms in Fig. P6.7(*b*), determine the output waveform with the following data in puts: $D_0 = 0$, $D_1 = 1$, $D_2 = 1$, $D_3 = 0$,

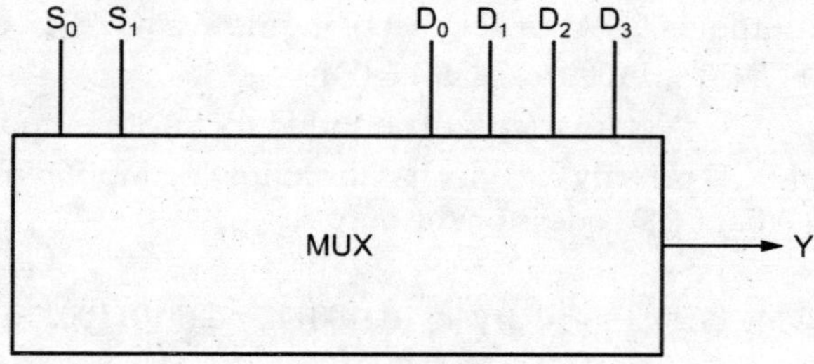

Fig. P6.7(*a*)

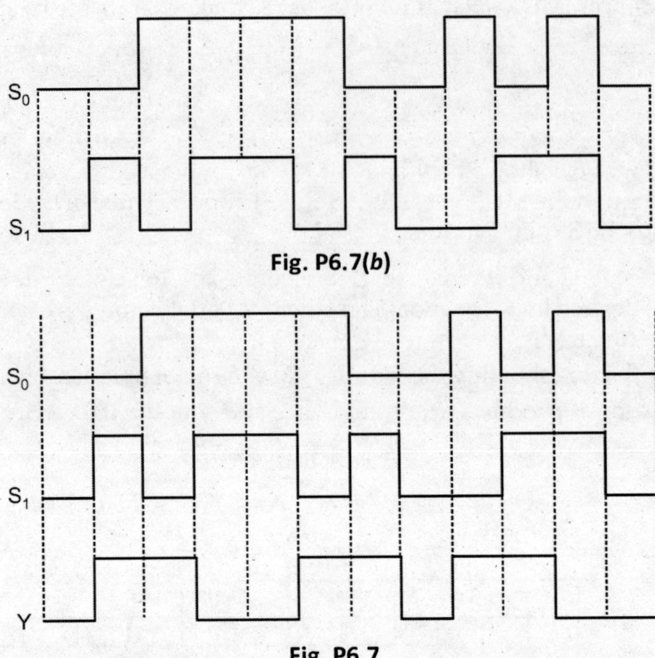

Fig. P6.7(*b*)

Fig. P6.7

8. Determine the logic required to decode the binary number 10112 by producing a HIGH indication on the output.

Ans: $Y = D\bar{C}BA$

9. A 3-to-8 decoder can be used for binary-to-octal decoding. When 101_2 is on the inputs, which output line is activated?

Ans: The 5 output

10. How many IC 74154—4-16 decoders are necessary to decode a six-bit binary number?

Ans: Four

11. Would you select a decoder/driver with active-HIGH or active-LOW outputs to drive a common anode seven-segment LED display?

Ans: Active-LOW

12. Determine which output is LOW for each set of inputs to a IC 74154 decoder, (*a*) *DCBA* = 1100 (*b*) *DCBA* = 1000 (*c*) *DCBA* = 0010.

Ans: (*a*) 12th output LOW (*b*) 8th output LOW (*c*) 2nd output LOW

13. IC 74147—decimal-BCD priority encoder has the following input pins LOW: 3, 4 and 7. The other inputs are HIGH. What BCD code is on the output?

Ans: The complement of BCD 7

14. Convert the following BCD into binary (*a*) 10010011 and (*b*) 01100111

Ans:

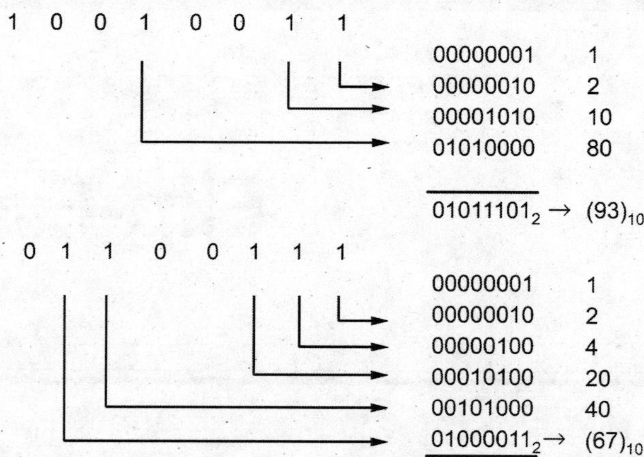

15. *ABCD*–seven segment decoder/driver is connected to an LED display. Which segments are illuminated for each of the following input codes?
(*a*) *DCBA* = 0001 (*b*) *DCBA* = 0111 (*c*) *DCBA* =0011.

Ans: (*a*) *b*, *c* (*b*) *a*, *b*, *c* (*c*) *a*, *b*, *c*, *d*, *g*

16. Show the logic required to convert a 10-bit binary number to Gray code and use the logic to convert the following binary code words to Gray code:
(*a*) 1010101010 (*b*) 1111100000 (*c*) 0000001110 (*d*) 1111111111

17. Show the logic required to convert a 10-bit Gray code to binary and use the logic to convert the following Gray code to binary:
(*a*) 1010000000 (*b*) 0011001100 (*c*) 1111000111 (*d*) 0000000001.

18. Convert the following decimal numbers first to BCD. Using the logical operation of the BCD-to-binary converter, convert the BCD to binary. Verify the result in each case.
(*a*) 2 (*b*) 8 (*c*) 13 (*d*) 26 (*e*) 33

Ans: (*a*) 0010 0010_2 (b) 1000 1000_2 (*c*) 00010011 1101_2
(*d*) 00100110 11010_2 (*e*) 00110011 100001_2

19. Use IC 7485 comparators to form a twelve-bit comparator. Illustrate with a diagram.

20. IC 7485—4-bit magnitude comparator has $A = 1011$ and $B = 1001$.
(*a*) Determine the outputs, (*b*) Show how to connect <, =, and > inputs if this is to be the least significant stage.

Ans: (*a*) $A > B = 1$, $A < B = 0$ and $A = B = 0$
(*b*) < and > inputs should be connected to LOW state and
= input should be connected to HIGH state

❑❑❑

7 Flip-Flops

7.1 INTRODUCTION

The logic circuits discussed in the previous chapters are examples of combinational circuits. The logic circuits whose outputs at any instant of time depend only on the input signals present at that time are known as combinational circuits. In particular, the output of combinational circuits does not depend upon any past inputs or outputs i.e. the output signals of combinational circuits are not fed back to the input of the circuit. Moreover, in a combinational circuit, for a change in the input, the output appears immediately, except for the propagation delay through circuit gates. The block diagram of a combinational circuit with *m* inputs and *n* outputs is shown in Fig.7.1 (*a*).

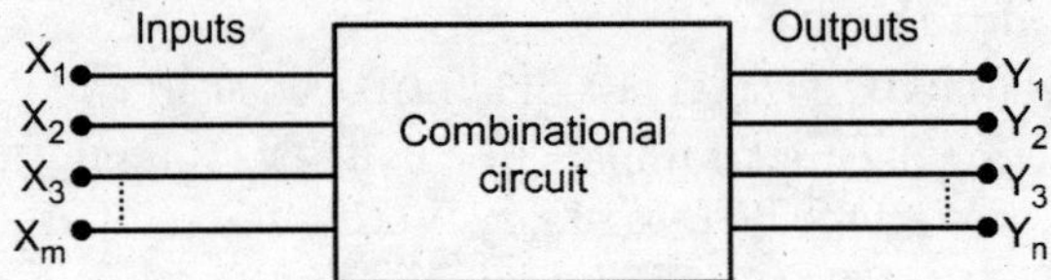

Fig. 7.1(*a*) Block Diagram of Combinational Circuit.

The logic circuits whose outputs at any instant of time depend not only on the present inputs but also on the past outputs are called sequential circuits. In sequential circuits, the output signals arc fed back to the input side. Thus, an output signal is a function of the present input signals and a sequence of the past input signals i.e. the past output signals. The block diagram of a sequential circuit is shown in Fig. 7.1(*b*). It consists of a combinational circuit to which memory elements are connected to form a feedback path. The circuit has *m* number of external inputs, denoted by x_1, x_2, ... x_m and *n* number of outputs, represented by z_1, z_2, ... z_n. The memory elements denoted by *M* are devices capable of storing binary information within them. The signal value at the output of memory elements, denoted by y_1, y_2, ... y_k, are referred to as the present state or simply the state of the sequential circuit.

The external inputs x_1, x_2,.........,x_m and the present state variables y_1, y_2,........ y_k are applied to the combinational circuit, which in turn produces the outputs z_1, z_2,........ z_n and the values Y_1, Y_2, ... Y_k. The values of *Y* which appear at the output of the combinational circuit at a particular time *t* are equal to the values of the present state variables y_s at the time (t + 1). So, Fs at the input of the memory elements are referred to as next state of the sequential circuit, i.e., the state, the circuit will assume next.

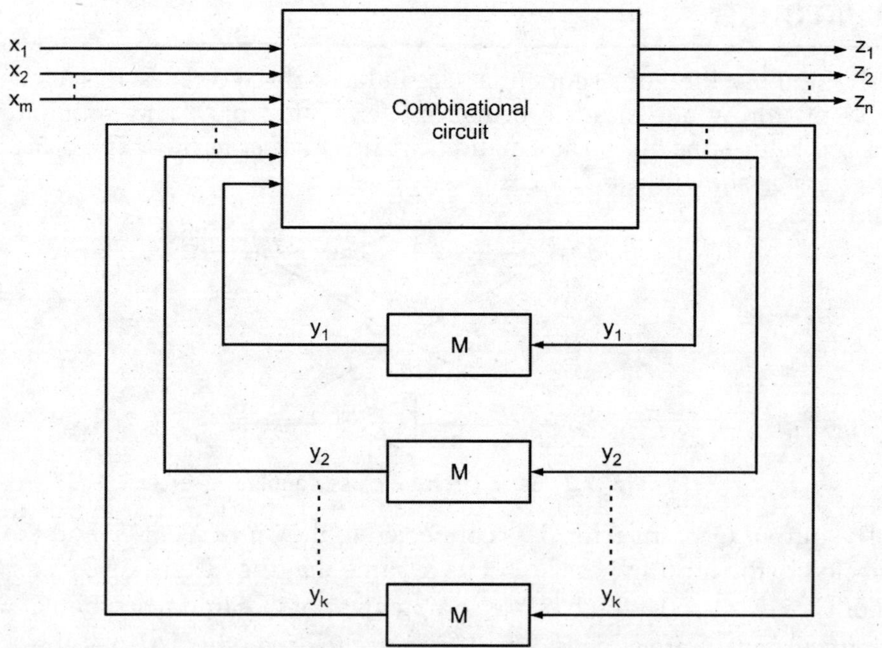

Fig. 7.1(*b*) Block diagram of sequential circuit

Combinational circuits are often faster than sequential circuits since the combinational circuits do not require memory elements whereas the sequential circuits need memory devices to perform their operations in sequence. However, modern digital computers, must have memories to function properly. Thus, sequential circuits are of prime importance in modern digital devices.

Sequential circuits are of two types, viz., (*i*) *synchronous or clocked* and (*ii*) *asynchronous or undocked*. In the synchronous sequential circuit, synchronisation is achieved by a timing device called a *master-clock generator*, which generates a periodic train of *clock pulses*. In practice, it may be achieved by applying clock pulses to various AND gates through which external inputs enter the sequential circuit. This ensures the gates to transmit input signals only, which coincide with the arrival of the clock pulses. The rate at which the master clock generates pulses must be slow enough to permit the slowest circuit to respond. This limits the speed of all circuits.

In an asynchronous sequential circuit, events can occur after one event is completed and there is no need to wait for a clock pulse. Therefore, in general, asynchronous circuits are considerably *faster* than synchronous sequential circuits. However, in an asynchronous circuit, events are allowed to occur without any synchronisation. In such a case, the system becomes unstable which results in difficulties.

Do You Know?	The flip-flop is known by other names, including latch and bistable multivibrator.

To have a sequential circuit, a storage device is required to know what has happened in the past. The basic unit of storage is the *flip-flop*.

7.2 LATCHES

The simplest kind of a sequential circuit has only two *states*. It is a memory cell, which is capable of storing one bit of information, i.e., logic 1 or 0. This sequential circuit is also called a *latch*, since one bit of information can be *locked or latched*. The basic latch consists of two inverters as shown in Fig. 7.2.

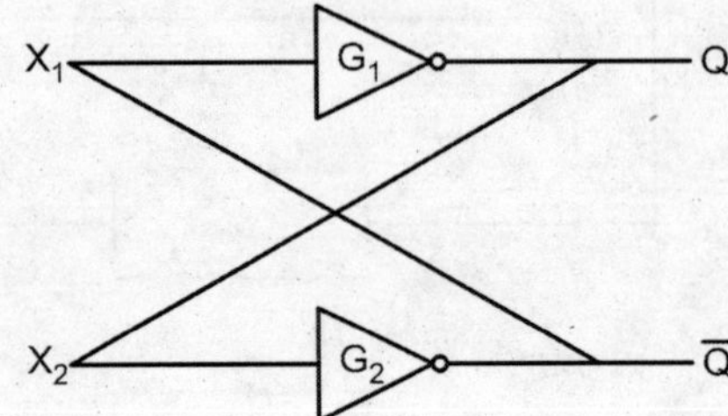

Fig. 7.2. Basic latch — cross coupled inverters.

The output Q of inverter G_1 is connected to the input X_2 of G_2 and the output $\bar{Q}$ of G_2 is connected to the input X_1 of G_1. Let us assume that the output of G_1, i.e., $Q = 1$. Then, the output of G_2, i.e., $\bar{Q} = 0$, which is the complement of Q. Similarly when $Q = 0$, $\bar{Q} = 1$. Thus the outputs Q and $\bar{Q}$ are always complementary to each other. Also, if the circuit is in state 1 or 0 at Q and $\bar{Q}$ respectively, it continues to remain latched in the same state. This property shows that it can store one bit of digital information. The basic latch shown in Fig. 7.2 has no provision to get any desired digital information stored. In fact, when the power is switched *on*, the circuit switches to one of the stable states (i.e., $Q = 1$ or 0) and it is not possible to predict the state. The inverters G_1 and G_2 can be replaced with two input NAND/NOR gates and the second input of the gates can be used to enter the desired digital information as discussed below.

Do You Know?	The term latch is used for certain types of flip-flops.

The general block schematic representation of a latch with provision to enter digital data is shown in Fig. 7.3. It has one or more inputs and two outputs Q and $\bar{Q}$. The two outputs are complement of each other; if $Q = 0$ i.e., *Reset*, then $\bar{Q} = 1$; if $Q = 1$ i.e., *Set*, then $\bar{Q} = 0$. When the latch output $Q = 0$ or 1, it will remain in the same state until one or more of the inputs are excited to effect a change in the output. Since the latch output will remain *set/reset* until the trigger pulse is given to change the state, it can be regarded as a *memory device* with the capability of storing one binary digit of information.

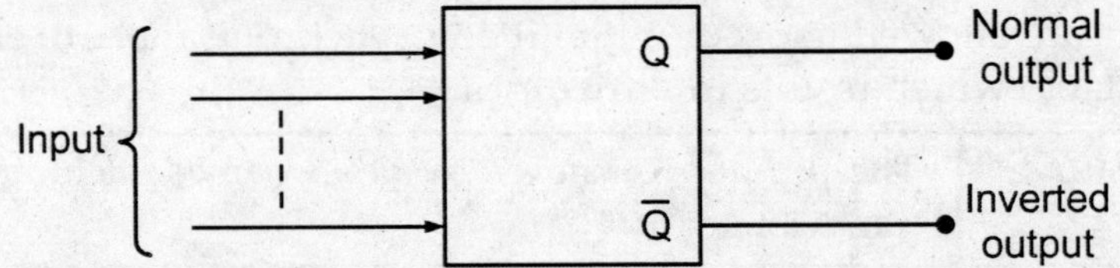

Fig. 7.3. Block diagram of a latch with provision to enter digital data

7.2.1 Set-Reset (S-R) Latch

The *S-R* latch has two inputs, namely SET (S) and RESET (R), and two outputs *Q* and $\bar{Q}$. The two outputs are complement to each other. The *S-R* latch can be easily immented using NOR gates or NAND gates as shown in the next section. The bl diagram of *S-R* latch is shown in Fig. 7.4(*a*).

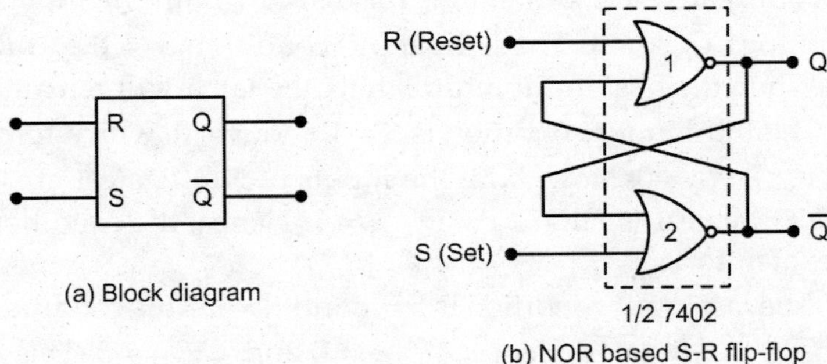

(a) Block diagram

(b) NOR based S-R flip-flop

Fig. 7.4. S-R latch.

7.2.2 NOR-based S-R Latch

The *S-R* latch can be easily constructed using two NOR gates connected back to back, as shown in Fig. 7.4(*b*). The cross-coupled connections from the output of one gal the input of the other gate constitute a feedback path. For this reason, the circuits classified as asynchronous sequential circuits. The truth table for the NOR-based *S-R* latch is shown in Table 7.1.

Table 7.1. Truth table of NOR-based *S-R* latch.

Inputs		Outputs		Action
S	*R*	Q_{n+1}	Q_{n+7}	
0	0	Q_n	$\bar{Q}_n$	No change
0	1	0	1	Reset
1	0	1	0	Set
1	1	?	?	Forbidden

To analyse the circuit of Fig. 7.4(*b*), one must remember that the output of a NOR gate is 0 if any input is 1 and the output is 1 only when all inputs are 0. From the truth table, it is evident that four possible input combinations exist for the *S-R* latch, outputs for these four possible input combinations are described below.

Case 1: For $S = 0$ and $R = 0$, the latch simply remains in its present state (Q_n). That is next state of the latch (Q_{n+1}) is just the present state. In this situation, the next stat the latch will be $Q_{n+1} = 0$ if $Q_n = 0$ and $Q_{n+1} = 1$ if $Q_n = 1$. First, let us assume that $Q_n = 0$ and $\bar{Q}_n = 1$. The inputs of NOR gate-1 are 1 and 0, and therefore its output $Q_{n+1} = 0$. This $Q_{n+1} = 0$ is fed back to NOR gate-2 input, thereby producing a 1 at its output; so $\bar{Q}_{n+1} = 1$, as originally assumed.

Next, let us assume that $Q_n = 1$ and $\overline{Q}_n = 0$. This 1 is applied to the input of NOR gate-2 and therefore the output becomes 0 (i.e. $\overline{Q}_{n+1} = 0$). This $\overline{Q}_{n+1} = 0$ is fed to the input of NOR gate-1, thereby producing a 1 at its output; so $Q_{n+1} = 1$, as originally assumed. Thus, the condition $S = 0$ and $R = 0$ will not affect the outputs of latch.

Case 2: The second input condition is $S = 0$ and $R = 1$. The 1 at the RESET input forces the output of NOR gate-1 LOW (i.e. $Q_{n+1} = 0$). Now both the inputs of NOR gate-2 are 0 and 0 and its output $\overline{Q}_{n+1} = 1$. Thus, the input condition $S = 0$ and $R = 1$ will always reset the latch to 0. When the *reset* input returns to 0, the latch will remain in the 0 state.

Case 3: The third jnput condition is $S = 1$ and $R = 0$, which forces the output of NOR gate-2 LOW, i.e. $\overline{Q}_{n+1} = 0$. Now, both the inputs of NOR gate-1 are 0 and 0, and therefore the output of NOR gate-1 is HIGH, i.e. $Q_{n+1} = 1$. Hence, the condition $S = 1$ and $R = 0$ will always set the latch to 1.

Case 4: The last input condition is $S = 1$ and $R = 1$. This condition will produce 0 at the output of both the NOR gates. Hence, $Q_{n+1} = 0$ and $\overline{Q}_{n+1} = 0$. This condition violates the fact that the outputs Q_{n+1} and $\overline{Q}_{n+1}$ are the complements of each other. In normal operation, this condition must be avoided by making sure that Is are not applied to both inputs simultaneously.

7.2.3 NAND-based $\overline{S}$ - $\overline{R}$ Latch

Another basic latch circuit constructed using cross-coupled NAND gates is shown in Fig. 7.5. The operation of NAND $\overline{S}$-$\overline{R}$ latch can be analysed in the same manner employed for the NOR latch. To understand the operation of NAND-based $\overline{S}$-$\overline{R}$ latch, one must remember that a low at any input of a NAND gate will force its output high. The truth table for the NAND-based $\overline{S}$-$\overline{R}$ latch is shown in Table 7.2 which is different from that of a NOR-based $\overline{S}$-$\overline{R}$ latch. This latch is called as $\overline{S}$-$\overline{R}$ latch, i.e., here $\overline{S} = 0$ and $\overline{R} = 1$ will set the latch.

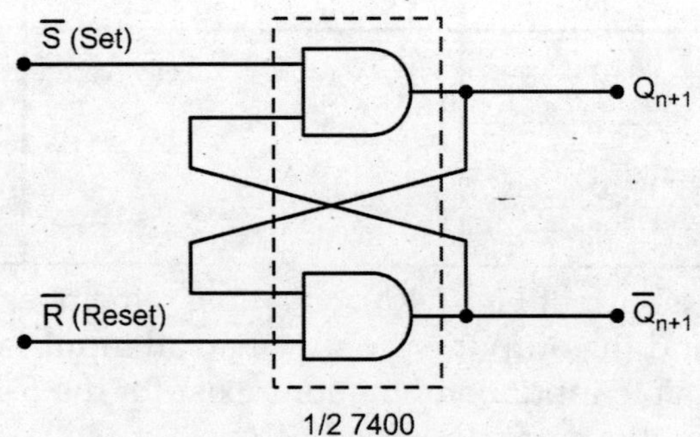

Fig. 7.5. NAND-based $\overline{S} - \overline{R}$ latch.

Case 1: The first condition is $\overline{S} = 0$ and $\overline{R} = 0$. When both inputs go to 0, both outputs go to 1, i.e., $Q_{n+1} = 1$ and $\overline{Q}_{n+1} = 1$. This condition is ambiguous and should not be used.

Case 2: The condition $\overline{S} = 0$ and $\overline{R} = 1$ always produces $Q_{n+1} = 1$ regardless of the present state of the latch output. This condition *sets* the state of the latch, i.e. as shown Q_{n+1} and $\overline{Q}_{n+1} = 0$.

Case 3: The condition $\overline{S}=1$ and $\overline{R}=0$ forces the lower NAND gate output to, i.e., $\overline{Q}_{n+1}=1$. Now, both the inputs of upper NAND gate are 1, and therefore the output of upper NAND gate is LOW, i.e. Q_{n+1} = 0, regardless of the prior state of the latch. This condition *resets* (clear) the latch, i.e. $Q_{n+1} = \overline{Q}_{n+1}=1$.

Case 4: The last condition $\overline{S}=1$ and $\overline{R}=1$ does not affect the state of the latch. It remains in its prior state.

Table 7.2. Truth table of NAND-based *S-R* latch.

Inputs		Outputs		Action
$\overline{S}$	$\overline{R}$	Q_{n+1}	$\overline{Q}_{n+1}$	
0	0	?	?	Forbidden
0	1	1	0	Set
1	0	0	1	Reset
1	1	Q_n	$\overline{Q}_n$	No change

Comparing the NAND latch and the NOR latch, we see that they operate basically in the same manner except for the following difference: The NOR latch inputs are normally 0 and must be pulsed to the 1 state (active HIGH) to change the state of the latch outputs; the NAND latch inputs are normally 1 and must be pulsed to the 0 state (active LOW) to change the latch output state.

One can convert a NAND based $\overline{S}$-$\overline{R}$ latch into an *S-R* latch by placing an inverter at each input of the latch. The resulting four NAND gate circuits, shown in Fig. 7.6, will behave in the same manner as an *S-R* latch.

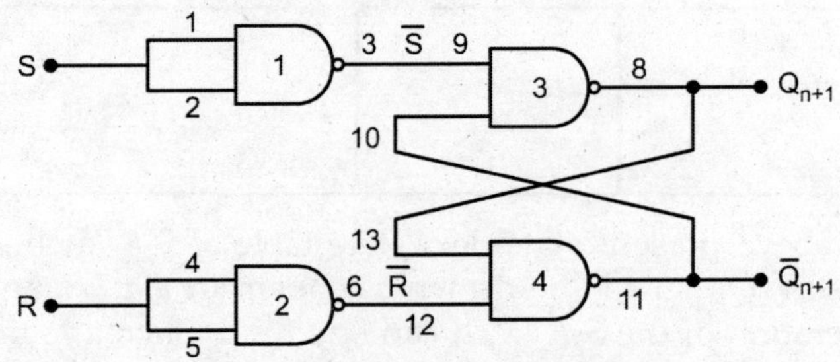

Fig. 7.6. S-R latch using 4 NAND gates.

7.2.4 State Diagram and Characteristic Equation of S-R Latch

The state transition diagram for *S-R* latch can be drawn as shown in Fig. 7.7.

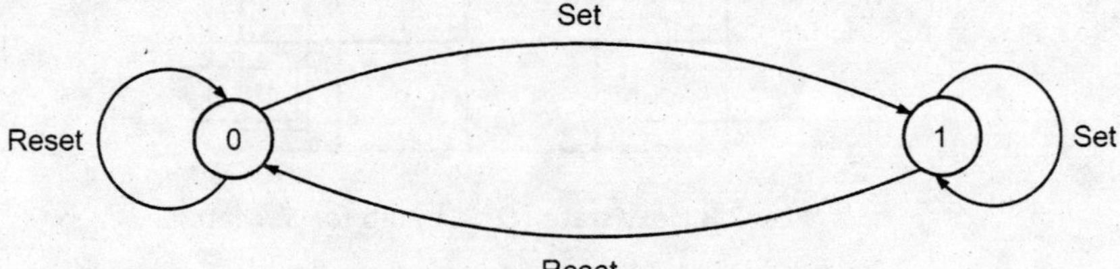

Fig. 7.7. State diagram of S-R latch.

From the state diagram, one can easily understand that an *S-R* latch is with two stable states 0 and 1. On applying SET input, the latch is set to 1; on applying RESET input, the latch is reset to 0 irrespective of its previous state.

From the above state diagram and truth table of *S-R* latch, (Table 7.1), the present state-next state table and application table or excitation table for *S-R* latch can be drawn as shown in Table 7.3 and Table 7.4 respectively.

Table 7.3. Present state-next state table for S-R latch.

Present state	SET input	RESET input	Next state
Q_n	S	R	Q_{n+1}
0	0	0	0
0	0	1	0
0	1	0	1
0	1	1	*d*
1	0	0	1
1	0	1	0
1	1	0	1
1	1	1	*d*

d: don't care

Table 7.4. Application or excitation table for S=R latch.

Q_n	Q_{n+1}	Excitation inputs	
		S	R
0	0	0	d
0	1	1	0
1	0	0	1
1	1	d	0

Using the abeve present state-next state table *of S-R* latch, a Karnaugh map for the next state transition (Q_{n+1}) can be drawn as shown in Fig.7.8, and a logic expression that describes the operation of the *S-R* latch can be found. Such a logic expression is called the *characteristic equation* of the *S-R* latch.

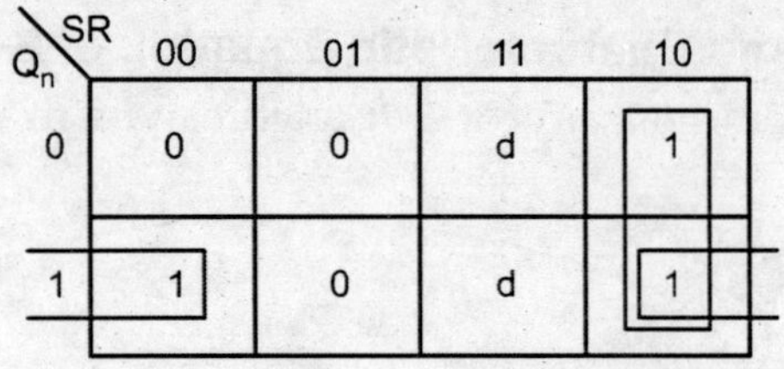

Q_n \ SR	00	01	11	10
0	0	0	d	1
1	1	0	d	1

Fig. 7.8 Next state (Q_{n+1}) map for *S-R* latch.

From the above *K*-map,

$$Q_{n+1} = S\overline{R} + \overline{R}\,Q_n$$

$$Q_{n+1} = (S + Q_n)\,\overline{R}$$

In the above *K*-map simplification, the don't care combinations are not considered because they correspond to invalid input combination of *S*-*R* latch.

7.3 FLIP-FLOPS

Synchronous circuits change their states only when clock pulses are present. The operation of the basic latch can be modified, by providing an additional control input that determines, when the state of the circuit is to be changed. The latch with the additional control input is called the Flip-Flop. The additional control input is either the clock or enable input.

7.3.1 Types of Flip-Flops

Flip-Flops are of different types depending on how their inputs and clock pulses cause transition between two states. There are four basic types, namely, *S-R, J-K, D* and Flip-Flops.

7.4 *S-R* FLIP-FLOP

The *S*-*R* flip-flop consists of two additional AND gates at the *S* and *R* inputs of 5 latch as shown in Fig. 7.9.

In this circuit, when the clock input is LOW, the output of both the AND gates are LOW and the changes in *S* and *R* inputs will not affect the output (*Q*) of the flip-flop. When the clock input becomes HIGH, the value at *S* and *R* inputs will be passed to the output of the AND gates and the output (*Q*) of the flip-flop will change according to the changes in *S* and *R* inputs as long as the clock input is HIGH. In this manner, one can strobe or clock the flip-flop so as to store either a 1 by applying *S* = 1, *R* = 0 (i.e. set) or a 0 by applying *S* = 0, *R* = 1 (i.e. reset) at any time and then hold that bit of information for any desired period of time by applying a LOW at the clock input. This flip-flop is called clocked *S*-*R* flip-flop.

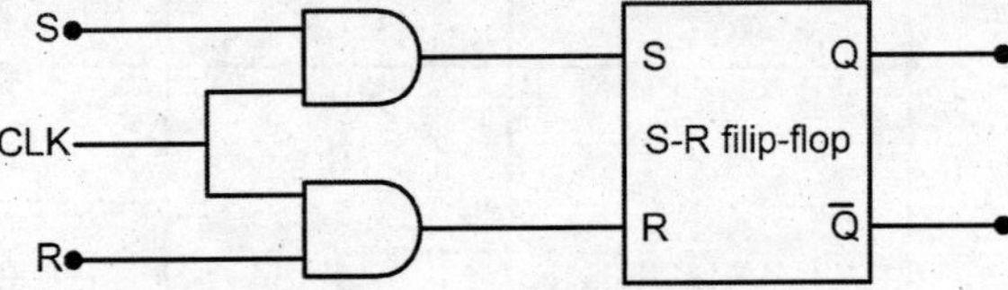

Fig. 7.9. Block diagram of *S*-*R* flip-flop

The *S*-*R* flip-flop which consists of the basic NOR latch and two AND gates is shown in Fig. 7.10.

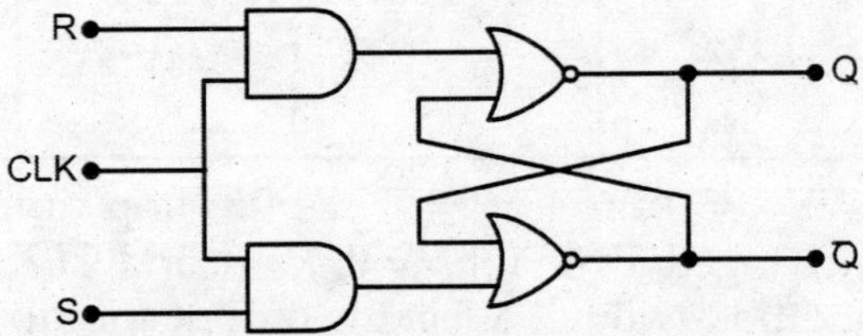

Fig. 7.10. Clocked NOR-based *S*-*R* flip-flop

The *S-R* flip-flop which consists of the basic NAND latch and two other NAND gates is shown in Fig. 7.11(*a*). The *S* and *R* inputs control the state of the flip-flop in the same manner as described earlier for the basic (unclocked) *S-R* latch. However, the flip-flop does not respond to these inputs until the rising edge of the clock signal occurs. The clock pulse input' acts as an enable signal for the other two inputs. The outputs of NAND gates 1 and 2 stay at the logic 1 level as long as the clock input remains at 0. This 1 level at the inputs of NAND-based basic *S-R* latch retains the present state, i.e. no change occurs. The characteristic table of the *S-R* flip-flop is shown in truth table of Table 7.5. This table shows the operation of the flip-flop in tabular form.

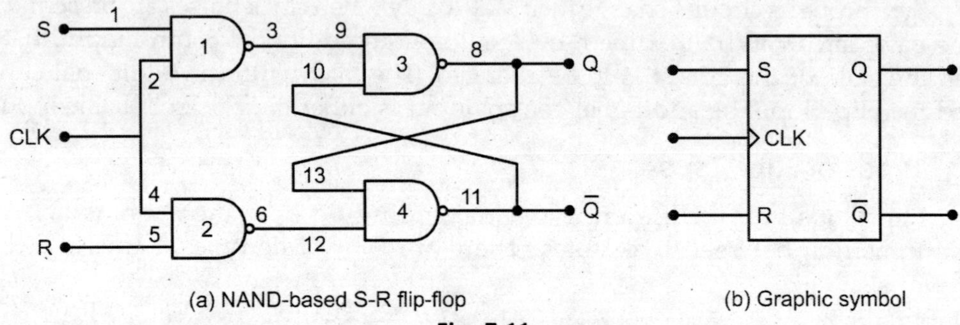

(a) NAND-based S-R flip-flop

(b) Graphic symbol

Fig. 7.11

Table 7.5. Characteristic table of *S-R* flip-flop

Present	**Clock pulse**	**Data inputs**	**Next state**	**Action**	
state Q_n	**CLK**	***S***	***R***	**Q_{n+1}**	
0	0	0	0	0	No change
1	0	0	0	1	No change
0	1	0	0	0	No change
1	1	0	0	1	No change
0	0	0	1	0	No change
1	0	0	1	1	No change
0	1	0	1	0	Reset
1	1	0	1	0	Reset
0	0	1	0	0	No change
1	0	1	0	1	No change
0	1	1	0	1	Set
1	1	1	0	1	Set
0	0	1	1	0	No change
1	0	1	1	1	No change
0	1	1	1	?	Forbidden
1	1	1	1	?	Forbidden

Case 1: For $S = 0$, $R = 0$ and CLK = 0, the flip-flop simply remains in its present state. That is, *Q* remains unchanged. Even for $S = 0$, $R = 0$ and CLK = 1, the flip flop remains in its present state. This condition will not affect the outputs of flip-flop. The first four rows of the truth table clearly indicate that the state of the flip-flop remains unchanged, i.e., $Q_{n+1} = Q_n$.

Case 2: For $S = 0$, $R = 1$ and CLK = 0, the flip-flop remains in its present state. But when CLK = 1, the NAND gate-1 output will go to 1 and the NAND gate-2 output will go to 0. Now, a 0 at NAND gate-4 input forces $\bar{Q}=1$ which in turn results in NAND gate-3 output $Q = 0$. Thus, for $S = 0$, $R = 1$ and CLK = 1, the flip-flop RESETS toll state.

Case 3: For $S = 1$, $R = 0$ and CLK = 0, the flip-flop remains in its present state. But for $S = 1$, $R = 0$ and CLK = 1, the set state of the flip-flop is reached. This causes NAND gate-1 output to go to 0 and the NAND gate-2 output to 1. Now, a 0 at NAND gate-3 input forces Q to 1 which in turn forces NAND gate-4 output $\bar{Q}$ to 0.

Case 4: An indeterminate condition occurs when all the inputs, namely CLK, S and R, are equal to 1. This condition results in 0's in the outputs of gate-1 and 2 and 1's in both outputs Q and $\bar{Q}$. When the CLK input goes back to 0 (while S and R remai 1), it is not possible to determine the next state, as it depends on whether the outpi gate-1 or gate-2 goes to 1 first.

The graphic symbol of the *S-R* flip-flop is shown in Fig. 7.11(*b*). It has 3 input *R* and CLK. The CLK input is marked with a small triangle. The triangle is a symh denote the fact that the circuit responds to a transition or edge at CLK input.

Do You Know?	The term bistable multivibrate is the more technical name for a flip-flop, but, it is too much of a mouthful to be used regularly.

The operation of *S-R* flip-flop is illustrated by the waveforms as show Fig. 7.12. This is analysed as follows:

1. Initially all inputs are 0 and the Q output is 0.
2. When the rising edge of the first clock pulse occurs (point a), the S and R inputs are both 0, so the flip-flop is not affected and it remains in the $Q = 0$ state.
3. At the occurrence of the rising edge of the second clock pulse (point c), $S = 1$ and $R = 0$. Thus, the flip-flop sets to the 1 state at the rising edge of this clock pulse.
4. When the third clock pulse makes its positive transition (point e), it finds that $S = 0$ and $R = 1$, which causes the flip-flop to reset to the 0 state.
5. The fourth pulse sets the flip-flop once again to the $Q = 1$ state (point g), because $S = 1$ and $R = 0$ when its positive edge occurs.
6. The fifth pulse finds that $S = 1$ and $R = 0$ when it makes its positive going transition. However, Q is already high, so it remains in that state.
7. The $R = 1$ and $S = 1$ condition should not be used as it results in an indetermi-nate condition.

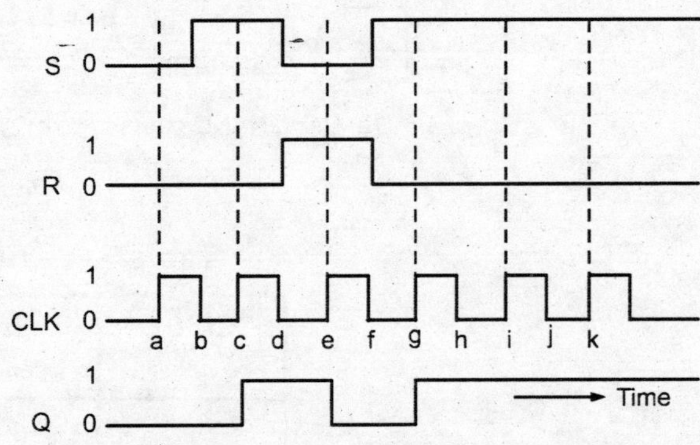

Fig. 7.12. Waveforms of *S-R* flip-flop.

7.5 *D* FLIP-FLOP

The *D* (delay) flip-flop has only one input called the Delay (D) input and two outputs *Q* and $\bar{Q}$. It can be constructed from an *S-R* flip-flop by inserting an inverter between *S* and *R* and assigning the symbol *D* to the *S* input. The structure of *D* flip-flop is shown in Fig. 7.13(*a*). Basically, it consists of a NAND flip-flop with a gating arrangement on its inputs. It operates as follows:

1. When the CLK input is LOW, the *D* input has no effect, since the set and reset inputs of the NAND flip-flop are kept HIGH.
2. When the CLK goes HIGH, the *Q* output will take on the value of the *D* input. If CLK =1 and *D* = 1, the NAND gate-1 output goes 0 which is the $\bar{S}$ input of the basic NAND-based *S-R* flip-flop and NAND gate-2 output goes 1 which is the $\bar{R}$ input of the basic NAND-based S-R flip-flop. Therefore, for $\bar{S}$ = 0 and $\bar{R}$ = 1, the flip-flop output will be 1, i.e. it follows D input. Similarly, for CLK=1 and *D* = 0, the flip-flop output will be 0. If D changes while the CLK is HIGH, *Q* will follow and change quickly.

Do You Know?	Most clockd, FFs also have asynchronous inputs that can set or clear the FF independently of the clock input.

The logic symbol for the *D* flip-flop is shown in Fig. 7.13(*b*). A simple way of building a delay *D* flip-flop is shown in Fig. 7.13(*c*). The truth table of *D* flip-flop is given in Table 7.6 from which it is clear that the next state of the flip-flop at time (Q_{n+1}) follows the value of the input *D* when the clock pulse is applied. As transfer of data from the input to the output is delayed, it is known as delay (*D*) flip-flop. The *D*-type flip-flop is either used as a delay device or as a latch to store 1 bit of bina information.

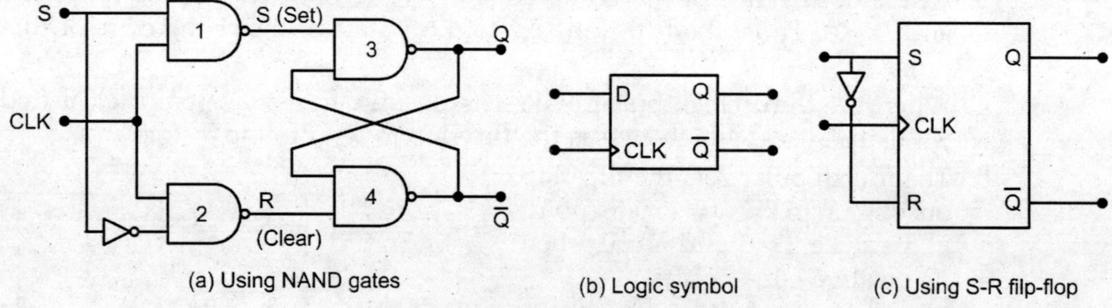

(a) Using NAND gates (b) Logic symbol (c) Using S-R filp-flop

Fig. 7.13. *D* flip-flop

Table 7.6. Truth table of *D* flip-flop.

CLK	Input *D* No change	Output Q_{n+1}
1	1	0
0	1	1
0	X	No change

7.5.1 State Diagram and Characteristic Equation of *D* Flip-flop

The state transition diagram for the delay flip-flop is shown in Fig.7.14.

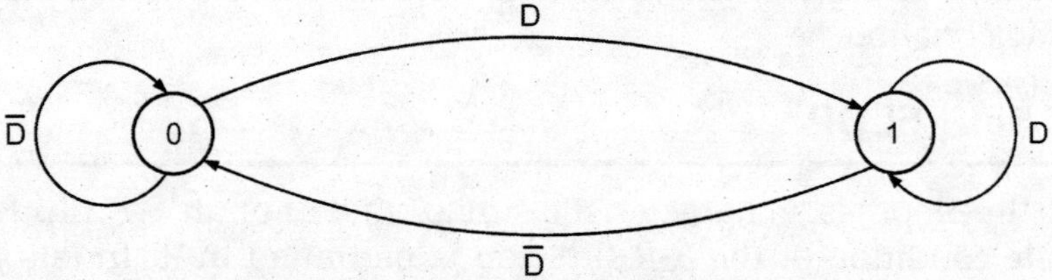

Fig. 7.14. State diagram of delay flip-flop.

From the above state diagram, it is clear that when D = 1, the next state will when D = 0, the next state will be 0, irrespective of its previous state. From the diagram, one can draw the present state-next state table and the application or tion table for the Delay flip-flop as shown in Table 7.7 and Table 7.8 respective)

Table 7.7. Present state-next state table for *D* flip-flop.

Present	Delay Input	Next state
Q_n	D	Q_{n+1}
0	0	0
0	1	1
1	0	0
1	1	1

Table 7.8. Application or excitation table for Dflip-flop.

Q_n	Q_{n+1}	Excitation Input D
0	0	0
0	1	1
1	0	0
1	1	1

Using the Present state-Next state table, the K-map for the next state (Q_{n+1}) of the Delay flip-flop can be drawn as shown in Fig. 7.15, and the simplified expression fore, Q_{n+1} can be found as follows.

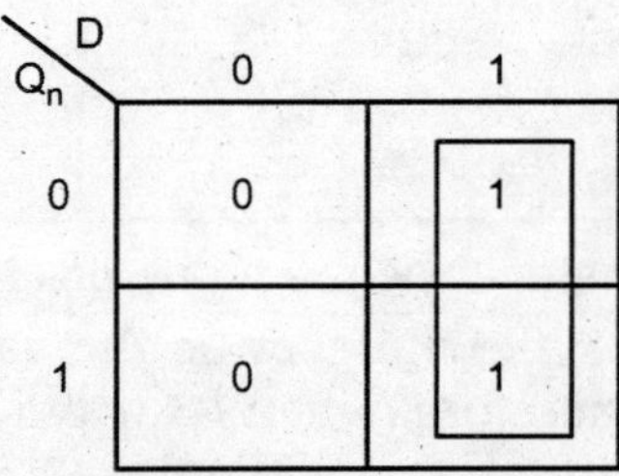

Fig. 7.15. Next state (Q_{n+1}) map for *D* flip-flop

From the above AT-map, the characteristic equation for Delay flip-flop is

$$Q_{n+1} = D$$

Hence, in a Delay flip-flop, the next state follows the Delay input as shown by the characterisitic equation.

7.6 J-K FLIP-FLOP

A *J-K* flip-flop has a characteristic similar to that of an *S-R* flip-flop. In addition, the indeterminate condition of the *S-R* flip-flop is permitted in it. Inputs *J* and *K* behave like inputs *S* and *R* to set and reset the flip-flop, respectively. When $J = K = 1$, the flip-flop output toggles, i.e. switches to its complement state; if $Q = 0$, it switches to $Q = 1$ and vice versa.

A *J-K* flip-flop can be obtained from the clocked *S-R* flip-flop by augmenting two AND gates as shown in Fig. 7. 1 6(*a*). The data input *J* and the output $\bar{Q}$ are applied to the first AND gate, and its output ($J\bar{Q}$) is applied to the *S* input of *S-R* flip-flop. Similarly, the data input AT and the output *Q* are connected to the second AND gate and its output (*KQ*) is applied to *R* input of *S-R* flip-flop. The graphic symbol of *J-K* flip-flop is shown in Fig. 7.16(*b*) and the truth table is shown in Table 7.9. The output for the four possible input sequences are described below.

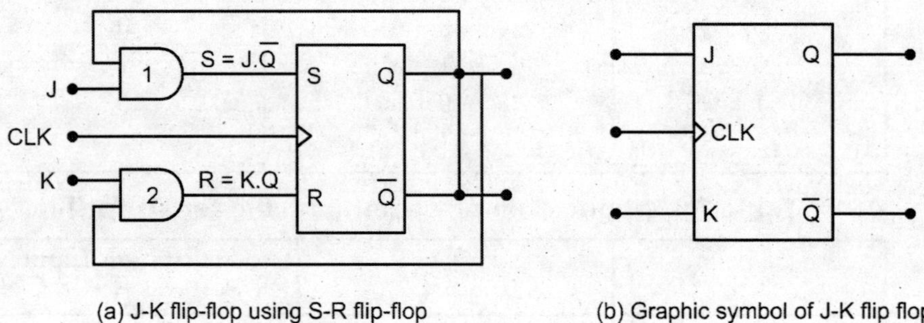

(a) J-K flip-flop using S-R flip-flop (b) Graphic symbol of J-K flip flop

Fig. 7.16

Table 7.9. Truth table of *J-K* flip-flop.

	Inputs		Output	Action
CLK	***J***	***K***	Q_{n+1}	
X	0	0	Q_n	No change
1	0	1	0	Reset
1	1	0	1	Set
1	1	1	$\bar{Q}_n$	Toggle

Case 1: When $J = 0$, whatever be the value of $\bar{Q}_n$ (0 or 1), the output of the first AND gate is 0. Similarly, when $K = 0$, whatever be the value of Q_n (0 or 1), the output of the second AND gate is also 0. Therefore, when $J = 0$ and $K = 0$, the inputs to the basic flip-flop are S = 0 and R = 0. This condition forces the flip-flop into the same state.

Case 2: When $J = 0$, $K = 1$ and the previous state of flip-flop is SET (i.e. $Q_n = \bar{Q}_n = 0$), then $S = J\bar{Q}_n = 0 \times 0 = 0$ and $R = KQ_n = 1 \times 1 = 1$. Since $S = 0$ and $R = 1$, the flip-flop RESETS on the application of a clock pulse. But, if the flip-flop is already RESET (i.e., $Q_n = 0$ and $\bar{Q}_n = 1$), then the application of $J = 0$ and $K = 1$ will not alter the state of the flip-flop ($S = J\bar{Q}_n = 0 \times 1 = 0$ and $R = KQ_n = 1 \times 0 = 0$), and it remains in the RESET state.

Case 3: When $J = 1$, $K = 0$ and the previous state of the flip-flop is RESET and ($Q_n = 0$), then $\bar{Q}_n = 1$), then $S = J\bar{Q}_n = 1 \times 1 = 1$ and $R = KQ_n = 0 \times 0 = 0$. Since $S = 1$ and $R = 0$, the flip-flop will SET on the application of a clock pulse. But, if the flip-flop is already set (i.e. $\bar{Q}_n = 1$, $\bar{Q}_n = 0$), then the inputs $J = 1$, $K = 0$ will make $S = 0$ and $R = 0$, and the flip-flop remains in the SET state.

Case 4: The condition $J = 1, K = 1$ deserves special attention. When $J = 1$, $K = 1$ and the previous state is a SET state (i.e. $Q_n = 1$, $\bar{Q}_n = 0$), then $S = J\bar{Q}_n = 1 \times 0 = 0$ and $R = KQ_n = 1 \times 1 = 1$. Since $S = 0$ and $R = 1$, the flip-flop RESETS on the applicatia of a clock pulse, i.e. the flip-flop toggles from SET to RESET state. Again for $J = 1$, $K = 1$, if the previous state is a RESET state (i.e. $Q_n = 0$, $\bar{Q}_n = 1$), then $S = J\bar{Q}_n = 1 \times 1 = 1$ and $R = K\bar{Q}_n = 1 \times 0 = 0$. Since $S = 1$ and $R = 0$, the flip-flop on tli application of a clock pulse will toggle to the SET state. Hence, when $J = 1$ and $K = 1$ the flip-flop toggles on the application of the next clock pulse. Toggle means to switch to the opposite state. The flip-flop will continuously change its state when $J = K = 1$ and CLK=HIGH, resulting in an unstable output.

Do You Know?	The D-latch is a modified NAND latch that operates like a D flip-flop except that it is not edge triggered.

It is very important to realise that because of the feedback connection in tin *J-K* flip-flop, a CLK pulse that remains in the 1 state while both *J* and *K* are equal to 1 will cause the output to complement again and again until the clock pulse goes bad to 0. To avoid this undesirable operation, the clock pulse must have a time duratioi that is shorter than the propagation delay time of the flip-flop. This is a restrictive requirement, since the operation of the circuit depends on the width of the pulse. Tin restriction on the pulse width can be eliminated with a master slave or edge triggerd construction which will be discussed shortly.

The operation of the *J-K* flip-flop is illustrated by the waveforms in Fig. 7.17, which can be analysed as follows:

1. Initially all inputs are 0 and the output *Q* is 1.
2. When the first clock pulse goes positive (point a), the $J = 0$, $K = 1$ condition exists. Thus, the flip-flop will be cleared to the $Q = 0$ state.
3. The second clock pulse finds $J = K = 1$ when it makes its positive transition (point c). This causes the flip-flop to toggle to its opposite state, $Q = 1$.
4. At point e on the clock waveform, *J* and *K* are both 0, and hence the flip-flop does not change state on this transition.
5. At point g, $J = 1$ and $K = 0$. This is the condition that sets *Q* to the 1 state. However, it is already 1, and it will remain in the same state.
6. At point i, $J = 1 = K$. The flip-flop toggles to its opposite state. The same toggling of state occurs at point k since $J = K = 1$.

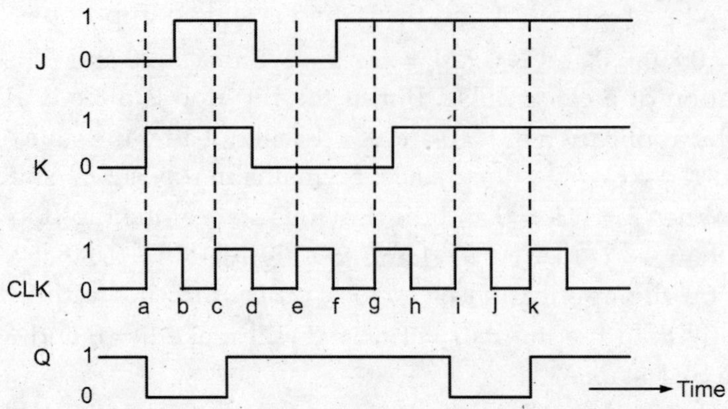

Fig. 7.17. Waveforms of the *J-K* flip-flop operation

7.6.1 State Diagram and Characteristic Equation of *J-K* Flip-flop

The state transition diagram for *J-K* flip-flop can be drawn as shown in Fig. 7.18.

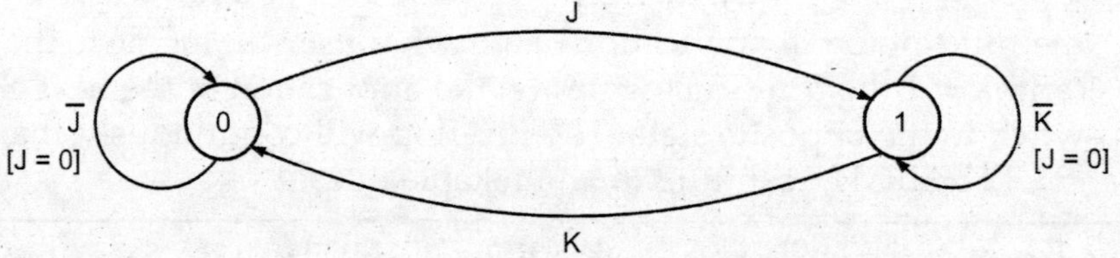

Fig. 7.18. State diagram of *J-K* flip-flop

From the above state diagram, one can easily understand that the state transition from 0 to 1 takes place whenever J is asserted (i.e. $J = 1$) irrespective of K value. Similarly, state transition from 1 to 0 takes place whenever K is asserted (i.e. $K = 1$) irrespective of the value of J. Also, the state transition from 0 to 0 occurs whenever $J = 0$, irrespective of the value of K, and the state transition from 1 to 1 occurs whenever $K = 0$, irrespective of J value.

From the above state diagram and truth table (Table 7.9) of *J-K* flip-flop, the present state-next state table and application table or excitation table for *J-K* flip-flop are shown in Table 7.10 and Table 7.11 respectively.

Table 7.10. Present state-next state table for *J-K* flip-flop.

Present state	Inputs		Next state
Q_n	J	K	Q_{n+1}
0	0	0	0
0	0	1	0
0	1	0	1
0	1	1	1
1	0	0	1
1	0	1	0
1	1	0	1
1	1	1	0

Table 7.11. Aplication or excitation table for *J-K* flip-flop.

Q_n	Q_{n+1}	Excitation inputs	
		J	*K*
0	0	0	d
0	1	1	d
1	0	d	1
1	1	d	0

From the Table 7.10, a Karnaugh map for the next state transition (Q_{n+1}) cant drawn as shown in Fig. 7.19, and the simplified logic expression which represents! characteristic equation of *J-K* flip-flop can be found as follows.

From the *K*-map shown in Fig. 7.19, the characteristic equation of *J-K* flip-flop can be written as

$$Q_{n+1} = J\bar{Q}_n + \bar{K}Q_n$$

Q_n \ JK	00	01	11	10
0	0	0	1	1
1	1	0	0	1

Fig. 7.19. Next state (Q_{n+1}) map for *J-K* flip-flop.

7.7 *T* FLIP-FLOP

Another basic flip-flop, called the *T* or trigger or toggle flip-flop, has only a single data (*T*) input, a clock input and two outputs *Q* and $\bar{Q}$. The *T*-type flip-flop is obtained from a *J-K* flip-flop by connecting its *J* and *K* inputs together. The designation *T* comes from the ability of the flip-flop to "toggle" or complement its state.

Do You Know?	The *T* flip-flop divides the clock frequency by 2. Hence, a *T* flip-flop can be used as a frequency divider.

The block diagram of a *T* flip-flop and its circuit implementation using a *J-K* flip-flop are shown in Fig. 7.20. The *J* and *K* inputs are wired together. The truth table for *T* flip-flop is shown in Table 7.12.

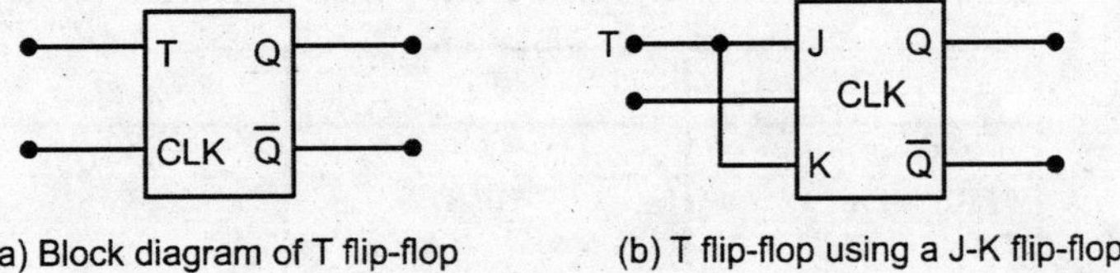

(a) Block diagram of T flip-flop (b) T flip-flop using a J-K flip-flop

Fig. 7.20

When the T input is in the 0 state (i.e. $J = K = 0$) prior to a clock pulse, the Q output will not change with clocking. When the T input is at a 1 (i.e. $J = K = 1$) level prior to clocking, the output will be in the Q state after clocking. In other words, if the T input is a logical 1 and the device is clocked, the output will change state regardless of what output was prior to clocking. This is called toggling hence the name T flip-flop.

Table 7.12. Truth table of *T* flip-flop.

Q_n	T	Q_{n+1}
0	0	0
0	1	1
1	0	1
1	1	0

The truth table shows that when $T = 0$, then $Q_{n+1} = Q_n$, i.e., the next state is the same as the present state and no change occurs. When $T = 1$, then $Q_{n+1} = \overline{Q}_n$, i.e. the state of the flip-flop is complemented.

Application of *T* flip-flop: T-type flip-flop is most often seen in counters and sequential counting networks because of its inherent divide-by-2 capability. When a clock pulse is applied, the output changes state once every input cycle, thus repeating one cycle for every two input cycles. This is the action required in many cases for binary counters.

7.7.1 State Diagram and Characteristic Equation of *T* Flip-flop

The state transition diagram for the Trigger flip-flop is shown in Fig. 7.21.

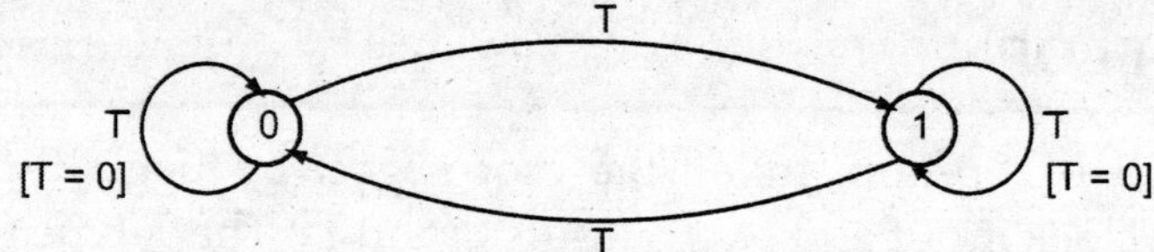

Fig. 7.21. State diagram of trigger flip-flop

From the above state diagram, it is clear that when $T = 1$, the flip-flop changes or toggles its state irrespective of its previous state; when $T = 1$ and $Q_n = 0$, the next state will be 1; when $T = 1$ and $Q_n = 1$, the next state will be 0. Similarly, one can understand that when $T = 0$, the flip-flop retains its previous state. From the above state diagram, one can draw the present state-next state table and application or excitation table for the Trigger flip-flop as shown in Table 7.13 and Table 7.14 respectively.

Table 7.13. Present state-next state table for *T* flip-flop.

Q_n	T	Q_{n+1}
0	0	0
0	1	1
1	0	1
1	1	0

Table 7.14. Application or excitation table for *T* flip-flop.

Q_n	Q_{n+1}	Excitation input *T*
0	0	0
0	1	1
1	0	1
1	1	0

From the Table 7.13, the *K*-map for the next state (Q_{n+1}) of trigger flip-flop can be drawn as shown in Fig. 7.22, and the simplified expression for Q_{n+1} can be found as follows.

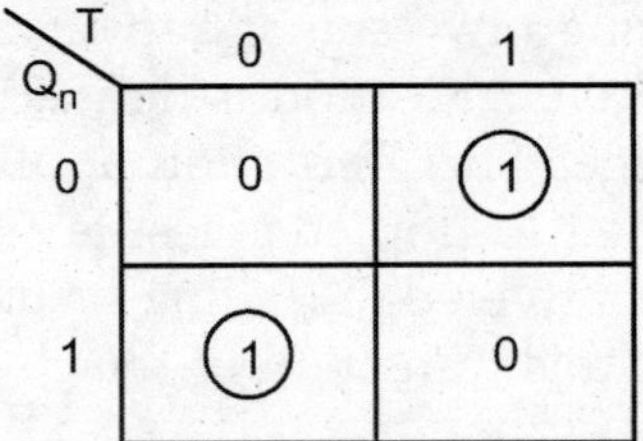

Fig. 7.22. Next state (Q_{n+1}) map for *T* flip-flop.

From the K-map shown in Fig. 7.22, the characteristic equation for trigger flip-flop is

$$Q_{n+1} = T\bar{Q}_n + \bar{T}Q_n$$

So, in a trigger flip-flop, the next state will be the complement of the previous state when $T = 1$.

7.8 TRIGGERING OF FLIP-FLOPS

Flip-flops are synchoronous bistable devices. The term synchronous means that the changes in the output occur at a specified point on a triggering input called the clock; that is, changes in the output occur in synchronisation (in time) with the clock. Based on the specfiec interval or point in the clock during or at which triggering of flip-flop takes place, it can be classified into two different types.

(*i*) Level triggering

(*ii*) Edge triggering

A clock pulse starts from an initial value of 0, goes momentarily to 1, and after a short time, returns to its initial 0 value. The time interval from the application of the pulse until the output transition occurs is a critical factor that needs further investigation.

7.8.1 Level Triggering in Flip-flops

While discussing clocked *S-R* flip-flop in the previous section, it was shown that the clock input triggers the flip-flop, i.e. enables the flip-flop, when the clock pulse goes HIGH, and die flip-flop is said to be level triggered flip-flop. Since the flip-flop changes its state when the clock is positive, it is termed as Positive level triggered flip-flop. If a NOT gate is introduced in between the clock input and die input of AND gate in Fig. 7.10, the flip-flop changes its state only when the clock is negative (i.e. when clock = LOW), and it is called Negative level triggered flip-flop.

The main drawback of level triggering is that, as long as the clock is positive or negative, the flip-flop changes its state more man once or many times for the change in inputs. If the inputs are made stable for the entire clock duration, then the output changes only once. On the other hand, if the frequency of input change is higher than die input clock frequency, the output of the flip-flop undergoes multiple changes when the clock is positive or negative. When the clock becomes unasserted (i.e. clock = 0), the output of the flip-flop reflects the last change in its inputs. This can be overcome in Master-Slave flip-flops and Edge-triggered flip-flops where die flip-flops change state only once for a clock.

7.8.2 Edge Triggering in Flip-flops

A clock pulse goes through two signal transitions from 0 to 1 and returns from 1 to 0. As shown in Fig. 7.23, a positive transition is defined as the positive edge and a nega-tive transition as the negative edge. This definition applies also to negative pulses. The term edge-triggered means that the flip-flop changes its state either at the positive edge (rising or leading edge) or at the negative edge (falling or trailing edge) of the clock pulse and is sensitive to its inputs only at this transition of the clock.

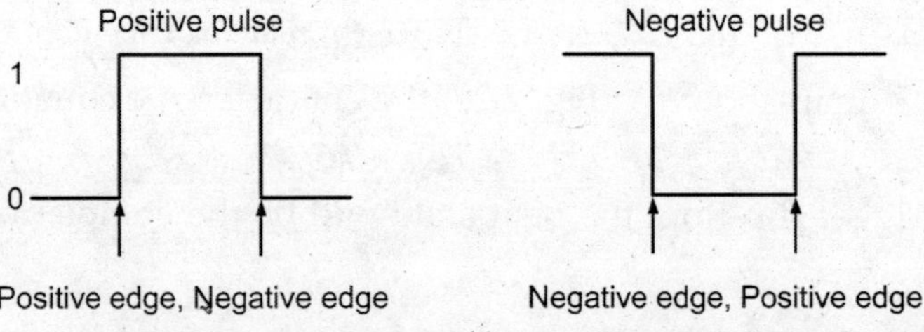

Fig. 7.23. Definition of clock pulse transition.

One way to make die flip-flop respond only to a pulse transition is to use capacitive coupling. An R-C (resistor-capacitor) circuit as shown in Fig. 7.24(*a*) must be inserted in the clock input of die flip-flop. By deliberate design, the RC time constant is much smaller than the clock's pulse width. Because of this, die capacitor can charge fully when the clock goes high; diis exponential charging produces a narrow positive voltage spike across the resistor. Later, die trailing edge of die pulse results in a narrow negative spike. The circuit that generates a spike in response to a momentary change of input signal is called R-C differentiator circuit. Edge triggering is achieved by designing the flip-flop to neglect one spike and trigger on die occurrence of the odier spike.

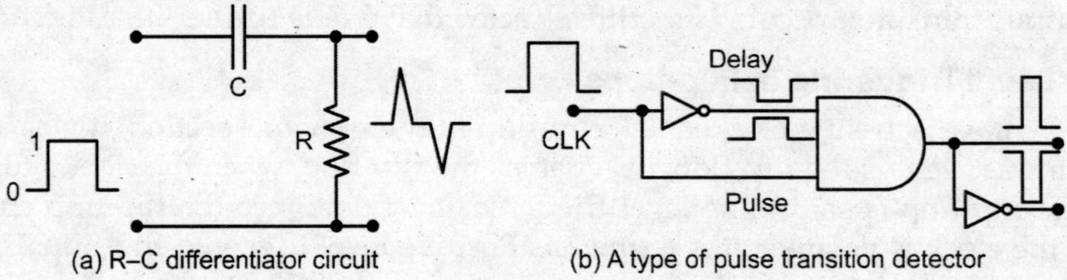

(a) R–C differentiator circuit

(b) A type of pulse transition detector

Fig. 7.24

Another type of a pulse transition detector is shown in Fig. 7.24(*b*). There is a small delay on one input of the AND gate due to propagation delay of the NOT gate. So, the inverted clock pulse arrives at the gate input a few nanoseconds after the true clock pulse. This produces an output spike with a time duration of only a few nanoseconds.

In certain applications, triggering on the negative edge is better suited. In this case, an inverter complements the clock pulse.

7.8.3 Edge-triggered *D* Flip-flop

The circuit of an edge-triggered *D* flip-flop is shown in Fig. 7.25. Here, a differentiating circuit comprising of a capacitor and a resistor is inserted between the clock and the input to the AND gates. The time constant (RC) is made much smaller than the clock's pulse width.

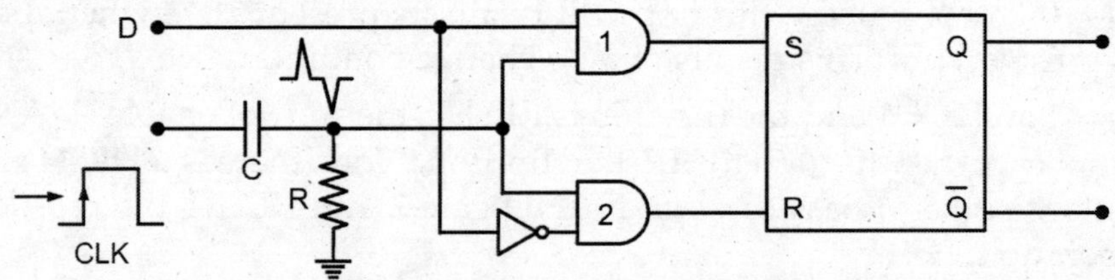

Fig. 7.25. Edge-triggered D flip-flop.

The narrow positive spike enables both the AND gates for an instant. During the time of narrow positive spike, *D* and its complement hit the inputs of the flip-flop, forcing *Q* to set or reset (unless *Q* already equals *D*).

The truth table given in Table 7.15 shows the action of a positive edge-triggen *D* flip-flop. When the CLK input is LOW, the flip-flop output retains its previous state, irrespective of whether $D = 0$ or 1. When the leading edge of the clock pulse designated by the upward arrow, is passed through the AND gates, the flip-flop output (*Q*) takes on the value of *D*. During the trailing edge of the CLK, designated by a downward arrow, *Q* remains in its last state, irrespective of the value of *D*.

Table 7.15. Truth table of the positive edge-triggered *D* flip-flop.

CLK	*D*	$Q_{(n+1)}$	Action
0	X	Q_n	Last state
↑	0	0	Reset
↑	1	1	Set
↓	X	Q_n	Last State

The logic diagram of a basic positive edge-triggered *J-K* flip-flop is shown in Fig. 7.26. Note that the *Q* output is connected back to the input of NAND gate-2, and the $\bar{Q}$ output is connected back to the input of NAND gate-1.

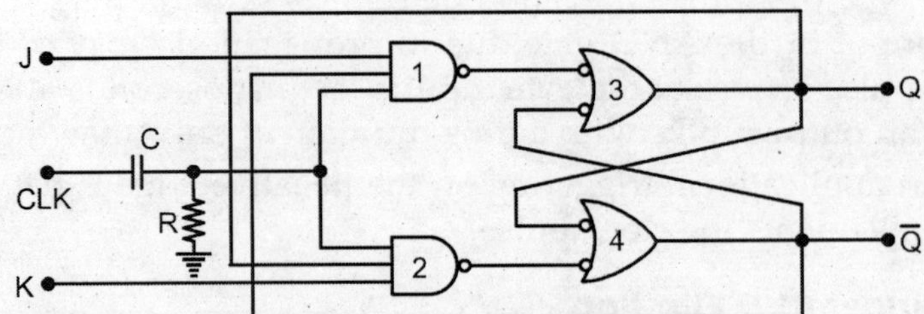

Fig. 7.26. Logic diagram for a positive edge-triggered *J-K* flip-flop.

Let us start by assuming that the flip-flop is RESET (i.e. $Q = 0$) and $J = 1$ and $K = 0$. When a CLK pulse occurs, a leading-edge spike is passed through NAND gate-1 because $\overline{Q} = 1$ and $J = 1$. This causes the latch portion of the flip-flop to change to the SET state.

If $J = 0$ and $K = 0$, the flip-flop will retain its present state when the leading edge of the clock pulse occurs. This results in a no change condition.

If $J = 1$ and $K = 1$ and the flip-flop is RESET, then $\overline{Q} = 1$ enables NAND gate-1, so that the clock spike passes through to SET the flip-flop. Now, there is a HIGH on Q which allows the next clock spike to pass through NAND gate-2 and RESET the flip-flop. This is called toggle operation.

The operation of the edge-triggered *J-K* flip-flop is summarised in truth table given in Table 7.16. The truth table for a negative edge-triggered device is the same except that it is triggered on the trailing edge of the CLK pulse.

Table 7.16. Truth table of edge-triggered *JK* flip-flop.

CLK	Inputs		Outputs		Action
	J	*K*	$Q_{(n+1)}$	$\overline{Q}_{(n+1)}$	
↑	0	0	Q_n	$\overline{Q}_n$	No change
↑	0	1	0	1	Reset
↑	1	0	1	0	Set
↑	1	1	$\overline{Q}_n$	Q_n	Toggle

7.9 ASYNCHRONOUS INPUTS IN FLIP-FLOPS

So far, in our discussion, we have encountered *S, R, J, K, D* and *T* inputs in the flip-flops. These inputs, normally referred to as control inputs, are also called as synchronous inputs. For such synchronous inputs, the flip-flop changes its state in synchronization with the clock input only. So, the synchronous control inputs must be used along with a clock signal to trigger a change in the flip-flop.

In addition to the above synchronous inputs, most clocked flip-flops have one or more asynchronous inputs whose effect on the flip-flop output is independent of the synchronous input and the clock input. Asynchronous inputs can be used to set the flip-flop to the 1 state, or clear the flip-flop to the 0 state at any time, regardless of the condition at the other inputs. The asynchronous inputs are override inputs, since they can be used to override all other inputs in order to place the flip-flop in one state or the other.

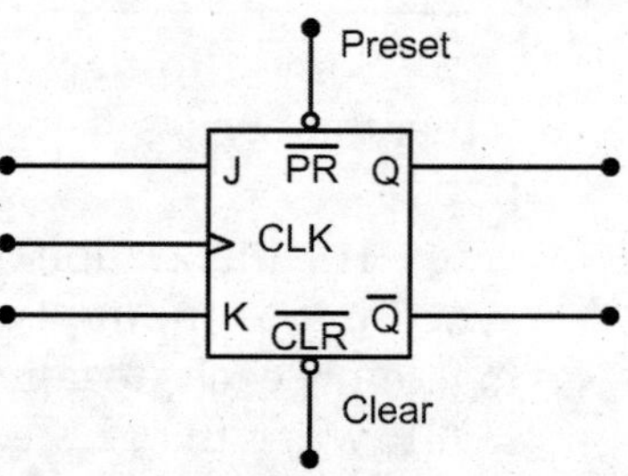

Fig. 7.27 Logic symbol for a J-K flip-flop with active LOW preset and clear inputs

Figure 7.27 shows a clocked *J-K* flip-flop with $\overline{\text{PRESET}}$ $\overline{\text{(PR)}}$ and $\overline{\text{CLEAR}}$ $\overline{\text{(CLR)}}$, or direct set and direct reset inputs. These are active LOW inputs. An active LOW level at the preset input will SET the flip-flop, and an active LOW level on the clear input will RESET it. These asynchronous inputs are activated by a 0 level, as indicated by the small bubbles on the flip-flop symbol. Both the preset and clear inputs must be kept HIGH for synchronous operation. Truth Table 7.17 shows how these inputs operate. Simultaneous LOW levels on preset and clear are forbidden since it leads to an ambiguous condition.

Table 7.17. Truth Table.

Inputs		**Output**		**Action**
Preset	**Clear**	Q_{n+1}	$\bar{Q}_{n+1}$	
0	0	?	?	Forbidden
0	1	1	0	Set
1	0	0	1	Reset
1	1	Q_n	$\bar{Q}_n$	No change

The logic diagram for an edge-triggered *J-K* flip-flop with preset $\overline{\text{PR}}$ and cle $\overline{\text{(CLR)}}$ inputs is shown in Fig. 7.28. The asynchronous inputs are directly connected! the latch portion of the flip-flop so that they override the effect of the synchrom inputs, *J*, *K* and the clock.

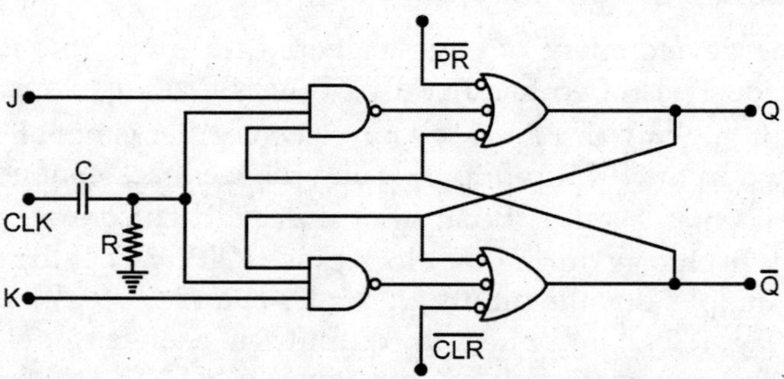

Fig. 7.28 Logic diagram for a basic *J-K* flip-flop with active-LOW preset and clear inpit.

From Fig. 7.28, it is clear that if a constant 0 is applied to the $\overline{\text{PR}}$ (preset) input, flip-flop output (*Q*) will become 1 regardless of the status of the other inputs.

Similarly, a constant LOW (0) at the $\overline{\text{CLR}}$ (clear) input holds the $\bar{Q}$ output of the flip-flop in the HIGH state, i.e. $Q = 0$. Thus, asynchronous inputs can be used-to hold the flip-flop in a particular state for any desired interval of time.

Some flip-flops have asynchronous inputs that are activated by 1's (active HIGH) rather than by 0's. The small bubble on the preset and clear input is not shown for these flip-flops.

The illustration of the operation of the asynchronous inputs is shown in Fig. 7.29. The synchronous inputs *J* and *K* are held HIGH as CLK pulses are applied. The flip-flop will toggle on each negative-going transition as long as the preset and clear inputs are both HIGH when the CLK edge occurs. This happens at points a, c, d and f. The output *Q* can also change states in response to a LOW at preset or clear. The low preset causes *Q* to go to the 1 state immediately, regardless of the CLK state (point b). Similarly, the LOW at clear causes *Q* to go to 0 immediately (point e). Note that this LOW at clear prevents the fourth CLK edge from toggling *Q*; this is the override feature.

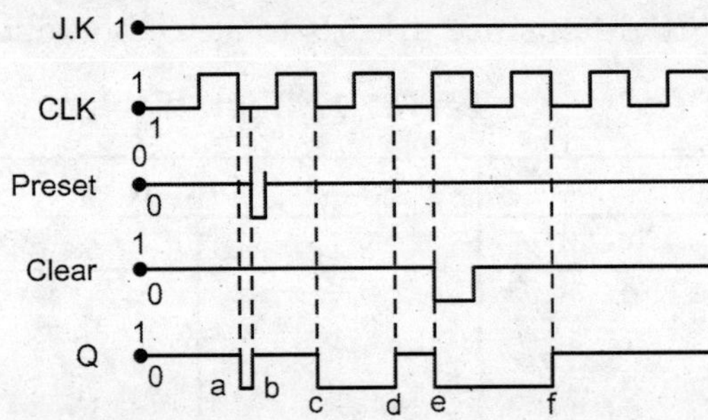

Fig. 7.29. Asynchronous inputs can cause flip-flop to change states immediately without waiting for clock transition.

7.10 MASTER-SLAVE FLIP-FLOPS

7.10.1 The Race-around Condition

Before the development of edge-triggered flip-flops, the timing problem in level-trig-gered flip-flops was often handled by Master-Slave flip-flops.

The condition *S* = 1 and *R* = 1 we have seen earlier is not allowed in an *S-R* flip-flop. This is eliminated in the *J-K* flip-flop by using the feedback connection from the outputs to the inputs of the gates 1 and 2. Because of the feedback connection $Q(\bar{Q})$ at the input to *K* (*J*), the input will change during the clock pulse (CLK = 1), if the output changes the state. Consider, for example, that the inputs are *J* = *K* = 1 and *Q* = 0. When a clock pulse of width t_p, as shown in Fig. 7.30(*a*) is applied, the output will change from 0 to 1 after a time interval Δt, where Δt *is* the propagation delay through two NAND gates in series. Now, after time Δt, we have *J* = *K* = 1 and *Q* = 1 and after, another interval of Δt, output *Q* will become 0. Hence, the output will oscillate back and forth between 0 and 1 in the duration t_p of the clock pulse width. So, at the end of the clock pulse, the value of *Q* is ambiguous. This situation is known as a race-around conditon.

The race-around condition can be avoided If $t_p < \Delta t$ as shown in Fig. 7.30(*b*). Lumped delay lines can be used in series with the feedback connections in order to increase the loop delay beyond, and hence to prevent the race-around difficulty. The above race-around condition can also be avoided in Master-Slave flip-flops.

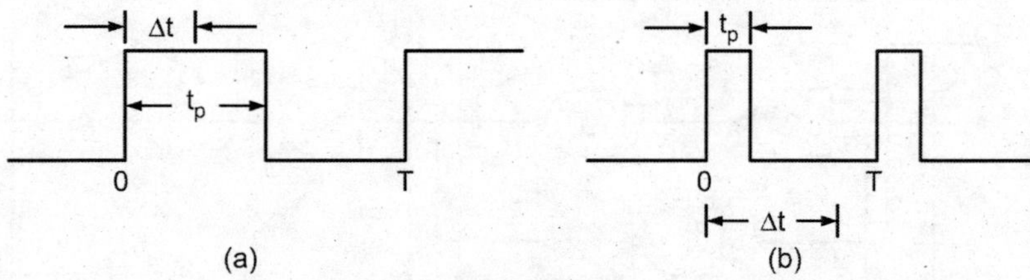

Fig. 7.30

A Master-Slave flip-flop is constructed using two separate flip-flops connected serially. The first flip-flop serves as the master and the second as a slave, and the overall circuit is referred to as a Master-Slave flip-flop.

7.10.2 Master-Slave *J-K* Flip-flop

A Master-Slave flip-flop can be constructed using two *J-K* flip-flops as shown in Fig. 7.31. The first flip-flop, called the master, is driven by the positive edge of the clock pulse; the second flip-flop, called the slave, is driven by the negative edge of the clock pulse. Therefore, when the clock input has a positive edge, the master acts according to its *J-K* inputs, but the slave does not respond since it requires a negative edge at the clock input. When the clock input has a negative edge, the slave flip-flop copies the master outputs. But the master does not respond to the feedback from Q and $\bar{Q}$, since it requires a positive edge at its clock input. Thus, the master-slave flip-flop does not have race-around problem.

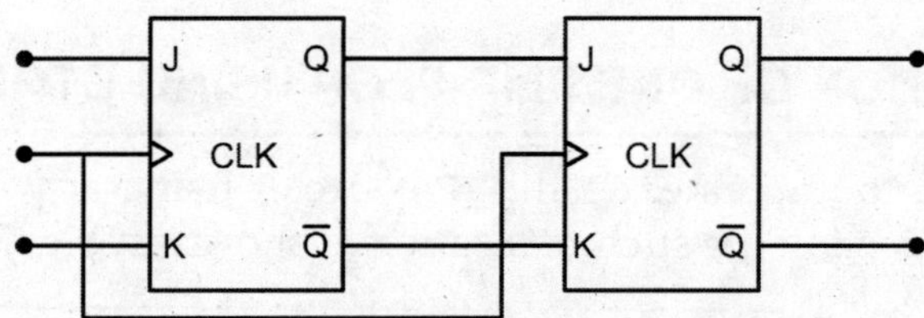

Fig. 7.31. Master-slave *J-K* flip-flop.

A master-slave *J-K* flip-flop constructed using NAND gates is shown in Fig 7.32. It consists of two flip-flops connected in series. NAND gates-1 through 4 form the master flip-flop and NAND gates-5 through 8 form the slave flip-flop. When tin clock is positive, a change in J and K inputs cause a change of state in the master flip-flop. During this period, the slave retains its previous state and serves as a buffet between the master and the output. When the clock goes negative, the master flip-flof does not respond, i.e. it maintains its previous state, while the slave flip-flop is enabled and changes its state to that of the master flip-flop. The new state of the slave then becomes the state of the entire master-slave flip-flop. The operation of master-slave *J-K* flip-flop for different *J-K* input combinations can be explained as follows:

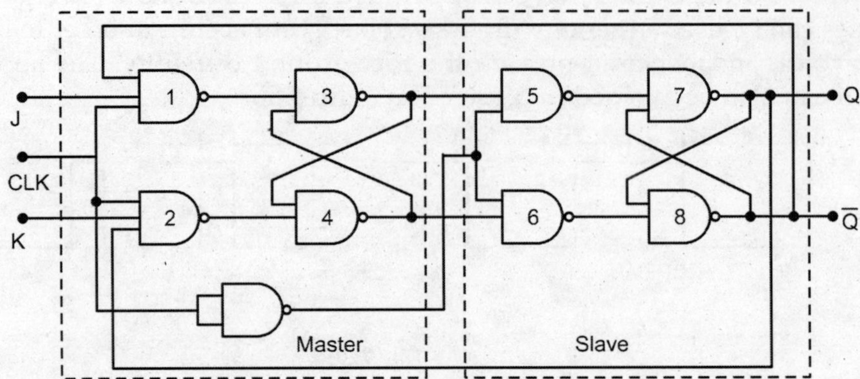

Fig. 7.32. Clocked master-slave *J-K* flip-flop using NAND gates.

If $J = 1$ and $K = 0$, the master flip-flop sets on the positive clock edge. The HIGH $Q(1)$ outputs of the master drives the input (J) of the slave. So, when the negative clock edge hits, the slave also sets. The slave flip-flop copies the action of the master flip-flop.

If $J = 0$ and $K = 1$, the master resets on the leading edge of the CLK pulse. The HIGH $\overline{Q}$ output of the master drives the input (K) of the slave flip-flop. Then, the slave flip-flop resets at the arrival of the trailing edge of the CLK pulse. Once again, the slave flip-flop copies the action of the master flip-flop.

If $J = K = 1$, the master flip-flop toggles on the positive clock edge and the slave toggles on the negative clock edge. The condition $J = K = 0$ input does not produce any change.

Master-Slave flip-flops operate from a complete clock pulse, and the outputs change on the negative transition.

7.11 REALISATION OF ONE FLIP-FLOP USING OTHER FLIP-FLOPS

It is possible to implement a flip-flop circuit using any other flip-flop. The general block diagram connection for such realisation is shown in Fig. 7.33.

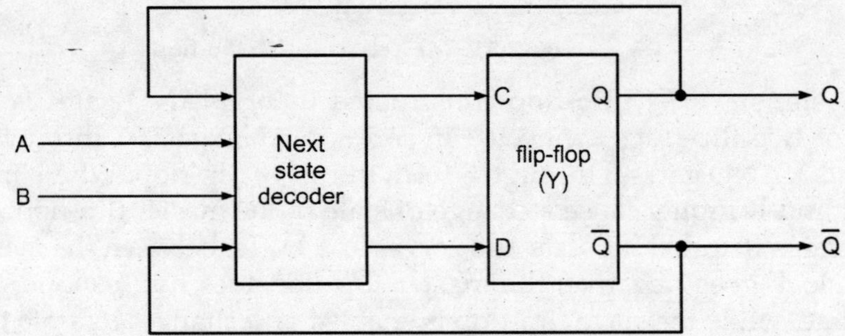

Fig. 7.33. Block diagram to realise flip-flop (X) using flip-flop (Y).

To realise a flip-flop (X) using another flip-flop (Y), the flip-flop (Y) along with a combinational circuit, called NEXT state decoder, is used which functions like flip flop (X).

For this to be realised, for each set of inputs of flip-flop (X), i.e. A & B and present state (Q_n), we have to find the inputs to the flip-flop (Y), i.e. C & D, that will cause the flip-flop (Y) to make transition into the proper next state (Q_{n+1}) of the flip flop (X). These inputs to flip-flop (Y) are called the next state code. The design procedure for such realisation can be summarised in the following steps:

(*i*) Obtain a clear word description of the desired flip-flop (X).

(*ii*) Obtain a Present State-Next State (PS-NS) table for the desired flip-flop (X),

(*iii*) Using the excitation table or application table of the chosen flip-flop (Y), append the next state code or the excitation input values to the above present state-next state table.

(*iv*) Using *K*-maps, simplify the logic expressions for excitation inputs of flip-flop (Y) and design the next state decoder logic,

(*v*) Draw a circuit for the desired flip-flop (X) using next state decoder logic an the chosen flip-flop (Y) as shown in the block diagram.

The above five steps can be clearly explained in the following sections.

7.11.1 Realisation of Delay Flip-flop using *S-R* Flip-flop

Consider the realisation of Delay flip-flop using *S-R* flip-flop. Here, the desired flip-flop (X) is Delay flip-flop and the chosen flip-flop(Y) is *S-R* flip-flop. The basic block diagram is shown in Fig. 7.34.

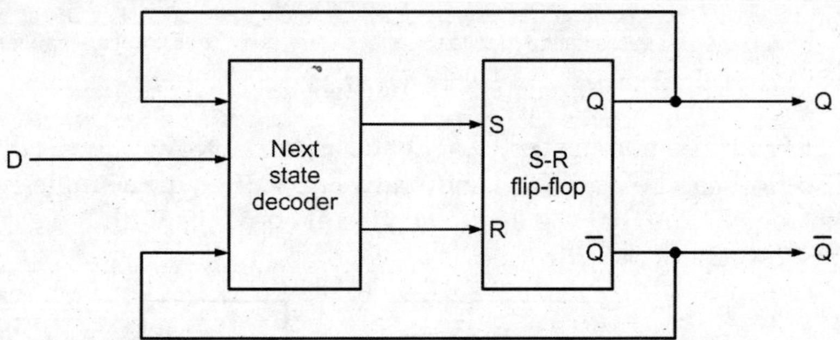

Fig. 7.34. Block diagram of *D*-flip-flop using *S-R* flip-flop.

Step 1: The operation of Delay flip-flop is well knwon, and hence ther is no need for word description.

Step 2: The Present State – Next State table for *D* flip-flop can be drawn as shown in Table 7.17.

Table 7.17. PS-NS table for *D* flip-flop.

Q_n	D	Q_{n+1}
0	0	0
0	1	1
1	0	0
1	1	1

Step 3: Using the application table of *S-R* flip-flop given in Table 7.4., the next state codes, i.e. *S* & *R* values, can be augmented in the above PS-NS table as shown in Table 7.18.

Table 7.18 Excitation table for *D* flip-flop realisation using *S-R* flip-flop.

Q_n	D	Q_{n+1}	Excitation inputs	
			S	R
0	0	0	0	d
0	1	1	1	0
1	0	0	0	1
1	1	1	d	0

Step 4: In this step, one can design the next state decoder, i.e. the simplified expression for *S* and *R*, from the excitation maps as shown in Fig. 7.35 (*a*) and (*b*).

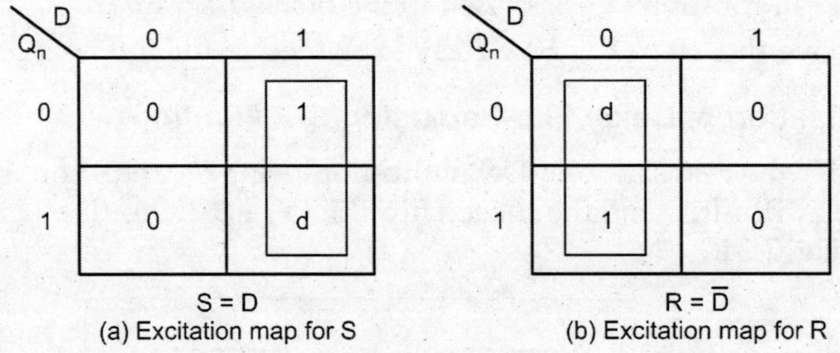

(a) Excitation map for S (b) Excitation map for R

Fig. 7.35

Step 5: From the above step, $S = D$ and $R = \overline{D}$. Now, the circuit for Delay flip-flop using *S-R* flip-flop can be drawn as shown in Fig. 7.36, with a single NOT gate in the next state decoder logic. Note that the designed circuit confirms with the earlier constructions of the delay flip-flop using *S-R* flip-flop.

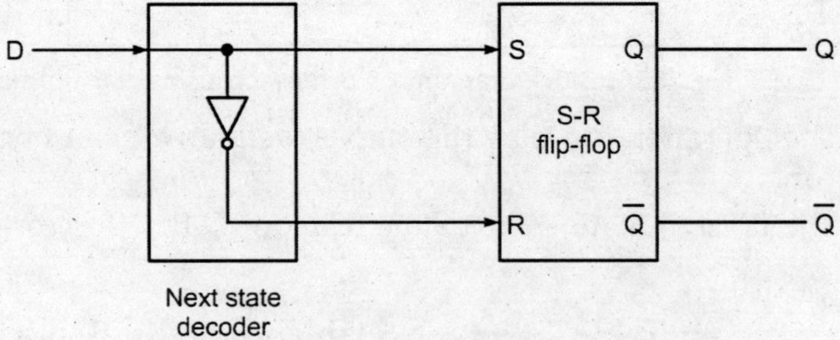

Fig. 7.36. *D* flip-flip using *S-R* flip-flop.

7.11.2 Realisation of *J-K* Flip-flop using *S-R* Flip-flop

In this section, let us consider the realisation of *J-K* flip-flop using *S-R* flip-flop. Here, the desired flip-flop (X) is a *J-K* flip-flop and the chosen flip-flop (Y) in an *S-R* flip-flop. The block diagram of *J-K* flip-flop using *S-R* flip-flop is shown in Fig. 7.37.

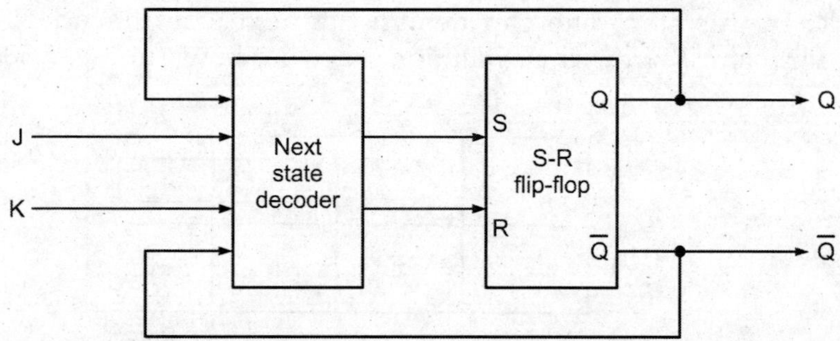

Fig. 7.37. Block diagram of *J-K* flip-flop using *S-R* flip-flop.

Step 1: The operation of a *J-K* flip-flop is well known; hence, dewscription is not given.

Step 2: The Present State-Next State table for *J-K* flip-flop can be drawn as shown in Table. 7.19.

Table 7.19 PS-NS table for *J-K* flip-flop.

Q_n	J	K	Q_{n+1}
0	0	0	0
0	0	1	0
0	1	0	1
0	1	1	1
1	0	0	1
1	0	1	0
1	1	0	1
1	1	1	0

Step 3: Using the application table of *S-R* flip-flop given in Table 7.4, the next state codes, i.e. *S* & *R* values, can be augmented in the above PS-NS table as shown in Table 7.20.

Table 7.20 Excitation table for *J-K* flip-flop realisation using *S-R* flip-flop.

Q_n	J	K	Q_{n+1}	Excitation inputs	
				S	R
0	0	0	0	0	d
0	0	1	0	0	d
0	1	0	1	1	0
0	1	1	1	1	0
1	0	0	1	d	0
1	0	1	0	0	1
1	1	0	1	d	0
1	1	1	0	0	1

Step 4: In this step, one can design the next state decoder, i.e. the simplified expressions for *S* and *R*, from the excitation maps as shown in Fig. 7.38(*a*) and (*b*).

Q_n \ JK	00	01	11	10
0	0	1	1	1
1	d	0	0	d

$S = J\overline{Q}_n$

Fig. 7.38(*a*) Excitation map for *S*.

Step 5: From the above step, $S = J\overline{Q}_n$ and $R = KQ_n$. Now, the circuit for *J-K* flip-flop using *S-R* flip-flop can be drawn as shown in Fig. 7.39 with two AND gates in the next state decoder logic.

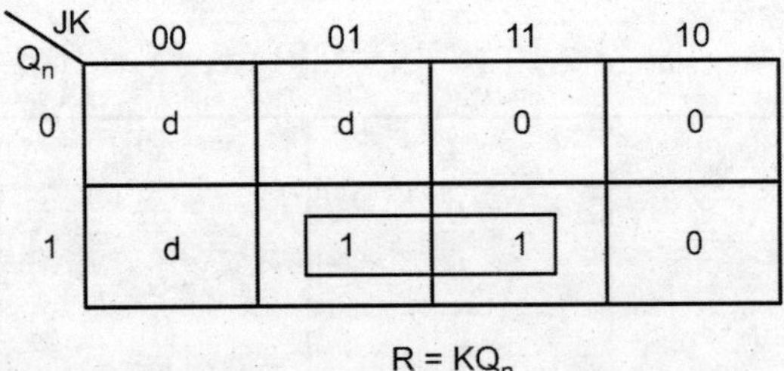

$R = KQ_n$

Fig. 7.38(*b*) Excitation map for *R*.

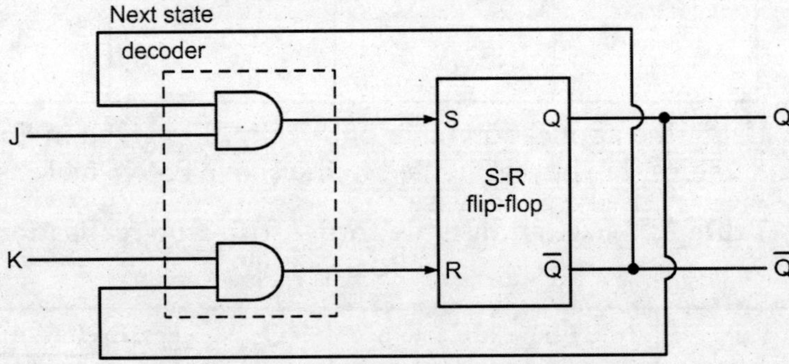

Fig. 7.39. *J-K* flip-flop using *S-R* flip-flop.

7.11.3 Realisation of Trigger Flip-flop using *S-R* Flip-flop

Consider the realisation of Trigger flip-flop using *S-R* flip-flop. Here, the desired flip-flop (X) is Trigger flip-flop and the chosen flip-flop (Y) is S-R flip-flop. The block diagram of Trigger flip-flop using *S-R* flip-flop is shown in Fig. 7.40.

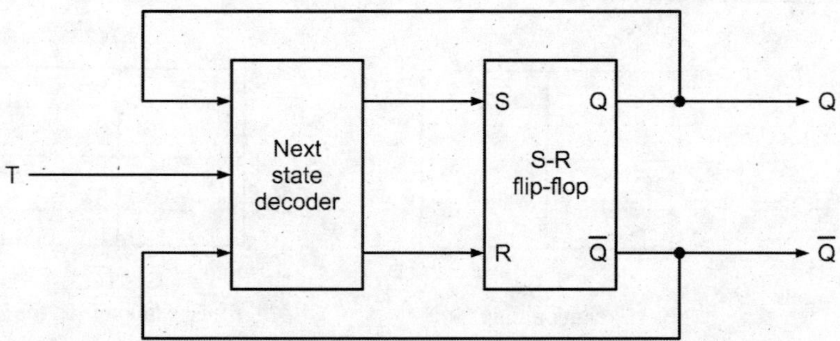

Fig. 7.40. Block diagram of Trigger flip-flop using *S-R* flip-flop.

Step 1: The operation of Trigger flip-flop was explained previously.

Step 2: The Present State-Next State table for Trigger flip-flop can be drawn as shown in Table 7.21.

Table 7.21. PS-NS table for *T* flip-flop.

Q_n	T	Q_{n+1}
0	0	0
0	1	1
1	0	1
1	1	0

Step 3: Using the application table of *S-R* flip-flop given in Table 7.4., the next state codes, i.e. *S* & *R* values, can be augmented in the above PS-NS table as shown in Table 7.22.

Table 7.22. Excitation table for realisation of *T* flip-flop using *S-R* flip-flop.

Q_n	T	Q_{n+1}	Excitation inputs	
			S	R
0	0	0	0	d
0	1	1	1	0
1	0	1	d	0
1	1	0	0	1

Step 4: In this step, one can design the next state recoder, i.e. the simplified expression for S and R, from the excitation maps as shown in Fig. 7.41(a) and (b).

Step 5: From the above step, $S = T\overline{Q}_n$ and $R = TQ_n$. Now, the circuit for Trigger flip-flop using *S-R* flip-flop can be drawn as shown in Fig. 7.42, with two AND gates in the next state decoder logic.

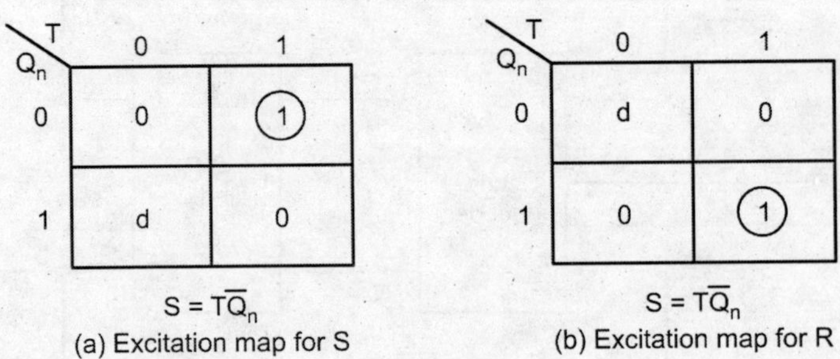

Fig. 7.41.

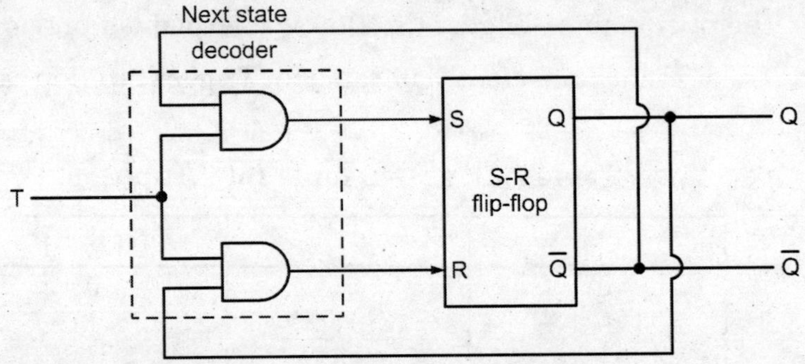

Fig. 7.42. ***T* flip-flop using *S-R* flip-flop.**

From the above realisation of Delay, *J-K* and Trigger flip-flop using *S-R* flip-flop, one can understand that the resultatn circuit is the same as the one discussed in the earlier sections.

Similarly, one can realise any flip-flop using another flip-flop. To get a better insight, that following two realisations are explained.

7.11.4 Realisation of Delay Flip-flop using *J-K* Flip-flop

Here, the desire flip-flop (X) is a Delay flip-flop and the chosen one (Y) is a *J-K* flip-flop. The simple block diagram of such a realisation is shown in Fig. 7.43.

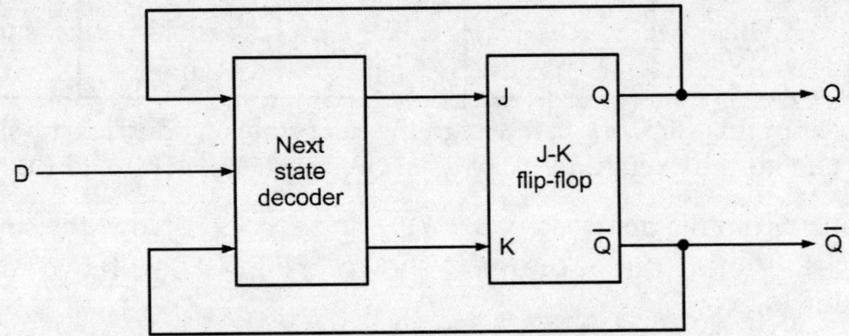

Fig. 7.43. Block Diagram of delay flip-flop using *J-K* flip-flop.

Step 1: The operation of Delay flip-flop is explained in the previous sections.

Step 2: The Present State-Next State table for *D* flip-flop can be drawn as shown in Table 7.23.

Table 7.23. PS-NS table for *D* flip-flop.

Q_n	D	Q_{n+1}
0	0	0
0	1	1
1	0	0
1	1	1

Step 3: Using the application table of *J-K* flip-flop given in Table 7.11, the next state codes, i.e. *J* & *K* values, can be augmented in the above PS-NS table as shown in Table. 7.24.

Table 7.24. Excitation table for *D* flip-flop realisation using *J-K* flip-flop.

Q_n	D	Q_{n+1}	Excitation inputs	
			J	K
0	0	0	0	d
0	1	1	1	d
1	0	0	d	1
1	1	1	d	0

Step 4: In this step, one can design the next state decoder, i.e. simplified expressions for *J* and *K*, from the excitation maps as shown in Fig. 7.44(*a*) and (*b*).

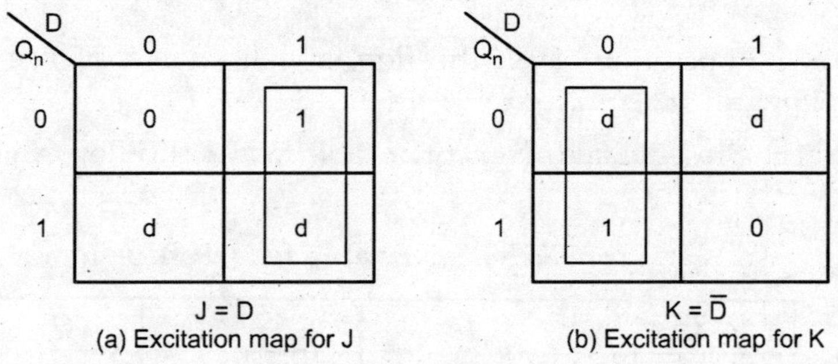

(a) Excitation map for J

(b) Excitation map for K

Fig. 7.44

Step 5: From the above step, it is seem that $J = D$ and $K = \overline{D}$. Now, the circuit for Delay flip-flop using a *J-K* flip-flop can be drawn as shown in Fig. 7.45, with a single NOT gate in the next state decoder logic.

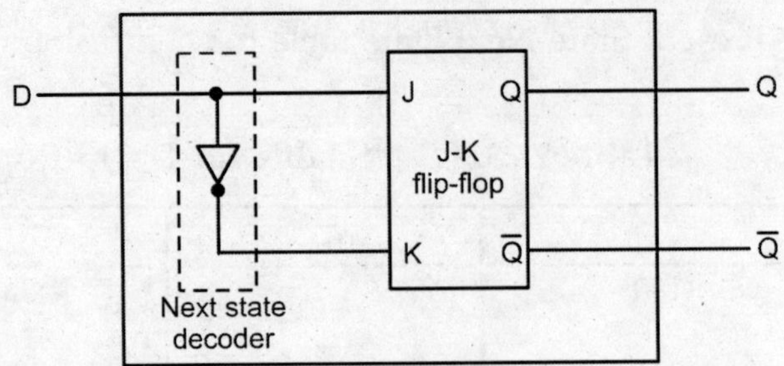

Fig. 7.45. *D* flip-flop using *J-K* flip-flop.

7.11.5 Realisation of *J-K* Flip-flop using *D* Flip-flop

Under this section, the realisation of *J-K* flip-flop using *D* flip-flop is taken into consideration. Here, the desired flip-flop (X) is a *J-K* flip-flop and the chosen one (Y) is a *D* flip-flop. The block diagram of this realisation is shown in Fig. 7.46.

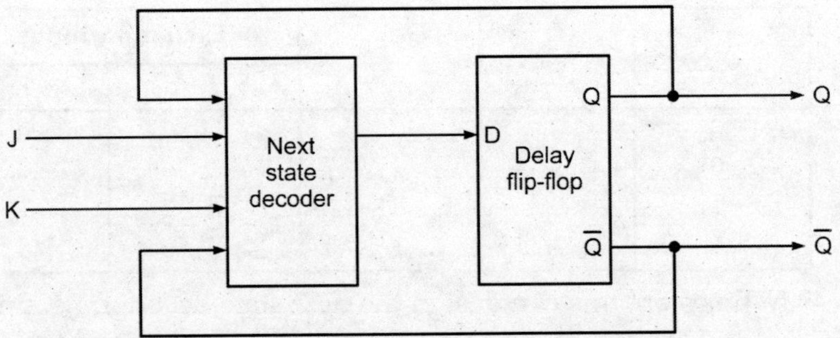

Fig. 7.46. Block diagram of *J-K* flip-flop using delay flip-flop.

Step 1: The operation of *J-K* flip-flop is well known. Hence, there is no need for word description.

Step 2: The Present State - Next State table for *J-K* flip-flop can be drawn as shown in Table. 7.25.

Table 7.25 PS-NS table for *J-K* flip-flop.

Q_n	J	K	Q_{n+1}
0	0	0	0
0	0	1	0
0	1	0	1
0	1	1	1
1	0	0	1
1	0	1	0
1	1	0	1
1	1	1	0

Step 3: Using the application table of *D* flip-flop given in Table. 7.8. the next state codes, i.e. *D* value, can be augmented in the above PS-NS table as shown in Table 7.26.

Table 7.26 Excitation table for realisation of *J-K* flip-flop using *D*-flip-flop

Q_n	J	K	Q_{n+1}	Excitation input D
0	0	0	0	0
0	0	1	0	0
0	1	0	1	1
0	1	1	1	1
1	0	0	1	1
1	0	1	0	0
1	1	0	1	1
1	1	1	0	0

Step 4: In this step, one can design the next state decoder, i.e. the simplified expression for *D*, from the excitation map as shown in Fig. 7.47.

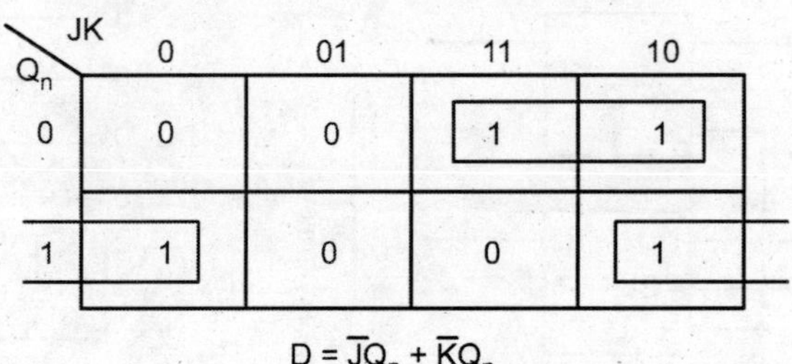

$D = \overline{J}Q_n + \overline{K}Q_n$

Fig. 7.47. Excitation map for *D*.

Step 5: From the above step, it can be seen that $D = J\overline{Q}_n + \overline{K}Q_n$. Now, the circuit for *J-K* Delay flip-flop using *D* flip-flop can be drawn as shown in Fig. 7.48, with two AND gates, one OR gate and one NOT gate in the next state decoder logic.

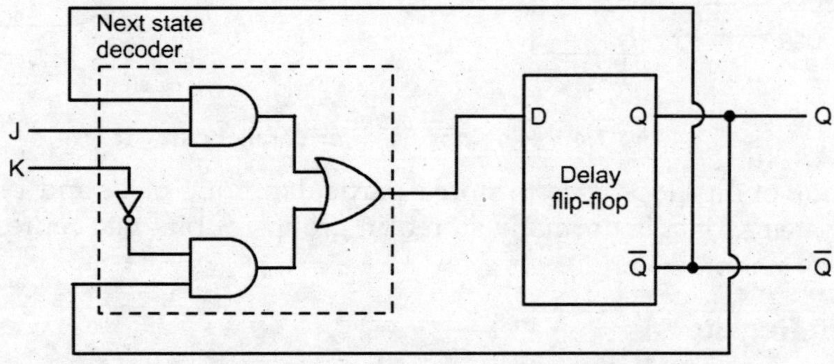

Fig. 7.48. *J-K* flip-flop using *D* flip-flop.

7.12 APPLICATIONS OF FLIP-FLOPS

Flip-flops find wide applications in counter circuits, frequency dividers, shift and storage registers. A wide variety of serial decoding, comparison and timing functions can be accomplished using flip-flops. Using flip-flops, we can generate a variety of one-shots.

An operation that occurs very frequencly in digital system is the transfer of information from one flip-flop or a ground of flip-flops to another flip-flop or another grounp of flip-flops.

7.12.1 Parallel Data Storage

Several bits of data can be stored simultaneously in a group of flip-flops. This is illustrated in Fig. 7.49. Each of the four parallel data lines is connected to the *D* input of a flip-flop. The clock inputs of all the flip-flps are connected to a common clock input, so that each flip-flop is triggered at the same time. As positive edge triggered flip-flops are used, the data on the *D* inputs are stored simultaneously be the flip-flops on the positive edge of the clock. The clear inputs are connected to a common $\overline{\text{CLR}}$ line, which resets all the flip-flops.

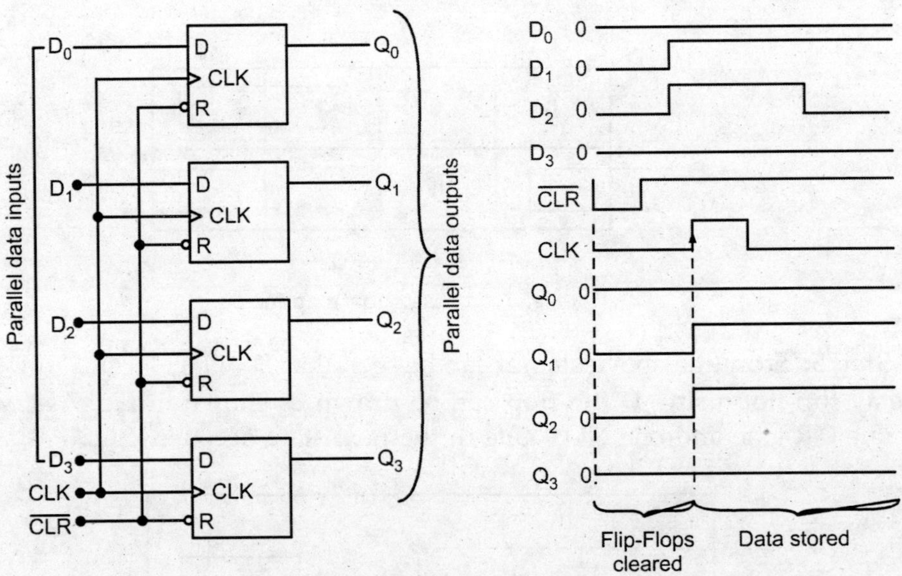

Fig. 7.49. Flip-flops used for parallel data storage.

A group of flip-flops used to store a particular group of 0's and 1's are called registers. In digital systems, data is normally stored in groups of bits that represent numbers, codes or other information.

7.12.2 Shift Registers

Shifts registers are used to transfer the contents of one register to another register, or within the same register one bit at a time. Serial shift registers may be implemented using *J-K*, *R-S* and *D*-type flip-flops. Fig. 7.50. shows a serial shift register implemented with *D*-type flip-flops.

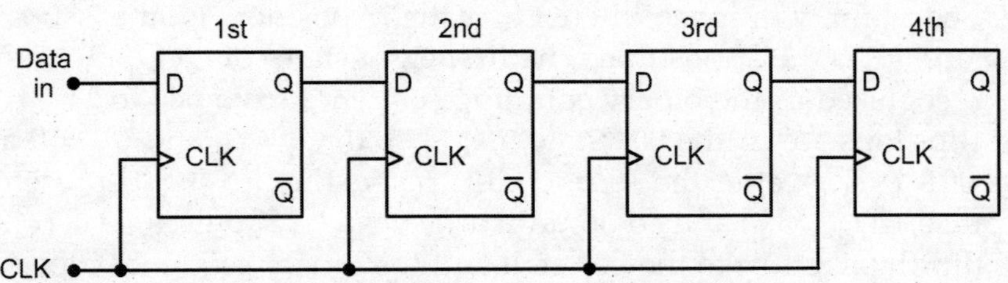

Fig. 7.50. Flip-flop used as shift registers.

7.12.3 Frequency Division

Flip-flops are also used to divide the frequency of a periodic waveform. When a pulse waveform is applied to the clock input of a *J-K* flip-flop whose *J* and *K* inputs are at 1 state, the *Q* output is a square wave with one-half the frequency of the clock input. Thus, a single flip-flop is a divide-by-2 device as shown in Fig. 7.51. The flip-flop changes state on each trigering clock edge.

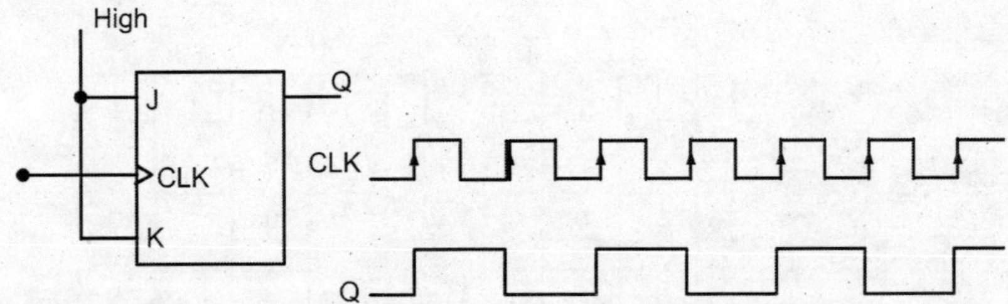

Fig. 7.51. *J-K* flip-flop as a divide-by-2 device

The division of a clock frequency can further be achieved by using the output of one flip-flop as the clock input to a second flip-flop, as shown in Fig. 7.52. The frequency of the Q_A output is divided by 2 by flip-flop *B*. The output, Q_B, is therefon one-fourth the frequency of the original clock input.

Similarly, a frequency division of 2^n is achieved by connecting *n* flip-flops in cascade For example, 4 flip-flops connected in cascade divide the clock frequency by 16 (2^4).

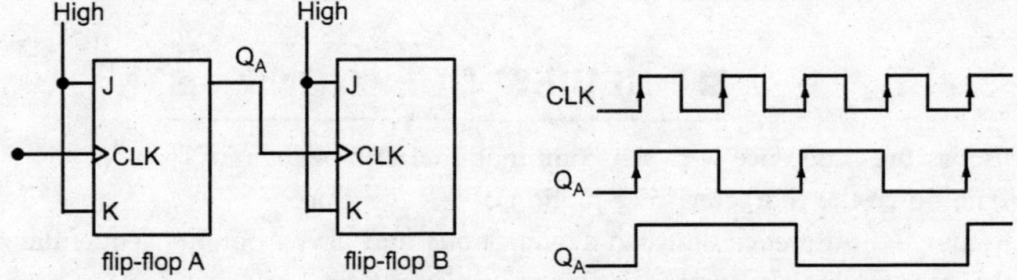

Fig. 7.52. Two *J-K* flip-flops used to divide the clock frequency by 4.

7.12.4 Counters

In addition to functioning as a frequency divider, the circuit of Fig. 7.53(*a*) operates as *a binary counter*. Here *J-K* flip-flops are negative edge-triggered. Flip-flops are intially RESET.

Let $Q_2Q_1Q_0$ be a binary number where Q_2, is the 2^2 position, Q_1 the 2^2 position and Q_1 is the 2^1 position and Q_0 is the 2^0 position. The first eight states of Q_2 Q_1 Q_0 in the timing diagram should be recognised as the binary counting sequence from 000 to 111. After the first clock pulse, the flip-flops are in the 001 state, i.e. $Q_2 = 0$, $Q_1 = 0$, and $Q_1 = 0$ and $Q_0 = 0$, which represents 001_2 (= 1_{10}); after the second CLK pulse, the flip-flop represent 010_2(= 2_{10}); after the third pulse, 011_2(= 3_{10}) and so on until after seva CLK pulses, 111_2 (= 7_{10}). On the eighth pulse, the flip-flops return to the 000 stall and the binary sequence repeats itself after every eight clock pulses as shown in tin timing diagram of Fig. 7.53(*b*). Thus, the flip-flops count in sequence from 0 to 7 an then recycle back to 0 to begin the sequence again.

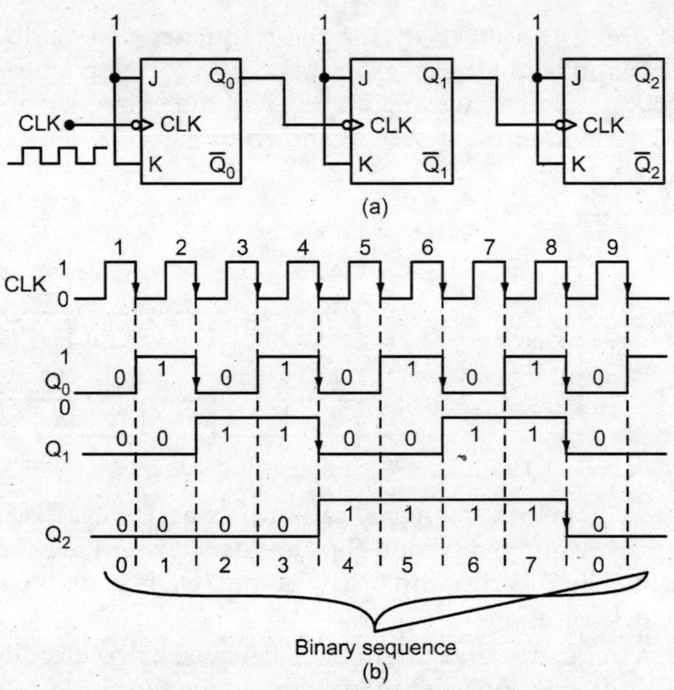

Fig. 7.53. *J-K* flip-flops wired as 3-bit binary counter.

■ REVIEW QUESTIONS ■

1. Discuss the difference between combinational and sequential logic.
2. Write down the characteristic equation of *S-R* flip-flop.
3. Discuss the difference between synchronous and asynchronous sequential circuits.
4. (*a*) Sketch the logic system for a clocked *S-R* flip-flop.
   (*b*) Verify that the state of the system does not change in between clock pulses.
   (*c*) Give the truth table and
   (*d*) Justify the entries in the truth table.
5. What is a *D* flip-flop?
6. Show how an *S-R* flip-flop can be converted into a *D* flip-flop.

7. Explain the function of a *D* flip-flop using a suitable diagram and discuss how it works as a latch.
8. Explain how a *J-K* flip-flop can be converted into a *D* flip-flop.
9. Name the two types of edge-triggered flip-flops classified by the method of triggering and explain the difference.
10. What advantage does a *J-K* flip-flop have over an *S-R* flip-flop?
11. What is the advantage of *D* flip-flop over an *S-R* flip-flop?
12. Show that a *J-K* flip-flop can be converted to a *D* flip-flop with an inverter between the *J* and *K* inputs.
13. Show how the *J-K* flip-flop can be operated as a toggle flip-flop.
14. How do *S-R*, *J-K* and *D* flip-flops differ?
15. (*a*) What is meant by edge triggering?
    (*b*) Give the difference between positive and negative edge triggering.
16. For a negative edge-triggered *J-K* flip-flop, $J = 0$, $K = 1$ and $Q = 0$. When the CLK input goes HIGH, what happens to *Q*? When the CLK input goes LOW, what hap-pens to *Q*?
17. Draw the logic diagram of a master-slave *D* flip-flop using NAND gates.
18. Explain the operation of master-slave flip-flop and show how the race around condition is eliminated in it.
19. What is the major restriction while operating a pulse-triggered flip-flop?
20. What is the major difference in the operation of edge-triggered flip-flops and master-slave flip-flops?
21. List four basic flip-flop applications.

■ PROBLEMS ■

1. If the waveforms in Fig. P7.1 are applied to an active-low input *R-S* latch, sketch the resulting *Q* output waveform in relation to the inputs. Assume that *Q* starts LOW.

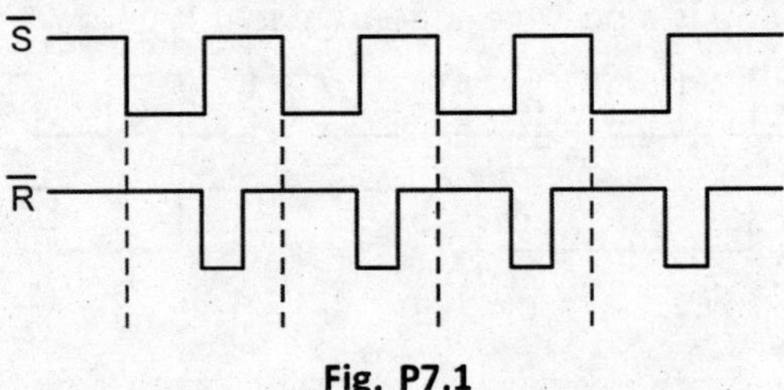

Fig. P7.1

Ans.

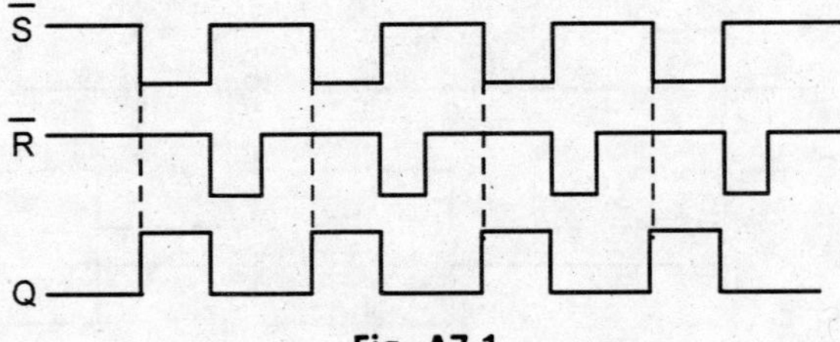

Fig. A7.1

2. Show how the *J-K* flip-flop can be operated as a toggle flip-flop. Apply a 10 KHz square wave to its input and determine the *Q* waveform.

Ans. 5 kHz square wave

3. For a clocked *R-S* latch, determine the *Q* and $\overline{Q}$ outputs for the given inputs in Fig. P7.3.

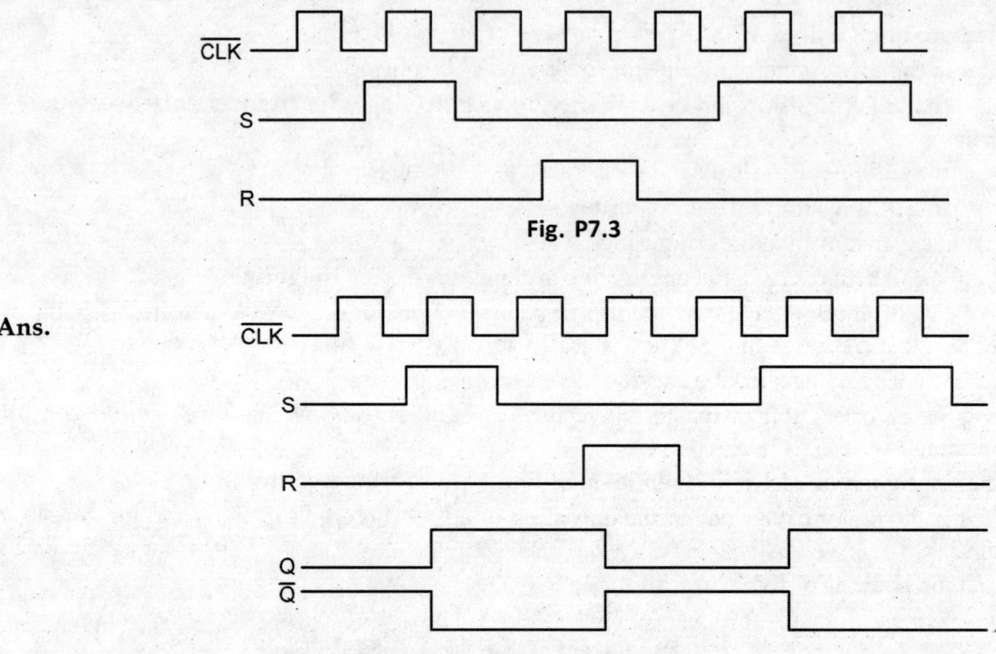

Fig. P7.3

Fig. A7.3

4. The *Q* output of the edge triggered *R-S* flip-flop in Fig. P7.4 is shown in relation to the clock signal. Determine the input waveforms on the S and R inputs that are required to produce this output if the flip-flop is a positive edge-triggered type.

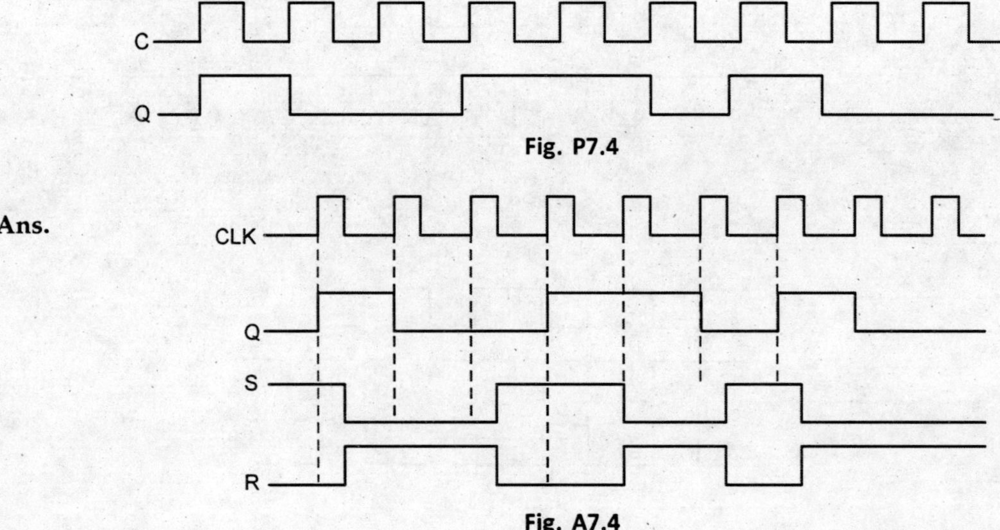

Fig. P7.4

Fig. A7.4

5. For a pulse-triggered (master-slave) *J-K* flip-flop with the inptus in Fig. P7.5, sketch the output *Q* waveforms. Assume that *Q* is initially LOW.

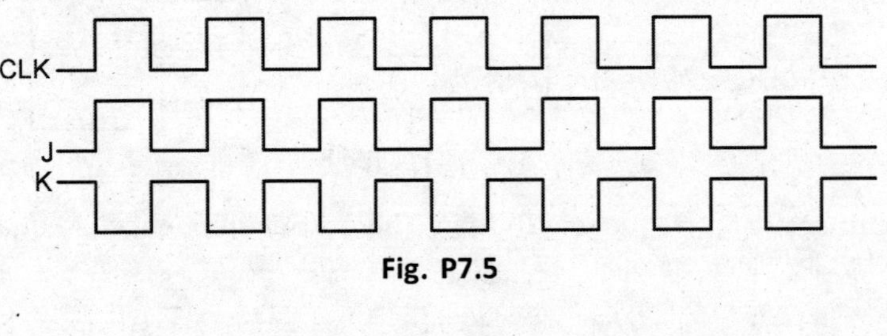

Fig. P7.5

Ans.

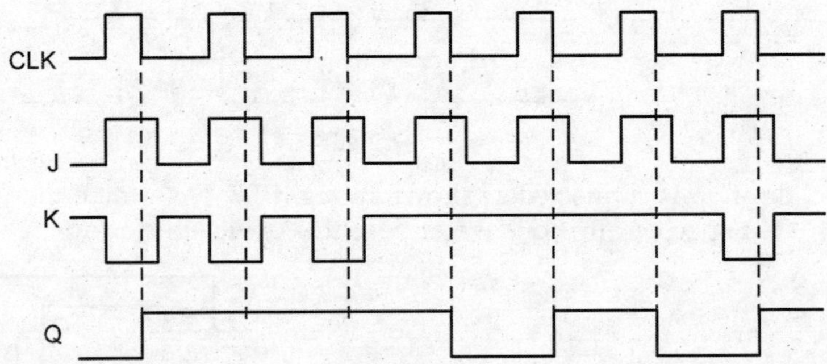

Fig. A7.5

6. Sketch the *Q* output of flip-flop B in Fig. P7.6 in proper relation to the clock. The flip-flops are initially RESET.

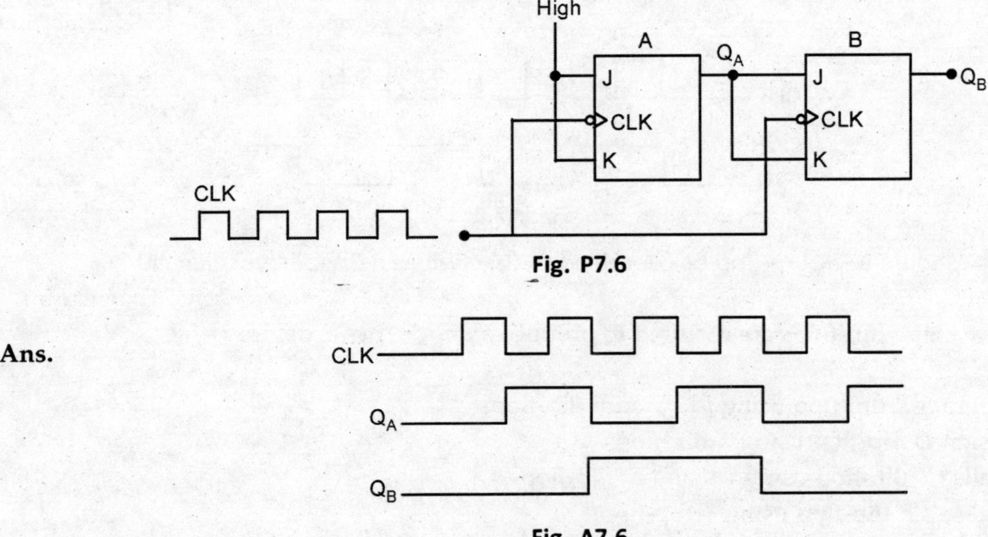

Fig. P7.6

Ans.

Fig. A7.6

7. The *J-K* master-slave flip-flop is Fig. P7.7 has its *J* and *K* inputs tied to + V_{CC}, and a series of pulses is applied to its CLK inputs. Describe the waveforms at *Q*.

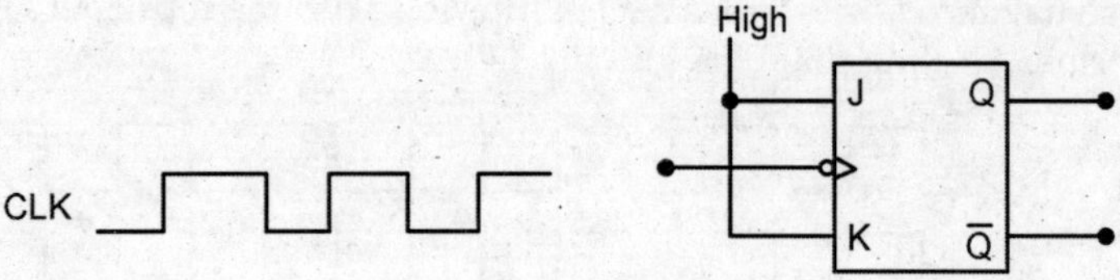

Fig. P7.7

Ans. The circuit acts as a frequency divider. The output frequency is equal to the input frequency divided by 2.

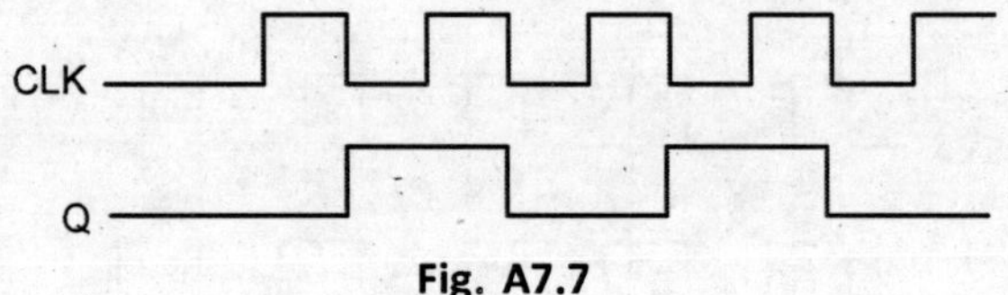

Fig. A7.7

8. As *R-S* flip-flop is connected as shown in Fig. P7.8. Determine the *Q* output in relation to the clock. What specific functions water does this device perform?

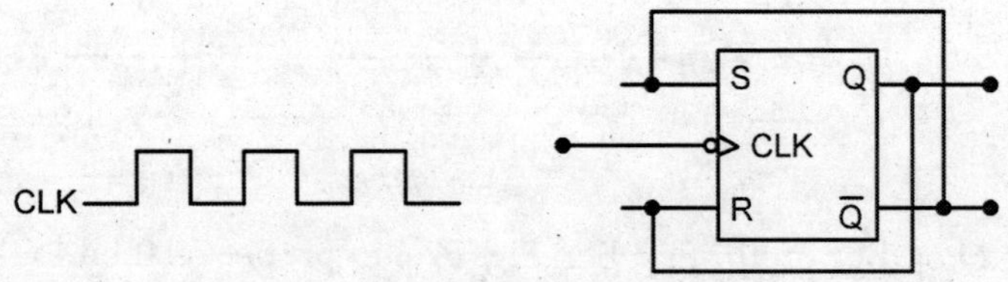

Fig. P7.8

Ans. 1.848 μs

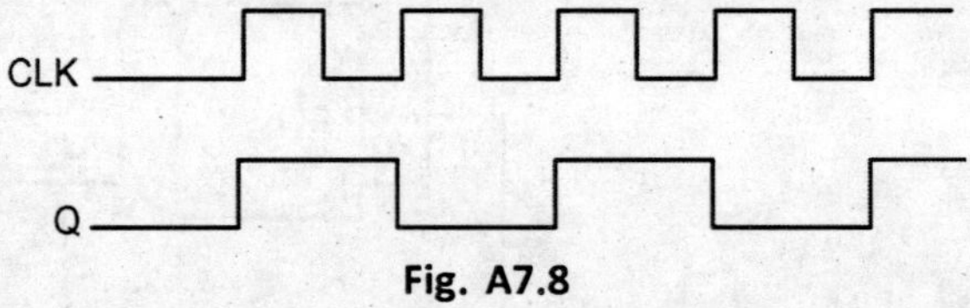

Fig. A7.8

9. How should a *J-K* flip-flop be connected to function as a divide-by-2 element ?

Ans. Toggle ($J = 1, K = 1$)

10. How many flip-flops are required to produce a divide- by-32 device ?

Ans. 5

11. Design *S-R* flip-flop using *J-K*, *T* and flip-flops.
12. Design *D* flip-flop using *T* flip-flops.
13. Realise *T* flip-flop using *J-K* and *D* flip-flops.
14. Realise *J-K* flip-flop using *T* flip-flop.

❑❑❑

8 Counters

8.1 INTRODUCTION

A counter, by function, is a sequential circuit consisting a set of flip-flops connected in a suitable manner to count the sequence of the input pulses presented to it in digital form. Counters can be broadly classified under 3 heads as follows:

(*i*) Asynchronous and Synchronous counters

(*ii*) Single and multi-mode counters

(*iii*) Modulus counters.

An *asynchronous* or *ripple counter* can be constructed using minimum hardware. In an asynchronous counter, each flip-flop is triggered by the output from the previous flip-flop which limits its speed of operation. The settling time in asynchronous counters, is the cumulative sum of the individual settling times of flip-flops. It is also called a *serial counter*. In *synchronous* counters, the speed limitation of ripple counters is overcome by applying clock pulses simultaneously to all the flip-flops which leads to the settling time of the counter being equal to the propagation delay of a single flip-flop. Hence synchronous counters are also called *parallel counters*. *Single mode counters* operate in a single mode, i.e. it counts either in the UP mode or in the DOWN mode, whereas *multimode counters* operate in both UP and DOWN modes. *Modulus counters* are defined based on the number of states they are capable of counting. For example, a MOD-10 counter has 10 states.

Counters are fundamental components of digital systems. Digital counters find wide applications like pulse counting, frequency division, time measurement and control and timing operations.

8.2 ASYNCHRONOUS (RIPPLE OR SERIAL) COUNTER

The asynchronous counter is the simplest in terms of logical operations, and is therefore the easiest to design. In this counter, all the flip-flops are not under the control of a single clock. Here, the clock pulse is applied to the first flip-flop, i.e. the least significant bit stage of the counter, and the successive flip-flop is triggered by the output of the previous flip-flop and thus the counter has a cumulative settling time. Hence, its speed of operation is limited. The first stage of the counter thus switches first on the application of a clock pulse to the first flip-flop and the successive stage change their states in turn causing a *ripple-through* effect of the count pulses. As the triggers move through the flip-flops like a ripple, it is called a *ripple counter*.

Do You Know?	An up/down counter can be commanded to count up or count down.

Fig. 8.1 shows a 4-bit binary ripple counter. As shown, a binary ripple counter is constructed using clocked *JK* flip-flops. The system clock, a square wave, drives flip flop *A*. The output of *A* drives flip-flop *B*, the output of *B* drives flip-flop *C*, and tl output of *C* drives flip-flop *D*. The overall propagation delay time of the counter is tl sum of the individual delays of flip-flops. All the *J* and *K* inputs are connected to V_{CC}, (1), which means that each flip-flop toggles on the negative edge of its clock input.

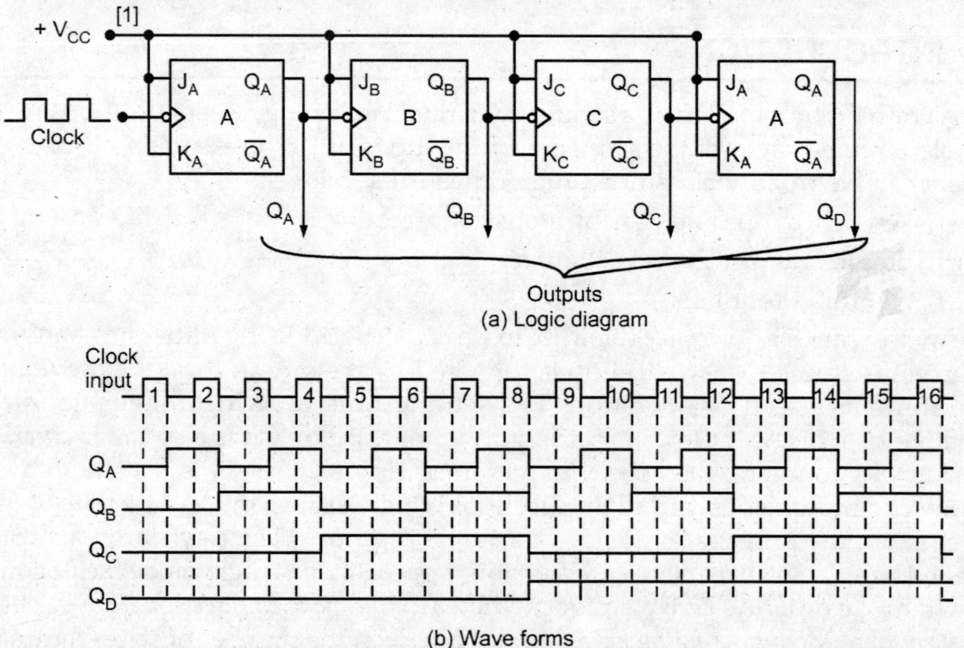

(a) Logic diagram

(b) Wave forms

Fig. 8.1 4-bit binary ripple counter.

Consider initially all flip-flops to be in the logical 0 state (i.e. $Q_A = Q_B = Q_C = Q_D = 0$). A negative transition (1 to 0) in clock input which drives flip-flop *A* causes Q_A to change from logical 0 to logical 1. Flip-flop *B* does not change its state since it also requires negative transition at its clock input, i.e. it requires its clock input (Q_A) change from logical 1 to logical 0. With the arrival of the second clock pulse to flip-flop *A*, Q_A goes from 1 to 0. This change of state creates the negative going edge needed to trigger flip-flop *B*, and thus Q_B goes from 0 to 1. Before the arrival of the sixteenth clock pulse, all the flip-flops are in the logical 1 state. Clock pulse 16 causes Q_t, Q_B, Q_D and Q_D to go to logical 0 state in turn.

Table 8.1 shows the sequence of binary states that the flip-flops will follow as clock pulses are applied continuously. An n-bit binary counter repeats the counting sequence for every 2^n (n = number of flip-flops) clock pulses and has discrete states from 0 to 2^n-l.

Table 8.1 Truth table of 4-bit binary ripple counter.

State	Q_D	Q_C	Q_B	Q_A
0	0	0	0	0
1	0	0	0	1
2	0	0	1	0
3	0	0	1	1
4	0	1	0	0
5	0	1	0	1
6	0	1	1	0
7	0	1	1	1
8	1	0	0	0
9	1	0	0	1
10	1	0	1	0
11	1	0	1	1
12	1	1	0	0
13	1	1	0	1
14	1	1	1	0
15	1	1	1	1
0	0	0	0	0

Mod-number or modulus: The counter in Fig. 8.1 has 16 different states. Thus, it is a MOD-16 ripple counter. The MOD-number (or the Modulus) of a counter is the total number of states it sequences through in each complete cycle.

$$\text{MOD-number}=2^n$$

where n = Number of flip-flops.

The maximum binary number counted by the counter is $2^n - 1$. Thus, a 4-flip-flop counter can count as high as $(1111)_2 = 2^4 - 1 = 15_{10}$.

Frequency division or frequency scaling: Consider the counter shown in Fig. 8.1. Here, the input consists of a sequence of pulses of frequency, f_0. As discussed in the previous section, Q_A changes only when the clock input makes a transition from 1 to 0. Thus, at the first negative transition of the clock, Q_A changes from 0 to 1 and at the second negative transition of the clock, Q_A shifts from 1 to 0. Therefore, two input pulses will result in a single *pulse* in Q_A. Hence the frequency of Q_A will be $f_0/2$. Similarly, the frequency of the Q_B signal will be half that *of* Q_A signal. Therefore, its frequency *is* $f_0/4$. Similarly, the frequency of Q_c and Q_D signals are $f_0/8$ and $f_0/16$ respectively. Thus, the circuit can be used to divide or scale down the input frequency. Such circuits are called *frequency dividers.* If n flip-flops are used, the frequency will be divided by 2^n.

8.3 RIPPLE COUNTER WITH DECODED OUTPUTS

Fig. 8.2 shows a 3-bit binary ripple counter with decoded outputs. It consists of 3 master-slave *J-K* flip-flops with decoding circuitry. In decoding the states of a ripple counter, pulses of one clock duration will occur at the decoding gate outputs as the flip-flops change their state.

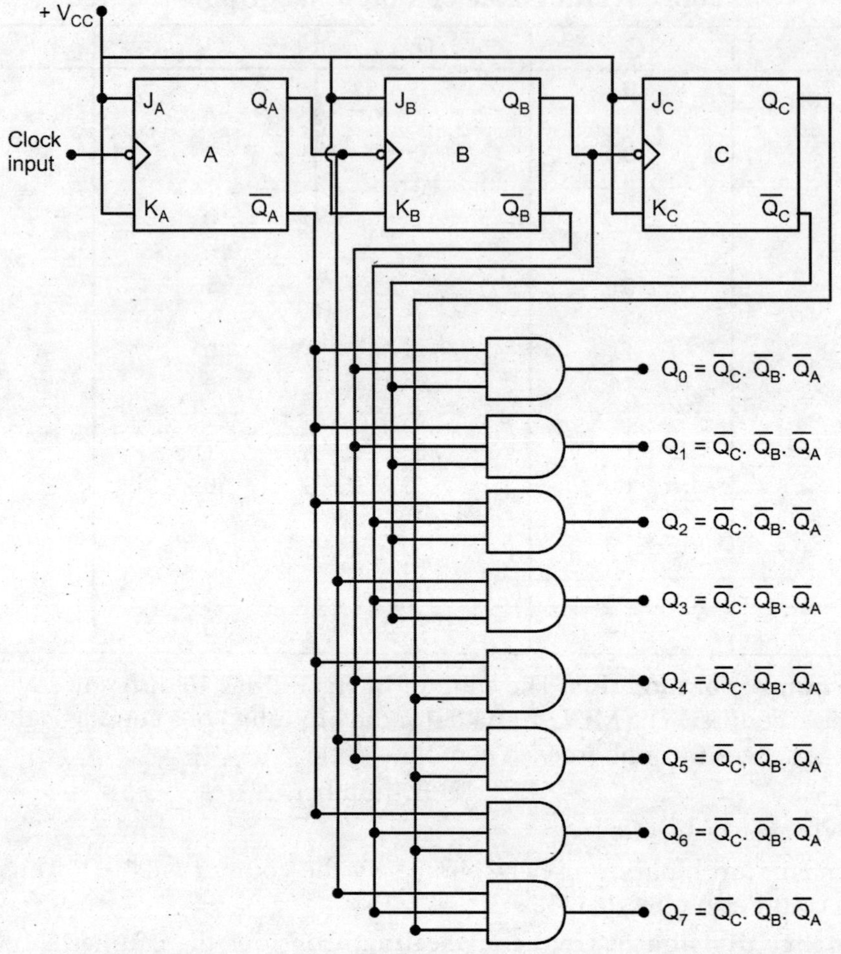

Fig. 8.2. 3-bit binary ripple counter with decoded outputs

A decoding gate is one which can be connected to the outputs of a counter and its output will be high only when the counter content is equal to the given state. For example, a decoding gate Q_5 connected inthe circuit will decode state 5 (i.e. $Q_C\,Q_B\,Q_A = 101$). Thus the gate output will be high only when $Q_C = 1$, $Q_B = 0$, and $Q_A = 1$. The logic expression for the decoding gate 5 is $Q_5 = Q_C = \bar{Q}_B\,Q_A$. The remaining seven states of the three-big couter can be decoded in a similar manner using AND gates as Q_0, Q_1, Q_2, Q_3, Q_4, Q_5, Q_6 and Q_7.

Theoretically, each decoding output will be high only when the counter content is equal to a given state, and this state occurs only once during a cycle of 2^n-states of the counter, where n is the number of flip-flops in the counter. But, pracically in an asynchronous counter, the decoding gate produces a high output more than once during a cycle of 2^n-states. Such undesired high or low pulses that appear at the decoding gate output at undesired time instants are called *glitches* or *false spikes* which are normally of very short duration. The reasion for these glitches is the cumulative propagation delay in the asynchornous counter.

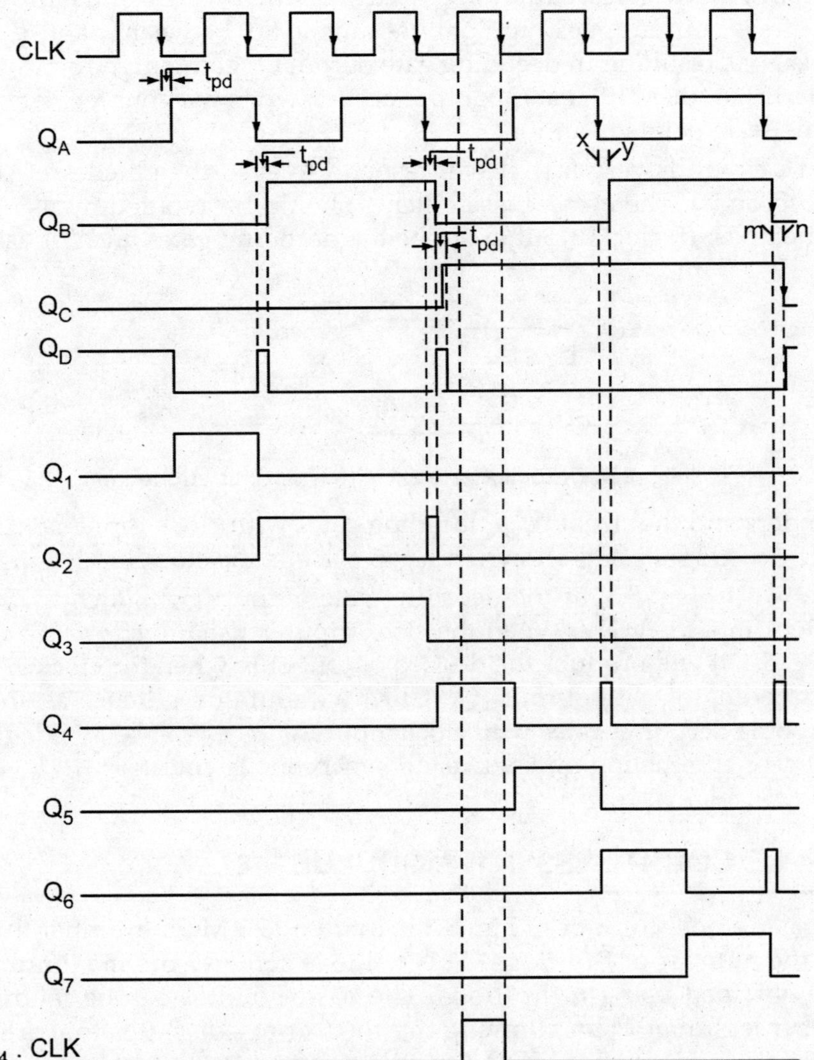

Fig. 8.3. Timing diagram of 3-bit ripple counter with decoded outputs.

A careful observation of the waveform given in Fig. 8.3 shows that the first flip-flip output (Q_A) is delayed by propagation delay (t_{pd}) of flip-flop from the negative clock trasition. Since Q_A acts as trigger for flip-flop *B*, Q_B is delayed by one flip-flop delay time from each negative transition of Q_A. Similarly, Q_C waveform is delayed by propagation delay (t_{pd}) from each negative transition of Q_B. Because of this cumulative propagation delay, glitches may appear at one or more decoding gate outputs as shown in Fig. 8.3.

Consider the gate (Q_4) used to decode state 4. Theoretically, its output waveform is high only when $Q_c = 1$, $Q_B = 0$ and $Q_A = 0$. But, glitches occur at the output of decoding gate

Q_4 when the counter progresses from state 5 to 6 as well as from state 7 to 0. When state transition from 5 to 6 occurs at time x, Q_A goes low. But, because of the flip-flop propagation delay time, the Q_B output goes high only at time y. So, between x and y, we have $Q_c = 1$, $Q_8 = 0$ and $Q_A = 0$, resulting in decoding gate output Q_4 going high for a small duration, i.e. *glitch*. Similarly, another glitch at Q_4 output occurs, when the counter state goes from 7 to 0, between time scale points m and n.

As TTL circuit is very fast, it will respond to even the glitches of smallest duration (a few nanoseconds). Therefore, these glitches should be avoided by using any one of the following methods: (*i*) clock input to strobe the decoding gates and (*ii*) using synchronous counters.

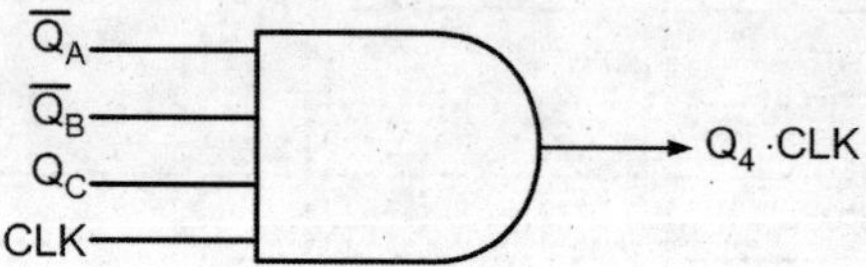

Fig. 8.4. Decoding gate Q4 with a clock as strobe input.

To understand the strobing of decoding gates with clock input, consider a four input AND gate to decode state 4 as shown in Fig. 8.4. Here, the clock input is used as the strobe.

As shown in Fig. 8.3, at the decoding gate output (Q_4), glitches are produced only when the clock input is LOW. When the clock input is used to strobe the decoding gate as shown in Fig. 8.4, it will produce the desired output only when the clock is HIGH, resulting in perfect decoding of gate output (Q_4. CLK) without any glitches as shown in Fig. 8.3. Thus, by strobing decoding gates with clock inputs, glitches can be avoided completely. The other method of eliminating glitches using synchronous counter will be discussed in the Section 8.9.

8.4 RIPPLE COUNTERS WITH MODULUS < 2^n

The basic ripple counter of Fig. 8.1 is limited to a MOD-number that is equal to 2^n, where n is the number of flip-flops. This value is actually the maximum MOD-number that can be obtained using n flip-flops. The basic counter can be modified to produce MOD-number less than 2^n by allowing the counter to *skip* states that are normallya part of the counting sequence. A MOD-6 ripple counter is shown in Fig. 8.5. Withod the NAND gate, this counter functions as a MOD-8 binary counter, which counts fron 000 to 111. However, the presence of the NAND gate shown in Fig. 8.5(*a*) alters this sequence as follows:

1. The NAND gate output is connected to the CLEAR inputs of each flip-flo As long as the NAND gate output is HIGH, it will have no effect on counter. When the NAND gate output goes LOW, it will clear all flip-flops and the counter immediately goes to the 000 state.
2. The outputs of the counter Q_B and Q_C are given as inputs to the NAND gate. The NAND gate output goes LOW whenever $Q_B = Q_C = 1$. This condition will occur when the counter goes from the 101 state to the 110 state (on the sixth input

pulse). The LOW at the NAND gate output will clear the counter to the 000 state. Once the flip-flops have been cleared, the NAND gate output goes back to HIGH, since $Q_B = Q_C = 1$ condition no longer exists.

3. Therefore, the counting sequence is

$000 \rightarrow 001 \rightarrow 010 \rightarrow 011 \rightarrow 100 \rightarrow 101 \rightarrow 000 \rightarrow$........

Although the counter does go to the 110 state, it remains there only for a few nano seconds before it recycles to 000. Thus, we can say that this counter counts from 000 to 101 and then recycles to 000. It essentially skips 110 and 111 states going only through six different states; thus, it is a MOD-6 counter.

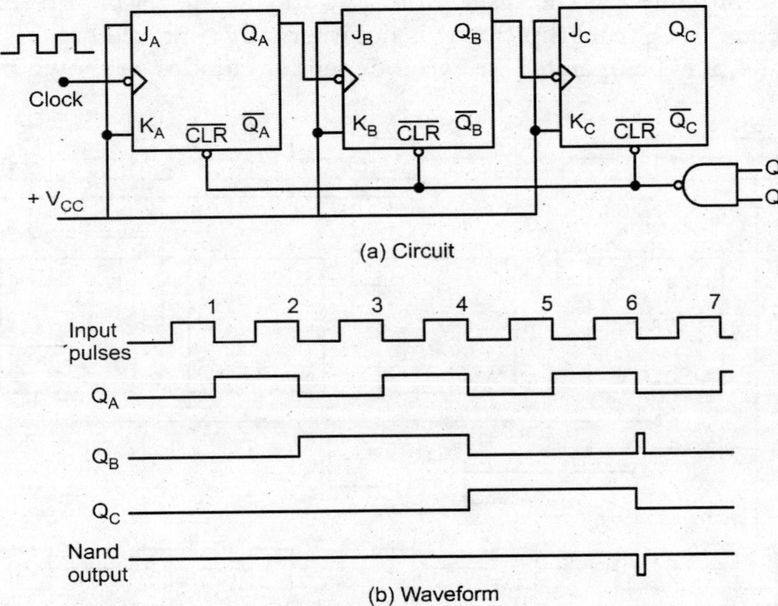

Fig. 8.5. Mod-6 ripple counter.

Note that the waveform at the Q_B output contains a *spike* or *glitch* caused by the momentary occurrence of the 110 state before clearning. This glitch is very narrow and so would not produce any visible indication. It should be noted that the Q_C output has a frequency equal to 1/6 of the input frequency; in other words, this MOD-6 counter has divide the input frequency by six.

To construct any MOD-N counter, the following method can be used.

1. Find the number of flip-flops (n) requiredd for the desired MOD number-N using the equation.

$$2^{n-1} \leq N \leq 2^n$$

2. Connect all the n flip-flops as a ripple counter.
3. Find the binary number for N.
4. Connect all flip-flop outputs, for which $Q = 1$ when the count is N, as inputs to NAND gate.

5. Connect the NAND gate output to thoe CLEAR input for each flip-flop.

When the counter reaches its N th state, the output of the NAND gate goes LOW, resetting all flip-flops to 0. So, the counter counts from 0 through $N - 1$, having N-states.

8.5 COUNTER ICs

8.5.1 IC 7493 – 4-bit Binary Ripple Counter

IC 7493 is a 4-bit binary ripple counter that consists of four master-slave flip-flops as shown in Fig. 8.6. These four flip-flops are internally connected to provide a divide by-2 and divide-by-8-bit counter. The reset inputs R_1 and R_2 are used to reset the counter to 0000. Since the output Q_4 from flip-flop A is not internally connected to the succeeding flip-flops, the counter may be operated in two independent modes as shown below.

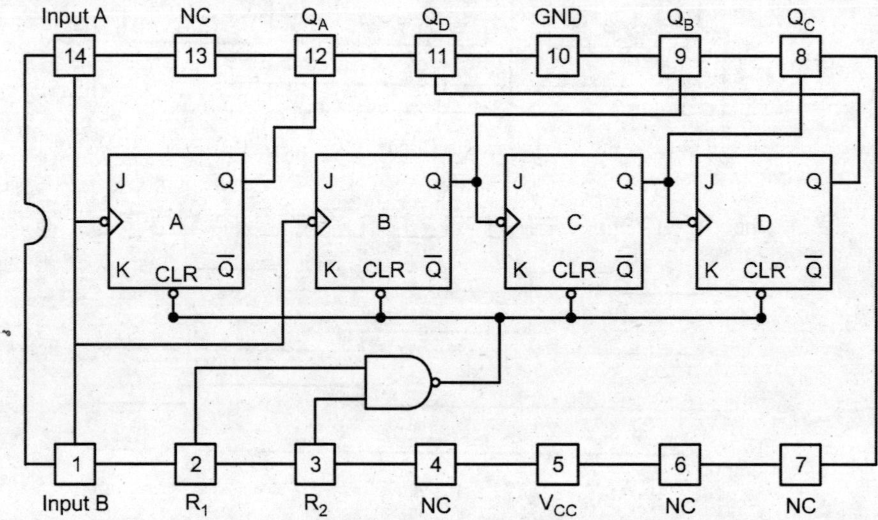

Fig. 8.6. 7493 – 4-bit binary ripple counter

1. When used as a 4-bit ripple counter, output Q_A must be extenally connectecd to input B. The input pulses are applied to input A. Simultaneous divisions of 2,4,8 and 16 are performed at the Q_A, Q_B, A_C, and Q_D outputs. The truth table for this connection is given in Table 8.2.
2. When used as a 3-bit ripple counter, the input count pulses are applied to input B. Simultaneous frequency divisions of 2, 4 and 8 are performed at Q_B, Q_C and Q_D outputs. Independent use of flip-flop A is available if the reset function conincides with the reset of the 3-bit ripple counter.

Table 8.2 Truth table for 7493—4-bit binary ripple counter.

Mode -1 (divide-by-16)				Mode-2 (divide-by-8)		
Q_D	Q_C	Q_B	Q_A	Q_D	Q_C	Q_B
0	0	0	0	0	0	0
0	0	0	1	0	0	1
0	0	1	0	0	1	0
0	0	1	1	0	1	1
0	1	0	0	1	0	0
0	1	0	1	1	0	1
0	1	1	0	1	1	0
0	1	1	1	1	1	1
1	0	0	0			
1	0	0	1			
1	0	1	0			
1	0	1	1			
1	1	0	0			
1	1	0	1			
1	1	1	0			
1	1	1	1			
0	0	0	0			

8.5.2 1C 7490—Decade Counter

IC 7490 is a decade counter (MOD-10), which consists of four master-slave flip-flops internally connected to provide a divide-by-2 counter and a divide-by-5 counter, as shown in Fig. 8.7. The reset inputs R_1 and R_2 are used to reset the counter to 0000 and the set inputs S_1 and S_2 are used to set the counter to 1001. Since the output Q_A from flip-flop A is not internally connected to the succeeding stages, the counter can be operated in two independent count modes.

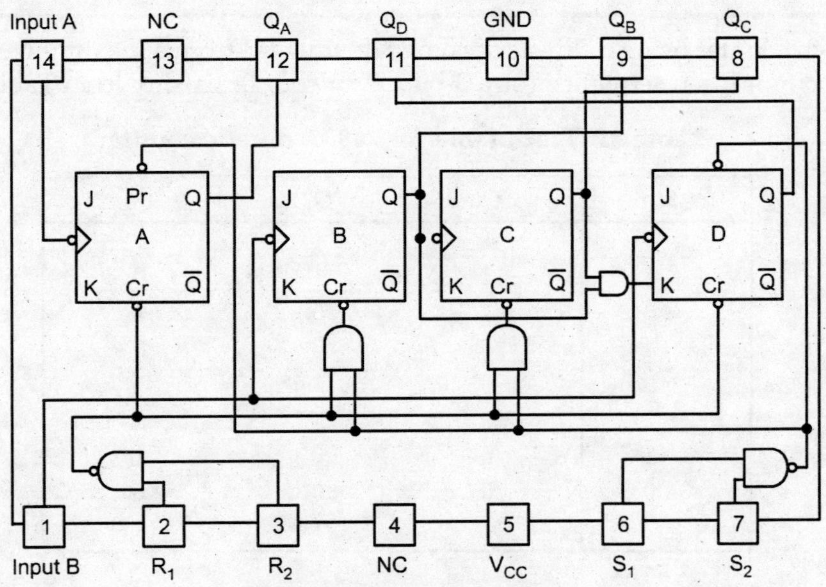

Fig. 8.7. 7490–decade counter.

1. When used as a BCD counter, the *B* input must be externally connected to the Q_A output. The input *A* receives the incoming pulses, and a count sequence is obtained in accordance with the BCD count sequence as shown in, Table 8.3. Two gated inputs are provided to reset the counter to 0. In addition, two more inputs are also provided to set to a BCD count of 9 for 9's complement decimal applications.
2. No external interconnections are necessary when it is required to function as a divide-by-2 counter and a divide-by-5 counter. Flip-flop *A* is used as a binary element for the divide-by-2 function. The *B* input is used to obtain binary divide-by-5 operation at the Q_B, Q_c and Q_D outputs. In this mode, the two counters operate independently; however, all four flip-flops are reset simultaneously.

Table 8.3 Truth table for 7490 – decade counter.

Mode -1 (divide-by-10)				*Mode-2 (divide-by-5)*		
Q_D	Q_C	Q_B	Q_A	Q_D	Q_C	Q_B
0	0	0	0	0	0	0
0	0	0	1	0	0	1
0	0	1	0	0	1	0
0	0	1	1	0	1	1
0	1	0	0	1	0	0
0	1	0	1			
0	1	1	0			
0	1	1	1			
1	0	0	0			
1	0	0	1			

8.6 ASYNCHRONOUS DOWN COUNTER

A down counter using *n* flip-flops counts downward from a maximum count of $(2^n - 1)$ to zero. The countdown sequence for a 3-bit down counter is shown in Table. 8.4.

Table 8.3 Truth table for 7490 – decade counter.

State	Q_C	Q_B	Q_A
7	1	1	1
6	1	1	0
5	1	0	1
4	1	0	0
3	0	1	1
2	0	1	0
1	0	0	1
0	0	0	0
7	1	1	1

Let Q_A, Q_B and Q_c represent the flip-flop output states as the counter goes through its sequence. The truth table shows that the output Q_A (LSB) changes its state (toggle) at each negative transition of clock as it does in the up-counter. The Q_B output changes state every time Q_A goes from LOW to HIGH, i.e. when $\bar{Q}_A$ goes from HIGH to LOW; Q_c changes state each time Q_B goes from LOW to HIGH, i.e. when $\bar{Q}_B$ goes from HIGH to LOW. Thus, in a down counter, each flip-flop, except the LSB flip-flop, must toggle when the inverted output ($\bar{Q}$) of the preceding flip-flop goes from HIGH to LOW. This is illustrated in Fig.8.8 for a MOD-8 down counter.

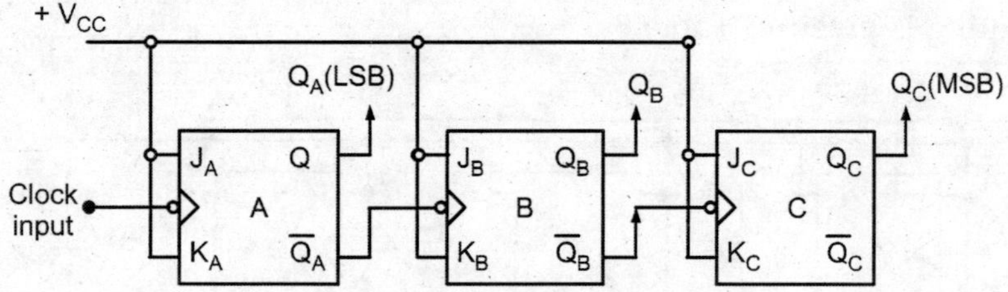

Fig. 8.8. Mod-8 down counter.

The input pulses are applied to the clock input of *A* flip-flop; the $\bar{Q}_A$ output serves as the clock input for the *B* flip-flop; the $\bar{Q}_B$ output serves as the clock input for the *C* flip-flop. The waveforms at Q_A, Q_B and Q_c of MOD-8 down counter is shown in Fig. 8.9. It shows that Q_A toggles at the negative transition of the clock, Q_B toggles whenever Q_A goes from LOW to HIGH, and Q_C toggles whenever Q_B goes from LOW to HIGH. Whenever Q_A and $\bar{Q}_B$ go HIGH, $\bar{Q}_A$ and $\bar{Q}_B$ go LOW, and it is this negative transition at $\bar{Q}_A$ and $\bar{Q}_B$ that triggers outputs Q_B and Q_c respectively.

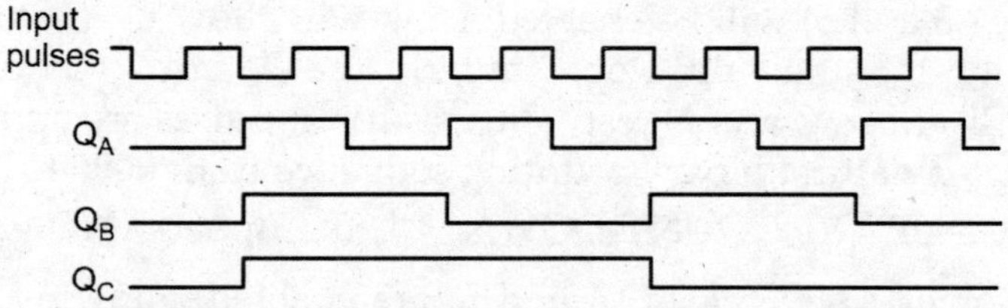

Fig. 8.9. Waveforms of MOD-8 down counter.

If the initial counter content is 000, at the first negative transition of the clock, the counter content becomes 111; at the second negative transition, the content becomes 110; at the third negative transition of clock, the content changes to 101, and so on. Thus, in the down counter, the counter content is decremented by one for every negative transition in clock.

The major application of down counters lies in situations where a desired number of input pulses that has occurred is to be found. In these situations, the down-counter is *preset* to the desired number and then allowed to countdown as the pulses are applied. When the counter reaches the *zero* state, it is detected by a logic gate whose output at that time indicates that the preset number of pulses are applied.

8.7 UP-DOWN COUNTER

The UP-DOWN counter is a combinaiton of the up-counter and the down-counter, the UP-DOWN counter has the capability of counting upwards as well as downwai it is also called *Multimode counter*. In an UP-counter, each flip-flop is triggered by the *normal* output of the preceding flip-flop; in a DOWN-counter, each flip-flop is trigered by the inverted output of the preceding flip-flop. In both the counters, the flip-flop is triggered by the input pulses.

A 4-bit UP-DOWN counter whose operation is controlled by the UP and DOWN control inputs is shown in Fig. 8.10. The counting sequence of UP/DOWN court the two modes of counting is given in Table 8.5.

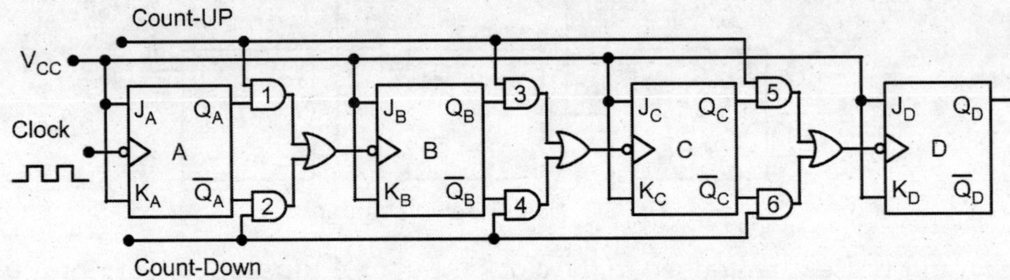

Fig. 8.10. Asynchronous 4-bit UP-DOWN counter.

Here, three logic gates per stage are required to switch the individual stages from COUNT-UP to COUNT-DOWN mode. The logic gates are used to allow either the non-inverted output or the inverted output of one flip-flop to the clock input of the following flip-flop, depending on the status of the control inputs. When the COUNT-UP line is held at 1 while the COUNT-DOWN line is at 0, the lower AND gates (2,4 and 6) will be disabled and their outputs are zero. So, it will have no effect on the outputs of OR gates. Also, the upper AND gates (1,3 and 5) will be enabled, i.e. it will allow Q_A pass through the OR gate and into the clock input of the *B* flip-flop. Similarly, the Q_B and Q_C outputs will be gated into the clock input of flip-flops *C* and *D* resp Thus, as input pulses are applied, the counter will count up and follow a natural bibary counting sequence from 0000 to 1111.

With COUNT-UP = 0, COUNT-DOWN = 1, the upper AND gates (1 ,_3 and 5) are disabled and the lower AND gates (2,4 and 6) are enabled, allowing $\overline{Q}_A$, $\overline{Q}_B$ and $\overline{Q}_C$ to pass through to the clock inputs of the following flip-flops. Thus, for this condl| the counter will count down as input pulses are applied.

When the control inputs are both 0 or 1, the counter will not count up or count down because the clock inputs of *B*, *C* and *D* will be held constant at either 0 or 1. The flip-flop *A* will keep toggling because it is always being clocked. These condition are not normally used.

Table 8.5 Truth Table of 4-bit up-down counter.

	COUNT-UP Mode					*COUNT-DOWN Mode*			
States	Q_D	Q_C	Q_B	Q_A	***State***	Q_D	Q_C	Q_B	Q_A
0	0	0	0	0	15	1	1	1	1
1	0	0	0	1	14	1	1	1	0
2	0	0	1	0	13	1	1	0	1
3	0	0	1	1	12	1	1	0	0
4	0	1	0	0	11	1	0	1	1
5	0	1	0	1	10	1	0	1	0
6	0	1	1	0	9	1	0	0	1
7	0	1	1	1	8	1	0	0	0
8	1	0	0	0	7	0	1	1	1
9	1	0	0	1	6	0	1	1	0
10	1	0	1	0	5	0	1	0	1
11	1	0	1	1	4	0	1	0	0
12	1	1	0	0	3	0	0	1	1
13	1	1	0	1	2	0	0	1	0
14	1	1	1	0	1	0	0	0	1
15	1	1	1	1	0	0	0	0	0
0	0	0	0	0	15	1	1	1	1

8.8 PROPAGATION DELAY IN RIPPLE COUNTER

The main drawback of a ripple counter is that it has a cumulative settling time. In ripple counters, each flip-flop is triggered by the transition at the output of the preceding flip-flop. Because of the inherent propagation delay time (t_{pd}) of each flip-flop, the first flip-flop will not respond for a period of t_{pd} even after receiving a clock pulse. Similarly, the second flip-flop will not respond for a period of $2t_{pd}$ after the input clock pulse occurs. Hence, the wth flip-flop cannot change states for a period of $n \times t_{pd}$ even after the clock pulse occurs. Therefore, to allow all the flip-flops in a w-bit counter to change states in response to a clock, the period of the clockr (T_{clock}) should be:

$$T_{clock} \geq n \times t_{pd}$$

Thus, the maximum frequency that can be used in an asynchronous counter is

$$\frac{1}{f_{clock}} \geq n \times t_{pd}$$

$$\frac{1}{n \times t_{pd}} \geq f_{clock}$$

Thus, the maxiumum frequency that can be used in an asynchronous counter is

$$\frac{1}{f_{clock}} \geq n \times t_{pd}$$

$$\frac{1}{n \times t_{pd}} \geq f_{clock}$$

Thus, f_{clock} should be less than or equal to $1/(n \times t_{pd})$. So, the maximum clock frequency that can be applied in an n-bit asynchronous counter is

$$f_{max} = \frac{1}{n \times t_{pd}}$$

For example, a 3-bit counter having flip-flops with identical tpd = 50ns will have a maximum input frequency limit of

$$f_{max} = \frac{1}{3 \times 50ns} = 6.67 \text{ MHz}$$

Obviously, as the number of flip-flops in the ripple counter increases, the total delay will increase and f_{max} will be lower.

8.9 SYNCHRONOUS (PARALLEL) COUNTER

Asynchronous (ripple) counter is the simplest type of binary counters as it requires less hardware. But its speed of operation is low because the propagation delay time of all flip-flops is cumulative and the total settling time is the product of the total numb of flip-flops and the propagation delay of a single flip-flop. Another problem encoutered with ripple counter is the glitches at the decoding gate outputs. These problem can be eliminated by applying clock pulses to all the flip-flops simultaneously, which is done in a synchronous counter. The speed of operation in a synchronous counter is limited py the propagation delay of control gating and a flip-flop. The design of a synchronous counter for any MOD-number other than the power of 2 is more difficult than that of asynchronous counter. This design can be simplified using Karnaugh m method.

A 4-bit (MOD-16) synchronous counter with parallel carry is shown in Fig. 8.11. In this counter, the clock inputs of all the flip-flops are connected together so that the input clock signal is applied simultaneously to each flip-flop. Also, only the LSB flip-flop *A* has its J and *K* inputs connected permanently to V_{cc} while the *J* and *K* inputs of the other flip-flops are driven by some combination of flip-flop outputs. The *J* and *K* inputs of the flip-flop *B* are connected with Q_A output of flip-flop *A*; the *J* and *K* inputs of flip-flop C are connected with AND operated output of Q_A and Q_B; similarly; the *J* and *K* inputs of *D* flip-flop are connected with AND operated output of Q_A, Q_B Q_c and Q_C. As shown in Fig. 8.11 and according to truth table (Table 8.6), flip-flop *A* changes its state with the occurrence of negative transition at each clock-pulse. The flip-flop *B* changes its state when $Q_A = 1$ and when there is negative transition at clock in Flip-flop C changes its state when $Q_A = Q_B = 1$ and when there is negative transition at clock input. Similarly, *D* flip-flop changes its state when $Q_A = Q_B = Q_c = 1$ and when there is negative transition at clock input.

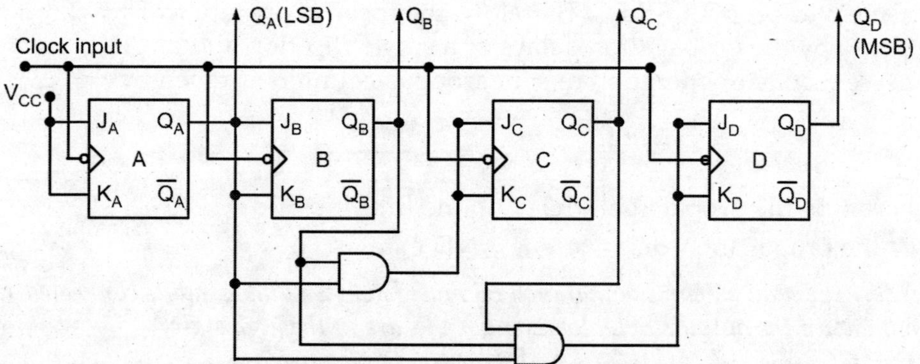

Fig. 8.11. 4-bit synchronous counter.

Table 8.6 Truth table of 4-bit binary synchronous counter.

State	Q_D	Q_C	Q_B	Q_A
0	0	0	0	0
1	0	0	0	1
2	0	0	1	0
3	0	0	1	1
4	0	1	0	0
5	0	1	0	1
6	0	1	1	0
7	0	1	1	1
8	1	0	0	0
9	1	0	0	1
10	1	0	1	0
11	1	0	1	1
12	1	1	0	0
13	1	1	0	1
14	1	1	1	0
15	1	1	1	1
0	0	0	0	0

In a parallel counter, all the flip-flops change their states simultaneously, i.e. they are all synchronised with the negative transition of the input clock signal. Thus, unlike asynchronous counters where the total propagation delay is cumulative, the total settling or response time of a synchronous counter is given as follows: The time taken by one flip-flop to toggle plus the time for the new logic levels to propagate through a single AND gate to reach the *J* and *K* inputs of the following flip-flop.

Total delay = Propagation delay of one flip-flop + Propagation delay of AND gate

The above total delay will be the same irrespective of the number of flip-flops present in the counter, and it will normally be much lower than that of an asynchronous counter with the same number of flip-flops. Therefore, the speed of operation of synchronous counters is limited only by the propagation delays of a single flip-flop and an AND gate. That is, the maximum frequency of operation of synchronous counters is given by

$$f_{max} = \frac{1}{t_p + t_g}$$

where t_p is the propagation delay of one flip-flop and

t_g is the propagation delay of one AND gate

Note: *Hence, in general we can say that each FF must have its J and K inputs connected such that they are high only when the outputs of all lower order FFs are in the high state.*

Also, because of common clocking of all the flip-flops, glitches can be avoided completely in synchronous counters. However, the synchronous counter has more complex circuitry than an asynchronous counter.

8.10 SYNCHRONOUS COUNTER WITH RIPPLE CARRY

The 4-bit synchronous counter shown in Fig. 8.11 is said to be a *Synchronous counter with parallel carry.* The maximum clock frequency for a 4-bit synchronous counter with parallel carry as discussed in the last section is

$$f_{max} = \frac{1}{t_p + t_g}$$

In this counter, as the number of stages in a synchronous counter with parallel carry increases, the flip-flops must drive an ever increasing number of AND gates. Similarly, the number of inputs per control gate also increases. The above problems of synchronous counter with parallel carry are eliminated in a *ripple carry synchronous counter* shown in Fig. 8.12, but the maximum clock frequency of the counter is reduced. Reduction of the maximum clock frequency is due to the delay through control logic which is now $2t_g$, instead of t_g which was achieved with parallel carry. The maximum clock frequency for an *n-bit* synchronous counter with ripple carry is given by

$$f_{max} = \frac{1}{t_p + (n-2)t_g}$$

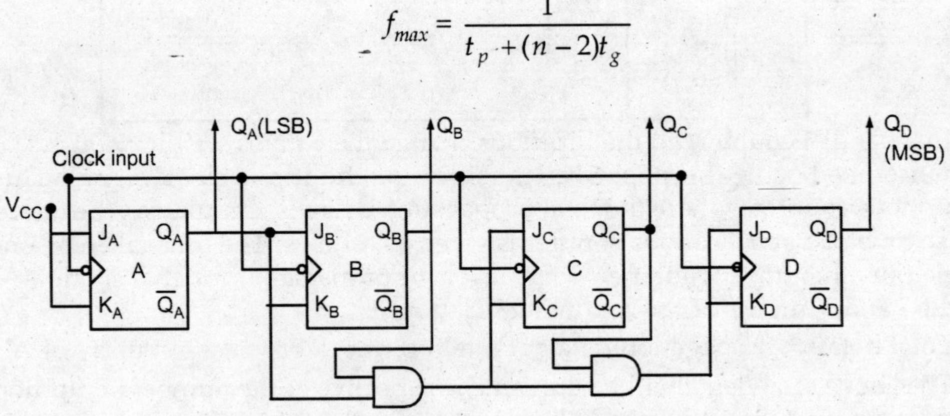

Fig. 8.12. 4-bit synchronous counter with ripple carry.

8.11 SYNCHRONOUS DOWN COUNTER

A parallel down counter can be made to count down by using the inverted outputs of flip-flops to feed the various logic gates. The parallel counter shown in Fig. 8.11 can be converted to a down counter by connecting the $\bar{Q}_A$, $\bar{Q}_B$, and ($\bar{Q}_C$) outputs to the AND gates in place of Q_A, Q_B and Q_C respectively as shown in Fig. 8.13. The counter will then proceed through the counting sequence (as shown in Table 8.7) when input pulses are applied.

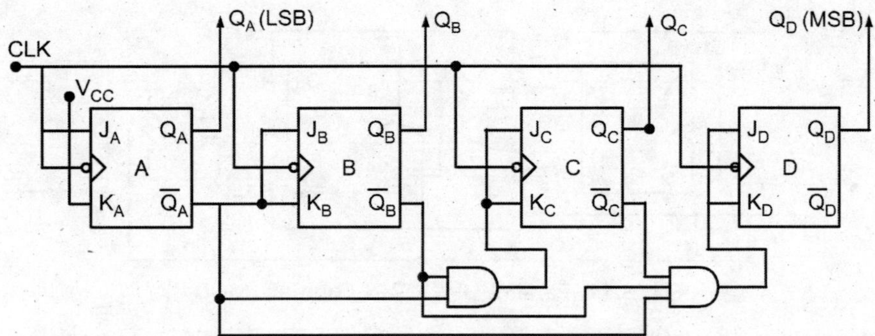

Fig. 8.13. 4-bit synchronous down counter.

Table 8.6 Truth table of 4-bit binary synchronous counter.

State	Q_D	Q_C	Q_B	Q_A
15	1	1	1	1
14	1	1	1	0
13	1	1	0	1
12	1	1	0	0
11	1	0	1	1
10	1	0	1	0
9	1	0	0	1
8	1	0	0	0
7	0	1	1	1
6	0	1	1	0
5	0	1	0	1
4	0	1	0	0
3	0	0	1	1
2	0	0	1	0
1	0	0	0	1
0	0	0	0	0
15	1	1	1	1

8.12 SYNCHRONOUS UP / DOWN COUNTER

To form a parallel UP/DOWN counter, the control inputs (COUNT-UP and COUNT-DOWN) are used to allow either the normal output or the inverted output of one flip-flop to the *J* and *K* inputs of the following flip-flop. A MOD-8 UP/DOWN counter which will count from 000 to 111 when the COUNT-UP = 1 and COUNT-DOWN = 0, or from 111 to 000 when the COUNT-DOWN = 1 and COUNT-UP=0, is shown in Fig. 8.14.

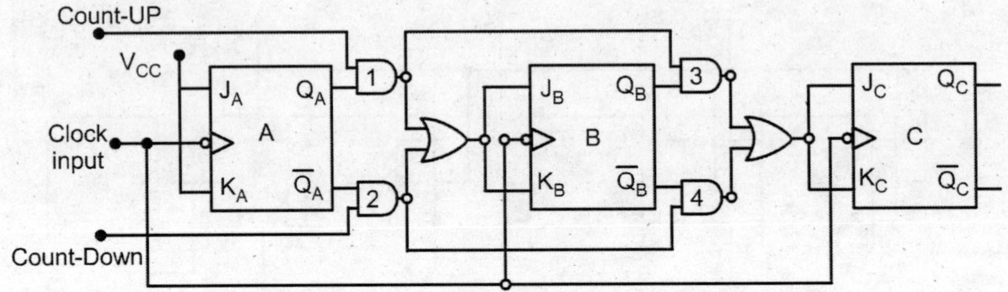

Fig. 8.14. Parallel UP/DOWN Counter (MOD-8)

A logical 1 on the COUNT-UP line while COUNT-DOWN = 0 enables AND gates 1 and 3 and disables gates 2 and 4. This allows the Q_A and Q_B outputs through the AND gates to the *J* and *K* inputs of the following flip-flops, so that the counter counts up as pulses are applied. The reverse action takes place when COUNT-UP = 0 and COUNT-DOWN=1.

8.13 SYNCHRONOUS/ASYNCHRONOUS COUNTER

The synchronous/asynchronous counter, constructed by combining the synchronous and asynchronous counters, represents a compromise between the speed of synchro-nous counters and the simplicity of asynchronous counters. In a decade [MOD-10] synchronous/asynchronous counter shown in Fig. 8.15, the input pulses are applied only to the flip-flop-A as in an asynchronous counter, whereas the Q_A output drives the clock inputs of both B and *D* flip-flops so that they will trigger simultaneously as in a synchronous counter. The sequence of operation of this counter is explained as follows:

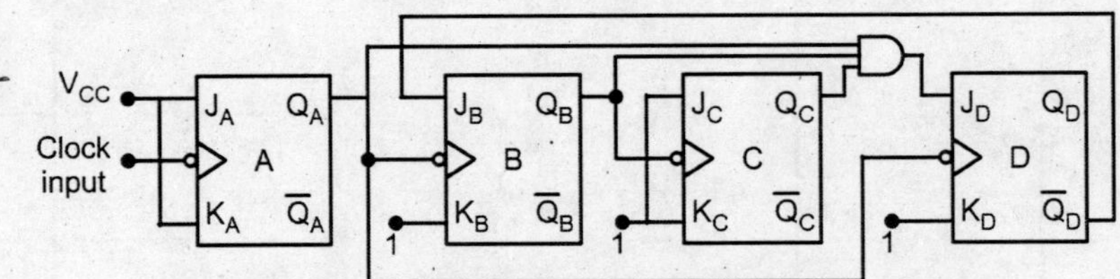

Fig. 8.15. Synchronous/asynchronous decade counter.

1. Assume that the counter is initially in the 0000 state. Thus, the *J* and *K* inputs of the flip-flops *A*, *B* and *C* are all HIGH. The *D* flip-flop has $J_D = 0$ and *K*= 1 and so it will not be affected by any clock transition.
2. Flip-flops *A*, *B* and *C* will function as a normal ripple counter for the first seven input pulses. The flip-flop *D* will remain LOW because its *J* input is at 0.

3. In the 0111 state, the AND gate output is at 1, so flip-flop *D* will have $J_D = K_D = 1$. When the *eighth* input pulse occurs, it will toggle Flip-flop *A* to the 0 state, which will in turn toggle flip-flop *B* LOW and flip-flop *D* HIGH. The Q_B output transition will toggle Q_c LOW. Thus, the counter is now in the 1000 state.
4. The *ninth* input pulse simply toggles Q_A high bringing the counter to the 1001 state.
5. In the 1001 state, the AND gate output is again 0, so the flip-flop *D* has $J_D = 0$, $K_D = 1$, which means it will go LOW on the next negative transition at its clock input. The flip-flop *B* also has $J_B = 0$, $K_B = 1$, so it will remain in the LOW state. Thus, when the *tenth* pulse occurs, flip-flop *A* will toggle LOW. which in turn will toggle Q_D LOW bringing the counter back to the 0000 state The operation then returns to step 1 and repeats the sequence.

8.14 PRESETTABLE (PROGRAMMABLE) COUNTERS

All the UP counters start from 00...0 state while DOWN counters start from 11... 1 state. This is accomplished by applying a momentary pulse to all the flip-flop's CLEAR inputs before the counting operation begins. A counter can also be made to start in any desired state through the use of appropriate logic circuitry. Counters that have the capability to start counting from any desired state are called *presettable or programmable* counters.

A presettable MOD-8 ripple UP counter is illustrated in Fig. 8.16. In this counts, the desired starting state is entered using the PRESET and CLEAR inputs irrespec of what is happening at the *J* and *K* or clock inputs. The desired preset count is determined by the preset inputs P_A,P_B and P_c whose values are transferred into the counter flip-flop when counter flip-flops when the 'PRESET LOAD' input is momentarily pulsed to the LOW level. When the 'PRESET LOAD' input returns HIGH, the NAND gates are disabled and the counter is free to count input clock pulses starting from the newly entered count that has been preset into the flip-flops.

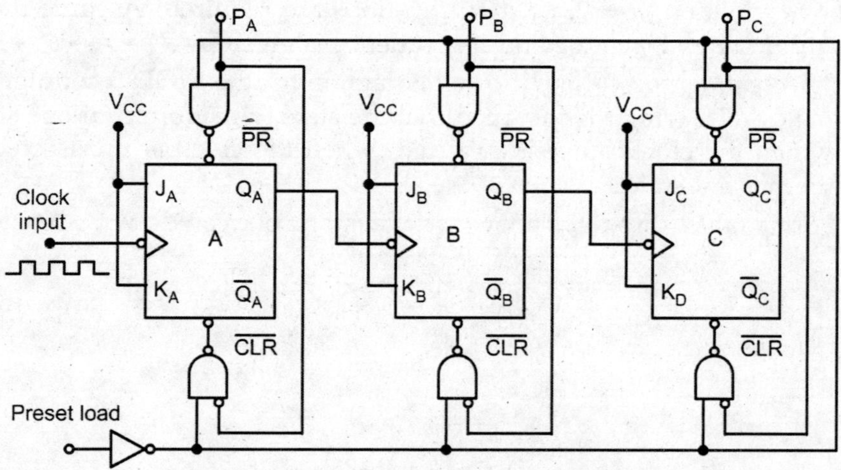

Fig. 8.16. Presettable MOD-8 UP counter.

8.15 DESIGN OF SYNCHRONOUS COUNTERS

In this section, the general method to design the different types of synchronous counters using various types of flip-flops is explained in detail.

The steps involved in the design of synchronous counter are listed below:

Step 1 From the word description of the problem, draw a state diagram which describes the operation of the counter.

Step 2 From the above state diagram, obtain Present State - Next State (PS-NS) table of the counter and check the same to ascertain whether it has any equivalent states. Any two states are said to be equivalent, if and only if their next states are one and the same. In such a case, one of the equivalent states can be eliminated from the state table. Thus, in this step, the state table is modified in such a way that there is no redundant state in it.

Step 3 Make a state assignment and document the same in the above state table.

Step 4 Decide the type of memory element to be used in the counter design and then obtain the excitation table from PS-NS table using the application table of the flip-flop.

Step 5 Draw the excitation maps for various excitation inputs of flip-flops and simplify the excitation functions.

Step 6 Draw the schematic diagram of the counter.

To understand the above design procedures, the following counter design problems can be considered, in which q_i represents the present state and Q_i represents the next state of the flip-flop where $i = 0$ to $n - 1$, and n is the number of flip-flops used.

8.15.1 Design of MOD-3 Counter

MOD-3 counter is one with three states. To design a counter with three states, the number of flip-flops required can be found using the equation $2^n \geq N \geq 2^{n-1}$, where n *is* the number of flip-flops required and N is the number of states present in the counter. For $N = 3$, from the above equation, $n = 2$, i.e. two flip-flops are required. Assume that the MOD-3 counter has three states a, b and c and its sequence is given by $a \rightarrow b \rightarrow c \rightarrow a$...

Step 1 State diagram Now, the state diagram for the MOD-3 counter can be drawn as shown in Fig. 8.17. Here, it is assumed that the state transition from one state to another takes place when the clock pulse is asserted; when the clock is unassorted, the counter remains in the present state.

Step 2 State table From the above state diagram, one can draw PS-NS table (Table 8 8).

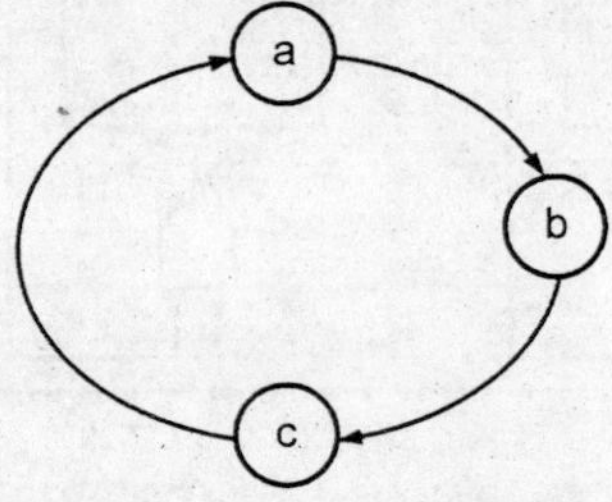

Fig. 8.17. State diagram of MOD-3 counter.

Table 8.8 PS-NS table for MOD-3 counter.

Present state (PS)	*Next State (NS)*
a	b
b	c
c	a

The state table given in Table 8.8 has no redundant state because no two states are equivalent. Hence, there is no modification required in the given state table.

Step 3 State assignment Let us assign two state variables to states *a*, *b* and *c* as follows: $a = 00$, $b = 01$ and c = 10. Then, the PS-NS table gets modified as shown in Table 8.9.

Table 8.9 PS-NS table for MOD-3 counter.

Present State (PS)		*Next State (NS)*	
q_1	q_0	q_1	Q_0
0	0	0	1
0	1	1	0
1	0	0	0
1	1	d	d

Step 4 Excitation table Although any one of the four flip-flops, i.e. *SR*, *JK*, *T* and *D*, can be used, the selection *of J-K* flip-flop will result in a simplified circuit for synchinous counters. The excitation table having entries for flip-flop inputs (J_1K_1 and J_0K_0) can be drawn from the above PS-NS table (and using the application table of *JK* Flip-flop given in Table 7.11) as shown in Table 8.10.

Table 8.10 Excitation table for MOD-3 counter.

PS		*NS*		*Excitation Inputs*			
q_1	q_0	Q_1	Q_0	J_1	K_1	J_0	K_0
0	0	0	1	0	*d*	1	*d*
0	1	1	0	1	*d*	*d*	1
1	0	0	0	*d*	1	0	*d*
1	1	*d*	*d*	*d*	*d*	*d*	*d*

In the first row of the above table, fir the flip-flop 2 of the counter to change from present state ($q_1 = 0$) to next state ($Q_1 = 0$), the J_1K_1 inputs required are 0*d*; for flip-flop 1 to change from $q_0 = 0$ to $Q_0 = 1$, the J_0K_0 inputs required are 1*d*. Similarly, other entries are also made using the appliation table.

Step 5 Excitation maps The excitation maps for J_1, K_1, J_0 and K_0 inputs of the counter can be drawn as shown in Fig. 8.18 from the excitation table (Table 8.10).

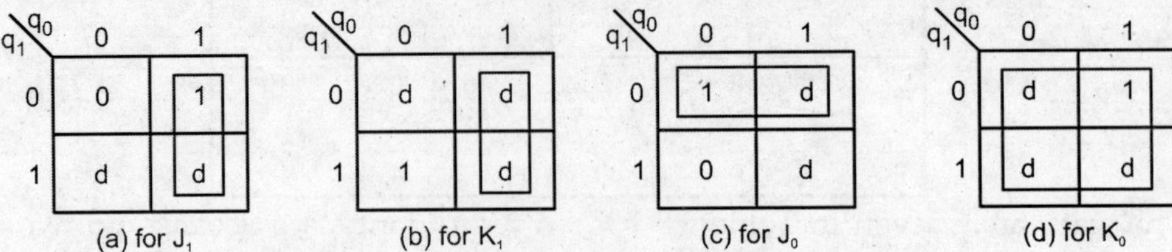

Fig. 8.18. Excitation Maps

From the above excitation maps, the simplified excitation funtions are:

$$J_1 = q_0,\ K_1 = 1;\ J_0 = \bar{q}_1 \text{ and } K_0 = 1$$

Step 6 Schematic diagram Using the above excitation equations, the circuit diagram for the MOD-3 counter can be drawn as shown in Fig. 8.19.

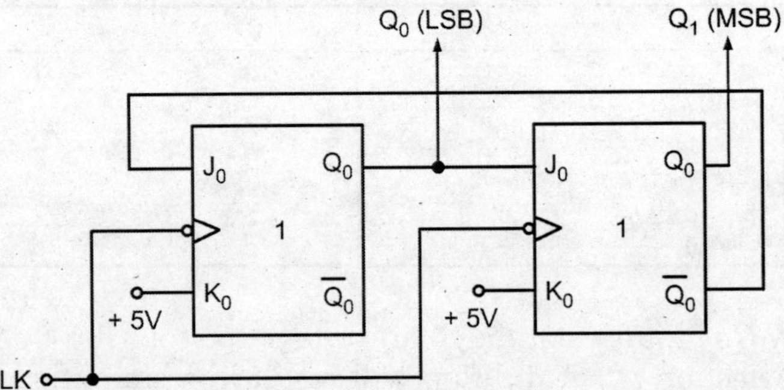

Fig. 8.19. Circuit diagram for MOD-3 synchronous counter.

8.15.2 Design of MOD-6 Counter

In order to design a MOD-6 counter with six states, the number of flip-flops required is three. This is found from the equation $2^n \leq N \leq 2^{n-1}$, where N = 6, the number of states present in the MOD-6 counter. Let us assume that the MOD-6 counter has six states, viz. *a*, *b*, *c*, *d*, *e* and *f*.

Step 1 State diagram The State diagram for the MOD-6 counter can be drawn as shown in Fig. 8.20. Here, it is assumed that the state transition from one state to another takes place when the clock pulse is asserted; when the clock is unasserted, the counter remains in the present state.

Step 2 State Table From the above state diagram, one can draw PS-NS table as shown in Table 8.11.

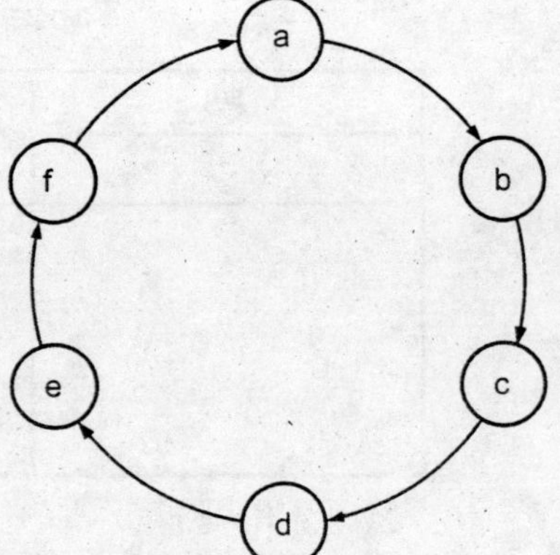

Fig. 8.20. State diagram of MOD-6 counter.

Table 8.11 PS-NS table for MOD-6 counter.

Present state (PS)	*Next State (NS)*
a	b
b	c
c	d
d	e
e	f
f	*a*

The above state table does not have any redundant state because no two states are equivalent. So, there is no modification in the above state table.

Step 3 State Assignment Let us assign three state variables to states *a, b, c, d, e* and *f* as follows; *a* = 000, *b* = 001, *c* = 010, *d* = 011, *e* = 100 and *f* = 101. Then the PS-NS table gets modified as shown in Table 8.12.

Table 8.12 PS-NS table for MOD-6 counter.

Present state (PS)			*Next State (NS)*		
q_2	q_1	q_0	Q_2	Q_1	Q_0
0	0	0	0	0	1
0	0	1	0	1	0
0	1	0	0	1	1
0	1	1	1	0	0
1	0	0	1	0	1
1	0	1	0	0	0
	....			....	
1	1	0	*d*	*d*	*d*
1	1	1	*d*	*d*	*d*

Step 4 Excitation table The JK flip-flop is selected for the counter design because it results in a simplified circuit. The excitation table having entries for flip-flop inputs (J_2 K_2, J_1 K_1 and J_0K_0) can be drawn from the above PS-NS table and using the application table of JK flip-flop given earlier. Table 8.13 gives the excitation values of MOD-6 counter.

Table 8.13 Excitation table for MOD-6 counter.

PS			*NS*			*Excitation inputs*					
q_2	q_1	q_0	Q_2	Q_1	Q_0	J_2	K_2	J_1	K_1	J_0	K_0
0	0	0	0	0	1	0	*d*	0	*d*	1	*d*
0	0	1	0	1	0	0	*d*	1	*d*	*d*	1
0	1	0	0	1	1	0	*d*	*d*	0	1	d
0	1	1	1	0	0	1	*d*	*d*	1	*d*	1
1	0	0	1	0	1	*d*	0	0	*d*	1	*d*
1	0	1	0	0	0	*d*	1	0	*d*	*d*	1
1	1	0	*d*	*d*	*d*	*d*	*d*	*d*	*d*	*d*	*d*
1	1	1	*d*	*d*	*d*	*d*	*d*	*d*	*d*	*d*	*d*

Step 4 Excitation maps The excitation maps for J_2, K_2, J_1, K_1, J_0 and K_0 inputs of the counter can be drawn as shown in Fig. 8.21. from the excitation table (Table 8.13.)

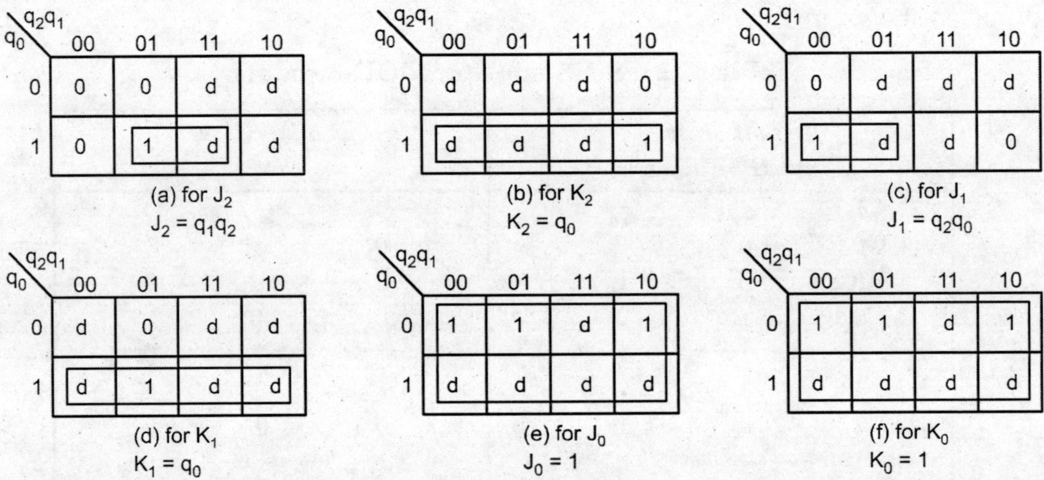

Fig. 8.21. Excitation maps for MOD-6 counter.

Step 6 Schematic diagram Using the above excitation equations, the circuit diagram for the MOD-6 counter can be drawn as shown in Fig. 8.22.

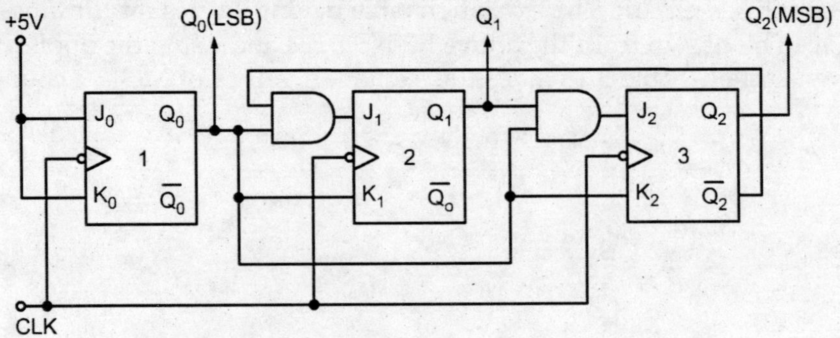

Fig. 8.22. Circuit diagram for MOD-6 synchronous counter.

8.15.3 Design of BCD of Decade (MOD-10) Counter

To design a BCD or Decade (MOD-10) counter that has ten states i.e., 0 to 9 the number of flip-flops required is four. Let us assume that the MOD-10 counter has ten states, viz. *a, b, c, d, e, f, g, h, i* and *j*.

Step 1 State diagram Now the state diagram for the MOD-10 counter can be drawn as shown in Fig. 8.23. Here, it is assume that the state transition from one state to another takes place when the clock pulse is asserted. When the clock is unasserted, the counter remains in the present state.

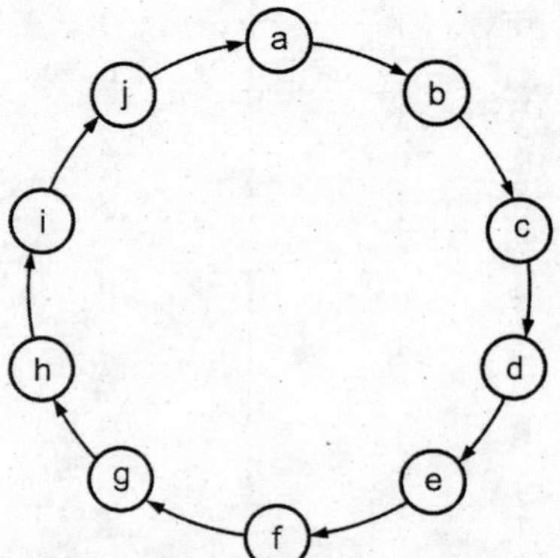

Fig. 8.23. State diagram of MOD-10 counter.

Step 2 State table From the above state diagram, one can draw the PS-NS table as shown in Table 8.14.

Table 8.14 PS-NS table for MOD-10 counter.

Present state (*PS*)	*Next State* (*NS*)
a	b
b	c
c	d
d	e
e	f
f	g
g	h
h	i
i	j
j	*a*

The above state table does not have any redundant state because no two states are equivalent. So, there is no modification required in the above state table.

Step 3 State assignment Let us assign four state variables to these states *a, b, c, d, e, f, g, h, i* and *j* as follows: *a* = 0000, *b* = 0001, *c* = 0010, *d* = 0011, *e* = 0100, *f* = 0101, *g* = 0110, *h* = 0111, *i* = 1000 and *j* = 1001. Then the above PS-NS table can be modified as shown in Table 8.15.

Table 8.15 PS-NS table for MOD-10 counter.

Present state (PS)				*Next State (NS)*			
0	0	0	0	0	0	0	1
0	0	0	1	0	0	1	0
0	0	1	0	0	0	1	1
0	0	1	1	0	1	0	0
0	1	0	0	0	1	0	1
0	1	0	1	0	1	1	0
0	1	1	0	0	1	1	1
0	1	1	1	1	0	0	1
1	0	0	0	1	0	0	1
1	0	0	1	0	0	0	0
....				....			
1	0	1	0	d	d	d	d
1	0	1	1	d	d	d	d
1	1	0	0	d	d	d	d
1	1	0	1	d	d	d	d
1	1	1	0	d	d	d	d
1	1	1	1	d	d	d	d

Step 4 Excitation table The excitation table having entries for flip-flop inputs (J_3K_3, J_2K_2, J_1K_1 and J_0K_0) can be drawn, from the above PS-NS table using the application table of JK flip-flop gien earlier, as shown in Table 8.16.

Table 8.16 Excitation table for MOD-10 counter.

PS				*NS*				*Excitation inputs*							
q_3	q_2	q_1	q_0	Q_3	Q_2	Q_1	Q_0	J_3	K_3	J_2	K_2	J_1	K_1	J_0	K_0
0	0	0	0	0	0	0	1	0	*d*	0	*d*	0	*d*	1	*d*
0	0	0	1	0	0	1	0	0	*d*	0	*d*	1	*d*	*d*	1
0	0	1	0	0	0	1	1	0	*d*	0	*d*	*d*	0	1	*d*
0	0	1	1	0	1	0	0	0	*d*	1	*d*	*d*	1	*d*	1
0	1	0	0	0	1	0	1	0	*d*	*d*	0	0	*d*	1	*d*
0	1	0	1	0	1	1	0	0	*d*	*d*	0	1	*d*	*d*	1
0	1	1	0	0	1	1	1	0	*d*	*d*	0	*d*	0	1	*d*
0	1	1	1	1	0	0	0	1	*d*	*d*	1	*d*	1	*d*	1
1	0	0	0	1	0	0	1	*d*	0	0	*d*	0	*d*	1	*d*
1	0	0	1	0	0	0	0	*d*	1	0	*d*	0	*d*	*d*	1
....				....				....				....			
1	0	1	0	*d*	*d*	*d*	*d*	*d*	*d*	*d*	*d*	*d*	*d*	*d*	*d*
1	0	1	1	*d*	*d*	*d*	*d*	*d*	*d*	*d*	*d*	*d*	*d*	*d*	*d*
1	1	0	0	*d*	*d*	*d*	*d*	*d*	*d*	*d*	*d*	*d*	*d*	*d*	*d*
1	1	0	1	*d*	*d*	*d*	*d*	*d*	*d*	*d*	*d*	*d*	*d*	*d*	*d*
1	1	1	0	*d*	*d*	*d*	*d*	*d*	*d*	*d*	*d*	*d*	*d*	*d*	*d*
1	1	1	1	*d*	*d*	*d*	*d*	*d*	*d*	*d*	*d*	*d*	*d*	*d*	*d*

Step 5 Excitation maps The excitation maps for J_3, K_3, J_2, K_2, J_1, K_1, J_0 and K_0 inputs of the counter can be drawn as shown in Fig. 8.24. from the Excitation Table 8.16.

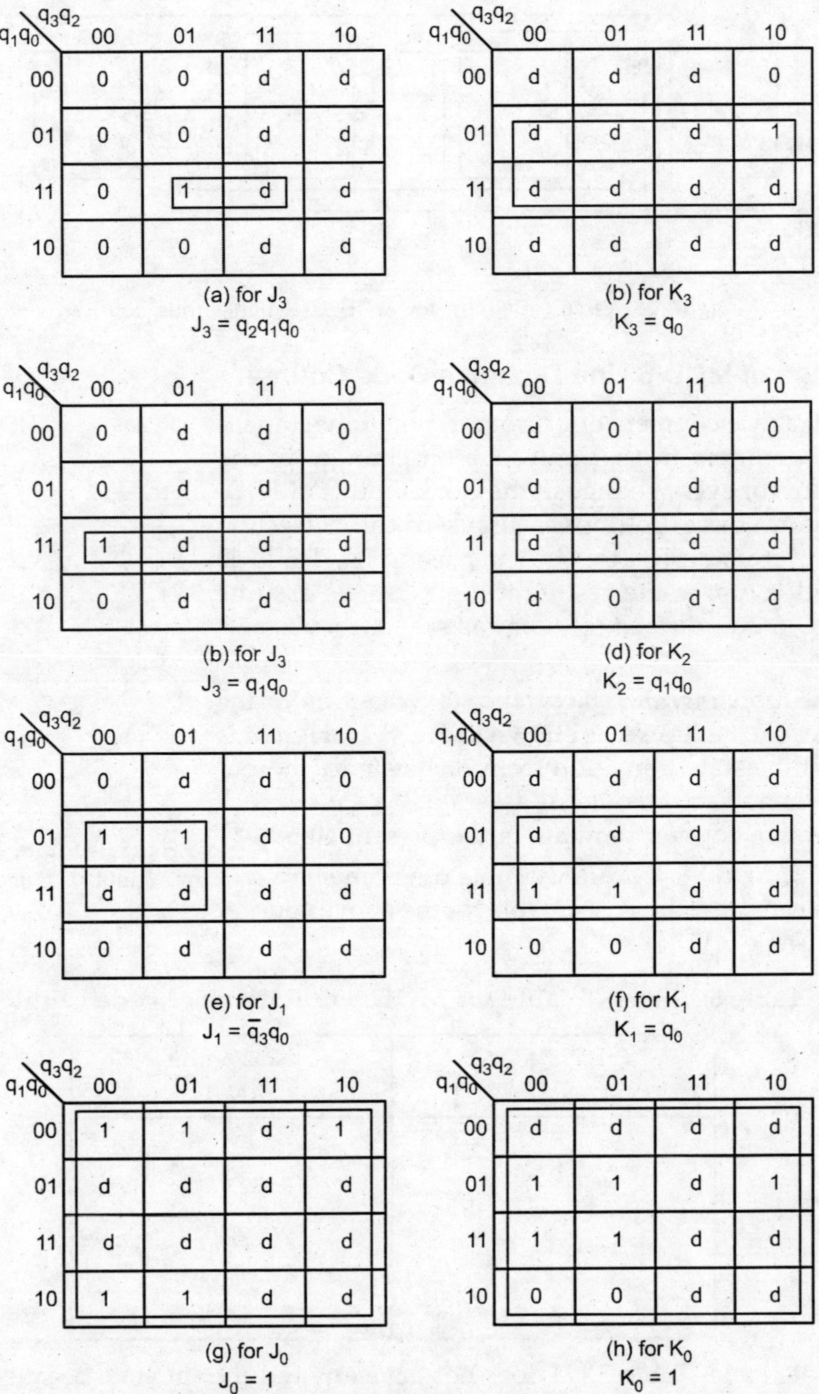

(a) for J_3
$J_3 = q_2q_1q_0$

(b) for K_3
$K_3 = q_0$

(b) for J_3
$J_3 = q_1q_0$

(d) for K_2
$K_2 = q_1q_0$

(e) for J_1
$J_1 = \bar{q}_3q_0$

(f) for K_1
$K_1 = q_0$

(g) for J_0
$J_0 = 1$

(h) for K_0
$K_0 = 1$

Fig. 8.24. Excitation maps for MOD-10 counter.

Step 6 Schematic diagram Using the above excitation equations, the circuit diagram for the MOD-10 counter can be drawn as shown in Fig. 8.25.

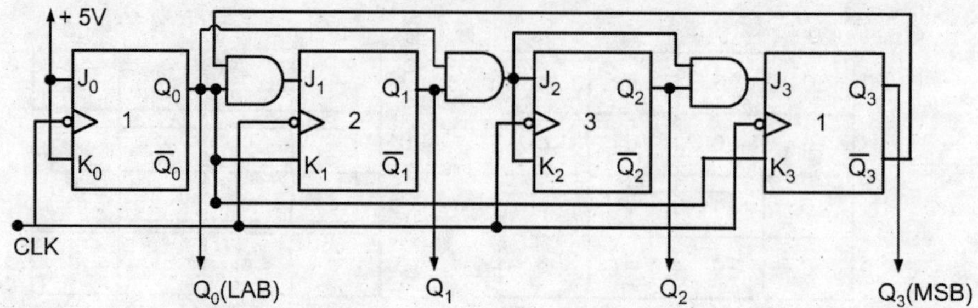

Fig. 8.25. Circuit diagram for MOD-10 synchronous counter.

8.15.4 Design of MOD-6 Unit Distance Code Counter

Unit distance counters offer from conventional binary and decimal counters in that only one bit changes in the counter output for every change in the clock input. A MOD-6 counter requires three flip-flops as discussed earlier (section 8.15.2). Now, let us design this counter using JK, T and D flip-flops and compare the resultant logic. Let us assume that the MOD-6 unit distace counter has six states, viz. *a*, *b*, *c*, *d*, *e* and *f*.

Step 1 State diagram Now, the state diagramfor the above counter can be drawn as shown in Fig. 8.26. Here it is assumed that the state transition from one state to another takes place when the clock pulse is asserted. When the clock is unasserted, the counter remains in the present state.

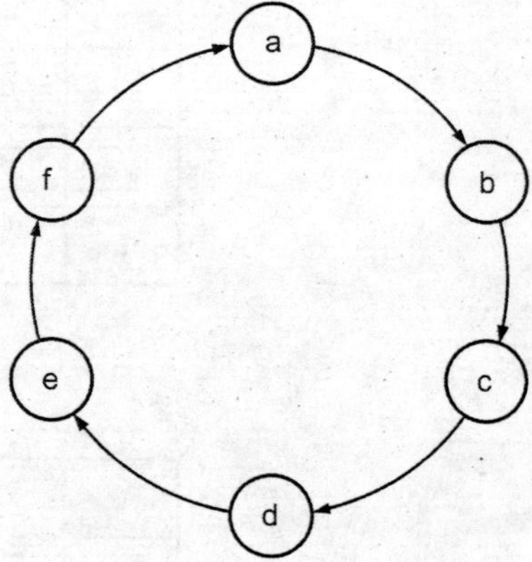

Fig. 8.26. State diagram of MOD-6 unit distance code counter.

Step 2 State table From the state diagram shown in Fig. 8.26, one can draw the PS-NS table for the above counter as shown in Table 8.17.

Table 8.17 PS-NS table for MOD-6 unit distance code counter.

Present state (*PS*)	*Next State* (*NS*)
a	b
b	c
c	d
d	e
e	f
f	*a*

The state table (Table 8.17) does not have any redundant state because no two states are equivalent. Hence, there i no modification required in the above state table.

Step 3 State assignment Let us assign three-bit unit distance state variables to the consecutive states *a*, *b*, *c*, *d*, *e* and *f* as follows: *a* = 000, *b* = 001, *c* = 011, *d* = 010, *e* = 110 and *f* = 100. Then, the above PS-NS table gets modified as shown in Table 8.18.

Table 8.18 PS-NS table for MOD-6 counter unit distance code counter.

Present state (PS)			*Next State (NS)*		
q_2	q_1	q_0	Q_2	Q_1	Q_0
0	0	0	0	0	1
0	0	1	0	1	1
0	1	1	0	1	0
0	1	0	1	1	0
1	1	0	1	0	0
1	0	0	0	0	0
....			....		
1	0	1	*d*	*d*	*d*
1	1	1	*d*	*d*	*d*

Step 4 Excitation table The JK flip-flop is selected for the counter design because it results in a simplified circuit. The excitation table having entries for flip-flop inputs (J_2 K_2, J_1 K_1 and J_0K_0) can be drawn from the above PS-NS table and using the application table of JK flip-flop given earlier. Table 8.13 gives the excitation values of MOD-6 counter.

Step 4 Excitation table So far, in the design of a synchronous counter, only *JK* flip-flop is considered, because it results in a simplified circuit. To verify this argument, in this design all the flip-flops (*JK*, Tand D) are considered and the resulting equation for excitation functions are compared. The excitation table having entries for *JK*, T and D flip-flop inputs can be written as shown in Table 8.19 from the above PS-NS table and using the application table of *JK*, *T* and *D* flip-flops given in Tables 7.8, 7.11 and 7.14 respectively.

Table 8.19 Excitation table for MOD-6 unit distance code counter.

Present state (PS)			*Next State (NS)*			*Excitation inputs* Using JK FF						*Using TFF*			*Using DFF*		
q_2	q_1	q_0	Q_2	Q_1	Q_0	J_2	K_2	J_1	K_1	J_0	K_0	T_2	T_1	T_0	D_2	D_1	D_0
0	0	0	0	0	1	0	d	0	d	1	d	0	0	1	0	0	1
0	0	1	0	1	1	0	d	1	d	d	0	0	1	0	0	1	1
0	1	1	0	1	0	1	d	d	0	d	1	0	0	1	0	1	0
0	1	0	1	1	0	1	d	d	0	0	d	1	0	0	1	1	0
1	1	0	1	0	0	d	0	d	1	0	d	0	1	0	1	0	0
1	0	0	0	0	0	d	1	0	d	0	d	1	0	0	0	0	0
...	...	...	...	...	...	...	...	...	...	...	...	...	...	...	...	...	...
1	0	1	*d*	*d*	*d*	*d*	*d*	*d*	*d*	*d*	*d*	*d*	*d*	*d*	*d*	*d*	*d*
1	1	1	*d*	*d*	*d*	*d*	*d*	*d*	*d*	*d*	*d*	*d*	*d*	*d*	*d*	*d*	*d*

Step 5 Excitation maps The excitation maps for J_2, K_2, J_1, K_1, J_0 and K_0 inputs for implementation using JK flip-flop can be drawn from the above excitation table as shown in Fig. 8.27.

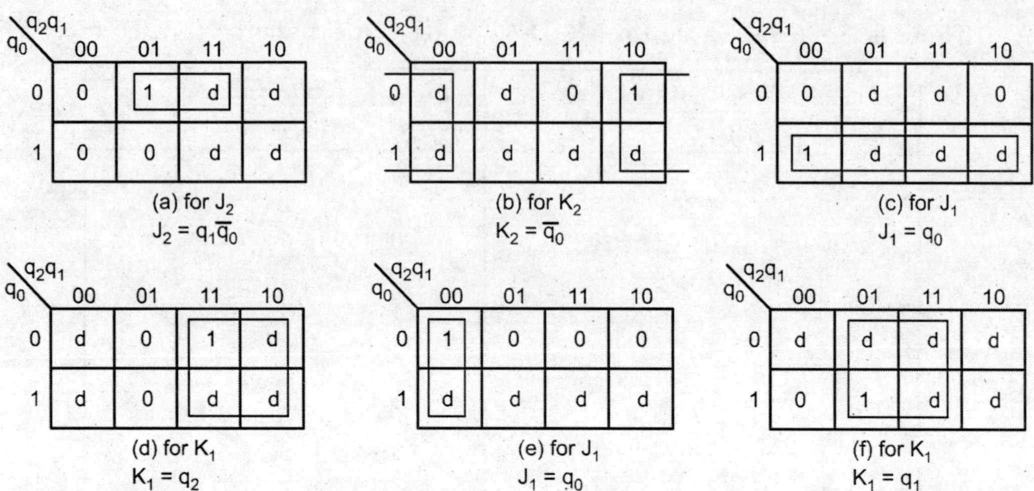

(a) for J_2 $J_2 = q_1\overline{q}_0$ (b) for K_2 $K_2 = \overline{q}_0$ (c) for J_1 $J_1 = q_0$

(d) for K_1 $K_1 = q_2$ (e) for J_1 $J_1 = q_0$ (f) for K_1 $K_1 = q_1$

Fig. 8.27. Excitation maps for MOD-6 unit distance code counter using Jk flip-flop.

The excitation maps for T_2, T_1 and T_0 inputs for implementaing the counter using a T flip-flop can be drawn from the above excitation table (Table 8.19) as shown in Fig. 8.28.

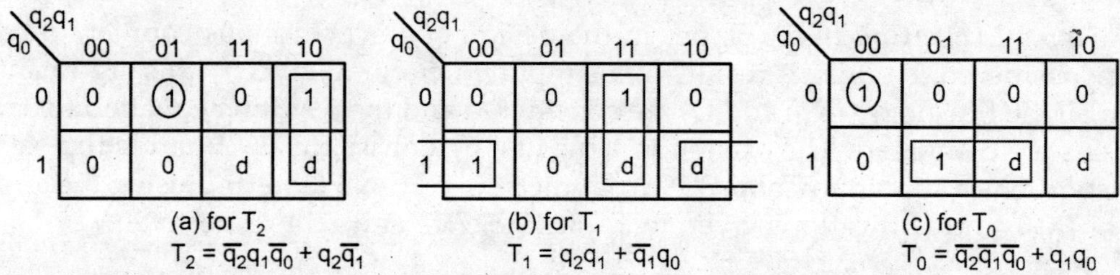

(a) for T_2 $T_2 = \overline{q}_2q_1\overline{q}_0 + q_2\overline{q}_1$ (b) for T_1 $T_1 = q_2q_1 + \overline{q}_1q_0$ (c) for T_0 $T_0 = \overline{q}_2\overline{q}_1\overline{q}_0 + q_1q_0$

Fig. 8.28. Excitation maps for MOD-6 unit distance code counter using T flip-flop.

Similarly, the excitation maps for D_2, D_1 and D_0 inputs for implementing the counter using a D flip-flop can be drawn from the excitation table (Table 8.19) as shown in Fig. 8.29.

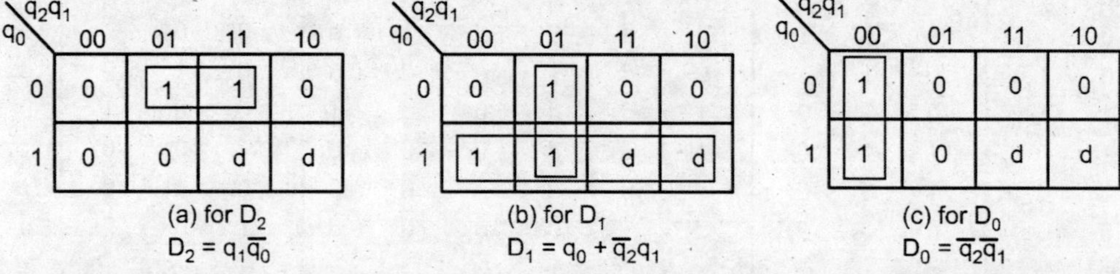

(a) for D_2 $D_2 = q_1\overline{q}_0$ (b) for D_1 $D_1 = q_0 + \overline{q}_2q_1$ (c) for D_0 $D_0 = \overline{q}_2\overline{q}_1$

Fig. 8.29. Excitation maps for MOD-6 unit distance code counter using D flip-flop.

Step 6 Schematic diagram From the above simplified equations for excitation inputs of JK, T and D flip-flops, it is found that implementation using *JK* flip-flop requires less number of logic gates. So, the circuit diagram for the MOD-6 unit distance code up counter using Jk flip-flop can be drawn as shown in Fig. 8.30.

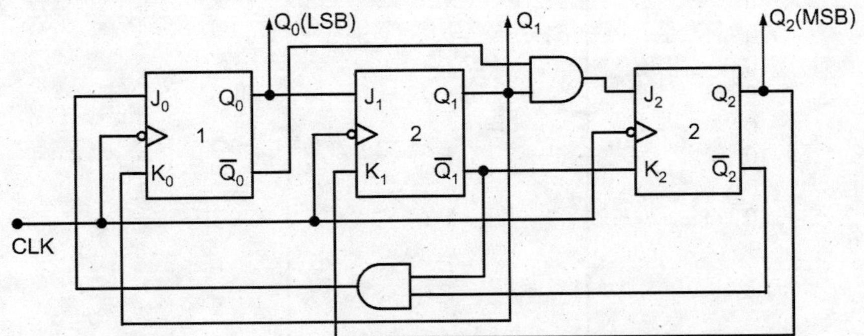

Fig. 8.30. Circuit diagram for MOD-6 unit distance code counter.

8.15.5 Design of MOD-8 UP-Down Counter

The block diagram of a 3-bit Up-Down counter is shown in Fig. 8.31.

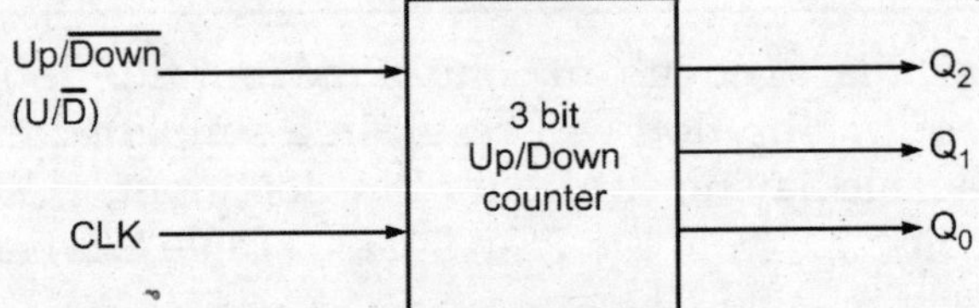

Fig. 8.31. Block diagram of 3-bit Up-Down counter.

From Fig. 8.31, when Up/$\overline{\text{Down}}$ = 1, the counter will work in the Up mode, i.e. it counts from state 000 to state 111. When Up/$\overline{\text{Down}}$ = 0, it will work in the Down mode, i.e. it counts from state 111 to state 000. Let us assume that the 3-bit Up-Down counter has eight states, viz. *a, b, c, d, e, f, g* and *h*.

Step 1 State diagram The state diagram for a MOD-8 up-down counter can be drawn as shown in Fig. 8.32. Here it is assumed that the state transition from one stae to another takes place when the clock pulse is asserted. When the clock is unasserted, the counter remains in the present state. Also, it is assumed that when Up/$\overline{\text{Down}}$ = 1, the state transition will take place in the forward directions; when Up/$\overline{\text{Down}}$ = 0, the state transition will take place in the reverse directions.

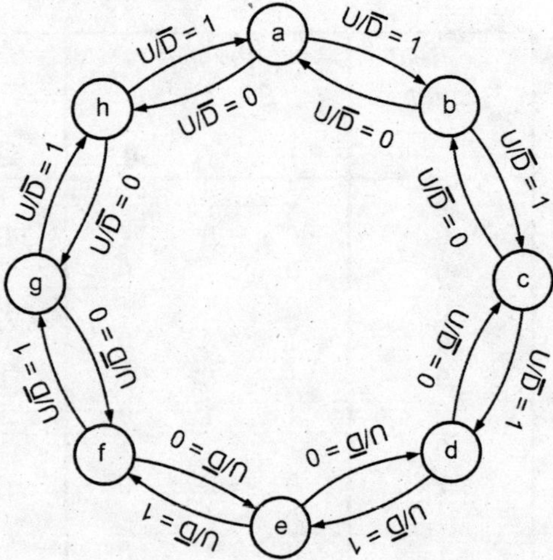

Fig. 8.32. State diagram of MOD-8 Up-Down counter.

Step 2 State table From the above state diagram, we can draw PS-NS table as shown in Table 8.20.

Table 8.20. PS-NS table for MOD-8 UP-Down counter.

Counter Input $Up/\overline{Down}$	*Present state (PS)*	*Next state (NS)*
0	*a*	*h*
0	*b*	*a*
0	*c*	*b*
0	*d*	*c*
0	*e*	*d*
0	*f*	*e*
0	*g*	*f*
0	*h*	*g*
1	*a*	*b*
1	*b*	*c*
1	*c*	*d*
1	*d*	*e*
1	*e*	*f*
1	*f*	*g*
1	*g*	*h*
1	*h*	*a*

Step 4 Excitation table Though any one of the four flip-flop, i.e. *SR*, *JK*, *T* and *D*, can be used, the selection of *J-K* flip-flop will result in a simplified circuit for a synchronous counter. The excitation table having entries for flip-flop inputs (J_2, K_2, J_1 K_1 and J_0K_0) can be drawn as shown in Table 8.22, from the PS-NS table (Table 8.21) and using the application table of *JK* flip-flop given in table 7.11.

Table 8.20. PS-NS table for MOD-8 Up-Down counter.

Control input	*Present State (PS)*			*Next state (NS)*			*Excitatino Inputs*					
$Up/\overline{Down}$	q_2	q_1	q_0	Q_2	Q_1	Q_0	J_2	K_2	J_1	K_1	J_0	K_0
0	0	0	0	1	1	1	1	d	1	d	1	d
0	0	0	1	0	0	0	0	d	0	d	d	1
0	0	1	0	0	0	1	0	d	d	1	1	d
0	0	1	1	0	1	0	0	d	d	0	d	1
0	1	0	0	0	1	1	d	1	1	d	1	d
0	1	0	1	1	0	0	d	0	0	d	d	1
0	1	1	0	1	0	1	d	0	d	1	1	d
0	1	1	1	1	1	0	d	0	d	0	d	1
1	0	0	0	0	0	1	0	d	0	d	1	d
1	0	0	1	0	1	0	0	d	1	d	d	1
1	0	1	0	0	1	1	0	d	d	0	1	d
1	0	1	1	1	0	0	1	d	d	1	d	1
1	1	0	0	1	0	1	d	0	0	d	1	d
1	1	0	1	1	1	0	d	0	1	d	d	1
1	1	1	0	1	1	1	d	0	d	0	1	d
1	1	1	1	0	0	0	d	1	d	1	d	1

Step 5 Excitation maps The excitation maps for J_2, K_2, J_1, K_1, J_0 and K_0 inputs of the Up-Down counter can be drawn from the above excitation table as shown in Fig. 8.33.

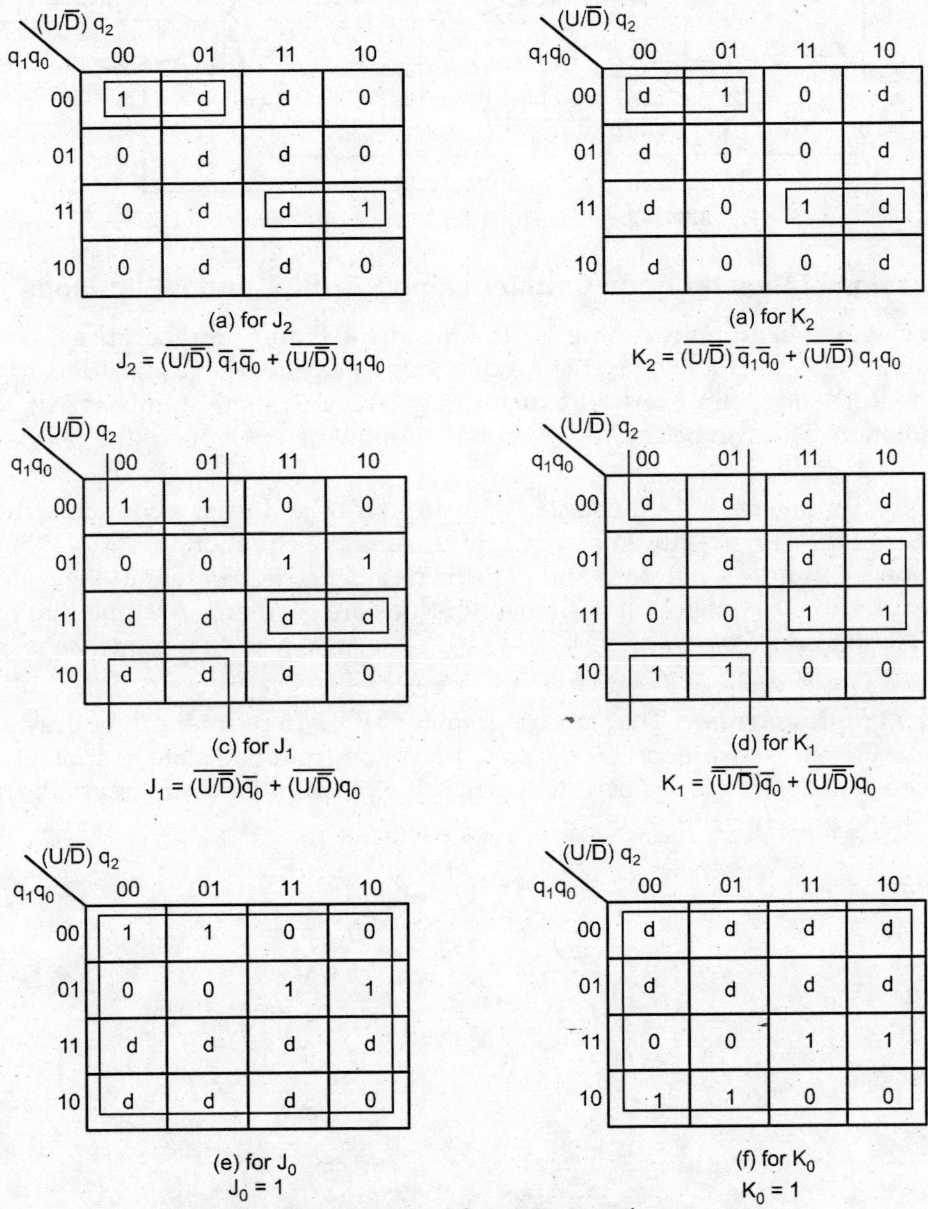

(a) for J_2

$J_2 = \overline{(U/\bar{D})}\,\bar{q}_1\bar{q}_0 + (U/\bar{D})\,q_1q_0$

(a) for K_2

$K_2 = \overline{(U/\bar{D})}\,\bar{q}_1\bar{q}_0 + (U/\bar{D})\,q_1q_0$

(c) for J_1

$J_1 = \overline{(U/\bar{D})}\bar{q}_0 + (U/\bar{D})q_0$

(d) for K_1

$K_1 = \overline{(U/\bar{D})}\bar{q}_0 + (U/\bar{D})q_0$

(e) for J_0

$J_0 = 1$

(f) for K_0

$K_0 = 1$

Fig. 8.33. Excitation maps for MOD-8 up-down counter.

Step 6 Schematic diagram Using the above excitation equations, the circuits diagram for the MOD-8 up=down counter can be drawn as shown in Fig. 8.34.

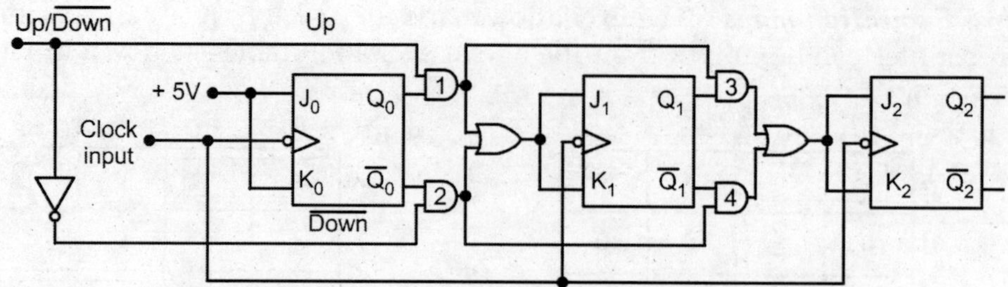

Fig. 8.34. Circuit diagram for MOD-8 Up-Down counter.

8.15.6 Design of Synchronous Counter using SR, JK, T and D Flip-Flops

Normally synchronous counter is designed usng Jk flio-flops because this will results in a simplified circuit. However, synchronous counter can also be designed using other flip-flops like SR, D and T flip-flops. But such design requires more number of logic gates for implementation. This can be explained with the following example of the MOD-16 counter edesign using all flip-flops.

Mod-16 counter is a 4-bit counter with 16 states. To design a counter with 16 states, the number of flip-flops required can be found using the equation $2^n \leq N \leq 2^{n-1}$, where n is the numbher of flip-flops required and N is the number states present in the counter. For N = 16, from the above equation, n = 4 i.e. 4 flip-flops are requierd. Assume that the Mod-16 counter has sixteen states $a, b, c, d, e, f\ g, h, i, j, k\ l, m, n, o, p$ and its sequence is given by

$$a \rightarrow b \rightarrow c \rightarrow d \rightarrow e \rightarrow f \rightarrow g \rightarrow h \rightarrow i \rightarrow j \rightarrow k \rightarrow l \rightarrow m \rightarrow n \rightarrow o \rightarrow p \rightarrow\ a \rightarrow \ldots\ldots$$

Step 1 State diagram The state diagram for MOD-16 counter can be drawn as shown in Fig. 8.35. Here it is assumed that the state transition from one state to another state takes place when the clock pulse is asserted; when the clock pulse is unasserted, the counter remains in the present state.

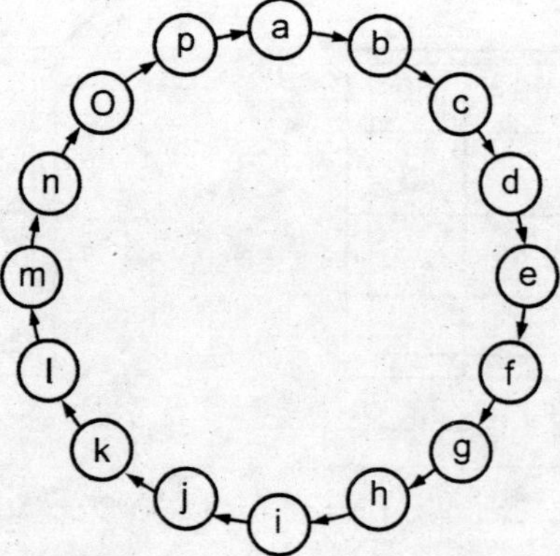

Fig. 8.35. State diagram of Mod-16 counter.

Step 2 From the state diagram given in Fig. 8.35, the Present State-Next State table can be drawn as shown in Table 8.23.

Table 8.23 PS-NS table for Mod-16 counter.

Present state (PS)	*Next State (NS)*
a	b
b	c
c	d
d	e
e	f
f	g
g	h
h	i
i	j
j	k
k	l
l	m
m	n
n	o
o	p
p	*a*

Step 3 State Assignment Let us assign four state variables to states *a*, *b*, ..., *p* as in PS-NS table, shown in Table 8.24.

Table 8.24 PS-NS table for Mod-16 counter.

Present State (PS)				*Next State (NS)*			
q_3	q_2	q_1	q_0	Q_3	Q_2	Q_1	Q_0
0	0	0	0	0	0	0	1
0	0	0	1	0	0	1	0
0	0	1	0	0	0	1	1
0	0	1	1	0	1	0	0
0	1	0	0	0	1	0	1
0	1	0	1	0	1	1	0
0	1	1	0	0	1	1	1
0	1	1	1	1	0	0	0
1	0	0	0	1	0	0	1
1	0	0	1	1	0	1	0
1	0	1	0	1	0	1	1
1	0	1	1	1	1	0	0
1	1	0	0	1	1	0	1
1	1	0	1	1	1	1	0
1	1	1	0	1	1	1	1
1	1	1	1	0	0	0	0

Table 8.25 Excitation table for Mod-16 counter using SR, JK, T and D flip-flops.

Present State (PS)				*Next State (NS)*				*Using SR flip-flop*				*Using JK flip-flop*				*Using T flip-flop*				*Using D flip-flop*			
q_3	q_2	q_1	q_0	Q_3	Q_2	Q_1	Q_0	S_3R_3	S_2R_2	S_1R_1	S_0R_0	J_3K_3	J_2K_2	J_1K_1	J_0K_0	T_3	T_2	T_1	T_0	D_3	D_2	D_1	D_0
0	0	0	0	0	0	0	1	0d	0d	0d	10	0d	0d	0d	1d	0	0	0	1	0	0	0	1
0	0	0	1	0	0	1	0	0d	0d	10	01	0d	0d	1d	d1	0	0	1	1	0	0	1	0
0	0	1	0	0	0	1	1	0d	0d	d0	10	0d	0d	d0	1d	0	0	0	1	0	0	1	1
0	0	1	1	0	1	0	0	0d	10	01	01	0d	0d	d1	d1	0	1	1	1	0	1	0	0
0	1	0	0	0	1	0	1	0d	d0	10	10	0d	1d	0d	d1	d	0	0	1	0	1	0	1
0	1	0	1	0	1	1	0	0d	d0	01	01	0d	d0	1d	d1	0	0	1	1	0	1	0	1
0	1	1	0	0	1	1	1	0d	d0	10	10	0d	d0	d0	1d	0	0	1	1	0	1	1	0
0	1	1	1	1	0	0	0	0d	01	01	01	1d	d0	d1	d1	0	0	0	1	0	1	1	1
1	0	0	0	1	0	0	1	10	0d	10	10	d1	d1	0d	1d	1	1	1	1	0	0	0	0
1	0	0	1	1	0	1	0	d0	0d	01	01	d0	0d	1d	d1	0	0	0	1	1	0	0	1
1	0	1	0	1	0	1	1	d0	0d	10	10	d0	0d	d0	1d	0	0	1	1	1	0	1	0
1	0	1	1	1	1	0	0	d0	10	01	01	d0	0d	d1	d1	0	1	1	1	1	1	0	0
1	1	0	0	1	1	0	1	d0	d0	10	10	d0	1d	0d	1d	0	0	0	1	1	1	0	1
1	1	0	1	1	1	1	0	d0	d0	01	01	d0	d0	1d	d1	0	0	1	1	1	1	1	0
1	1	1	0	1	1	1	1	d0	d0	10	10	d0	d0	d0	1d	0	0	0	1	1	1	1	1
1	1	1	1	0	0	0	0	01	01	01	01	d1	d1	d1	d1	1	1	1	1	0	0	0	0

Step 4 Excitation Table Here, all the four flip-flops i.e. SR, JK, T and D are used for the counter design. The excitation table having entreies for flip-flop inputs, S_3, R_3, S_2, R_2 S_1 R_1 and S_0 R_0 (for design using SR flip-flops); J_3K_3, J_2K_2, J_1K_1, and J_0K_0 (for design using JK flip-flops); T_3,T_2,T_1,T_0 (for design using T flip-flops) and D_3, D_2, D_0 (for design using D flip-flops) can be drawn from the above PS-NS table shown in Table 8.24 using application table of respective flip-flop as shown in Table 8.25.

Step 5 Excitation Maps The excitation maps for S_3R_3, S_2R_2, S_1R_1 and S_0R_0 (for design using SR flip-flops); J_3K_3, J_2K_2, J_1K_1, and J_0K_0 (for design using JK flip-flops); T_3, T_2,T_1,T_0 (for design using T flip-flops) and D_3, D_2, D_1, D_0 (for design using D flip-flops) can be drawn as shown in Fig. 8.36, 8.37, 8.38 and 8.39 respectively from the excitation table (Table 8.24).

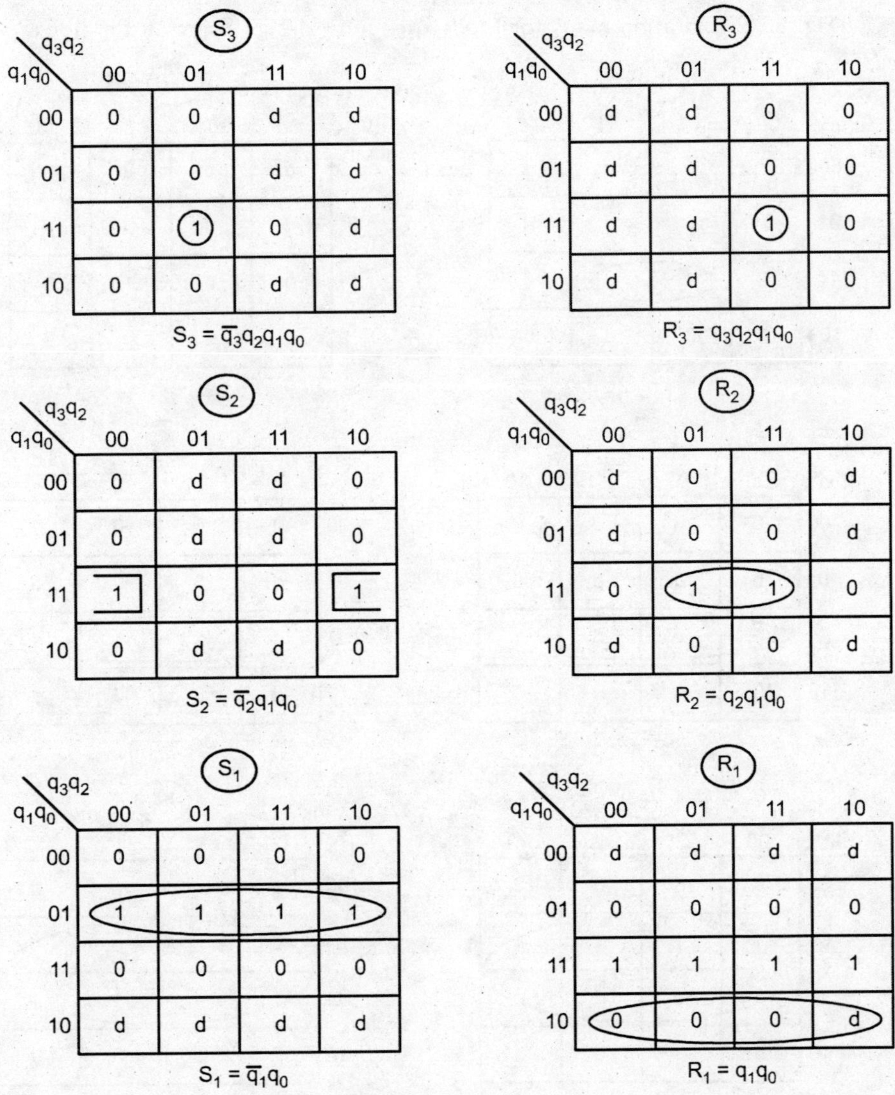

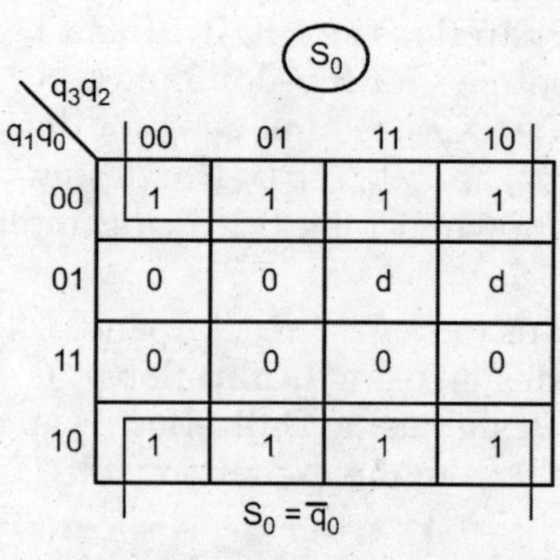

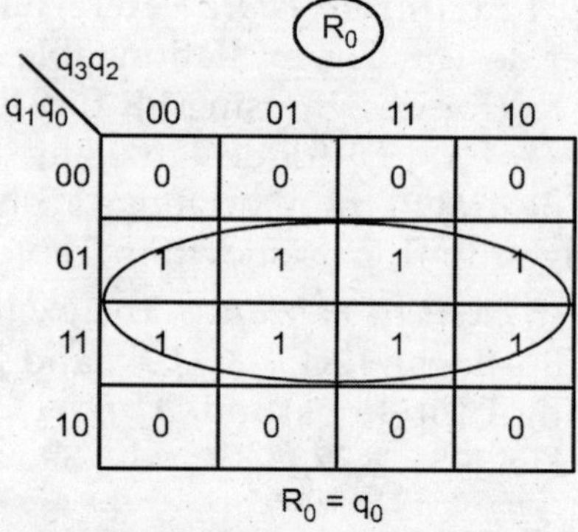

Fig. 8.36. Excitation maps for Mod-16 counter design using SR flip-flops.

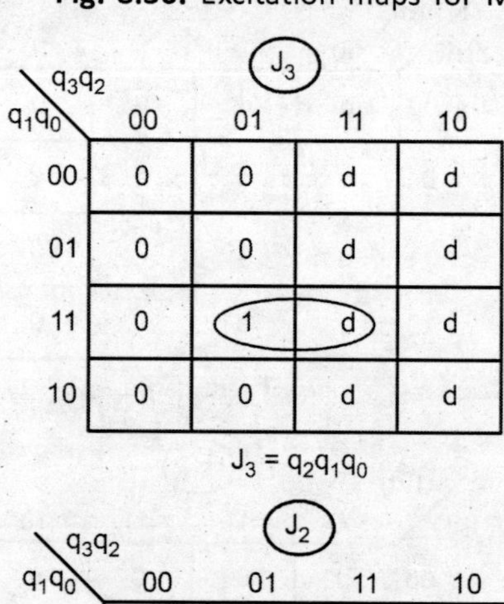

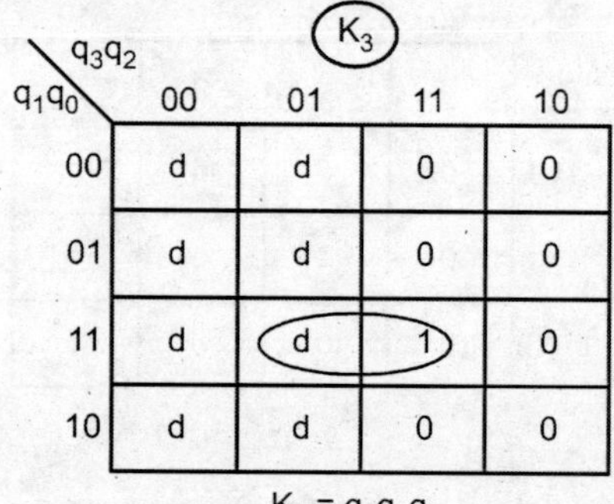

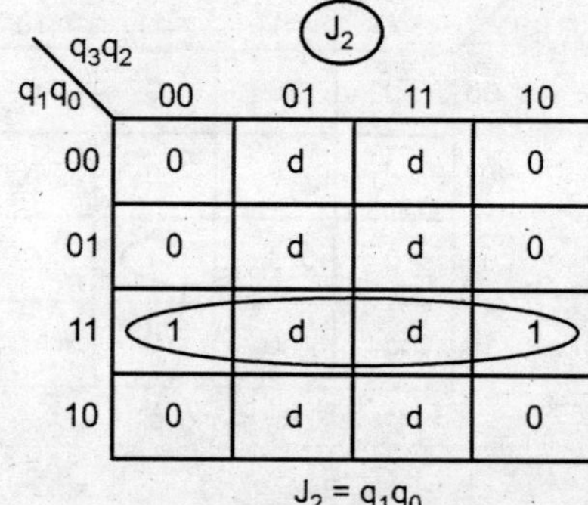

K_2

q_1q_0 \ q_3q_2	00	01	11	10
00	d	0	0	d
01	d	0	0	d
11	d	1	1	d
10	d	0	0	d

$K_2 = q_1q_0$

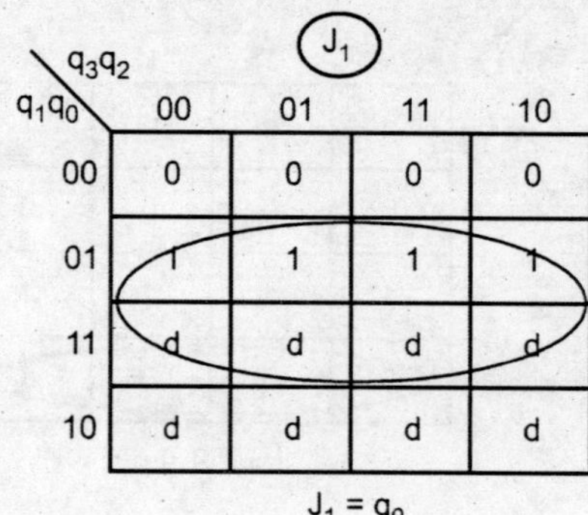

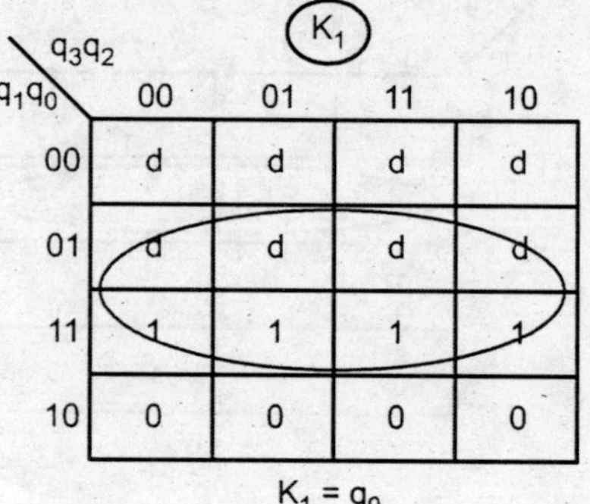

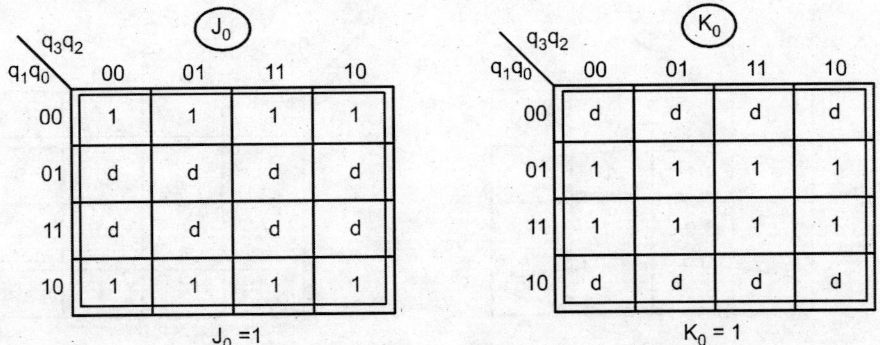

Fig. 8.37. Excitation maps for Mod-16 counter design using SR flip-flops.

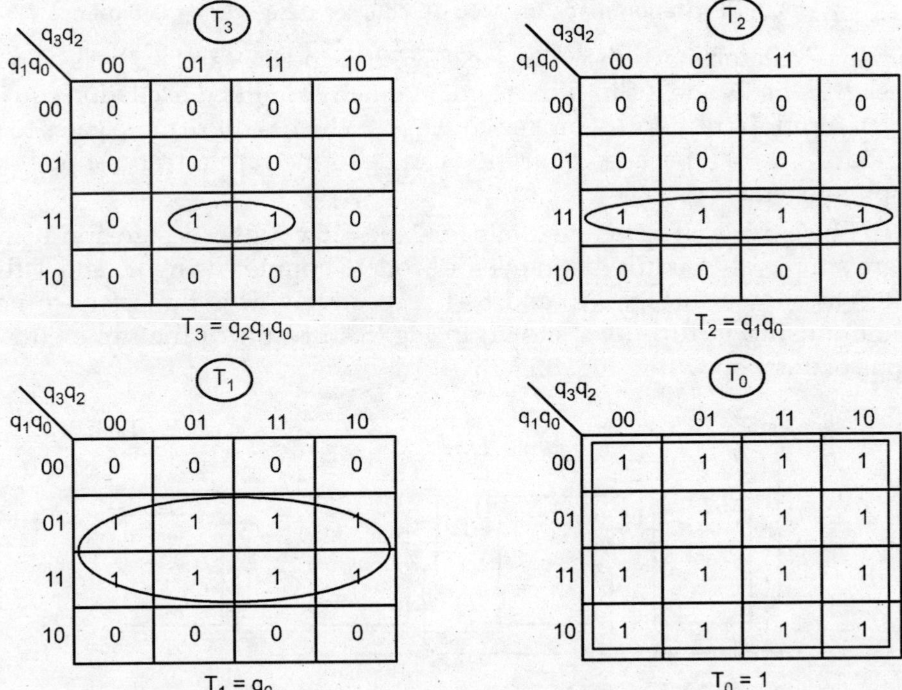

Fig. 8.38. Excitation maps for Mod-16 counter design using T flip-flops.

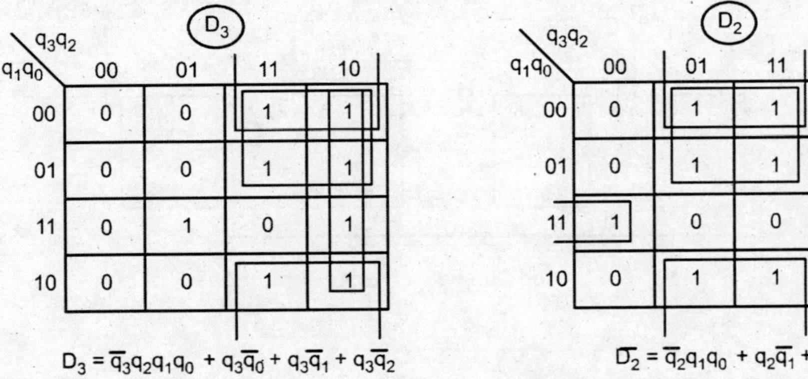

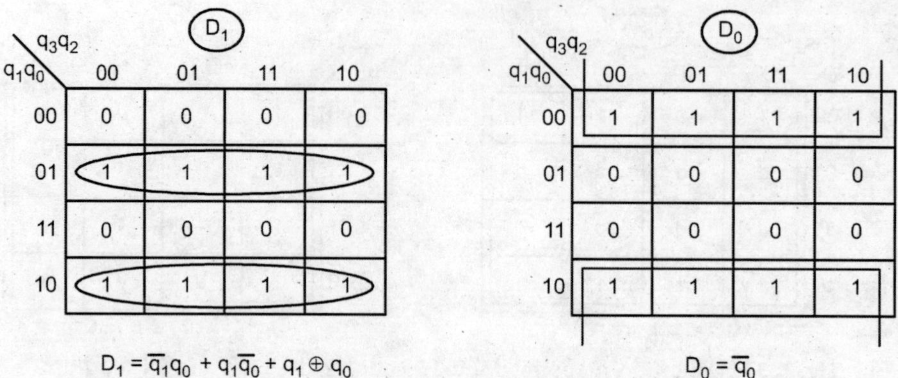

Fig. 8.39. Excitation maps for Mod-16 counter design using D flip-flops.

From the excitaion maps shown in Fig. 8.36, 8.37 and 8.38 and 8.39, one can understand that the design using JK and T flip-flops result in much simplified excitation equations while SR and Delay flip-flops result in compkex equations which require more gates to implement. Also, note that $J_i = K_i = T_i$ because T flip-flop can be formed from JK flip-flop by connecting J and K inputs together.

Step 6 Since the excitation equations for Mod=16 counter design and using JK and T flip-flops are simple, the circuit diagrams for Mod-16 counter using JK and T flip-flops are implemented as shown in FIg. 8.40 and 8.41 respectively. It is important to note that the implementationusing Jk flip-flops shown in Fig. 8.40 result in the same circuits as 4-bit synchronous ocunter shown in Fig. 8.13.

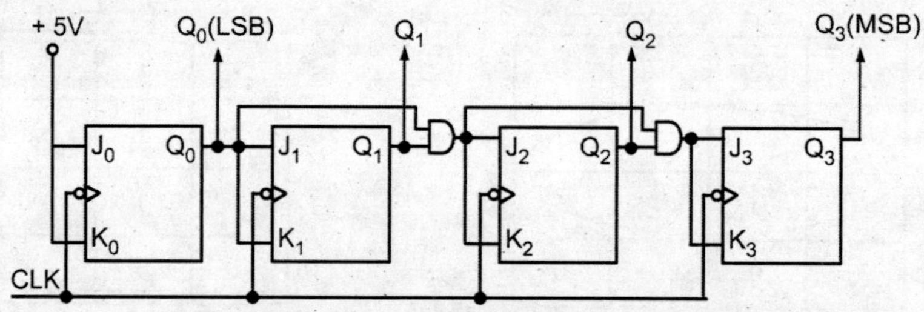

Fig. 8.40. Mod-16 counter using JK flip-flops

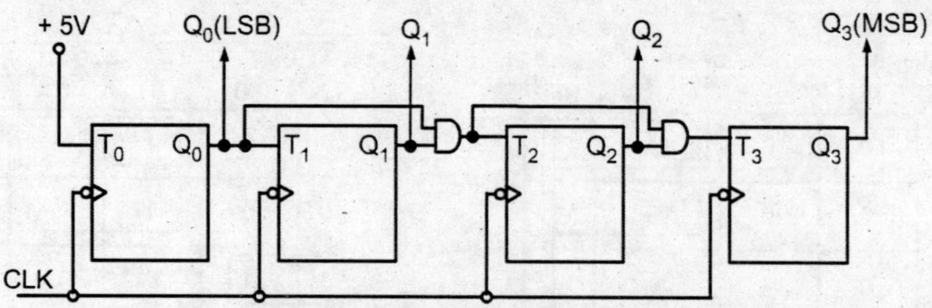

Fig. 8.41. Mod-16 counter using T flip-flops

8.16 COUNTER IMPLEMENTATION AND APPLICATIONS

8.16.1 Frequency Counter

The frequency of a pulse signal can be measured and displayed using a counter. This is illustrated in Fig. 8.42. The counter is driven by the output of an AND gate. The inputs to the AND gate are (i) the pulse signal whose frequencyis to be measured and (ii) a Sample pulse that is high between time t_1 and t_2.

The output of the AND gate will be held LOW except during the time interval (t_1 to t_2) called the *sample interval*. During this time, the pulse signal of the unknown frequency will appear at the output of the AND gate and will be counted by the counter. After t_2, the output of the AND gate stays LOW, and hence the counter stops countng. Thus, the counter counts the number of pulses that occur during the sampling interval. It is a direct measure of the frequency of the pulse waveform.

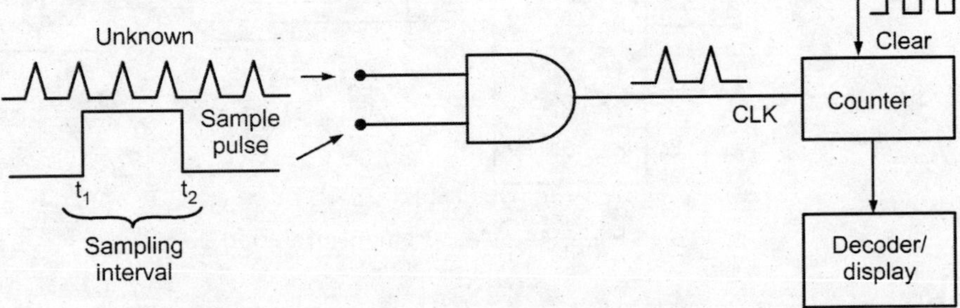

Fig. 8.42. Basic frequency counter arrangement.

The accuracy of this method depends on the duration of the sampling interval, which must be very accurately controlled. A commonly used method for obtaining very accurate sample pulses in shown in Fig. 8.43. A crystal controlled oscillator generatesa very accurate waveform of frequency equal to 100KHz. This waeform is shaped into square pulses and fed to a series of decade counters, Which are being used to successivelydivide this 100 KHz frequency by 10.

This switch is used to select one of the decade counter outputs which will be fed to the CLK inputs of the flip-flop to the divided by 2.

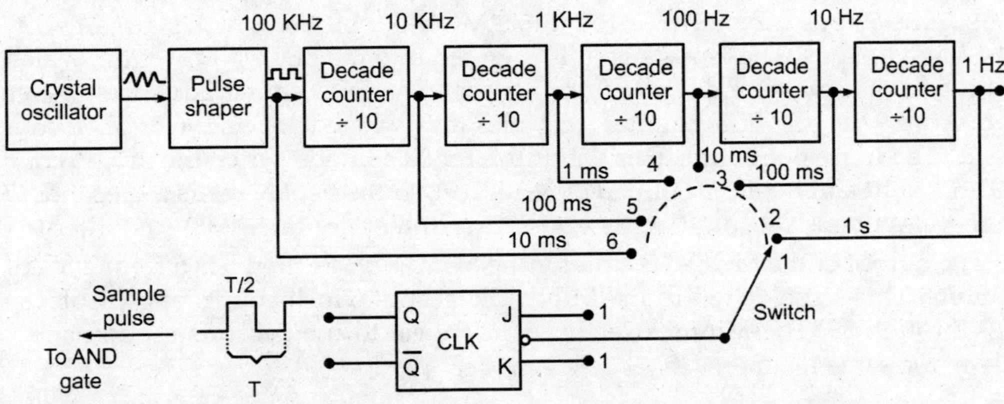

Fig. 8.43. Method for obtaining accurate sampling intervals for frequency counter.

8.16.2 Measurement of Period

The operating principle of the frequency counter can be modified and applied tothe measurement of period rather than frequency. The basic idea is illustrated in FIg. 8.44. An acurate 1 MHz referece frequency is gated into the counter/display for a time durationequal to T_x, the period of the signal being measured. The counter will count and display the value of T_x in units of 1 μs. For example, if T_x is 1.19 ms, the gate will allow 1190 pulses into the counter.

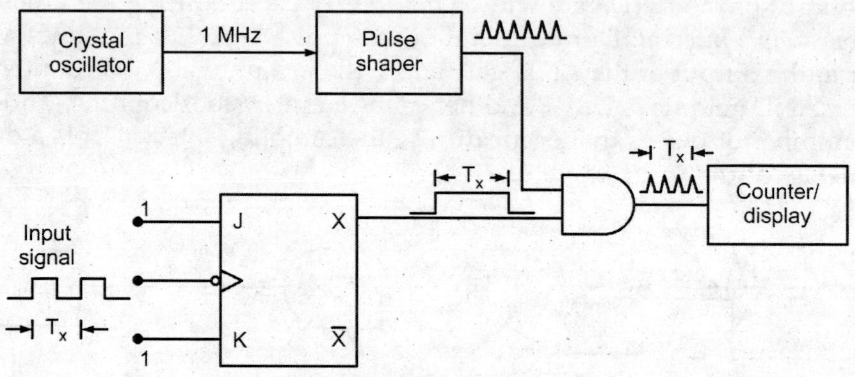

Fig. 8.44. Measurement of Period.

8.16.3 Digital Clock

One of the most popular application of counters is the digital clock – a time clock which displays the time of days in hours, minutes and seconds. In order to construct an accurate digital clock, a highly accurate beais clock frequency isrequired. For battery-operated digital clocks (or watches), the basic frequency is normally obtained from a quartz-crystal oscilltor. DIgital clocks operated from the a.c. power line can use the 50 Hz power frequency as the basic clock frequency. In either case, the basic frequency has to be divided down to a frequency of 1 Hz or 1 pulse per second (pps) The basic block diagram for a digital clock operating from 50 Hz is shownin Fig. 8.45.

The 50 Hz signal is sent through a shaping circuit to produe square pulses at the rate of 50 pps. The 50-pps waveforms is fed into a MOD-50 counter which is used to divided the 50 pps down to 1pps. The 1-pps signal is then fed into the SECONDS sctino, which is used to coutn and display seconds from 0 through 59. The BCD couter advances one count per second. After 9 sec, the *BCD* counter recyclles to 0, which triggers the MOD-6 counter and causes it to advance one count. This continues for 59 seconds, when the mod-6 counter is in the 101(5) count and the *BCD* counter is at 1001(9), so thedisplay reads 59 seconds. The next pulse recycles the *BCD* counter to 0, which in turn recycles the MOD-6 counter to 0.

The output of the MOD-6 counter inthe SECONDS Section has a frequency of 1 pulse per minute. This signal is fed to the MINUTES sectin, which counts and displays minutes from 0 through 59. The Minutes section is identical to the Seconds section and operates exactly in the same manner.

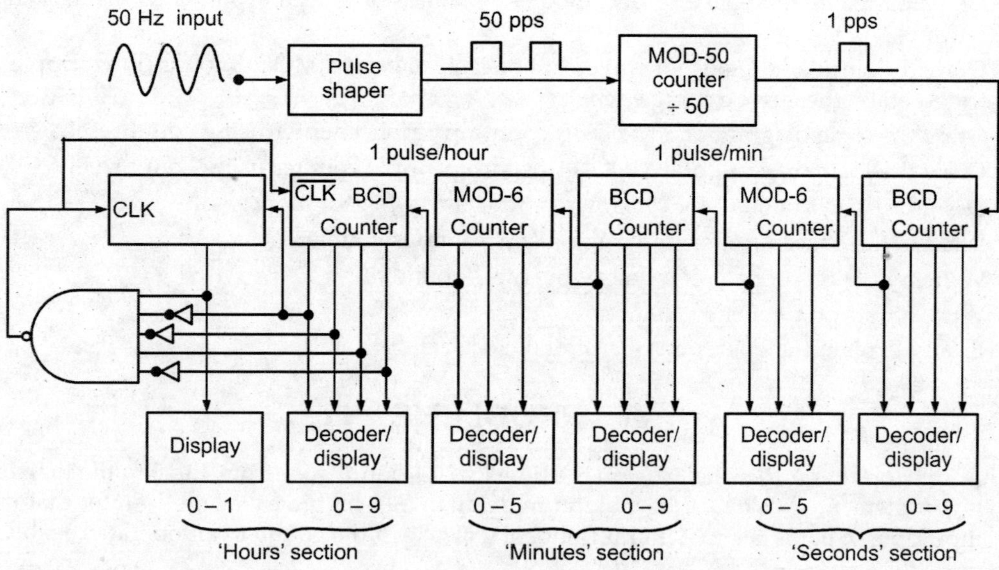

Fig. 8.45. Block dkagram for digial clock.

The output of the MOD-6 counter in the Minutes section has a frequency of 1 pulse per hour, This signal is then fed to the Hours section, which counts and displays hours from 1 through 12. When the hours counters reaches 12, it will be reset to zero by the NAND gate.

■ REVEIW QUESTIONS ■

1. What is a ripple counter?
2. Explain the difference between the performance of asychronous of synchronous counters.
3. What is the primary disadvantage of an asynchronous counter.?
4. What factors determine whether a cunter operates as a count-up or count-down counter ?
5. What is a modulus counter?
6. How is a modulus ocunter built using ocunt reset ?
7. What is a synchronous counter ?
8. Draw the logic diagram of a 4-bit binary ripple counter using fligp-flops that trigger on the positive-edge transition.
9. Draw the logic diagram of a 4-bit binary ripple down-counter using the followings:
   (a) Flip-flops that trigger on the positive-edge transition of the clock.
   (b) Flip-flops that trigger on thenegative-edge transition of the clock.
10. A decade counter does not use its maximum possible modulus and so ther are several invalid states. List the states
11. A 4=bit binary up/down couter (modulus-16) is in the binary state of 0. What is its next state in the up mode and in this down mode ?
12. Draw the logic diagram of a binary ripple counter using toggle flip-flops.

13. Draw the circuit diagram of a MOD-64 parallel counter.
14. Draw the gates necessary to decode all the stages of a MOD-16 counter using active-LOW outputs.
15. Draw the clock and *Q*-output waveform of each stage of a MOD-8 count-down ripple couinter triggered by the clock trailing edge.
16. Draw the logic diagram of a MOD-10 count-up ripple counter using count reset.
17. Draw the logic diagram of a MOD-14 count-up ripple counter using count reset.
18. Design a synchronous Mod-12 down-couter using J-K flip-flops.
19. What is the procedure for designing a synchronous counter ?
20. Write a note on propagation deleay in ripple counters.
21. Write a note on frequency counters.
22. Briefly explain the operation of a digital clock.

■ PROBLEMS ■

1. A flip-flops has a 10 ns delay from the time its CLK input goes from 1 to 0 to the time its output is complemented. What is the maximum delay in a 10-bit binary ripple counter that uses these flip-flops? What is the maximum frequency at which the counter can operate reliably ?

   Ans. 100 ns; 10 MHz

2. How many flip-flops willbe complemented in a 10-bit binary ripple counter to reach the next count after the following count: (a) 1001100111, (b) 0011111111.

   Ans. (a) 4; (b) 9

3. Determine the next state for each of the six unused states in the BCD ripple counter shown in Fig. P8.3. Determine whether the counter is self-correcting.

   Ans. 1010 – 1011 – 0100 1100 – 1101 – 0100 – 1110 – 1111 – 0000

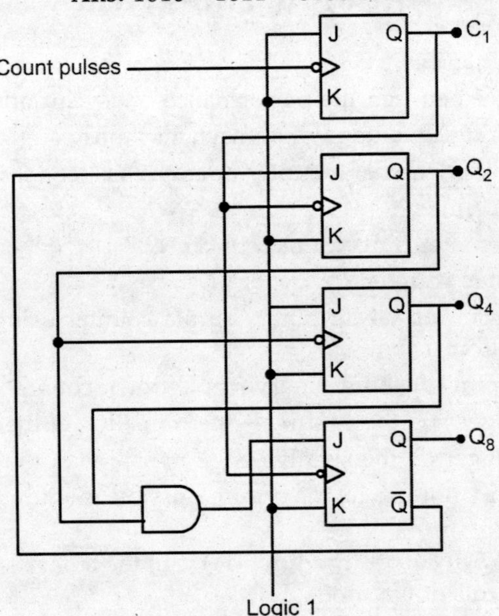

Fig. P8.3. BCD ripple counter.

4. Determine the number of flip-flops that would be required to build the following counters: (a) Mod-6 (b) Mod-11 (c) Mod-15 (d) Mod-19 (e) Mod-31
5. Draw the waveform of a 10 flip-flop ripple counter. What difficulties do you enconter ?

   Ans: 1111, 0000, 1111
6. How many states does a modulus-14 counter have ?

   Ans: 14
7. A 4-bit up/down binary counter is in the down mode and in the 1010 states. On the next clock pulse, to what state does the counter go?

   Ans: 1001
8. What is the terminal count of a 4-bit counter in the up mode and in the down mode ? What is the next state after the terminal count in the down mode ?

   Ans: 1111, 0000, 1111
9. A binary counter is in the 1010_2 state (a) what is its next state? (b) what condition must exist on each flip-flop input to ensure that it goes to the *proper next* state on the clok pulse ?

   Ans: 1011_2 (b) D (MSB): NC or SET; C : NC or Reset: B: NC or SET; A (LSB) = SET or toggle.
10. For the ripple counter shown in FIg. P8.10., draw the complete timing diagram from eight clock pulses showing the clock, Q_A, and Q_B waveforms?

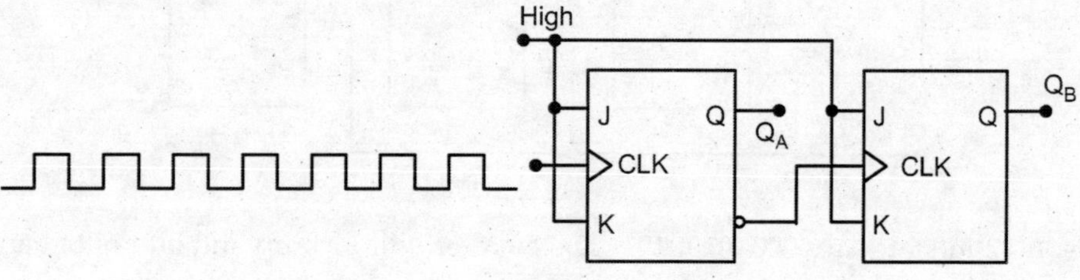

Fig. P8.10.

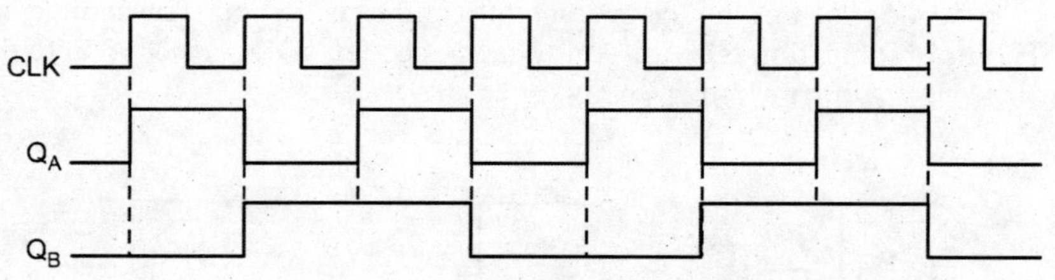

Fig. A8.10

11. Design a 4-bit binary syunchronous counter with D flip-flops.

    Ans: $DA_1 = A_1, DA_2 = A_2 \oplus A_1, DA_3 = A_3 \oplus (A_1A_2),$
    $DA_4 = A_4 \oplus (A_1A_2A_3)$
12. Design a synchronous BCD counter with *J-K* flip-flops.

    Ans: $JA_1 = KA_1 = 1; JA_2 = KA_2 = A_1\bar{A}_8$ (or $= A_1$);
    $JA_4 = KA_4 = A_1A_2, JA_8 = A_1A_2A_4; KA_8 = A_1$

13. An 8-MHz square wave clocks a 5-bit ripple counter. What is the frequency of the last flip-flop? What is the duty cycle of this output waveform? **Ans:** 250 kHz, 50 per cent
14. Repeat, Problem 8.13 if the input has a 20 per cent duty cycle. **Ans:** Same as Problem 8.13
15. Use *J-K* flip-flops and other necessary logic to construct a MOD-24 asynchronous counter.

    Ans: 5 flip-flop are required: $Q_0 - Q_4$ with Q_4 as the MSB. Connect Q_3 and Q_4 outputs to a NAND gate whose output is connected to all CLR or CLK.
16. A 4-bit ripple counter is driven by a 20-MHz clock signal. Draw the waveforms at the output of each flip-flop if each flip-flop has t_{pd} = 20 ns. Determine which counter states, if any, will not occur because of propagation delays.

    Ans: 1000 and 0000 states never occur
17. Analyse the counter in Fig. P8.17 for a "lock-up" condition in which the counter cannot escape from an invalid state or states. An invalid state is one that is not in the counter's normal sequence.

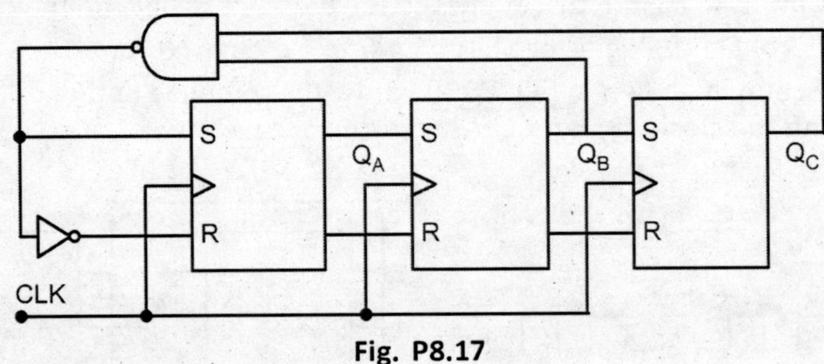

Fig. P8.17

Ans: Invalid states are 000, 010, 101. The counter will lock up and alternate between 010 and 101 if it gets into any of the invalid states.

18. Given a BCD decade counter with only the *Q* outputs available, show what decoding logic is required to decode each of the following states and how it should be connected to the counter. A HIGH output indication is required for each decoded state. The MSB is to the left.

    (*a*) 0001 (*b*) 0011 (*c*) 0101 (*d*) 0111 (*e*) 1000

    Ans: Refer Fig. A8.18

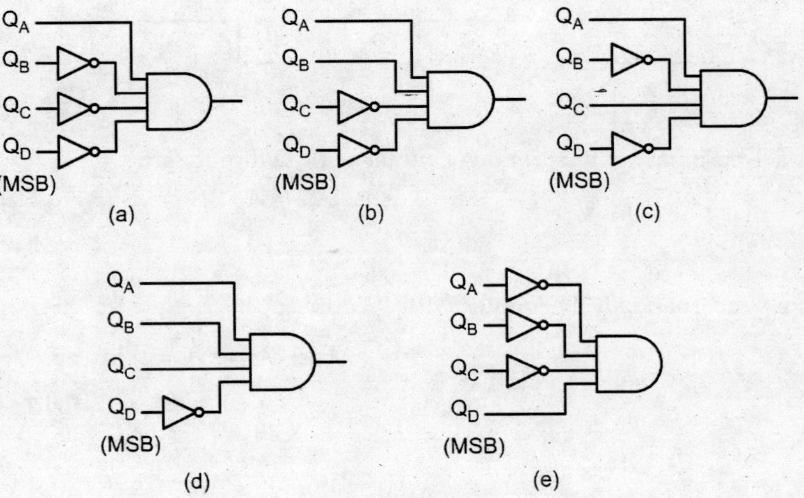

Fig. A8.18

19. How many decade counters are necessary to implement a divide-by-1000 counter and divide-by-10,000 counter.

Ans: $n = 3$; $n = 4$

20. Show with general block diagram how to achieve each of the following using a flip-flop, a decade counter, and a four-bit binary counter, or any combination of these:
(*a*) Divide-by-20 counter (*b*) Divide-by-32 counter (*c*) Divide-by-160 counter (*d*) Divide-by-320 counter.

Ans: (*a*) flip-flop and DIV 10 (*b*) flip-flop and DIV 16
(*c*) DIV 16 and DIV 10 (*d*) DIV 16 and DIV 10 and flip-flop

21. (*a*) Determine f_{max} for MOD-64 parallel counter if each flip-flop has t_{pd} = 20 ns and each gate has t_{pd} = 10 ns; (*b*) A more efficient gating arrangement can be used for this counter by deriving the inputs of each AND gate from the output of the proceeding gate (See Fig. P8.21).

Ans: (*a*) Add two flip-flops (E and F) to Fig. P8.21, Connect AND gates below to appropriate flip-flop inputs, 33 MHz (*b*) 16.7 MHz

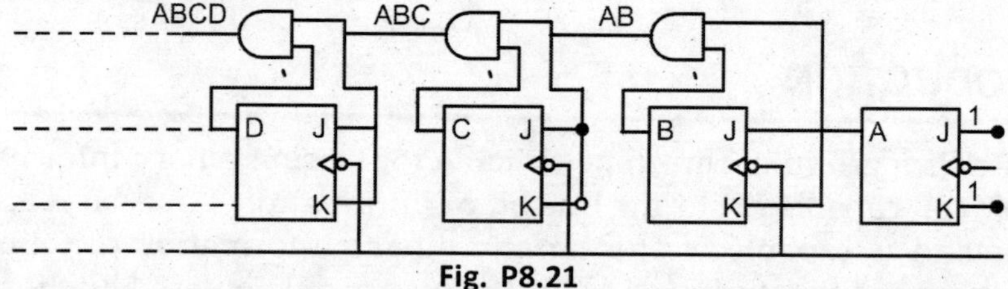

Fig. P8.21

22. Design a synchronous counter that has the following sequence: 000, 010, 101, 110 and repeat. The undesired states 001, 011, 100 and 111 must always go to 000 on the next clock pulse.

Ans: $J_A = B\overline{C}, K_A = 1, J_B = CA + \overline{C}\overline{A}, K_B = 1, J_C = A\overline{B}, K_C = B + \overline{A}$

23. Redesign the counter part of Problem 8.22 without any requirement of the unused states; that is, their NEXT states can be don't cares. Compare with design from Problem 8.22.

Ans: $J_A = B\overline{C}, K_A = 1, J_B = K_B = 1, J_C = K_C = B$

24. Use the synchronous counter design procedure to design a 4-bit synchronous down counter that counts through all states from 1111 down to 0000.

Ans: $J_A = K_A = 1, J_B = K_B = \overline{A}, J_C = K_C = \overline{A}\overline{B}, J_D = K_D = \overline{A}\overline{B}\overline{C}$

25. Design a synchrnous counter with the following sequence:
0000 → 0010 → 0100 → 0110 → 1000 → 1010 → 1100 → 1110 → 0000...
26. Design a synchronous counter with the following sequence:
0001 → 0011 → 0101 → 0111 → 1001 → 1011 → 1101 → 1111 → 0001...
27. Design a 4-bit unit distance Up-down counter using *J-K*, *T* and *D* flip-flops.
28. Design MOD-10 counter with states 0, 1, 2, 3, 4, 8, 9, 10, 11, 12.
29. Design a 3-bit counter with two control signals S_1 and S_0. The counter should operate as:
(a) binary Up counter when S_1S_0 = 00,
(b) binary Down counter when S_1S_0 = 01,
(c) unit distance Up counter when S_1S_0 = 10,
(d) unit distance down counter when S_1S_0 = 11.
30. Design a MOD-16 synchronous Count-Up counter using *J-K* flip-flops.

❑❑❑

9 Registers

9.1 INTRODUCTION

A register is a group of flip-flops suitable for storing binary information. Each flip-flop is a binary cell capable of storing one bit of information. An n-bit register has a group of n flip-flops and is capable of storing any binary information containing n bits. The register is mainly used for storing and shifting binary data entered into it from an external source.

The register is a type of sequential circuit and an important building block used in digital systems like multipliers, dividers, memories, microprocessors, etc. The main difference between a register and a counter is that a register has no specific sequence of states except in certain specialised applications.

9.1.1 4-bit Shift Register

A number of registers are available in MST circuits. A simple register constructed using D flip-flops without any external gates is shown in Fig. 9.1. The register has a common clock-pulse input. This clock pulse input enables all the flip-flops, so that the information available at the four inputs I_1, I_2, I_3 and I_4 can be transferred into the 4-bit register. The four outputs can be read to obtain the information presently stored in the register.

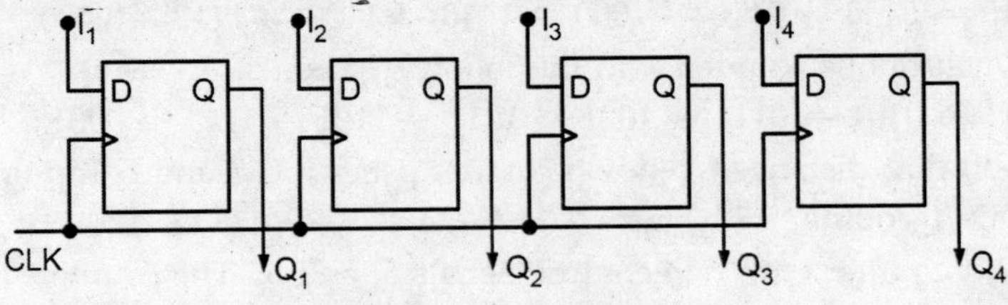

Fig. 9.1. 4-bit register.

All the flip-flops in a register should respond to the clock pulse transition. Hence they should be either of the edge triggered type or the master-slave type. A group of flip-flops sensitive to pulse duration is called a *latch*.

9.2 SHIFT REGISTERS

A register that is used to store binary information is known as a *memory register*. A register capable of shifting binary information either to the right or to the left is called a *shift register*. The shift register permits the stored data to move from a particular location to some other location within the register. In a shift register, the flip-flops are connected in such a way that the bits of a binary number are entered into the shift register, shifted from one position to another and finally shifted out.

There are two methods of shifting the data viz., (*i*) serial shifting and (*ii*) parallel shifting. The *serial shifting* method shifts one bit at a time for each clock pulse in a serial fashion, beginning with either MSB or the LSB. For example, a 4-bit register required four clock pulses to shift a bit from the input to the output. In *parallel shifting* operation, all the data (input or output) get shifted simultaneously during a single clock pulse. Hence, the parallel shifting method is much faster than the serial shifting method. These two method can be used to shift data into a register and out of the register.

Shift register are classified into the following four types based on how binary information is entered or shifted out :

1. Serial-in Serial-out (SISO)
2. Serial-in Parallel-out (SIPO)
3. Parallel-in Serial-out (PISO)
4. Parallel-in Parallel-out (PIPO)

The block diagrams of the four basic register types are shown in Fig. 9.2. Registers can be designed using discrete flip-flop (*S-R*, *J-K* and *D*-type). An *n*-bit shift register consists of *n* flip-flops and the required gates to control the shift operation. Registers are also available as MSI devices.

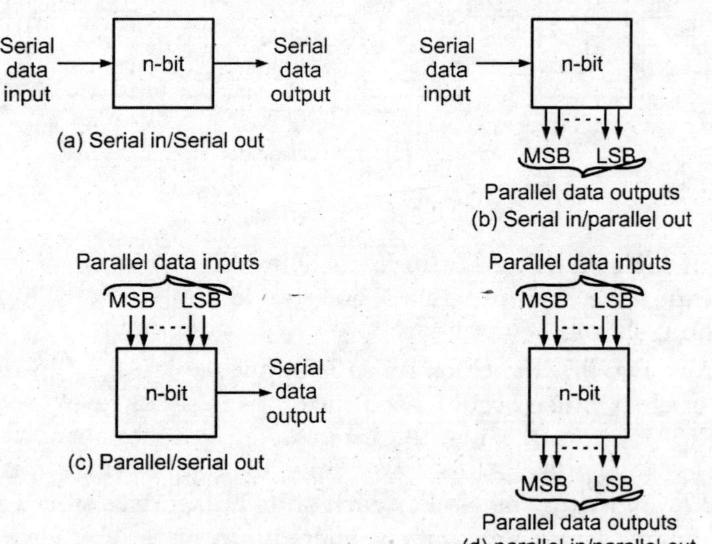

Fig. 9.2. Shift register types.

Shift registers are used in digital systems for temporary storage of information, data manipulation and transferring. In addition, they are used in counting circuits, such as simple counters, variable modulo counters, Up/Down counters and increment counters.

9.2.1 Serial-in-Serial-out Shift Register

This type of shift register accepts data serially, i.e. one bit at a time on a single input line. It produces the stored information on its single output also in serial form. Data may be shifted left (from low to high order bits) using *shift-left register* or shifted right (from high to low order bits) using *shift-right register*.

Do You Know?	A sequential logic system uses FFs, counters and registers, along with logic gates. It outputs and sequence of operations depend on present and past inputs.

Shift-left register: A shift-left register can be built using *J-K* flip-flops or *D* flip-flops as shown in Fig. 9.3(*a*) and (*b*) respectively. A *J-K* flip-flop based shift register requires connection of both *J* and *K* inputs. Input data are connected to the *J* and *K* inputs of the rightmost (lowest order) flip-flop. To input a 1, one should apply a 1 at *J* input, i.e. $J = 1$ and $K = 0$; to input a 0, a 0 at a *J* input, i.e. $J = 0$ and $K = 1$, should be applied. When the clock pulse is applied, the data will be shifted bit by bit to the left.

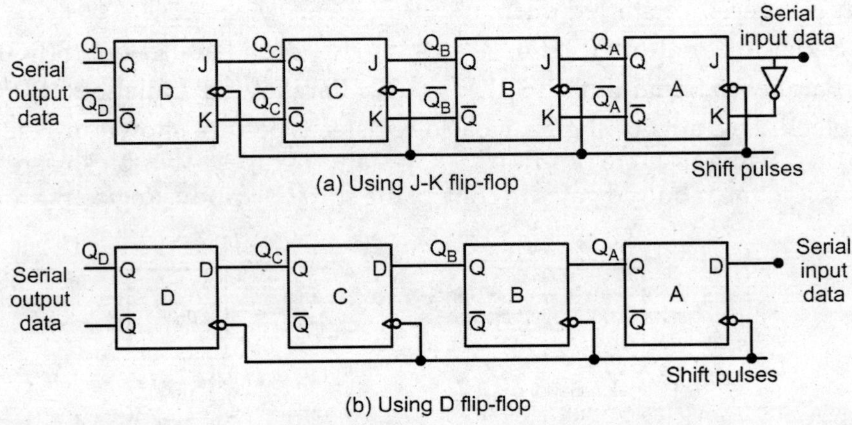

(a) Using J-K flip-flop

(b) Using D flip-flop

Fig. 9.3. Shift-left register.

In the shift register using *D* flip-flops delay (*D*) input of the rightmost flip-flop is used as a serial input line. To input data 1, one should apply a 1 at *D* input and to input data 0, a 0 at the *D* input should be applied.

As shown in Fig. 9.3, the clock pulse is applied to all the flip-flops simultaneously. When the shift or clock pulse occurs, each flip-flops is set or reset according to the data at the respective flip-flop input. Thus, the input data bit at serial input line is entered into stage *A* by the first shift pulse. At the same time, the data of stage *A* is shifted into stage *B* and so on for the following stages. For each shift pulse, data stored in the register stage shifts to the left by one stage. New data is entered into stage *A*, whereas the data present in stage *D* are shifted out (to the left) for use by some other shift register or other building block of digital system.

For example, in case 1, consider that all stages are reset and a steady logical-1 is applied at the serial input line connected to stage *A*. The data in each stage after each of the four shift pulses is shown in Table 9.1. Note that the logical-1 input enters into stage *A* and then shifts left to stage *D* after four shift pulses.

In case 2, consider shifting of alternate 0 and 1 data into stage *A* starting with all stages set. Table 9.2 shows the data in each stage after each of four shift pulses.

In cases 3, consider starting with the count 5 (0101) as shown in Table 9.3 and applying four more shift pulses while placing a steady logical-0 at serial input to stage *A*.

Table 9.1. Operation of shift-left register

Shift pulse	Q_D	Q_C	Q_B	Q_A
0	0	0	0	0
1	0	0	0	1
2	0	0	1	1
3	0	1	1	1
4	1	1	1	1

Table 9.2

Shift pulse	Q_D	Q_C	Q_B	Q_A
0	1	1	1	1
1	1	1	1	0
2	1	1	0	1
3	1	0	1	0
4	0	1	0	1

Table 9.3

Shift pulse	Q_D	Q_C	Q_B	Q_A
0	0	1	0	1
1	1	0	1	0
2	0	1	0	0
3	1	0	0	0
4	0	0	0	0

Shift-right register: A shift-right register can also be built using *D* flip-flops or *J-K* flip-flops as shown in Fig. 9.4(*a*) and (*b*) respectively. Let us illustrate the entry of the 4-bit binary number 1101 into the register, beginning with the right most bit. A 1 is applied at the serial input line i.e. at the delay (*D*) input of the first flip-flop. When the first clock pulse is applied, flip-flop *A* is SET, thus storing the 1. Next, a 0 is applied to the serial input, making $D = 0$ for flip-flop *A* and $D = 1$ for flip-flop *B* because *D* input of flip-flop *B* is connected to the Q_A output.

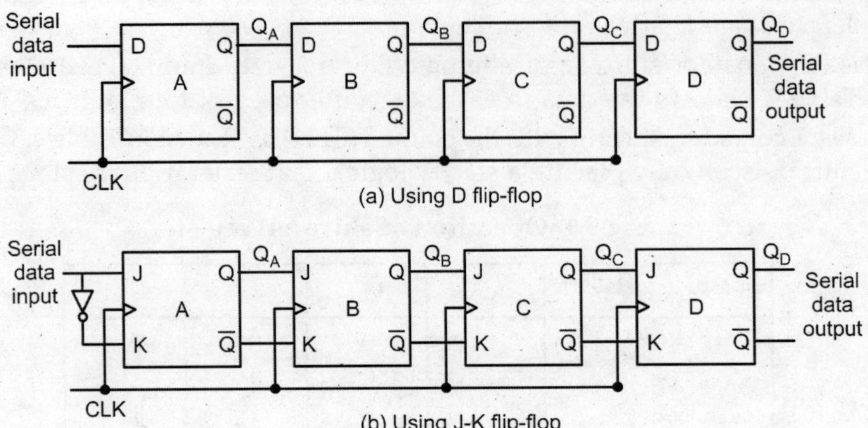

Fig. 9.4. Shift-right register.

When the second clock pulse occurs, the 0 on the data input is "shifted" to flip-flop *A* and the 1 in flip-flop *A* is "shifted" to flip-flop *B*. The next 1 in the binary number is now applied at the serial input line, and a clock pulse is applied. This 1 is entered into flip-flop *A*, and the 0 stored in flip-flop *A* is shifted to flip-flop *B* and the 1 stored in flip-flop *B* is shifted to flip-flop *C*. the last bit in the binary number, i.e. 1, is now applied at the serial input and a clock pulse is applied. Now, this 1 is entered into flip-flop *A*, and the 1 stored in flip-flop *A* is shifted to flip-flop *B*, the 0 stored in flip-flop *B* is shifted to flip-flop *C*, and the 1 stored in flip-flop *C* is shifted to flip-flop *D*. This completes the serial entry of the 4-bit binary number into the shift register, where it can be stored for an indefinite time. Table 9.4 shows the register operation for the entry of 1101 and Table 9.5 show the action of shifting all logical-1 inputs into an initially reset register.

Table 9.4 Operation of shift-right register

Shift pulse	Q_A	Q_B	Q_C	Q_D
0	0	0	0	0
1	1	0	0	0
2	0	1	0	0
3	1	0	1	0
4	1	1	0	1

Table 9.5

Shift pulse	Q_A	Q_B	Q_C	Q_D
0	0	0	0	0
1	1	0	0	0
2	1	1	0	0
3	1	1	1	0
4	1	1	1	1

The waveforms shown in Fig. 9.5 illustrate the entry of a 4-bit number 0100. For a *J-K* flip-flop, the data bit to be shifted into the flip-flop must be present at the *J* and *K* inputs when the clock transition from low to high occurs. Since the data bit is either a 1 or a 0, there are two cases :

1. To shift a 0 into the flip-flop, $J = 0$ and $K = 1$,
2. To shift a 1 into the flip-flop, $J = 1$ and $K = 0$.

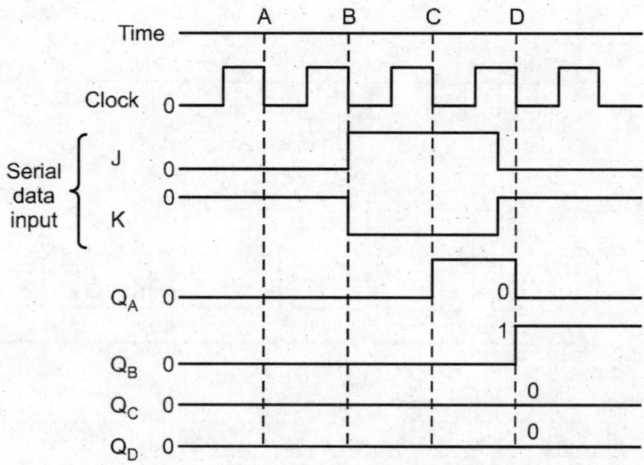

Fig. 9.5. Waveforms of 4-bit serial input shift register.

At time *A* : All the flip-flops are reset. The output of flip-flop just after time *A* are $Q_AQ_BQ_CQ_D = 0000$.

At time *B* : All the flip-flops contain 0s. The flip-flop outputs are $Q_AQ_BQ_CQ_D = 0000$.

At time *C* : All the flip-flops still contain 0s. The flip-flop outputs after time *C* are $Q_AQ_BQ_CQ_D = 1000$.

At time *D* : The flip-flop outputs are $Q_AQ_BQ_CQ_D = 0100$.

9.2.2 IC 7491-8-bit Serial-in-Serial-out Shift Register

IC 7491 is an example of a serial-in-serial-out register. The pinout and logic diagram of IC 7491 shift register is shown in Fig. 9.6. There are eight *S-R* flip-flop connected to provide a serial input as well as a serial output. The clock input at each flip-flop is sensitive to the negative edge. The applied clock signal is passed through an inverter. Hence, the data will be shifted on the positive edges of the input clock pulses. The inverter connected between *S* and *R* on the first flip-flop means that this circuit functions as a *D*-type flip-flop. So, the input to the register is a single line on which the data to be shifted into the register has to be applied. The inputs *A* and *B* of NAND gate are two gated data-input lines for serial data entry. When the data are entered on *A*, the *B* input must be HIGH, and vice versa. Therefore, either *A* or *B* can be used as serial data input and the other input can be used as a control line.

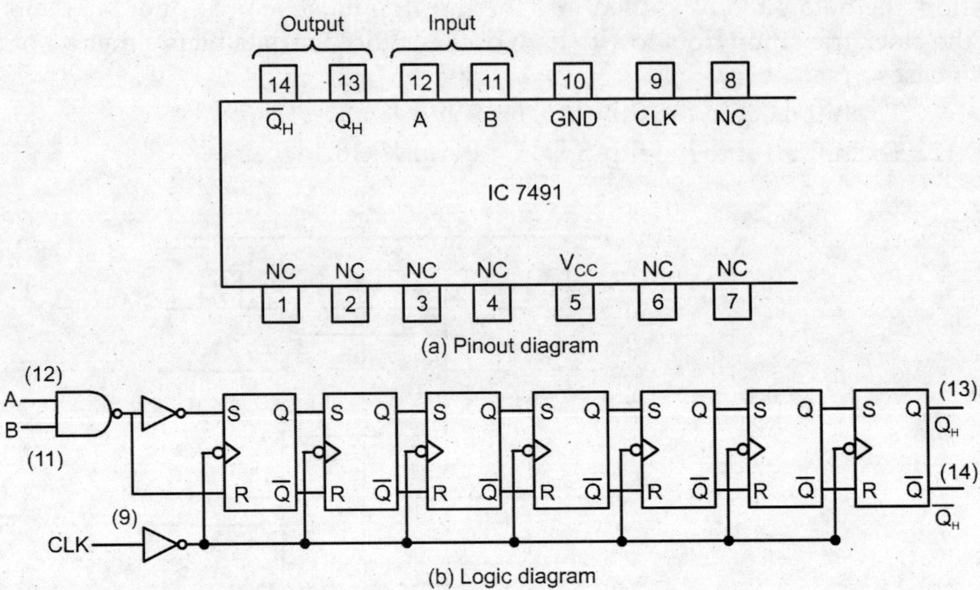

Fig. 9.6. IC 7491—8-bit serial-in-serial-out shift register

9.2.3 Serial-in–Parallel-out Shift Register

A 4-bit serial-in–parallel-out shift register is shown Fig. 9.7. It consists of one serial input, and outputs are taken from all the flip-flops parallel. In this register, data is shifted in serially but shifted out in parallel. In order to shift the data out in parallel, it is necessary to have all the data available at the outputs at the same time. Once the data is stored, each bit appears on its respective output line and all the bits are available simultaneously, rather than on a bit-by-bit basis as with the serial output.

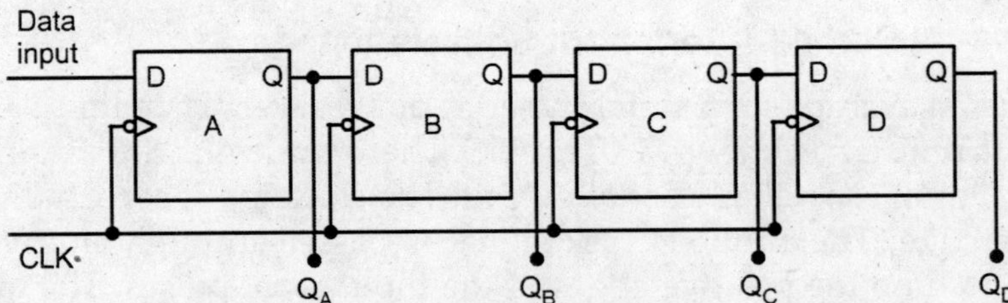

Fig. 9.7. 4-bit serial-in-parallel-out shift register.

9.2.4 IC 74164—8-bit Serial-in-Parallel-out Shift Register

The pinout and logic diagram of an IC 74164 eight-bit serial-in-parallel-out shift register is shown in Fig. 9.8. The logic diagram is constructed using *S-R* flip-flops having clock inputs that are sensitive to negative clock transitions. Since the input clock signal is passed through an inverter, the data will be shifted in, on the arrival of positive edges of input clock pulses. This device has two gated serial inputs, *A* and *B*. Let Q_A through Q_H be the parallel outputs. Each flip-flop has an asynchronous CLEAR input. A low level at the CLR input i.e. at pin 9, will reset every flip-flop. Here, data is shifted into the register serially as given in the previous section.

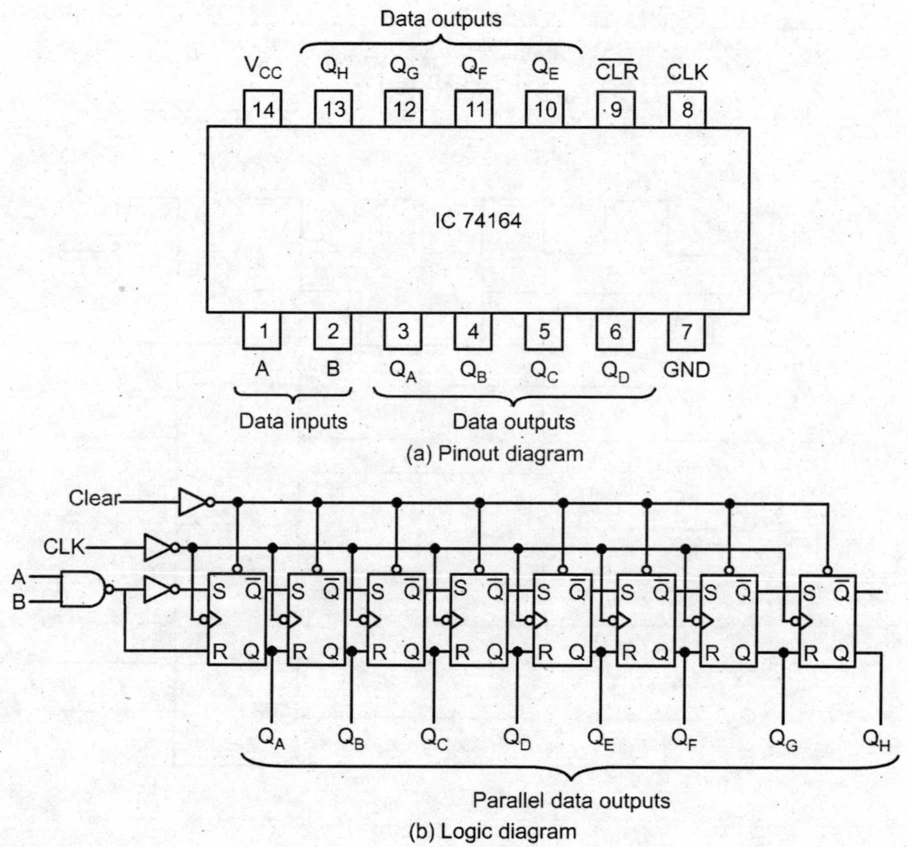

Fig. 9.8. 74164—8-bit serial-in-parallel-out shift register

In IC 74164, data bits are shifted into the register in the same manner as previously discussed for IC 7491. Data at the serial inputs may be changed when the clock is either low or high.

Let the serial data be connected to *A* and terminal *B* be used as a control line. When *B* is held high, the NAND gate is enabled and the serial input data that passes through the NAND gate gets inverted, resulting in shifting of serial data into the first flip-flop. When the input line *B* is held low, the NAND gate output is high, the input data is inhibited, and the next positive clock transition will shift a 0 into the first flip-flop.

Example 9.1. *Show the states of the five-bit register shown in Fig. E9.1(a) for the specified data input and clock waveforms. Assume that the register in initially cleared (all 0s).*

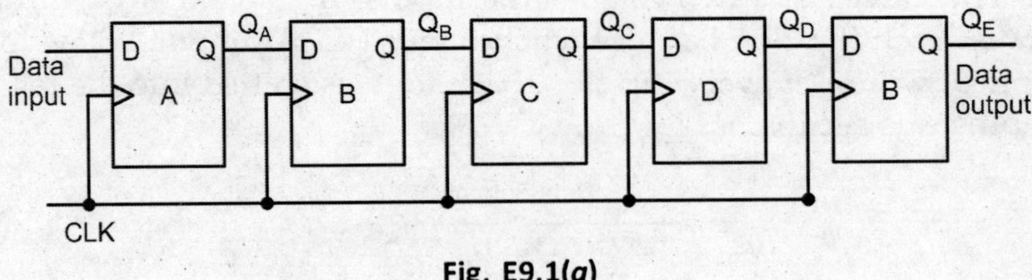

Fig. E9.1(*a*)

Solution:

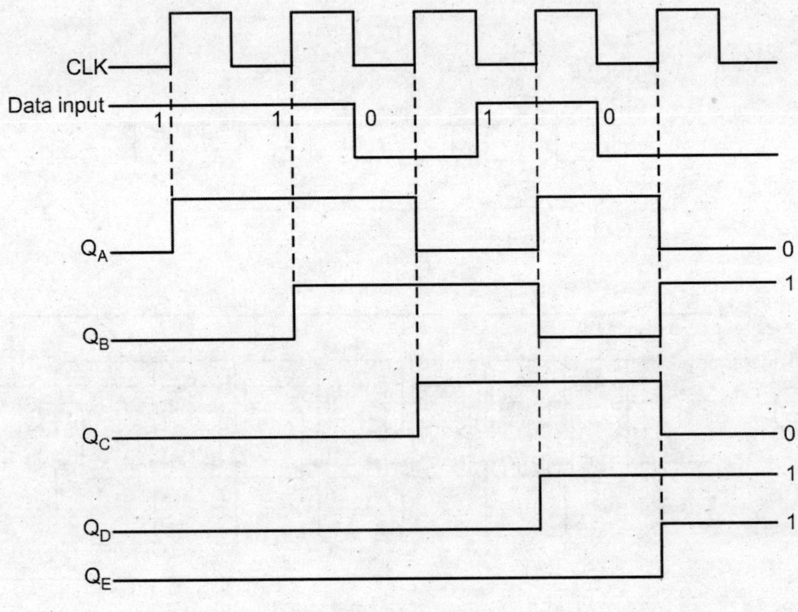

Fig. E9.1(*b*)

Example 9.2. *How long will it take to shift an 8-bit number into a 74164 shift register if the clock is set at 10 MHz?*

Solution: Eight clock pulses will be required since the data is entered serially. Since the clock period is 100 ns, this shift register requires a minimum of 800 ns.

Timing diagram: The timing diagram for the IC 74164 is shown in Fig. 9.9. The first CLEAR pulse occurs at time *A* and simply resets all flip-flop to 0. The clock begins at *B*, but the first positive clock does not do anything since the control line (*B*) is LOW. At *C*, the control is HIGH and hence the first data bit (0) is shifted into the register at *D*. The next 7 data bits are shifted in at timing points, *E, F, G, H, I, J* and *K* in the given order. The clock remains HIGH after time *K*, and the 8- bit number 0010 1100 now resides in the register and is available on the eight output lines. The clock must be stopped after its positive transition at *K* as otherwise shifting will continue and the data will be lost.

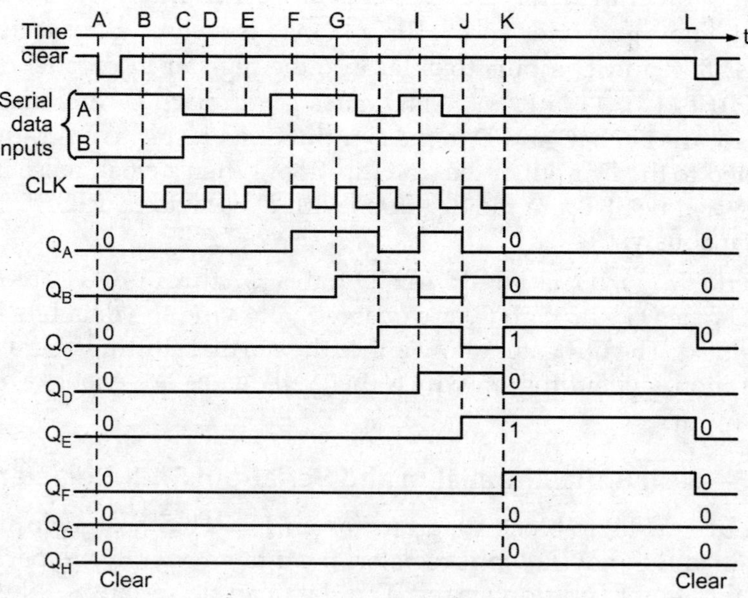

Fig. 9.9. Timing diagram for an IC 74164

9.2.5 Parallel-in-Serial-out Shift Register

For a register with parallel data inputs, the bits are entered simultaneously into their respective flip-flops rather than a bit-by-bit basis on one line.

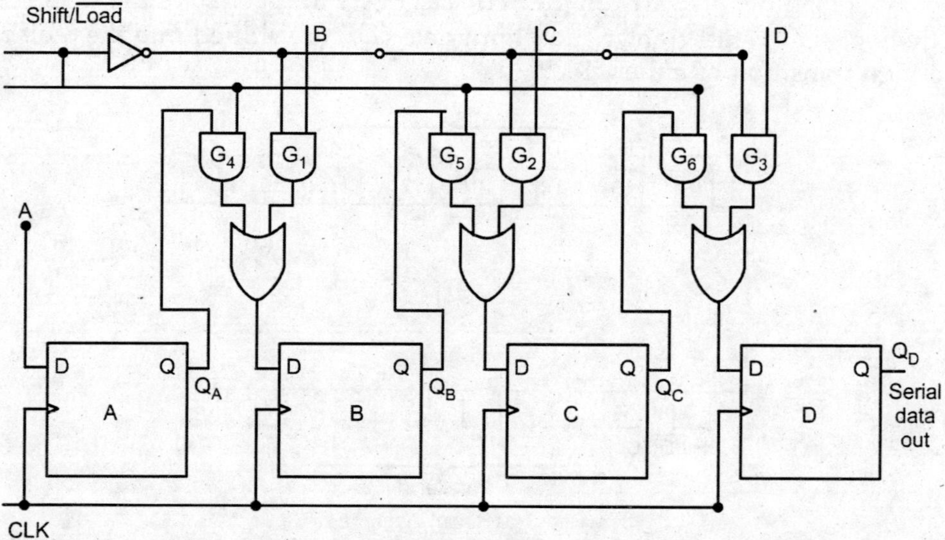

Fig. 9.10. A 4-bit parallel-in-serial-out shift register

A four-bit parallel-in serial-out register is illustrated in Fig. 9.10. Let *A, B, C* and *D* be the four parallel data-input lines and SHIFT/$\overline{\text{LOAD}}$ is a control input that allows the four bits of data at *A, B, C* and *D* inputs to enter into the register in parallel or shift the data in serial. When SHIFT/$\overline{\text{LOAD}}$ is LOW, AND gates G_1 through G_3 are enabled, allowing the data at parallel inputs i.e. *B, C* and *D* to the *D* input of its respective flip-flop. The A input is directly connected to the *D* input of the first flip-flop. When a clock pulse is applied, the flip-flops with $D = 1$ will be SET and the flip-flops with $D = 0$ will be RESET, thereby storing all four bits simultaneously.

When SHIFT/$\overline{\text{LOAD}}$ is HIGH, AND gates G_1 through G_3 are disabled and the remaining AND gates G_4 through G_6 are enabled, allowing the data bits to shift right from one stage to the next. The OR gates allow either the normal shifting operation or the parallel data-entry operation, depending on which the AND gates are enabled by the level on the SHIFT/$\overline{\text{LOAD}}$ input.

9.2.6 IC 74165—8-bit Serial/Parallel-in and Serial-out Shift Register

The pinout and logic block diagram for an IC 74165 are given in Fig. 9.11. It is a Serial/Parallel-in and Serial-out shift register. It contains circuitry to parallel transfer an 8-bit data (*A-H*) into eight flip-flop stages. The data can be serially shifted out at Q_H using either of two clocks (CLK or CLK inhibit). It also contains a serial input, D_S, so that data can also be serially shifted in.

When the input SHIFT/$\overline{\text{LOAD}}$ (SH / $\overline{\text{LD}}$) is LOW, it enables all the. NAND gates for parallel loading. When an input data bit is a 1, the flip-flop is asynchronously SET by a LOW output of the upper NAND gate. Similarly, when an input data bit is a 0, the flip-flop is asynchronously RESET by a LOW output of the lower NAND gate. The clock is inhibited during parallel loading. A HIGH on the SHIFT/$\overline{\text{LOAD}}$ input enables the clock, causing the data in the register to shift right. Serial input data (D_S) are shifted into the 8-bit register by a low-to-high transition of either clock.

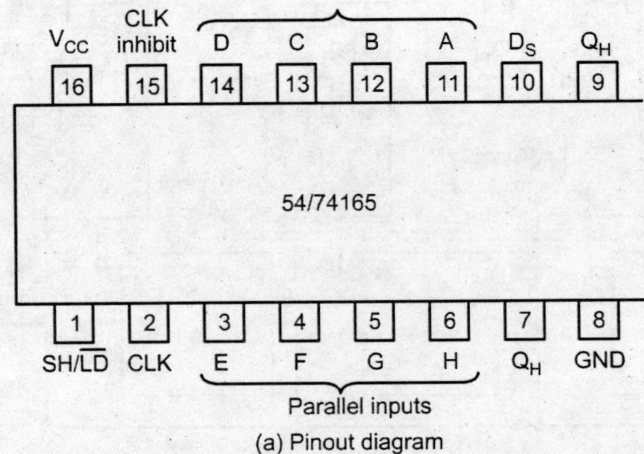

(a) Pinout diagram

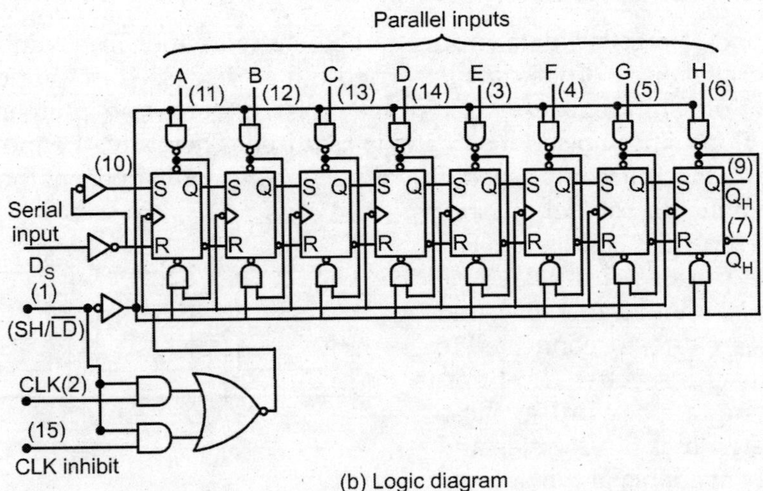

(b) Logic diagram

Fig. 9.11. IC 74165 — 8-bit serial/parallel-in and serial out shift register

With proper control signals at SHIFT/$\overline{\text{LOAD}}$, the IC an be used to load the data parallel into the shift register at one time and then transfer the data serially out at a later time. Fig. 9.12 is a timing diagram that shows the operation of a 74165 shift register.

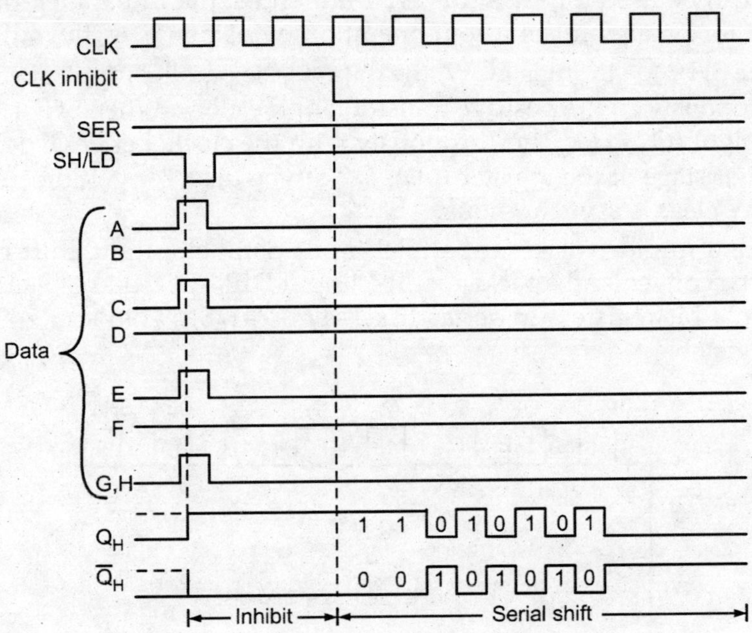

Fig. 9.12. Timing diagram for an IC 74165 shift register

9.2.7 Parallel-in–Parallel-out Register

In this type of register, data inputs can be shifted either in or out of the register in parallel. The parallel entry of the data is carried out as discussed in the previous section of parallel-in serial-out shift register. Also, in this register, there is no interconnection between successive flip-flops since no serial shifting is required. Therefore, the moment the parallel entry of the input data is accomplished, the respective bits will appear at the parallel outputs. A simple 4-bit parallel-in-parallel-out shift register using *D* flip-flops is shown in Fig. 9.13. Here the parallel inputs to be entered should be applied at *A*, *B*, *C* and *D* inputs which are directly connected to delay (*D*) inputs of respective flip-flops. Now on applying a clock pulse, these inputs are entered into the register and are immediately available at the outputs Q_1, Q_2, Q_3 and Q_4.

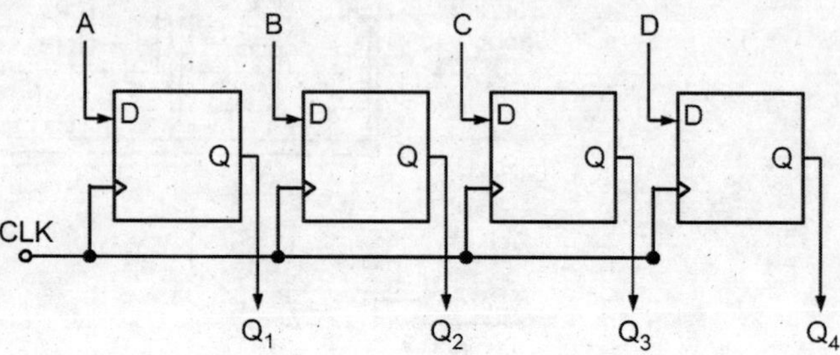

Fig. 9.13. A 4-bit parallel-in-parallel-out shift register.

9.2.8 IC 74195—4-bit Serial/Parallel-in and Serial/Parallel-out Shift Register

IC 74 I 95 is a 4-bit TTL MSI having both serial/parallel-in and serial/parallel-out capability. The pinout diagram of IC 74195 is shown in Fig. 9.14. Since this IC also has a serial input, it can be used for serial-in-serial-out and-serial in-parallel -out operation. This IC can be used for parallel-in serial-out operation by using Q_D as the output.

When the SH / $\overline{\text{LD}}$ input is LOW, the data on the parallel inputs, i.e. *A*, *B*, *C* and *D*, are entered synchronously on the positive transition of the clock. When SH/ $\overline{\text{LD}}$ is HIGH, stored data will shift right (Q_A to Q_D) synchronously with the clock. Let *J* and $\overline{\text{K}}$ be the serial data inputs to the first stage of the register (Q_A); Q_D can be used for getting a serial output data. The active-LOW clear is asynchronous.

There are a number of 4-bit, parallel-input-parallel-output shift registers available since they can be conveniently packaged in a 16-pin DIP. An 8-bit register can be created by connecting two 4-bit registers in series ICs 74174, 74178, 74198 and 7495 are parallel-in-parallel-out registers.

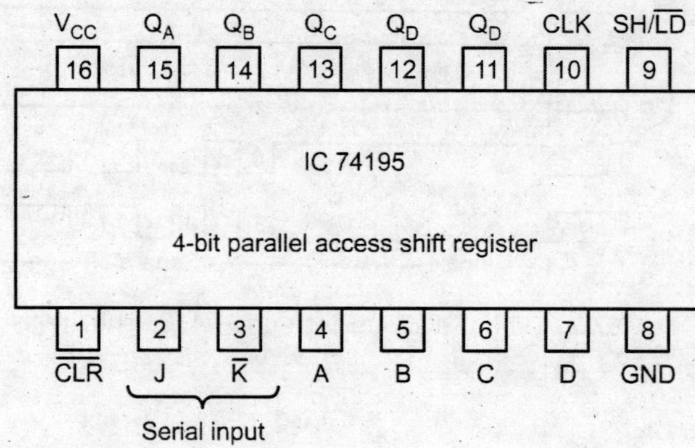

Fig. 9.14. Pinout diagram of IC 74195 shift register.

9.3 UNIVERSAL SHIFT REGISTERS

A register which is capable of shifting data both to the right and left is called a *bi-directional shift register*. A register that can shift in only one direction is called a *uni-directional shift register*. If the register has shift and parallel load capabilities, then it is called a *shift register with parallel load or Universal Shift Register*.

Shift registers can be used for converting serial data to parallel data, and vice-versa. If a parallel load capability is added to a shift register, then data entered in parallel can be taken out in serial fashion by shifting the data stored in the register.

Some shift registers have necessary input and output terminals and also have both shift-right and shift-left capabilities. The most general shift register has the capabilities listed below. Others may have only some of these functions, with at least one shift operation.

1. A clear control to clear the register to 0.
2. A CLK input for clock pulses to synchronise all operations.
3. A *shift-right* control to enable the shift-right operation and *serial input* and *output* lines associated with the shift right.
4. A *shift-left* control to enable the shift-left operation and the serial input and output lines associated with the shift left.
5. A *parallel load* control to enable a parallel transfer and the n input lines associated with the parallel transfer.
6. *n* parallel output lines.
7. A control line that leaves the information in the register unchanged even though clock pulses are continuously applied.

The diagram of a universal shift register that has all the capabilities listed above is shown in Fig. 9.15. It consists of four *D* flip-flops and four 4-input multiplexers (MUX). Let S_0 and S_1 be the two selection inputs connected to all the four multiplexers. These two selection inputs are used to select one of the four inputs of each multiplexer. Input 0 in each MUX is selected when $S_1S_0 = 00$ and input 1 is selected when $S_1S_0 = 01$. Similarly, inputs 2 and 3 are selected when $S_1S_0 = 10$ and $S_1S_0 = 11$ respectively. The inputs S_1 and S_0 control the mode of operation of the register. When $S_1S_0 = 00$, the present value of the register is applied to the *D* inputs of the flip-flops. This is done by connecting the output of each flip-flop to the 0 input of the respective multiplexer. The next clock pulse transfers into each flip-flops, the binary value it held previously, and hence no change of state occurs. When $S_1S_0 = 01$, terminal 1 of the multiplexer inputs has a path to the *D* inputs of the flip-flops. This causes a shift-right operation, with the left serial input transferred into flip-flop A_4. When $S_1S_0 = 10$, a shift-left operation results, with the right serial input going into flip-flop A_1. Finally, when $S_1S_0 = 11$, the binary information on the parallel input lines (I_1, I_2, I_3 and I_4) are transferred into the register simultaneously during the next clock pulse. The function table of bidirectional shift register with parallel inputs and parallel outputs is shown in Table 9.6.

Table 9.6 Function table of bidirectional shift register.

Mode control *S1*	*S0*	Register operation
0	0	No change
0	I	Shift-right
I	0	Shift-left
I	I	Parallel Load

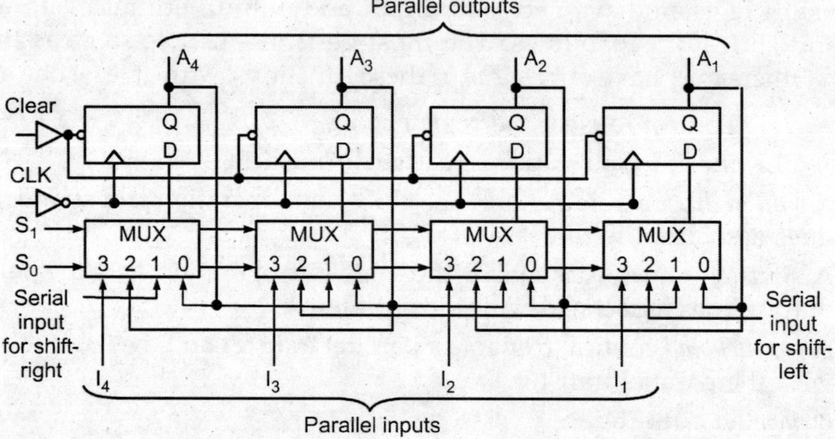

Fig. 9.15. A 4-bit Universal shift register.

A 4-bit bidirectional shift register is shown in Fig. 9.16. When mode control $M = 1$, the AND gates G_1 though G_4 are enabled and the data at D_R is shifted to the right when the clock pulses are applied, and thus it acts as a shift-right register. When $M = 0$, the AND G_5 through G_8 are enabled allowing the date at D_L to be shifted to the left, and thus it acts as a shift-left register. M should be changed only when CLK = 0, otherwise the data stored in the register may be altered.

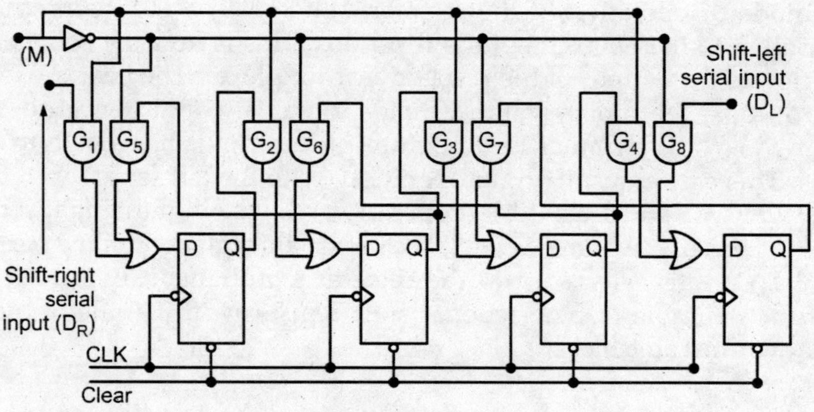

Fig. 9.16. A 4-bit bidirectional shift register

9.3.1 IC 74194—4-bit Bidirectional Shift Register

IC 74194 is a four-bit bidirectional shift register. The pinout diagram of IC 74194 is shown in Fig. 9.17. Parallel loading, which is synchronous with a positive transition of the clock, is accomplished by applying the four bits of data to the parallel inputs and a HIGH to the S_0 and S_1 inputs.

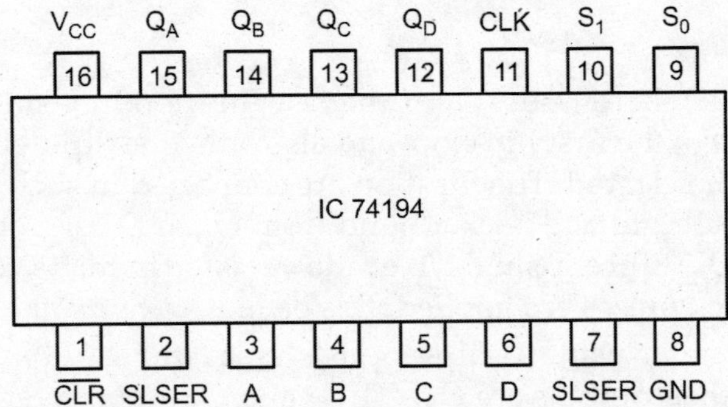

Fig. 9.17. Pinout diagram of IC 74194.

Shift right is accomplished synchronously with the positive edge of the clock when S_0 is HIGH and S_1 is LOW. Serial data in this mode is entered at the shift-right serial input. When S_0 is LOW and S_1 is HIGH, data bits shift left synchronously with the clock and new data is entered at the shift-left serial input. The timing diagram of IC 74194 is shown is Fig. 9.18.

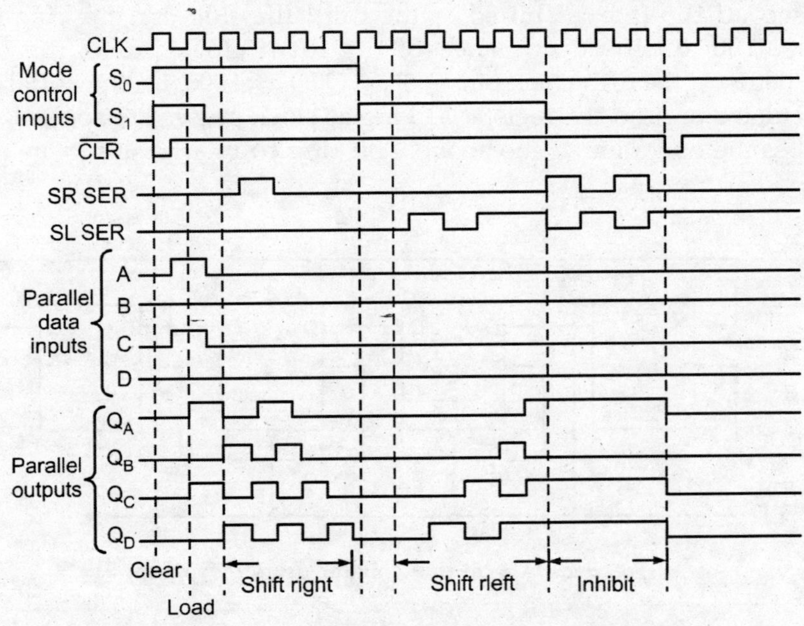

Fig. 9.18. Timing diagram for IC 74194 shift register.

9.4 SHIFT REGISTER COUNTERS

Shift registers can be arranged to form types of counters. All shift-register counters use *feedback*, whereby the output of the last flip-flop in the shift register is connected back to the first flip-flop. Based on the types of feedback connection, the shift register counters are classified as (*i*) standard ring or ring counter and (*ii*) twisted ring or Johnson or Shift counter.

9.4.1 Ring Counter

In a ring counter, the output (Q) of the last flip-flop in a shift register is connected back to the serial input of the first flip-flop, and also only one flip-flop is set at any particular time while all others are cleared. The flip-flop are connected in such a way that information shifts either from left to right and back around from Q_D to Q_A of from right to left and back around from Q_A to Q_D. Since a single 1 in the register is made to circulate around the register as long as clock pulses are applied, it is called a *ring counter*.

A 4-big ring counter using D flip-flop is shown in Fig. 9.19. This circuit consists of four D flip-flops and their outputs are Q_A, Q_B, Q_C and Q_D respectively. The PRESET input of first flip-flop and CLEAR inputs of the other three flip-flop are connected together and brought out as INIT input. Now, on applying a LOW pulse at this INIT input, the first flip-flop is SET to 1 and the other three flip-flops are cleared to 0, i.e. $Q_AQ_BQ_CQ_D$ = 1000. Now, from this circuit it is clear that $D_A = 0$, $D_B = 0$, $D_C = 0$ and $D_D = 0$. Therefore, when a clock pulse is applied, the second flip-flop are reset to 0 i.e. the output of the ring counter $Q_AQ_BQ_CQ_D$ = 0100. As a result, on the occurrence of the first clock pulse, the 1 in the first flip-flop is shifted to the second flip-flop. Similarly, when the second clock pulse is applied, the 1 in the second flip-flop is shifted to the third flip-flop and the ring counter output $Q_AQ_BQ_CQ_D$ = 0100; on the occurrence of the forth clock pulse, the output will be $Q_AQ_BQ_CQ_D$ = 0001; on the fifth clock pulse, $Q_AQ_BQ_CQ_D$ = 1000, i.e. the initial state. Thus, 1 is shifted or circulated around the register as long as clock pulses are applied. The truth table which describes the operation of the above 4-bit ring counter is shown in Table 9.7.

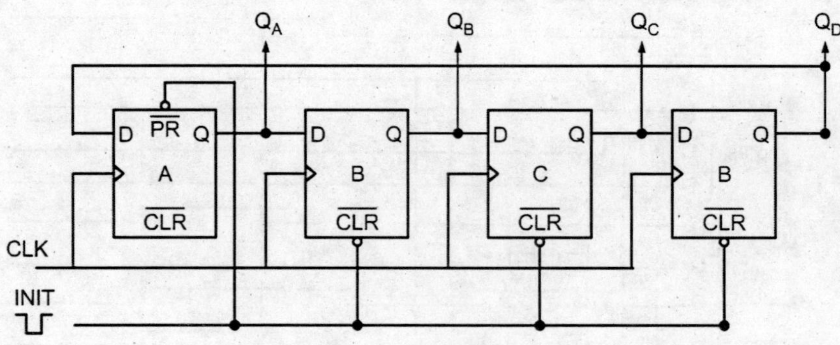

Fig. 9.19. A 4-bit ring counter using *D* flip-flop.

Table 9.7 Truth table for 4-bit ring counter.

INIT	CLK	Q_A	Q_B	Q_C	Q_D
L	X	1	0	0	0
H	↑	0	1	0	0
H	↑	0	0	1	0
H	↑	0	0	0	1
H	↑	1	0	0	0

As shown in the above truth table, the ring counter has only 4 valid states, i.e. 1000, 0100, 0010 and 0001. The ring counter can *hang* or enter into any one of the invalid state due to noise or any other condition without returning to the main countering sequence. Hence, it is a must to design ring counters which are self-correcting and capable of recovering from invalid states to valid states.

The design of a self-correcting ring counter can be accomplished by completely avoiding don't care entries (*d*) in one of the excitation map simplification, most in the excitation map corresponding to MSB. The design of a 4-bit self-correcting ring counter is described in the flowing section.

9.4.2 Design of 4-bit Self-correcting Ring Counter

Step 1: State diagram: Assuming that the ring counter is initially cleared to 0000, the state diagram for the 4-big ring counter is drawn as shown in Fig. 9.20.

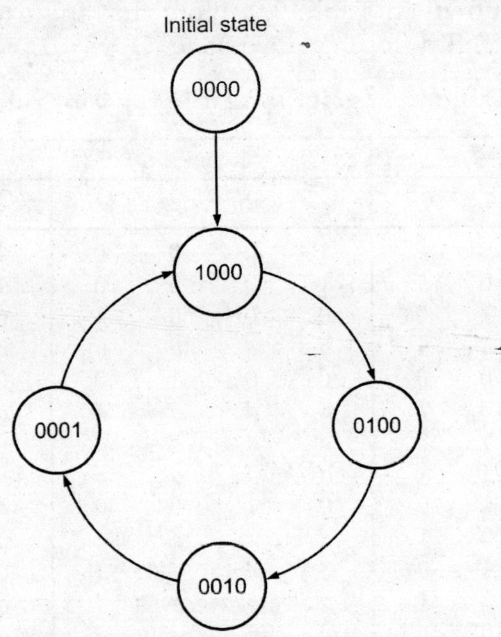

Fig. 9.20. State diagram of 4-bit ring counter.

Step 2: State table: From the above state diagram, the PS-NS table of ring counter can be drawn as shown in Table 9.8.

Table 9.8. PS-NS table for 4-bit standard ring counter.

PS				*NS*			
q_A	q_B	q_C	q_D	Q_A	Q_B	Q_C	Q_D
0	0	0	0	1	0	0	0
0	0	0	1	1	0	0	0
0	0	1	0	0	0	0	1
0	0	1	1	*d*	*d*	*d*	*d*
0	1	0	0	0	0	1	0
0	1	0	1	*d*	*d*	*d*	*d*
0	1	1	0	*d*	*d*	*d*	*d*
0	1	1	1	*d*	*d*	*d*	*d*
1	0	0	0	0	1	0	0
1	0	0	1	*d*	*d*	*d*	*d*
1	0	1	0	*d*	*d*	*d*	*d*
1	0	1	1	*d*	*d*	*d*	*d*
1	1	0	0	*d*	*d*	*d*	*d*
1	1	0	1	*d*	*d*	*d*	*d*
1	1	1	0	*d*	*d*	*d*	*d*
1	1	1	1	*d*	*d*	*d*	*d*

Step 3: Excitation table: Ring counters can be easily implemented using *D* flip-flop. So, from the above PS-NS table and the application table of *D* flip-flop, one can draw the excitation table as shown in Table 9.9. Note that the excitation inputs $D_A D_B D_C D_D = Q_A Q_B Q_C Q_D$, since for *D* flip-flop, the next state, $Q_{n+1} = D$.

Table 9.9. Excitation table for 4-bit ring counter.

PS				*NS*				**Excitation inputs**			
q_A	q_B	q_C	q_D	Q_A	Q_B	Q_C	Q_D	Q_A	Q_B	Q_C	Q_D
0	0	0	0	1	0	0	0	1	0	0	0
0	0	0	1	1	0	0	0	1	0	0	0
0	0	1	0	0	0	0	1	0	0	0	1
0	0	1	1	*d*	*d*	*d*	*d*	*d*	*d*	*d*	*d*
0	1	0	0	0	0	1	0	0	0	1	0
0	1	0	1	*d*	*d*	*d*	*d*	*d*	*d*	*d*	*d*
0	1	1	0	*d*	*d*	*d*	*d*	0	0	1	0
0	1	1	1	*d*	*d*	*d*	*d*	*d*	*d*	*d*	*d*
1	0	0	0	0	1	0	0	*d*	*d*	*d*	*d*
1	0	0	1	*d*	*d*	*d*	*d*	*d*	*d*	*d*	*d*
1	0	1	0	*d*	*d*	*d*	*d*	0	1	0	0
1	0	1	1	*d*	*d*	*d*	*d*	*d*	*d*	*d*	*d*
1	1	0	0	*d*	*d*	*d*	*d*	*d*	*d*	*d*	*d*
1	1	0	1	*d*	*d*	*d*	*d*	*d*	*d*	*d*	*d*
1	1	1	0	*d*	*d*	*d*	*d*	*d*	*d*	*d*	*d*
1	1	1	1	*d*	*d*	*d*	*d*	*d*	*d*	*d*	*d*

Step 4: Excitation maps: From Table 9.9, the excitation maps for D_A, D_B, D_C and D_D inputs of the self-correcting ring counter are obtained as shown in Fig. 9.21.

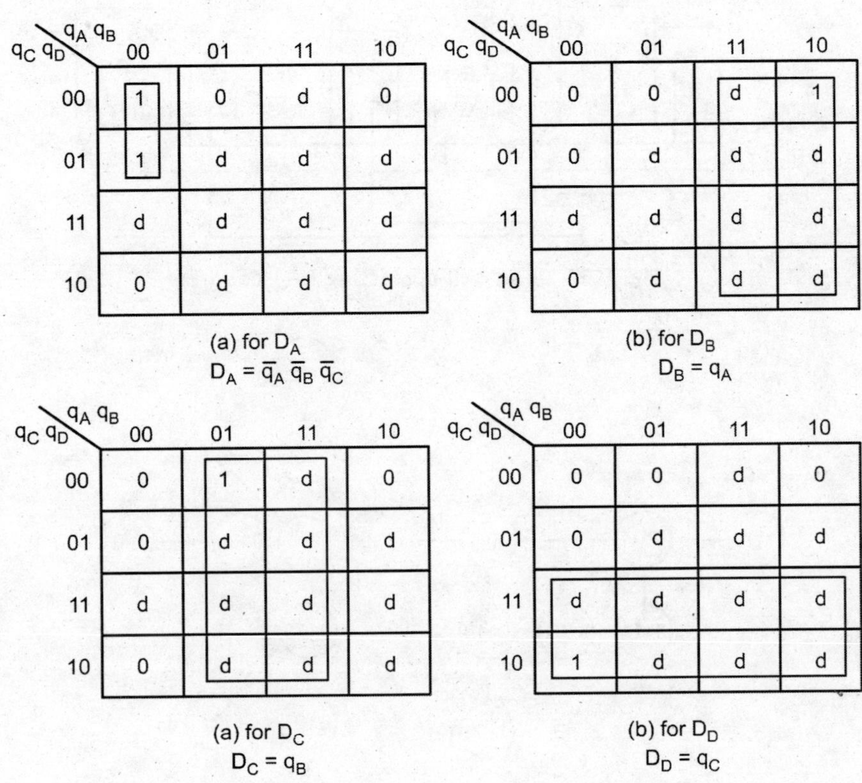

Fig. 9.21. Excitation maps for 4-bit self-correcting ring counter

In Fig. 9.21, the don't care combinations are completely avoided in D_A simplification, i.e don't care = 0. But, in excitation map simplifications for D_B, D_C and D_D, the don't care combination are considered for grouping. Note that, wherever don't care combinations are used for grouping, the assume value1; wherever not considered, they assume value 0.

Using the above simplified excitation equations, the circuit for a 4-bit self-correcting standard ring counter is drawn as shown in Fig. 9.22.

In this circuit, let us assume that due to some conditions, the ring counter has an invalid state 1111 initially. Now, by referring to the above excitation maps, the next state of 1111 is found to be 0111; the next state of 0111 is 0011 and the next state of 0011 is 0001, which is a valid state. Similarly, if other invalid states occur, this ring counter is capable of self-correcting and entering into the valid counting sequence after the occurrence of a maximum number of three clock pulses, as shown in the state diagram given in Fig. 9.23.

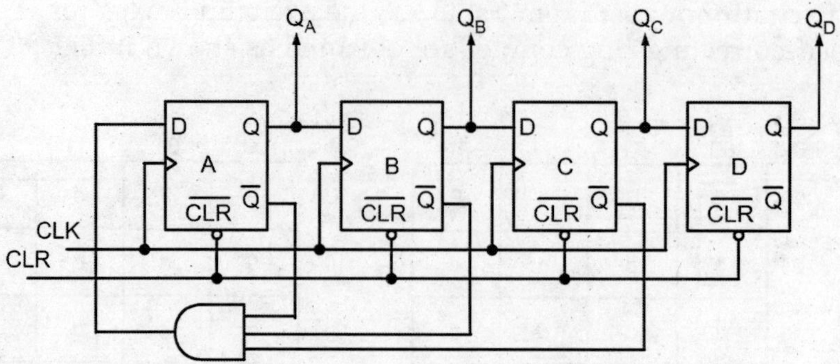

Fig. 9.22. 4-bit self-correcting ring counter

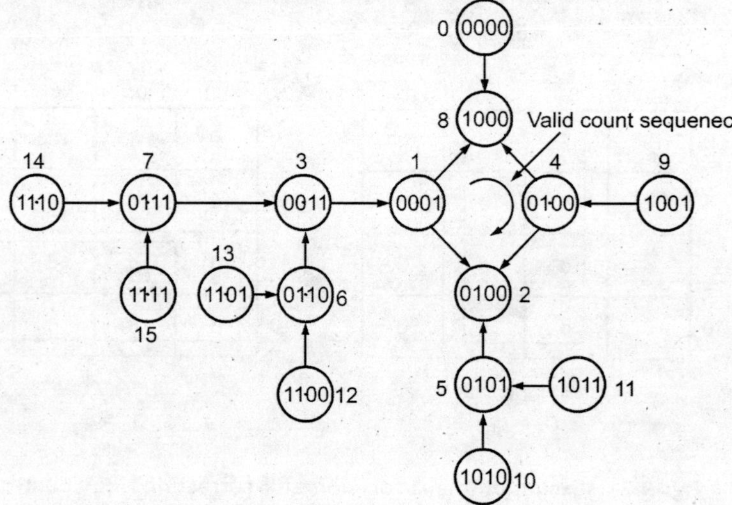

Fig. 9.23. State diagram of 4-bit self-correcting ring counter

9.4.3 Ring Counter using IC 74164 Serial Shift Register

IC 74164 is a simple serial shift register shown in Fig. 9.24(*a*). One of the most logical applications of feedback might be to connect the output of the last flip-flop Q_H back to the *D* input of the first flip-flop *A*. Here, the data inputs *A* and *B* are connected together. Now, suppose that all flip-flops are reset and the clock is applied to run. Since *S* = 0 and *R* = 1 at the first flip-flop, the output remains the same. Therefore, every time the clock goes high, the zero in each flip-flop will be shifted into the next flip-flop, which the zero in the last flip-flop *H* will travel around the feedback loop and shift into the first flip-flop *H* will travel around the feedback loop and shift into the first flip-flop *A*. In other words, all the flip-flops are in a reset state and each clock transition resets them again, so that each flip-flop output simply remains low. Consider the register as a tube full of zeros that shift round and round the register, moving ahead one flip-flop with each clock transition.

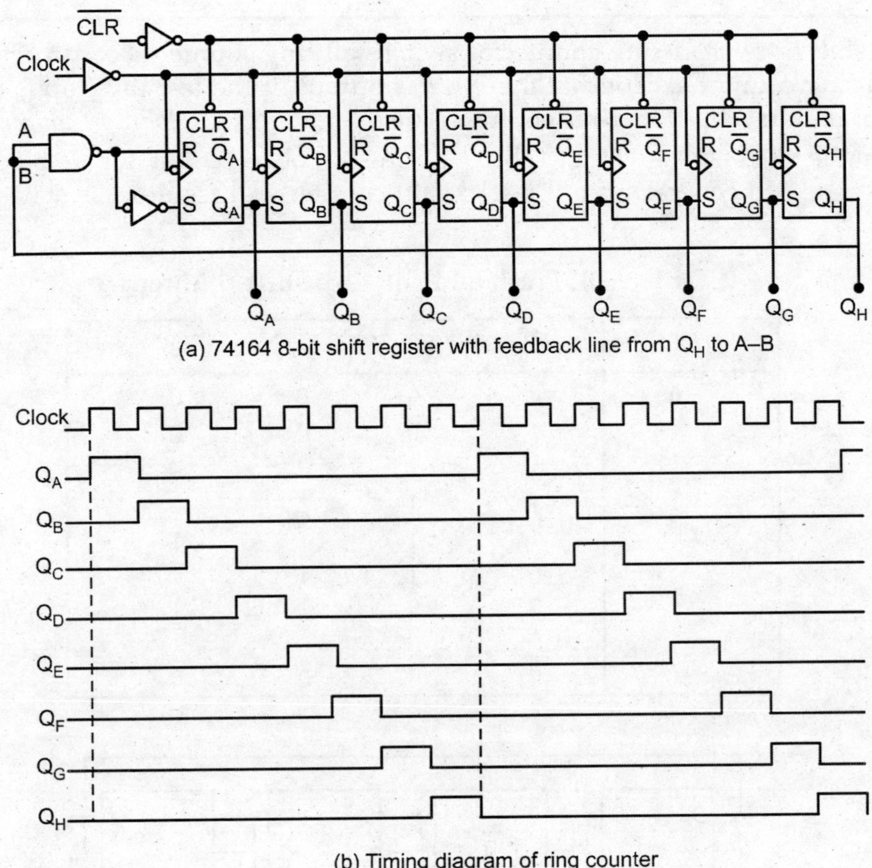

(a) 74164 8-bit shift register with feedback line from Q_H to A–B

(b) Timing diagram of ring counter

Fig. 9.24

Suppose the Q_A is high and all other flip-flops are low, and then the clock pulse is applied. When the clock goes high, the 1 in A will shift to B and A will reset, since the 0 in H will shift into A. All other flip-flops will still contain 0s. The second clock pulse will shift the 1 from B to C, while B resets. The third clock pulse will shift the 1 from C to D and so on. Thus, this single 1 will shift down the register, travelling from one flip-flop to the next, each time the clock goes high. When it reaches flip-flop H, the next clock will shift it into flip-flop A by means of the feedback connection. Again, consider the register as a tube of seven os and a single 1, which circulate the register in a clockwise direction, moving ahead by one flip-flop with each clock transition. This configuration is known as a circulating register or a ring counter. The waveforms for this counter are shown in Fig. 9.24(*b*).

The waveforms are ideal for controlling events that must occur in a strict time sequence, i.e. event A, B, then C, and so on.

9.5 SHIFT COUNTER/JOHNSON COUNTER

The shift or twisted-ring counter or switch-tail ring counter is constructed similar to a standard ring counter except that the *inverted* output of the last flip-flop is connected to the input of the first flip-flop. This counter is also known as the *Johnson counter*.

From the truth table given in Table 9.10, one can observe that, for a 4-bit shift counter, there are 8 states. In general, an *n*-flip-flop shift counter will result in $2n$ states or Modulo-$2n$ counter.

Table 9.10 Truth table of 4-bit shift counter.

CLK	Q_A	Q_B	Q_C	Q_D
0	0	0	0	0
↑	1	0	0	0
↑	1	1	0	0
↑	1	1	1	0
↑	1	1	1	1
↑	0	1	1	1
↑	0	0	1	1
↑	0	0	0	1
↑	0	0	0	0

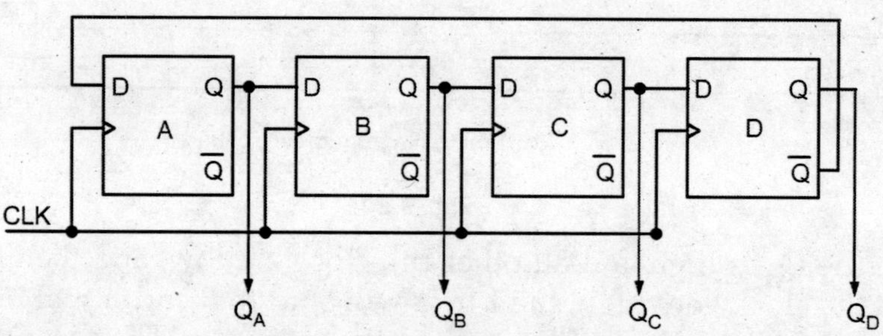

Fig. 9.25. 4-bit shift counter.

A 4-bit Johnson counter using *D* flip-flop is shown in Fig. 9.25. The counter of each flip-flop (*Q*) is connected to the *D* input of the next stage. However, the inverted output of the last flip-flop, i.e. $\overline{Q}_D$ is connected to the *D* input of the first flip-flop *A*. As shown in Table 9.10, assume that initially the counter is reset to 0, i.e. $Q_AQ_BQ_CQ_D = 0000$, and the Value at $D_B = D_C = D_D = 0$ whereas $D_A = 1$, since $\overline{Q}_D$ is connected to the D_A input of the first flip-flop. Now, when the first clock pulse is applied, the first flip-flop is set to 1 and the other flip-flops are reset to 0, i.e. $Q_AQ_BQ_CQ_D = 1000$. Now, $D_A = 1$ as $\overline{Q}_D = 1$; $D_B = 1$ as $Q_A = 1$; $D_C = D_D = 0$ as $Q_B = Q_C = 0$. Therefore, on the arrival of the second clock pulse, the first and second flip-flop are set to 1 while the third and fourth flip-flops are reset to 0. Similarly, after the third clock pulse, the counter content $Q_AQ_BQ_CQ_D = 1110$; after the fourth clock pulse, $Q_AQ_BQ_CQ_D$ = 1000. Now, $D_A = 0$ as $\overline{Q}_D = 0$ and $D_B = D_C = D_D = 1$ as $Q_A = Q_B = Q_C = 1$. Therefore, on the occurrence of the fifth clock pulse, the first flip-flop is reset to 0 and the other three flip-flops are set to 1, i.e. $Q_AQ_BQ_CQ_D$ = 0111. Now, a 0 is shifted from left to right until the counter $Q_AQ_BQ_CQ_D$ = 0000, i.e. the initial state.

9.5.1 Design of 4-bit Self-correcting Shift Counter

A Johnson counter has $2^n - 2n$ unused states; thus for $n > 2$, there will be undesired sequences of states i.e. invalid states. The counter can enter an invalid state either when the power is turned ON or because of noise. In the case of system turn-on, the counter can be cleared or preset or both, thus starting in a valid state. On the other hand, the shift counter can *hang* or enter into any invalid state due to noise or any other conditions without returning to the main counting sequence. Hence, it is a must to design shift counters which should be self-correcting and capable of recovering from invalid states.

The design of self-correcting shift counters can be accomplished by completely avoiding don't are entries (*d*) in one of the excitation map simplification, mostly in the excitation map corresponding to MSB. The design of 4-bit/Modulo-8 self-correcting shift counter is explained in the following section.

Step 1: State diagram: The state diagram for the 4-bit shift counter is shown in Fig. 9.26.

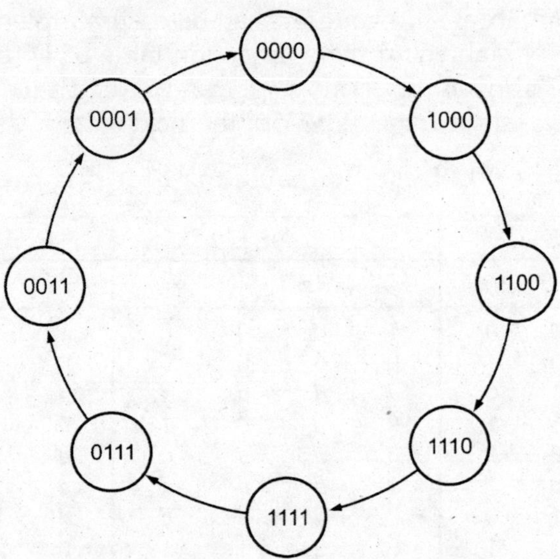

Fig. 9.26. State diagram of 4-bit shift counter.

Step 2: State table: From the state diagram given in Fig. 9.26, the PS–NS table of the shift counter is drawn as shown in Table 9.11.

Table 9.11 PS–NS table for 4-bit shift counter.

PS				*NS*			
q_A	q_B	q_C	q_D	Q_A	Q_B	Q_C	Q_D
0	0	0	0	1	0	0	0
0	0	0	1	0	0	0	0
0	0	1	0	d	d	d	d
0	0	1	1	0	0	0	1
0	1	0	0	d	d	d	d
0	1	0	1	d	d	d	d
0	1	1	0	d	d	d	d
0	1	1	1	0	0	1	1
1	0	0	0	1	1	0	0
1	0	0	1	d	d	d	d
1	0	1	0	d	d	d	d
1	0	1	1	d	d	d	d
1	1	0	0	1	1	1	0
1	1	0	1	d	d	d	d
1	1	1	0	1	1	1	1
1	1	1	1	0	1	1	1

Step 3: Excitation table: Shift counters can be easily implemented using *D* flip-flops. So, from the above PS–NS table and the application table of *D* flip-flop, one can draw the excitation table as shown in Table 9.12. Note that the excitation inputs $D_AD_BD_CD_C = Q_AQ_BQ_CQ_D$, since for *D* flip-flop, the next state, $Q_{n+1} = D$.

Table 9.12 Excitation table for 4-bit ring counter.

PS				**NS**				**Excitation inputs**			
q_A	q_B	q_C	q_D	Q_A	Q_B	Q_C	Q_D	Q_A	Q_B	Q_C	Q_D
0	0	0	0	1	0	0	0	1	0	0	0
0	0	0	1	0	0	0	0	0	0	0	0
0	0	1	0	*d*	*d*	*d*	*d*	*d*	*d*	*d*	*d*
0	0	1	1	0	0	0	1	0	0	0	1
0	1	0	0	*d*	*d*	*d*	*d*	*d*	*d*	*d*	*d*
0	1	0	1	*d*	*d*	*d*	*d*	*d*	*d*	*d*	*d*
0	1	1	0	*d*	*d*	*d*	*d*	*d*	*d*	*d*	*d*
0	1	1	1	0	0	1	1	0	0	1	1
1	0	0	0	1	1	0	0	1	1	0	0
1	0	0	1	*d*	*d*	*d*	*d*	*d*	*d*	*d*	*d*
1	0	1	0	*d*	*d*	*d*	*d*	*d*	*d*	*d*	*d*
1	0	1	1	*d*	*d*	*d*	*d*	*d*	*d*	*d*	*d*
1	1	0	0	1	1	1	0	1	1	1	0
1	1	0	1	*d*	*d*	*d*	*d*	*d*	*d*	*d*	*d*
1	1	1	0	1	1	1	1	1	1	1	1
1	1	1	1	0	1	1	1	0	1	1	1

Step 4: Excitation maps: From Table 9.12, the excitation maps for D_A, D_B, D_C and D_D inputs of the self-correcting shift counter are obtained as shown in Fig. 9.27.

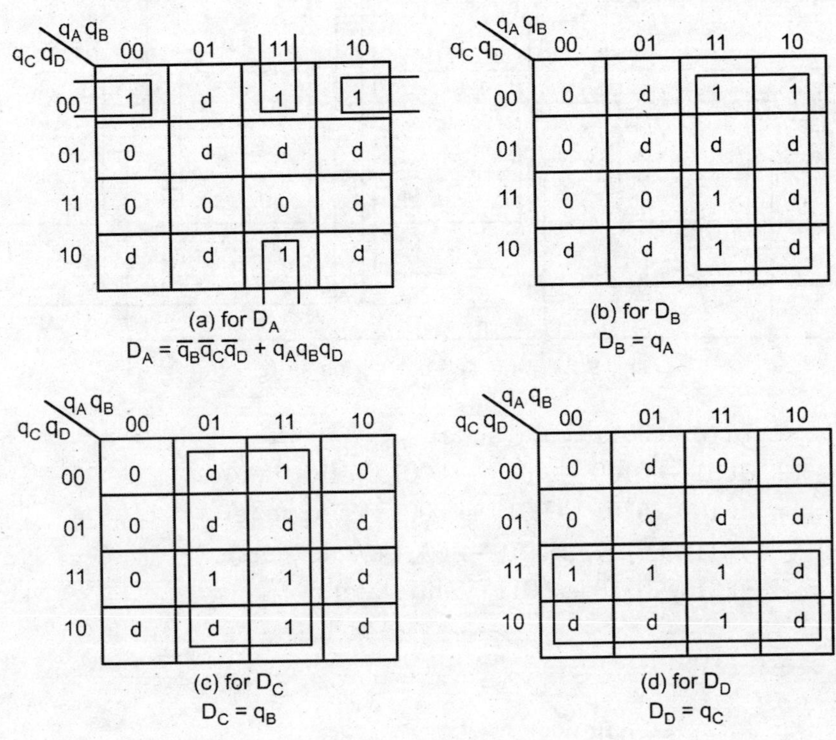

Fig. 9.27 Excitation maps for 4-bit self-correcting shift counter.

In Fig. 9.27, the don't care combinations are completely avoided in D_A simplification, i.e. don't care = 0. But, in other excitation map simplification, i.e. for D_B, D_C and D_D, the don't care combinations are considered for grouping. Note that, whereever don't care combinations are used for grouping, they assume value 1; wherever not considered, they assume value 0.

Using the above simplified excitation equations, the circuit for 4-bit self-correcting shift counter is drawn as shown in Fig. 9.28.

In this circuit, let us assume that due to some conditions, the shift counter has invalid state 1001 initially. By referring to the above excitation maps for D_A, D_B D_C and D_D, the next state of 1001 is found to be 0100; the next state of 0100 is 0010; the next state of 0010 is 0001, which is a valid state. Similarly, if any other invalid state occurs, this shift counter is capable of entering into valid state after the occurrence of a maximum number of three clock pulses as shown in the state diagram given in Fig. 9.29.

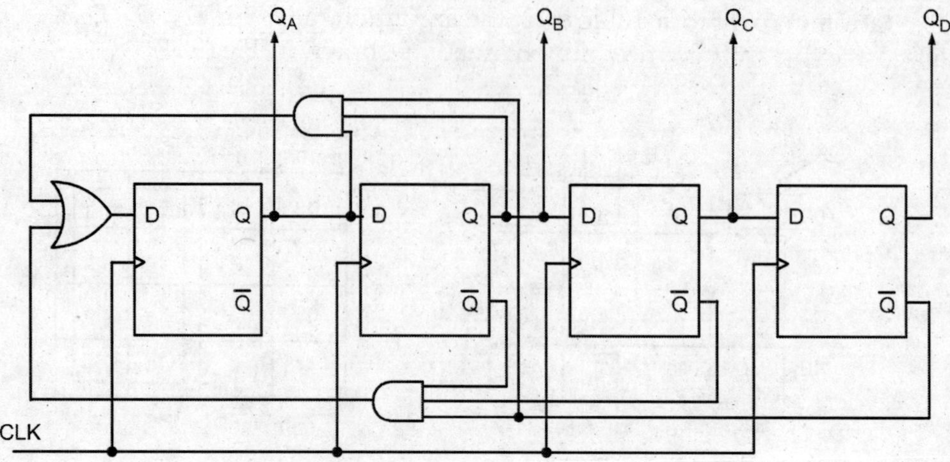

Fig. 9.28 4-bit self-correcting shift counter

For invalid 0110 → 0011 (Valid state)
1001 → 0100 → 0010 → 0001 (Valid state)
1010 → 0101 → 0010 → 0001 (Valid state)
1011 → 0101 → 0010 → 0001 (Valid state)
1101 → 0110 → 0011 (Valid state)

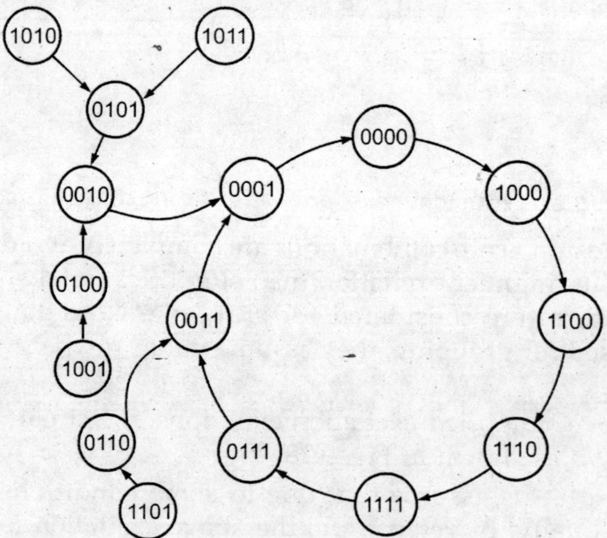

Fig. 9.29 State diagram of 4-bit self-correcting shift counter.

9.6 SEQUENCE GENERATOR

Sequence generator is a circuit that generates a desired sequence of bits in synchronization with a clock. Such a sequence generator can be used as a random bit generator, code generator and prescribed period generator. The block diagram of a sequence generator is shown in Fig. 9.30.

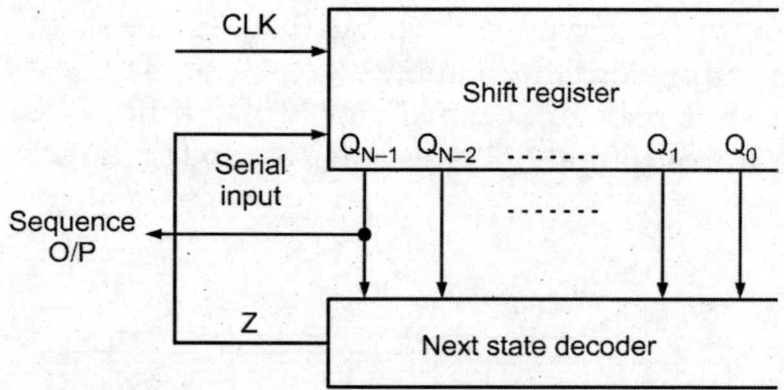

Fig. 9.30 Block diagram of a sequence generator.

The sequence generator can be constructed using shift register and next state decoder. The output of the next state decoder (Z) is a function of Q_{N-1}, Q_{N-2},....., Q_1, Q_0 and is connected to the serial input of the shift register. This sequence generator is similar to ring counter / shift counter in which $Q_0/\overline{Q}_0$ is connected to the serial input of the shift register and hence they are special cases of sequence generators. The design of sequence generators is discussed in the following sections.

9.6.1 Design of 5-bit Sequence Generator

Consider the design of sequence generator to generate a sequence 11001. The minimum number of flip-flops (n) required to generate a sequence of length N is given by

$$N \leq 2^n - 1$$

Here $N = 5$ and hence the minimum value of n to satisfy the above condition is 3, i.e., three flip-flops are required to generate the given sequence. If the given sequence does not lead to five distinct states, then more than three flip-flops are required. The states of the given sequence generator are given in Table 9.13.

Table 9.13 State table for 5 bit (11001) sequence generator.

CLK	Flip-Flop outputs			Serial Input
	Q_2	Q_1	Q_0	Z
1	1	1	0	1
2	1	1	1	0
3	0	1	1	0
4	0	0	1	1
5	1	0	0	1
1	1	1	0	1
2	1	1	1	0
X	0	1	1	0
X	0	0	1	1
X	1	0	0	1

In Table 9.13, the given sequence (11001) is listed under Q_2 and the sequence under Q_1 and Q_0 are the same sequence delayed by one and two clock pulses respectively as indicated by arrow marks. Also, it is observed that all the 5 states are distinct and hence three flip-flops are sufficient to implement the sequence generator. The last column gives the serial input required at the shift register (i.e, at D_2 of MSB flipflop), assuming D flip-flops are used and considering the output at Q_2. Now, the K-map for the serial input (Z) is shown in Fig. 9.31.

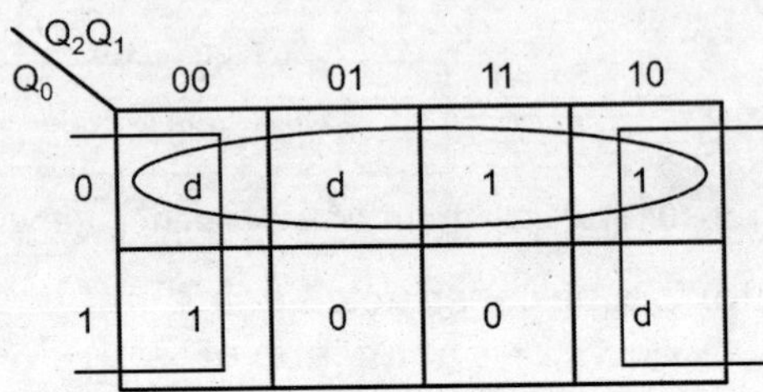

Fig. 9.31. *K*-map of serial input (*Z*) for 5 bit (11001) sequence generator

From the K-map shown in Fig. 9.31, the simplified expression for serial input Z can be written as

$$Z = \overline{Q}_0 + \overline{Q}_1$$

Therefore, using the simplified expression for Z, the logic diagram of given 5-bit sequence generator can be drawn as shown in Fig. 9.32.

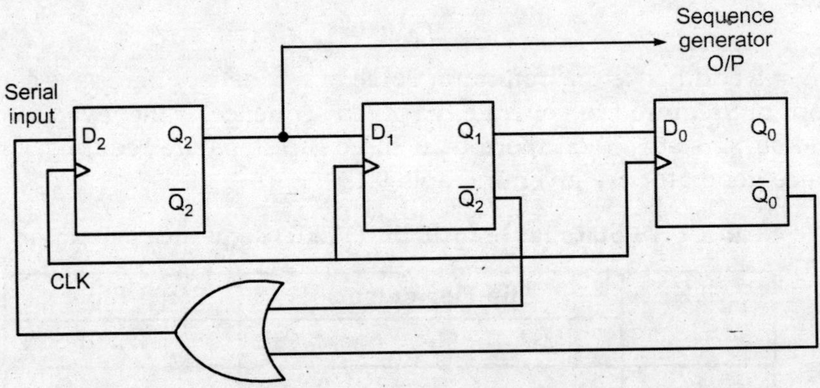

Fig. 9.32 Logic diagram of 5 bit (11001) sequence generator

9.6.2 Design of 6-bit Sequence Generator

Consider the design of 6-bit sequence generator to generate the sequence (101011). To generate the sequence of 6 bits (N), the minimum number of flip-flops (n) required can be found using the equation $N \leq 2^n - 1$.

For $N = 6$, the minimum number of flip-flops (n) required is 3. More flip-flops can be used if the given sequence does not lead to six distinct states. The state table for the given sequence generator is shown in Table 9.14.

Table 9.14 State Table for 6-bit (101011) sequence generator.

CLK	Flip-Flop outputs		
	Q_2	Q_1	Q_0
1	1	1	1
2	0	1	1
3	1	0	1
4	0	1	0
5	1	0	1
6	1	I	0
1		I	1
2			1

As explained in the previous section, the given sequence (101011) is listed under Q_2 and the sequence listed under Q_1 and Q_0 are the same sequence delayed by one and two clock pulses respectively as indicated by arrow marks. From the Table 9.14, it is observed that all the six states are not distinct, i.e., 101 state occurs twice. Hence, three flip-flops are not sufficient to generate the given sequence. Next, assuming $n = 4$, the modified state table for the given sequence generator can be drawn as shown in Table 9.15.

Table 9.15 Modified state table far 6-bit (101011) sequence generator

CLK	Flip-Flop outputs				Serial Input
	Q_3	Q_2	Q_1	Q_0	Z
1	1	1	1	0	0
2	0	1	1	1	1
3	1	0	1	1	0
4	0	1	0	1	0
5	1	0	1	0	1
6	1	1	0	1	1
1	1	1	1	0	0
2	0	1	1	1	1
3	1	0	1	1	0
X	0	1	0	1	0
X	1	0	1	0	1
X	1	1	0	1	1

From the Table 9.15, it is observed that all the six states are distinct and hence the four flip-flops are sufficient to implement the sequence generator. The last column gives the serial input required at the shift register (i.e., at D_3 input of MSB flipflop), assuming D flip-flops are used and considering Q_3 outputs. The logic expression for Z can be simplified using a *K*-map shown in Fig. 9.33.

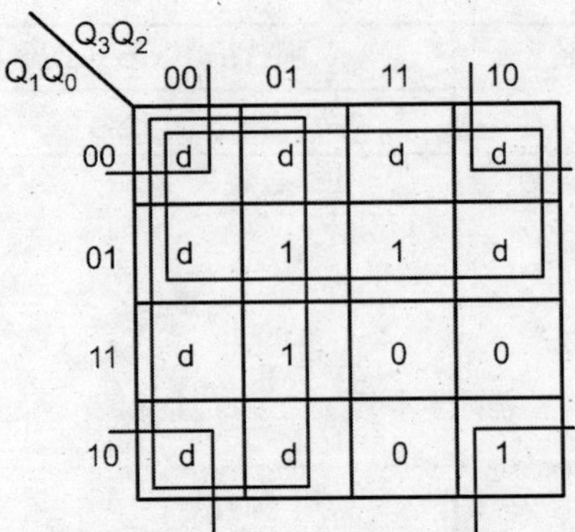

Fig. 9.33. *K*-map of serial input (*Z*) for 6-bit (101011) sequence generator.

From the *K*-map shown in Fig. 9.33, the simplified expression for serial input *Z* is given by

$$Z = \overline{Q}_3 + \overline{Q}_1 + (\overline{Q}_2\,\overline{Q}_0)$$

Therefore, using the simplified expression for *Z*, the logic diagram of given six bit sequence generator can be drawn as shown in Fig. 9.34.

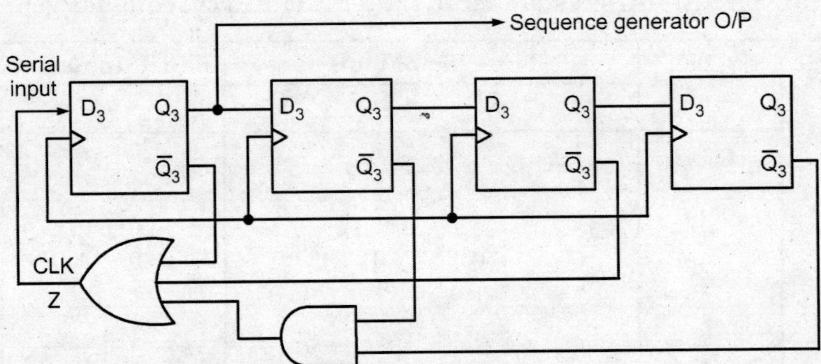

Fig. 9.34. Logic diagram of 6-bit (101011) sequence generator.

9.6.3 Design of 7-bit Sequence Generator

Consider the design of sequence generator to generate the sequence 1011011. To generate the sequence of 7 bits (*N*), the minimum number of flip-flops (*n*) required can be found using the equation

$$N \leq 2^n - 1$$

The minimum value of n to satisfy the above condition is 3, i.e., three flip-flops are required to generate the given sequence. If the given sequence does not lead to seven distinct states, then more than four flip-flops are required. The states of given sequence generator circuit are given in Table 9.16.

In the state table shown in Table 9.16, the given sequence (1011011) is listed under Q_2 and the sequence listed under Q_1 and Q_0 are the same sequence delayed by one and two clock pulses respectively as indicated by arrow marks. From the Table 9.16, it is seen that all the states are not distinct and hence $n = 3$ i.e., 3 flip-flops are not sufficient.

Table 9.16 State table for 7-bit (1011011) sequence generator.

Number of clock pulse	Flip-Flop outputs		
	Q_2	Q_1	Q_0
1	1	1	1
2	0	1	1
3	1	0	1
4	1	1	0
5	0	1	1
6	1	0	1
7	1	1	0
1		1	1
2			1

For $n = 4$, the modified state table of the given sequence generator is given in Table 9.17 in a similar manner. From Table 9.17, it is also observed that the states 1011 and 1101 occur twice and hence 4 flip-flops are not sufficient.

Table 9.17 Modified state table for 7-bit (1011011) sequence generator.

CLK	Flip-Flop outputs			
	Q_3	Q_2	Q_1	Q_0
1	1	1	1	0
2	0	1	1	1
3	1	0	1	1
4	1	1	0	1
5	0	1	1	0
6	1	0	1	1
7	1	1	0	1
1		1	1	0
2			1	1
3				1

Then assuming $n = 5$, the state table for the given sequence generator can be modified as shown in Table 9.18. Here, the state 11011 occurs twice and hence five flip-flops are not sufficient.

Table 9.18 Modified state table for 7-bit (1011011) sequence generator.

CLK	Flip-Flop outputs				
	Q_4	Q_3	Q_2	Q_1	Q_0
1	1	1	1	0	1
2	0	1	1	1	0
3	1	0	1	1	1
4	1	1	0	1	1
5	0	1	1	0	1
6	1	0	1	1	0
7	1	1	0	1	1
1	1	1	1	0	1
2	0	1	1	1	0
3	1	0	1	1	1
4	1	1	0	1	1
X	0	1	1	0	1
X	1	0	1	1	0
X	1	1	0	1	1

Now, let us assume that n = 6, the state table can be again modified as shown in Table 9.19. Here, all the 7 states are distinct. The last column gives the serial input required at the input of the shift register, assuming D flip-flops are used and considering the output at Q_5. Now, the K-map for the Table 9.19 is shown in Fig. 9.35.

Table 9.19. Modified state table for 7-bit (1011011) sequence generator.

CLK	Flip-Flop outputs						Serial Input
	Q_5	Q_4	Q_3	Q_2	Q_1	Q_0	
1	1	1	1	0	1	1	0
2	0	1	1	1	0	1	1
3	1	1	1	1	1	0	1
4	1	1	0	1	1	1	0
5	0	1	1	0	1	1	1
6	1	0	1	1	0	1	1
7	1	1	0	1	1	0	1
1	1	1	1	0	1	1	0
2	0	1	1	1	0	1	1
3	1	0	1	1	1	0	1
4	1	1	0	1	1	1	0
X	0	1	1	0	1	1	1
X	1	0	1	1	0	1	1
X	1	1	0	1	1	0	1

$Q_2Q_1Q_0$ \ $Q_5Q_4Q_3$	000	001	011	010	110	111	101	100
000	d	d	d	d	d	d	d	d
001	d	d	d	d	d	d	d	d
011	d	d	1	d	d	0	d	d
010	d	d	d	d	d	d	d	d
110	d	d	d	d	1	d	1	d
111	d	d	d	d	0	d	d	d
101	d	d	d	d	d	d	1	d
100	d	d	d	d	d	d	d	d

Fig. 9.35 *K*-map of *Z* for 7-bit (1011011) sequence generator

From the *K*-map shown in Fig. 9.35, the simplified expression for serial input *Z* can be written as

$$Z = \overline{Q_5} + \overline{Q_4} + \overline{Q_0}$$

Hence, using the simplified expression for Z, the logic diagram of 7-bit sequence generator is shown in Fig. 9.36.

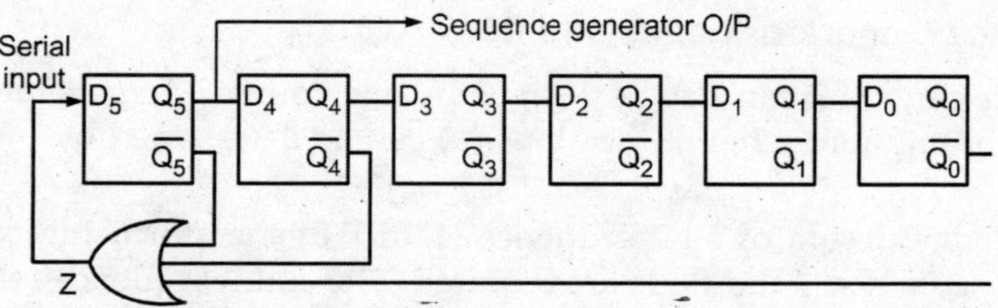

Fig. 9.36. Logic diagram of 7-bit (1011011) sequence generator.

9.6.4 Sequence Generator using Multiplexer and Counter

Sequence generator can also be designed using multiplexer and counter. An *N*-bit sequence can be generated using a multiplexer with at least *N* inputs and a modulo-*N* counter. Consider the design of 6-bit sequence generator (110101) using multiplexer and counter. To design the given 6-bit sequence generator, a multiplexer with at least six inputs and a modulo-6 counter are required. The given 6-bit generator can be constructed using an 8 to 1 multiplexer and a modulo-6 counter as shown in Fig. 9.37.

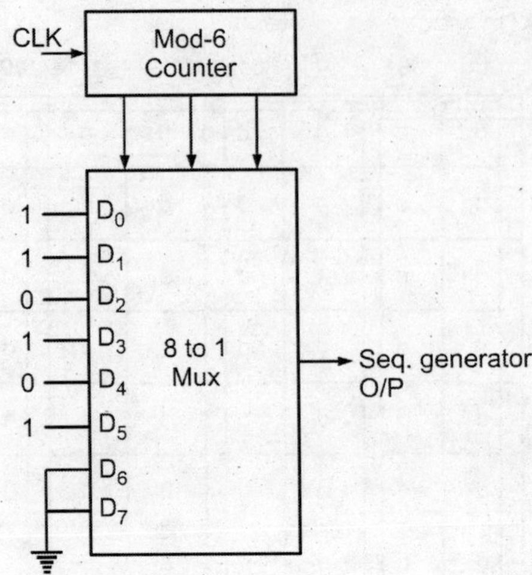

Fig. 9.37. 6-bit sequence (110101) generator using multiplexer and counter.

As shown in Fig. 9.37, the D_0, D_1, D_3 and D_5 inputs of multiplexer are connected with logic '1' (+5V) while D_2 and D_4 inputs are connected with logic '0' (0V). It is important to note that the MSB of the given sequence (i.e., 1) is connected with D_0; the next bit (i.e., 1) is connected with D_1; the third bit (i.e., 0) with D_2 and so on. Since a 6-bit sequence has to be generated, the remaining two inputs D_6 and D_7 of multiplexer are not required. The select lines (*A*, *B* and *C*) of the multiplexer are connected with the outputs of the modulo-6 counter that counts from 000 to 101. Now, when continuous clock is applied to modulo-6 counter, the counter from 000 to 101 in cyclic manner and subsequently the data at multiplexer inputs ($D_0 - D_5$) are available in serial manner at the multiplexer output as 110101.

9.6.5 Sequence Generator Using Counter

Sequence generator can also be designed using counter. To generator N-bit sequence, the required minimum number of flip-flops (*n*) can be determined by

$$N \leq 2^n - 1$$

Consider the design of 6-bit sequence (110101) generator using counter. Hence, for the given 6-bit sequence, 3 flip-flops are required. The count sequence of the 3-bit generator to generate the given sequence can be drawn, with the given sequence in one of the three flip-flops, preferably in the LSB flip-flop, as shown in Table 9.20.

Table 9.20 Count Sequence of Sequence (110101) Generator Using Counter.

CLK	Flip-Flop outputs		
	Q_2	Q_1	Q_0
1	0	0	1
2	0	1	1
3	1	0	0
4	1	0	1
5	1	1	0
6	1	1	1

Here, the input sequence 110101 is given under Q_0. Then the values for Q2 and Q1 are assigned in such a way that the six states obtained are distinct or valid. Now for the count sequence a counter can be designed by using the counter design procedure as follows.

Table 9.21 PS–NS and Excitation Table.

PS			NS			Excitation inputs		
q_2	q_1	q_0	Q_2	Q_1	Q_0	J_2K_2	J_1K_1	J_0K_0
0	0	0	*d*	*d*	*d*	*dd*	*dd*	*dd*
0	0	1	0	1	1	0*d*	1*d*	*d*0
0	1	0	*d*	*d*	*d*	*dd*	*dd*	*dd*
0	1	1	1	0	0	1*d*	*d*1	*d*1
1	0	0	1	0	1	*d*0	1*d*	*d*1
1	0	1	1	1	0	*d*0	1*d*	*d*1
1	1	0	1	1	1	*d*0	*d*0	1*d*
1	1	1	0	0	1	*d*1	*d*1	*d*0

From the Table 9.21, the excitation maps for J_2, K_2, J_1, K_1, J_0 and K_0 can be drawn as shown in Fig. 9.38.

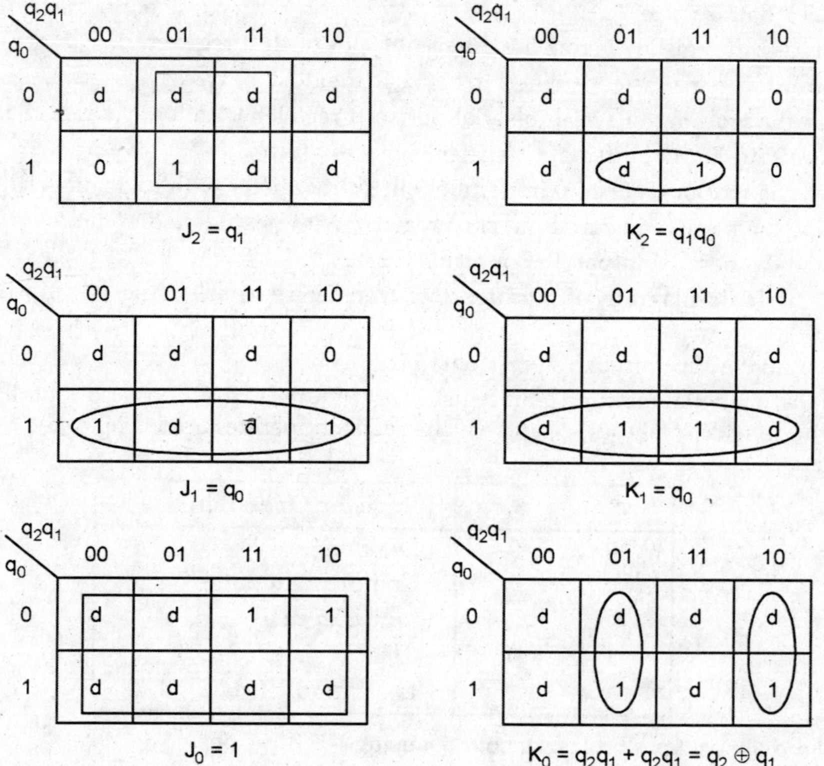

Fig. 9.38

Now, using the simplified expression for excitation inputs shown in Fig. 9.38, the 3-bit counter circuit to generate the given sequence 110101 can be drawn as shown in Fig. 9.39.

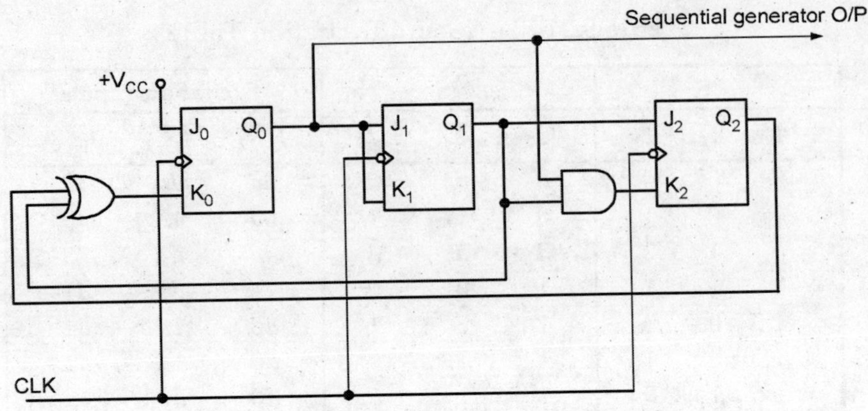

Fig. 9.39 Sequence (110101) Generator Using Counter

■ REVEIW QUESTIONS ■

1. What is a data register?
2. What is a shift register? Name the different types of shift register.
3. Why are shift registers considered to be basic memory devices?
4. Explain the working of serial in-serial out shift register with logic diagram and waveforms.
5. How are shift-left or shift-right transfer registers built?
6. Explain the working of serial in-parallel out shift register with logic diagram and waveforms.
7. Describe the parallel in-serial out shift register with neat logic diagram.
8. Describe the parallel in-parallel out shift register.
9. Draw the logic diagram of parallel data transfer registers using *J-K* flip-flop synchronous inputs.
10. Explain how a flip-flop can store a data bit.
11. Draw the logic diagram of a 4-bit register with four 4 D flip-flops and four 4-to-1 multiplexers with mode selection inputs S_1 and S_0. The register operates according to the following function table :

S_1	S_0	**Register Operation**
0	0	No change
		Complement of the four outputs
1	0	Clear register to 0
1	1	Load parallel data

12. Draw a diagram for a 5-bit ring counter using *J-K* flip-flops.
13. Write a note on Johnson counter.
14. Construct a Johnson counter for ten timing signals.

15. What is the basic difference between a Johnson counter and a ring counter?
16. Write the sequence of states for a 4-bit Johnson counter.
17. Draw the logic diagram for a divide-by-18 Johnson counter. Sketch the timing diagram and write the sequence in tabular form.
18. What are the needs for sequence generator?
19. Explain how ring and shift counters can be used as sequence generators.
20. Design a sequence generator to generate the sequence 111101.
21. Design a sequence generator to generate the sequence 1001.
22. Design a sequence generator using multiplexes and counter to generate the sequence 1101011110100011.

■ PROBLEMS ■

1. Draw the diagram for a MOD-10 Johnson counter using *J-K* flip-flops and determine its counting sequence. Draw the decoding circuit needed to decode each of the 10 states.

Ans. Counting sequence is: 00000, 10000, 11000, 11100, 11110, 11111, 01111, 00111, 00011, 00001 and repeats.

2. Determine the frequency of the pulses at points *w*, *x*, *y* and *z* in the circuit of Fig. P9.2

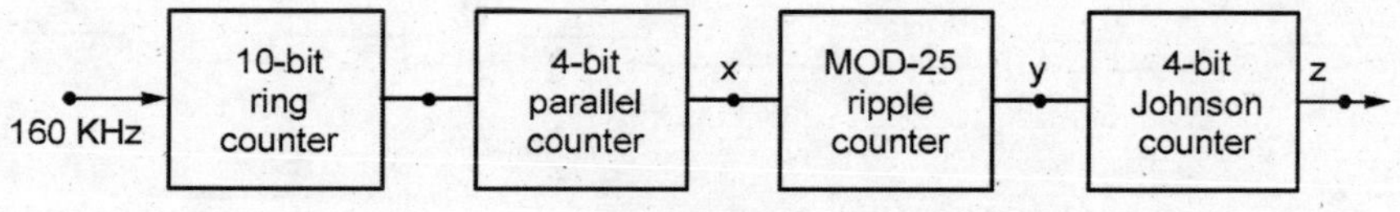

Fig. P9.2

Ans: Frequency of Pulses at w =16 KHz; x = 1 KHz; y = 40 Hz; z = 5 Hz.

3. Design a shift register with parallel load that operates according to the following function table:

Shift	*Load*	*Register operation*
0	0	No change
0	1	Load parallel data
1	X	Shift right

4. The content of a 4-bit register is initially 1101. The register is shifted 6 times to the right with the serial input being 101101. What is the content of the register after each shift?

Ans: 1110; 0111; 1011; 1101; 0110; 1011

5. What is the state of the register in Fig P9.5 after each clock pulse if it starts in the 101001111000 state?

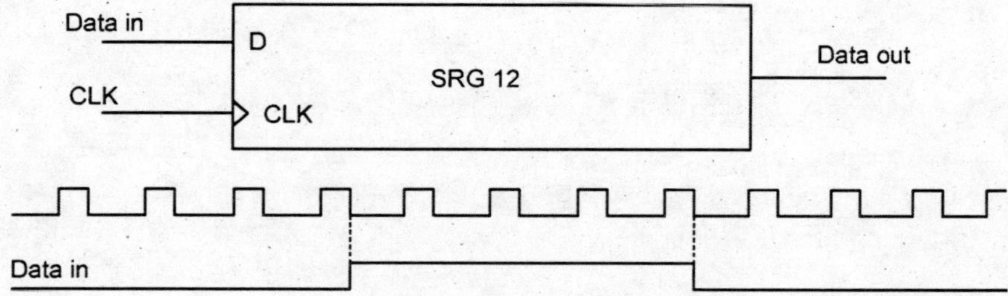

Fig. P9.5

Ans: Initially	101001111000
CLK_1	010100111100
CLK_2	001010011110
CLK_3	000101001111
CLK_4	000010100111
CLK_5	100001010011
CLK_6	110000101001
CLK_7	111000010100
CLK_8	111100001010
CLK_9	011110000101
CLK_{10}	001111000010
CLK_{11}	000111100001
CLK_{12}	000001110000

6. For the aerial-in-serial-out shift register, determine the output waveform for the data input and clock waveforms in Fig P9.6.

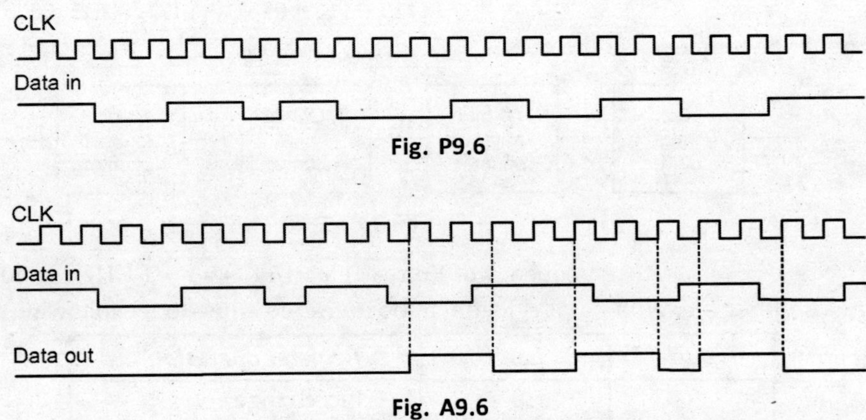

Fig. P9.6

Fig. A9.6

7. Design a MOD-7 Johnson counter using *J-K* flip-flop.
8. Design a self-correcting MOD-9 shift counter using *D* flip-flops.
9. Design a 4-bit self-correcting ring counter which is capable of rotating 1 in left direction.
10. Design a 5-bit self-correcting ring counter.

❑❑❑

Memory and Programmable Logic Devices

10.1 INTRODUCTION

Memory is an indispensable part of computer and microprocessor based systems. The data used in a program as well as the instructions for executing the program are stored in the memory. Hence, digital systems require memory facilities for temporary as well as permanent storage of data to perform their functions. A flip-flop stores one bit of information; a *register* is able to hold a word; a *register file* holds a modest number of words of information.

The very first computer memory consisted of a minute magnetic toroid, called *core memory*, which required large, bulky circuit boards stored in large cabinets. Semi conductor memory, on the other hand, is very compact and can be accessed at very high speeds and is capable of storing data in extremely high densities. All modern computers and microprocessor systems have been made possible by the development of inexpensive and reliable VLSI semiconductor memory chips utilising NMOS, CMOS, BJT and BiCMOS technologies.

Do You Know?	All memory devices store binary logic levels (1s and 0s) in an array structure. The size of each binary word (number of bits) that is stored varies depending on the memory device. These binary values are referred to as data.

Information from magnetic and optical storage devices such as hard disk, floppy disk, Compact Disk-Read Only Memory (CD-ROM) and digital tape must be accessed sequentially, starting at the beginning of a data file or track. In contrast, data stored in an electronic memory cell can be accessed at random and on demand using *direct addressing*. Direct addressing eliminates the need to process a large stream of irrelevant data in order to find the desired data word.

The evolution of *Programmable Logic Device* (PLD) began with Programmable Read Only Memory (PROM). A Read Only Memory (ROM) is a memory device that consists of AND and OR arrays which can be programmed by the user to implement combinational and sequential functions. For reprogrammability, PLDs use Erasable PROM (EPROM) or Electrically Erasable and Programmable ROM (EEPROM) like cells. Generally, PLDs may be classified depending upon the programmability of the AND and OR arrays. PLDs with programmable AND and fixed OR arrays are called *Programmable Array Logic* (PAL) devices. When both the AND and OR arrays are programmable, such PLDs are known as *Programmable Logic Arrays* (PLA).

The programmability and high density of PLDs make them useful in the design of Application-Specific Integrated Circuits (ASICs) where design changes can be made rapidly and inexpensively.

A Field Programmable Gate Array (FPGA) is a reprogrammable gate array that uses antifuse. The differences between FPGA and PLD is that FPGA incorporates logic blocks instead of fixed AND-OR gates and is faster with low power dissipation.

10.2 CLASSIFICATION OF MEMORIES

Memories are classified as:

(*i*) Registers, Main memory and Secondary memory

(*ii*) Sequential Access Memory and Random Access Memory

(*iii*) Static and Dynamic Memory

(*iv*) Volatile and Non-volatile Memory

(*v*) Magnetic and Semiconductor Memory.

10.2.1 Registers, Main Memory and Secondary Memory

Though memories are scattered throughout the computer, those based on the location and usage are called Registers, Main memory and Secondary memory. Registers are available within the CPU to store data temporarily during arithmetic and logical operations like addition, subtraction, AND, OR, etc. They have very low access time, as they are available inside the CPU. Main memories of a computer, usually of semiconductor type, are available external to the CPU to store program and data during the execution of a program. In the main memory, each memory location is identified by an unique address and is accessed for read/write operation in a lesser speed than registers. As the storage capacity of main memory is inadequate, secondary or auxiliary memories are added to enhance storage capabilities. This secondary memory operates at a lesser speed when compared to registers and main memory. Normally, secondary memories are of magnetic memory type (Magnetic tape, Magnetic drum, Floppy disk and Hard disk) that are used to store large quantities of data.

10.2.2 Sequential Access Memory and Random Access Memory

Based on the method of access, memory devices can be classified as Sequential Access and Random Access Memories (RAM). A sequential access memory is one in which a particular memory location is accessed sequentially, i.e. the *i* th memory location is accessed only after sequencing through previous (*i*-1) memory locations. Therefore, the access time of a sequential memory varies depending on the location to be accessed. An example of sequential access memory is the magnetic tape memory.

On the other hand, a random access memory is one in which any location can be accessed in a random manner and thus has equal access time for all memory locations. An example of random access memory is the semiconductor RAM.

10.2.3 Static and Dynamic Memory

In static memory, the content does not change with time; in dynamic memory, its content changes with time. Dynamic memory cells use the capacitance of a transistor as the storage device. Only one transistor is needed to store one bit of information. The capacitor must be *refreshed* periodically without being discharged in order to prevent loss of information. Static memory devices require no refreshing, and hold data as long as d.c. power is applied. Examples for static memory are register and MOS cell; semiconductor dynamic RAM and circulating registers using Charge-coupled Devices (CCD) are examples of dynamic memory.

10.2.4 Volatile and Non-volatile Memory

Volatile memory loses its stored data when power to the memory circuit is removed; a non-volatile memory retains stored data permanently even after the power supply is turned OFF. Magnetic Core Memory and Read Only Memory (ROM) are examples of non-volatile memory devices.

10.2.5 Magnetic and Semiconductor Memory

These memories are classified based on the material used for construction. The magnetic memories are constructed using magnetic material, e.g. magnetic tape, floppy and compact disks. Magnetic recording is the process of storing data magnetically on the surface of a tape, disk or drum. Magnetic tape is a storage medium using the surface of a magnetic tape to hold data. Magnetic disk is a storage medium using the surfaces of a disk to hold magnetically stored data. Magnetic drum is a storage medium using the surface of a rotating magnetic drum to hold data. Magnetic core is the digital memory in which data is stored magnetically in individual cores operated by row and column select wires, with data obtained from sense wire.

Semiconductor memories are constructed out of semiconductor material using LSI and VLSI technologies. The examples of this type are Random Access Memory (RAM) and Read Only Memory (ROM) that are discussed in detail in the later sections of this chapter.

10.3 BASIC MEMORY STRUCTURE

A memory unit is a collection of storage cells with associated circuits needed to transfer information in and out of the device. It stores data or binary information in groups of bits called *words*. A word is an entity of bits that move in and out of the memory as a unit. Each word consists of a sequence of 0s and 1s. A word may represent a number, an instruction, one or more alphanumeric characters, or any other binary-coded information. Each word stored in a memory location is represented by an *address.* We know that a group of eight bits is called a *byte.* The capacity of a memory unit is the total number of bytes that can be stored.

Do You Know?	The place (location) in the memory device where any data values is stored is identified by another binary number referred to as an address. Each memory location has a unique address.

The communication between a memory and its environment is achieved through data input/output lines, address selection lines, and control lines that specify the direction of transfer. Fig. 10.1 shows a block diagram of the memory unit. It has *n* address lines to address or access 2^n locations in the memory. Also, it has *m* data input/output lines through which data is transferred in and out of the memory. Also, it has two control inputs called Read and Write. The write input causes binary data on I/O line to be transferred into the memory whereas the read input causes the binary data in a memory location to be transferred out of memory.

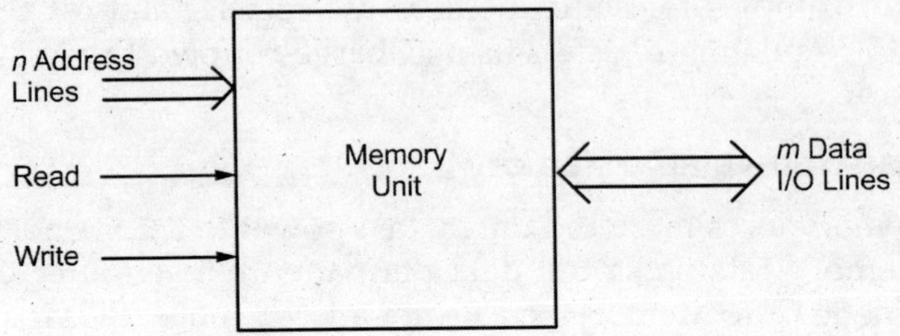

Fig. 10.1 Block diagram of memory unit

The memory unit is specified by the total number of words in it and the number of bits in each word. The address lines select one particular word. Each word in memory is assigned an identification number, called an *address*, starting from 0 and continuing with 1, 2, 3, up to $2^n - 1$, where *n* is the number of address lines. The selection of a specific word inside the memory unit is done by applying the *n*-bit binary address to the address lines. A decoder inside the memory accepts the address and opens the paths needed to select the specified word.

10.4 READ ONLY MEMORY (ROM)

A read only memory (ROM) is a semiconductor memory device used to store the information permanently. It performs only read operation and does not have a write capability. A ROM is programmed for a particular purpose during the manufacturing process and the user cannot alter its function. ROM circuits are typically used to provide the computer with resident programs and key operating functions needed to *boot* the operating system of the computer.

The ROM is a combinational logic circuit. It includes both the decoder and the OR gates within a single IC package. In order to minimize the number of address lines, decoders are used. The address of the desired line is given in binary. The connections between the output of the decoder and the input of the OR gates can be specified for each particular configuration. The ROM is used to implement complex combinational circuits within one IC package or as permanent storage package for binary information. The binary information must be specified by the designer and is then embedded in the unit to form the required interconnection pattern.

10.4.1 Architecture of ROM

The block diagram of a ROM is shown in Fig. 10.2. It consists of *n* address lines and *m* output lines. Each bit combination of the address variables is called an *address*. Each bit combination that comes out of the output lines is called a *data word*. Hence, the number of bits per word is equal to the number of output lines, *m*; an address is essentially a binary number that denotes one of the 2^n memory locations. An output word can be selected by a unique address; since there are 2^n distinct addresses in a ROM, there are 2^n distinct words that are said to be stored in the unit. The word available on the output lines at any instant depends on the address value applied to the input lines. A ROM is characterized by the number of words 2^n and the number of bits per word m; this terminology is used because of the similarity between the read only memory and the random access memory.

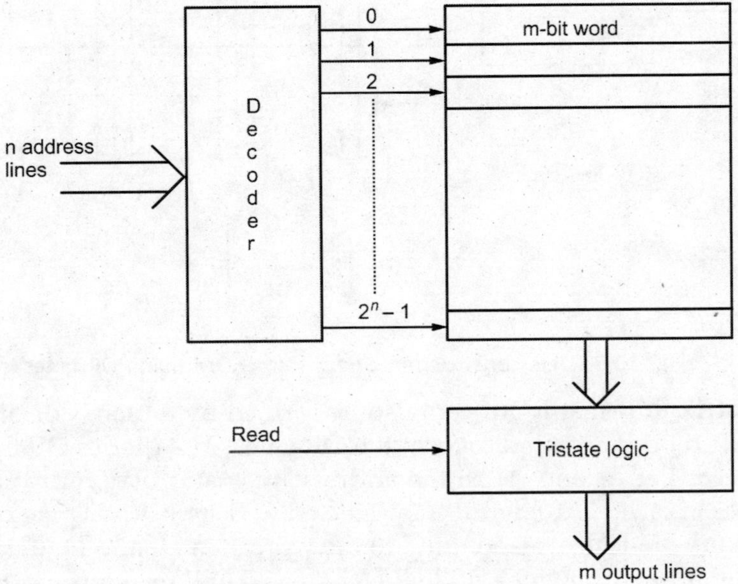

Fig. 10.2 Block diagram of ROM

Consider a 32 × 8 ROM. This unit consists of 32 words of 8 bits each. This means that there are eight (m) output lines and 32 distinct words in the unit, each of which may be applied to the output lines. There are only five (n) address lines in a 32 × 8 ROM because $2^5 = 32$; with five address variables, we can specify 32 memory locations. For each address, a unique word is selected. Thus, if the input address is 00010, word number 2 is selected and it appears on the output lines. If the input address is 10011, word number 19 is selected and applied to the output lines. Totally 32 addresses can select 32 words.

A ROM is sometimes specified by the total number of bits it contains, which is $2^n \times m$. For example, a 2048-bit ROM may be organised as 512 words of 4 bits each. This means that the unit has four (m) output lines and nine (n) address lines to specify 512 ($= 2^9$) words. The number of bits stored in the unit is 512 × 4 = 2048. The capacity of ROM varies from 256 bits to 256 KB. They are well suited to LSI manufacturing process.

32 × 4 ROM using OR gates: Internally, the ROM is a combinational circuit with AND gates connected as *decoders*, and the number of OR gates is equal to the number of output lines in the unit. The internal logic construction of a 32 × 4 ROM is shown in Fig. 10.3. The five input variables are decoded into 32 ($= 2^5$) lines by means of 32 AND gates and 5 inverters. Each one of the 32 addresses selects one and only one output of the decoder. The 32 outputs of the decoder are connected through fuses to each OR gate. Only four of these fuses are shown in Fig. 10.3, but actually each OR gate has 32 inputs and each input of the OR gate goes through a fuse that can be blown as desired.

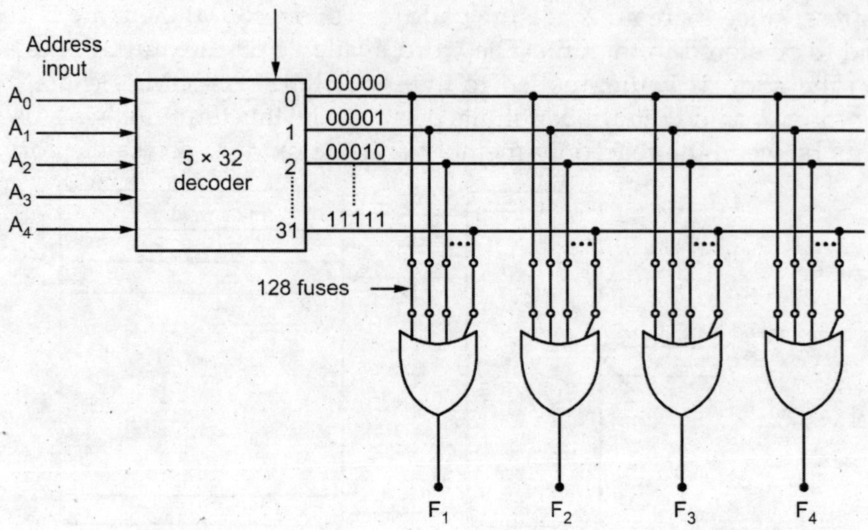

Fig. 10.3 Logic construction of a 32 × 4 ROM using OR gates

Diode matrix ROM: A ROM can also be formed by a matrix of diodes as shown in Fig. 10.4. Here the diodes are used to form the OR gates. The diodes OR the inputs to which they are connected. Let A_0 and A_1 be the address lines and they form the two-bit address bus. For example, if A_1A_0 is 10, word 2 is selected and its NAND gate output goes LOW. This will enable the diodes corresponding to the word 2 in D_2 and D_2 data lines to conduct, resulting in 001 in the column lines, and when inverted will result in a 110 at the output data bus. Thus, memory location 2 contains a 6.

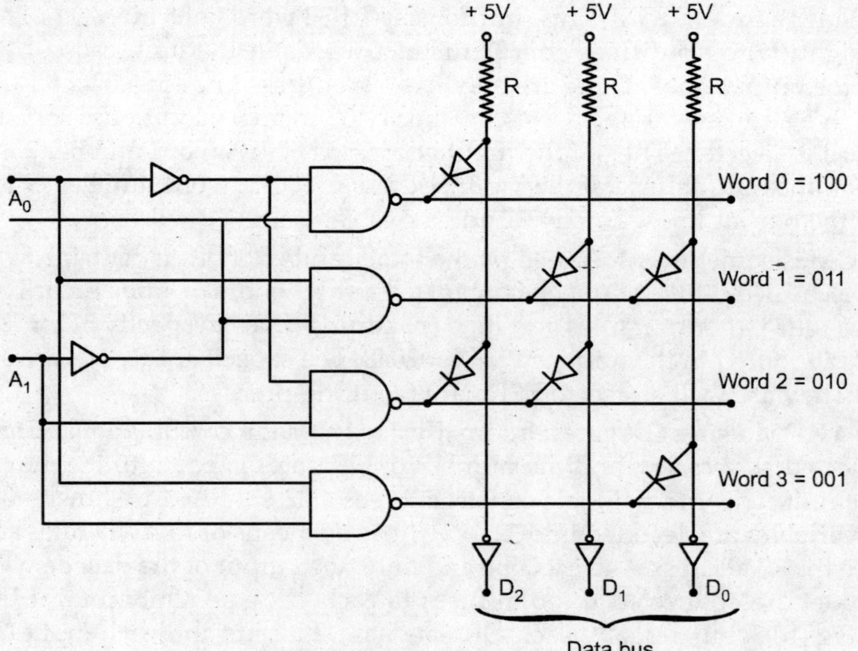

Fig. 10.4 Diode matrix ROM

10.4.2 Types of ROM

Semiconductor ROMs are manufactured with *bipolar technology* or with *MOS technology*. ROMs can be classified into several types, which differ as to how information is written or *programmed* into the memory storage locations. Fig. 10.5 shows how ROMs are categorised. The bipolar ROMs can be subdivided into Mask ROMs and Programmable ROMs (PROMs). The MOS ROMs can be divided into Mask ROMs, Programmable ROMs and Electrically Erasable Programmable ROMs (EEPROMs).

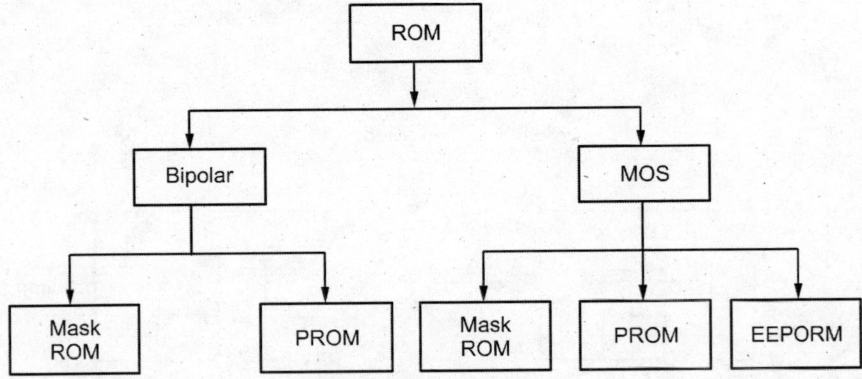

Fig. 10.5 The ROM family

Bipolar ROMs use bipolar transistor within an integrated circuit much in the same way as the diode matrix. The bipolar ROMs can be programmed. MOS ROM is constructed using MOSFETs. In a MOS cell, the silicon gate is completely insulated from the *n*-type substrate. The MOS can also be programmed in two ways. One is by placing charge on the gate at the time of manufacture.

10.4.3 Programming Mechanisms

ROMs may be programmed in two different ways. The first is called *mask programming* and is done by the manufacturer during the last fabrication process of the unit. The second type of ROM is, *programmable read only memory* (PROM).

Mask-programmed ROM: Mask ROM is permanently programmed during the manufacturing process. It is actually programmed by the manufacturer according to the customer's applications. Once the memory is programmed, it cannot be changed. Most IC ROMs utilize the presence or absence of a transistor connection at a ROW/COLUMN junction to represent 1 or 0. The particular 1s or 0s are obtained by providing a mask in the last fabrication step. A photographic negative called a mask is used to control the electrical interconnections on the chip.

Fig. 10.6(*a*) shows bipolar ROM cells. The presence of a connection from a ROW line to the base of a transistor represents a 1 at that location because, when ROW line is taken HIGH, all transistors with base connection to that ROW line turn ON and connect the HIGH to the associated COLUMN lines. When there are no base connections at ROW or COLUMN junctions, the COLUMN lines remain LOW (0) when the ROW is addressed.

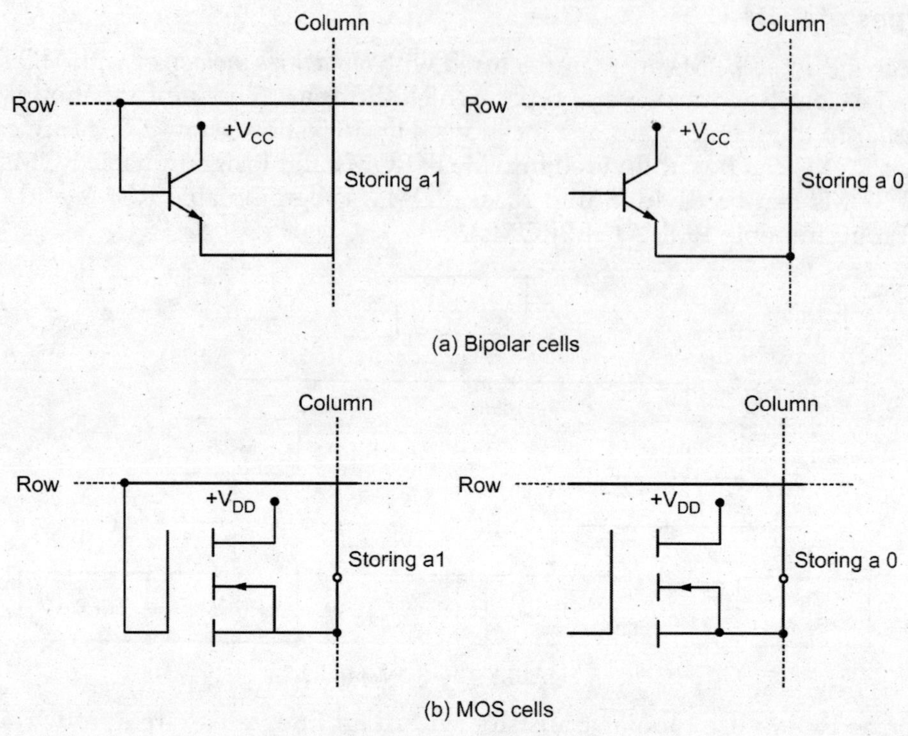

Fig. 10.6 ROM cells

The illustration of MOS ROM cells is given in Fig.10.6(*b*). The presence or absence of a gate of MOSFET connection at a junction stores 1 or 0 permanently.

The manufacturer makes the corresponding mask for the paths to produce the 1s and 0s according to the customer's truth table (specifications). This process is called custom or mask programming. Mask programmed ROMs are economical only if large quantities of the same ROM configuration are to be manufactured.

A major disadvantage of this type of ROM is the fact that it cannot be reprogrammed in the event of a design change requiring a modification of the stored program. Several types of user-programmable ROMs have been developed to overcome this disadvantage.

10.4.4 Organization of a Simple ROM

A 16 × 8-bit simplified ROM array is shown in Fig. 10.7. This ROM is organized into 16 addresses, each of which stores 8 data bits. The total capacity of this ROM is 128 bits. The dark squares in this diagram represent stored 1s by means of base-connected transistor or gate-connected MOSFET, and the light squares represent stored 0s. When a 4-bit binary address is applied to the address inputs, the corresponding ROW line becomes HIGH. This HIGH is connected to the COLUMN line through the transistor at each junction where 1 is stored. The column line stays LOW at each cell where 0 is stored because of the terminating resistor. The COLUMN lines form the data output. Thus, the eight data bits stored in the selected ROW appear on the output lines.

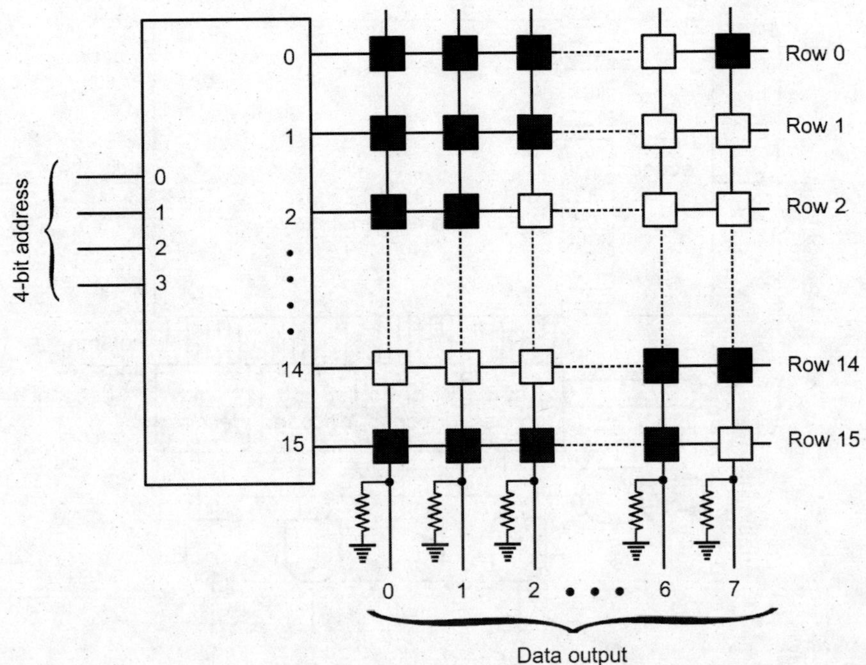

Fig. 10.7 A 16 × 8-bit ROM array

10.4.5 ROM ICs

Some of the ROM ICs are discussed in the following sections.

IC 74187-1024-bit ROM: A typical MSI/TTL ROM IC 74187 is shown in Fig. 10.8. It is a 16-pin, 1024-bit, 256 words of 4 bits each ROM. Although the 256 × 4 organization of this device implies that there are 256 rows and 4 columns in the memory, the memory cells are organized in a 32 × 32 matrix. Word selection of this memory is accomplished by means of an 8-bit address or select lines with the *A* being the least significant bit, progressing alphabetically through the select inputs to *H*, which is the most significant bit. The word-select table of this ROM is described in Table 10.1.

The working of ROM is as follows: Five of the eight address lines (*H* to *D*) are decoded by the ROW (*Y*) decoder to select one of the 32 rows. Three of the eight address lines (*C* to *A*) are decoded by the COLUMN (*X*) decoders to select four of the 32 columns. The *X* decoder consists of 1-of-8 decoders as shown in Fig. 10.8(*a*). The pinout diagram of 74187 is shown in Fig. 10.8 (*b*).

When an eight-bit address code (*A* to *H*) is applied, a four-bit word appears on the data outputs (Y_4, Y_3, Y_2, Y_1), when the ENABLE inputs $\overline{S}_1$ and $\overline{S}_2$ are LOW.

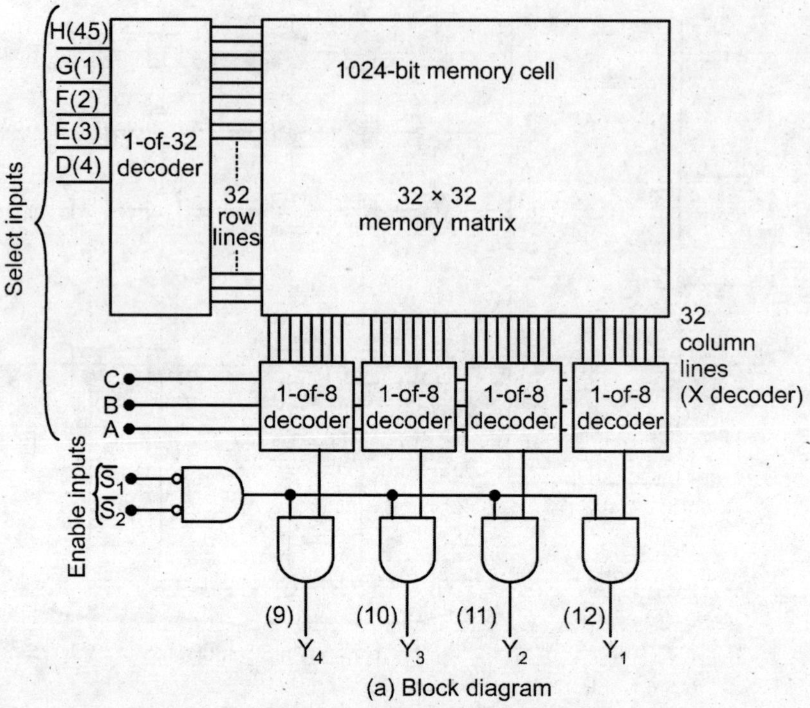

(a) Block diagram

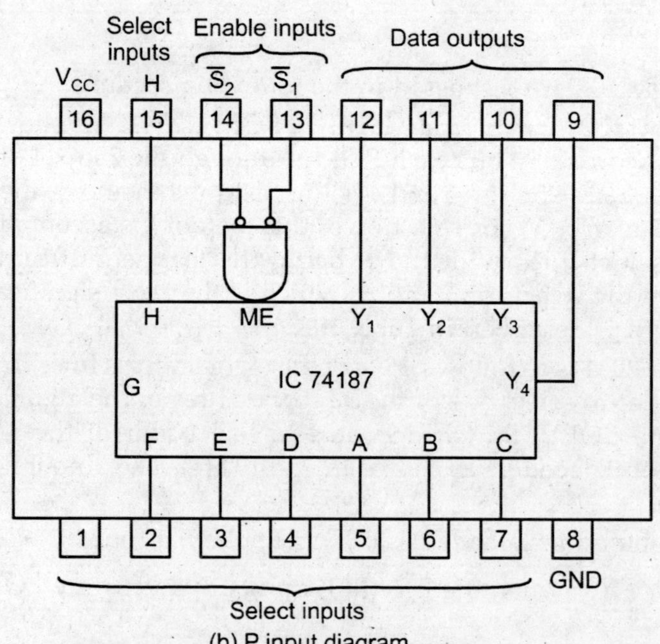

(b) P input diagram

Fig. 10.8 IC 74187-1024-bit ROM

Table 10.1 Word-select table for IC 74187

Word	Inputs							
	H	*G*	*F*	*E*	*D*	*C*	*B*	*A*
0	0	0	0	0	0	0	0	0
1	0	0	0	0	0	0	0	1
2	0	0	0	0	0	0	1	0
3	0	0	0	0	0	0	1	1
4	0	0	0	0	0	1	0	0
5	0	0	0	0	0	1	0	1
6	0	0	0	0	0	1	1	0
7	0	0	0	0	0	1	1	1
8	0	0	0	0	1	0	0	0
⋮	Words 9 through 250 are omitted for simplicity							
251	1	1	1	1	1	0	1	1
252	1	1	1	1	1	1	0	0
253	1	1	1	1	1	0	0	1
254	1	1	1	1	1	1	1	0
255	1	1	1	1	1	1	1	1

IC 7488-256-bit bipolar ROM: The logic symbol of an IC 7488 bipolar ROM is shown in Fig. 10.9. It is a 256-bit ROM and is organised as a 32 × 8 memory. It has 32 address locations, each of which has eight bits of storage. A five-bit address on inputs *A* through *E* selects one of the 32 (0 through 31) locations, when the enable input ($\overline{S}$) is asserted low and places the selected data byte on the 8 outputs (Q_1 through Q_8).

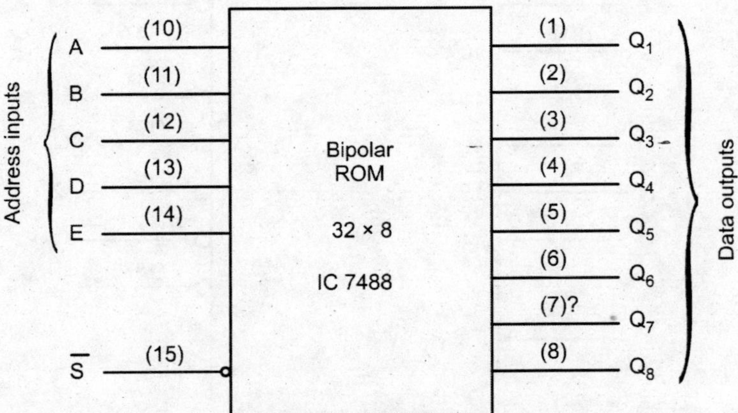

Fig. 10.9 Logic symbol for IC 7488-256-bit bipolar ROM

TMS 47256-32 KB MOS static ROM: The IMS 47256 is a large-capacity MOS static ROM. It is a 2,62,144-bit MOS ROM and is organised as a 32768 × 8(32 K × 8) memory. It has

32,768 address locations with eight bits at each location. The logic symbol of this device is shown in Fig. 10.10. Since $2^{15} = 32768$, a 15-bit address on input *A* through *O* selects one of the 32768 (0 through 32767) memory locations. There is an active-LOW output enable input ($\overline{S}$) and a chip enable/power down input ($\overline{E}$). $\overline{S}$ and $\overline{E}$ must be LOW to enable the memory input. When $\overline{E} = 1$, the device is put into a low power standby mode, which reduces the current drain on the d.c. power supply.

10.4.6 ROM Access Time

The ROM access time, t_a, is the time from the application of a valid address code on the address inputs until the appearance of valid output data. It can also be measured from the activation of *chip select* ($\overline{CS}$) to the occurrence of valid output data when a valid address is already on the inputs. A typical diagram that illustrates ROM access time is shown in Fig. 10.11.

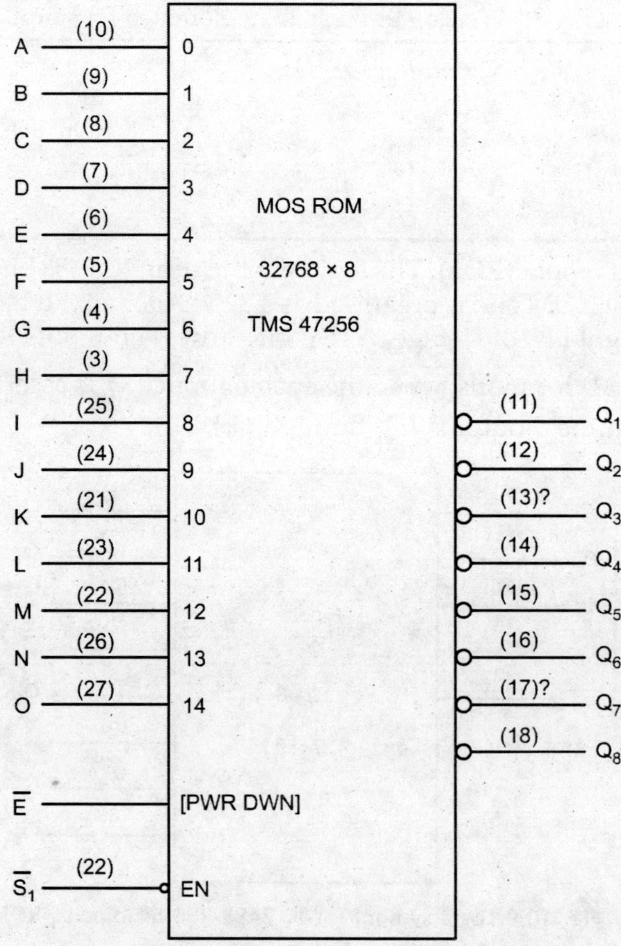

Fig. 10.10 Logic symbol of TMS 47256-32 KB MOS ROM

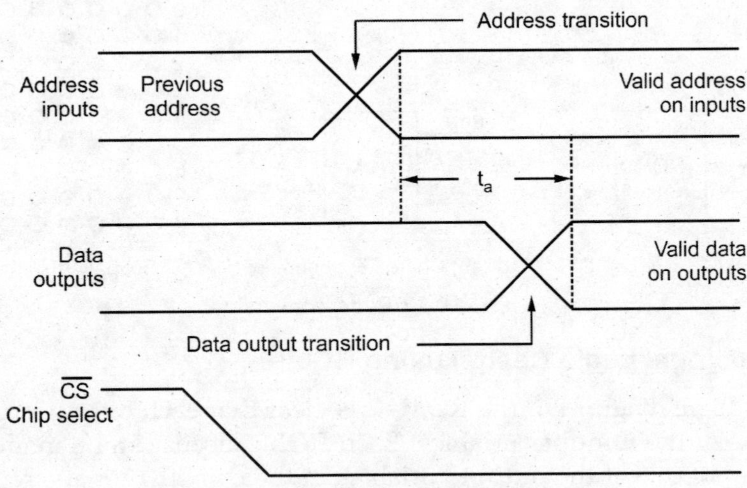

Fig. 10.11 Diagram of ROM access time

10.4.7 Applications of ROM

1. ROMs are used for a variety of tasks within a digital system. They can be used as a direct substitute for any random logic of AND, OR and NOT gates.
2. ROMs are used to store bootstrap program that loads operating system program available in secondary memory and language interpreters in personal and business computers and to store, monitor or control programs in microcomputer and microprocessor based systems like electronic games, electronic cash registers, electronic scales and microcomputer controlled automobile fuel injection.
3. A very significant application of MOS ROM is for character generation. This includes display control for moving billboards and light-emitting diode arrays. The TMS 4100 is a family of MOS ROMs organised to function primarily as a character generator. TMS 4100 family ROMs with a capacity of 2,240 bits, mounted in a 28-pin package with the following two different organisations, are available:
   (*i*) 64 words of 35 bits (5 by 7)
   (*ii*) 32 words of 70 bits (5 by 14)
4. ROMs can also be used for code conversion. In a ROM for ASCII to EBCDIC conversion as shown in Fig. 10.12(*a*), if ASCII data is placed on the address bus, the data bus will contain extended binary-coded decimal interchange code (EBCDIC). ROMs can also convert an ASCII data and row number into a dot pattern as shown in Fig. 10.12(*b*) & (*c*). Address lines A_0 to A_3 are used to select the row of dots, and address lines $A_4 - A_{10}$ are used to select the character that is to be displayed.
5. A ROM and a DAC can be used to generate sine waves, sawtooth waves, triangular waves and square waves.
6. A ROM can be used to implement any or a set of logic expressions and is therefore helpful in the design of combinational circuits.

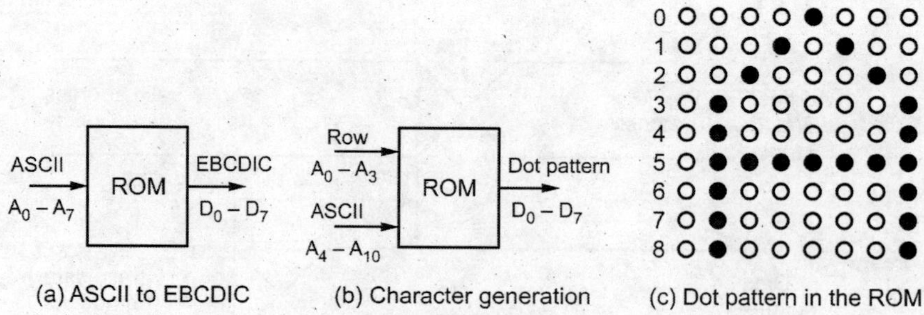

(a) ASCII to EBCDIC (b) Character generation (c) Dot pattern in the ROM

Fig. 10.12 Code conversion

10.4.8 Combinational Logic Design using ROM

From the logic diagram of the ROM, it is clear that each output provides the sum of all the minterms of the *n* input variables. Each ROM output can be made to represent the Boolean function of one of the output variables in the combinational circuit.

When a combinational circuit is implemented by means of a ROM, the output functions must be expressed in sum of minterms from its truth table. The advantage of a ROM is in the implementation of complex combinational circuits.

When the ROM array is used to implement combinational logic, the gate reduction of the design is not important. This is clearly explained in the following examples.

Example 10.1. *Design a Binary-to-Gray code converter similar to basic ROM structure shown in Fig. E10.7.*

Solution: The conversion of binary code into gray code is shown in Table E10.1.

Table E 10.1 Binary-to-gray code conversion

Binary				Gray			
B_3	B_2	B_1	B_0	G_3	G_2	G_1	G_0
0	0	0	0	0	0	0	0
0	0	0	1	0	0	0	1
0	0	1	0	0	0	1	1
0	0	1	1	0	0	1	0
0	1	0	0	0	1	1	0
0	1	0	1	0	1	1	1
0	1	1	0	0	1	0	1
0	1	1	1	0	1	0	0
1	0	0	0	1	1	0	0
1	0	0	1	1	1	0	1
1	0	1	0	1	1	1	1
1	0	1	1	1	1	1	0
1	1	0	0	1	0	1	0
1	1	0	1	1	0	1	1
1	1	1	0	1	0	0	1
1	1	1	1	1	0	0	0

The ROM programmed as a binary-to-gray-code converter is shown in Fig. E 10.1. In this figure, the black and white cells represent the presence of 1 or 0 in it respectively. A binary code placed on $B_3B_2B_1B_0$ produces the corresponding gray code $G_3G_2G_1G_0$ on the output (COLUMNS) lines.

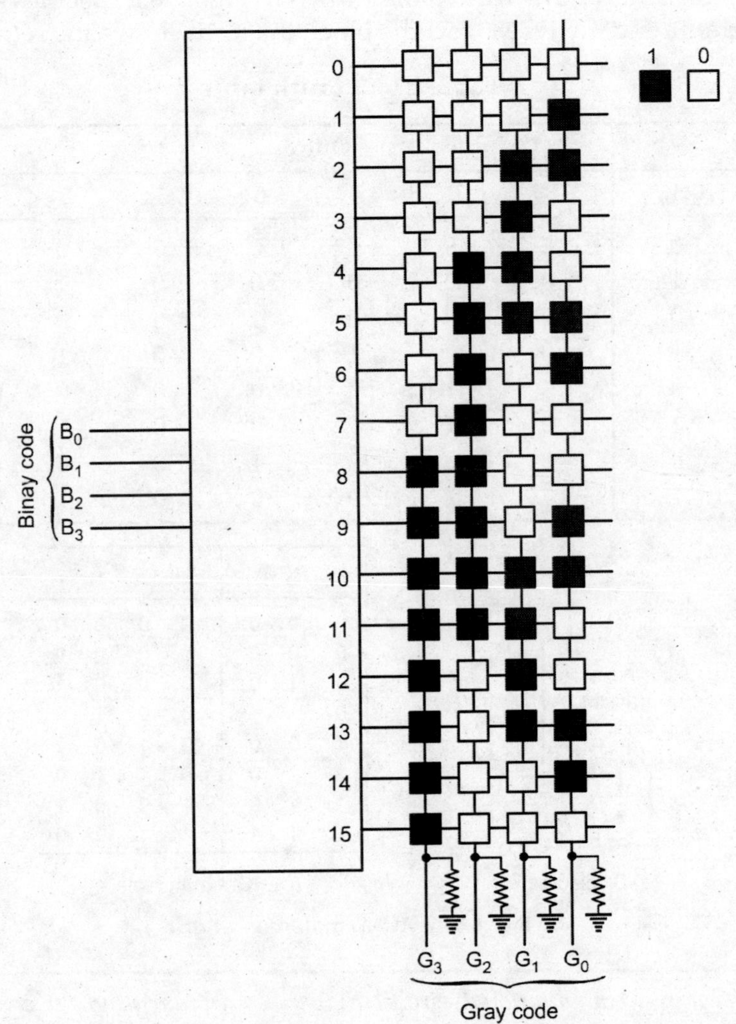

Fig. E 10.1 ROM programmed as a binary-to-gray code converter

Example 10.2. *Design a combinational circuit that accepts a 3-bit number as input and generates an output binary number equal to the square of the input number using ROM.*

Solution: Let us first derive the truth table for the combinational circuit. Table E10.2 is the truth table for the combinational circuit. We require 3 inputs and 6 outputs to accommodate all possible numbers. Note that output B_0 is always equal to input A_0. Therefore, there is no need to generate B_0 with a ROM since it is equal to an input variable. Further, as the output

B_1 is always 0, this output is always known. Hence, only four outputs ($B_2B_3B_4B_5$) are to be generated. Therefore, the minimum-size ROM needed must have three inputs and four outputs. As three inputs specify eight words of 4 bits each, the ROM size must be 8 × 4. The implementation of ROM is shown in Fig. E10.2. The other 2 outputs of the combinational circuit are equal to 0 and A_0. The truth table shown in Fig. E10.2 specifies all the information needed for programming the ROM, and the block diagram shows the required connections.

Table E10.2 Truth table

Inputs			Outputs						Decimal
A_2	A_1	A_0	B_5	B_4	B_3	B_2	B_1	B_0	
0	0	0	0	0	0	0	0	0	0
0	0	1	0	0	0	0	0	1	1
0	1	0	0	0	0	1	0	0	4
0	1	1	0	0	1	0	0	1	9
1	0	0	0	1	0	0	0	0	16
1	0	1	0	1	1	0	0	1	25
1	1	0	1	0	0	1	0	0	36
1	1	1	1	1	0	0	0	1	49

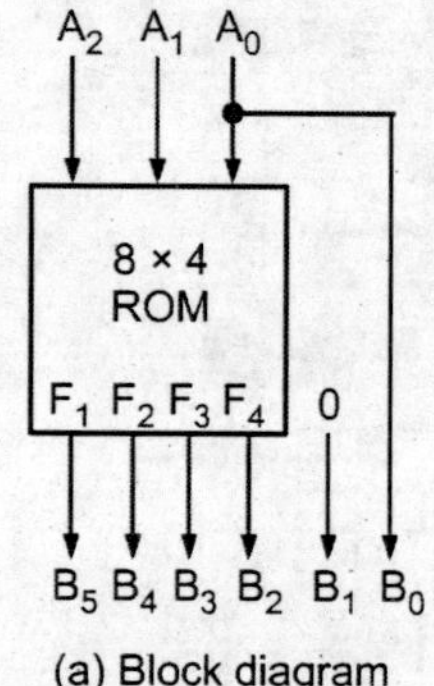

(a) Block diagram

A_2	A_1	A_0	F_1	F_2	F_3	F_4
0	0	0	0	0	0	0
0	0	0	0	0	0	0
0	1	0	0	0	0	1
0	1	1	0	0	1	0
1	0	0	0	1	0	0
1	1	0	0	1	1	0
1	1	0	1	0	0	1
1	1	1	1	1	0	0

(b) ROM truth table

Fig. E10.2 ROM implementation

Example 10.3. *Design a combinational circuit that gives a binary output equal to the square of binary-coded decimal numbers 0 through 9 using diode matrix ROM.*

Solution: The truth table for the given square function is shown in Table E10.3, and this can be implemented using diode matrix ROM with four inputs and seven outputs as shown in Fig. E10.3. Here, the presence of the diode indicates 1, and its absence indicates 0.

From the above truth table, the outputs Y_6 to Y_0 can be expressed as SOP expressions as shown below:

$$Y_6 = \Sigma_m (8, 9);\ Y_5 = \Sigma_m (6, 7);\ Y_4 = \Sigma_m (4, 5, 7, 9);$$
$$Y_3 = \Sigma_m (3, 5);\ Y_2 = \Sigma_m (2, 6);\ Y_1 = 0;\ Y_0 = \Sigma_m(1, 3, 5, 7, 9)$$

Here, the output Y_6 is the sum of minterms corresponding to decimal codes 8 and 9. This summation is done by means of two diodes in the Y_6 column shown in Fig. E10.3. Similarly, output Y_0 is the sum of minterms corresponding to decimal codes 1, 3, 5, 7 and 9, and this is implemented by the presence of five diodes at the intersection of Y_0 column and the corresponding rows. Also, it is interesting to note that $Y_1 = 0$, and hence there is no diode in Y_1 column.

Table E10.3 Truth table

Decimal number	BCD code				Square of number						
	D	C	B	A	Y_6	Y_5	Y_4	Y_3	Y_2	Y_1	Y_0
0	0	0	0	0	0	0	0	0	0	0	0
1	0	0	0	1	0	0	0	0	0	0	1
2	0	0	1	0	0	0	0	0	1	0	0
3	0	0	1	1	0	0	0	1	0	0	1
4	0	1	0	0	0	0	1	0	0	0	0
5	0	1	0	1	0	0	1	1	0	0	1
6	0	1	1	0	0	1	0	0	1	0	0
7	0	1	1	1	0	1	1	0	0	0	1
8	1	0	0	0	1	0	0	0	0	0	0
9	1	0	0	1	1	0	1	0	0	0	1

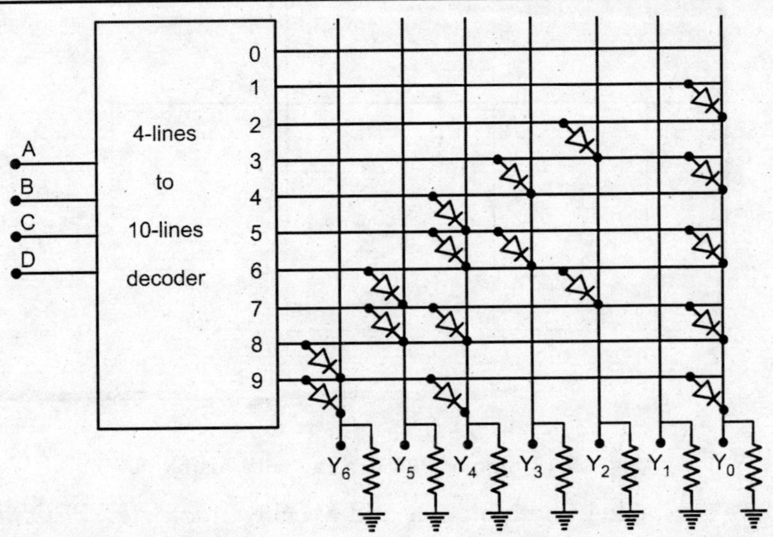

Fig. E10.3 Implementation of square function using diode matrix ROM

10.4.9 Programmable ROM (PROM)

In order to provide some flexibility in the possible applications of ROM, programmable ROMs (PROMs) have been introduced. The PROM can be programmed electrically by the user but cannot be reprogrammed. In a PROM chip, the manufacturer includes a connection at every intersection of the grid of address and data lines. PROMs are widely used in the control of electrical equipment such as washing machines and electric ovens.

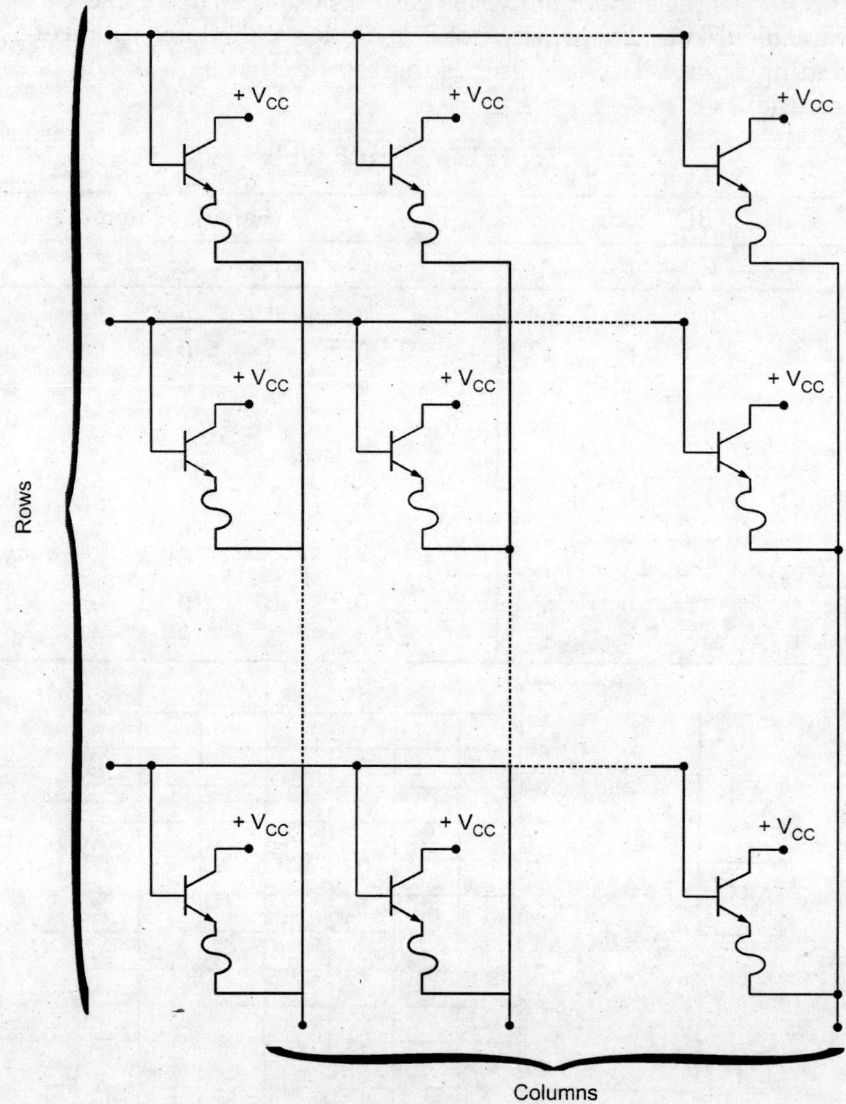

Fig. 10.13 Bipolar PROM array with fusible links

PROMs are available in both bipolar and MOS technologies. PROMs have 4-bit or 8-bit output word formats with capacities ranging in excess of 2,50,000 bits. A PROM is manufactured as a generalised integrated circuit with all the matrix intersections linked by *fusible* diodes or transistors. In series with each connection, the manufacturer includes a fusible link which can be melted, and thereby opened, by passing a large current through it. A memory link is fused open or left intact to represent a binary number 0 or 1. The user can selectively *burnout* some of the fuses by passing enough current through them and can thereby program the PROM according to the required truth table. Once a PROM is programmed, it cannot be changed, and therefore it has to be done carefully and correctly in the first time itself. Hence, the fusing process is irreversible.

Fuse technology used in PROM: A bipolar PROM array with *fusible links* is illustrated in Fig. 10.13. In PROM, the fuse links are placed between the emitter of each cell's transistor and its column line. The cell type is called a *fuse cell.*

The following are the three basic fuse technologies used in PROMs:

(*i*) metal links

(*ii*) silicon links

(*iii*) *p-n* junctions.

Metal links are made of a Nichrome material. Each bit in the memory array is represented by a separate link. The metal link is either "blown" open or left intact during programming. This is done basically by addressing a given fuse cell and then forcing sufficient amount of current through the link to make it open.

Silicon links are formed by narrow notched strips of polycrystalline silicon. Programming of silicon links requires melting of the links by passing enough amount of current (20-30 mA) through them. The current causes a high temperature (1400°C) at the fuse location, and the silicon gets oxidized and forms an insulation around the new open link.

***p-n* junctions:** This technology, also referred to as shorted junction or avalanche-induced migration, consists of two *p-n* junctions arranged back to back as shown in Fig. 10.14(*a*). During programming, the *p-n* junction of diode D_1 is avalanche reverse-biased. The sudden heavy flow of electrons in the reverse direction and heat cause aluminium ions to migrate and short the junction of the emitter-base. The current is of the order of 200-300 mA. The other *p-n* junction is forward-biased, and this diode can be used to represent a data bit.

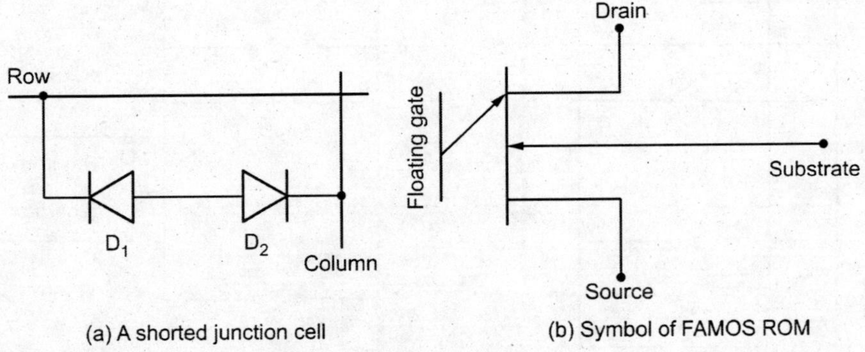

(a) A shorted junction cell

(b) Symbol of FAMOS ROM

Fig. 10.14

The above fuse technologies are irreversible. These technologies do not work with MOS memory devices where high current levels required for the fusing process are incompatible with MOS impedance levels. The following are the MOS technologies used for the fabrication of programmable memories:

(*i*) Floating Gate Avalanche Injection MOS (FAMOS) and

(*ii*) MAOS.

FAMOS ROM: This is the storage device used in silicon gate MOSFET with no electrical connection to the gate, i.e. the gate is electrically floating in an insulating layer of silicon dioxide. The symbol of FAMOS ROM is shown in Fig. 10.14(*b*).

MAOS PROM: A MOS memory cell using alumina (Al_2O_3) as gate dielectric is called a MAOS memory element. This gate dielectric is used for charge storage and provide a reprogramming feature. The MAOS element is programmed by the application of a positive or negative gate voltage pulse above a threshold level, as per the specifications of that PROM.

10.4.10 PROM Programming

A PROM is programmed by plugging it into a special device called *PROM Programmer*. The programming is accomplished by using a simplified set up shown in Fig. 10.15. An address is selected by the switch settings on the address lines (0 through *m*) and then a pulse is applied to those output lines corresponding to bit locations where 1s are to be stored. The PROM actually starts out with all 0s. These current pulses blow the fuse links, thus creating the desired bit pattern. The next address is then selected and the process is repeated. This sequence is done automatically by the PROM programmer.

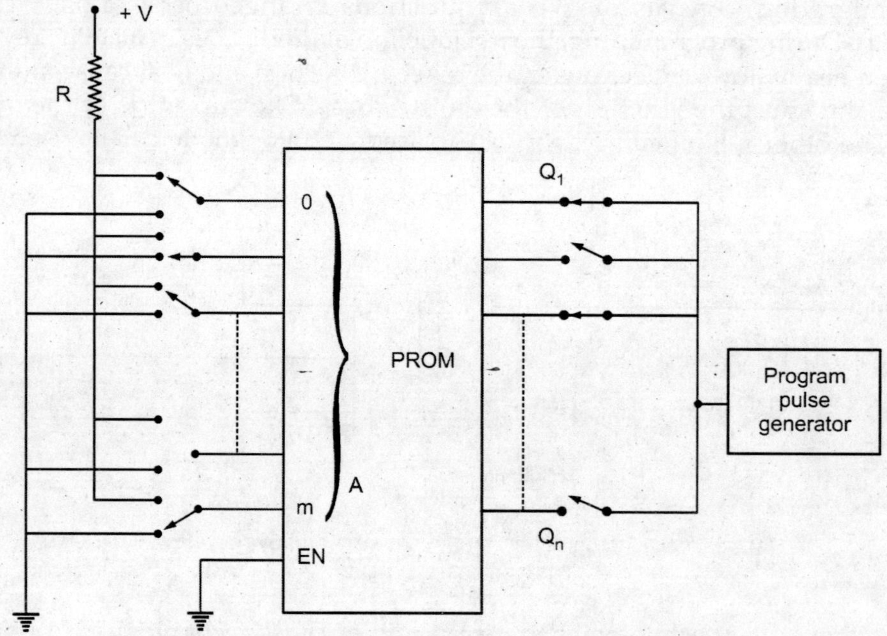

Fig. 10.15 Simplified PROM programming set up

IC 74186-512-bit PROM: The IC 74186 is a TTL LSI 512-bit PROM. It is organised as 64 words of 8 bits each. The logic diagram of this IC PROM is shown in Fig. 10.16(*a*). The 6 bits of address (*A, B, C, D, E* and *F*) are decoded to uniquely select one of the 64, 8-bit words in a linear addressing scheme.

The chip is initially stored with full of 0s, i.e. the contents of any 8-bit word is 0000 0000. We know that the programming is done by applying current pulse to each output terminal where a logic 1 appears. The IC 74186 PROM is programmed using the following procedure as shown in Fig. 10.16(*b*).

1. First of all, apply the correct address (*ABCDEF*) for the word to be programmed. For instance, if the desired contents of word 29 is 1001 1011, the address would be 0001 1101. A 0 is a closed switch and a 1 is an open switch.
2. Apply a current pulse to each bit to store a 1.
3. Repeat the above steps for all words to be stored in the memory.

Specific information such as current pulse limits, voltage levels and so on for chip (IC) programming must be taken from the manufacturer's specification sheets.

Once the field programming is done, it cannot be altered.

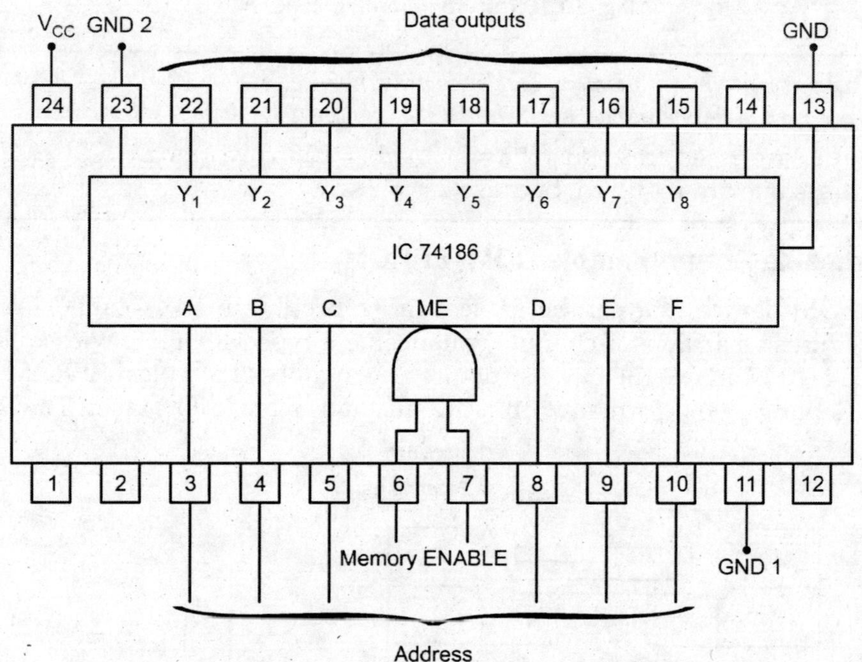

Fig. 10.16 (*a*) Logic and pinout diagrams

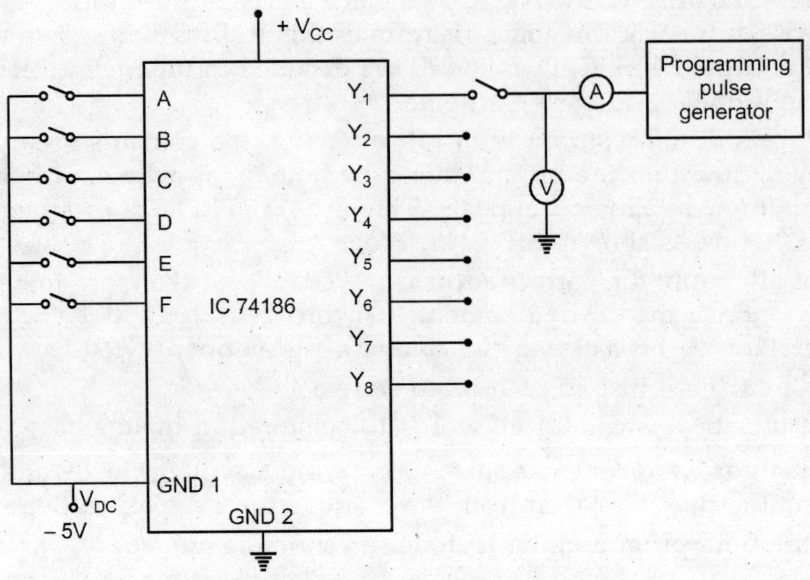

Fig. 10.16 (*b*) Programming of IC 74186

Example 10.4 (*a*) *How many memory locations are addressed using 18 address bits?* (b) *How many address bits are needed to operate a 2K × 8-bit ROM memory?*

Solution: (*a*) Using 18 address bits, $2^{18} = 262{,}144$ (= 256K) words are addressed. (*b*) For 2K, only 11 address bits are required, because 2^{11} = 2K.

10.4.11 Erasable Programmable ROM (EPROM)

A PROM device that can be erased and reprogrammed is called Erasable PROM (EPROM). It uses an array of *n*-channel enhancement type MOSFETs with an insulated gate structure. Fig. 10.17 shows the basic structure and symbol of a typical EPROM cell. Here, an additional floating gate is formed within the silicon dioxide (SiO_2) layer. The floating gate is

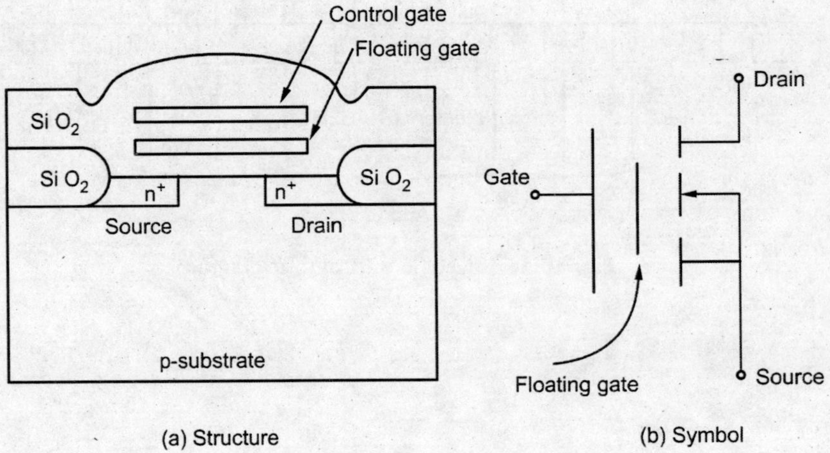

Fig. 10.17 EPROM cell

left unconnected while the normal control gate is connected to the row decoder output of EPROM. The data bits are represented by the presence or absence of a stored charge. The initial values of unprogrammed EPROM cells may be all 0s or all 1s.

10.4.12 Programming of EPROM

Consider the programming of an EPROM with 1s as initial values in all the cells. To program or store a 0 in such a cell, the floating gate must be charged. For this, a high voltage of about 16 to 20 V is applied between the source and drain, and a voltage of about 25 to 50 V is applied to the control gate for a specified amount of time (typically 50 ms per address location). Due to the high electric field established by the positive control gate voltage, the high energy electrons penetrate the thin insulating SiO_2 and reach the floating gate. Thus, the charge is stored on the floating gate. Since more negative charge accumulates on the floating gate, the electric field strength is reduced, and thereby further accumulation is inhibited. Programming actually involves selecting the desired cell gate and repeatedly injecting charge onto the floating gate until a sufficient amount of charge is trapped. Since the gate is surrounded by SiO_2, there is no discharge path available. Therefore, the charge remains trapped on the floating gate for an indefinite period of time. Now, the cell is programmed for a logic 0.

To program a different data, all cells in the EPROM must be erased. This is done by illuminating the cells by a strong ultraviolet (UV) light having a wavelength typically 253.7 nm for about 20 minutes. Now, the electrons trapped on the floating gate acquire sufficient energy from the UV light and escape through the SiO_2 layer to the substrate.

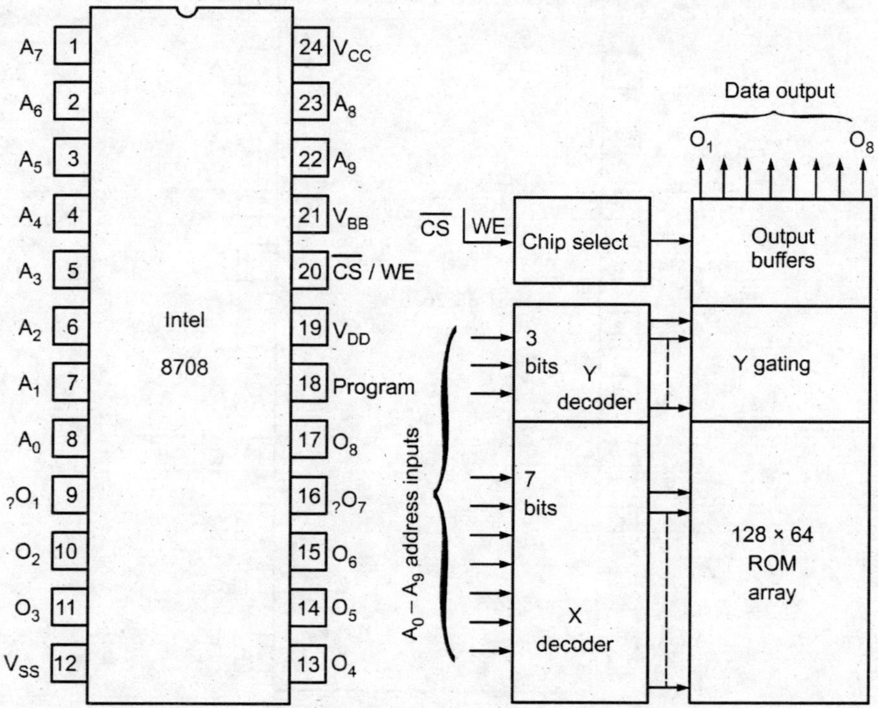

Fig. 10.18 IC 8708-1 KB EPROM

EPROMs are provided with a transparent quartz window on the top of the chip to allow the UV rays for erasing the data. All the contents of the memory locations in EPROM become erased by exposure to ultraviolet rays. The EPROM window should be covered with an opaque sticker to protect the memory from unwanted exposure to UV rays from sunlight and fluorescent lamps. The new data can be stored by programming electrically. EPROMs can be erased and reprogrammed as often as desired. Thus, the EPROM is ultraviolet-erasable and electrically programmable. An EPROM must be programmed with a desired data before it is connected in a circuit. Programming is done byte by byte using a set of switches for setting data bits as well as address bits.

EPROM ICs: Manufacturers fabricate ROMs, PROMs and EPROMs that can store thousands of words. The IC 2764 is a 8 KB or 65536-bit EPROM organised as 8192 words of 8 bits each. It has 13 address lines and 8 data lines.

IC 8708 -1 KB EPROM: The IC Intel 8708 is a MOS EPROM. It has 8192 bits that are arranged as 1024 words of 8 bits each. The logic diagram of Intel 8708 is shown in Fig. 10.18. The memory is arranged as a rectangular array of 8,192 cells having 128 rows and 64 columns. The *X* decoder uses 7 address bits to select one of the 128 rows, while the *Y* decoder uses 3 address bits to select one of the eight groups of 8 bits from the 64 columns. The 8 data bits are available at the outputs through three-state buffer amplifiers.

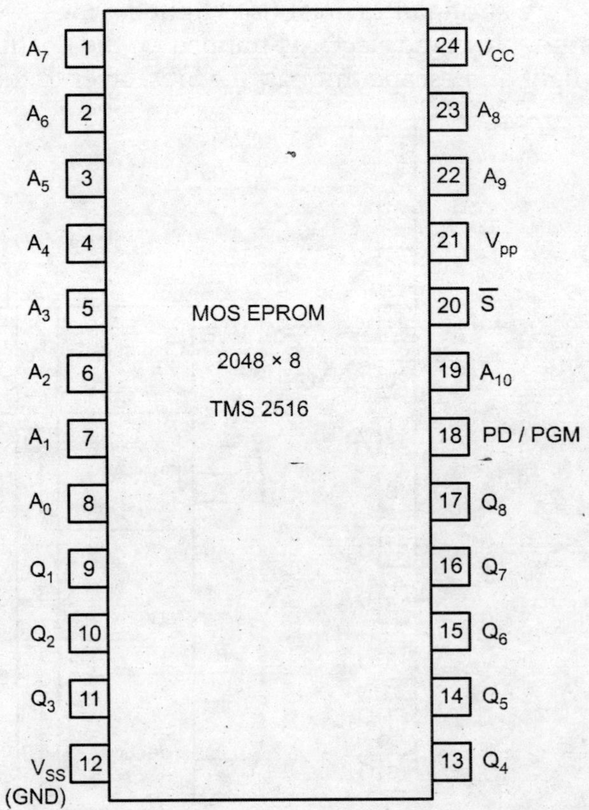

Fig. 10.19 Pin diagram of TMS 2516-2 KB MOS EPROM

IC TMS 2516-2 KB MOS EPROM: The TMS 2516 is a MOS EPROM device and its pin diagram is shown in Fig. 10.19. This device has 2048 addresses ($2^{11} = 2048$), each with 8 bits (three-state).

To *read* from the memory, the select input ($\overline{S}$) and the power down program (*PD/PGM*) input must be LOW. To program the device, + 25V dc is applied to V_{PP} and $\overline{S}$ is made HIGH. The 8 data bits to be programmed into a given address are applied to the outputs (Q_1 through Q_8) and the address is selected on inputs A_0 through A_{10}. A HIGH level pulse of width 10 ms to 55 ms is applied to the *PD/PGM* input. To *erase* the stored data, the device is exposed to a 12 mW/cm^2 filterless UV lamp through the transparent lid.

By adding three-state switches to the data lines of a memory, we can get a three-state output. Most of the commercially available ROMs, PROMs and EPROMS have three-state outputs. ROMs, PROMs and EPROMs are *non-volatile* memories.

Access time: The access time is the time taken to read a stored word after applying the address bits. Bipolar memories have faster access time than MOS memories. The Intel 8708 is a MOS EPROM with an access time of 450 ns. The Intel 2716 is also a MOS EPROM with an access time of 450 ns. The 3636 is a bipolar PROM with an access time of 80 ns. The Intel 2732 is a 32K (4K × 8) EPROM that is pin-compatible with the 2716—it simply has twice the memory storage. The 2764 is a 64K (8K × 8) EPROM.

Disadvantages of EPROM:

(1) Changes in the selected memory locations cannot be made in the reprogramming. The entire memory should be erased before reprogramming.

(2) The process of reprogramming cannot take place with the IC in the circuit. The EPROM IC must be removed from the circuit and the stored program can be erased by exposing the memory cells to ultraviolet light through a "window" on the IC package. This process takes about half an hour.

10.4.13 Electrically Erasable Programmable ROM (EEPROM)

Another type of reprogrammable ROM device is EEPROM (Electrically Erasable Programmable ROM), which is also known as Electrically Alterable Programmable ROM (EAPROM). The EEPROM overcomes the disadvantages of EPROM. EEPROM can be erased and programmed by the application of controlled electric pulses to the IC in the circuit, and thereby changes can be made in the selected memory locations without disturbing the correct data in other memory locations. EEPROM is non-volatile like EPROM but does not require ultraviolet light to be erased. The non-volatility of EEPROM permits a system to be immune to power interruptions.

EEPROM is a rugged, low power semiconductor device and it occupies less space. It has the advantages of program flexibility, small size and semiconductor memory ruggedness, i.e. low voltages and no mechanical parts. The requirement of low power supports field programming in portable devices for communication encoding, data formatting and conversion, and program storage. With EEPROM, the programs can be altered remotely, possibly by telephone.

10.5 RANDOM ACCESS MEMORY (RAM)

A Random Access Memory (RAM) is a volatile chip memory in which both read and write operations can be performed. Any random memory location can be accessed for information transfer to or from the memory. It is also called *Read-Write Memory* (RWM).

A block diagram of a read-write memory is shown in Fig. 10.20(*a*). It has *n* address input lines to access 2^n memory locations and *m* data lines to read/write data from/to memory. The *n*-bit address is placed in the *Memory Address Register* (MAR) to select one of the 2^n memory locations. Read and Write are control signals that are used to enable memory for read and write operations respectively. When a read is asserted, the data from the memory location, selected by the *m*-bit address, is placed in Memory Buffer Register (MBR) which is a m-bit register. To perform write operation, the data to be written into a particular memory location is placed in the MBR, and then the write line is asserted. When the write line is asserted, the data in the MBR is written into the memory location, selected by the *n*-bit address.

Thus, any computer communicates with the memory by means of the memory address register, the memory buffer register and the READ and WRITE inputs. Computers invariably use RAMs for their high-speed main memory and then use backup or slower-speed memories to hold auxiliary data.

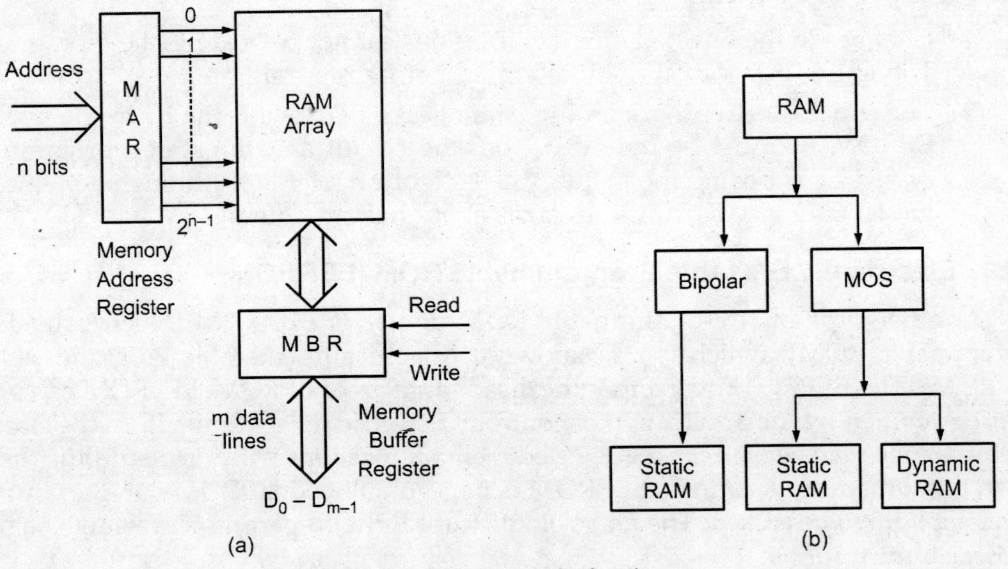

Fig. 10.20 The RAM family

10.5.1 Types of RAM

Semiconductor RAMs may be static or dynamic. The static RAM employs bipolar or MOS flip-flops, and the dynamic RAM uses MOSFETs and capacitors that store data. In either case, RAMs are volatile because the stored data will be lost once the d.c. power applied to the flip-flops is removed.

Semiconductor RAMs are available with large storage capacities and have replaced magnetic core memories in most of the computer circuits. The different categories of RAMs are shown in Fig. 10.20(*b*).

10.5.2 Static RAM

A static RAM essentially contains an array of flip-flops, one for each stored bit. Data written into a flip-flop remains stored as long as a d.c. power is maintained. The memory capacity of a static RAM varies from 64 bits to 1 Mega bit.

Static RAM cell: The logic diagram of a static RAM cell is shown in Fig. 10.21. The cell (or a group of cells) is selected by HIGH values on the ROW and COLUMN lines. The input data bit (1 or 0) is written into the cell by setting the flip-flop for a 1 and resetting the flip-flop for a 0 when the READ/ $\overline{\text{WRITE}}$ line is LOW (i.e. write). When the READ/ $\overline{\text{WRITE}}$ line is HIGH, the flip-flop is unaffected. It means that the stored bit (data) is gated to the *data out* line.

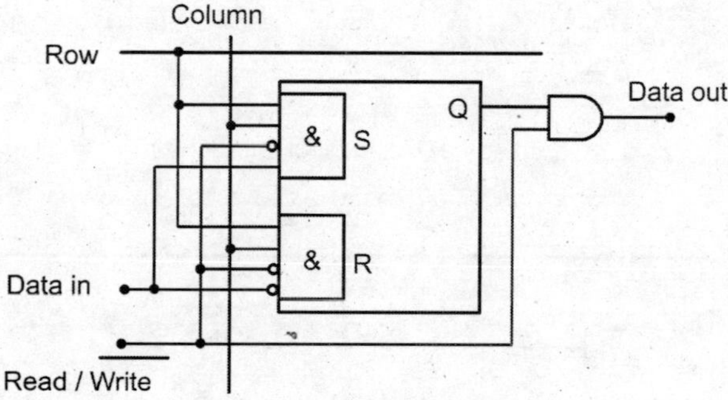

Fig. 10.21 Logic diagram of a static RAM cell

The flip-flop in static memory cell can be constructed using Bipolar Junction Transistor (BJT) and MOSFETs that are shown in Fig. 10.22(*a*) and (*b*) respectively.

In a bipolar static RAM cell shown in Fig. 10.22(*a*), two BJTs Q_1 and Q_2 are cross-coupled to form a flip-flop. Here, each transistor has three emitters, namely Row Select input, Column Select input, and Write input. To select the cell, both the row and column select lines must be held HIGH. When selected, a data-bit can be stored in the cell (Write operation) or the content of the cell can be read (Read operation). If either row or column select line is LOW, then the memory cell is disabled.

In a MOS static RAM cell shown in Fig. 10.22(*b*), Q_1 and Q_2 act like switches while Q_3 and Q_4 act as active load resistors. The transistor Q_1 conducts and Q_2 is cut off or vice versa. As a static RAM uses a flip-flop as the basic memory cell, it consists of thousands of flip-flops.

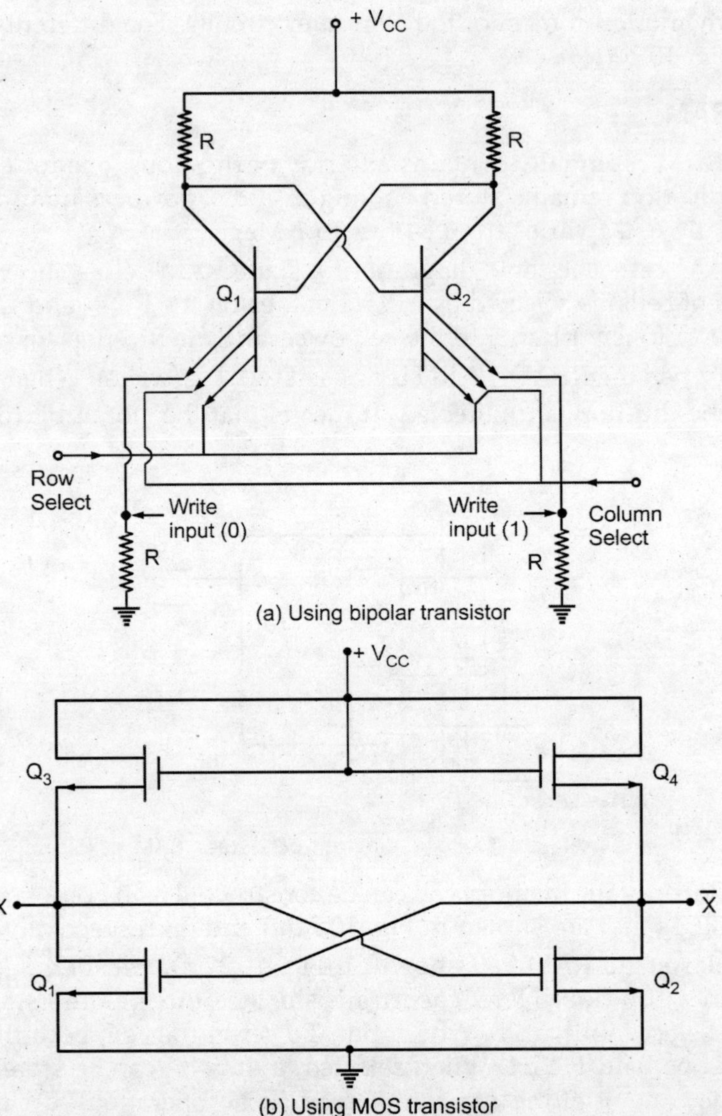

Fig. 10.22 Static RAM cell

Basic structure of a static RAM: Basically, RAM is addressed in the same way as ROM. RAM has address inputs, data inputs and a READ/ $\overline{\text{WRITE}}$ control. The logic diagram for a 1024-bit device with 256 × 4 organisation is shown in Fig. 10.23.

When READ/ $\overline{\text{WRITE}}$ is HIGH (Read mode) and chip select ($\overline{CS}$) is LOW, four data bits from the selected address appear on the data outputs. When READ/ $\overline{\text{WRITE}}$ is LOW (write mode), the four data bits that are applied to the data inputs are stored at the selected address.

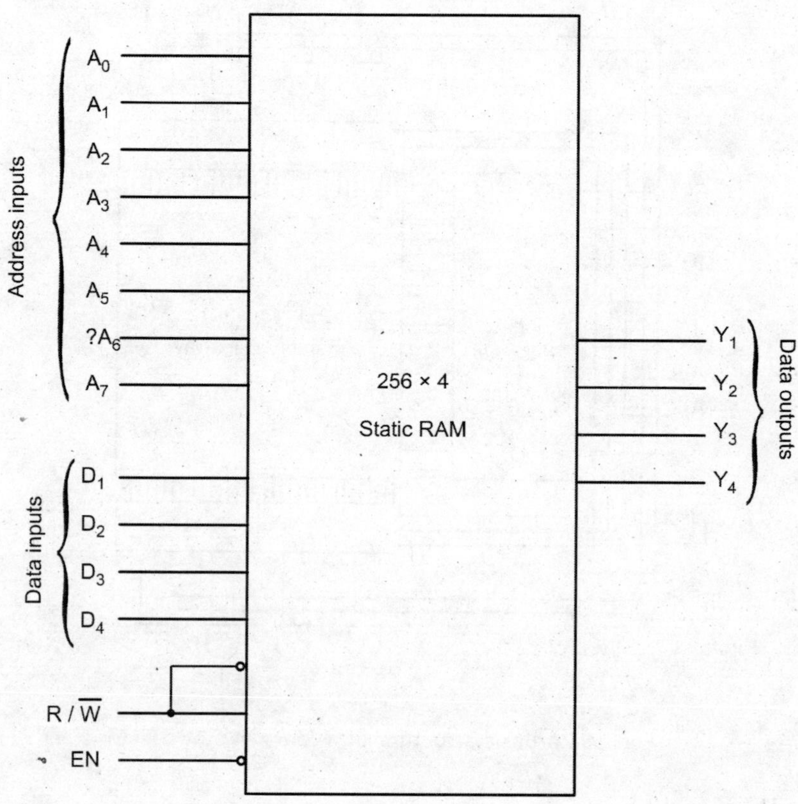

Fig. 10.23 Logic diagram for a 256 × 4 static RAM

A 1024-bit device with a 256 × 4 organisation implies that there are 256 rows and 4 columns. Here, the memory cells are arranged in an array of 32 × 32 matrix as shown in Fig. 10.24.

The operation of this memory is explained as follows: Five of the eight address lines (A_0 through A_4) are decoded by the ROW decoder to select one of the 32 rows. The remaining 3 address lines (A_5 through A_7) are used by the column decoder to select 4 of the 32 column lines as shown in Fig. 10.24. In the READ mode, the output buffers are enabled, and the four data bits from the selected memory location appear on the outputs ($Y_4\ Y_3\ Y_2\ Y_1$). In the WRITE mode, the input data buffers are enabled, and the four input data bits are routed through the input data selector by the address bits, A_5 through A_7, to the selected address for storage. During READ and WRITE, the chip select must be LOW.

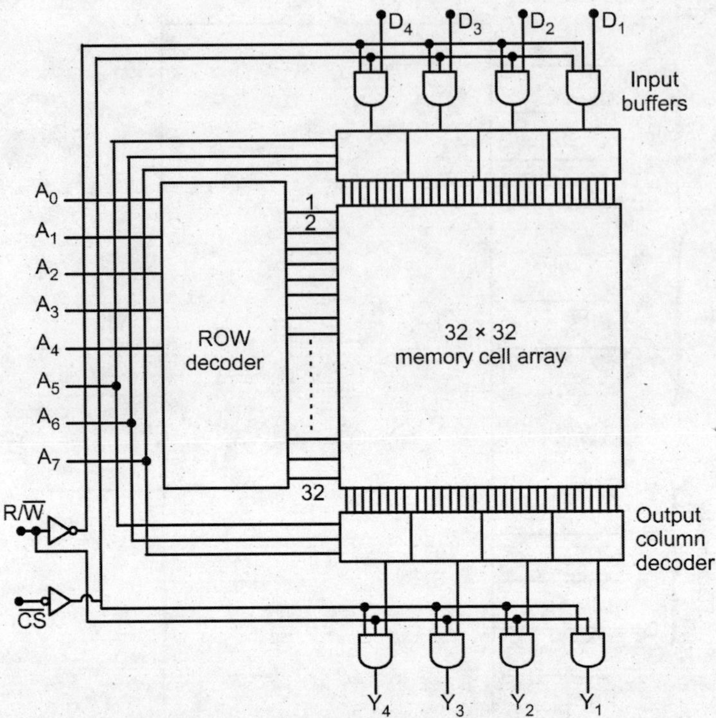

Fig. 10.24 Basic structure of a 256 × 4 static RAM

10.5.3 Static RAM ICs

Some of the static RAM ICs are discussed in the following sections.

IC 7489-64-bit Static RAM: The IC 7489, a TTL, 64-bit Static RAM, arranged as 16 words of 4 bits each, is shown in Fig. 10.25. Data can be written into the memory via the data inputs by supplying an address to the SELECT inputs and holding both the memory enable and write enable LOW. As the internal output of the data input gate is common to the input of the sense amplifier, the sense output will assume the opposite state of the information at the data input when the write enable is LOW.

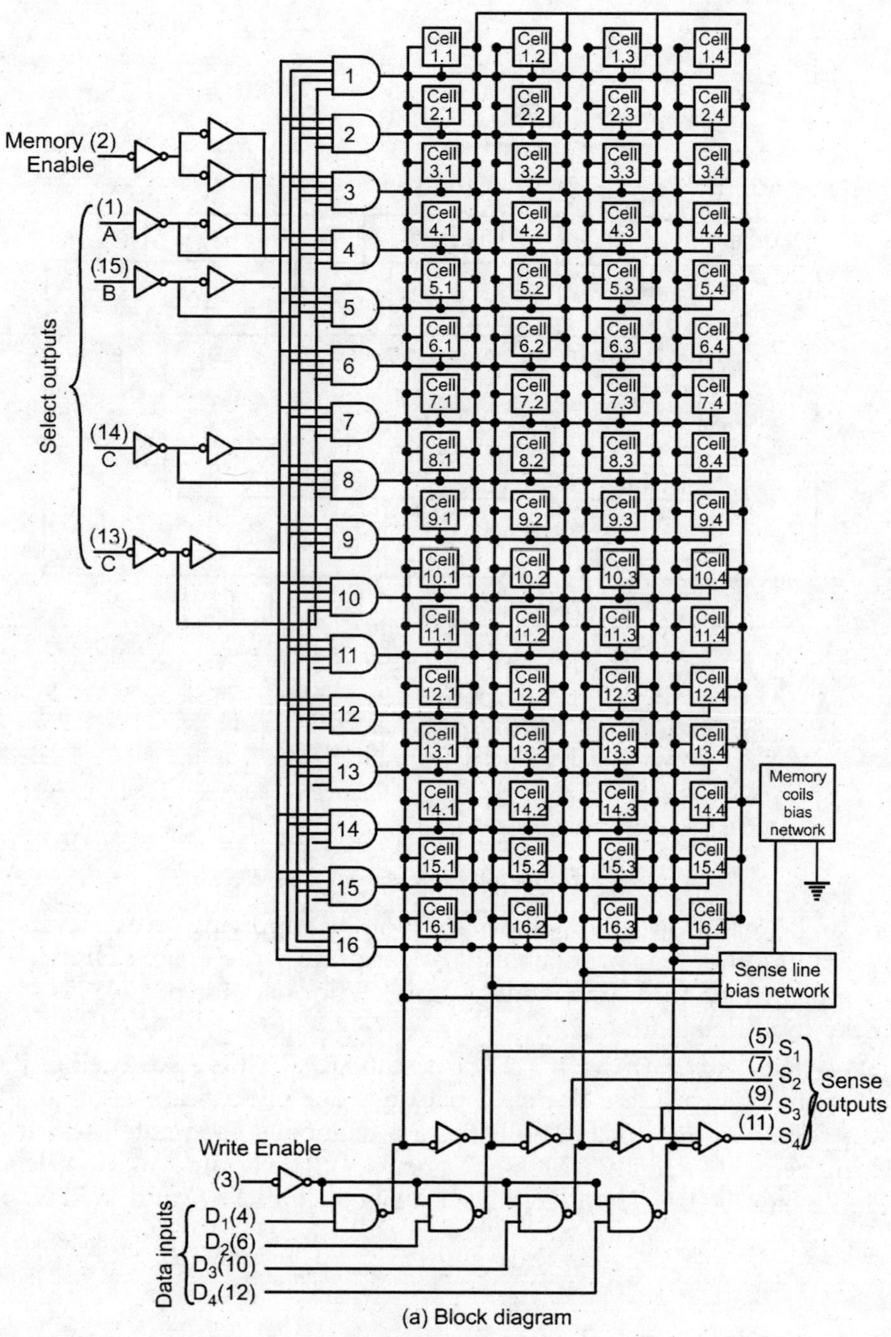

(a) Block diagram

Fig. 10.25(*a*) 7489-64-bit Static RAM

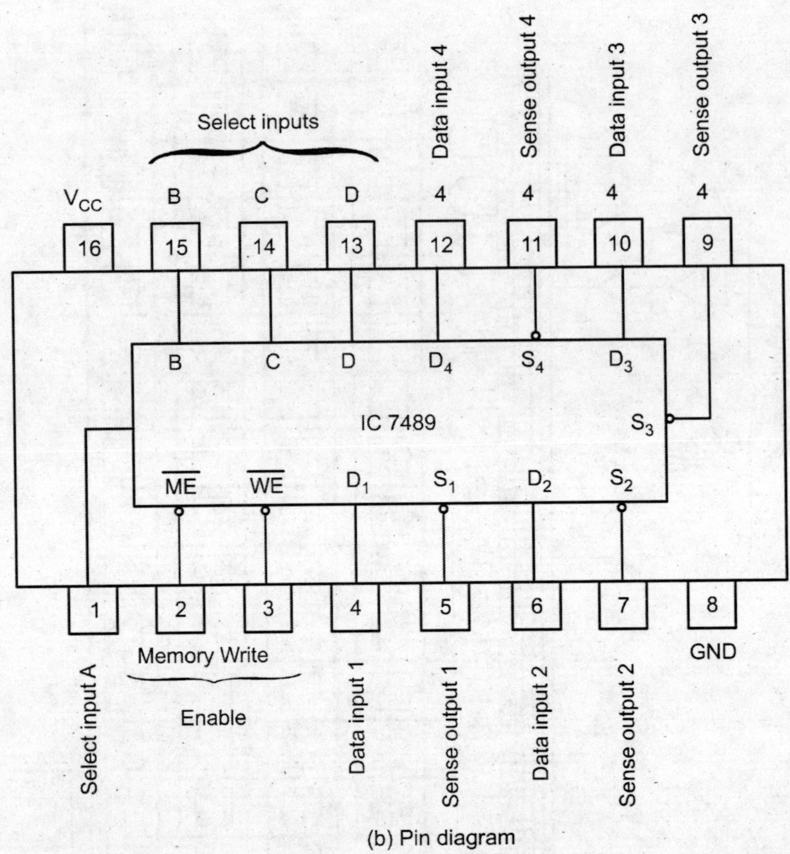

(b) Pin diagram

Fig. 10.25(*b*) 7489-64-bit DRAM

Data can be read from any memory location by supplying an address to the SELECT inputs and holding the memory enable LOW, and the write enable HIGH. The data will appear in complemented form at the sense outputs. The operation of a 7489 is straightforward and very easy to understand.

IC 2114-4K bit static RAM: IC 2114 is a static RAM. It is a 4096 cell unit organised as 1024 × 4-bit available in a single 18-pin IC package. The pin configuration and logic symbol of IC 2114 is shown in Fig. 10.26. The IC uses *n*-channel silicon gate MOS transistors in a static cell arrangement (actually 6 MOSFETs per cell). The *CS* line allows selection of the IC for expanded memory size. The *WE* line selects between READ and WRITE operations.

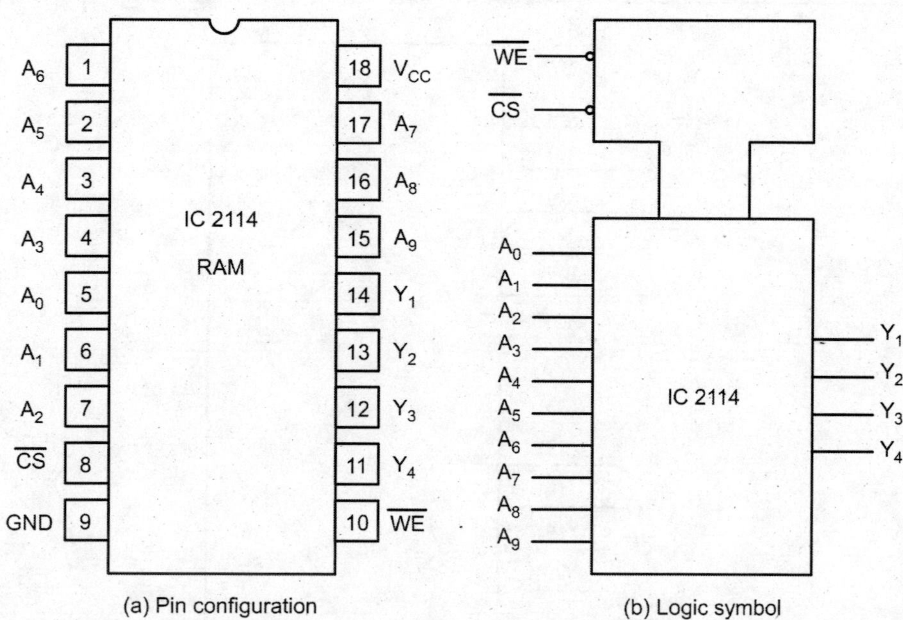

Fig. 10.26 IC 2114-1K × 4-bit static RAM

The 4 data pins operate as input lines on a WRITE operation and as output lines on a READ operation. The common input-output data lines using tri-state outputs allow connection of a number of ICs to a common data bus.

TMS 4016-2 KB static MOS RAM: The IC TMS 4016 is a 2 KB (2,048 × 8) static MOS RAM. Its pin out diagram is shown in Fig. 10.27. It is a byte organised static RAM. There are eleven address inputs to select the 2048 locations. The data in/data out (DY_1 through DY_8) are bidirectional terminals. These terminals provide for both data input during WRITE operation and data output during READ operation. The operation of TMS 4016 is given in Table 10.2.

Table 10.2 Truth table of TMS 4016

$\overline{WR}$	$\overline{CS}$	$\overline{RD}$	DY_1–DY_8	Mode
0	0	X	Data input	Write
1	0	0	Data output	Read
X	1	X	High impedance	Device disabled
1	0	1	High impedance	Output disabled

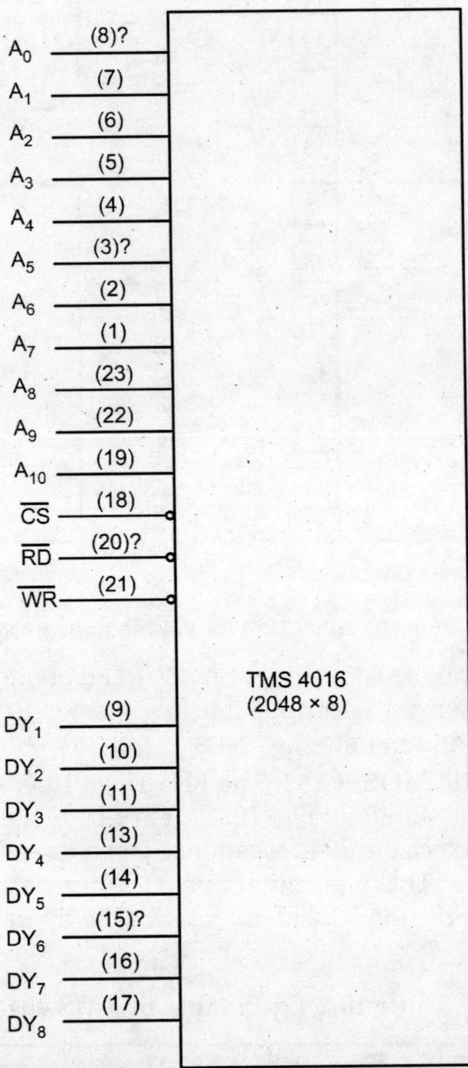

Fig. 10.27 TMS 4016—2 KB MOS static RAM

10.5.4 Dynamic RAM (DRAM)

The Dynamic Random Access Memory (DRAM) is the lowest cost, highest density random access memory available. Nowadays, computers use DRAM for main memory storage with the memory sizes ranging from 16 to 256 Mega bytes.

The DRAM stores its binary information in the form of electric charges on capacitors. Data are stored as charge on every capacitor, which must be *recharged* or *refreshed* thousands of times every second in order to retain the stored charge. These memory devices make use of an integrated MOS capacitor as basic memory cell instead of a flip-flop. The advantage of this cell is that it allows very large memory arrays to be constructed on a chip at a lower cost per bit than in static memories. The disadvantage is that the MOS capacitor cannot hold the stored charge over an extended period of time and it has to be refreshed every few milliseconds. This requires more circuitry and complicates the design problem. Static RAMs are simpler than dynamic RAMs.

Do You Know?	Dynamic RAM (DRAM) uses capacitors to store data by charging or discharging them.

A typical dynamic RAM cell consisting of a single MOSFET and a capacitor is shown in Fig. 10.28. A dynamic RAM consists of an array of such memory cells. In this type of cell, the transistor acts as a switch. The memory cell also requires MOSFETs for READ and WRITE gating to operate the cell. Data input is connected for storage by a WRITE control signal.

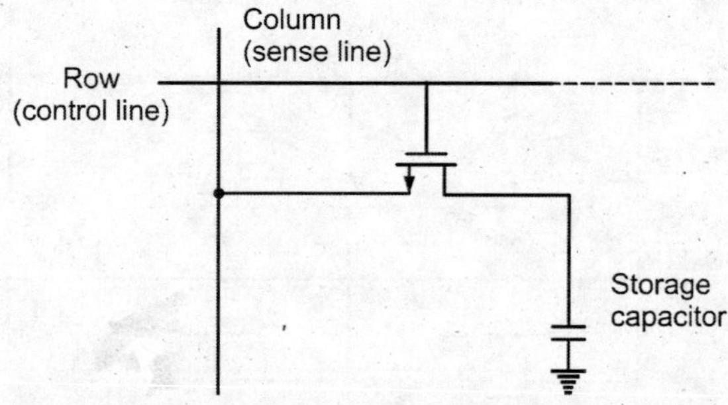

Fig. 10.28 A dynamic MOS RAM cell

The Dynamic RAM offers reduced power consumption and large storage capacity in a single memory chip. With the availability of such high packing density memory ICs, the capacity of memory will continue to grow.

Dynamic memory organisation: The organisation of a dynamic memory allows many cells to be accessed by a minimum amount of circuitry. Each memory cell is gated onto a data input/output line when a ROW and a COLUMN select MOSFETs are driven ON as shown in Fig. 10.29. For example, to select the DRAM cell marked 1-1, a 1 (HIGH) has to be applied at ROW 1 and COLUMN 1 select inputs. A HIGH at ROW 1 input selects all 128 transistors connected to the ROW 1 select line while the HIGH at the COLUMN 1 input selects one of the 128 possible COLUMNs, thereby the cell marked 1-1 being read onto the I/O data line. A sense amplifier for each column is necessary to convert from the low voltage and low energy to a sufficient level voltage on the I/O data line. The circuitry interposed between a bit-storage capacitor and external data line is called a *sense amplifier*. A matrix of 64 × 128 accounts for 8192 individual memory cells, and only one cell is connected onto the data line at any time.

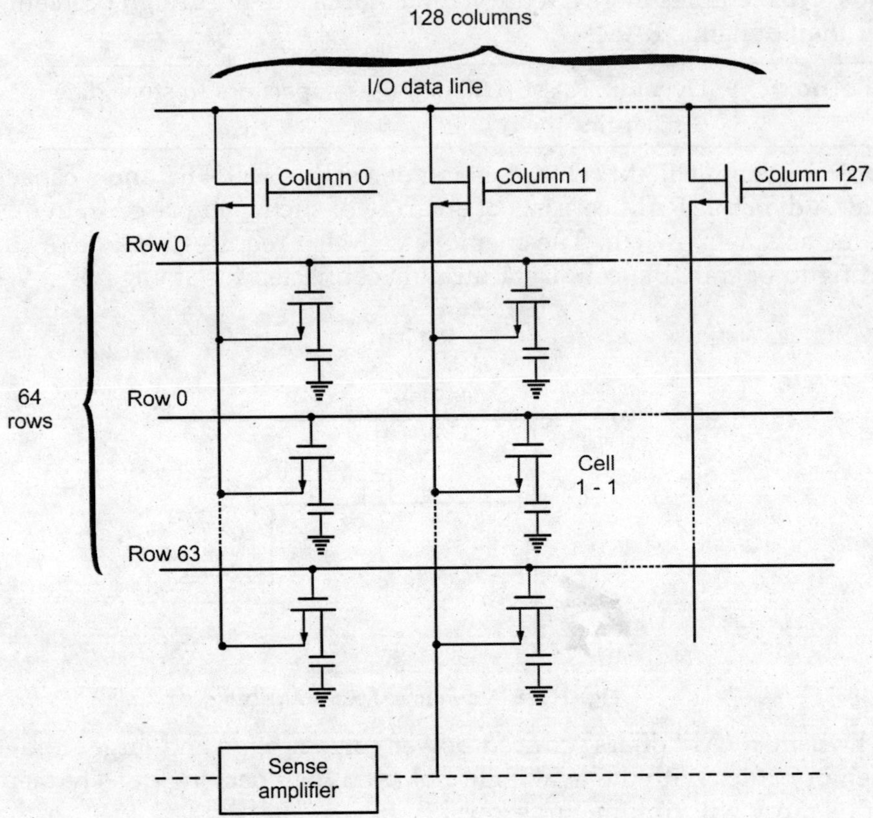

Fig. 10.29 ROW and COLUMN selection of a DRAM cell

The column and row decoders together selects a single capacitor and provide access to that 1-bit storage site for reading and writing. All the capacitors in an entire row can be refreshed simultaneously. There is a provision on the chip for activating all the sense amplifiers simultaneously and closing simultaneously all the appropriate switches in the bit lines. Such simultaneous activation will refresh all the capacitors in the ROW selected by the ROW decoder.

Basic structure of a dynamic RAM: A Dynamic RAM usually employs a technique called *address multiplexing* to reduce the number of address lines and thus the number of input/output pins on the IC package.

A block diagram of a 16 Kbit (16,384 bits) dynamic RAM that has been simplified to illustrate address multiplexing is shown in Fig. 10.30. Since 2^{14} = 16,384, the fourteen-bit address is applied (seven bits at a time) to the address inputs. First of all, the seven-bit ROW address has to be applied and the $\overline{RAS}$ (row address strobe) latches the seven bits into the ROW *address latch.* Next, the seven-bit COLUMN address is applied to the address inputs, and the $\overline{CAS}$ (column address strobe) latches the remaining seven bits into the

COLUMN *address latch*. Then, the seven-bit ROW address and the seven-bit COLUMN address are decoded to select the appropriate memory cell in the 128 × 128 dynamic memory array for a READ or WRITE operation. The timing diagram of the address multiplexing operation is shown in Fig. 10.31.

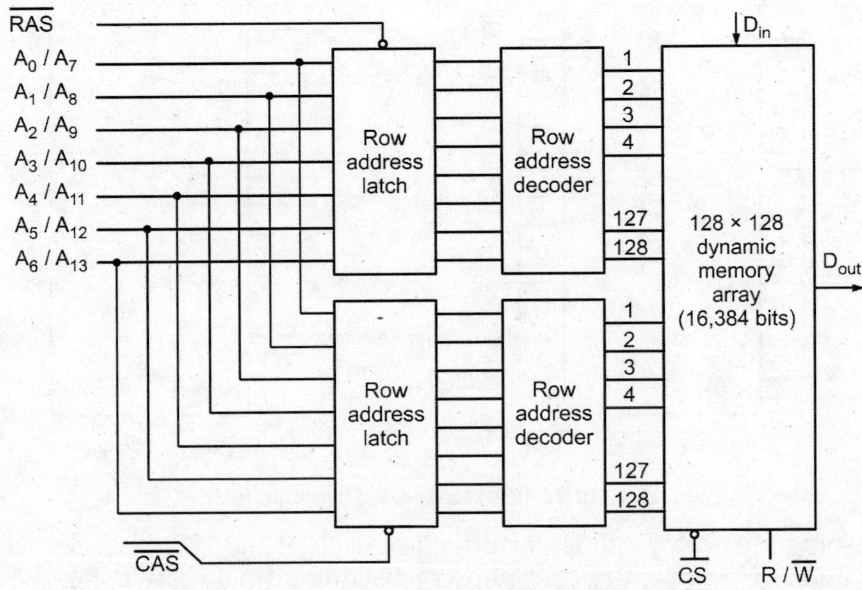

Fig. 10.30 A block diagram of a 16 K dynamic RAM

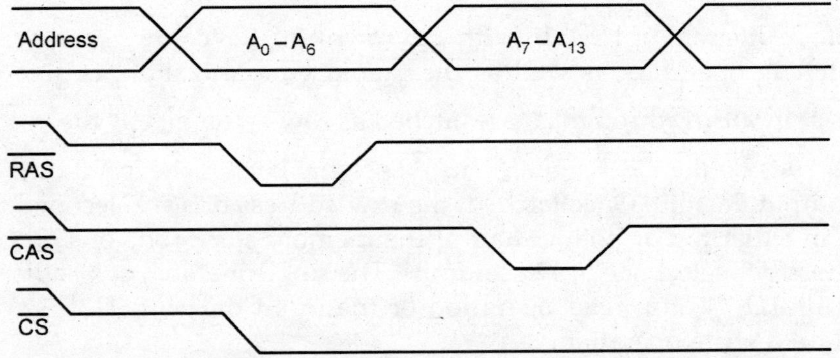

Fig. 10.31 Timing diagrams for address multiplexing in DRAM

IC 4116-16 K bit dynamic RAM: The pin configuration and logic symbol of IC 4116,16K × 1-bit dynamic memory is shown in Fig. 10.32. It has a single data input (D_{in}) and a single data output (D_{out}). The LOW at $\overline{WE}$ input enables writing the data input bit into the selected memory cell. The HIGH at $\overline{WE}$ input allows for reading the data out of the selected cell. The 4116 requires three supply voltages (+5 V, –5V, and +12V) to operate the IC unit.

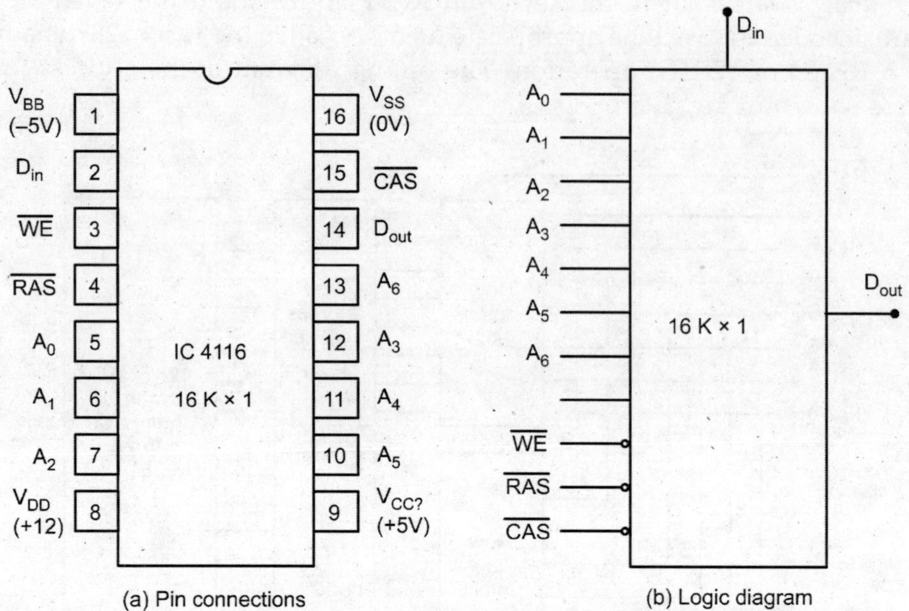

(a) Pin connections (b) Logic diagram

Fig. 10.32 IC 4116-16K bit dynamic RAM

Refreshing a memory cell is performed at each of the 128 row address every 2 ms. The organization of the memory cells allows selecting a single row using the row-address bits. A read operation for that row also refreshes all memory cells in that row. With 128 rows, the memory unit must be cycled to read from each of the 128 rows and it is repeated at least every 2 ms.

The block diagram of IC 4116 with 128 rows and 128 columns, showing the address and data latching operation as well as the memory organization, is given in Fig. 10.33. Address bits at inputs A_0 through A_6 are latched as row-select bits by the $\overline{RAS}$ signal and as column-select bits by the $\overline{CAS}$ signal. The $\overline{WE}$ signal latches the input data bit while $\overline{CAS}$ latches the output data bit. Once latched, the row address bits are decoded to select one of the 64 rows in either top or bottom half of the memory stack (one of 128 rows), while the column address bits select one of 128 columns. The single memory cell is then latched as the output data bit (D_{out}) on a read operation or the input data bit (D_{in}) is written into the selected cell on a write operation.

10.5.5 Pseudo Static RAM (PSRAM)

Pseudo static RAM is basically a dynamic RAM. It has a built-in-refresh logic. Hence, no external refreshing circuit is required and thereby PSRAM can be used as a static RAM device. When this device is performing internal refresh, it cannot be accessed for read or write operation. They are fabricated with capacities of 64K bits and more.

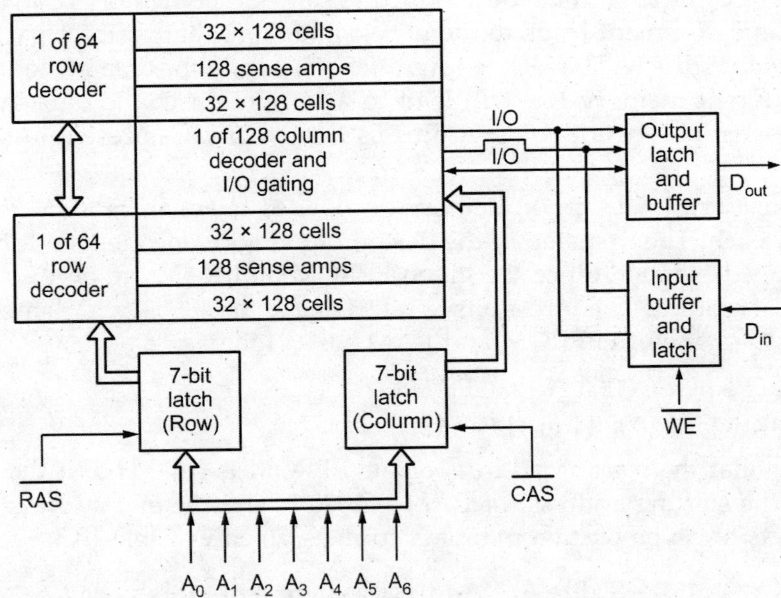

Fig. 10.33 Block diagram of IC 4116-16K × 1 dynamic memory

10.5.6 Advantages of RAM

RAMs are utilized in the computer as scratch-pad, buffer and main memories. The following are the advantages of MSI/LSI RAMs over magnetic devices:

1. *Non-destructive read out:* Read out of a RAM does not affect the content stored.
2. *Fast operating speed:* Access time can be as low as 150 ns, with on-chip decoding.
3. *Low power dissipation:* It is typically less than 1 mW per bit for static RAM and less than 0.5 mW per bit for dynamic RAM.
4. *Compatibility:* As semiconductor memories enjoy common interface and technology between sensing and decoding circuitry and the storage element itself, they are self-compatible.
5. *Economy:* MOS memories are more economical than magnetic core for small and medium-sized systems.

10.6 MEMORY DECODING

The Memory IC used in a digital system is selected or enabled only for the range of addresses assigned to it and this process is called memory decoding. The memory IC is disabled for the range of addresses that are not assigned to it.

10.6.1 Address Assignment

For the design of memory decoder, each memory IC used in the system should be assigned with a range of addresses according to its capacity. The address range assigned to a particular memory IC should not be assigned to some other memory IC in the same digital system. Such an assignment leads to memory contention, that is, for a particular address, two or more memory ICs will be selected simultaneously and the data in the selected memory location of different memory ICs will lead to a contention in the data bus. Hence, it is important to assign non-overlapping addresses to memory ICs present in the same digital system.

Next consider the design of memory decoder to select an EPROM and a RAM of 4 Kbyte capacity each. The first step in the design of memory decoder is address assignment in non-overlapped manner. Since the given EPROM and RAM are of 4 Kbyte (4 × 1024 = 4096) capacity, it requires 12 address bits to select one of the 4096 (i.e., 2^{12}) memory locations. Let the address assigned to EPROM and RAM are as follows:

EPROM (4 Kbyte): 0000 H to 0FFF H

RAM (4 Kbyte) : 1000 H to 1FFF H

It means that the starting address of the EPROM is 0000 H and the end address is 0FFF H while the starting address for RAM is 1000 H and the end address is 1FFF H. The range of addresses can be written in binary form as given in Table 10.3.

Table 10.1 Address assignment table

Memory	Address in Hex	Address in Binary							
		A_{15}	A_{14}	A_{13}	A_{12}	A_{11}	A_{10}	A_9	A_8
	0000	0	0	0	0	0	0	0	0
EPROM	:	:	:	:	:	:	:	:	:
(4 Kbyte)	:	:	:	:	:	:	:	:	:
	0FFF	0	0	0	0	1	1	1	1
	1000	0	0	0	1	0	0	0	0
RAM	:	:	:	:	:	:	:	:	:
(4 Kbyte)	:	:	:	:	:	:	:	:	:
	1FFF	0	0	0	1	1	1	1	1

Memory	Address	Address in Binary							
	in Hex	A_7	A_6	A_5	A_4	A_3	A_2	A_1	A_0
	0000	0	0	0	0	0	0	0	0
EPROM	:	:	:	:	:	:	:	:	:
(4 Kbyte)	:	:	:	:	:	:	:	:	:
	0FFF	1	1	1	1	1	1	1	1
	1000	0	0	0	1	0	0	0	0
RAM	:	:	:	:	:	:	:	:	:
(4 Kbyte)	:	:	:	:	:	:	:	:	:
	1FFF	1	1	1	1	1	1	1	1

The least significant 12-bit addresses ($A_n - A_0$) are changing from 0s to 1s while the most significant 4-bit addresses ($A_{15} - A_{12}$) are constant i.e., 0000 for EPROM and 0001 for RAM. Hence, the least significant 12-bit addresses can be used to access one of the memory location in the respective 4 Kbyte EPROM/RAM while the 4 most significant bit addresses can be used to select EPROM or RAM at a time. The memory decoder for the given address assignment can be designed using a 3 to 8 decoder IC (IC 74138) as shown in Fig. 10.34.

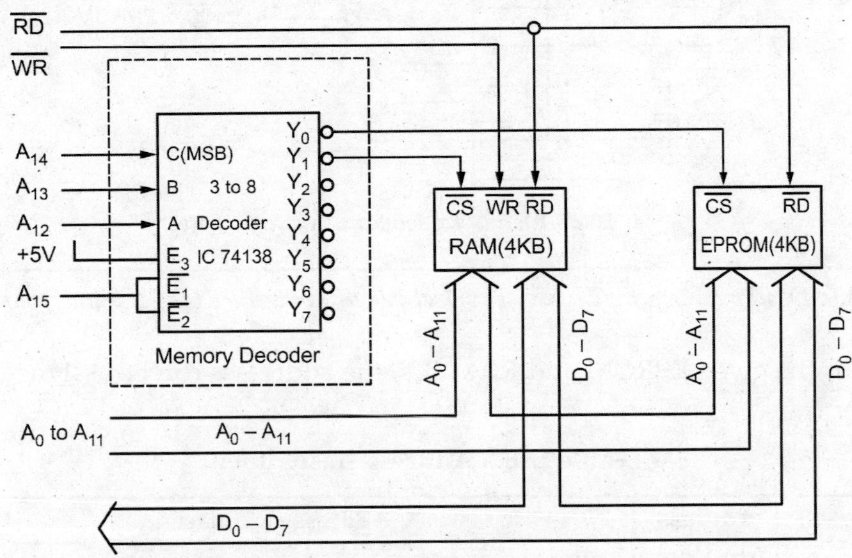

Fig. 10.34 Memory decoder for 4 Kbyte EPROM and RAM

The IC 74138 is a 3 to 8 decoder with three inputs (*C*, *B* and *A*), eight active low outputs ($Y_0 - Y_7$), two active low enable inputs ($\overline{E}_1$ and $\overline{E}_2$) and one active high enable input (E_1). The address lines ($A_0 - A_{11}$) are connected to ($A_8 - A_{11}$) pins of EPROM and RAM ICs to select one of the 4096 memory locations. The address lines ($A_{14} - A_{12}$) are connected with three inputs of the decoder. Address line A_{15}, which is 0 for both EPROM and RAM range of addresses, is connected to $\overline{E}_1$ and $\overline{E}_2$ of IC 74138 and enable input E_3 is permanently connected to +5 V. Therefore, IC 74138 will be enabled for both EPROM and RAM address ranges. From the Table 10.3, it is observed that $A_{14}A_{13}A_{12}$ = 000 for EPROM and $A_{14}A_{13}A_{12}$ = 001 for RAM. Now, based on the $A_{14}A_{13}A_{12}$ address bits, either the output Y_0 or output Y_1 will be selected. In other words, for EPROM addresses, Y_0 will be selected which is connected to chip select ($\overline{CS}$) input of EPROM whereas for RAM addresses, Y_1 will be selected which is connected to chip select ($\overline{CS}$) input of RAM.

Therefore, for any EPROM address, Y_0 will be low and hence EPROM will be selected and one memory location in EPROM will be accessed based on ($A_0 - A_{11}$) address bits. But at the same time RAM is disabled since Y_1 is in high state. Similarly, for RAM addresses, only RAM is selected and EPROM is not selected.

Alternatively, the memory decoder, for the selection of given 4 Kbyte EPROM and RAM can be designed using two 4-input NAND gates and 4 NOT gates as shown in Fig. 10.35. The NAND gate 1 output becomes low for $A_{15}A_{14}A_{13}A_{12}$ = 0000 and hence EPROM is selected, while for $A_{15}A_{14}A_{13}A_{12}$ = 0001, the NAND gate 2 output becomes low and hence RAM is selected.

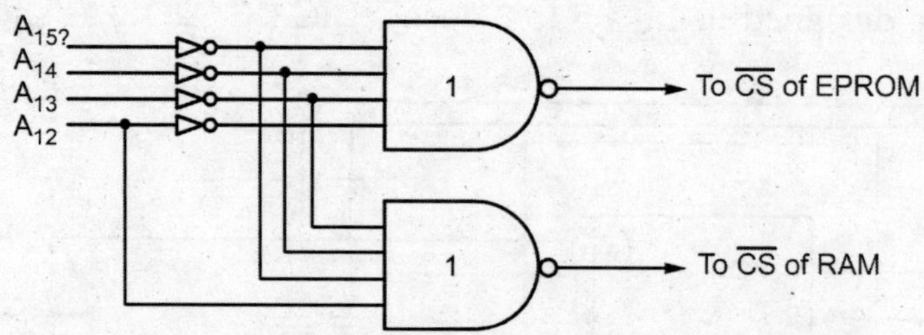

Fig. 10.35 Memory decoder using logic gates

Example 10.5. *Design a memory decoder to select two 8 Kbyte EPROM ICs and two 8 Kbyte RAM ICs.*

Solution: For the given EPROM and RAM ICs, the addresses can be assigned as shown in Table E10.5.

Table E10.5 Address Assignment

Memory	Address in Hex	Address in Binary							
		A_{15}	A_{14}	A_{13}	A_{12}	A_{11}	A_{10}	A_9	A_8
EPROM1 (8 Kbyte)	0000	0	0	0	0	0	0	0	0
	:	:	:	:	:	:	:	:	:
	:	:	:	:	:	:	:	:	:
	1FFF	0	0	0	1	1	1	1	1
EPROM2 (8 Kbyte)	2000	0	0	1	0	0	0	0	0
	:	:	:	:	:	:	:	:	:
	:	:	:	:	:	:	:	:	:
	3FFF	0	0	1	1	1	1	1	1
RAM1 (8 Kbyte)	4000	0	1	0	0	0	0	0	0
	:	:	:	:	:	:	:	:	:
	:	:	:	:	:	:	:	:	:
	5FFF	0	1	0	1	1	1	1	1
RAM2 (8 Kbyte)	6000	0	1	1	0	0	0	0	0
	:	:	:	:	:	:	:	:	:
	:	:	:	:	:	:	:	:	:
	7FFF	0	1	1	1	1	1	1	1

Memory	Address in Hex	Address in Binary							
		A_7	A_6	A_5	A_4	A_3	A_2	A_1	A_0
EPROM1	0000	0	0	0	0	0	0	0	0
(8 Kbyte)	:	:	:	:	:	:	:	:	:
	:	:	:	:	:	:	:	:	:
	1FFF	1	1	1	1	1	1	1	1
EPROM2	2000	0	0	1	0	0	0	0	0
(8 Kbyte)	:	:	:	:	:	:	:	:	:
	:	:	:	:	:	:	:	:	:
	3FFF	1	1	1	1	1	1	1	1
RAM1	4000	0	0	0	0	0	0	0	0
(8 Kbyte)	:	:	:	:	:	:	:	:	:
	:	:	:	:	:	:	:	:	:
	5FFF	1	1	1	1	1	1	1	1
RAM2	6000	0	0	0	0	0	0	0	0
(8 Kbyte)	:	:	:	:	:	:	:	:	:
	:	:	:	:	:	:	:	:	:
	7FFF	1	1	1	1	1	1	1	1

To select one of the 8 Kbyte (8192) memory locations in EPROM/RAM 13 address bits ($A_0 - A_{12}$) are required and this ($A_0 - A_{12}$) are changing from 0s to 1s for respective range of addresses. The next two bits A_{14} A_{13} are 00 for EPROM1; 01 for EPROM2; 10 for RAM1 and 11 for RAM2 and the MSB A15 is 0 for all addresses. Now, a memory decoder to select the given two EPROMs and two RAMs can be designed using a 2 to 4 decoder (IC 74139) as shown in Fig. E10.5. Here, A_{14} and A_{13} are connected to two inputs (B and A) and A_{15} is connected to enable input ($\overline{E}$) of IC 74139. Therefore, for all ranges of assigned addresses A = 0 and hence IC 74139 is enabled. Based on A_{14} and A_{13}, either EPROM1 or EPROM2 or RAM1 or RAM2 is selected and based on ($A_{12} - A_0$), one of the 8 Kbyte memory locations in respective memory IC is selected.

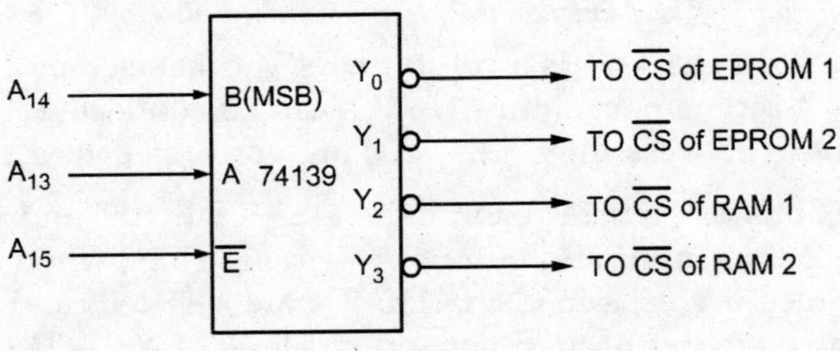

Fig. E10.5

Example 10.6. *Construct a 16 KB RAM using two 6264 RAM ICs.*

Solution: IC 6264 is a static RAM with 8 KB capacity. It has 13 address lines (A_0 to A_{12}) to access any one of 8 K memory locations. The memory capacity can be expanded to 16 KB by connecting two 6264 RAM ICs. But, to access 16 KB memory locations, 14 address lines (A_0 to A_{13}) are required. Of these, 14 address lines, A_0 to A_{12}, are used to select one of the 8 K memory locations either in RAM1 or in RAM2, while address line A_{13} is used to select RAM1 or RAM2. When A_{13} = 0, the RAM1 will be selected; when A_{13} = 1, the RAM2 will be selected. The connection diagram of two 6264 ICs is shown in Fig. E10.6.

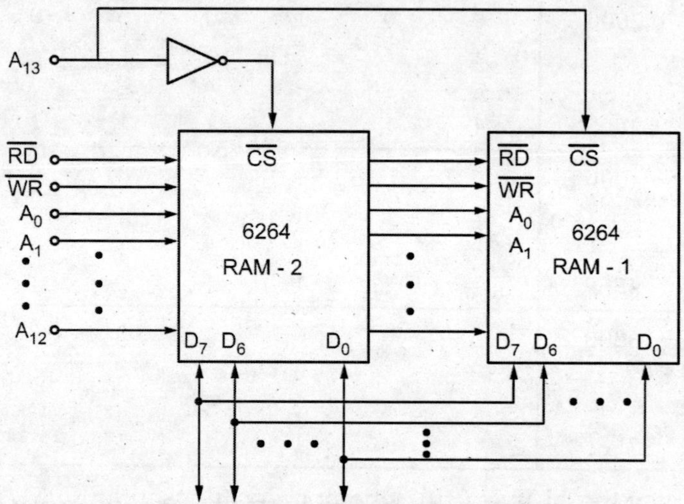

Fig. E10.6 Connection of two 6264 ICs as 16 KB memory

In the above expansion, the memory capacity is expanded while the number of bits per word in the memory remains the same.

10.7 MEMORY EXPANSION

Memory ICs can be connected together to expand the number of memory words or the number of bits per word. A few examples are given below in order to understand the memory expansion.

Example 10.7. *Construct 16K × 4-bit memory using four number of 4116 ICs.*

Solution: It is known that IC 4116 is a 16K bit DRAM organised as 16 K × 1 bit. Therefore, to construct 16 K × 4-bit memory, four 4116 ICs can be connected in parallel as shown in Fig. E10.7. Here, the 7 address lines ($A_0 - A_6$) are connected to a common address bus.

Address bits on this bus are transferred into the 4116 ICs by the and $\overline{CAS}$ signals. Separate data bits are connected as input and output of each IC. Thus, when an address is applied, the corresponding memory cells in each 4116 ICs are selected, and the 4-bit data will be either read from the memory through data output lines [$D_0(0) - D_0(3)$] or written into the memory through data input lines $D_i(0) - D_i(3)$ depending upon write signal.

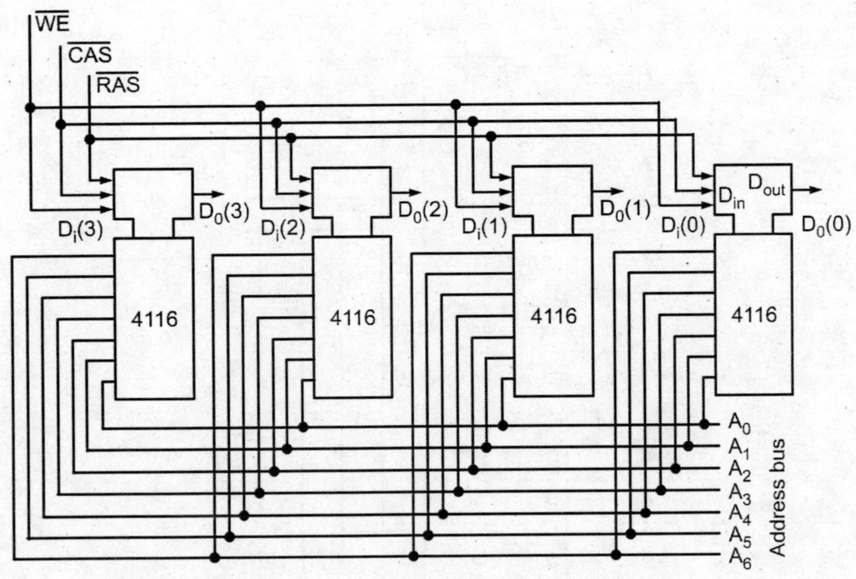

Fig. E10.7 Connection of four 4116 ICs as 16K × 4-bit memory

Example 10.8. *How many 1024 × 1 RAM chips are needed to construct a 1024 × 8 memory system?*

Solution: Eight chips are required, with each chip storing 1-bit of each of the 1024, 8-bit words.

Example 10.9. *Construct 512 × 4-bit memory using two 256 × 4-bit PROM.*

Solution: We know that Memory ICs can be organized using the chip enable inputs and open-collector or tri-state outputs to expand the number of memory words or the number of bits per second.

Fig. E10.9 shows a connection of two 256 × 4-bit ICs for operation as a 512 × 4-bit memory. The Programmed PROMs can be operated with open-collector (or tri-state) outputs connected on the same data signal line. The data bits ($D_1 - D_4$), from each PROM are tied to the others. When address bit A_8 is LOW, A_8 provides a low input to CS_2 of lower PROM, enabling the lower PROM unit. Then, the address bits A_8 through A_7 select one word from the lower PROM. Since $\overline{A}_8$ is HIGH, the upper PROM is disabled and its data outputs are in tri-stated condition. When the address bit A_8 is HIGH, the lower PROM outputs are in tri-stated condition; since $\overline{A}_8$ is LOW, the upper PROM is enabled. In this way the eight address bits select among 512 different locations, reading 4 bits of data from each location.

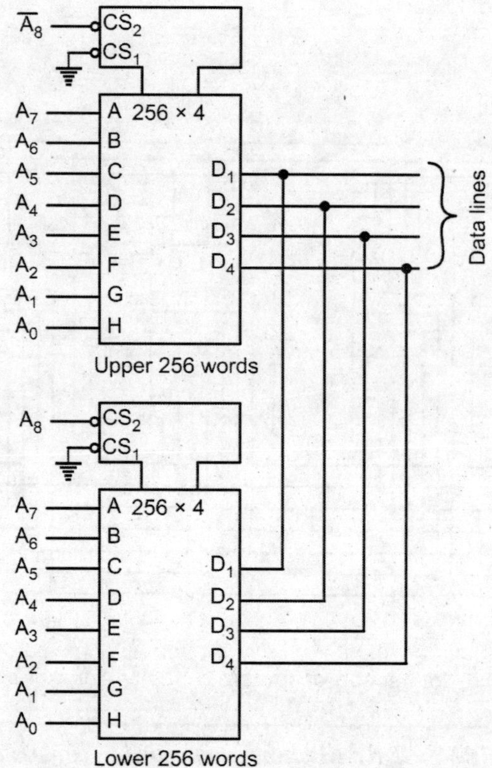

Fig. E10.9 Connection of two 256 × 4-bit PROMs as 512 × 4-bit memory

Example 10.10. *How many 16K × 4 RAMs are required to achieve a memory with a capacity of 64K and a word length of 8 bits?*

Solution: Eight IC chips are required.

10.8 PROGRAMMABLE LOGIC DEVICES (PLD)

Various combinational and sequential logic circuits are designed using discrete logic gates and flip-flops. To implement such combinational and sequential circuits, the designer has to interconnect several SSI and MSI chips by making connections external to the IC package. Logic circuits can also be designed using *Programmable Logic Devices* (PLDs) that have all the gates necessary for a logic circuit design in a single package. In such devices, there are provisions to perform the interconnections of the gates internally so that the desired logic can be implemented.

A PLD is an IC that contains large number of gates, flip-flops and registers that are interconnected on the chip. The IC is said to be programmable because the specific function of the IC for a given application is determined by the selective breaking of some of the interconnections while leaving others intact. Programming is accomplished by using fusible links, as in the field programmable ROMs. Alternatively- PLDs may be fabricated with *antifuses* at the cross points. A dielectric of multilayer oxygen-nitrogen-oxygen (ONO) formed between n^+ diffusion and polysilicon, or a layer of amorphous silicon between metal layers, forms an *antifuse*. Normally, the antifuse has a very high resistance between its terminals. When a high voltage (10 to 20 V) is applied across its terminals, sufficient current of about 5 mA to 15 mA is passed through the antifuse and melts the dielectric material, thereby creating a low resistance of 600 Ω to 100 Ω link permanently.

Do You Know?	A programmable logic device (PLD) is an IC that contains a large number of logic gates whose interconnections can be programmed by the user to generate the desired logic relationship between inputs and outputs.

There are two types of programmable logic devices, viz. (*i*) devices with fixed architecture, (*ii*) devices with flexible architecture. Programmable Array Logic (PAL) devices and Programmable Logic Array (PLA) devices are programmable logic devices with fixed architecture. PAL and PLA devices differ in the way the gate connections can be made internally. The best example for second type is Field Programmable Gate Arrays (FPGAs). FPGAs have flexible architecture and logic circuits can be designed using FPGA in the same way as logic circuit design using discrete gates.

The basic idea used by all programmable logic ICs is demonstrated in Fig. 10.36. It shows an array of AND gates and an array of OR gates that can be interconnected to generate four outputs, each of which can be any logic function of the two input variables A and B.

Each input feeds both a non-inverting buffer and an inverting buffer to produce the true and inverted forms of each variable. These are the *input lines* to the array of AND gates. The AND outputs are called the *product lines*. Each one of the product lines is connected to one of the four inputs of each OR gate through a fusible link. With all the links initially intact, the output of each OR gate will be a constant 1. Let Y_1, Y_2, Y_3 and Y_4 be the outputs. Each one of the four outputs can be programmed to any function of A and B by selectively blowing the appropriate fuses. A blown OR input acts as a logic 0. If all the fuses in the OR gate 1 are intact, then its output 7, is

$$Y_1 = \overline{AB} + \bar{A}B + A\bar{B} + AB$$

If we blow fuses 1 and 4 of OR gate 1, then the output Y_1 is given by

$$Y_1 = 0 + \bar{A}B + A\bar{B} + 0 = \bar{A}B + A\bar{B}$$

Thus, one can program any of the OR outputs to any desired function in a similar manner.

PLD symbology: The symbology for a 4-input AND gate is shown in Fig. 10.37. The input buffers are represented as a single buffer with two outputs. A single line is shown going into the AND gate to represent the four inputs. The connections from the input variable lines to the AND gate are indicated as either an X or a dot. An X represents an intact fuse. A dot represents a hard-wired connection. The absence of either of these indicates no connection.

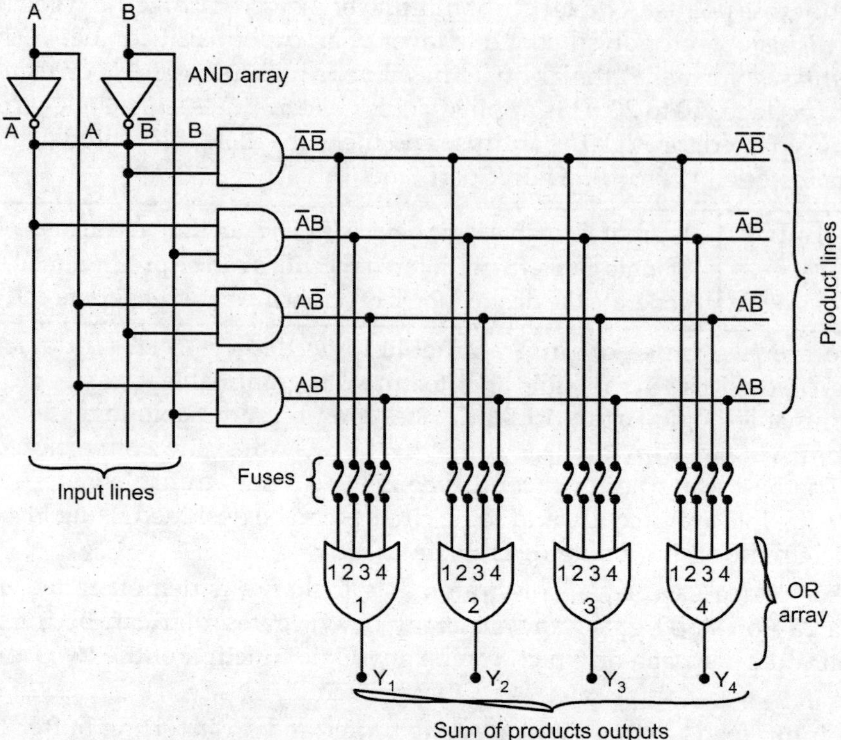

Fig. 10.36 A programmable logic device

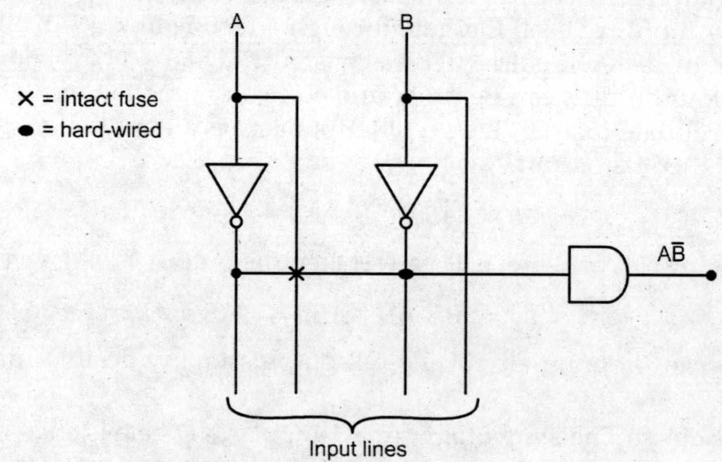

Fig. 10.37 Simplified symbology for PLDs

10.8.1 Programmable Logic Array (PLA)

Programmable Logic Array (PLA) is a type of fixed architecture logic devices with programmable AND gates followed by programmable OR gates. The PLA is used to implement a complex combinational circuit. In VLSI design, PLAs are used because the area required by the regular AND and OR arrays is less than the area required by randomly interconnected gates. Several PLAs include storage elements such as flip-flops on the silicon chip, so that they can be used for sequential applications.

Do You Know?	Programmable logic devices (PLDs) are the key technology in the future of digital systems.

A PLA is similar to a ROM in concept except that it does not provide full decoding of the variables and does not generate all the minterms as in the ROM. Thus, in a PLA, the decoder is replaced by a group of AND gates, each of which can be programmed to produce a product (AND) term of the input variables. The AND and OR gates inside the PLA are initially fabricated with fuses among them. The specific Boolean functions are implemented in sum-of-products (SOP) form by blowing appropriate fuses and leaving the desired connections. It is similar to the reprogramming of ROMs. For this reason, the logic array is called a programmable logic array. If a manufacturer makes a provision to allow programming by the user, the PLA is referred to as a FPLA, a *field-programmable logic array*. The FPLA can be programmed by the user by following certain recommended procedures.

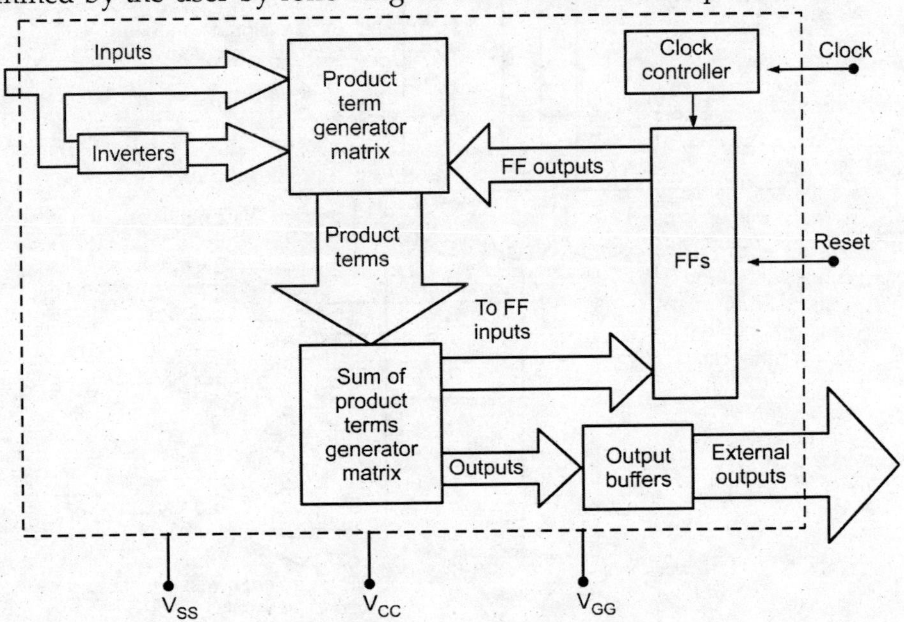

Fig. 10.38 Block diagram of PLA

The block diagram of PLA is shown in Fig. 10.38. It contains two matrices namely product term generator matrix and sum term generator matrix with clocked flip-flops in feedback path. Since each matrix may contain from 1,000 to 20,000 summing nodes, one can readily note that a great logical complexity is possible using the PLA circuit system. In this system, the feedback loop must be clocked carefully. Clocked flip-flops must be connected into all feedback paths from the summing matrix to the product matrix. The reset of the flip-flops to initialize the logic can be controlled.

IC 7575-PLA with 14 inputs and 8 outputs: A typical commercial PLA IC 7575 having 14 inputs and 8 outputs is shown in Fig. 10.38. Let I_0, I_1, ... I_{13} be the inputs applied to inverters to make available the complements $\overline{I}_0, \overline{I}_1, \ldots \overline{I}_{13}$ as well as the uncomplemented inputs. The chip provides 96 AND gates. Each AND gate has a potential fan-in of 14. The AND gate inputs are established during chip fabrication by the sites at which connections are made in the crossed array of I and $\overline{I}$ lines and input lines to the AND gates. If an AND gate has each of the I variables as inputs, then the AND gate output is a minterm. Otherwise, the output is a product term. Each OR gate has a potential fan-in of 96. Therefore, it has 96 product terms. At the OR gate outputs, inverters are provided which can be bypassed if required. We can therefore select at each output a function F or its complement $\overline{F}$.

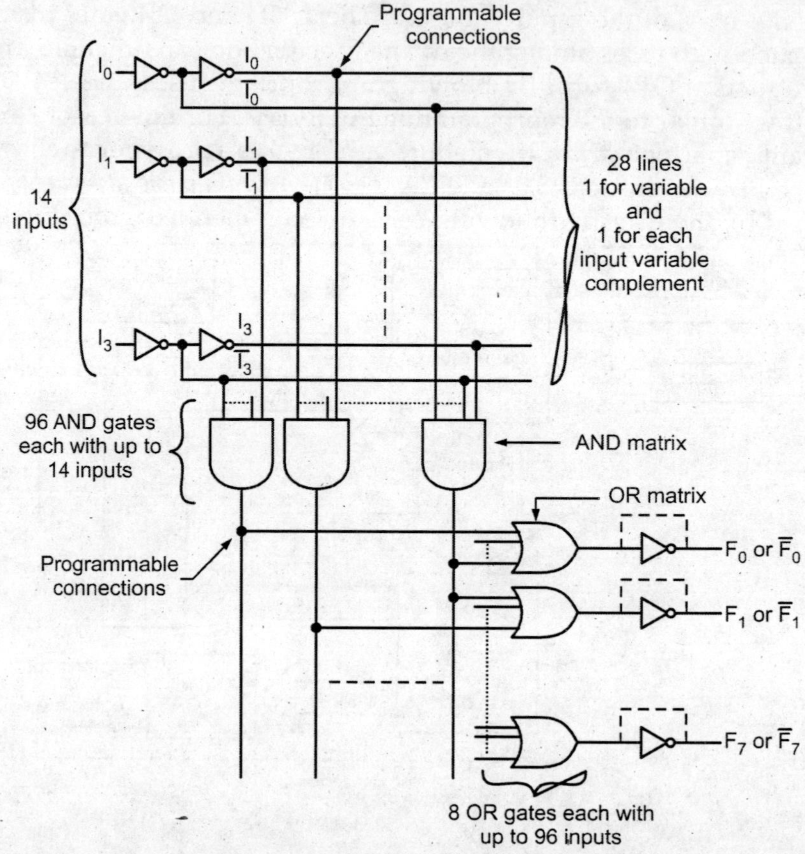

Fig. 10.39 Logic structure of an IC 7575 PLA

The size of the PLA is specified by the number of inputs, the number of product terms and the number of outputs.

In the manufacturing technique of PLA, the desired connections are made during manufacture. The manufacturer originally makes the IC array layout so that any desired connections can be made. The manufacturer creates a *mask,* which generates the desired connections when layers of metallisation are added to the chip during manufacture.

The AND gate which generates $\overline{A}\overline{B}$, shown in Fig. 10.40, has its output connected to two OR gates. Larger PLAs contain several hundred gates. 15 to 25 inputs, and 5 to 15 outputs. This offers the logic designer a greater flexibility. To design these large arrays, a simplifying symbology has proved useful. Fig. 10.41 shows this for the array in Fig. 10.40. The crosses drawn on the function indicate ANDs, and the squares indicate ORs. The figure also shows that the AND can be realized by a single semiconductor junction (called a diode), and the OR can be realized by a single junction pointed the other way. In practice, the manufacturer lays out the chip with junctions at every intersection of the lines, and only the desired diode connections are made. For field-programmable logic arrays (FPLAs), the diodes are fused so that they can be blown by an instrument called an array programmer. This sets the FPLA as desired.

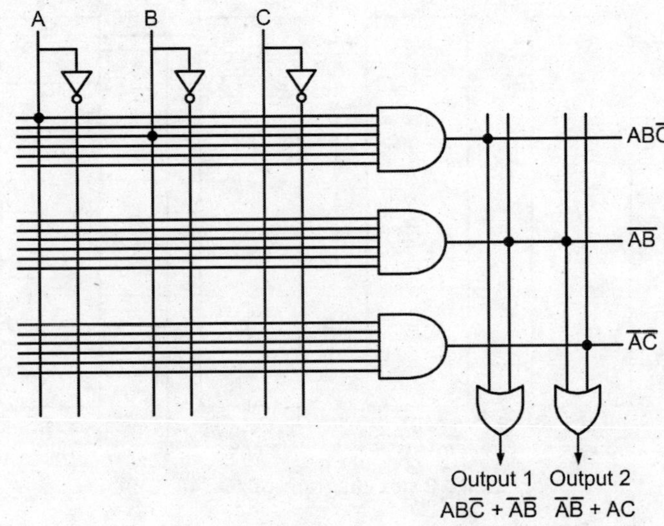

Fig. 10.40 Connection diagram for 3-input 2-output PLA

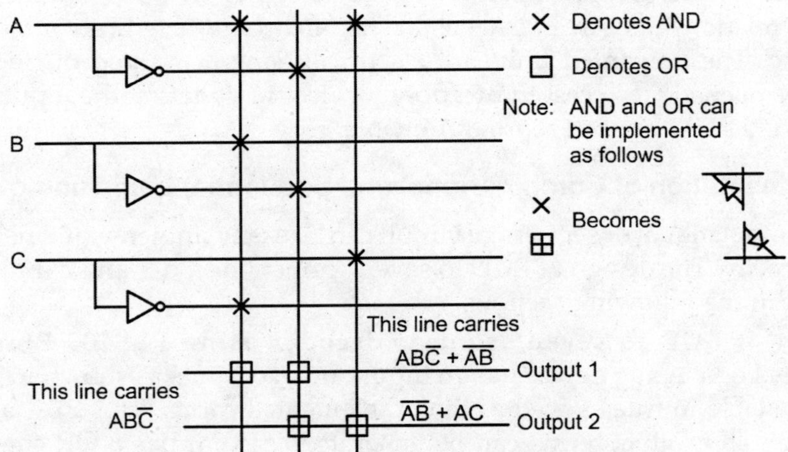

Fig. 10.41 Common method of PLA-design

IC 74839-PLA with 14 inputs and 6 outputs: The block diagram of IC 74839 PLA is shown in Fig. 10.42. It has fourteen inputs. The input buffer (▷) provides the input variables and their complements on its outputs. The AND matrix (&) consists of 32 product terms (AND), and the OR matrix (≥) consists of six OR terms. The AND and OR matrices can be programmed for the desired SOP output functions. The fused inputs (~) to the AND matrix can be programmed. The exclusive-OR array (⊕) is for programming inversions of the output functions and is accomplished by programming the fused input (~). A fused input left connected to ground (0) results in the input bit remaining uncomplemented ($1 \oplus 0 = 1, 0 \oplus 0 = 0$). A fused input blown open (1) results in inversion of the input ($1 \oplus 1 = 0, 0 \oplus 1 = 1$)

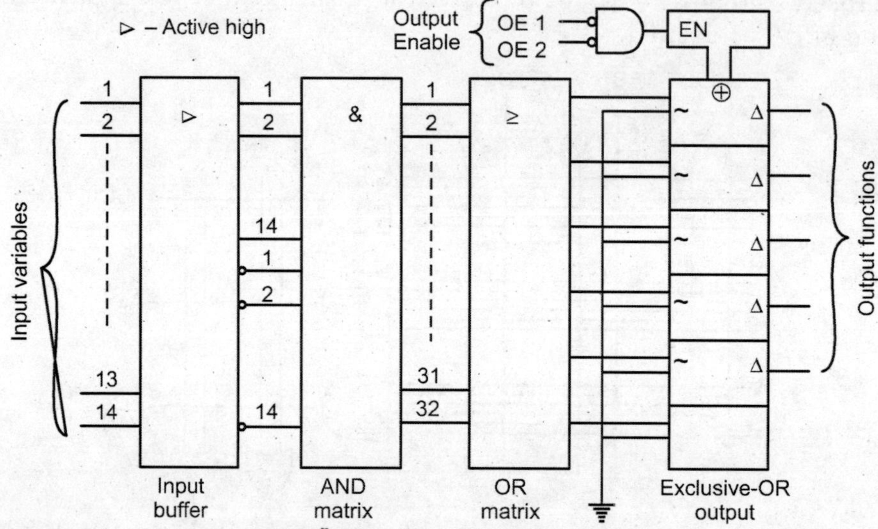

Fig. 10.42 Block diagram of IC 74839 PLA

Programming the PLA: Programming the PLA means that the interconnection paths are specified in its AND-OR-NOT pattern. A PLA device is to be programmed for the desired input-output relationship. An FPLA has all its Nichrome fuse links intact at the time of manufacturing. The unwanted links are electrically open circuited during programming. The links to be opened are accessed by applying electric signals at the inputs and outputs of the device. The FPLAs are not reprogrammable.

10.8.2 Implementation of Combinational and Sequential Logic using PLA

Any combinational or sequential circuit can be easily implemented using PLA from its truth table directly. The design of BCD to seven segment decoder and a multi-mode counter are explained in the following sections.

Design of BCD to seven segment decoder using PLS100 PLA: The PLS 100 Programmable Logic Array device has 16 inputs and 8 outputs as shown in Fig. 10.43. These inputs are available in true as well as in complemented forms. This PLA also has 48 AND gates by which 48 product terms can be generated. Also, it has 8-OR gates, each with 48 inputs, i.e., outputs of 48 AND gates. The outputs of 8 OR gates are passed through EX-OR gates to get true/complemented outputs of the EX-OR gates are passed through tri-state buffers that will be enabled when hen a low signal is applied at pin 19 of PLA.

The design of *BCD* to seven segment decoder using logic gates has been discussed in section 6.5.9. The truth table of the *BCD* to seven segment decoder provided in Table 6.12 shows *ABCD* as *BCD* inputs and a, b, c, d, e, f and g as seven segment outputs. The logic expressions for the seven segment coder outputs without considering don't care combinations are as follows:

$$a = \sum_m(0,2,3,5,6,7,8,9)$$

$$b = \sum_m(0,1,2,3,4,7,8,9)$$

$$c = \sum_m(0,1,3,4,5,6,7,8,9)$$

$$d = \sum_m(0,2,3,5,6,8,9)$$

$$e = \sum_m(0,2,6,8)$$

$$f = \sum_m(0,4,5,6,8,9)$$

$$g = \sum_m(2,3,4,5,6,8,9)$$

The outputs of seven segment decoder are given in the form of sum of minterms. The *BCD* inputs A, B, C and D are applied at input pins 9, 8, 7 and 6 respectively. From these four inputs, 10 minterms (0 to 9) are generated using first 10 AND gates. The seven segment decoder outputs a, b, c, d, e, f and g are taken from pins 18, 17, 16, 15, 13, 12 and 11 respectively. The fuses at the inputs of the respective OR gates are left intact or blown as per the corresponding logic expressions. The fuses connected to the one input of EX-OR gates are left intact so that true output is available. The structure of PLS 100 PLA programmed for the given *BCD* to seven segment decoder is shown hi Fig. 10.43.

Design of 3-bit multimode counter using PLS 105A PLA: Consider the design of 3-bit multimode counter that works as per function table shown in Table 10.3.

Table 10.3 Function table of multimode counter

Mode	Mode Inputs		Counter Operation
	M_1	M_2	
0	0	0	Normal UP counter
1	0	1	Normal DOWN counter
2	1	0	Unit distance UP counter
3	1	1	Unit distance DOWN counter

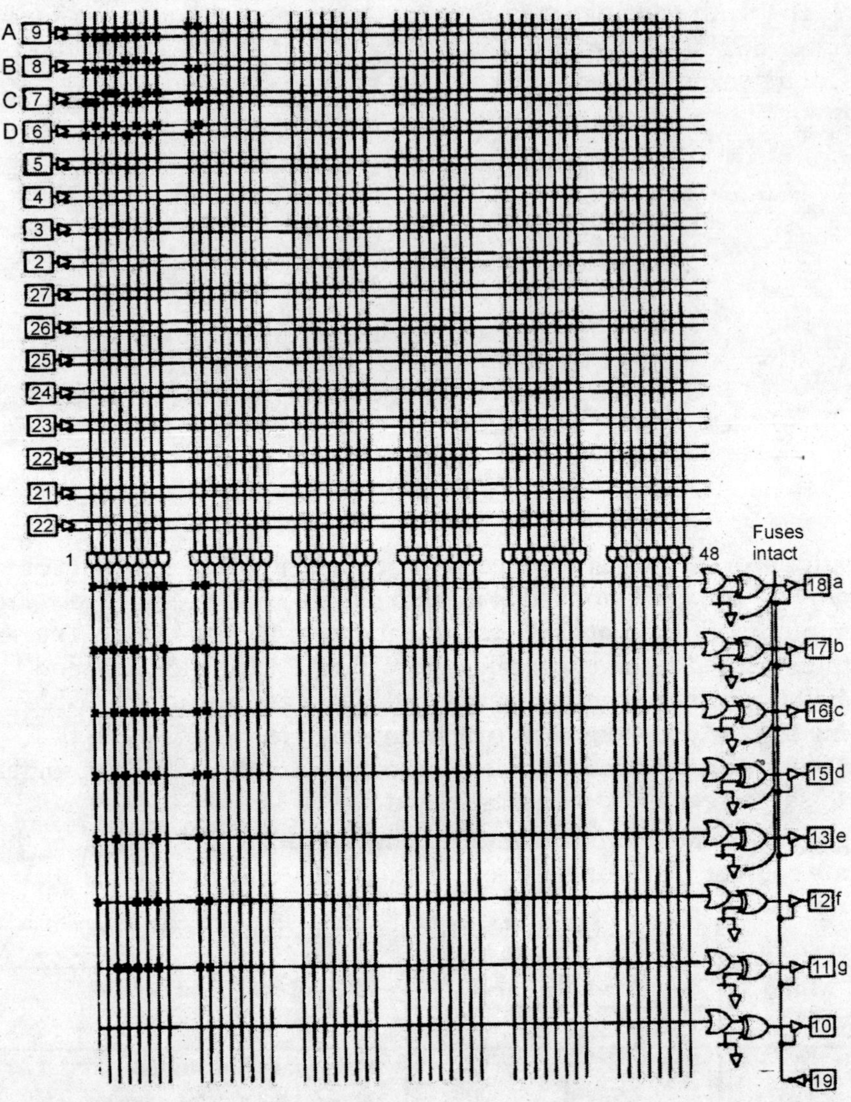

Fig. 10.43 Structure of PLS 100 PLA,
Programmed for BCD to seven segment decoder

As shown in Table 10.3, the multimode counter functions as a 3-bit normal sequence UP counter when mode inputs $M_1M_0 = 0\ 0$; when $M_1M_0 = 0\ 1$, it functions as a 3-bit normal sequence DOWN counter; when $M_1M_0 = 1\ 0$, it functions as a 3-bit unit distance UP counter and for $M_1M_0 = 1\ 1$, it functions as a 3-bit unit distance DOWN counter.

The given multimode counter can be implemented using a sequential PLA PLS 105 A. It consists of 14 SR flip-flops, of which the output of six flip-flops are fed back to the input side of AND arrays as shown in Fig. 10.44. These 6 flip-flop outputs are not available to the external world while the remaining 8 flip-flop outputs are available as 8 outputs. Clock input is given at pin 1 while pin 19 acts as preset input (P) of flip-flops as well as enable input to the output buffers. Also, it has 16 inputs and 48 AND gates to generate 48 product terms. In addition, one combinational (i.e., OR gate) output is fed back to the input side of AND arrays. Therefore, to each of 48 AND gates, 16 external inputs, 6 inputs fed back from 6 flip-flops and one input fed back from combinational output are available both in true and complemented forms i.e., a total of 46 inputs are available to each AND gate.

The complete truth table for the given multimode counter is shown in Table 10.4.

Table 10.4 Truth table for the given multimode counter

Mode	Mode input		Present State			Next State			Excitation Inputs					
	M_1	M_0	q_2	q_1	q_0	Q_2	Q_1	Q_0	S_2	R_2	S_1	R_1	S_0	R_0
	0	0	0	0	0	0	0	1	0	d	0	d	1	0
0	0	0	0	0	1	0	1	0	0	D	1	0	0	1
	0	0	0	1	0	0	1	1	0	d	d	0	1	0
(Normal	0	0	0	1	1	1	0	0	1	0	0	1	0	1
UP)	0	0	1	0	0	1	0	1	d	0	0	d	1	0
	0	0	1	0	0	1	1	0	d	0	1	0	0	1
	0	0	1	1	1	1	1	1	d	0	d	0	1	0
	0	0	1	1	0	0	0	0	0	1	0	1	0	1
	0	1	0	0	1	1	1	1	1	0	1	0	1	0
1	0	1	0	0	0	0	0	0	0	d	0	d	0	1
	0	1	0	1	1	0	0	1	0	d	0	1	1	0
(Normal	0	1	0	1	0	0	1	0	0	d	d	0	0	1
distance	0	1	1	0	1	0	1	1	0	1	1	0	1	0
UP)	0	1	1	0	0	1	0	0	d	0	0	d	0	1
	0	1	1	1	1	1	0	1	d	0	0	1	1	0
	0	1	1	1	0	1	1	0	d	0	d	0	0	1
	1	0	0	0	1	0	0	1	0	d	0	d	1	0
2	1	0	0	0	0	0	1	1	0	d	1	0	d	0
	1	0	0	1	1	1	1	0	1	0	d	0	0	d
(Unit	1	0	0	1	0	0	1	0	0	d	d	0	0	1
distance	1	0	1	0	1	0	0	0	0	1	0	d	0	d
UP)	1	0	1	0	0	1	0	0	d	0	0	d	0	1
	1	0	1	1	1	1	1	1	d	0	d	0	1	0
	1	0	1	1	0	1	0	1	d	0	0	1	d	0
	1	1	0	0	1	1	0	0	0	1	0	d	0	d
3	1	1	0	0	0	0	0	0	0	d	0	d	0	1
	1	1	0	1	1	0	1	1	0	d	0	0	1	0
(Unit	1	1	0	1	0	0	0	1	0	d	0	1	d	0
distance	1	1	1	0	1	1	0	1	d	0	d	d	1	0
DOWN)	1	1	1	0	0	1	1	1	d	0	d	0	d	0
	1	1	1	1	1	0	1	0	d	0	d	0	0	d
	1	1	1	1	0	1	1	0	d	0	d	0	0	1

Assuming don't care, $d = 0$, the expressions for S_2, R_2, S_1, R_1, S_0 and R_0 can be written from the truth table 10.4, as function of M_1, M_0, q_2, q_1 and q_0.

$$S_2 = \sum_m (3, 8, 18)$$

$$R_2 = \sum_m (7, 12, 20, 24)$$

$$S_1 = \sum_m (1, 5, 8, 12, 17, 29)$$

$$R_1 = \sum_m (3, 7, 10, 14, 23, 27)$$

$$S_0 = \sum_m (0, 2, 4, 6, 8, 10, 12, 14, 16, 22, 26, 28)$$

$$R_0 = \sum_m (1, 3, 5, 7, 9, 11, 13, 15, 19, 21, 25, 31)$$

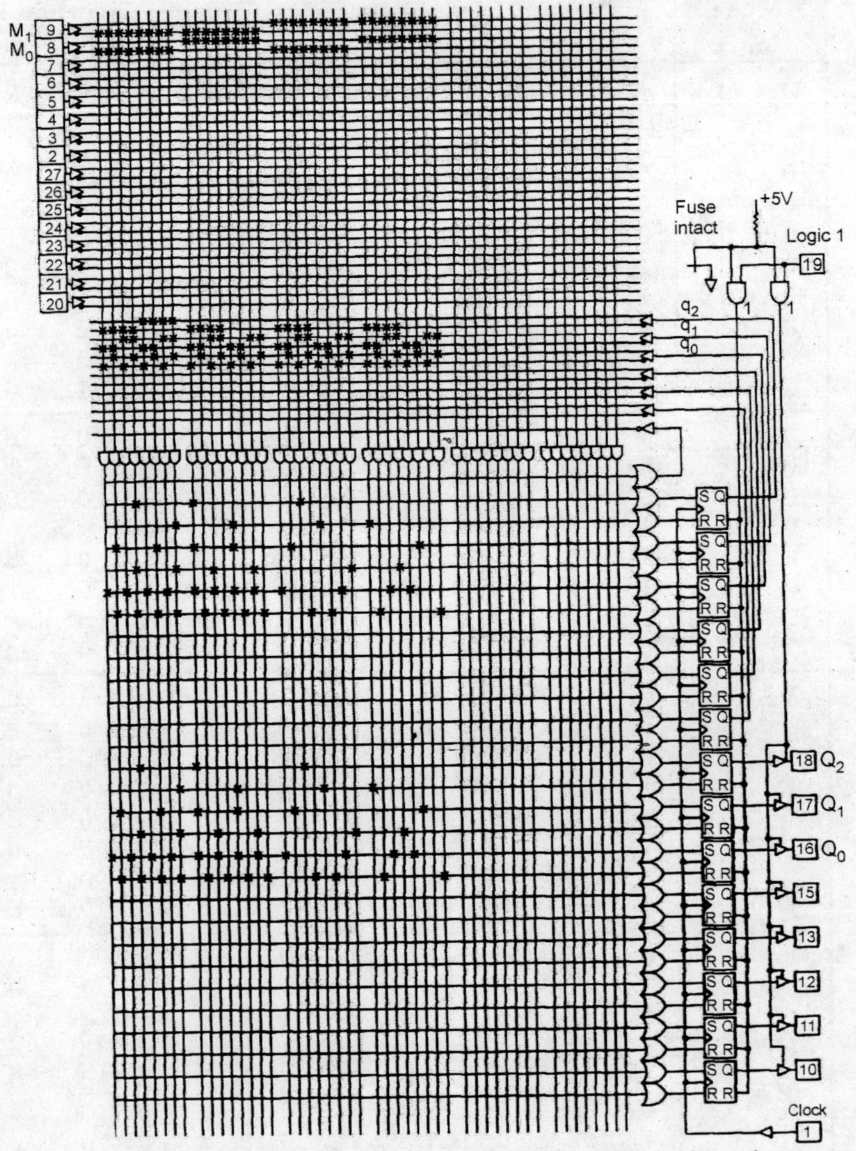

Fig. 10.44 Structure of PLS 105A PLA programmed for 3-bit multimode counter

To generate the SOP expressions for S_2, R_2, S_1, R_1, S_0 and R_0, 32 minterms are first generated using 5 variables (i.e., M_1, M_0, q_2, q_1 and q_0) using first 32 AND gates and sum of minterms are obtained by six OR gates as shown in Fig. 10.44. Here, first 3 flip-flops are used to give the feedback inputs while flip-flops (7 – 9) are used to give the outputs ($Q_2 - Q_0$) of 3-bit multimode counter.

10.8.3 Application of PLAs

The PLAs are used to replace ROMs in many applications. They are used for implementing combinational logic functions, and this results in compact circuitry and high switching speed. The following are the steps used for implementing the combinational logic functions:

1. Prepare the truth table.
2. Write the Boolean relations in SOP form.
3. Simplify the equations to obtain minimum SOP form.
4. Determine the input connections of AND matrix to generate the required product terms.
5. Determine the input connections of OR matrix to generate the required sum-of-product terms.
6. Determine the connections of Ex-OR matrix required for invert/non-invert matrix to set the active logic level of the outputs.
7. Program the PLA.

The PLAs can also be programmed as *conditionally addressable* ROMs.

Comparison of PLA and ROM: PLAs are similar to ROMs. Both of them have a first level of AND gates as followed by a second level of OR gates. A two-level AND-OR gate structure will serve to generate the most general logic function of the inputs. When the switching functions are complex and more in number, then the ROM and the associated decoders require complicated logic circuitry. But, the complex combinational logic circuit can be implemented easily using programmable AND and OR gates in PLA.

The ROMs completely decode the input variables to minterms, whereas PLAs do not. The number of gates in a PLA is comparable with a five or six-input ROM, while the input variable size is greater for PLAs. Functions of fewer input variables that can be described with fewer product terms can be realized with one PLA rather than a network of ROMs or other combinational logic blocks. Simplification of multiple output switching function is necessary in the programming of PLAs, whereas simplification is not so essential in the programming of ROMs.

10.8.4 Programmable Array Logic (PAL)

Programmable array logic (PAL) is a type of fixed architecture logic devices with programmable AND gates followed by fixed OR gates. Because only the AND gates are programmable, the PAL is easier to program, but is not as flexible as the PLA. The PAL has the same AND and OR arrays. In PAL, the inputs to the AND gates are programmable, while the inputs to the OR gates are hard-wired. This means that every AND gate can be programmed to generate any desired product of the input variables and their complements. Each OR gate is hard-wired to only selected AND gate outputs. A. PAL with 4 inputs and 4 outputs is shown in Fig. 10.45. Xs on the input side of AND gates represent the fusible links and Xs on the output side of AND gate are fixed connections. This limits each output function to sum of four product terms. If a function requires more than four product terms, it cannot be implemented with this PAL; one having more OR inputs would have to be used. If fewer than four product terms are required, the unneeded 1s can be made 0.

Though ROM and PLA are very versatile and simplify the logic design task greatly, their higher cost and larger size make them unattractive. An intermediate form of programmable logic device called Programmable Array Logic (PAL) integrated circuit has emerged that provides high performance at lower cost and size.

The names assigned to these ICs usually include a two-digit number followed by a letter and then followed by a digit, in which the number before the letter indicates the number of input lines and number after the letter represents the number of output lines. If the output gates are OR gates, the letter in the part number is *H* (active HIGH output); *L* is used when the output gates are NORs. For example, (*i*) 10H8 PAL provides 10 input and 8 output AND-OR gates and (*ii*) 14L4 PAL provides 14 input and 4 output AND-NOR gates.

Registered PALS: In addition to combinational circuits, Flip-Flops are required for the design of sequential logic circuits. PALs with flip-flops in the output to satisfy this requirement are called registered PALs. 16R4, 16R6 and 16R8 are some of the available registered PALs. The internal organization of 16R6 registered PAL is shown in Fig. 10.49. It consists of 16 inputs such as 8 external inputs ($I_1 - I_8$), 2 inputs fed back from combinational outputs (IO_2 and IO_8) and 6 inputs fed back from 6 flip-flop outputs ($O_2 - O_7$); 6 delay flip-flops and 8 outputs such as 2 from combinational logic and 6 from flip-flop outputs.

Configurable PAL: Configurable PALs are PALs with configurable (i.e., programmable) outputs which enhance the output capabilities of such devices. They are also called Generic PALs. In configurable PALs, macro cell is available in each output. A macrocell has special circuitry with fuses that can be configured for a variety of options.

The internal architecture of a configurable PAL IC 22V10 is shown in Fig. 10.50. It consists of 22 inputs, of which 12 are external inputs ($I_1 - I_{12}$) and the remaining 10 are inputs/outputs ($IO_1 - IO_{10}$). It also consists of 10 outputs in multiplexed with 10 inputs and 120 product terms. The I_1 input acts as the common clock input for *D* flip-flops, available in output macro cells. Here, the number of inputs (i.e., product terms) associated with various OR gates are not the same. All the 10 outputs are passed through configurable macrocells and tristate inverters. The macrocells cab is configured hi 4 different ways, i.e., it allows this configurable PAL to function as registered PAL or combinational PAL, in which the output of each case is in inverted or non-inverted form.

PAL programming: Consider the following Boolean functions to be generated:

$$Y_3 = \bar{A}B\bar{C}D + \bar{A}BC\bar{D} + \bar{A}BCD + ABC\bar{D}$$

$$Y_2 = \bar{A}BC\bar{D} + \bar{A}BCD + ABCD$$

$$Y_1 = \bar{A}B\bar{C} + \bar{A}BC + AC + AB\bar{C}$$

$$Y_0 = ABCD$$

On the top input line of Fig. 10.45, remove the first X, the fourth X, the fifth X and the eighth X. The top AND gate then has an Output of $\bar{A}B\bar{C}D$. By removing Xs on the next three input lines, we can make the top four AND gates produce the fundamental products of equation Y_3. The fixed OR connections on the output side imply that the first OR gate produces an output given by the relation

$$Y_3 = \bar{A}B\bar{C}D + \bar{A}BC\bar{D} + \bar{A}BCD + ABC\bar{D}$$

Similarly, we can remove Xs to generate the outputs Y_2, Y_1 and Y_0. The diagram of programmed PAL after having removed the unnecessary Xs is shown in Fig. 10.46.

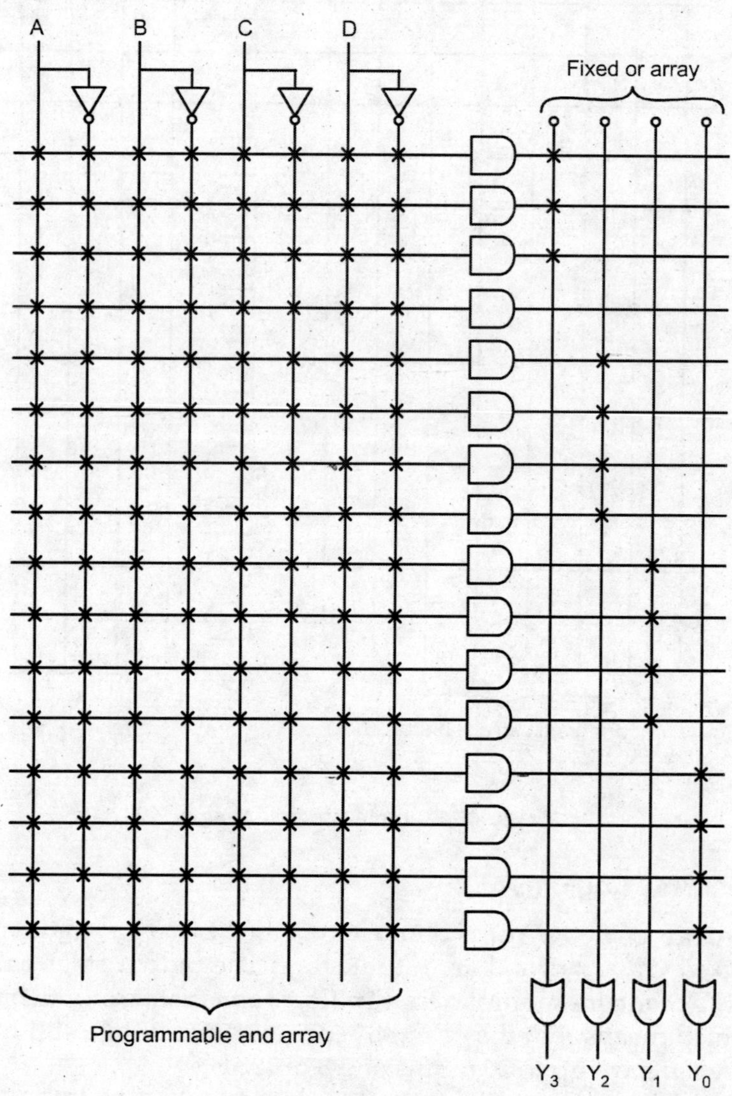

Fig. 10.45 Structure of PAL

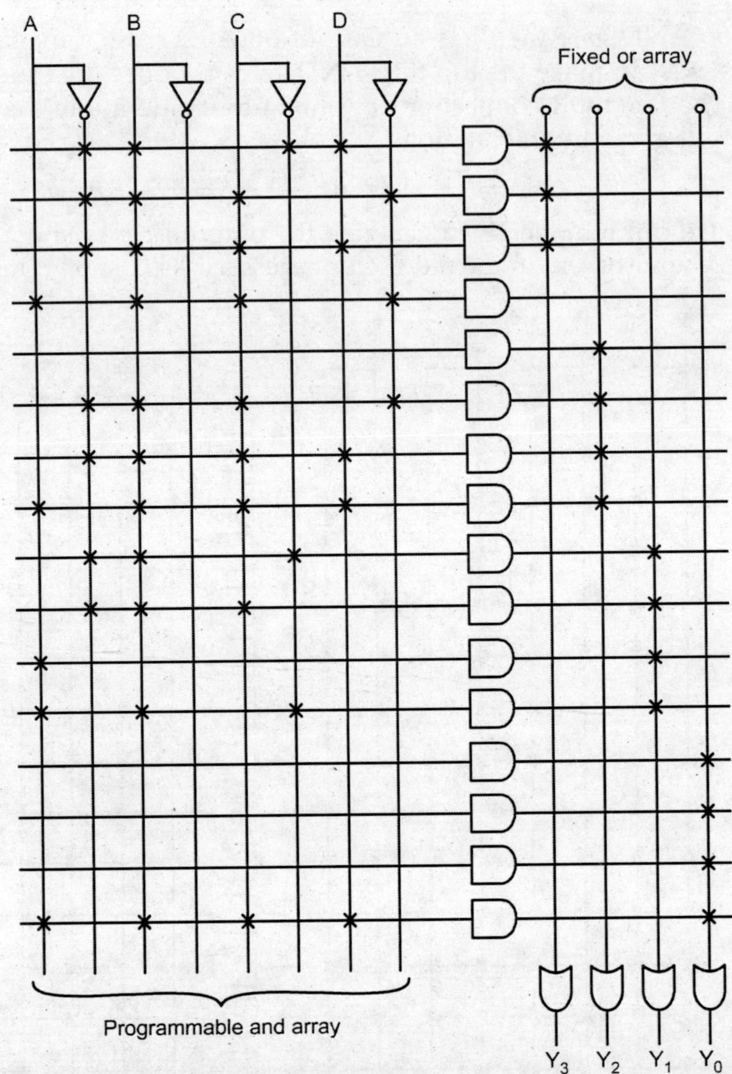

Fig. 10.46 Example programming a PAL

10.8.5 Generic Array Logic (GAL)

Generic Array Logic (GAL) is another type of PLD. This contains a programmable AND array, a fixed OR array, and an output stage. The output state, called an *output logic* macro cell (OLMC), contains multiplexers, flip-flops, and tristate output buffers. GAL devices are EEPLDs, which means that they can be reprogrammed while still in the circuit board. The GAL 16V8A can have up to 20 inputs and 8 outputs.

The logic diagram for the GAL16V8A is shown in Fig. 10.47. The OLMC block that receives the p-terms from the inputs or from flip-flops and output has tristate buffers.

The OLMC unit has two input multiplexers such as tristate mux product term MUX (PT MUX), one feedback MUX and one output multiplexer, one tristate flip-flop, and a fixed OR array. Figure 10.48 shows the block diagram of the cell.

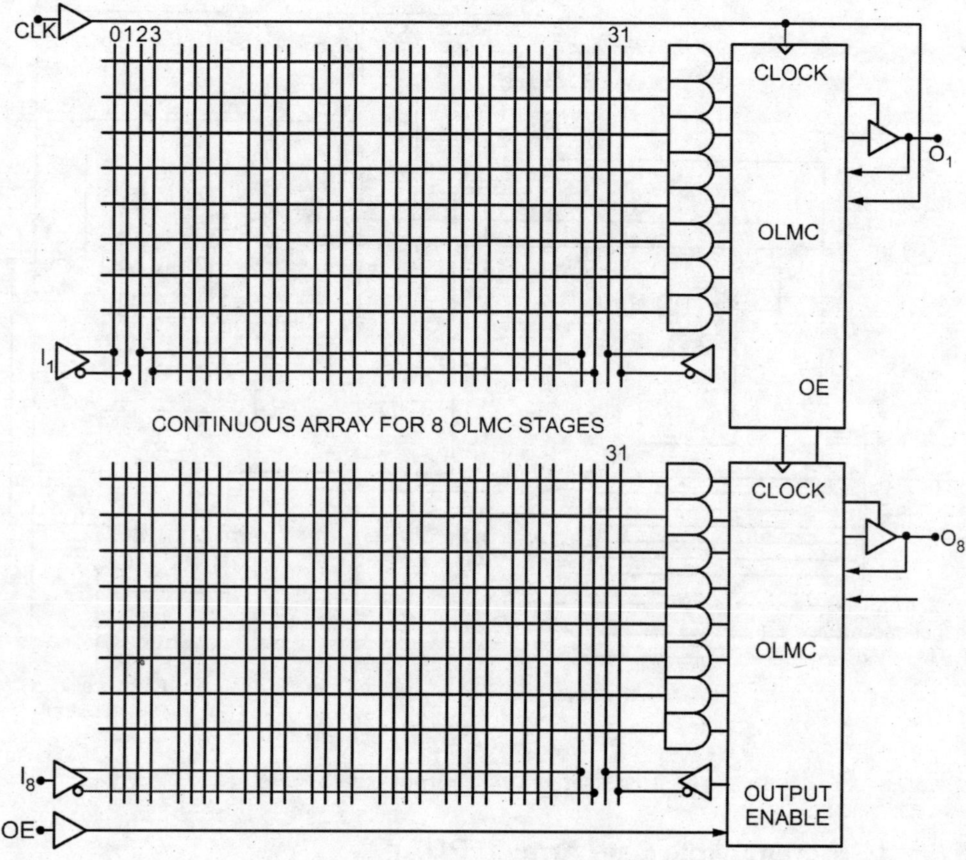

Fig. 10.47 GAL16V8A logic diagram

Inputs from the programmable AND array are given to the fixed OR array. The output of the OR array provides one input to a XOR gate. One of the inputs to the OR array comes from the PTMUX output. Two control inputs C_0 and C_1 control the PTMUX and TSMUX. The XOR output provides the input to the D flip-flop. The excitation level can be active-high or low, depending on the XOR (*n*) signal. The flip-flop output and the XOR output are inputs to the OUTMUX. This means that the cell can be configured as a combinational function or a registered function, depending on the values of C_0 and $C_1(n)$.

Registered feedback to the AND array is provided by the flip-flop Q', the output of tristate output of the OMUX or from an adjacent cell, under the control of C_0 and C_1 (*n*). A common positive edge clock provides flip-flop triggering. The tristate output can be controlled by a single output enable or by a p-term output steered through the Tristate MUX (TSMUX), to the output tristate enable.

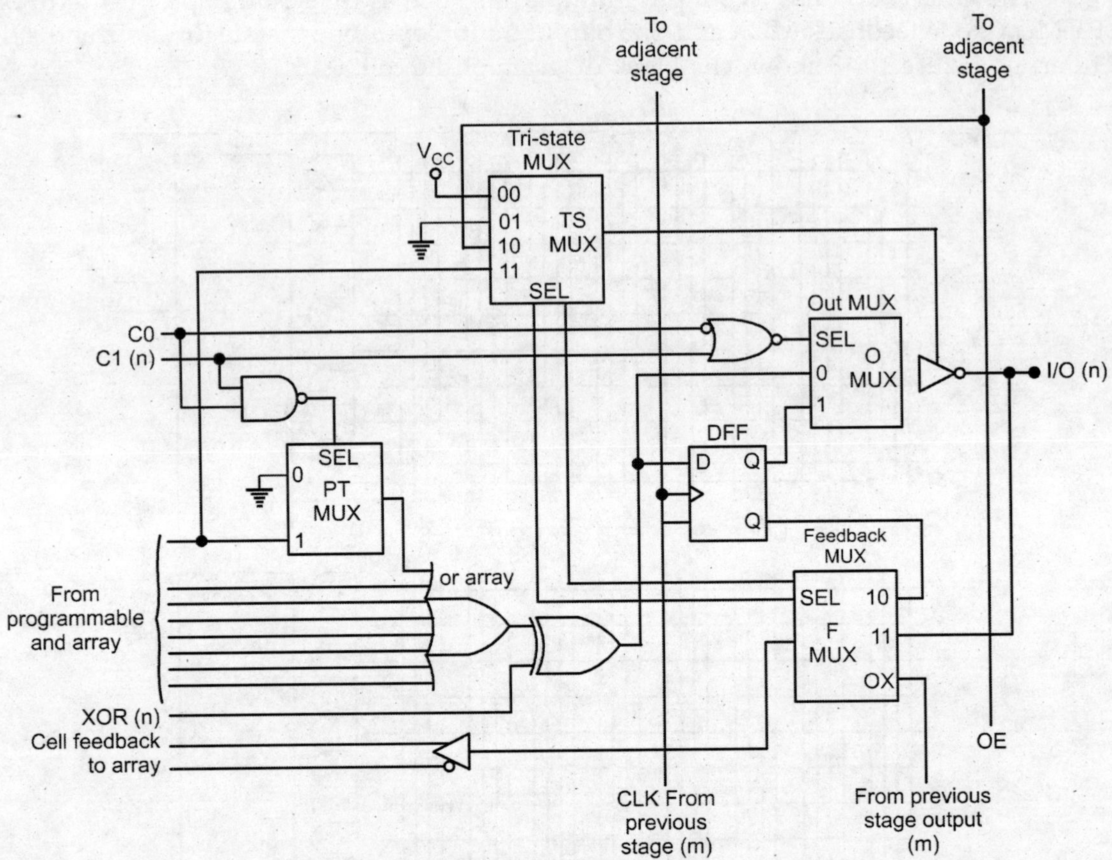

Fig. 10.48 GAL 16V8A output logic macro cell

10.8.6 Field Programmable Gate Array (FPGA)

Field Programmable Gate Array (FPGA) is a flexible architecture programmable logic device. It is a single Very Large Scale Integrated (VLSI) circuit constructed on a single piece of silicon. It consists of identical individually programmable rectangular modules as shown in Fig. 10.49. The modules are separated in both horizontal and vertical directions by horizontal and vertical metallic conductors called *channels.* In addition, each module has vertical and horizontal conductors at its input and output that cross one or more of the channels. Each intersection between the horizontal and vertical conductors, marked as a ⊕ in the figure, is a programmable link. These programmable links are used to interconnect the modules and also to program the individual modules.

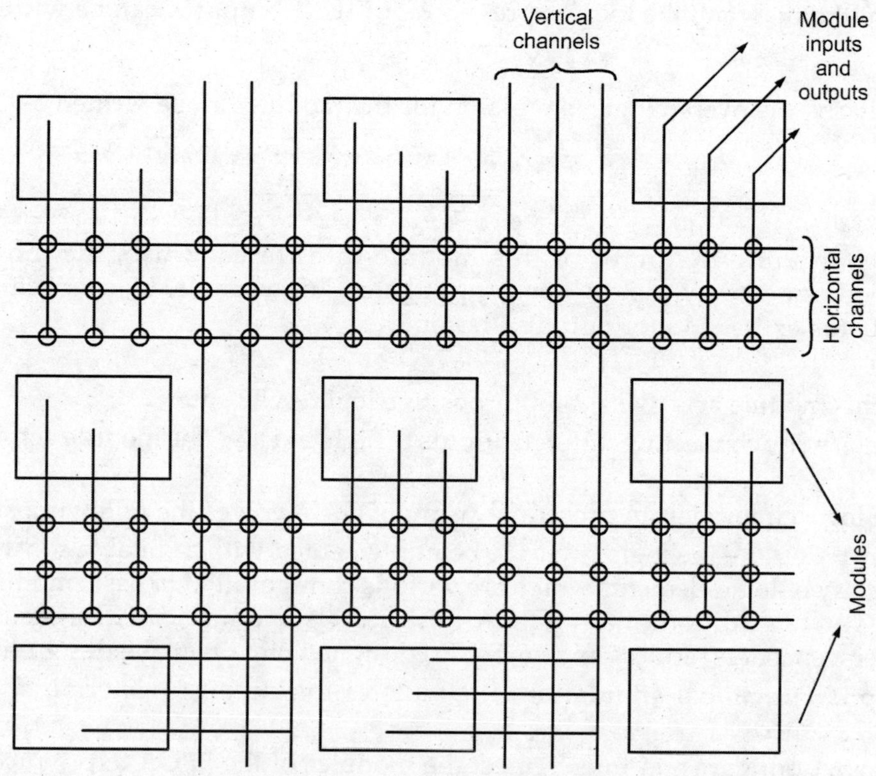

Fig. 10.49 Architecture of FPGA

The content of the modules depends on the type of FPGA. For easy use, the modules need to be programmable into the gates and sequential elements. A module may have both combinational and sequential components. On the other hand, there may be two different types of modules, namely combinational and sequential modules. Fig. 10.50 shows the circuit of a sample module used by one manufacturer in one type of FPGA. Other FPGA use different modules, but the concept of an array of programmable modules is common. The module shown in Fig. 10.50 consists of 3 number of interconnected 2-to-1 multiplexers (MUX1, MUX2 and MUX3) and an OR gate whose output selects the MUX3.

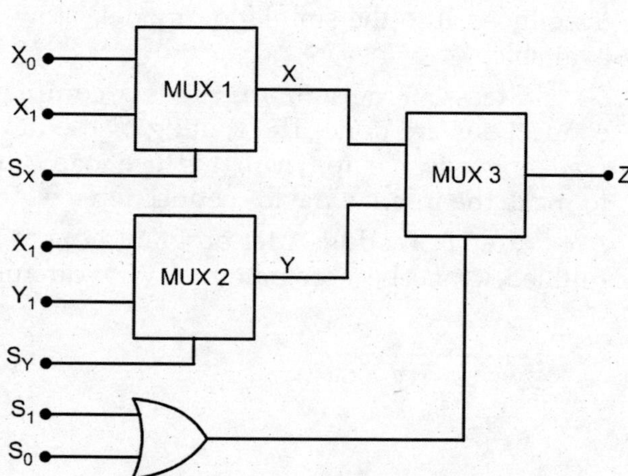

Fig. 10.50 Circuit of a sample module used in an FPGA

From the diagram, the logic expression for MUX1 output X can be written as

$$X = X_0\overline{S_X} + X_1S_X$$

Similarly, the overall logic expression for the module can be written as

$$Z = (X_0\overline{S_X} + X_1S_X)(\overline{S_0 + S_1}) + (Y_0\overline{S_Y} + Y_1S_Y)(S_0 + S_1)$$

$$= X_0\overline{S_X}\,\overline{S_0}\,\overline{S_1} + X_1S_X\overline{S_0}\,\overline{S_1} + Y_0\overline{S_Y}\,S_0 + Y_1S_YS_0 + Y_1S_YS_1$$

By connecting the input of this module in different ways, the module can be programmed as a variety of 2 or 3 or 4 input gates. For example, by connecting X_0, X_1, Y_0 and S_1 to 0, the expression for output Z becomes

$$OUT = Y_1\,S_Y\,S_0$$

and now this module acts like a 3-input positive logic AND gate.

Similarly, by connecting different inputs to high and low, the module acts like different gates.

The logic circuit design procedure using FPGA involves the following steps:

Step 1: Capture the logic circuit to be implemented with a suitable software package, using a library of logic elements which are various configuration of basic modules available in the FPGA. In addition, many FPGA libraries also contain predesigned circuits for multiplexers, encoders, adders and so on. Predesigned circuits make design much easier.

Step 2: Functional simulation: It simulates the circuit to determine whether it is functioning properly.

Step 3: Configure and interconnect the modules of the FPGA to produce the desired logic circuit. This may be done automatically by a routing software called *router*. Once the routing is over, it is now possible to determine the actual circuit delays which can now be introduced into the simulation model. Now, an accurate simulation of the circuit can be available.

Step 4: Programming: It is a completely automated step in which FPGA interconnections are done. The routing of the devices determined in the previous step is now made into a fuse map. Then, this fuse map is used in conjunction with a device programmer to make the internal device connections.

Step 5: Testing: After programming, it must be tested. If the designed function is not fulfilled, it must be reprogrammed. With careful simulation, reprogramming can be minimized.

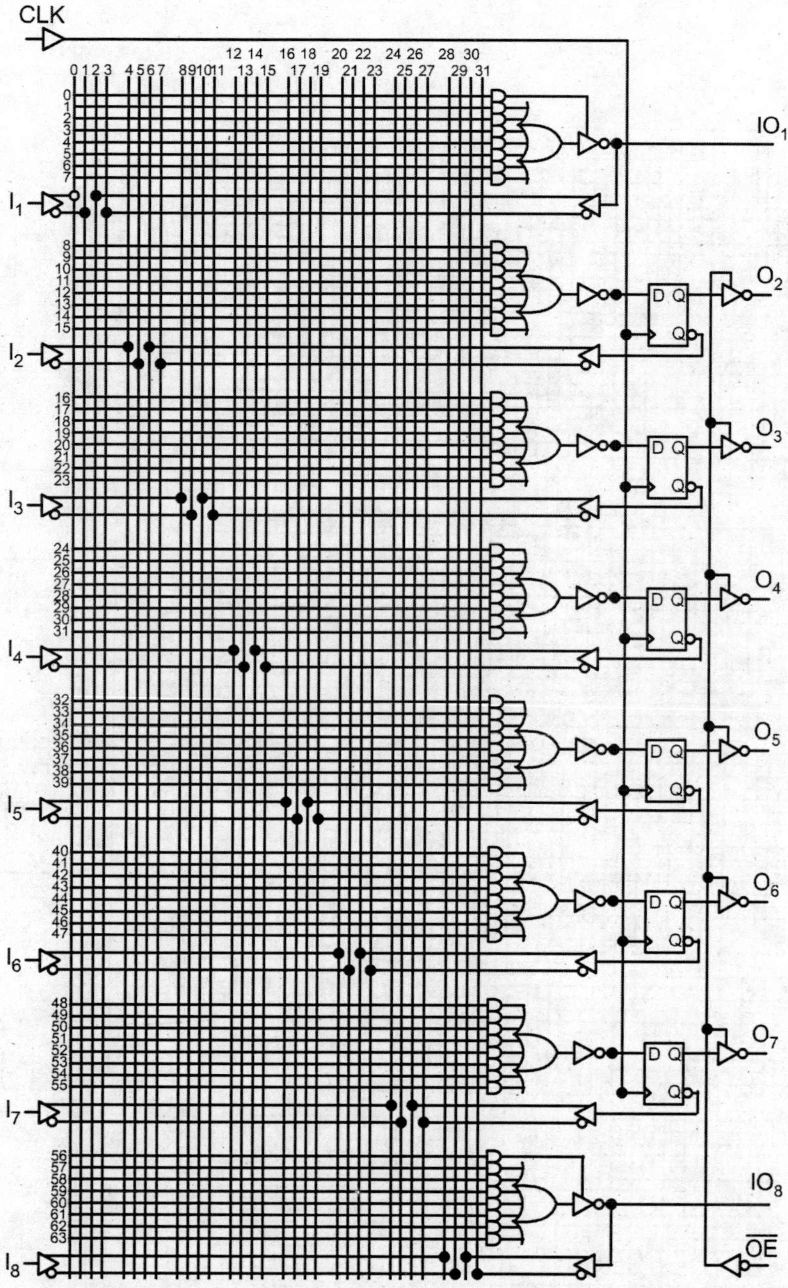

Fig. 10.51 Internal organisation of Registered PAL 16R6

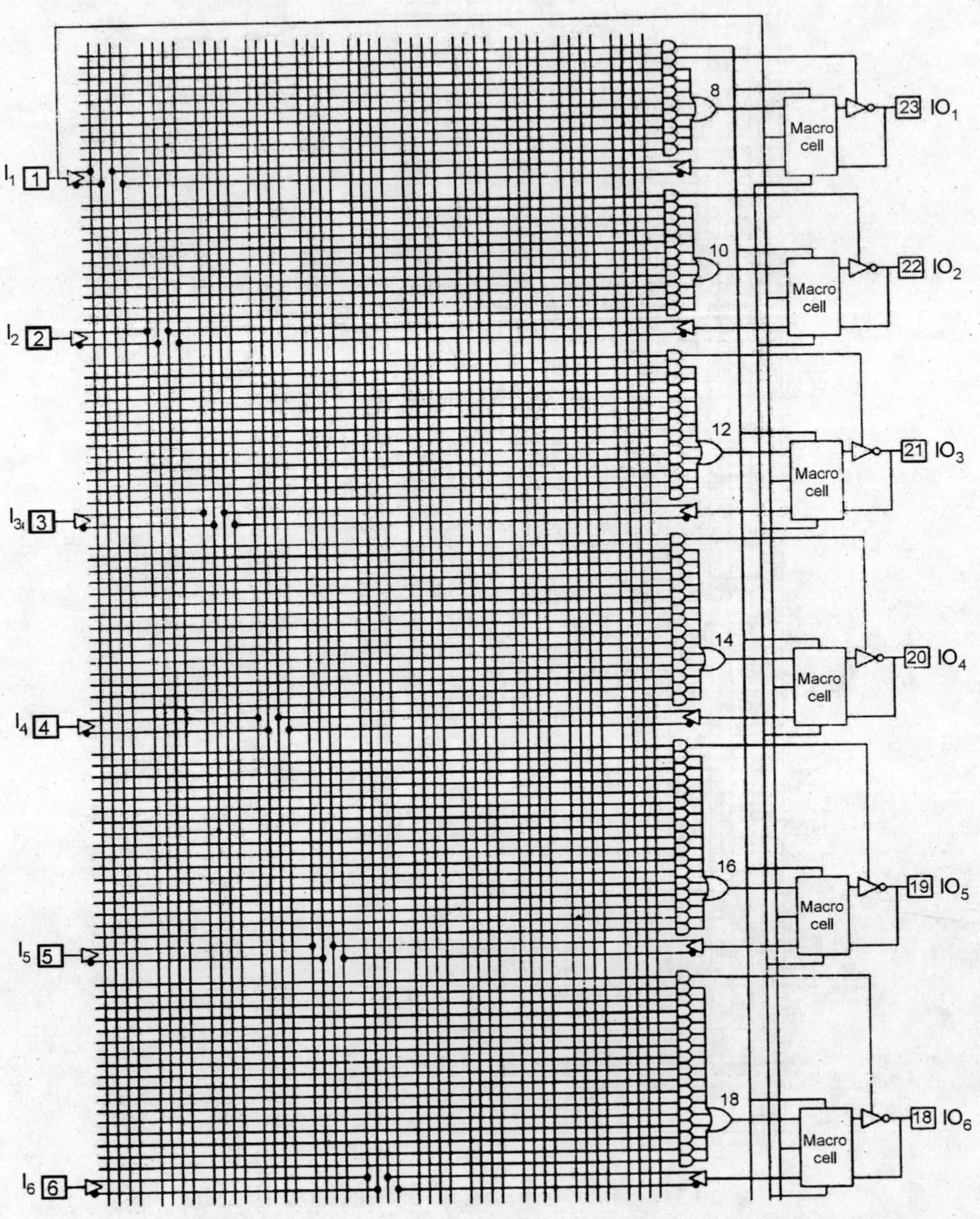

I1 1
I2 2
I3 3
I4 4
I5 5
I6 6
8
10
12
14
16
18
Macro cell
Macro cell
Macro cell
Macro cell
Macro cell
Macro cell
23 IO1
22 IO2
21 IO3
20 IO4
19 IO5
18 IO6

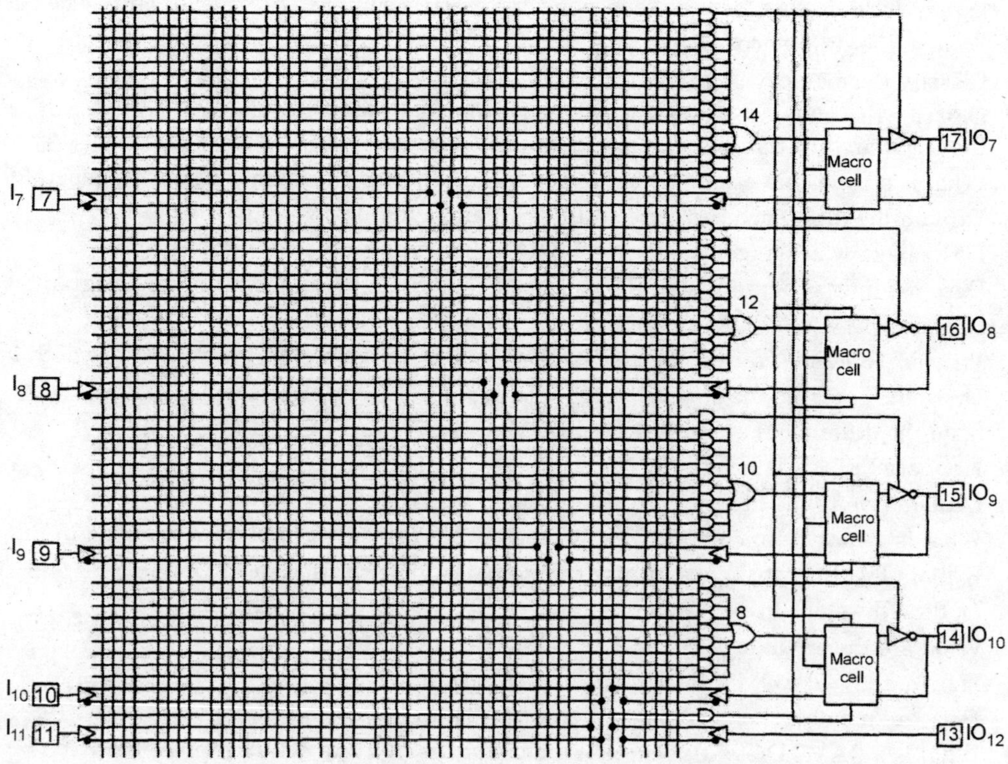

Fig. 10.52 Internal architecture of a configurable PAL IC 22V10

■ REVIEW QUESTIONS ■

1. Define the following terms: (*a*) memory cell (*b*) memory word (*c*) address (*d*) byte (*e*) access time.
2. Explain the difference between read (fetch) and write (store) operations.
3. Explain the terms: (*a*) volatile memory (*b*) non-volatile memory.
4. Explain the difference between static and dynamic memories.
5. What are the primary advantages of bipolar memory over MOS memory?
6. Explain the difference between primary and secondary memories.
7. Describe and compare sequential access memories, random access memories and read only memories.
8. Describe a word stored in a memory and discuss its address.
9. What is a ROM?
10. How are memory cells organized?
11. What is row-column selection?
12. How many address bits are needed for a given number of memory locations?

13. Explain the difference between semicnductor memories and memories that use magnetic materials.
14. Explain the difference between RAM and ROM.
15. Classify the memory devices.
16. Describe the function of the row-select decoder, column-select decoder and output buffers in the ROM architecture.
17. What is a code converter? What kind of information is stored in a character generator ROM?
18. Explain the difference between an EPROM and PROM.
19. Describe how a semiconductor read only memory is built and used.
20. Describe how programmable ROM ICs are constructed and how they are programmed.
21. Explain why an EPROM is not a volatile memory.
22. Why are ROMs used in computer systems? What sort of memory elements are used in ROMs?
23. Draw the basic circuit of a ROM cell and explain its working.
24. State the differences among ROM, PROM, EPROM, EAPROM and EEPROM.
25. How are PROM data written?
26. Explain briefly the different types of ROMs.
27. What is meant by fusible link? How many types of fuse technologies are used in PROM?
28. Explain the different types of fuse technologies that are used in PROM.
29. Outline the steps in programming and verifying a 74186 PROM.
30. What are the advantages of an EEPROM over an EPROM?
31. How are UV-EPROM data written and erased?
32. How are memory units connected onto a data bus?
33. What is a RAM? Draw and explain the circuit of a typical cell of bipolar RAM.
34. Describe the input conditions needed to read a word from a specific RAM address location.
35. How does a static RAM cell differ from a dynamic RAM cell?
36. Draw the logic diagrams of a static RAM cell and a dynamic RAM cell.
37. What are the RAM timing parameters that will determine its operating speed?
38. What is the reason for the refresh operation in dynamic RAMs?
39. What are the advantages of dynamic RAM over static RAM?
40. Define the term mask programming.
41. What are the ways of achieving nonvolatile RAM storage?
42. Explain the advantages and disadvantages of dynamic RAM.
43. What is meant by PSRAM?
44. What is a PLD? What do a dot and an represent on a PLD diagram?
45. How does the architecture of a PAL differ from a PROM?
46. What functions does a PLD programmer perform?
47. List the applications of PLA.
48. What are the steps used for implementing combinational circuit using PLA?
49. How does the architecture of PLA differ from ROM and PAL?
50. Explain why RAM and ROM are both random access memories.
51. Explain FPGA with architecture.
52. What are the advantages of FPGA?
53. Explain the steps involved in programming of FPGA?

54. Design a memory decoder to select 4 numbers of 4 KB EPROM ICs and two numbers of 4KBRAMICs.
55. Design a memory decoder to select 1 number of 16 KB EPROM IC and one number of 32 KB RAM IC.
56. Design a suitable circuit for the selection of one number of 8 KB EPROM, one number of 4 KB EPROM, one 16 KB RAM and one 8 KB RAM.
57. Design a 5 bit UP/DOWN counter using PLA.
58. Design Binary to Gray and Gray to Binary converters in a single PLA.

■ PROBLEMS ■

1. A certain memory has capacity of 16K × 32. How many words does it store?

   [*Ans: 16,384 words; 5,24,388 bits*]

2. How many different addresses are required by the memory that contain 16 K words?

   [*Ans: 16,384*]

3. What is the bit storage capacity of a ROM with a 512 × 4 organization?

   [*Ans: 2048 bits*]

4. How many address bits are required for a 2048 × 1-bit memory? [*Ans: 11 bits*]
5. For the ROM array in Fig. P10.5, determine the outputs for all possible input combinations, and summarize them in tabular form (light cell is a 1, dark cell is a 0).

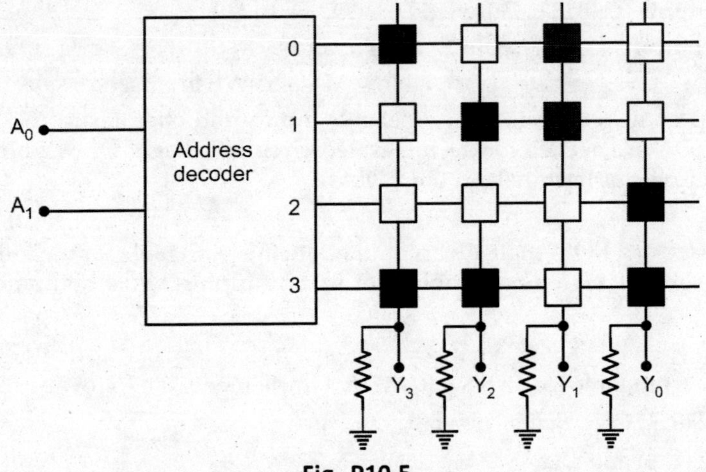

Fig. P10.5

Ans:

Inputs		Outputs			
A_1	A_0	Y_3	Y_2	Y_1	Y_0
0	0	0	1	0	1
0	1	1	0	0	1
1	0	1	1	1	0
1	1	0	0	1	0

6. How many 16K × 1 RAMs are required to achieve a memory with word capacity of 16K and a word length of eight bits? *[Ans: 8]*
7. To expand the 16K × 8 memory obtained in Problem 6 to a 32K × 8 organization, how many more 16K × 1 RAMs are required? *[Ans: 8]*
8. What ranges of hex address values are used in a 64K memory? *[Ans: 0000 to FFFF]*
9. How many memory locations are there for address values form 0000 to 07FFF? *[Ans: 2048_{10} locations]*
10. How many memory locations are there from C000 to C3FF? *[Ans: 1024 locations]*
11. How many memory locations are there from A000 to BFFF? *[Ans: 8192 memory locations]*
12. What is the hex address range for a 4K × 8 ROM with hex addresses starting from $6000)_{16}$? *[Ans: $6FFF_{16}$]*
13. How many 2114 ICs are required for a 4K × 8-bit memory? *[Ans: 8 ICs]*
14. A 64K × 1 dynamic memory chip uses multiplexed address inputs. How many address inputs does this chip have? *[Ans: 8]*
15. A certain memory stores 8K sixteen-bit words. How many data input and data output lines does it have? How many address lines does it have? What is its capacity in bytes? *[Ans: 16; 16; 13; 16,384]*
16. Show how to connect 7489s in series to construct a memory that has thirty-two 4-bit words. *[Ans: Two identical chips are connected together as follows:*
    (*a*) Select inputs (*A* to *A*, *B* to *B*, *C* to *C*, *D* to *D*)
    (*b*) Data inputs (D_1 to D_1, D_2 to D_2, D_3 to D_3, D_4 to D_4)
    (*c*) Sense outputs (S_1 to S_1, S_2 to S_2, S_3 to S_3 and S_4 to S_4).
    Now ME and WE are used to select one chip or the other
17. A ROM chip of 4096 × 8 bits has two enable inputs and operates from a 5-volt power supply. How many pins are needed for the integrated-circuit package? Draw a block diagram and label all input and output terminals in the ROM. *[Ans: 24 Pins]*
18. Specify the size of a ROM that will accommodate the truth table for the following combinational circuit components: (*a*) a binary multiplier that multiplies two 4-bit numbers, (*b*) a 4-bit adder-subtracter. *[Ans: (a) 256 × 8; (b) 512 × 5]*
19. Tabulate the truth table for an 8 × 4 ROM that implements the following four Boolean functions:
    $A(x, y, z) = \Sigma(0, 1, 6, 7)$, $B(x, y, z) = \Sigma(0, 1, 6, 7)$,
    $C(x, y, z) = \Sigma(2, 6)$ and $D(x, y, z) = \Sigma(1, 2, 3, 5, 7)$
20. Tabulate the PLA programming table for the four Boolean functions listed in Example Problem 19. Minimize the number of product terms. *[Ans: Seven product terms: $y\bar{z}, x\bar{z}, \bar{x}\bar{y}z, \bar{x}\bar{y}, xy, \bar{x}y, z$]*
21. Show the main connections of 4116 ICs used in a 16K × 4-bit memory.

❑❑❑

Synchronous Sequential Circuits

11.1 INTRODUCTION

Digital circuits are of two types, namely *combinational* and *sequential* circuits. In combinational circuits, the value of the outputs at any instant of time is dependent only on the present values of the inputs. In sequential circuits, the outputs depend not only on the present inputs but also on the sequence of all the past inputs, i.e. previous output state of the circuit. Sequential circuits require memory elements to store the previous output/state of the machine to determine the present output. Unlike combinational circuits, it is not practicable to prepare an input-output table of all possible sequences of inputs and outputs.

A sequential circuit can be classified into synchronous and asynchronous circuits. Sequential circuits can be viewed as machines that proceed through an orderly set of conditions, called *states*. If the transitions of the sequential circuit from one state to the next state are controlled by a clock, then the circuit is called a *synchronous sequential circuit*. When the circuit is not controlled by the clock, the transition from one state to the next occurs whenever there is a change in the input to the circuit at any time, and hence this circuit is called an *asynchronous sequential circuit*. Asynchronous sequential circuits give high speeds, as they do not use a clock. They are useful in applications where the input signals to the system may change at any time, independently of an internal clock. They are also classified *into fundamental mode asynchronous sequential circuits* and *pulse mode asynchronous sequential circuits*.

All the state variables in sequential circuits are binary in nature. If the number of state variables is 'n', then the sequential circuit has 2^n possible states. Even for larger values of 'n', the number of possible states is still finite. Therefore, sequential circuits are referred to *as finite state machines*. This chapter deals with the finite state circuits.

11.2 GENERAL SEQUENTIAL CIRCUIT MODEL

The model for a general sequential circuit is shown in Fig. 11.1. The present state (PS) of the circuit is stored in the memory element. The memory can be any device capable of storing enough information to specify the state of the circuit. For example, if a sequential circuit has 'n' states, then the memory could be any device that can store codes representing these 'n' states.

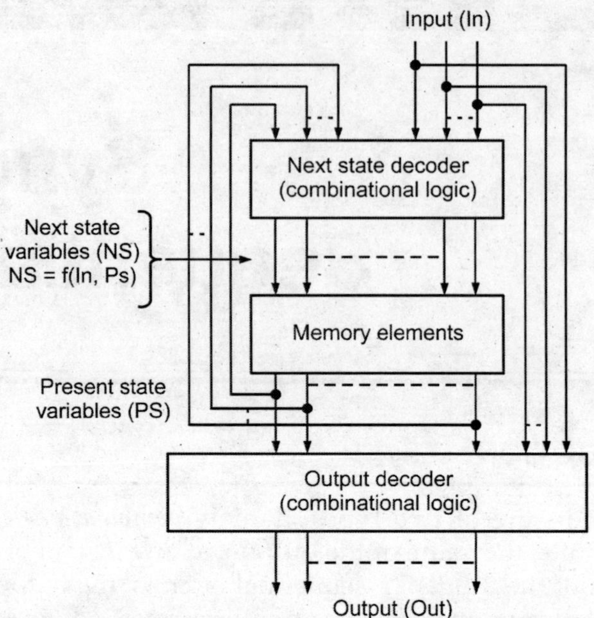

Fig. 11.1 General sequential circuit model

The next state (NS) of the circuit is determined by the present state (PS) of the circuit and by the inputs (In). The function of the NEXT STATE decoder logic is to decode the external inputs (In) and the present state (PS) of the circuit (stored in the memory) and to generate at its output a code called NEXT STATE variable. The next state variables will become the present state variables when the memory loads and stores them. This process is called a *state change.* State changing is a continuous process with new set of inputs and each new state being decoded to form the new next state variables. Thus, sequential circuit is a feedback system in which the present state of the circuit is fed back to the next state decoder and used along with the input to determine the next state.

The output (Out) of the circuit is determined by the present state of the machine and also possibly by the input to the circuit. The function of the output decoder is to decode the present state of the machine and the present inputs for the purpose of generating the desired outputs. The output of a synchronous machine may be clocked, just as the state transition is clocked.

11.3 CLASSIFICATION OF SEQUENTIAL CIRCUITS

Sequential circuits are generally classified into five different classes:

(*i*) Class A circuit
(*ii*) Class B circuit
(*iii*) Class C circuit
(*iv*) Class D circuit
(*v*) Class E circuit

The sequential circuit models for the above classes can be derived from the general sequential circuit model shown in Fig. 11.1.

The class A circuit is defined as a MEALY circuit named after G.H. Mealy, one of the leading personalities in sequential circuit design. The basic property of Mealy circuit is that the output is a function of the present input conditions and the present state (PS) of the circuit. The class A model is shown in Fig. 11.2.

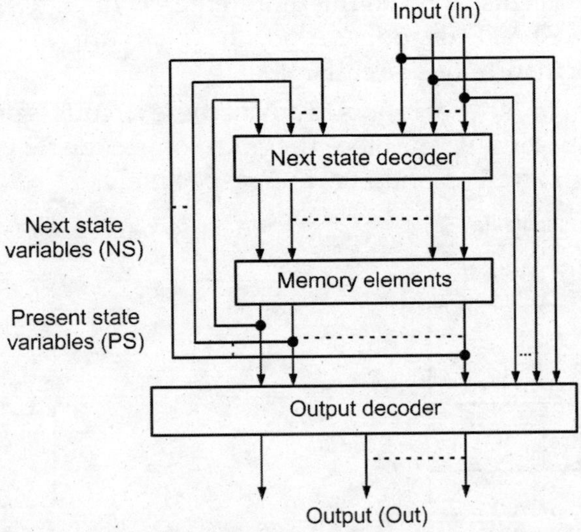

Fig. 11.2 Class A circuit

The class B and class C circuits are generally defined as MOORE circuits, named after E.F. Moore, another leading personality in sequential circuit design. The basic property of a Moore circuit is that its output is strictly a function of present state (PS) of the circuit inputs. The block diagram of class B and class C circuits are shown in Figs. 11.3 and 11.4 respectively.

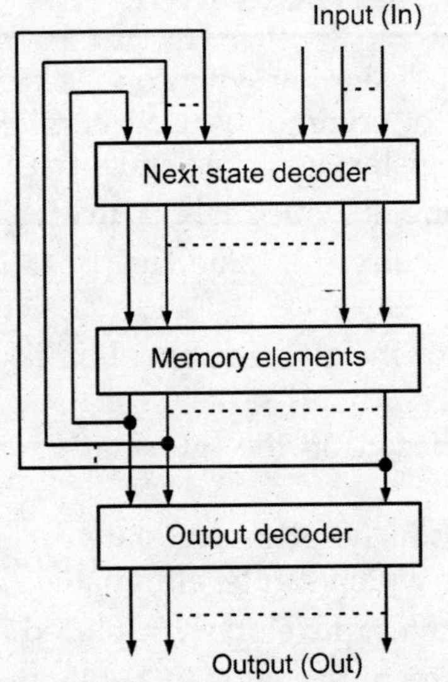

Fig. 11.3 Class B circuit (MOORE circuit with output decoder)

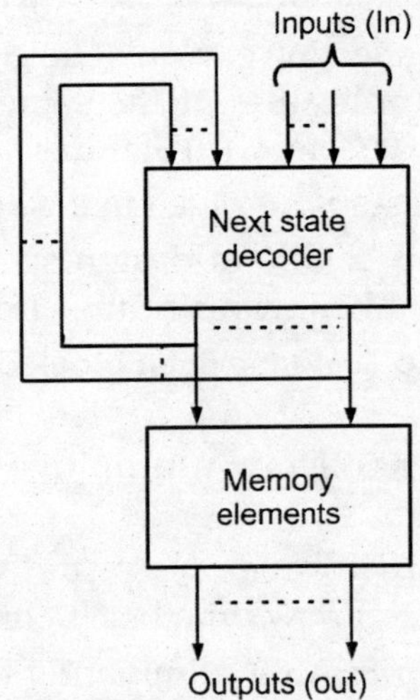

Fig. 11.4 Class C circuit (MOORE circuit without output decoder)

Both Mealy and Moore circuits are widely used; in some cases, one can get circuits that are combinations of both, i.e. some outputs depend on both the inputs and present state of the machine and other outputs depend only on present state of the circuit.

Class A, B and C circuits with a single input form the general model for a *counter circuit* in which the events to be counted are entered directly into the memory element or through the NEXT STATE decoder. Also, these circuits are equally applicable to both synchronous and asynchronous circuits.

The minimum number of inputs to any of these circuits is one. For synchronous circuits, the single input is clock input. The block diagram connection for class D and class E sequential circuits are shown in Fig. 11.5 and Fig. 11.6 respectively.

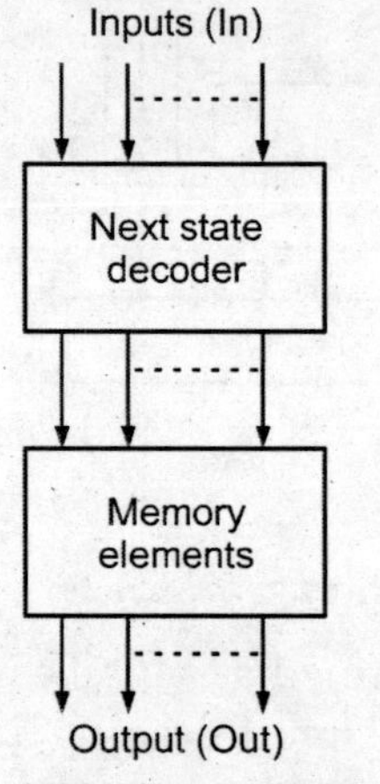

Fig. 11.5 Class D circuit

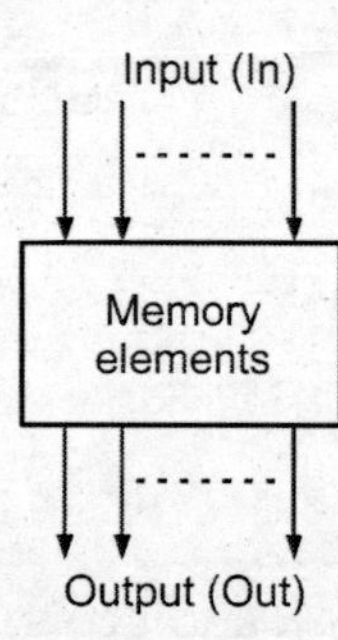

Fig. 11.6 Class E circuit

11.4 DESIGN OF SYNCHRONOUS SEQUENTIAL CIRCUITS

This section explains the general method of designing synchronous sequential circuits and the application of the same for the design of commonly used circuits, viz. (*i*) Code sequence detectors, (*ii*) Standard counters, (*iii*) Registers and (*iv*) Sequential code generators.

The steps involved in the design of synchronous sequential circuit are given as follows:

Step 1: Obtain design specifications and study the same clearly to understand the operational behaviour of the circuit.

Step 2: Construct a block diagram model for the given design. Identify all inputs and outputs.

Step 3: Draw a primitive state diagram based on the information obtained in the above steps.

Step 4: Develop a primitive state table from the primitive state diagram and check this table to know whether it contains any states which can be eliminated (i.e. redundant).

Step 5: After eliminating the redundant states, develop a simplified state table.

Step 6: Make a state assignment and document the same in a state map.

Step 7: Develop a Present state/Next state (PS/NS) table using the above state assignments and output table.

Step 8: Decide the type of memory elements to be used in the design and then obtain the excitation table from PS/NS table using the application table of the flip-flop.

Step 9: Draw excitation and output maps from step 7 and step 8 and simplify the excitation and output function, i.e. derive the next state decoder and output decoder logic.

Step 10: Draw the schematic diagram.

To understand the above design procedure in detail, the following examples can be considered.

11.4.1 Design of Serial Binary Adder

The serial binary adder is capable of adding two binary numbers serially, i.e., bit by bit. The design of serial binary adder is explained in Example 11.1.

Example 11.1. *Design a serial binary adder using delay flip-flop.*

Solution: The serial binary adder is capable of adding two binary numbers serially, i.e. bit by bit. Before proceeding with the design process, consider the addition of two 4-bit binary numbers *A* and *B*,

$$A = A_3A_2A_1A_0 = \quad 1011$$
$$B = B_3B_2B_1B_0 = \quad \underline{1011}$$
$$10001$$

Step 1: In the above addition process, the output at a time instant t_i depends not only on A_i and B_i bits but also on carry (C_{i-1}) generated in the previous addition of A_{i-1} and B_{i-1} bits. Now, one can understand that the C_{i-1} must be memorised and fed back to the input-side at time t_i, i.e. after a time delay of 1 unit. Therefore, the adder must be able to preserve or memorise the carry, generated at any instant of time (say t_{i-1}) upto time t_i.

Step 2: From the above word description, the block diagram of the serial binary adder can be drawn as shown in Fig. E11.1 (*a*). In the case of serial adder, the carry generated at any instant of time will either be '0' or '1'. Let us designate the carry as the state of the adder. Let *X* be the state of the adder at time t_i if a carry 0 is generated at time t_{i-1} and *Y* be the state if '1' is generated as carry at time t_{i-1}. The state of the adder at the time (t_i) when the present inputs (A_i and B_i) are applied is referred to as *Next State* because this is the state to which the adder goes due to the new carry. The output of the memory element is called *Present State* of the adder. The output Z at time t_i depends on inputs at that time (A_i and B_i) and the state of the adder at that time.

Step 3: Primitive state diagram: Now, a primitive state diagram of serial binary adder can be constructed as shown in Fig. E11.1 (*b*).

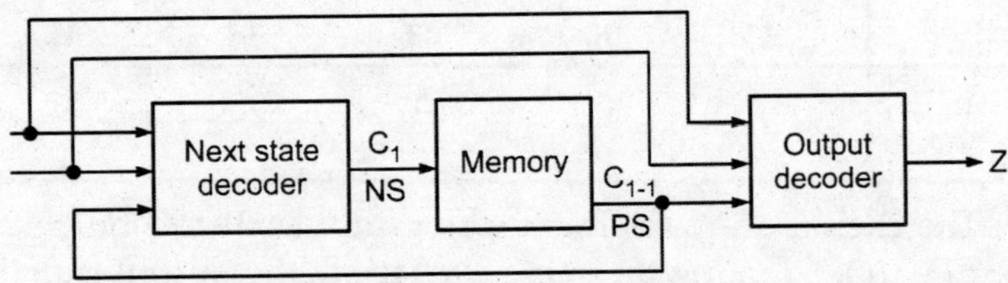

Fig. E11.1(*a*) Block diagram of serial binary adder

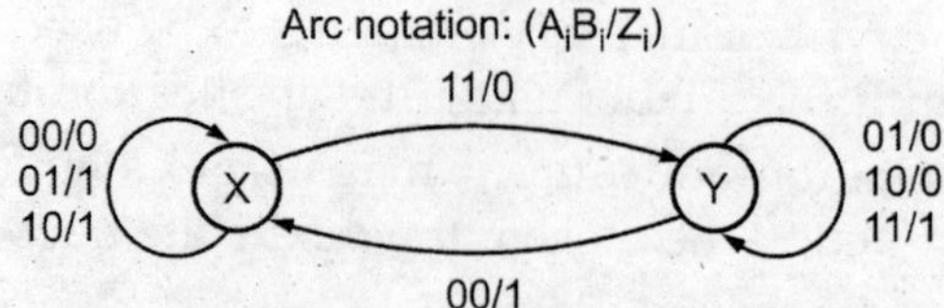

Fig. E11.1(*b*) State diagram of serial binary adder

Step 4: Primitive state table: From the primitive state diagram, the primitive state table can be constructed as shown in Table E11.1(*a*).

Table E11.1(*a*) State table of serial binary adder

Present state (PS)	Next state, Output (NS, Z) A_iB_i = 00	01	11	10
X	X, 0	X, 1	Y, 0	X, 1
Y	X, 1	Y, 0	Y, 1	Y, 0

Step 5: Now, there are only two states X and Y which are not redundant since states X and Y are not equivalent states.

Step 6: State assignment: Since there are only two state variables, '0' can be assigned to X and '1' to Y.

Step 7: Now, using the above state assignment, the modified Present State/Next State and output table can be constructed as shown in Table E11.1(*b*).

Table E11.1(*b*) PS/NS and output table

PS C_{i-1}	Next state (C_1) A_iB_i =				Output (Z)			
	00	01	11	10	00	01	11	10
0	0	0	1	0	0	1	0	1
1	0	1	1	1	1	0	1	0

Step 8: Here, the type of memory element, i.e. flip-flop, to be used has to be decided. If D flip-flop is used as a memory element in serial adder, then the excitation table can be obtained from PS/NS table using application table of D flip-flop as shown in Table E.11.1(*c*).

Table E.11.1(*c*) Excitation table of serial binary adder

C_{i-1}	A_iB_i =	*D* 00	01	11	10
0		0	0	1	0
1		0	1	1	1

Step 9: The excitation maps for flip-flop input '*D*' and output map can be drawn as shown in Fig. E11.1(*c*). By using the same excitation, the output functions can be simplified.

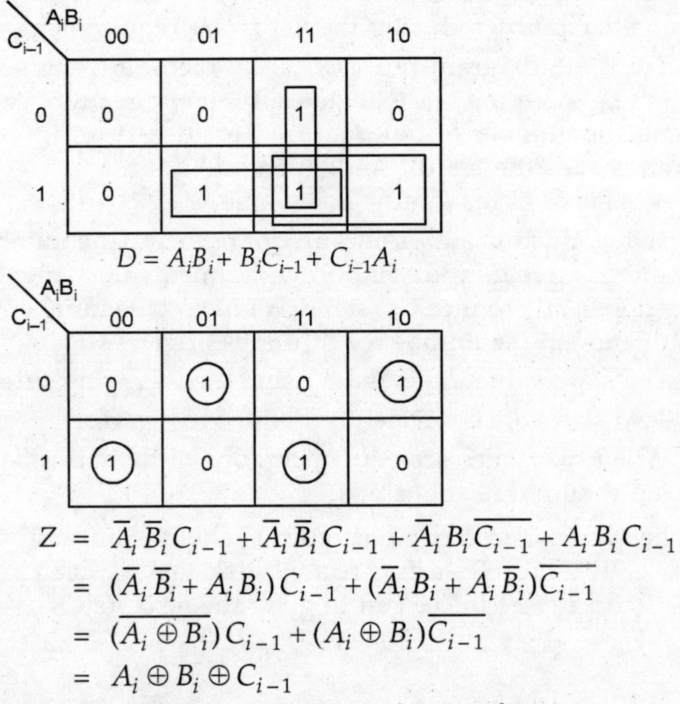

Fig. E11.1(*c*) *K*-map for simplifying excitation and output functions

Step 10: Using the simplified excitation and output functions, the circuit diagram for serial adder can be implemented as shown in Fig. E11.1(*d*).

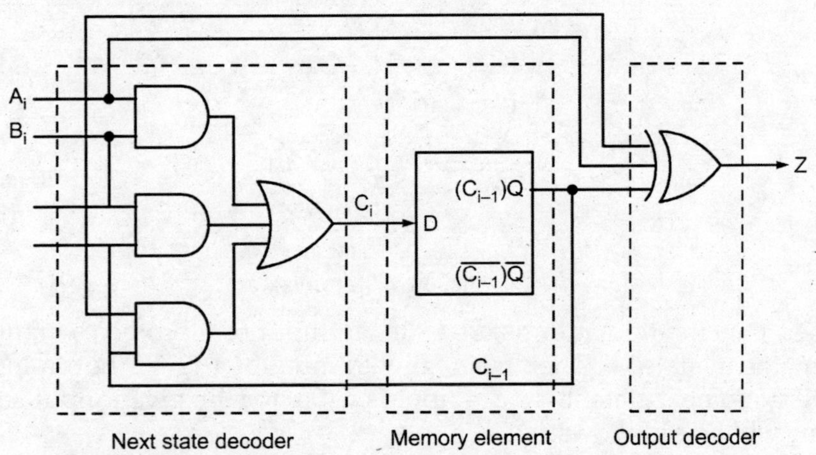

Fig. E11.1(*d*) Circuit diagram of serial binary adder

11.4.2 State Reduction and Assignment

The synchronous sequential circuits are usually represented by state diagram and state table. When the state diagram or state table is directly used to implement the sequential circuit, more hardware is needed. Hence, the state reduction may be used to optimize the sequential circuit by reducing the hardware requirement.

The state diagram represents the pictorial representation of inputs, outputs and transition between the internal states. When large number of states are present in the state diagram, the number of bits required for binary assignment is more and the hardware requirement for the circuit implementation is more. Hence, it is essential to reduce the number of states as maximum as possible.

State reduction algorithms are for reducing the number of states in a state table while keeping the external input-output requirements unchanged. The number of state variables (or flip-flops) 'm' produces 2^m states. Reducing the number of states may (or may not) result in a reduction in the number of flip-flops or gates.

The state reduction is done using the following rules:

- Two states are equivalent if both have the same next states and outputs.
- When two states are equivalent, one of them can be eliminated without altering the input-output relationships.

Let us explain the input-output relationship with an example. Consider the input sequence 01010110100 starting from initial state 'a'. Each input of 0 or 1 produces an output of 0 or 1 and causes the circuit to go to the next state.

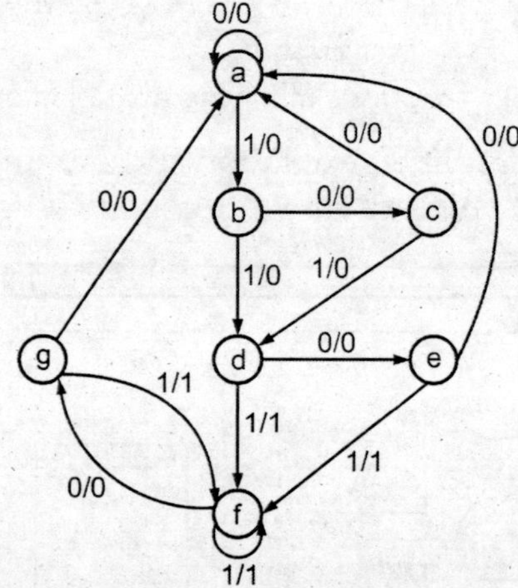

Fig. 11.7 State diagram

When the circuit is in initial state 'a', an input of 0 produces an output of 0 and the circuit remains in state 'a'. Then from 'a' with input of 1 (2nd input in input sequence), the circuit will go to next state 'b' and output is 0. So, for the given input sequence, the states and output will be as follows:

State :	a	a	b	c	d	e	f	f	g	f	g	a
Input :	0	1	0	1	0	1	1	0	1	0	0	
State :	0	0	0	0	0	1	1	0	1	0	0	

The next step is to develop the state table corresponding to the state diagram shown in Fig. 11.7. The state table of circuit is prepared directly from the state diagram as shown in Table 11.1. In this table, each row represents the state of the circuit and each column stands for every combination of the input signals. Here, there is only one input signal. Hence, there are only two columns in the next state; one for $x = 0$ and other for $x = 1$.

Table. 11.1 State Table of State Diagram

	Next State		Output	
Present State	$x = 0$	$x = 1$	$x = 0$	$x = 1$
a	a	b	0	0
b	c	d	0	0
c	a	d	0	0
d	e	f	0	1
e	a	f	0	1
f	g	f	0	1
g	a	f	0	1

Here state 'e' and 'g' are equivalent and one of these states can be removed. Hence, the 'g' can be removed, and state 'g' is replaced by state 'e', each time it occurs in the next-state columns.

The partially reduced state table after removing state 'g' is shown in Table 11.2.

Table. 11.2 Partially reduced state table

	Next State		*Output*	
Present State	$x = 0$	$x = 1$	$x = 0$	$x = 1$
a	a	b	0	0
b	c	d	0	0
c	a	d	0	0
d	e	f	0	1
e	a	f	0	1
f	e	f	0	1

In this state table also, two equivalent states 'd' and 'f' are there. Hence, 'f' can be removed. The reduced state table is shown in Table 11.3.

Table. 11.3 Reduced state table

	Next State		Output	
Present State	$x = 0$	$x = 1$	$x = 0$	$x = 1$
a	a	b	0	0
b	c	d	0	0
c	a	d	0	0
d	e	d	0	1
e	a	d	0	1

Again check for the equivalency. But here no equivalent states were found. So, we get this state table with only five states. The state diagram for the reduced state table can be drawn as shown in Fig. 11.8.

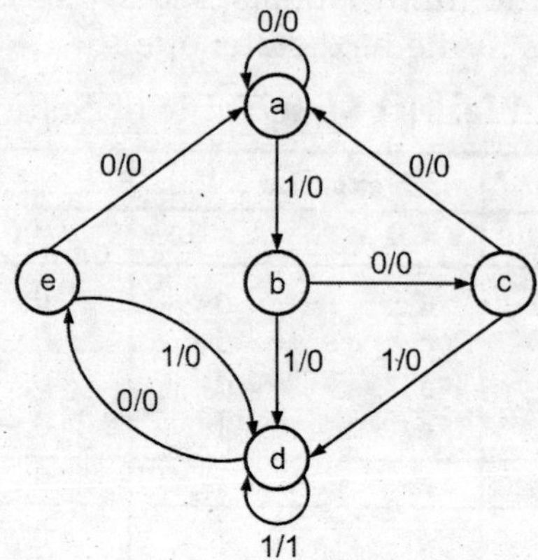

Fig. 11.8 Reduced State Diagram

This state diagram satisfies the original input-output specifications and will produce the required output sequence for any given input sequence. The following is the input-output relationship of the reduced state table.

State	:	a	a	b	c	d	e	d	d	e	d	e	a
Input	:	0	1	0	1	0	1	1	0	1	0	0	
Output	:	0	0	0	0	0	1	1	0	1	0	0	

Note that the output remains unaltered for the same input sequence.

The reduced state diagram and state table will have fewer states and hence will need lesser hardware. However, the fact is that a state table has been reduced to fewer states does not guarantee a saving in the number of flip-flops or gates.

State Assignment: After getting the minimized state table, the next step is to assign the unique binary codes to the state variables of a sequential circuit. This process of assigning the binary values to the states of the sequential machine is called *state assignment.*

State assignment is required for obtaining simplified equation for implementation. The simplest way to assign the states is to use binary numbers in ascending order. The second method is using the gray code where only one bit changes while going from one number to the next. This code makes it easier for the Boolean functions to be placed in the map for simplification. The third is one-hot assignment, which uses as many bits as there are states in the circuit. Here, at any given time, only one bit is equal to 1 while all others are kept at 0. This type of assignment uses one flip-flop per state. The first two methods can be used whenever we need the minimum hardware. The third method of assignment may be preferred for easy implementation but needs more flip-flops. These three possible binary assignments are shown in Table 11.4.

Table 11.4 Three possible binary assignments

State	Binary	Gray Code	One hot
a	000	000	00001
b	001	001	00010
c	010	011	00100
d	011	010	01000
e	100	110	10000

Example 11.2. *Reduce the number of states in the following state table:*

Table E11.2 (*a*)

	Next State		Output	
Present State	$x = 0$	$x = 1$	$x = 0$	$x = 1$
a	f	b	0	0
b	d	c	0	0
c	f	e	0	0
d	g	a	1	0
e	d	c	0	0
f	f	b	1	1
g	g	h	0	1
h	g	a	1	0

Solution:

From the Table E11.2(*a*), we observe that present states 'd' and 'h' have the same next states and outputs for the same input conditions. Hence states 'd' and 'h' are said to be equivalent. Similarly, states 'b' and 'e' are found to be equivalent. So, we can remove one of the states (let us remove state 'h' and 'e' for example) and replace states 'h' and 'e' with 'd' and 'b' respectively wherever it is present in the next state as shown in Table E11.2 (*b*).

Table E11.2 (*b*)

	Next State		Output	
Present State	$x = 0$	$x = 1$	$x = 0$	$x = 1$
a	f	b	0	0
b	d	c	0	0
c	f	b	0	0
d	g	a	1	0
f	f	b	1	1
g	g	d	0	1

From Table E11.2(*b*), we observe that the present states 'a' and 'f' have same next states but the outputs are not same. Hence they are not equivalent states. This procedure shall be repeated until no further reduction is possible.

On further comparison, it is found that states 'a' and 'c' are equivalent. Hence, state 'c' can be eliminated.

Table E11.2 (*c*)

Present State	Next State		Output	
	$x = 0$	$x = 1$	$x = 0$	$x = 1$
a	f	b	0	0
b	d	a	0	0
d	g	a	1	0
f	f	a	1	1
g	g	d	0	1

Table E11.2(*c*) is the final reduced state table. In this example, the states were reduced from 8 to 5.

11.4.3 Design of Sequence Detectors

The sequence detector is a single input circuit that will accept a stream of bits and generate an output '1' whenever the particular sequence is detected. The sequence detector can detect the given sequence either in non-overlapped manner or in overlapped manner. The design of non-overlapping sequence detector is explained in Example 11.3 and overlapping sequence detector is explained in Example 11.4.

Example 11.3. *Design a sequence detector that produces an output '1' whenever the non-overlapping sequence 1011 is detected.*

Solution:

Step 1: From the word description of the problem, one can understand that the circuit will accept a stream of bits and generate an output '1' whenever the sequence 1011 is detected. Then, the circuit will go back to the initial state and wait for the next 1011 sequence to generate the output. For example, if the input is 1011011011011, the output generated will be 0001000001000. But in the case of overlapping sequence detector, the output will be 0001001001001 i.e., additional two 1's are generated due to overlapping sequence.

Step 2: From the above description of the problem, the sequence detector is a block with an input (x) and output (Z) as shown in Fig. E11.3(*a*).

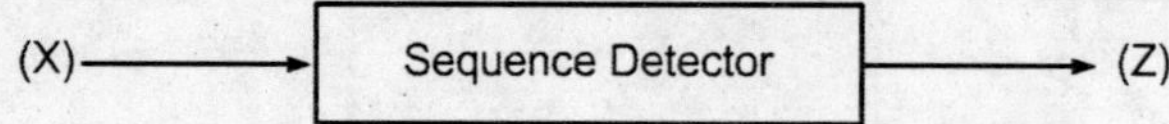

Fig. E11.3(*a*) Block diagram of sequence detector

Step 3: Primitive state diagram: Let A be the initial state; B be the state when the last received one symbol in the input sequence is 1; C be the state when the last received two symbols in the input sequence is 10 and D be the state when the last received three symbols in the input sequence is 101. The Primitive state diagram for the given sequence detector is shown in Fig. E11.3(*b*).

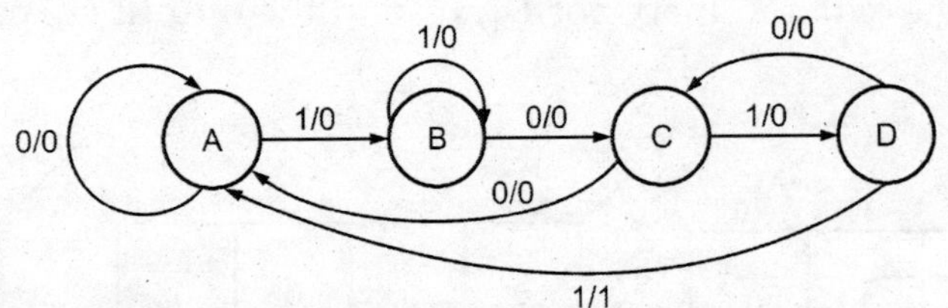

Fig. E11.3(*b*) State diagram for the given sequence detector

Step 4: Primitive state table: From the above primitive state diagram, primitive state and output table can be drawn as shown in Table E11.3(*a*).

Table E11.3(*a*) Primitive Present State/Next State and output table

PS	NS	Z
	$x = 0$	$x = 1$
A	A, 0	B, 0
B	C, 0	B, 0
C	A, 0	D, 0
D	C, 0	A, 1

Step 5: From the primitive state table, one can understand that no state is redundant since no two states are equivalent states. Hence, the state table cannot be minimised further.

Step 6: State Assignment: As there are 4-states, two state variables are required for state assignment. The state assignment is as follows:

$$A \rightarrow 00; B \rightarrow 01; C \rightarrow 10 \text{ and } D \rightarrow 11$$

Step 7: Using the state assignment, the above PS/NS table can be modified as shown in Table E11.3(*b*).

Table E11.3(*b*) Present State/Next State and Output Table

PS	NS (Y_1 Y_0)		Z	
(y_1 y_0)	$x = 0$	$x = 1$	$x = 0$	$x = 1$
0 0	0 0	0 1	0	0
0 1	1 0	0 1	0	0
1 0	0 0	1 1	0	0
1 1	1 0	0 0	0	1

Step 8: Let us assume that delay flip-flops are used for the implementation. Then the excitation table can be drawn as shown in Table E11.3(*c*).

Table E11.3(*c*) Excitation table of sequence detector

PS	D_1 D_0	
(y_1 y_2)	$x = 1$	$x = 1$
0 0	0 0	0 1
0 1	1 0	0 1
1 0	0 0	1 1
1 1	1 0	0 0

Step 9: The excitation maps for D_1, D_0 and the output (Z) can be drawn as shown in Fig. E11.3(*c*).

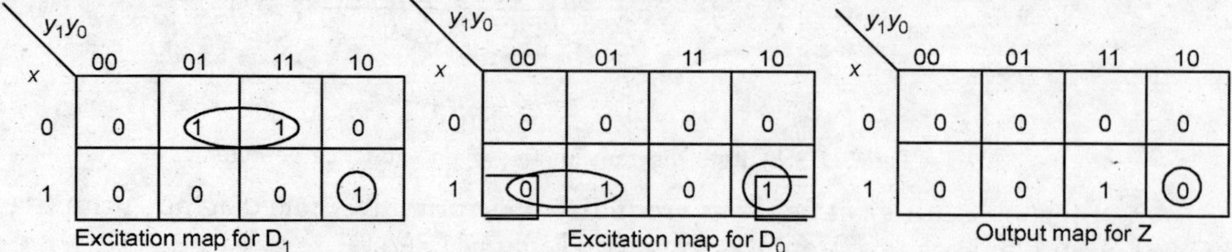

Fig. E11.3(*c*) Excitation and Output Maps for Non-overlapping sequence detector

From the excitation and output maps, the simplified expressions for D_1, D_0 and Z are given by

$$D_1 = y_0\bar{x} + y_1\overline{y_0}x$$
$$D_0 = \overline{y_1}x + \overline{y_0}x = x(\overline{y_0} + \overline{y_1})$$
$$Z = xy_1y_0$$

Step 10: Using the simplified expressions for D_1, D_0 and Z, the logic diagram of non-overlapping sequence detector can be drawn as shown in Fig. E11.3(*d*).

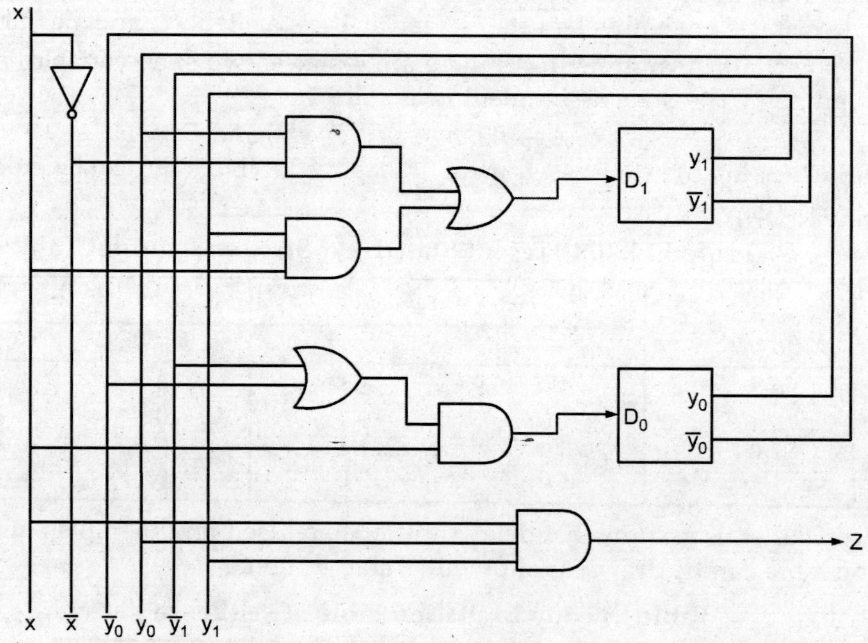

Fig. E11.3(*d*) Logic diagram of non-overlapping sequence detector

---

Example 11.4. *Design a sequence detector that produces an output 1 whenever the sequence 101101 is detected.*

Solution: Step 1: From the word description of the problem, one can understand that the sequence detector is a single input circuit that will accept a stream of bits and generate an output '1' whenever the sequence 101101 is detected. Otherwise, output '0' is generated. For example, for the input 0101101101, the output 0000001001 will be generated, and for the input 10110101101 the output is 00000100001.

Step 2: From the above description of the problem, the sequence detector is a block with one input (x) and output (Z) as shown in Fig. E11.4(a).

Fig. E11.4(*a*) Block diagram of sequence detector

Step 3: Primitive state diagram: Considering the input sequence 101101 to be detected, let

A be the initial arbitrary state,

B be the state when the last received one symbol in the input sequence is 1,

C be the state when the last received two symbols in the input sequence is 10,

D be the state when the last received three symbols in the input sequence is 101,

E be the state when the last received four symbols in the input sequence is 1011 and

F be the state when the last received five symbols in the input sequence is '10110'.

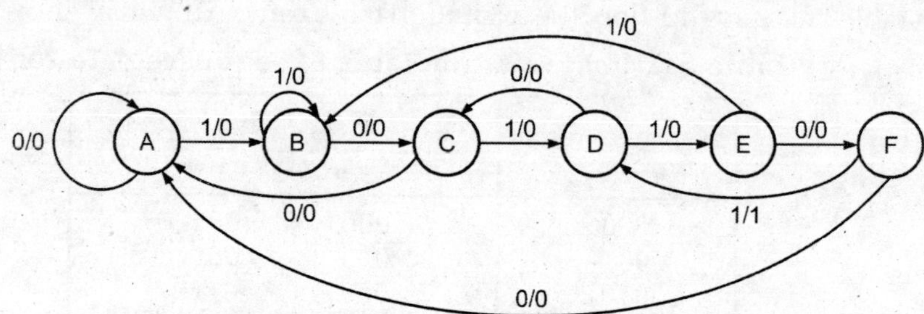

Fig. E11.4(*b*) State diagram of sequence detector

Step 4: Primitive state table: From the above primitive state diagram, the primitive Present state/Next state (PS/NS) and output table can be drawn as shown in Table E11.4(a).

Table E11.4(*a*) Primitive PS/NS and output table

PS	NS, Z (o/p)	
	$x = 0$	$x = 1$
A	*A*, 0	*B*, 0
B	*C*, 0	*B*, 0
C	*A*, 0	*D*, 0
D	*C*, 0	*E*, 0
E	*F*, 0	*B*, 0
F	*A*, 0	*D*, 1

Step 5: From the primitive state table, one can understand that no state is redundant since no two states are equivalent states. Therefore, the state table cannot be reduced further.

Step 6: State assignment: In this step, the following state assignments can be made arbitrarily to the states A to F. Since there are six states, at least three state variables are required:

$$A - 000;\ B - 001;\ C - 010;\ D - 011;\ E - 100;\ F - 101$$

Step 7: Now, using the above state assignment, the Present state/Next state and output table can be modified as shown in Table E11.4(*b*).

Table E11.4(*b*) PS/NS and output table

PS	NS $Y_3Y_2Y_1$		Z	
$y_3y_2y_1$	$x = 0$	$x = 1$	$x = 0$	$x = 1$
000	000	001	0	0
001	010	001	0	0
010	000	011	0	0
011	010	100	0	0
100	101	001	0	0
101	000	011	0	0

Step 8: In this step, one has to decide the type of memory element (i.e. flip-flop) to be used. If three Delay flip-flops are used as memory elements in this design, then, using the application table of Delay flip-flop, the excitation table can be drawn as shown in Table E11.4(*c*).

Table E11.4(*c*) Excitation table of sequence detector

$y_3y_2y_1$	$D_3D_2D_1$ $x = 0$	$x = 1$
000	000	001
001	010	001
010	000	011
011	010	100
100	101	001
101	000	011

Step 9: The excitation maps for flip-flop inputs D_3, D_2, D_1 and output map can be drawn as shown in Fig. E11.4(*c*) and the excitation functions and output function can be simplified.

y_1x \ y_3y_2	00	01	11	10
00	0	0	*d*	1
01	0	0	*d*	0
11	0	1	*d*	0
10	0	0	*d*	0

$D_3 = y_3\,\overline{y}_1\overline{x} + y_2y_1x$

Excitaton map for D_3

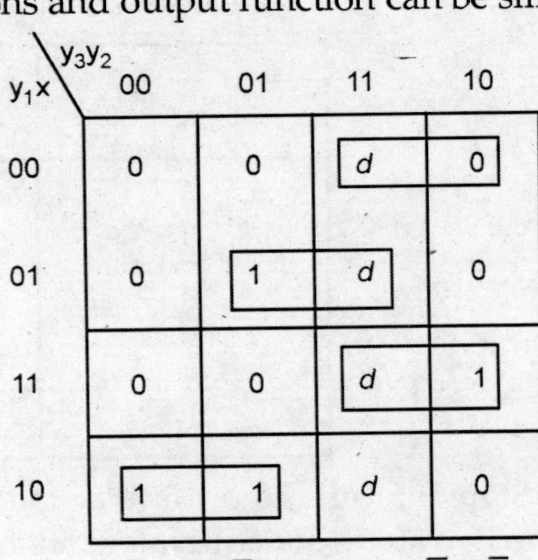

$D_2 = y_2\,\overline{y}_1x + y_3\,y_1x + \overline{y}_3\,y_1\overline{x}$

Excitaton map for D_2

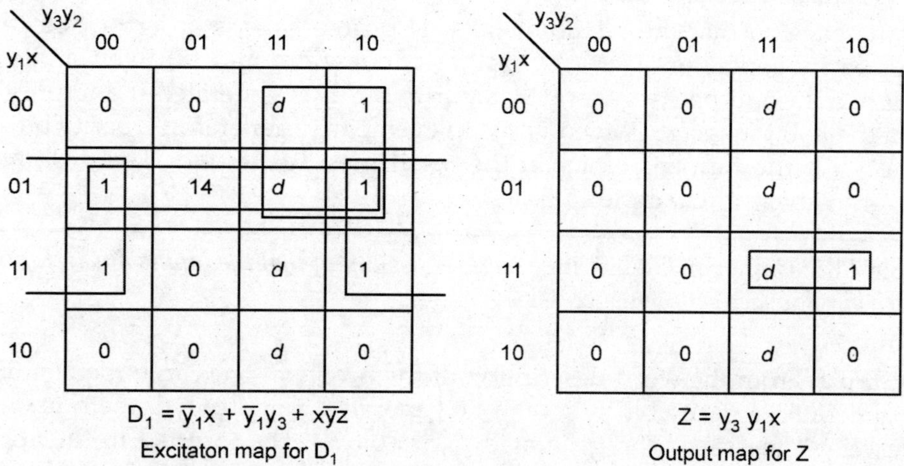

Fig. E11.4(*c*) K-map simplification for excitation and output functions

Step 10: Using the simplified expressions for excitation and output functions, the logic diagram of sequence detector can be drawn as shown in Fig. E11.4(*d*).

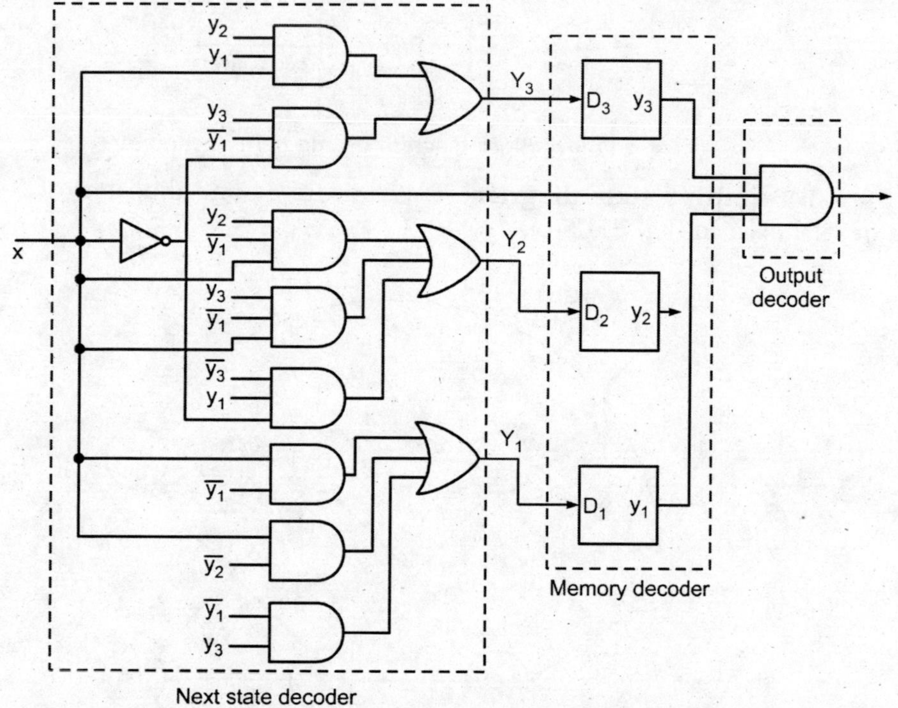

Fig. E11.4(*d*) Circuit diagram of sequence detector

11.4.4 Design of Odd/ Even Parity Generator

A parity generator is a two terminal circuit (i.e., with one input and one output), which receives coded messages in serial format at its input and adds a parity bit to every *m*-bits message so that the resultant ($m + 1$) bit message is an error detecting code. Two types of parity generators viz., (*i*) Odd parity generator and (*ii*) Even parity generator can be designed. In odd parity generator, a parity bit '1' is generated if and only if the number of 1's in the *m*-bit message is even while in even parity generator, a parity bit '1' is generated if and only if the number of 1's in the *m*-bit message is odd. The design of odd parity generator is given in Example 11.5.

Example 11.5. *Design an Odd parity generator assuming that the inputs are arriving in strings of 3-bit with a single space between successive strings.*

Solution:

Step 1: From the word description of the problem, one can understand that in an odd parity generator, a parity bit '1' should be generated if and only if the number of 1's is even for every 3-bit message. The generated parity bit is to be inserted in the appropriate space between successive strings so that the resultant one is a continuous stream of bits without spaces.

Step 2: From the above description of the problem, the block diagram of odd parity generator with one input and one output can be drawn as shown in Fig. E11.5(*a*).

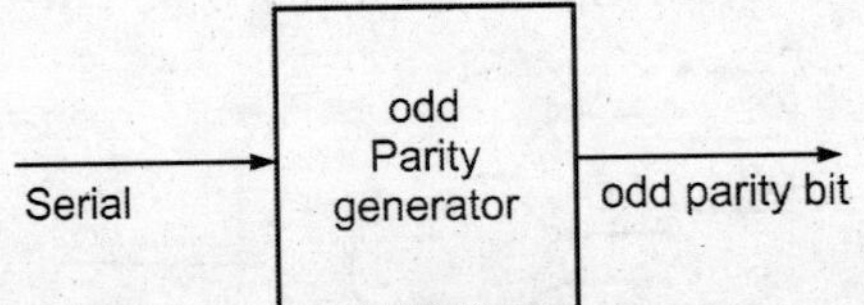

Fig. E11.5(*a*) Block diagram of odd parity generator

Step 3: Primitive state diagram: The primitive state diagram of 3-bit odd parity generator is shown in Fig. E11.5(*b*).

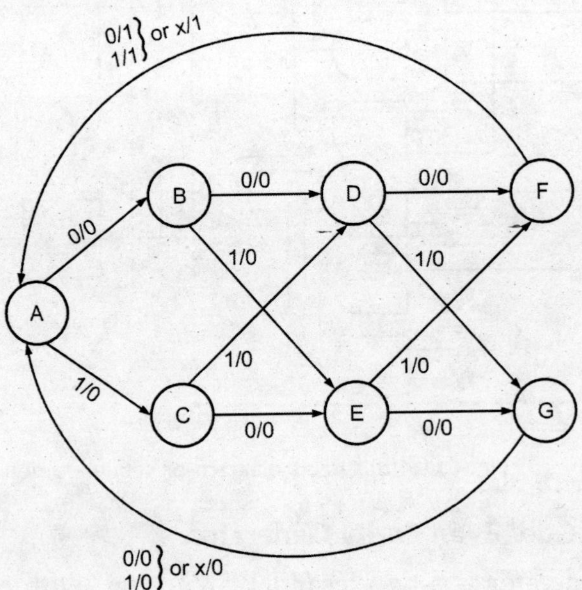

Fig. E11.5(*b*) State diagram of 3-bit odd parity generator

Here, the states *B*, *D* and *F* correspond to even number of 1's out of one, two and three incoming input bits respectively while the states *C*, *E* and G correspond to odd number of 1's out of one, two and three incoming input bits respectively. From state *F*, it returns to state *A* with 1 as output i.e., a parity bit '1' is generated since the number of 1's in the last 3-bit message is even. Similarly, from state *G*, it returns to state *A* with 0 as output i.e., a parity bit '0' is generated since the number of 1's in the last 3-bit message is already odd. It is important to note that from either state *F* or state *G*, it returns to state *A* irrespective of the input; since the fourth input is a space, it is marked as a don't care in the state diagram.

Step 4: Primitive state table: The primitive state table corresponding to the state diagram shown in Fig. E11.5(*b*) is given in Table E11.5(*a*).

Table E11.5(*a*) State table of 3-bit odd parity generator

Present State (PS)	Next state, output NS, *Z*	
	$x = 0$	$x = 1$
A	*B*, 0	*C*, 0
B	*D*, 0	*E*, 0
C	*E*, 0	*D*, 0
D	*F*, 0	*G*, 0
E	*G*, 0	*F*, 0
F	*A*, 1	*A*, 1
G	*A*, 0	*A*, 0

Step 5: From the primitive state table, one can understand that no state is redundant since no two states are equivalent states. Hence, the state table cannot be minimized further.

Step 6: State assignment: As there are seven states, three variables are required for state assignments. Let us assign 000 for state A; 001 for state B; 010 for state C; 011 for state D; 100 for state E; 101 for state F and 110 for state G.

Step 7: The state table with the above state assignment is shown in Table E11.5(*b*).

Table E11.5(*b*) State table of 3-bit odd parity generator (with state assignment)

Present State (PS)			Next State (NS) $Y_2Y_1Y_0$		Output *Z*	
y_2	y_1	y_0	$x = 0$	$x = 1$	$x = 0$	$x = 11$
0	0	0	001	010	0	0
0	0	1	011	100	0	0
0	1	0	100	011	0	0
0	1	1	101	110	0	0
1	0	0	110	101	0	0
1	0	1	000	000	1	1
1	1	0	000	000	0	0

Step 8: If D flip-flops are used for the design of odd parity generator, then the excitation inputs $D_2D_1D_0$ will have similar entries as $Y_2Y_1Y_0$.

Step 9: Excitation maps: The logic equation for D_2, D_1 and D_0 can be simplified by using 4 variable K-maps as shown in Fig. E11.5(c). As there are only 7 valid states for the remaining 1 state, the entry is don't care as in the excitation maps shown in Fig. E11.5(c).

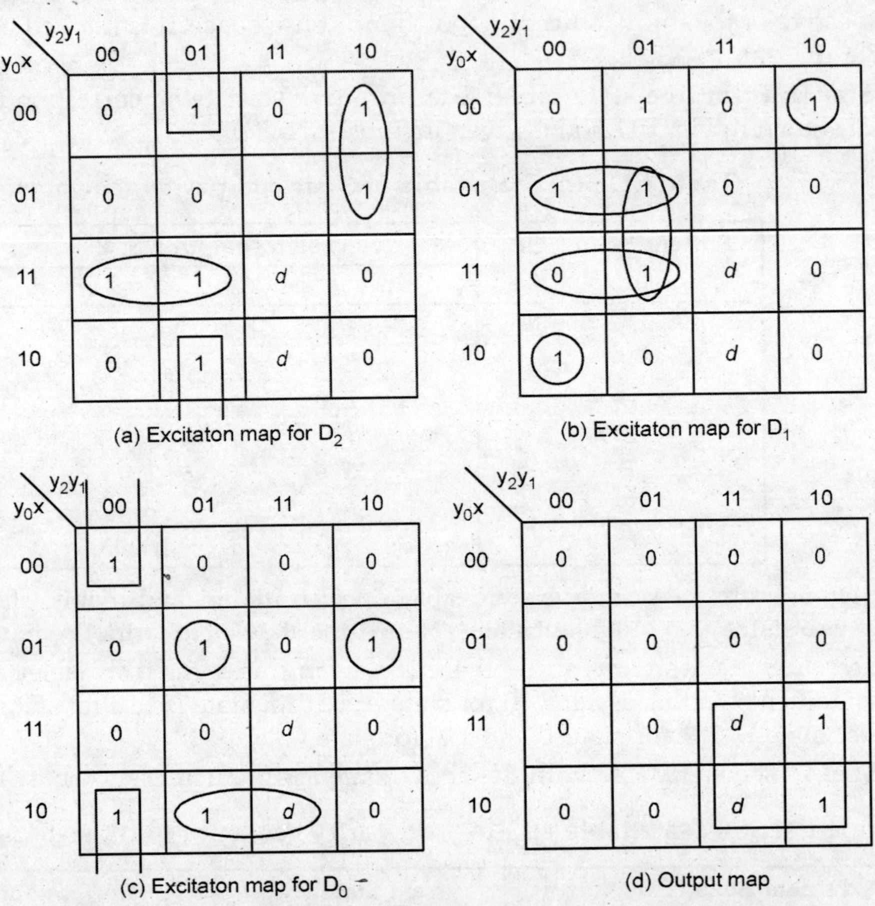

Fig. 11.5(c) Excitation and output maps of 3-bit odd parity generator

From the excitation and output maps, the simplified expressions for D_2, D_1, D_0 and Z are given by

$$D_2 = \overline{y_2}\, y_1\, \overline{x} + y_2\, \overline{y_1}\, \overline{y_0} + \overline{y_2}\, y_0\, x$$

$$D_1 = \overline{y_2}\, \overline{y_0}\, x + \overline{y_2}\, y_1\, x + y_2\, \overline{y_1}\, \overline{y_0}\, \overline{x} + \overline{y_2}\, \overline{y_1}\, y_0\, \overline{x} = \overline{y_2}\, x(\overline{y_0} + y_1) + \overline{y_1}\, x(y_2 \oplus y_0)$$

$$D_0 = \overline{y_2}\, \overline{y_1}\, \overline{x} + y_1 y_0\, \overline{x} + y_2\, \overline{y_1}\, \overline{y_0}\, x + \overline{y_2}\, y_1\, \overline{y_0}\, x = \overline{y_2}\, \overline{y_1}\, \overline{x} + y_1 y_0 \overline{x} + \overline{y_0}\, x\, (y_2 \oplus y_1)$$

$$Z = y_2 y_0$$

Step 10: Using the simplified expressions for D_2, D_1, D_0 and Z, the logic diagram of 3-bit odd parity generator is shown in Fig. E11.5(d).

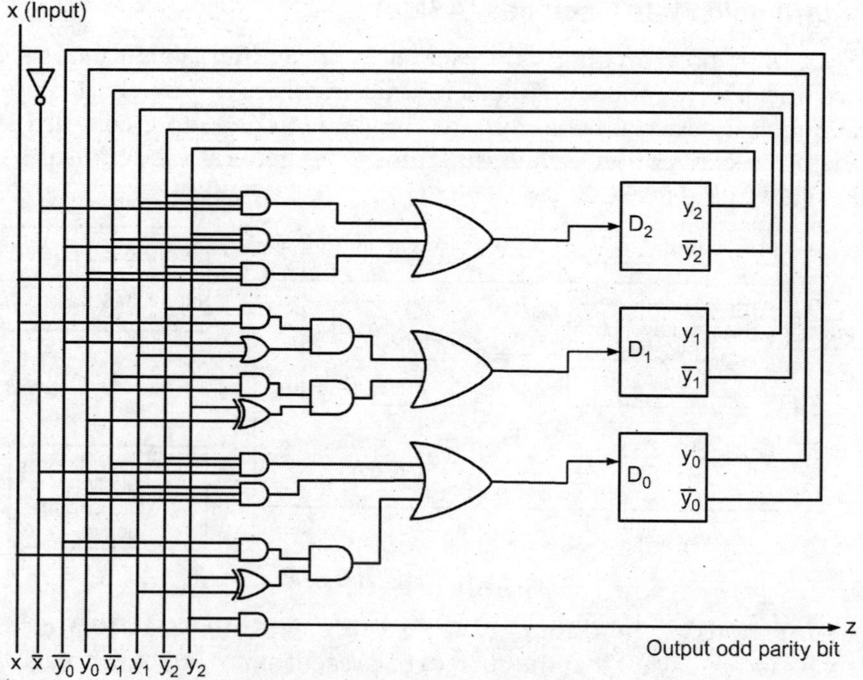

Fig. E11.5(*d*) Logic diagram of 3-bit odd parity generator

11.5 SYNCHRONOUS SEQUENTIAL CIRCUIT DESIGN USING ALGORITHMIC STATE MACHINE (ASM)

Any digital system can be viewed as having a controller and controlled architecture as shown in Fig. 11.9. The controller has to supply a time sequence of commands to the controlled architecture, which is otherwise called as a data processor. It normally comprises components such as flip-flops, shift registers, counters, adders/subtracters and comparators for manipulating the data.

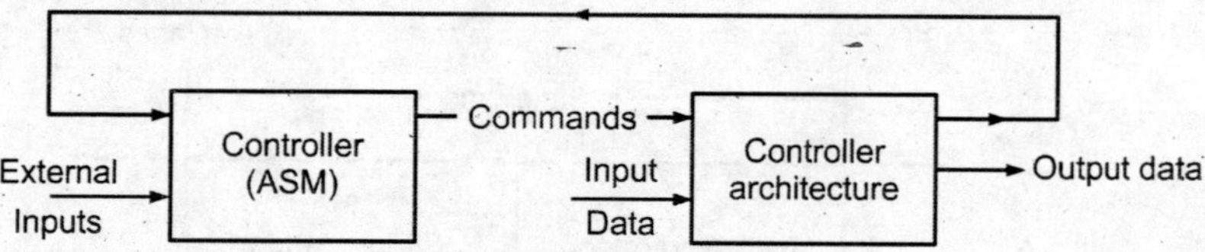

Fig. 11.9 A digital system

The controller in Fig. 11.9 can be designed to function as a clocked synchronous sequential network having several states, with each state associated with the generation of a set of commands. Arrangement of various states in a sequence controls the data processor. Hence, the controller can be regarded as a hardware algorithm operating following a finite number of prescribed states. It is referred to as Algorithmic State Machine, or simply ASM.

An algorithmic state machine can also serve as a stand-alone sequential network model.

11.5.1 Algorithmic State Machines (ASMs)

ASM can be used an alternative way of representing synchronous sequential circuits like Moore and Mealy models. This technique involves a flow-chart representation of the desired sequential network behavior and can characterize more complex systems easily as compared to the conventional state diagrams. The general model of an Algorithmic state machine is shown in Fig. 11.10.

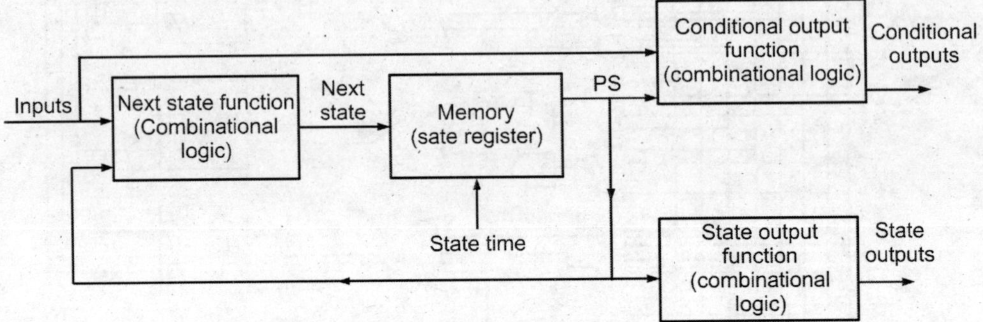

Fig. 11.10 General model of ASM

The functions of the different blocks presented in Fig. 11.10 are similar to Moore's and Mealy's models except that the current or present state (PS) of the machine is represented by the memory block. Each state lasts for one state time, and the next state replaces present state at the end of a state time, which marks the beginning of the next statetime.

Do You Know?	ASM can be used as an alternative way of representing synchronous sequential circuits like Moore and Mealy Models.

Timing of an ASM: An algorithmic state machine experiences a series of states as illustrated in Fig. 11.11 with each state lasting for one state time. A transition period and a stable period divide each state time into two slots. Transition period is the time duration when the next state and output variables change. The time slot over which these values are usable is called the stable period. For efficient operation of an ASM, the state time must be greater than the transition period.

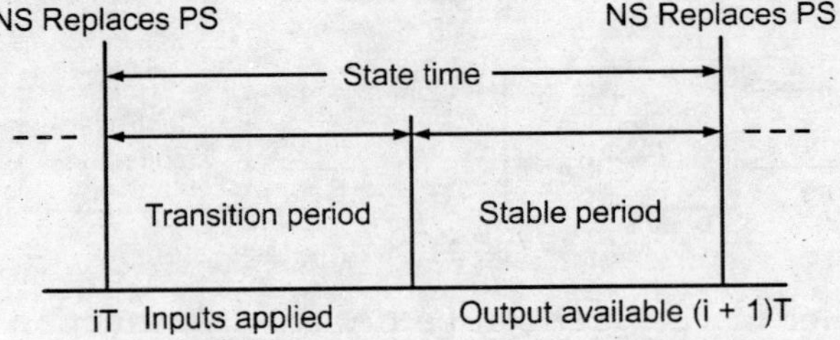

Fig. 11.11 Timing of an ASM

Differences between Moore/Mealy models and ASM: Mealy/Moore models define the sequential network behavior only at sampling instants, whereas ASM model defines it for the entire stable period of the state time. Further, both conditional and state outputs can occur in the same ASM.

11.5.2 ASM Charts

ASM charts resemble the flow-charts conventionally used in software design in many aspects. The difference lies in their time interpretation. The conventional software flow-charts give a sequence of events that must occur one after the other. But an ASM chart will use a collection of box-like symbols to represent each event occurring in a particular state time. Hence, ASM chart describes a sequence of time intervals.

Do You Know?	The ASM chart resembles a conceptional flowchart, but is interpret somewhat differently.

ASM charts have three basic components namely, the state box, the decision box and the conditional output box. All these comprise an ASM block, which corresponds to a single state time, i.e., events within an ASM block occur simultaneously. Hence, the overall behavior of an algorithmic state machine can be described by a collection of ASM blocks forming an ASM chart. The standard representations of the state box, decision box and conditional output box are shown in Fig. 11.12 to Fig. 11.14.

The state box: The state box of the ASM machine is shown in Fig. 11.12.

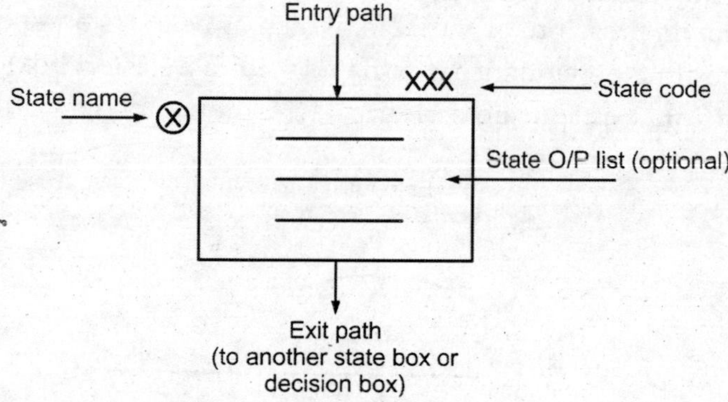

Fig. 11.12 State box

Features of the state box are:

- State box represents one state of ASM.
- There is a single entry path and an exit path.
- Name assigned to the box appears encircled to the left of it.
- It is the only basic component in an ASM chart that is time-dependent, i.e., system resides in a state box for one full state time.
- State assignment, which is made at a later instant, may be included as the state code appearing in the top right corner.
- Name of the output variables that are asserted or active, while the system is in the given state are listed as entries within the state box.
- If the single exit path leads to a state box, then there is an unconditional state transition.
- If it leads to a decision box, the next state and outputs depend on the status of inputs.

The decision box: The decision box is shown in Fig. 11.13.

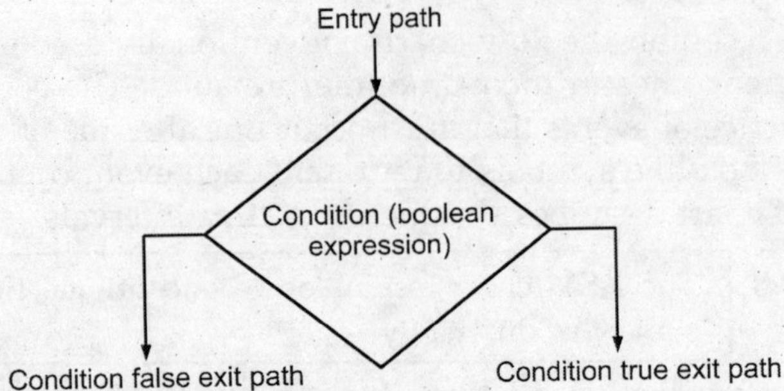

Fig. 11.13 Decision box

Features of decision box are:

- This is the second basic ASM component.
- There is a single entry path and two exit paths (condition true exit path, condition false exit path)
- Basic function is to provide alternative paths in an ASM chart based on the logic value of a Boolean expression, or variable, which appears as an entry in the decision box.
- There is no time dependence associated with a decision box.

The conditional output box: Figure 11.14 shows the conditional output box.

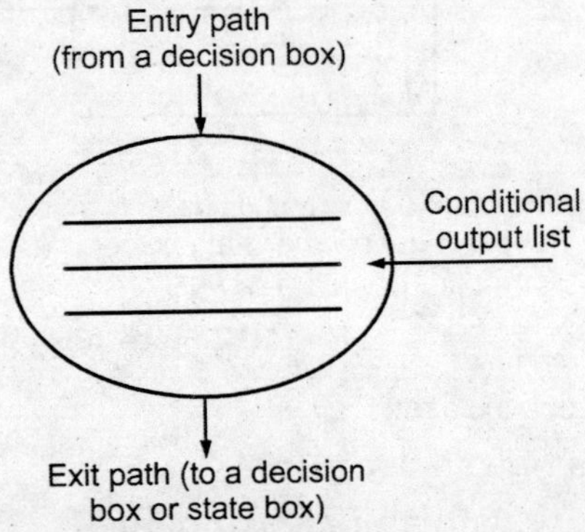

Fig. 11.14 Conditional output box

Features of the conditional output box are:

- This is the final component of the ASM block.
- There is a single entry path and an exit path.
- Those output variables being asserted should be included in the output list present in the box.
- There is no time dependence associated with them.
- These boxes are always associated with a state box.
- A decision box must occur in a path between the associated state box and the conditional output box.

Rules to be observed while constructing ASM blocks: For any valid combination of values to the decision-box variables, all simultaneously selected link paths must lead to the same exit path, i.e., next state behavior of the system should be uniquely defined.

There can be no closed loop existing, which do not contain at least one state box, since a state box is the only component that is time-dependent.

11.5.3 Examples of ASM Charts

An ASM chart is a collection of ASM blocks interconnected for the purpose of describing the behavior of a clocked synchronous sequential network. The ASM chart for mod-8 binary up counter and 3-bit up/$\overline{\text{down}}$ counter are discussed in the next section.

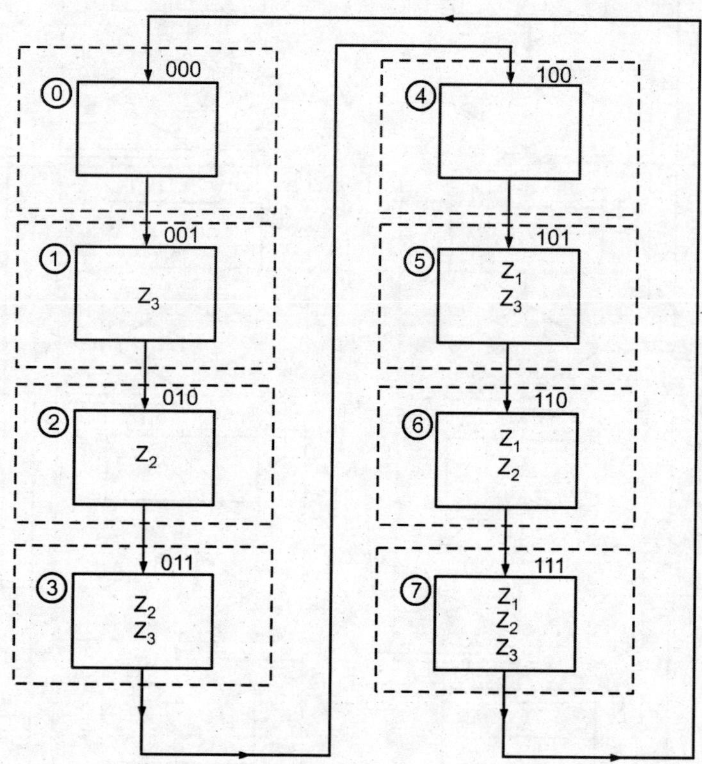

Fig. 11.15 ASM for Mod-8 binary counter

ASM chart for mod-8 binary counter: Figure 11.15 shows an ASM chart describing the working of mod-8 binary counter. The circled entries to the left of each state box represent the state name while the state assignment is given in top right corner. Variables which are to be asserted high are present as the state output list in each state box. Every dashed box indicates one state of the synchronous sequential machine, i.e. one of the eight states of the mod-8 counter.

ASM chart for 3-bit up/$\overline{\text{down}}$ counter: The ASM chart for a Mod-8 binary up/$\overline{\text{down}}$ counter can be constructed as shown in Fig. 11.16.

Counter is incremented or decremented depending on the value of the control variable *I*. If *I* = 0, counter works in countdown mode and vice versa. Each ASM block has a state box followed by a decision box to describe the behavior. The output variables to be asserted for that particular state are highlighted in the state output list of each state box.

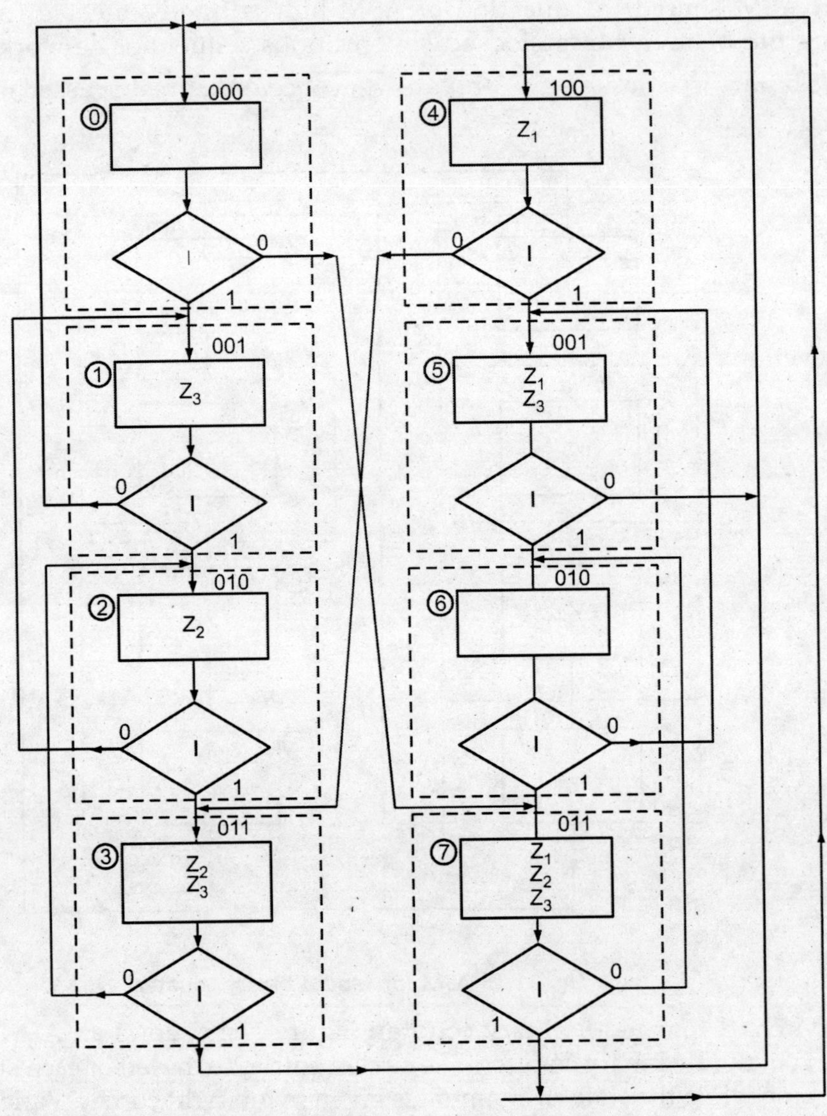

Fig. 11.16 ASM Chart for 3-bit up/$\overline{\text{down}}$ counter

11.5.4 Relationship Between State Diagrams and ASM Charts

In a Moore sequential network, outputs are dependent only on the state variables. Hence, to construct an equivalent ASM chart for a network described by Moore model, there should be as many state boxes and ASM blocks as the number of nodes in the state diagram. This can be illustrated as follows: Consider the state diagram of a Moore's model, shown in Fig 11.17(*a*) where the arc label represents the external input

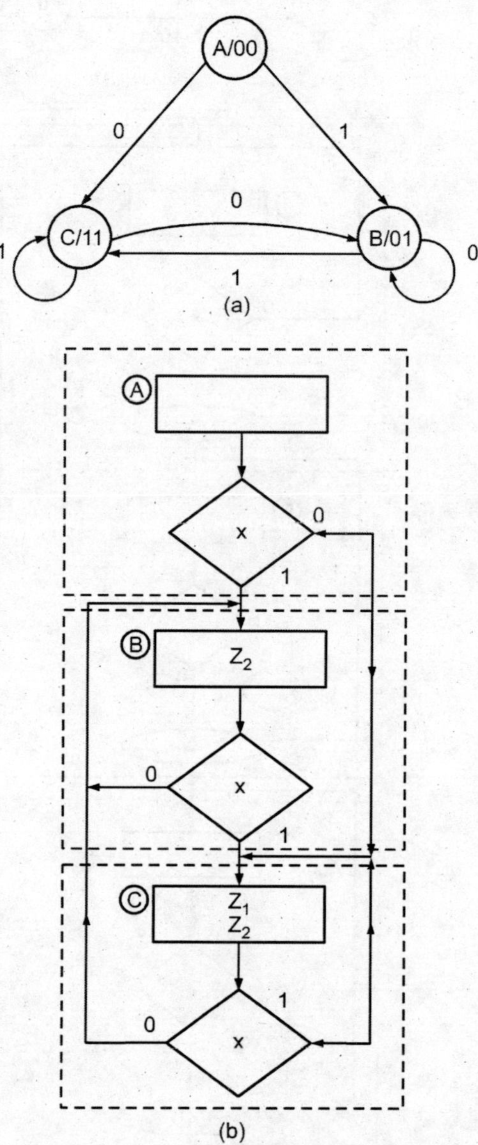

Fig. 11.17 Moore's Machine (*a*) State diagram (*b*) Equivalent ASM Chart

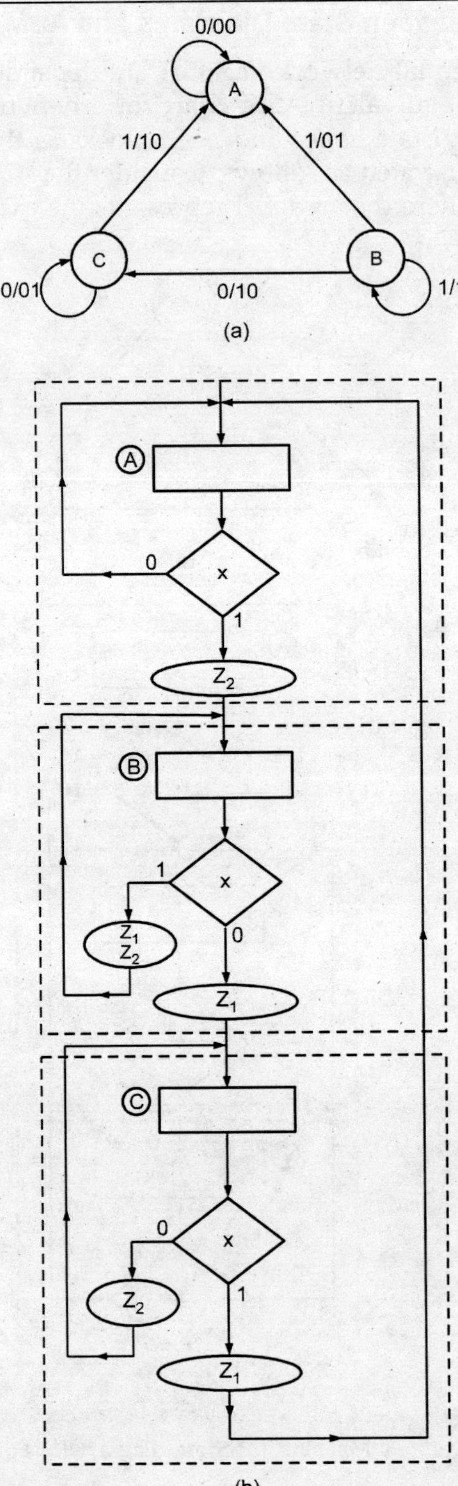

Fig. 11.18 Mealy's Machine (*a*) State diagram (*b*) Equivalent ASM Chart

x. The equivalent ASM chart is shown in Fig 11.17(*b*). For a state diagram represented by Moore model, each ASM block will have a state box and a decision box. There is no need for a conditional output box as the outputs are not dependent on any external input. The output variables that need to be asserted can be comfortably enlisted in the state box itself as shown in Fig. 11.17(*b*).

In the case of a Mealy's machine, the outputs depend on state variables as well as external inputs. So each ASM block has to include all the three basic components namely, the state box, the decision box and the conditional output box to represent one particular state of a Mealy machine. One such Mealy machine is shown in Fig. 11.18(*a*), where the arc label represents external input/state variables (i.e., x/Z_1Z_2) and its equivalent ASM chart will be as in Fig. 11.18(*b*). If $Z_1 = Z_2 = 0$, the condition need not be represented by an output box. The illustration shows one such condition, i.e., the state A being maintained when the external input is 0.

11.5.5 Procedure for Design using ASM Charts

The various steps involved in the design using ASM charts are as follows:

Step 1: Framing the equivalent ASM chart

Step 2: Making state assignment

Step 3: Forming ASM transition table

Step 4: Obtaining ASM excitation table

Step 5: ASM realization

To understand the above design procedure in detail, the following examples can be considered.

11.5.6 Design of Sequence Detector Using ASM Chart

The sequence detector problem discussed in example 11.4 is revisited here to explain the concept of designing using an ASM chart.

Example 11.6. *Design a sequence detector that produces an output 1 whenever the sequence 101101 is detected using ASM chart.*

Solution: The conventional state diagram of sequence detector to detect a sequence of 101101 is shown in Fig. E11.6(*a*).

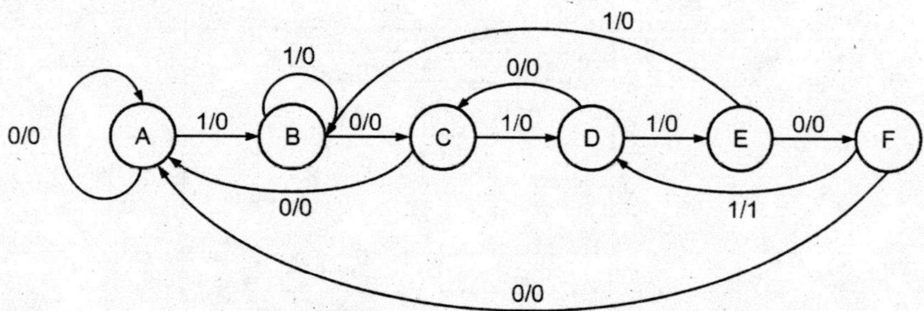

Fig. E11.6(*a*) State diagram of a sequence detector

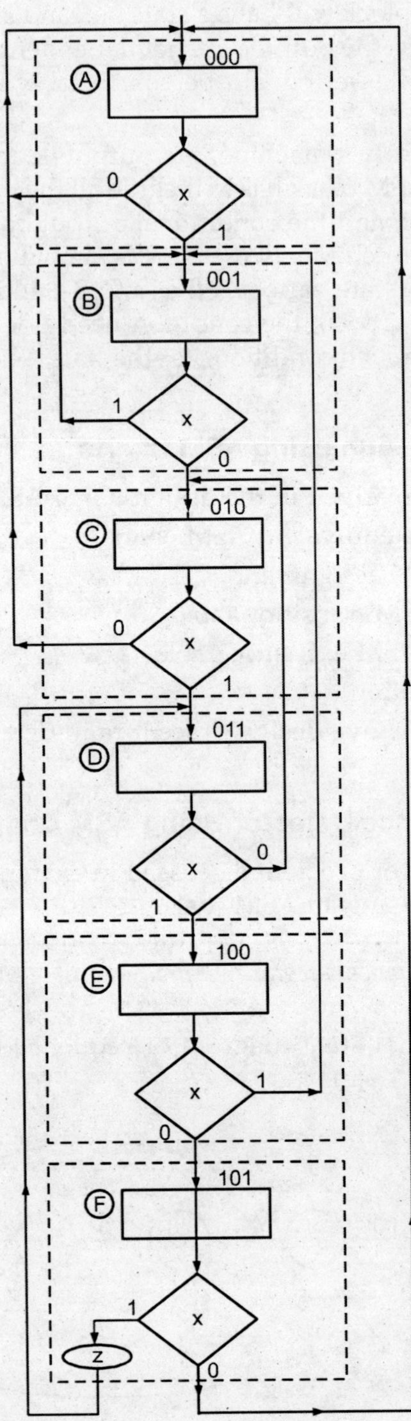

Fig. E11.6(*b*) Equivalent ASM Chart

Step 1: Equivalent ASM Chart: The equivalent ASM Chart corresponding to the conventional state diagram is shown in Fig. E11.6(*b*). The six states have been shown encircled next to the left of each state box.

Step 2: State assignment: The State assignments have been indicated in the top right corner of each state box, which completes step 2.

Step 3: ASM transition table: Information available in the ASM chart can be converted into tabular form, which is known as ASM transition table. It consists of five sections namely link path, present state, inputs, next state and outputs. For the problem being considered, ASM transition table can be framed from the ASM chart as shown in Table E11.6(*a*).

Table E11.6(*a*) ASM transition table

Link path	PS	I/Ps	NS	O/P
		x		Z
L1	A	0	A	0
L2	A	1	B	0
L3	B	0	C	0
L4	B	1	B	0
L5	C	0	A	0
L6	C	1	D	0
L7	D	0	C	0
L8	D	1	E	0
L9	E	0	F	0
L10	E	1	B	0
L11	F	0	A	0
L12	F	1	D	1

Data within horizontal dashed lines pertain to one particular state box. Each link path in the ASM chart is associated with a row in the ASM transition table. In Fig. E11.6(*b*), for state A there are two link paths. Link path $L1 = \bar{x}$ corresponds to state A being maintained, while link path $L2 = x$ indicates state transition from A to B. The other link paths (L3-L12) can also be derived extending the same concept to all other ASM blocks.

A simple variation of the ASM transition table is the Assigned ASM transition table in which the state assignments are done. The assigned ASM transition table is shown in Table E11.6(*b*).

Table E11.6(*b*) Assigned ASM transition table

Link path	State name	PS State code			I/Ps	State name	NS State code			O/P
		y_2	y_1	y_0	x		Y_2	Y_1	Y_0	Z
L1	A	0	0	0	0	A	0	0	0	0
L2	A	0	0	0	1	B	0	0	1	0
L3	B	0	0	1	0	C	0	1	0	0
L4	B	0	0	1	1	B	0	0	1	0
L5	C	0	1	0	0	A	0	0	0	0
L6	C	0	1	0	1	D	0	1	1	0
L7	D	0	1	1	0	C	0	1	0	0
L8	D	0	1	1	1	E	1	0	0	0
L9	E	1	0	0	0	F	1	0	1	0
L10	E	1	0	0	1	B	0	0	1	0
L11	F	1	0	1	0	A	0	0	0	0
L12	F	1	0	1	1	D	0	1	1	1

Step 4: ASM excitation table: This is similar to deriving a conventional excitation table from PS-NS table. Table E11.6(*c*) has been derived for an implementation with *JK* flip-flops using application table of *JK FF*.

Table E11.6(*c*) ASM excitation table

Link path	State name	PS State code			I/Ps	State name	NS State code			Excitations			O/P
		y_2	y_1	y_0	x		Y_2	Y_1	Y_0	J_2K_2	J_1K_1	J_0K_0	Z
L1	A	0	0	0	0	A	0	0	0	0d	0d	0d	0
L2	A	0	0	0	1	B	0	0	1	0d	0d	1d	0
L3	B	0	0	1	0	C	0	1	0	0d	1d	d1	0
L4	B	0	0	1	1	B	0	0	1	0d	0d	d0	0
L5	C	0	1	0	0	A	0	0	0	0d	d1	0d	0
L6	C	0	1	0	1	D	0	1	1	0d	d0	1d	0
L7	D	0	1	1	0	C	0	1	0	0d	d0	d1	0
L8	D	0	1	1	1	E	1	0	0	1d	d1	d1	0
L9	E	1	0	0	0	F	1	0	1	d0	0d	1d	0
L10	E	1	0	0	1	B	0	0	1	d1	0d	1d	0
L11	F	1	0	1	0	A	0	0	0	d1	0d	d1	0
L12	F	1	0	1	1	D	0	1	1	d1	1d	d0	1

Step 5: ASM realization: This step includes deriving excitation and output expressions and implementing those Boolean expressions. The hardware implementation can be done using logic gates, programmable logic devices.

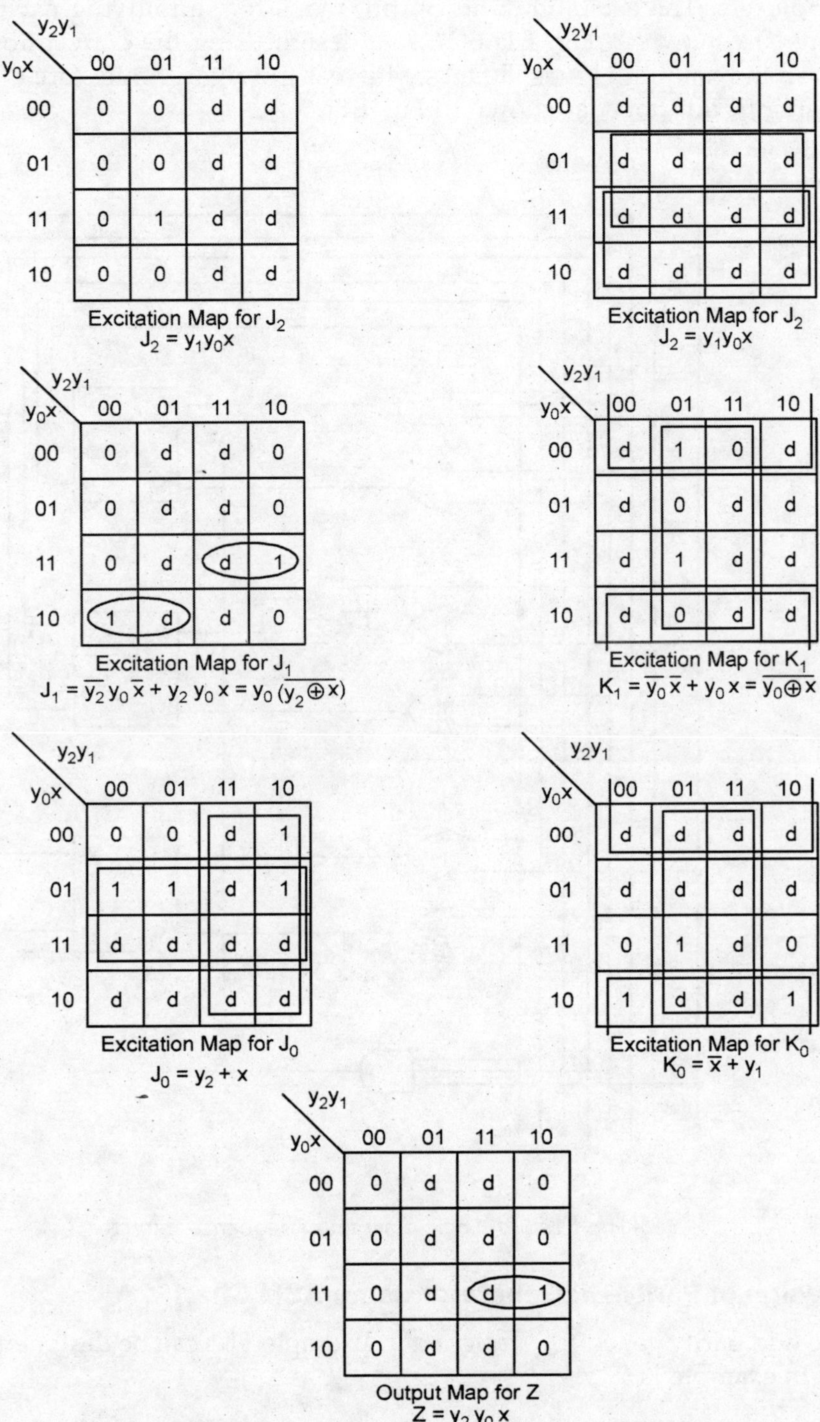

Fig. E11.6(*c*) Excitation and output maps of separate detector and multiplexers. The excitation and output maps for simplifying excitation and output expressions are shown in Fig. E11.6(*c*). The design using the conventional state diagram model given in Example 11.4 used DFFs, whereas the algorithmic state machine realization is done using *JK* flip-flops, as shown in Fig. E11.6(*d*).

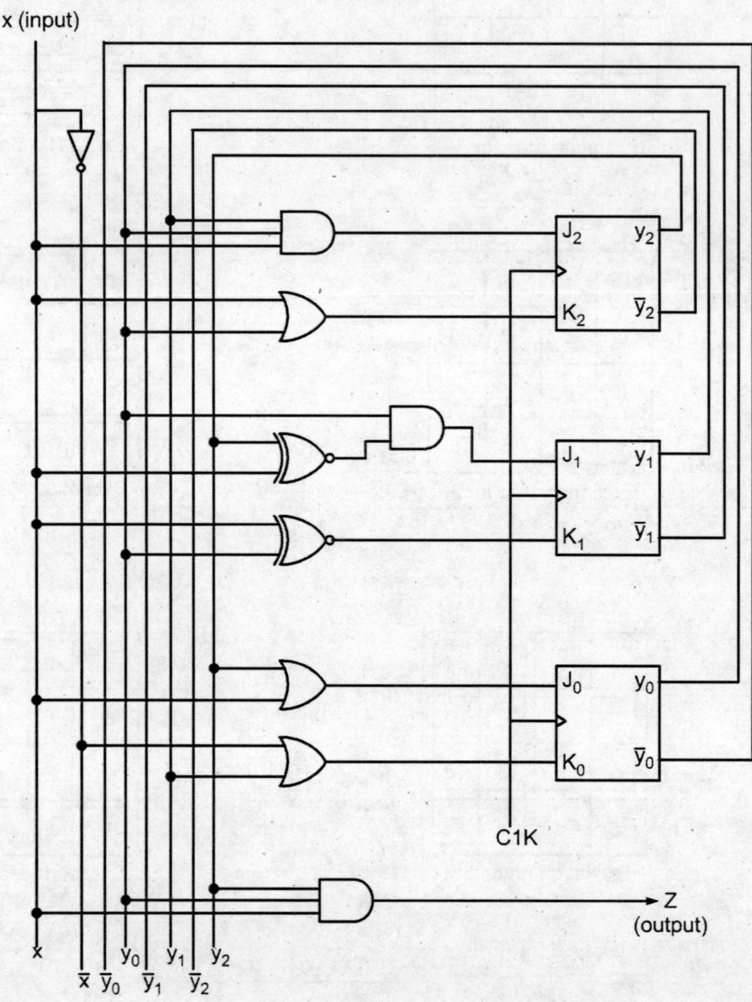

Fig. E11.6(*d*) Logic diagram of sequence detector

11.5.7 Design of Parity-Bit Generator using ASM Chart

The odd parity generator, designed in Example 11.5 can be designed using ASM chart as shown in example 11.7.

Example 11.7. *Design an odd parity generator assuming that inputs are arriving in strings of 3-bit with a single space between successive strings.*

Solution: The conventional state diagram of the 3-bit odd parity generator is redrawn as shown in Fig. E11.7(*a*).

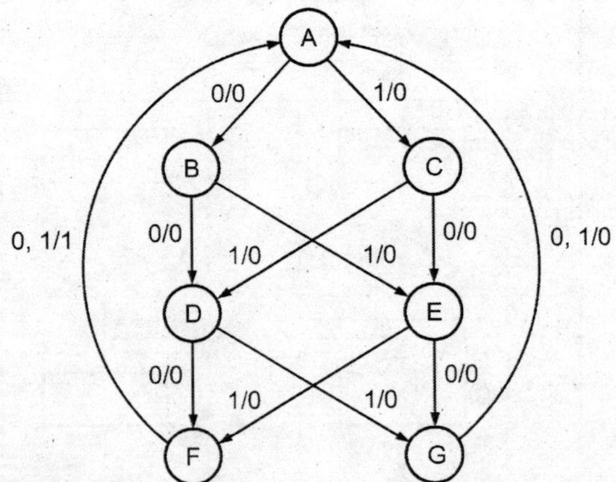

Fig. E11.7(*a*) State diagram of a 3-bit odd-parity generator

Step 1: Equivalent ASM chart: The equivalent ASM chart can be realized as shown in Fig. E11.7(*b*). States F and G have a single link path as they go to the same next state A irrespective of the external input x. This has been illustrated in Fig. E11.7(*b*), and in ASM transition table, shown in Table E11.7(*a*).

Table E11.7(*a*) ASM transition table

Link path	PS	I/Ps x	NS	O/P Z
L1	A	0	B	0
L2	A	1	C	0
L3	B	0	D	0
L4	B	1	E	0
L5	C	0	E	0
L6	C	1	D	0
L7	D	0	F	0
L8	D	1	G	0
L9	E	0	G	0
L10	E	1	F	0
L11	F	0, 1	A	1
L12	G	0, 1	A	0

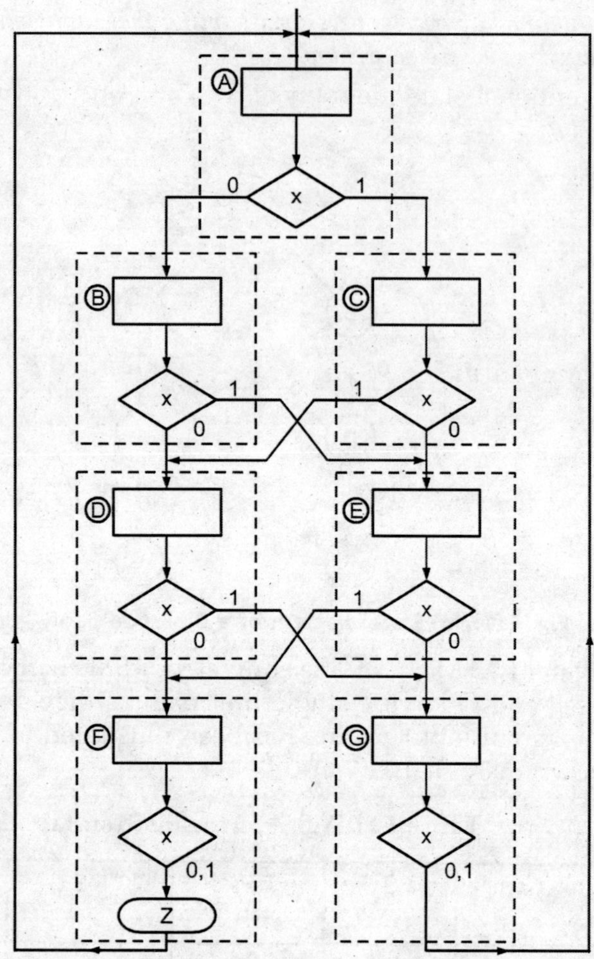

Fig. E11.7(*b*) Equivalent ASM chart

Step 2: State assignment: There are 7 states and they can be given a 3-bit state assignment as follows:

A → 000, B → 001, C → 010, D → 011, E → 100, F → 101 and G → 110

Step 3: ASM transition table: The Assigned ASM transition table is derived as shown in Table E11.7(*b*).

Table E11.7(*b*) Assigned ASM transition table

Link path	State name	PS State code			I/Ps x	State name	NS State code			O/P Z
		y_2	y_1	y_0			Y_2	Y_1	Y_0	
L1	A	0	0	0	0	B	0	0	1	0
L2	A	0	0	0	1	C	0	1	0	0
L3	B	0	0	1	0	D	0	1	1	0
L4	B	0	0	1	1	E	1	0	0	0
L5	C	0	1	0	0	E	1	0	0	0
L6	C	0	1	0	1	D	0	1	1	0
L7	D	0	1	1	0	F	1	0	1	0
L8	D	0	1	1	1	G	1	1	0	0
L9	E	1	0	0	0	G	1	1	0	0
L10	E	1	0	0	1	F	1	0	1	0
L11	F	1	0	1	0,1	A	0	0	0	1
L12	F	1	1	0	0,1	A	0	0	0	0

Step 4: ASM excitation table: Choosing an implementation using D-FFs, the ASM excitation table will be as shown in Table E11.7(*c*).

Table E11.7(*c*) ASM excitation table

Link path	State name	PS State code			I/Ps x	State name	NS State code			Excitations			O/P Z
		y_2	y_1	y_0			Y_2	Y_1	Y_0	D_2	D_1	D_0	
L1	A	0	0	0	0	B	0	0	1	0	0	1	0
L2	A	0	0	0	1	C	0	1	0	0	1	0	0
L3	B	0	0	1	0	D	0	1	1	0	1	1	0
L4	B	0	0	1	1	E	1	0	0	1	0	0	0
L5	C	0	1	0	0	E	1	0	0	1	0	0	0
L6	C	0	1	0	1	D	0	1	1	0	1	1	0
L7	D	0	1	1	0	F	1	0	1	1	0	1	0
L8	D	0	1	1	1	G	1	1	0	1	1	0	0
L9	E	1	0	0	0	G	1	1	0	1	1	0	0
L10	E	1	0	0	1	F	1	0	1	1	0	1	0
L11	F	1	0	1	0,1	A	0	0	0	0	0	0	1
L12	G	1	1	0	0,1	A	0	0	0	0	0	0	0

Step 5: ASM realization: The simplified expressions for excitation inputs (D_2, D_1 and D_0) and output (Z) can be obtained and odd parity generator can be realized as discussed in step 9 and step 10 of example 11.5.

11.6 ANALYSIS OF SYNCHRONOUS SEQUENTIAL CIRCUITS

The behaviour of a sequential circuit can be determined from the inputs, outputs and the states of its flip-flops. Both the output and next state are a function of the inputs and the present state. The suggested analysis procedure of a sequential circuit is given below:

Step 1: Draw the given logic schematic diagram.

Step 2: From the logic schematic diagram, obtain the expressions for the flip-flop excitation inputs and the outputs.

Step 3: Draw the excitation and output maps corresponding to flip-flop excitation and output expressions.

Step 4: Obtain excitation and output table from excitation output maps.

Step 5: Obtain Present State/Next State (PS/NS) and Output Table.

Step 6: From the PS/NS Table, draw the state diagram.

Step 7: From the state diagram, give a word description about the given sequential circuit.

The analysis of the sequential circuits and the steps involved in it are explained in the following four examples 11.8 to 11.11.

Example 11.8. *Analyze the sequential circuit shown in Fig. E11.8(a).*

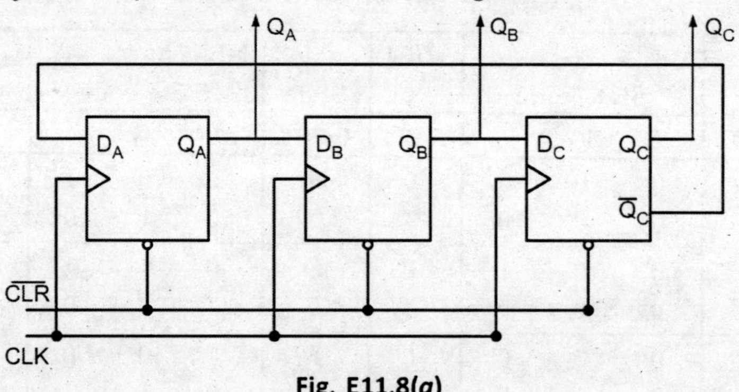

Fig. E11.8(*a*)

Solution:

Step 1: Draw the given circuit diagram.

Step 2: Flip-flop excitation input expressions: From the given circuit diagram shown in Fig. E11.8(*a*), the expressions for the flip-flop excitation inputs, i.e., D_A , D_B and D_C are given by

$$D_A = \overline{q_C}; \quad D_B = q_A; \quad D_C = q_B$$

Step 3: Excitation maps: Using the excitation expressions for D_A, D_B and D_C, the excitation maps, i.e., *K*-maps corresponding to D_A, D_B and D_C can be drawn as shown in Fig. E11.8(*b*).

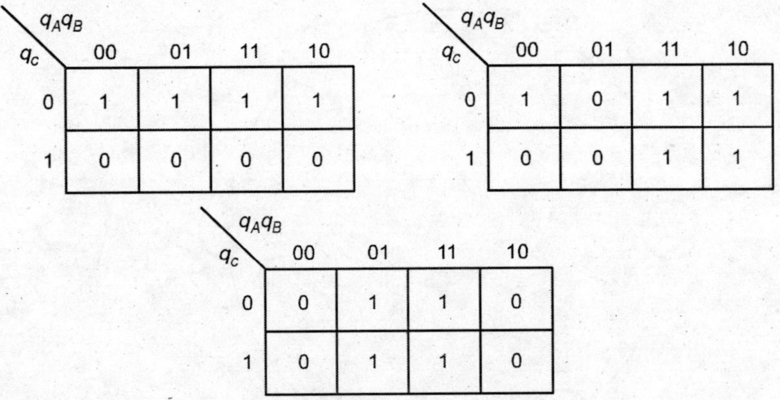

Fig. E11.8(*b*) Excitation Table

Step 4: Excitation table: From the excitation maps shown in Fig. E11.8(*b*), the excitation table can be drawn as shown in Table E11.8(*a*).

Table E11.8(*a*) Excitation Table

Present State			Excitation Inputs		
q_A	q_B	q_C	D_A	D_B	D_B
0	0	0	1	0	0
0	0	1	0	0	0
0	1	0	1	0	1
0	1	1	0	0	1
1	0	0	1	1	0
1	0	1	0	1	0
1	1	0	1	1	1
1	1	1	0	1	1

Step 5: Present state and next state table: From the excitation table given in Table E11.8(*a*), the Present state and Next state table can be drawn as shown in Table E11.8(*b*).

Table E11.8(*b*) Present state and Next state table

Present State			Next State		
q_A	q_B	q_C	Q_A	Q_B	Q_B
0	0	0	1	0	0
0	0	1	0	0	0
0	1	0	1	0	1
0	1	1	0	0	1
1	0	0	1	1	0
1	0	1	0	1	0
1	1	0	1	1	1
1	1	1	0	1	1

Step 6: State diagram: From the Present State and Next State Table shown in Table E11.8(*b*), the state diagram can be drawn as shown in Fig. E.11.8(*c*).

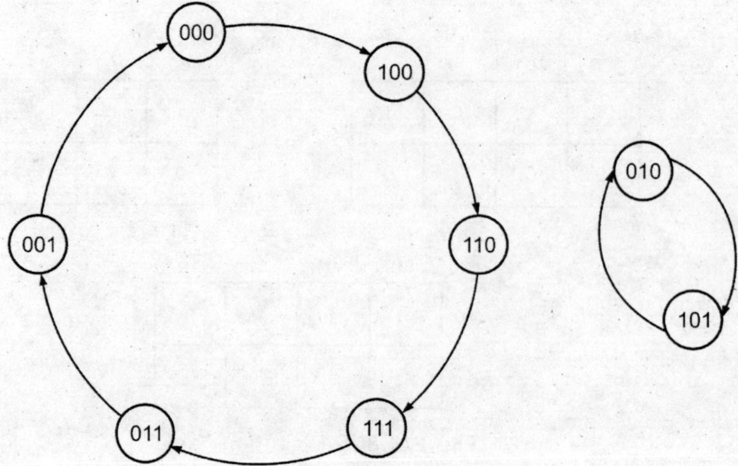

Fig. E11.8(c) State diagram

Step 7: Word description: From the state diagram shown in Fig. E11.8(*c*), it is understood that the states 010 and 101 are isolated from the main sequence. The main sequence shows that the given sequential circuit behaves as a 3-bit shift counter or twisted ring counter.

Example 11.9. *Analyse the synchronous sequential circuit shown in Fig. E11.9(a).*

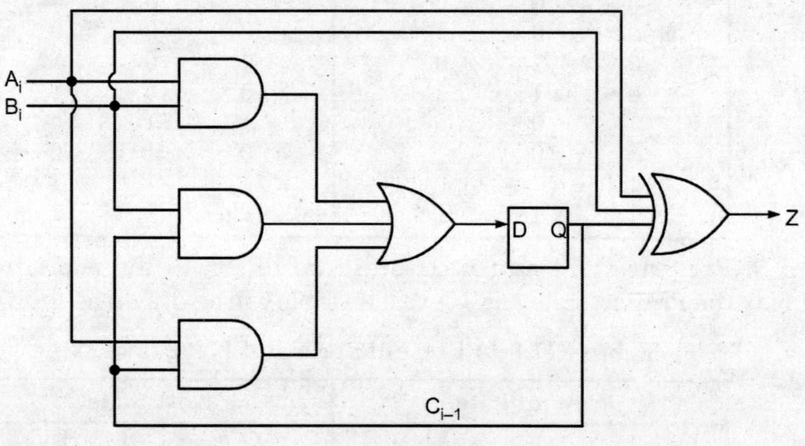

Fig. E11.9(a)

Solution:

Step 1: Draw the given circuit diagram.

Step 2: Flip-flop excitation input and output expressions: From the given circuit diagram shown in Fig. E11.9(*a*), the expression for the flip-flop excitation input (*D*) and output (*Z*) can be written as follow:

$$D = A_iB_i + B_iC_{i-1} + A_iC_{i-1}$$

$$Z = A_i \oplus B_i \oplus C_{i-1} = A_i\overline{B_i C_{i-1}} + \overline{A_i} B_i \overline{C}_{i-1} \overline{A_i B_i} C_{i-1} + A_i B_i C_{i-1}$$

Step 3: Excitation map: Using the excitation and output expression for *D* and *Z*, the excitation map i.e., the *K*-map corresponding to *D* input and the output map for *Z* can be drawn as shown in Fig. E11.9(*b*).

C_{i-1} \ A_iB_i	00	01	11	10
0	0	0	1	0
1	0	1	1	1

C_{i-1} \ A_iB_i	00	01	11	10
0	0	1	0	1
1	1	0	1	0

Fig. E11.9(*b*) Excitation and output maps

Step 4: From the excitation and output maps shown in Fig. E11.9(*b*), the excitation and output table can be drawn as shown in Table E11.9(*a*).

Table E11.9(*a*) Excitation and output table

Present State	*D, Z*			
	$A_i B_i$ =			
C_{i-1}	00	01	11	10
0	0,0	0,1	1,0	0,1
1	0,1	1,0	1,1	1,0

Step 5: From the excitation and output table shown in Table E11.9(*a*), the Present State-Next State (PS-NS) table and output table can be drawn separately as shown in Table E11.9(*b*) and Table E11.9(*c*) respectively.

Table E11.9(*b*)

Present State	Next State			
	A_iB_i =			
C_{i-1}	00	01	11	10
0	0	0	1	0
1	0	1	1	1

Table E11.9(*c*)

Present State	Output (*Z*)			
	A_iB_i =			
C_{i-1}	00	01	11	10
0	0	1	0	1
1	1	0	1	0

Step 6: From the PS-NS Table, shown in Table E11.9(*b*), it is understood that there are only two states such as 0 and 1. The state diagram can be drawn as shown in Fig. E11.9(*c*).

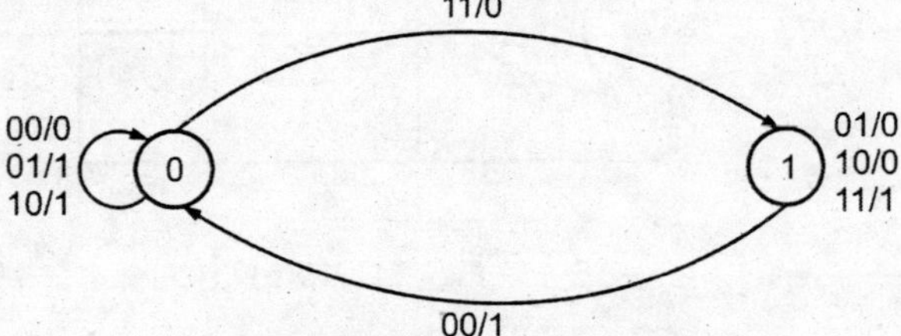

Fig. E11.9(*c*) State diagram

Step 7: In the state diagram, one can observe that for 00, 01 and 10 inputs, there is no state transition from state 0. Also, for 01 and 10 transition, the output is 1. When 11 input occurs, there is a transition from 0 to 1 state with '0' as output. Similarly, when the present state is 1 for 01, 10 and 11 input combinations, there is no state transition. But for 00 input, the next state is 0 with 1 as output. Then, using the state diagram and PS-NS Table, the Truth table of the given circuit can be drawn as shown in Table E11.9(*b*).

Table E11.9(*d*)

Inputs	Present State	Next State	Output	
A	B	C_{i-1}	C_i	Z
0	0	0	0	0
0	0	1	0	1
0	1	0	0	1
0	1	1	1	0
1	0	0	0	1
1	0	1	1	0
1	1	0	1	0
1	1	1	1	1

The truth table shown in E11.9(*d*) describes the operation of full adder. Since the output of the flip-flop (i.e. C_i) is fed back to the input as C_{i-1}, it act as serial adder.

Example 11.10. *Analysis the given sequential circuit shown in Fig. E11.10(a).*

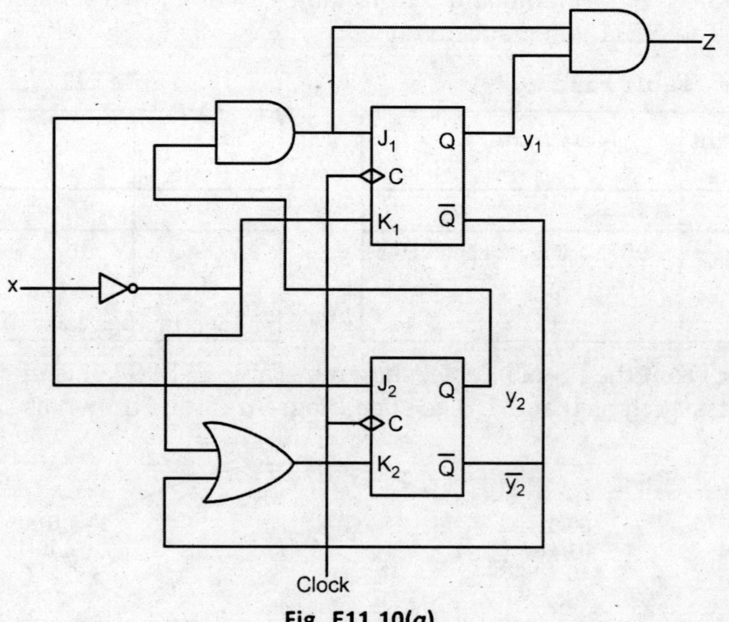

Fig. E11.10(*a*)

Solution:

Step 1: Draw the given circuit diagram.

Step 2: Excitation and output expression: From the given circuit diagram shown in Fig. E11.10(*a*), the expression for the flip-flop excitation input (J_1, K_1, J_2 and K_2) output (Z) can be written as follows:

$$J_1 = xy_2; \qquad K_1 = \overline{x}$$

$$J_2 = x; \qquad K_2 = \overline{x} + \overline{y}_1$$

$$Z = xy_1y_2$$

Step 3: Excitation and output maps: Using the excitation and output expressions given above, the excitation and output maps can be drawn as shown in Fig. E11.10(*b*).

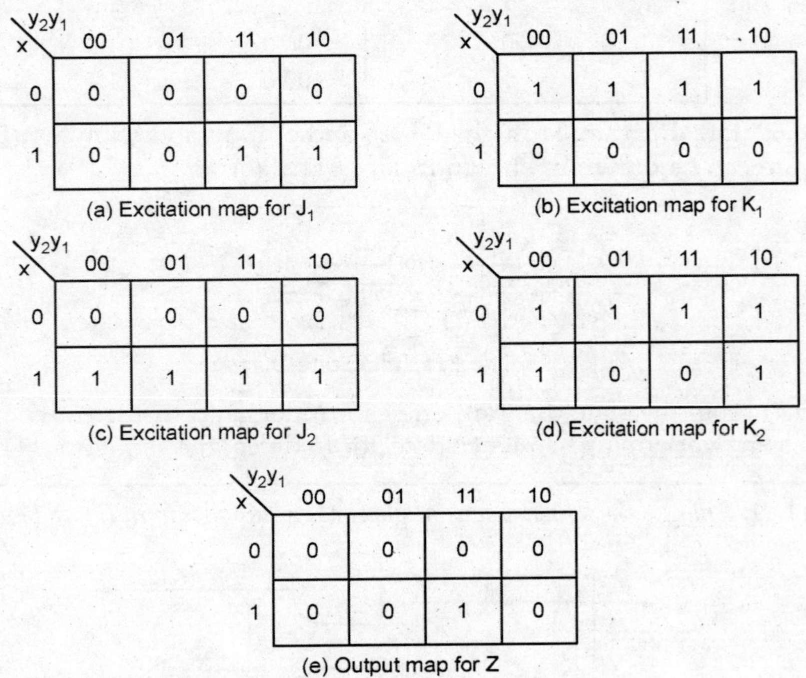

x \ y_2y_1	00	01	11	10
0	0	0	0	0
1	0	0	1	1

(a) Excitation map for J_1

x \ y_2y_1	00	01	11	10
0	1	1	1	1
1	0	0	0	0

(b) Excitation map for K_1

x \ y_2y_1	00	01	11	10
0	0	0	0	0
1	1	1	1	1

(c) Excitation map for J_2

x \ y_2y_1	00	01	11	10
0	1	1	1	1
1	1	0	0	1

(d) Excitation map for K_2

x \ y_2y_1	00	01	11	10
0	0	0	0	0
1	0	0	1	0

(e) Output map for Z

Fig. E11.10(*b*) Excitation and Output maps

Step 4: Excitation table: From the excitation and output maps shown in Fig. E11.10(*b*), the excitation and output table can be drawn as shown in Table E11.10(*a*).

Table E11.10(*a*) Excitation and output table

Present State		Excitation Inputs (J_2K_2, J_1K_1)		Output (Z)	
y_2	y_1	$x = 0$	$x = 1$	$x = 0$	$x = 1$
0	0	01,01	11,00	0	0
0	1	01,01	10,00	0	0
1	0	01,01	11,10	0	0
1	1	01,01	10,10	0	1

Step 5: PS/NS and output table: From the excitation and output table shown in Table E11.10(*a*), the Present State-Next State (PS-NS) and output table can be drawn as shown in Table E11.10(*b*).

E11.10(*b*) Present State-Next State (PS-NS) and Output Table

Present State		Next State, Output $Y_2\ Y_1\ Y_0$	
y_2	y_1	$x = 0$	$x = 1$
0	0	00,0	10,0
0	1	00,0	11,0
1	0	00,0	01,0
1	1	00,0	11,1

Step 6: State diagram: From the PS-NS and output table shown in Table E11.10(*b*), the state diagram can be drawn as shown in Fig. E11.10(*c*).

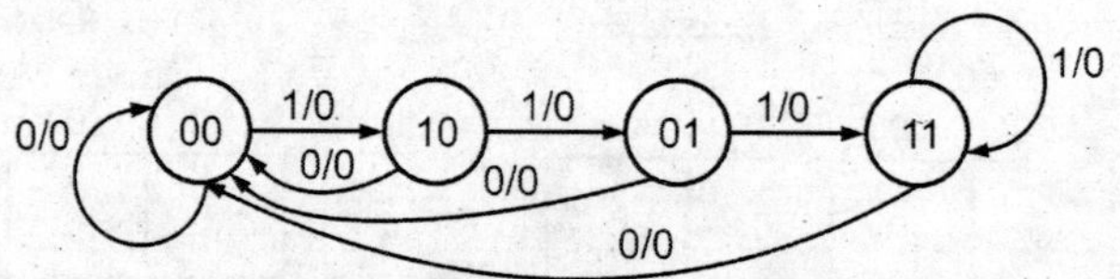

Fig. E11.10(*c*) State Diagram

Step 7: From the state diagram one can understand that the given sequential circuit functions as an overlapping sequence detector to detect the sequence 1111.

Example 11.11. *Analyse the synchronous sequential circuit shown in Fig. E11.11(a).*

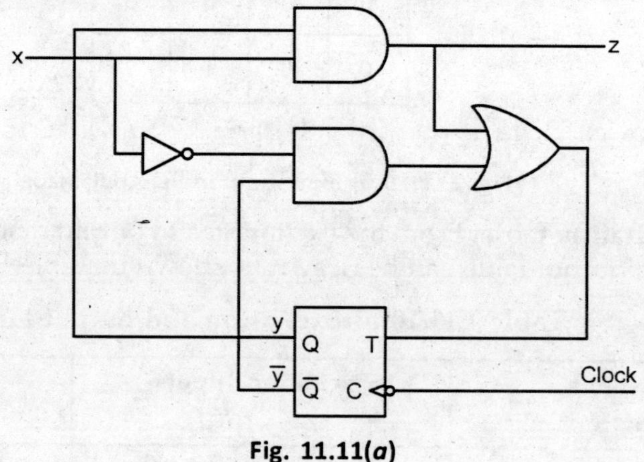

Fig. 11.11(*a*)

Solution:

Step 1: Draw the given circuit diagram.

Step 2: Excitation and output expression: From the given circuit diagram shown in Fig. E11.11(*a*), the expression for the flip-flop excitation inputs (*T*) and output (*Z*) can be written as follows:

$$T = xy + \overline{x}\,\overline{y}$$

$$Z = x\,y$$

Step 3: Excitation and output maps: Using the excitation and output expressions given above, the excitation and output maps can be drawn as shown in Fig. E11.11(b).

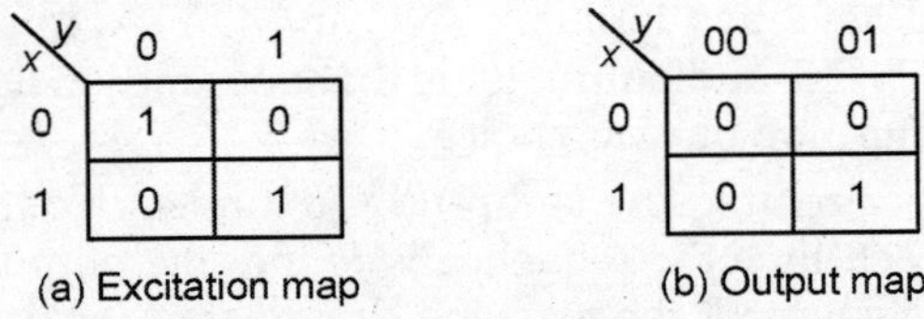

(a) Excitation map (b) Output map

Fig. E11.11(b)

Step 4: Excitation and output table: From the excitation and output maps shown in Fig. E11.11(b), the excitation and output table can drawn as shown in Table E11.11(a).

Table E11.11(a) Excitation and output table

Present State y	Excitation Input T		Output Z	
y	$x = 0$	$x = 1$	$x = 0$	$x = 1$
0	1	0	0	0
1	0	1	0	1

Step 5: PS/NS and output table: From the excitation and output table shown in Table E11.11(a), the Present State-Next State (PS-NS) and output table can be drawn as shown in Table E11.11(a).

Table E11.11(b) Present-State-Next State (PS-NS) and output table

Present State	Next State, Output (Y, Z)			
y	$x = 0$		$x = 1$	
0	1	0	0	0
1	1	0	0	1

Step 6: State diagram: From the PS-NS and output table shown in Table E11.11(b), the state diagram of the given sequential circuit can be drawn as shown in Fig. E11.11(c).

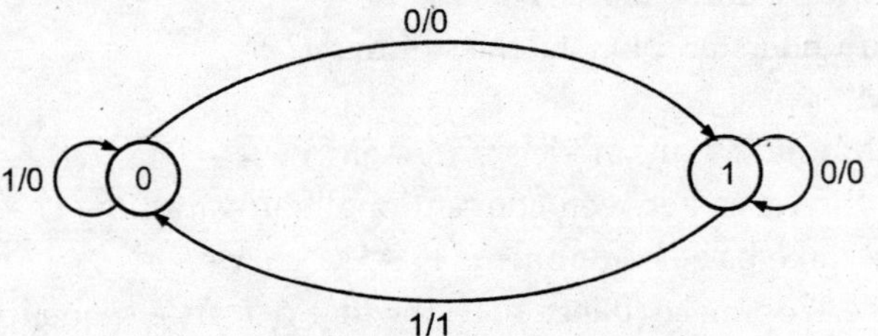

Fig. E11.11(c) State Diagram

■ REVIEW QUESTIONS ■

1. Differentiate combinational and sequential circuits.
2. Write down the steps involved in the design of synchronous sequential circuits.
3. Design a clocked synchronous sequential circuit that checks a serial data line for even parity. The circuit should have two inputs, SYNC and DATA, in addition to CLK, and one output, ERROR. Devise a PS/NS and output table that does the job using just four states and design the logic circuit with flip-flop of your choice.
4. Design a sequential circuit, using D flip-flops to find each of the following sequences in a serial input signal (x). Each time the sequence is found, the output (Z) should be asserted. When a complete sequence is found, the machine should start looking for the sequence again without missing a bit. Changes in the signal (x) are synchronous with the negative going edge of the clock.
   (*a*) 000 (*b*) 010 (*c*) 110 (*d*) 0001 (*e*) 0011 (*f*) 1010
5. What is the need for state table reduction?
6. Explain, with an example, the state table reduction and binary assignment procedures?
7. Reduce the number of states in the following state table and assign the binary values.

	Next State		Output	
Present State	$x = 0$	$x = 1$	$x = 0$	$x = 1$
a	f	b	0	0
b	d	c	0	0
c	f	e	0	0
d	g	a	1	0
e	d	c	0	0
f	f	b	1	1
g	g	h	0	1
h	g	a	1	0

8. Design a sequence detector that detects a sequence of 1011101 in the given binary input stream in non-overlapped manner.
9. Design an overlapping sequence detector that detects a sequence of 110011.
10. Design an even parity generator that generates an even parity bit for every input string of 3-bits.
11. Give the step-by-step procedure to analyse a given synchronous sequential circuit.
12. How are sequential circuits classified?
13. Distinguish Mealy and Moore machines.
14. What do you understand by state assignment?
15. What is ASM?
16. Distinguish Moore/Mealy models and ASM.
17. What is the difference between conventional software flow charts and ASM charts?
18. What are the three basic components is ASM charts?
19. Design an overlapping sequence detector that detects a sequence of 101101 using ASM charts.
20. What are the features of decision box in ASM charts?
21. List the steps involved in the design using ASM charts.

❑❑❑

Asynchronous Sequential Circuits

12.1 INTRODUCTION

Sequential circuits without clock pulses are called *Asynchronous Sequential Circuits.* It can be classified into two types, viz. (*i*) fundamental mode asynchronous sequential circuits and (*ii*) pulse mode asynchronous sequential circuits. In fundamental mode circuits, the inputs and outputs are represented by levels rather than pulses; in pulse mode circuits, the inputs and outputs are represented by pulses. In the design of both types of circuits, it is assumed that a change occurs in only one of the inputs and no change occurs in any other inputs until the circuit enters a stable state. A block diagram of an asynchronous sequential circuit is shown in Fig. 12.1.

Do You Know?	To ensure proper operation, a synchronous sequential circuits must be allowed to attain a stable state before the input is changed to a new value.

The level signals applied at the input lines, namely a_1, a_2, ... a_m, and the output of memory devices, namely y_1, y_2, ..., y_k are combinely called total state of the circuit. The input signals at a_1, a_2... a_m are called input state while the signals y_1, y_2, ... y_k are called secondary or internal state or present state variables of the circuit. The level signals that appear at the outputs of the combinational circuit, namely a_1, b_2, ... b_n, are the outputs of the entire circuit, and the signals that appear at the inputs of memory devices Y_1, Y_2,... Y_k are called excitation or next state variables.

Fig. 12.1 Block diagram of asynchronous sequential circuit

In an asynchronous sequential circuit, the circuit is said to be in a stable state if and only if $y_i = Y_i$ for $i = 1, 2, \dots k$ for a given set of inputs. If there is a change in the inputs, then the combinational circuit produces a new set of excitation variables, namely Y_1, Y_2, ... Y_k, at the inputs of memory devices and the circuit enters into unstable state. When the outputs of memory devices become equal to their inputs, the circuit is said to enter the next stable state.

In fundamental mode asynchronous sequential circuit, it is also assumed that the time difference between two successive input changes is larger than the duration of internal changes.

12.2 DESIGN OF FUNDAMENTAL MODE ASYNCHRONOUS SEQUENTIAL CIRCUITS

The steps involved in the design of fundamental mode asynchronous sequential circuits are as follows:

Step 1: From the verbal description of the problem, formulate precisely what the circuit has to do and develop a state diagram specifying the Next state, Output (NS, Z).

Step 2: Develop a primitive state table with Present State (PS), Next State (NS) and Output (Z) from the state diagram obtained in step 1.

Step 3: Identify the redundant states by using the implicant chart and minimize the primitive state table by merging process. Minimisation is the process of identifying those states that are related by input sequences and can be combined together into a single state, i.e., finding the states that can be put together in the same row of the excitation map. At the end of this merge process, the number of rows will be reduced.

Step 4: State assignment Assign state variables (i.e. secondary variables) to the rows of merged primitive state table and obtain Present State/Next State and output table. The outputs assigned to the unassigned to the unstable states vary according to the design requirements.

Step 5: Excitation and output table Decide the flip-flop to be used and obtain excitation and output table.

Step 6: Excitation and output maps Obtain the simplified expression for the excitation and output function by using K-map.

Step 7: Circuit Diagram Draw the schematic diagram.

Example 12.1. *Design an asynchronous sequential circuit with two inputs x_1 and x_2, and one output Z. Initially, both inputs are equal to zero. When x_1 or x_2 becomes '1', the output Z becomes 1. When the second input also becomes 1, the output changes to 0. The output stays at 0 until the circuit goes back to the initial state.*

Solution:

Step 1: State diagram From the word description of the problem, the operation of the circuit can be easily understood. Now, from this, a state diagram can be drawn, as shown in Fig. E12.1(*a*).

Step 2: Primitive state table: The primitive state table for the given requirement can be developed with Present State (PS), Next State (NS) and output (Z) as shown in Table E12.1(*a*).

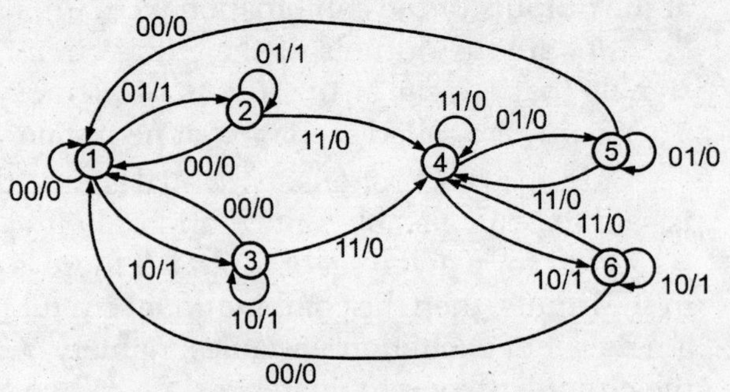

Fig. E12.1(*a*) State diagram

Table E12.1(*a*) Primitive state table

PS	NS, Z			
	$x_1, x_2 =$ 00	01	11	10
1	1, 0	2, 1	-, 0	3, 1
2	1, 0	2, 1	4, 0	-, -
3	1, 0	-, -	4, 0	3, 1
4	-, 0	5, 0	4, 0	6, 0
5	1, 0	5, 0	4, 0	-, -
6	1, 0	-,-	4, 0	6, 0

From the state diagram shown in Fig. E12.1 (*a*), all the rows of primitive state table shown in Table E12.1 (*a*) can be entered. Initially when both the inputs are equal to zero (i.e. $x_i = x_2 = 0$), the state is designated by 1. Also, when $x_1 = x_2 = 0$, the output should be zero which is indicated by '0' to the right of 1 in the first row and column $x_1, x_2 = 00$ of the primitive state table. Now, if x_1 remains at zero and x_2 becomes '1' the circuit enters a state 2 and its, output is '1'. This is entered as 2, '1' in the first row and column $x_1, x_2 = 01$. Similarly, from the initial input state $x_1, x_2 = 00$, if x_1 becomes '1' while x_2 remains '0', then the circuit enters a state 3 with the output '1', indicated as 3, '1' in the first row and column $x_1, x_2 = 10$. Since two inputs are not allowed to change simultaneously, a dash is entered in the first row and column $x_1, x_2 = 11$ with 0 as the output. To complete the primitive state table, the transitions from every state for each change in the input should be specified as in the state diagram.

Step 3: Minimization of primitive state: The minimization of primitive state table has two functions, viz., (*i*) eliminating redundant states and (*ii*) merging those states which are distinguishable by the input states. The redundant states are identified using implicant chart shown in Fig. E12.1(*b*).

Compare the NS, Z for row 1 and 2 of Table E12.1(*a*) for all possibilities of x_1 and x_2. If they are same, put a '✓' else, put an '×' in the box belonging to first row and first column of implicant chart shown in Fig. E12.1(*b*). Similarly, compare row 1 with all other rows and enter the appropriate symbol (tick or ×) in the corresponding space allocated

2	✓				
3	✓	✓			
4	×	×	×		
5	×	×	×	✓	
6	×	×	×	✓	✓
	1	2	3	4	5

Fig. E12.1(*b*) Implicant chart

for indicating row comparison result in the implicant chart shown Fig. E12.1(*b*). Repeat this for all rows of the state table to complete the Fig. E12.1(*b*). From the Fig. E12.1(*b*), it is clear that states 1, 2 and 3 are equivalent and similarly 4, 5 and 6 are equivalent states. Therefore, states 1, 2 and 3 can be mentioned as 1 and states 4, 5 and 6 can be mentioned as 4 in the Table E12.1(*b*). The minimized state table as shown in Table E12.1(*b*).

Table E12.1(b) Minimized state table

PS	NS, Z			
	$x_1\ x_2 =$ 0 0	0 1	1 1	1 0
1	1, 0	1, 1	4, 0	1, 1
4	1, 0	4, 0	4, 0	4, 0

Step 4: State Assignment: Since there are only two rows in the merged state table, a one-bit variable '0' and '1' can be assigned to row 1 and 2 respectively, i.e. state 1 and 4 are replaced by 0 and 1 respectively. The PS/NS and output table after the state assignment is shown in Table E12.1(*c*).

Table E12.1(*c*) PS/NS and output table

PS	NS, Z			
	$x_1\ x_2 =$ 0 0	0 1	1 1	1 0
1	0, 0	0, 1	1, 0	0, 1
1	0, 0	1, 0	1, 0	1, 0

Step 5: Excitation and output table: If Delay Flip-Flop is used, one can obtain the excitation and output table as shown in Table E12.1(*d*).

Table E12.1(*d*) Excitation and output table

PS	NS, Z			
	$x_1\ x_2 =$ 0 0	0 1	1 1	1 0
1	0, 0	0, 1	1, 0	0, 1
1	0, 0	1, 0	1, 0	1, 0

Step 6: Excitation and output maps: The K-map for simplifying the excitation function (*D*) and output (*Z*) are shown in Fig. 12.1(*c*).

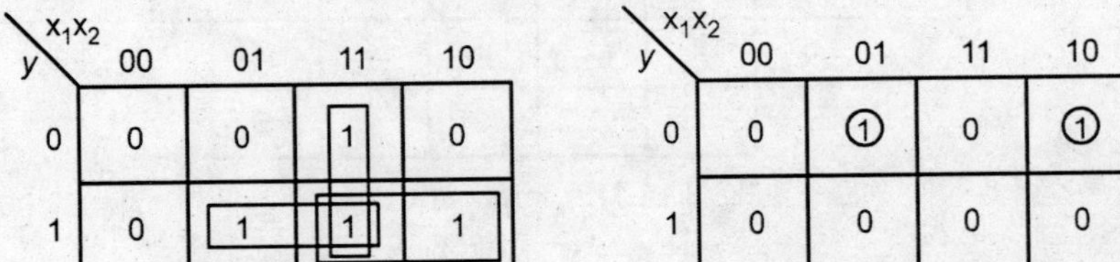

y \ x_1x_2	00	01	11	10
0	0	0	1	0
1	0	1	1	1

y \ x_1x_2	00	01	11	10
0	0	1	0	1
1	0	0	0	0

Fig. E12.1(*c*) *K*-map simplification for excitation and output function

From the *K*-map shown in Fig. E12.1(*c*), the simplified expression for *D* and *Z* are:

$$D = x_1x_2 + x_2y + yx_1$$

$$Z = \overline{x}_1x_2\overline{y} + x_1\overline{x}_2\overline{y}$$

$$= \overline{y}(\overline{x}_1x_2 + x_1\overline{x}_2)$$

$$= \overline{y}(x_1 \oplus x_2)$$

Step 7: Circuit diagram: Now, using the simplified expressions for *D* and *Z*, the circuit diagram for the given asynchronous sequential circuit can be drawn as shown in Fig. E12.1(*d*).

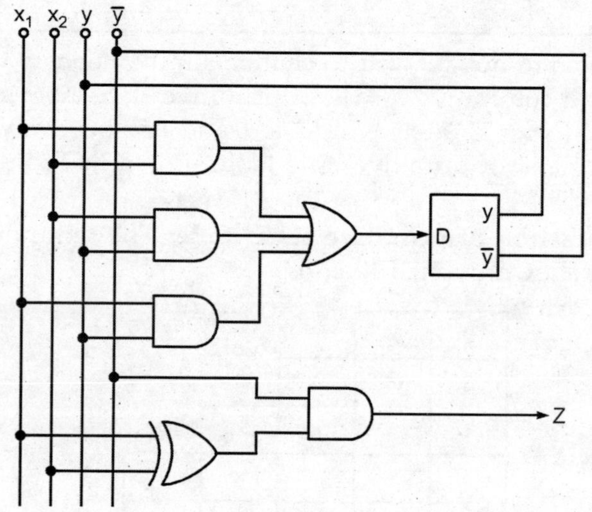

Fig. E12.1(*d*) Circuit diagram

Example 12.2. *Design an asynchronous sequential circuit with two inputs x_1 and x_2 and one output Z. The output Z = 1 if x_1 changes from 0 to 1, Z = 0 if x_2 changes from 0 to 1, and Z = 0 otherwise. Realize the circuit using (a) D flip-flops and (b) JK flip-flops.*

Solution:

Step 1: State diagram: From the word description of the problem, the operation of the circuit can be easily understood. Now from this, a state diagram can be drawn, as shown in Fig. E12.2(*a*).

Step 2: Primitive state table: The primitive state table for the given requirement can be developed with Present State (PS), Next State (NS) and output (Z) as shown in Table E12.2(*a*).

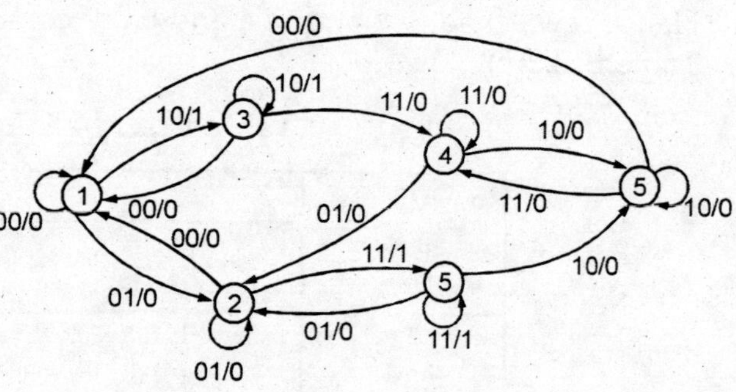

Fig. E12.2(*a*) State diagram

Table E12.2(*a*) Primitive state table

PS	NS, Z			
	$x_1 x_2 =$ 0 0	0 1	1 1	1 0
1	1, 0	2, 0	–, –	3, 1
2	1, 0	2, 0	5, 1	–, –
3	1, 0	–, 0	4, 0	3, 1
4	–, 0	2, 0	4, 0	6, 0
5	–, 0	2, 0	5, 1	6, 0
6	1, 0	–, 0	4, 0	6, 0

Since two inputs are not allowed to change simultaneously, a dash is entered as the NS in the first row and column $x_1\ x_2$ = 11 of primitive state table shown in Table E12.2(*a*). State 4 (NS = 11) and 5 (NS = 11) are having Z as 0 and 1 respectively. Hence the Z in the first row and column $x_1\ x_2$ = 11 of primitive state table shown in Table E12.2(*a*) is indicated as dash.

Step 3: Minimization of primitive state table: The redundant states are identified using implicant chart shown in Fig. E12.2(*b*).

2	✓				
3	✓	×			
4	×	×	×		
5	×	✓	×	×	
6	×	×	×	✓	×
	1	2	3	4	5

Fig. E12.2(*b*) Implicant chart

From the implicant chart shown in Fig. E12.2(*b*), it is clear that states 1 and 2 are equal, states 1 and 3 are equal, states 2 and 5 are equal and states 4 and 6 are equal. But it cannot be concluded that states 1, 2 and 3 are equal because states 2 and 3 are not equal. Hence row 1 and 3, 2 and 5, 4 and 6 are combined. The minimized state table is shown in Table E12.2(*b*).

Table E12.2(*b*) Minimised state table

PS	NS, Z			
	$x_1 x_2 =$ 0 0	0 1	1 1	1 0
1	1, 0	2, 0	4, 0	1, 1
2	1, 0	2, 0	2, 1	4, 0
4	1, 0	2, 0	4, 0	4, 0

Step 4: State assignment: Since there are 3 rows in the merged state table, the state variables 00,01,10 can be assigned to the rows 1, 2 and 3 respectively, i.e. State 1, 2 and 4 are replaced by variables 00, 01 and 10 respectively. The PS/NS and output table after the state assignment is shown in Table E12.2(*c*).

Table E12.2(*c*) PS/NS and output table

y_1, y_2	$Y_1\,Y_2, Z$			
	$x_1\,x_2 =$ 0 0	0 1	1 1	1 0
0 0	00, 0	01, 0	10, 0	00, 1
0 1	00, 0	01, 0	01, 1	10, 0
1 0	00, 0	01, 0	10, 0	10, 0

12.2.1 (*a*) Realization using *D* flip-flop

Step 5: Excitation and output table: If Delay (*D*) flip-flop is used as memory element, then the excitation and output table can be obtained as shown in Table E12.2(*d*).

Table E12.2(*d*) PS/NS and output table

y_1, y_2	$D_1\,D_2, Z$			
	$x_1\,x_2 =$ 0 0	0 1	1 1	1 0
0 0	00, 0	01, 0	10, 0	00, 1
0 1	00, 0	01, 0	01, 1	10, 0
1 0	00, 0	01, 0	10, 0	10, 0

Step 6: Excitation and output maps: The *K*-map for simplifying the excitation functions D_1 and D_2 and output Z are shown in Fig. E12.2(*c*). In the K-map simplification for $y_1, y_2 = 11$ state, the next state and outputs are assumed as zero and not don't care.

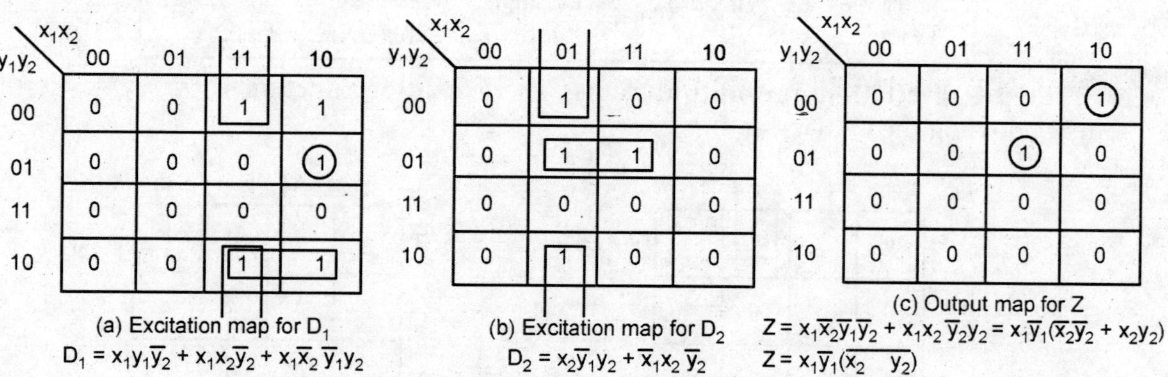

(a) Excitation map for D_1

$D_1 = x_1y_1\bar{y}_2 + x_1x_2\bar{y}_2 + x_1\bar{x}_2\,\bar{y}_1y_2$

(b) Excitation map for D_2

$D_2 = x_2\bar{y}_1y_2 + \bar{x}_1x_2\,\bar{y}_2$

(c) Output map for Z

$Z = x_1\bar{x}_2\bar{y}_1\bar{y}_2 + x_1x_2\,\bar{y}_2y_2 = x_1\bar{y}_1(\bar{x}_2\bar{y}_2 + x_2y_2)$

$Z = x_1\bar{y}_1(\overline{x_2 \quad y_2})$

Fig. E12.2(*c*) K-map simplification for excitation and output function

Now using the simplified excitation functions D_1, D_2 and output function Z, the logic diagram for the given asynchronous sequential circuit using Delay flip-flop can be drawn as shown in Fig. E12.2(*d*).

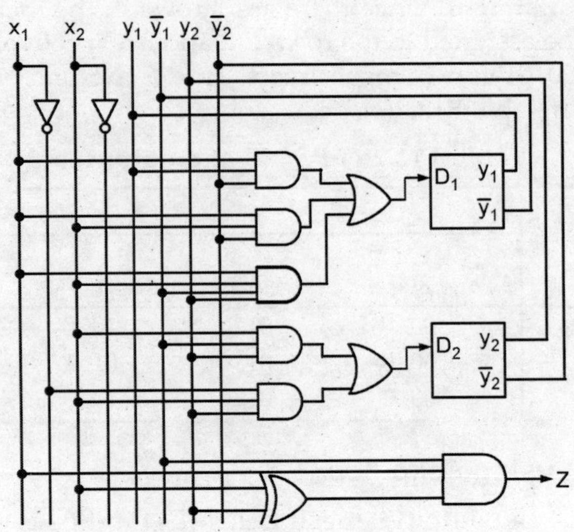

Fig. E12.2(*d*) Circuit diagram

12.2.1(*b*) Realisation using *JK* Flip-Flops

Step 7: Excitation and output table: If *JK* flip-flops are used as memory elements, then the excitation and output table can be drawn using application table of *JK* flip-flop, as shown in Table E12.2(*e*).

Table E12.2(*e*) Excitation and output table

y_1y_2	J_1K_1, J_2K_2, Z			
	x_1x_2 = **0 0**	**0 1**	**1 1**	**1 0**
00	0*d*, 0*d*, 0	0*d*, 1*d*, 0	1*d*, 0*d*, 0	0*d*, 0*d*, 1
01	0*d*, *d*1, 0	0*d*, *d*0, 0	0*d*, *d*0, 1	1*d*, *d*1, 0
10	*d*1, 0*d*, 0	*d*1, 1*d*, 0	*d*0, 0*d*, 0	*d*0, 0*d*, 0

Step 8: Excitation and output maps: The excitation function J_1, K_1, J_1, K_1 can be simplified using the K-maps as shown in Fig. E12.2(e).

y_1y_2 \ x_1x_2	00	01	11	10
00	0	0	1	0
01	0	0	0	1
11	0	0	0	0
10	d	d	d	d

Excitation map for J_1

$J_1 = x_1x_2\overline{y}_2 + x_1\overline{x}_2\,\overline{y}_1y_2$

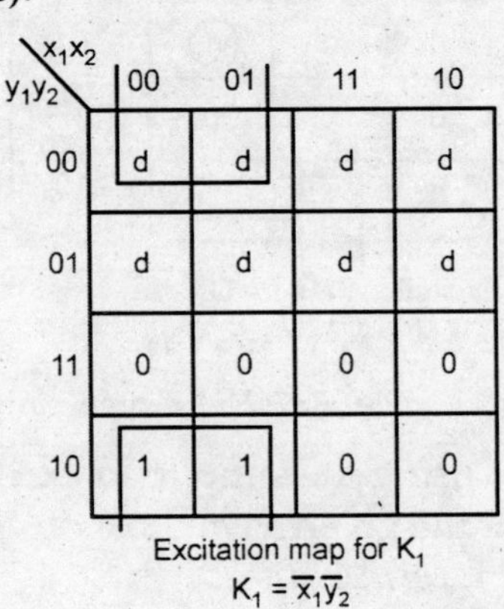

Excitation map for K_1

$K_1 = \overline{x}_1\overline{y}_2$

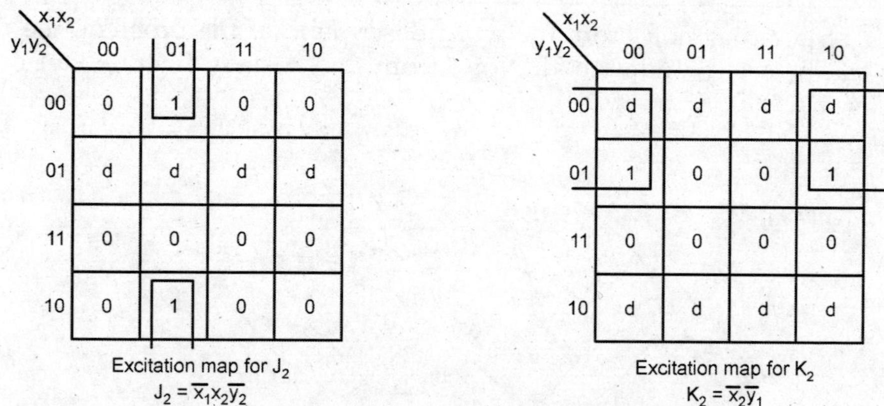

Fig. E12.2(*e*) K-map simplification for excitation function

Step 9: Now using simplified expressions for excitation function J_1, K_1, J_2, K_2 and output function Z, derived in the Fig. E12.2(*e*) and (*c*) respectively. The circuit diagram for the given asynchronous circuit using *JK* flip-flop can be drawn as shown in Fig. E12.2(*f*).

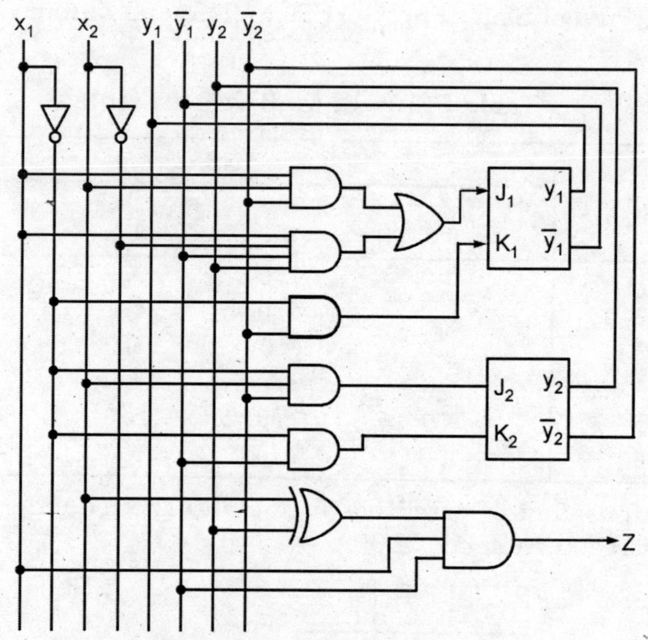

Fig. E12.2(*f*) Circuit diagram

Example 12.3. *Design an asynchronous circuit that has two inputs x_1 and x_2 and one output Z. The circuit is required to give an output wherever the input sequence (0, 0) (0, 1) and (1, 1) received but only in that order.*

Solution:

Step 1: State diagram: From the word description of the problem, the operation of the circuit can be easily understood. Now, from this, a state diagram can be drawn, as shown in Fig. E12.3(*a*).

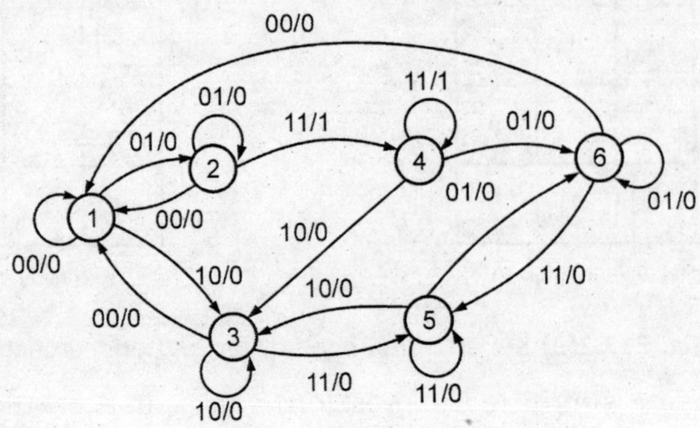

Fig. E12.3(*a*) State diagram

Step 2: Primitive state table: The primitive state table for the given requirement can be developed with Present State (PS), Next State (NS) and output (Z) as shown in Table E12.3(*a*).

Table E12.3(*a*) Primitive state table

PS	NS, Z			
	$x_1 x_2 =$ 0 0	0 1	1 1	1 0
1	1, 0	2, 0	-, -	3, 0
2	1, 0	2, 0	4, 1	-, 0
3	1, 0	-, 0	5, 0	3, 0
4	-, 0	6, 0	4, 0	3, 0
5	-, 0	6, 0	5, 0	3, 0
6	1, 0	6, 0	5, 0	-, 0

Step 3: Minimisation of primitive state table: The redundant states are identified using implicant chart shown in Fig. 12.3(*b*).

2	✓				
3	✓	×			
4	×	×	×		
5	×	×	✓	×	
6	×	×	✓	×	×
	1	2	3	4	5

Fig. E12.3(*b*) Implicant chart

From the implicant chart shown in Fig. E12.3(*b*), it is obvious that row 1 and 2; row 3, 5 and 6 can be combined and row 4 can be kept as it is. The minimized state table is shown in Table E12.3(*b*).

Table E12.3(*b*) Minimized state table

PS	NS, Z			
	$x_1 x_2 =$ 0 0	0 1	1 1	1 0
1	1, 0	1, 0	4, 1	3, 0
3	1, 0	3, 0	3, 0	3, 0
4	-, 0	3, 0	4, 1	3, 0

Step 4: State assignment: Since there are 3 rows in the merged state table, the state variables 00, 01, 11 can be assigned to the rows 1, 2 and 3 respectively, i.e. State 1, 3 and 4 are replaced by variable 00, 01, 11 can be assigned to the rows 1, 2 and 3 respectively, i.e. State 1, 3 and 4 are replaced by variable 00, 01 and 11 respectively. The PS/NS and output table after the state assignment is shown in Table 12.3(*c*).

Table E12.3(*c*) PS/NS and output table

Row $y_1 y_2$	$Y_1 Y_2, Z$			
	$x_1 x_2 =$ *0 0*	*0 1*	*1 1*	*1 0*
0 0	00, 0	00, 0	11, 1	01, 0
0 1	00, 0	01, 0	01, 0	01, 0
1 1	00, 0	01, 0	11, 1	01, 0

Step 5: Excitation and output table: If Delay (*D*) flip flop is used as memory element, then the excitation and output table can be obtained as shown in the Table E12.3(*d*).

Table E12.3(*d*) Excitation and output table

$y_1 y_2$	$D_1 D_2, Z$			
	$x_1 x_2 =$ 0 0	0 1	1 1	1 0
0 0	00, 0	00, 0	11, 1	01, 0
0 1	00, 0	01, 0	01, 0	01, 0
1 1	00, 0	01, 0	11, 1	01, 0

Step 6: Excitation and output maps: The excitation functions (D_1 and D_2) and the output function (*Z*) can be simplified using K-map method as shown in Fig. E12.3(*c*).

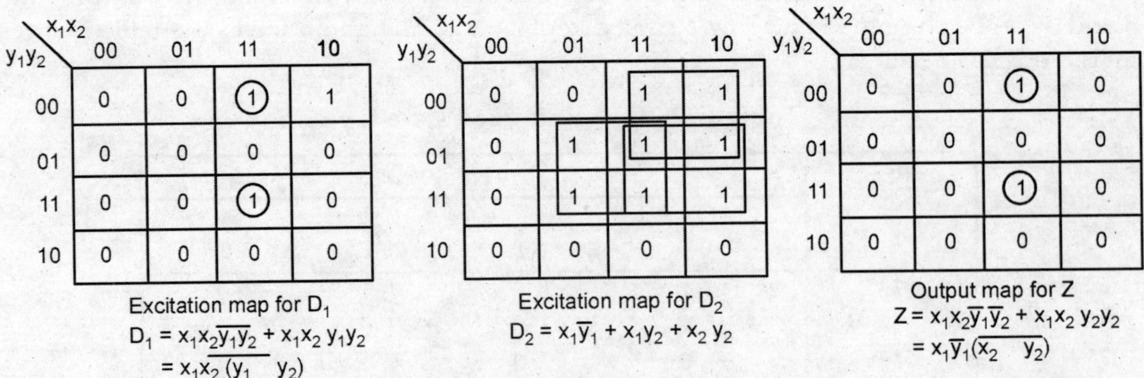

Fig. E12.3(*c*) *K*-map simplification for excitation and output function

Step 7: Using the simplified excitation and output functions, the circuit diagram of the given asynchronous circuit can be drawn as shown in Fig. E12.3(*d*).

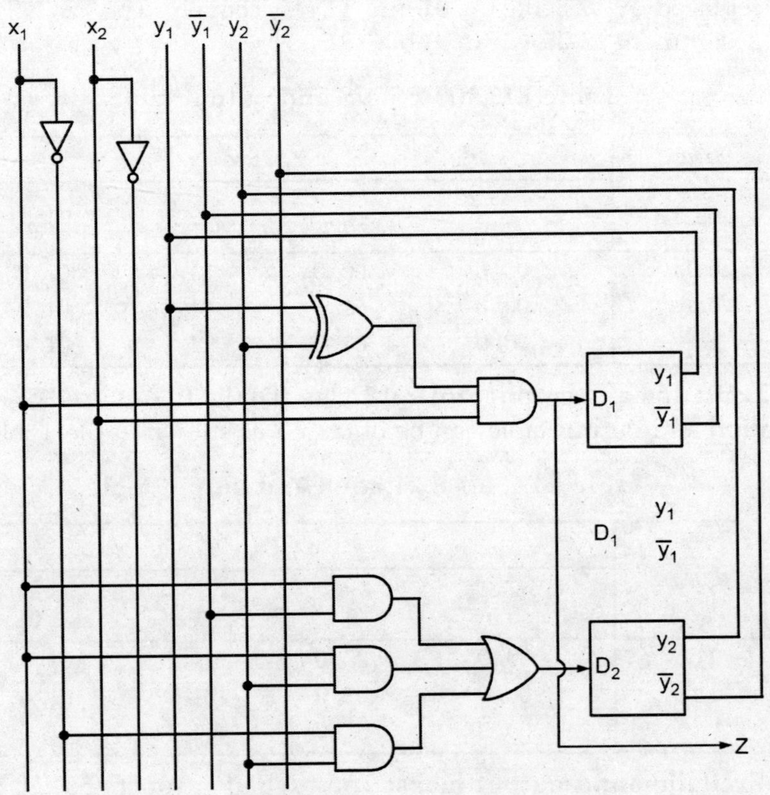

Fig. E12.3(*d*) Circuit diagram

12.3 DESIGN OF PULSE MODE ASYNCHRONOUS SEQUENTIAL CIRCUITS

In pulse mode sequential circuits, the inputs are pulses. In this circuit, it is assumed that no two pulses will arrive at the same time. Also, it is assumed that the duration of the pulses is long enough to cause state transition and is short enough so that there will not be more than one state transition for a single pulses.

The design procedure employed for fundamental mode circuits are also applicable to pulse mode asynchronous circuits. The design of pulse mode asynchronous circuits is explained with the following examples.

Example 12.4. *Design an asynchronous circuit that will give output when only the first pulse received whenever a control input is asserted from LOW to HIGH state. Any further pulses will be ignored.*

Solution:

Step 1: State diagram: As per the given design requirement, only the first pulse received in pulse input (*P*), after the control input is asserted from LOW to HIGH state, will be passed through the circuit, i.e. available at the output (Z). Any further input pulses will be ignored. To output any pulses, thereafter, the control input must once again go to HIGH state from a LOW state. Then only, the next first pulse will be transmitted. The block diagram of this circuit is shown in Fig. E12.4(*a*). To understand the operation of the circuit, consider the timing diagram shown in Fig. E12.4(*b*).

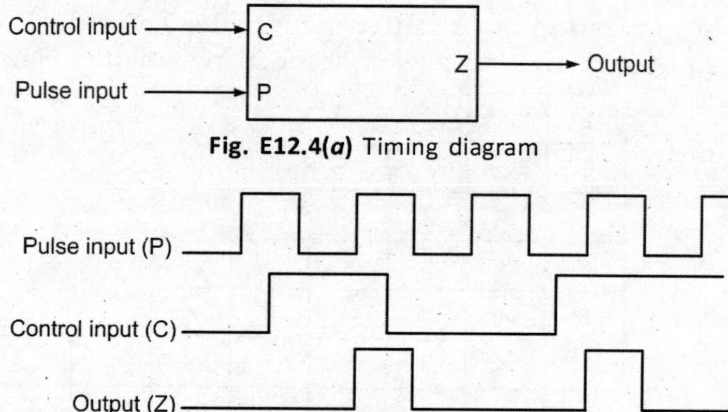

Fig. E12.4(*a*) Timing diagram

Fig. E12.4(*b*) Timing diagram

Now from this, a state diagram can be drawn, as shown in Fig. E12.4(*c*).

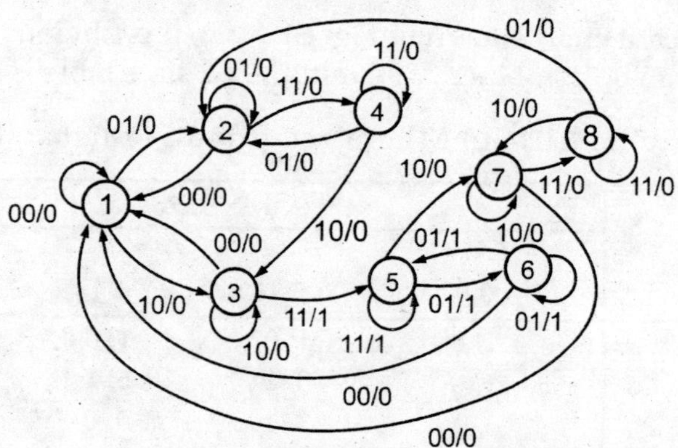

Fig. E12.4(*c*) State diagram

Step 2: Primitive state table: The primitive state table for the given requirement can be developed as shown in Table E12.4(*a*).

Table E12.3(*a*) Primitive state table

PS	NS, Z			
	CP = 0 0	0 1	1 1	1 0
1	1, 0	2, 0	-, -	3, 0
2	1, 0	2, 0	4, 0	-, 0
3	1, 0	-, -	5, 1	3, 0
4	-, 0	2, 0	4, 0	3, 0
5	-, 0	6, 1	5, 1	7, 0
6	1, 0	6, 1	5, 1	-, 0
7	1, 0	-, 0	8, 0	7, 0
8	-, 0	2, 0	8, 0	7, 0

Step 3: Minimization of primitive state table: The redundant states of primitive state table can be found out using the implicant chart shown in Fig. E12.4(*d*).

2	✓						
3	✓	✓					
4	✓	✓	×				
5	×	×	×	×			
6	×	×	✓	×	✓		
7	×	×	×	×	×	×	
8	×	×	×	×	×	×	✓
	1	2	3	4	5	6	7

Fig. E12.4(*d*) Implicant chart

From the implicant chart shown in Fig. E12.4(*b*), it is obvious that row 1, 2 and 4; row 3 and 6; row 7 and 8 can be combined. The minimized state table is shown in Table E12.4(*b*).

Table E12.4(*b*) Minimized primitive state table

PS	NS, Z			
	CP = 0 0	0 1	1 1	1 0
1	1, 0	1, 0	1, 0	3, 0
3	1, 0	3, 1	5, 1	3, 0
5	-, 0	3, 1	5, 0	7, 0
7	1, 0	1, 0	7, 0	7, 0

Step 4: State assignment: There are four rows in the merged primitive state table. Therefore, two secondary variables are required for the state assignment. Let 00 be the state variables assigned to first row, 01 to 2nd row, 11 to third row and 10 to 4th row. The entry '-, 0' in the 3rd row of minimized state table is assigned with 00. Now, the PS/NS and output table is as shown in Table E12.4(*c*).

Table E12.4(*c*) PS/NS and output table

$y_1\ y_2$	$Y_1\ Y_2, Z$			
	CP = 0 0	0 1	1 1	1 0
0 0	00, 0	00, 0	00, 0	01, 0
0 1	00, 0	01, 1	11, 1	01, 0
1 1	00, 0	01, 1	11, 1	10, 0
1 0	00, 0	00, 0	10, 0	10, 0

Step 5: If *D* flip-flops are used as the memory elements, then the excitation and output table is as shown in Table E12.4(*d*).

Table E12.4(*d*) Excitation and output table

$y_1\ y_2$	$D_1\ D_2, Z$			
	CP = 0 0	0 1	1 1	1 0
0 0	00, 0	00, 0	00, 0	01, 0
0 1	00, 0	01, 1	11, 1	01, 0
1 1	00, 0	01, 1	11, 1	10, 0
1 0	00, 0	00, 0	10, 0	10, 0

Step 6: Excitation and output maps: The *K*-maps for simplifying the excitation functions D_1, D_2 and output *Z* are shown in Fig. E12.4(*e*).

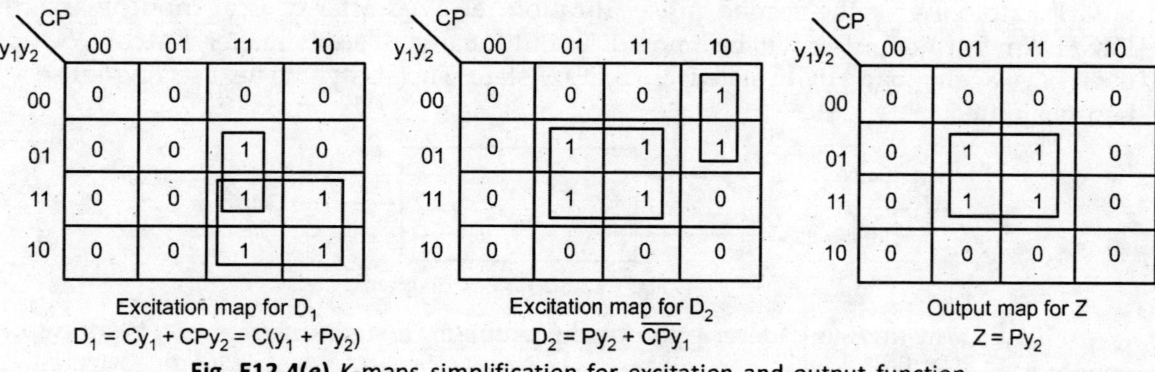

Excitation map for D_1
$D_1 = Cy_1 + CPy_2 = C(y_1 + Py_2)$

Excitation map for D_2
$D_2 = Py_2 + \overline{CPy_1}$

Output map for Z
$Z = Py_2$

Fig. E12.4(*e*) *K*-maps simplification for excitation and output function

Step 7: Using the above simplified expressions for excitation and output functions, the circuit diagram for the asynchronous sequential circuit can be drawn as shown in Fig. E12.4(*f*).

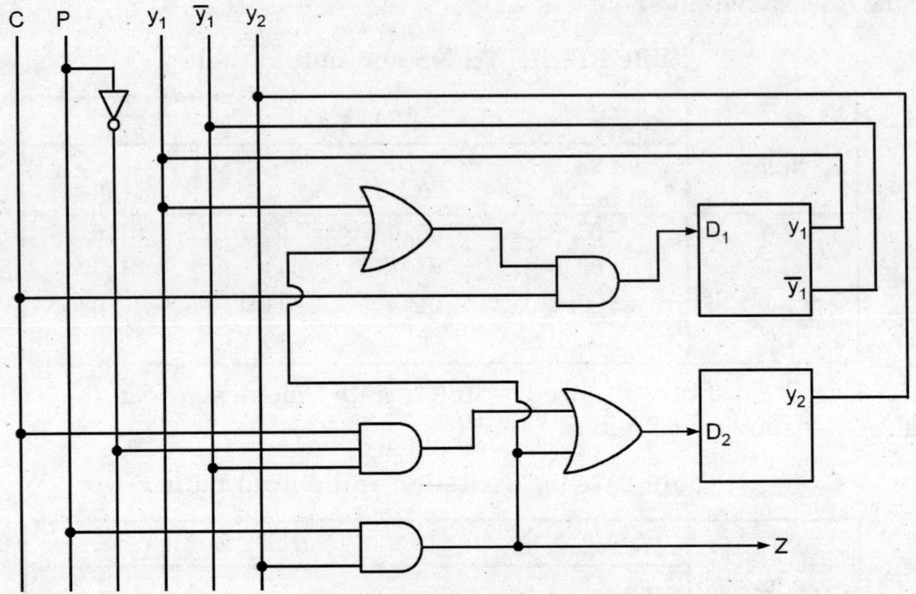

Fig. E12.4(*f*) Circuit diagram

Example 12.5. *Design an asynchronous circuit that will output only the second pulse received whenever a control input is asserted from LOW to HIGH state and will ignore any other pulse.*

Solution:

Step 1: Let us assume that the circuit for the given requirement consists of two input (*i*) Control Input (*C*) and (*ii*) Pulse Input (*P*) as shown in Fig. E12.5(*a*). The output goes to HIGH state only for the second pulse duration received after control input is asserted HIGH. Any further pulses will be ignored. To output any pulse, thereafter, the control input must once again go to HIGH state from a LOW state. Then only, the next second pulse will be transmitted.

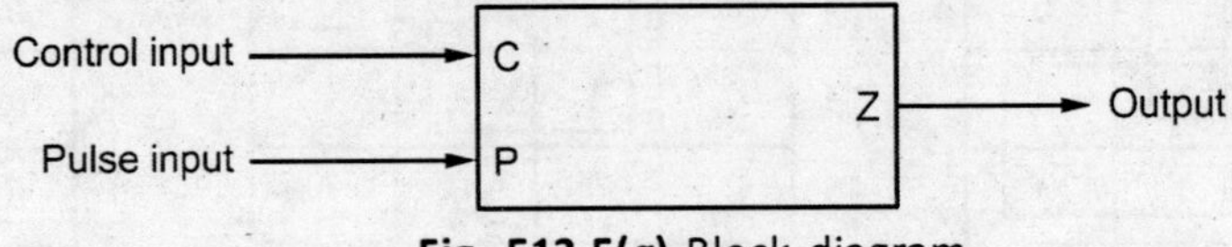

Fig. E12.5(*a*) Block diagram

Now from the word description of the problem, a state diagram can be drawn, as shown in Fig. E12.5(*b*).

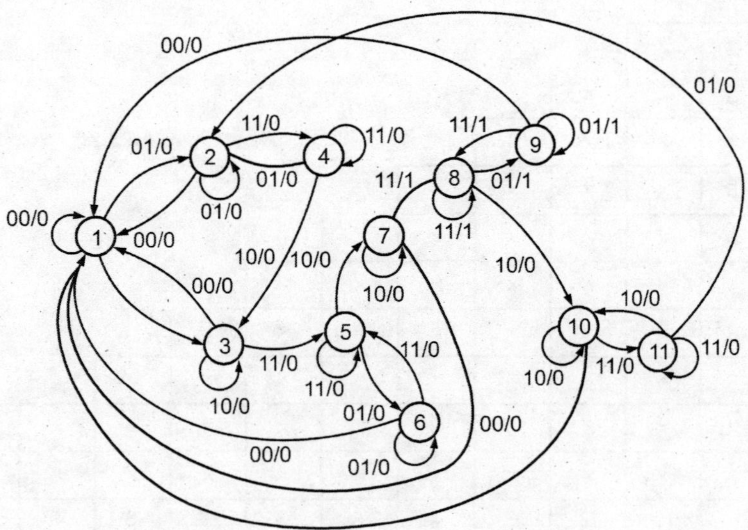

Fig. E12.5(*b*) State diagram

Step 2: Primitive state table: The primitive state table for the given requirement is shown in Table E12.5(*a*).

Table E12.5(*a*) Primitive state table

PS	NS, Z			
	CP = 0 0	0 1	1 1	1 0
1	1, 0	2, 0	-, -	3, 0
2	1, 0	2, 0	4, 0	-, 0
3	1, 0	-, 0	5, 0	3, 0
4	-, 0	2, 0	4, 0	3, 0
5	-, 0	6, 0	5, 0	7, 0
6	1, 0	6, 0	5, 0	-, 0
7	1, 0	-, 0	8, 1	7, 0
8	-, 0	9, 1	8, 1	10, 0
9	1, 0	9, 1	8, 1	-, 0
10	1, 0	-, 0	11, 0	10, 0
11	-, 0	2, 0	11, 0	10, 0

Step 3: Minimisation of primitive state table: The redundant states of primitive state table can be found out using the implicant chart shown in Fig. E12.4(*c*).

	1	2	3	4	5	6	7	8	9	10
2	✓									
3	✓	×								
4	✓	✓	×							
5	×	×	×	×						
6	×	×	×	×	✓					
7	×	×	×	×	×	×				
8	×	×	×	×	×	×	×			
9	×	×	×	×	×	×	×	✓		
10	×	×	×	×	×	×	×	×	×	
11	×	×	×	×	×	×	×	×	×	✓

Fig. E12.5(*c*) Implicant chart

From the implicant chart shown in Fig. E12.5(*c*), it is obvious that row 1, 2 and 4; row 5 and 6; row 8 and 9; row 10 and 11 can be combined and row 3 and 7 can be maintained as it is. The minimized state table is shown in Table E12.5(*b*).

Table E12.5(*b*) Primitive state table

PS	NS, Z			
	CP = 0 0	0 1	1 1	1 0
1	1, 0	1, 0	1, 0	3, 0
3	1, 0	-, 0	5, 0	3, 0
5	1, 0	5, 0	5, 0	7, 0
7	1, 0	-, 0	8, 1	7, 0
8	1, 0	8, 1	8, 1	10, 0
10	1, 0	1, 0	10, 0	10, 0

Step 4: The state variable assigned to the rows of the merged state table as follows 000 to 1st row, 001 to 2nd row, 011 to 3rd row, 010 to 4th row, 110 to 5th row and 111 to 6th row. Here the state variable is assigned with unit distance codes to get the simplified excitation function. Now the PS/NS and the output table is shown in Table E12.5(*c*).

Table E12.5(c) PS/NS and output table

$y_1y_2y_3$	$Y_1Y_2Y_3$, Z			
	CP = **0 0**	**0 1**	**1 1**	**1 0**
000	000, 0	000, 0	000, 0	001, 0
001	000, 0	000, 0	011, 1	001, 0
011	000, 0	011, 0	011, 1	010, 0
010	000, 0	000, 0	110, 1	010, 0
110	000, 0	110, 0	110, 1	111, 0
111	000, 0	000, 0	111, 0	111, 1

Step 5: Excitation and output table: If Delay Flip-Flops are used as memory elements, then the excitation and output table can be drawn as shown in Table E12.5(*d*).

Table E12.5(*d*) Excitation and output table

$y_1y_2y_3$	$Y_1Y_2Y_3$, Z			
	CP = **0 0**	**0 1**	**1 1**	**1 0**
000	000, 0	000, 0	000, 0	001, 0
001	000, 0	000, 0	011, 1	001, 0
011	000, 0	011, 0	011, 1	010, 0
010	000, 0	000, 0	110, 1	010, 0
110	000, 0	110, 0	110, 1	111, 0
111	000, 0	000, 0	111, 0	111, 1

Step 6: Excitation and output maps: The *K*-maps for simplifying excitation function D_1, D_2 and D_3 and output function Z are shown in Fig. E12.5(*d*).

CP \ $y_1y_2y_3$	000	011	011	010	110	111	101	100
00	0	0	0	0	0	0	0	0
01	0	0	0	0	1	0	0	0
11	0	0	0	1	1	1	0	0
10	0	0	0	0	1	1	0	0

Excitation map for D_1 $D_1 = y_1y_2C + y_1y_2\bar{y}_3P + y_2\bar{y}_3CP$

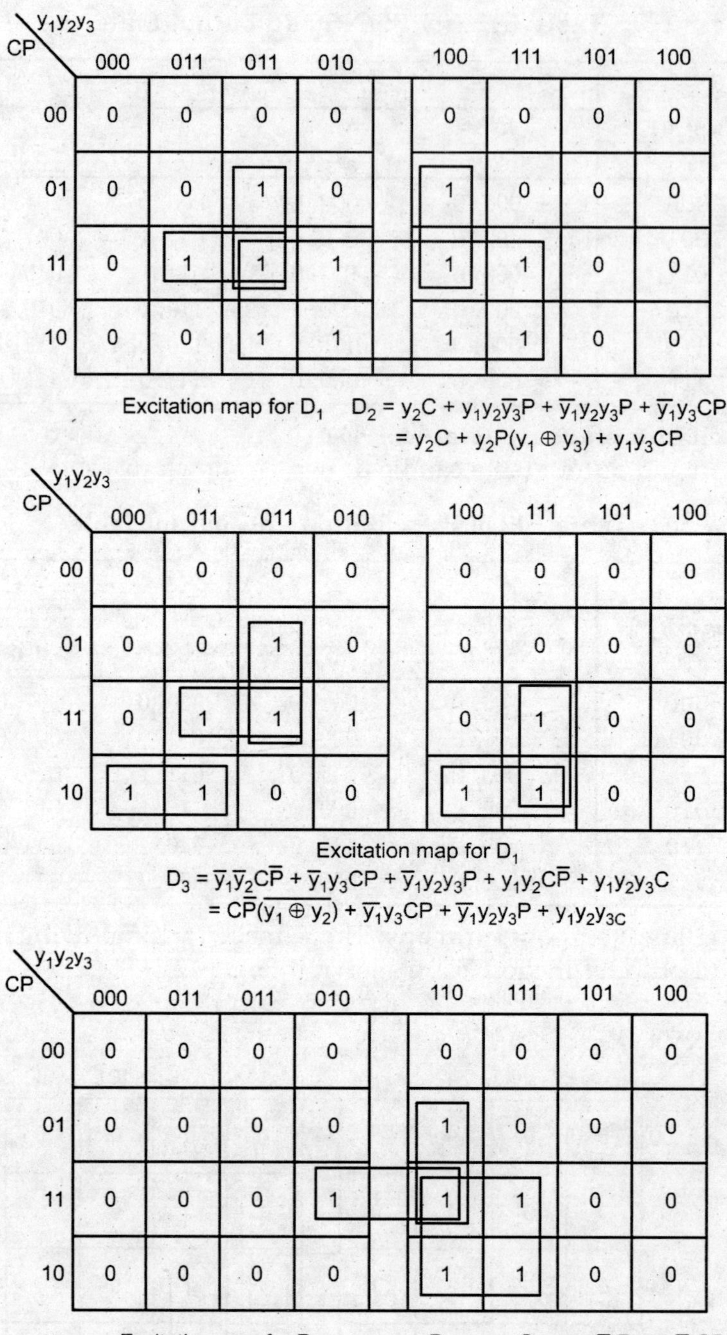

Fig. E12.5(*d*) *K*-maps for simplifying excitation and output functions

Step 7: Using the above simplified expressions for excitation and output function, the circuit diagram for the asynchronous sequential circuit can be drawn as shown in Fig. E12.5(*e*).

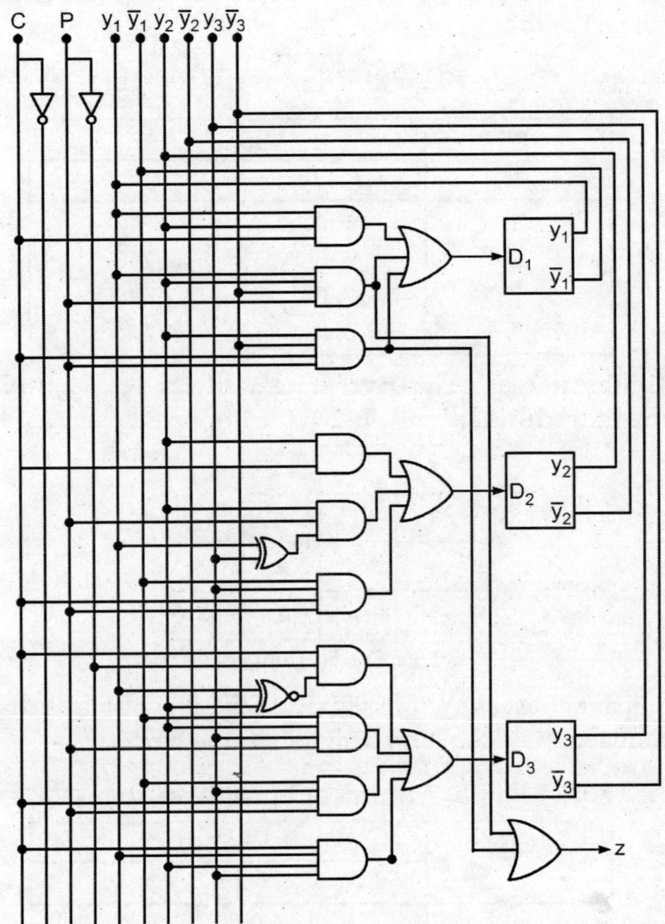

Fig. E12.5(*e*) Circuit diagram

Example 12.6. *Design an asynchronous circuits that will output only the first pulse received and ignore any other pulses.*

Solution:

Step 1: State diagram: As per the given design requirement, the circuit to be designed should consist of only one input called pulse input (*F*) and one output (*Z*). The state diagram for the given design requirement is shown in Fig. E12.6(*a*).

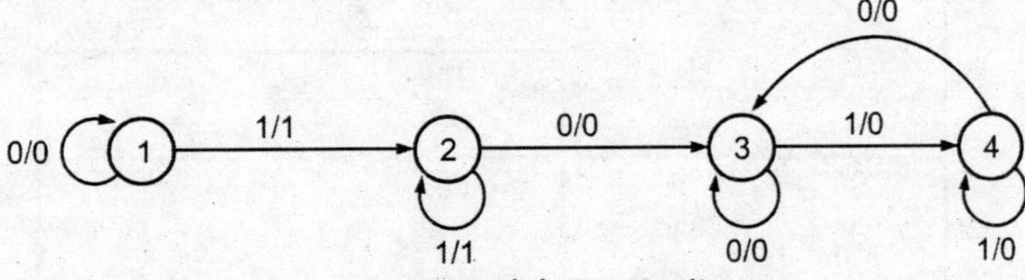

Fig. E12.6(*a*) Circuit diagram

Step 2: Primitive state table: The primitive state table for this circuit can be constructed as shown in the Table E12.6(*a*).

Table E12.6(*a*) Primitive state table

PS	NS, Z	
	$P = 0$	$P = 1$
1	1, 0	2, 0
2	3, 0	2, 0
3	3, 0	4, 0
4	3, 0	4, 0

Step 3: Minimization of primitive state table: The primitive state table can be minimized by finding the redundant state using the implicant chart as shown in Fig. E12.6(*b*).

2	×		
3	×	×	
4	×	×	✓
	1	2	3

Fig. E12.6(*b*) Implicant chart

From the implicant chart shown in Fig. E12.6(*b*), it is obvious that row 3 and 4 can be combined. The minimized state table is shown in Table E12.6(*b*).

Table E12.6(*b*) Minimized primitive state table

PS	NS, Z	
	$P = 0$	$P = 1$
1	1, 0	2, 1
2	3, 0	2, 1
3	3, 0	3, 0

Step 4: State assignment: There are three rows in merged primitive state table. Therefore the secondary variables are required for the state assignment. Let 00 be the state variables assigned to first row, 01 to second row and 11 to third row. The PS/NS and output table is as shown in Table E12.6(*c*).

Table E12.6(*c*) PS/NS and output table

		Y_1Y_2, Z	
		$P =$	
y_1	y_2	*0*	*1*
0	0	00, 0	01, 1
0	1	11, 0	01, 1
1	1	11, 0	11, 0

Step 5: If *D* flip-flops are used as the memory element, the excitation and output table is as shown in Table E12.6(*d*).

Table E12.6(*d*) Excitation and output table

		D_1D_2, Z	
		P =	
y_1	y_2	**0**	**1**
0	0	00, 0	01, 1
0	1	11, 0	01, 1
1	1	11, 0	11, 0

Step 6: Excitation and output maps: The *K*-maps for simplifying the excitation functions D_1, D_2 and output function *Z* are shown in Fig. E12.6(*c*).

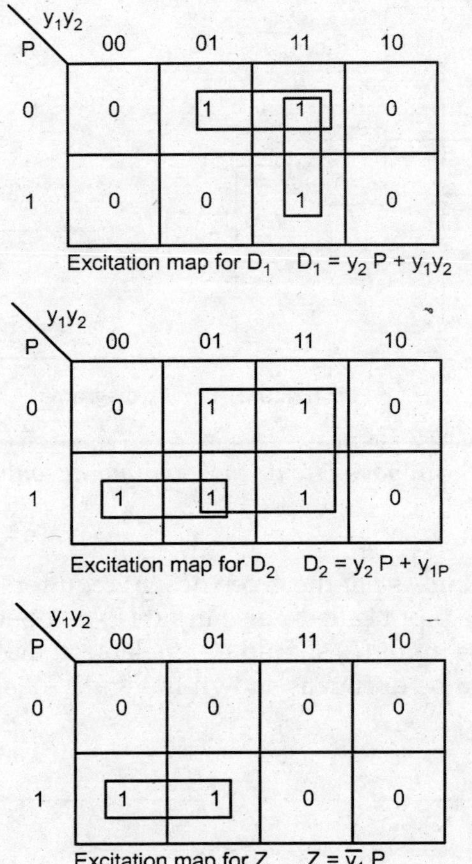

Fig. E12.6 (*c*) *K*-maps simplification for excitation and output function

Step 7: Using the above simplified expression for excitation and output function, the circuit diagram for the asynchronous sequential circuit can be drawn as shown in Fig. E12.6(*d*).

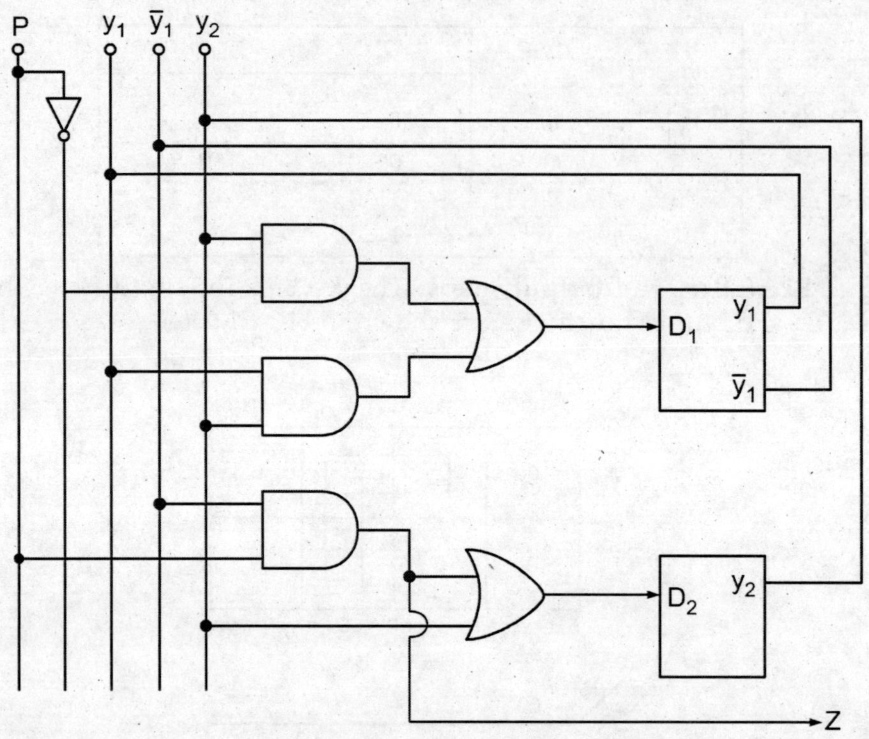

Fig. E12.6(*d*) Circuit diagram

Example 12.7. *Design asynchronous circuit that will output only the second pulse received and ignore any other pulses.*

Solution:

Step 1: State diagram: As per the given design requirement, the circuit to be designed should consist of only one input called pulse input (*P*) and one output (*Z*). Only the second pulse received in the pulse input (*P*) should be available at the output. The state diagram for the given requirement can be drawn as shown in Fig. E12.7(*a*).

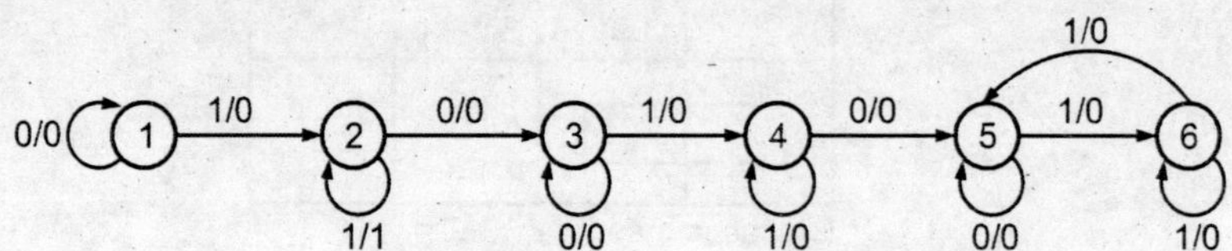

Fig. E12.7(*a*) Circuit diagram

Step 2: Primitive state table: The primitive state table for this state diagram can be constructed as shown in the Table E12.7(*a*).

Table E12.7(*a*) Primitive state table

PS	NS, Z	
	$P = 0$	1
1	1, 0	2, 0
2	3, 0	2, 0
3	3, 0	4, 1
4	5, 0	4, 1
5	5, 0	6, 0
6	5, 0	6, 0

Step 3: Minimization of primitive state table: The primitive state table can be minimized by using the implicant chart as shown in Fig. E12.7(*b*).

2	×				
3	×	×			
4	×	×	×		
5	×	×	×	×	
6	×	×	×	×	✓
	1	2	3	4	5

Fig. E12.7(*b*) Implicant chart

From the implicant chart shown in Fig. E12.7 (*b*), it is obvious that row 5 and 6 can be combined. The minimized state table is shown in Table E12.7 (*b*).

Table E12.7(b) Minimised state table

PS	NS, Z	
	$P = 0$	$P = 1$
1	1, 0	2, 0
2	3, 0	2, 0
3	3, 0	4, 1
4	5, 0	4, 1
5	5, 0	5, 0

Step 4: State assignment: The state variables assigned to the rows of the merged state table are as follows: 000 to 1st row, 001 to 2nd row, 011 to 3rd row, 010 to 4th row and 110 to 5th row. Here, the state variables are assigned with graycode to get the simplified excitation function. Now, the PS/NS and output table is as shown in Table E12.7 (*c*).

Table E12.7(*c*) PS/NS and output table

$y_1y_2y_3$	$Y_1Y_2Y_3, Z$	
	$P = 0$	$P = 1$
000	000, 0	001, 0
001	011, 0	001, 0
011	011, 0	010, 1
010	110, 0	010, 1
110	110, 0	110, 0

Step 5: The *D* Flip-flop used as the memory element, the excitation and the output table as shown in the Table E12.7(*d*).

Table E12.7(*d*) Excitation and output table

$y_1y_2y_3$	$D_1D_2D_3, Z$	
	$P = 0$	1
000	000, 0	001, 0
001	011, 0	001, 0
011	011, 0	010, 1
010	110, 0	010, 1
110	110, 0	110, 0

Step 6: Excitation and output table: The *K*-maps for simplifying excitation function D_1, D_2, D_3 and output function *Z* are shown in Fig. E.12.7(*c*).

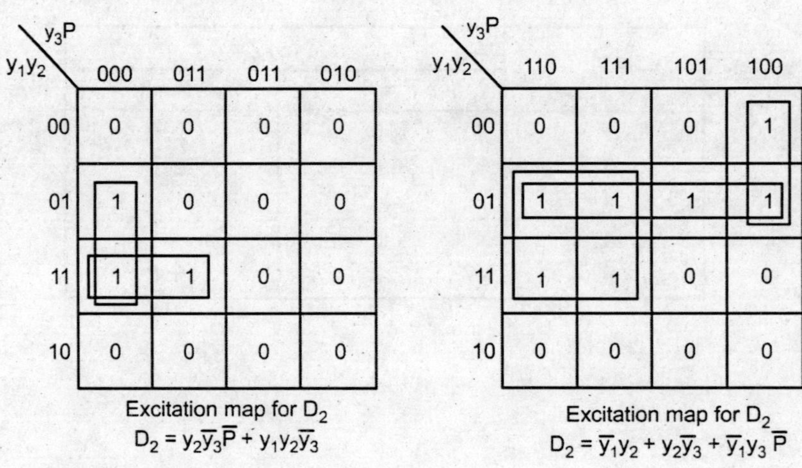

Excitation map for D_2
$D_2 = y_2\bar{y}_3\bar{P} + y_1y_2\bar{y}_3$

Excitation map for D_2
$D_2 = \bar{y}_1y_2 + y_2\bar{y}_3 + \bar{y}_1y_3\bar{P}$

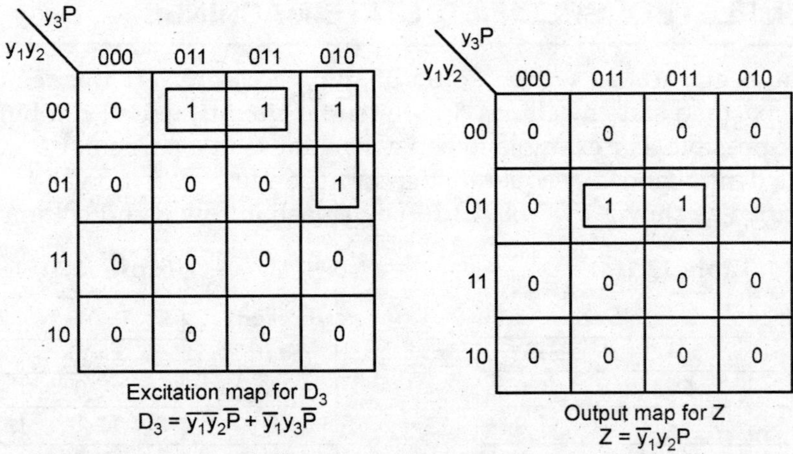

Excitation map for D_3
$D_3 = \overline{y}_1\overline{y}_2\overline{P} + \overline{y}_1 y_3\overline{P}$

Output map for Z
$Z = \overline{y}_1 y_2 P$

Fig. E12.7(*c*) *K*-maps simplification for excitation and output function

Step 7: Using the above simplified expression for excitation and output function, the circuit diagram for the asynchronous sequential circuit can be drawn as shown in Fig. E12.7(*d*).

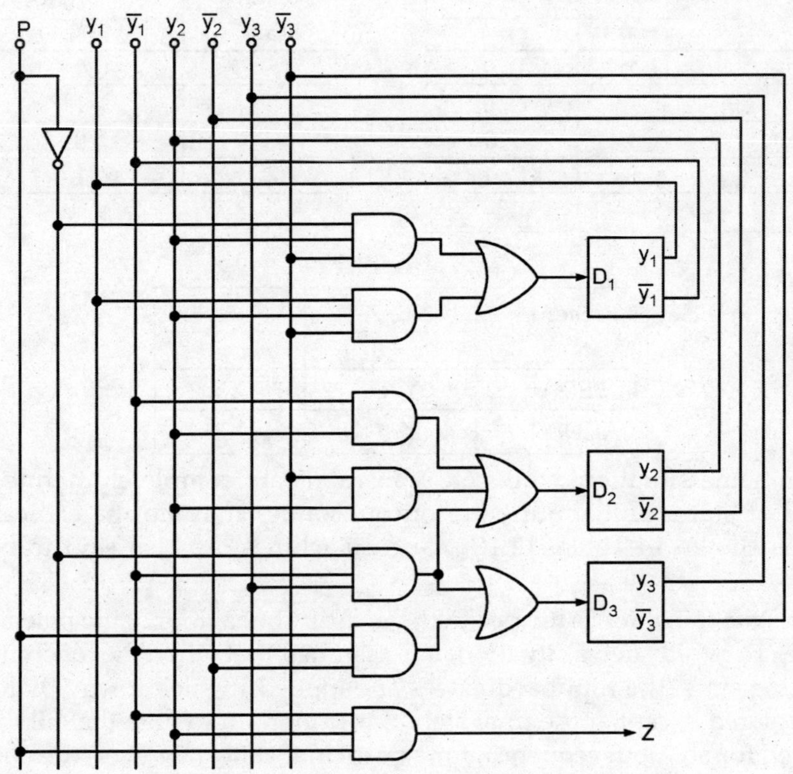

Fig. E12.7(*d*) Circuit diagram

12.4 INCOMPLETELY SPECIFIED STATE MACHINES

Sequential circuits in which some of the states are left unspecified are called Incompletely specified state machines. In sequential circuits, not all combinations of states and inputs are possible. For example, when a machine is in a particular state (say *B*), it will never receive a 0 input and consequently the corresponding transition and output map may be left unspecified as shown in Table 12.1(*a*) by a dash in row *B* and column $x = 0$.

Table 12.1(*a*)

Present State	NS, Z	
	$x = 0$	$x = 1$
A	D, 1	B, 0
B	-	C, 0
C	A, 1	-,0
D	A, 0	D, 1

Table 12.1(*b*)

Present State	NS, Z	
	$x = 0$	$x = 1$
A	A, 0	B, 1
B	D, 1	B, 0
C	D, 0	E, 1
E	C, 1	E, -

Table 12.1(*c*)

Present State	NS, Z	
	$x = 0$	$x = 1$
A	D, 1	B, 0
B	T, -	C, 0
C	A, 1	T, 0
D	A, 0	D, 1
T	T, -	T, -

Table 12.1(*d*)

Present State	NS, Z	
	$x = 0$	$x = 1$
A	A, 0	B, -
B	D, 1	B, 0
C	D, 0	D, 1
E	C, 1	D, 1

Table 12.1(*e*)

Present state	NS, Z	
	$x = 0$	$x = 1$
BE(X)	Y, 1	X, 0
ACD(Y)	Y, 0	X, 1

Also, in some situations, the state transitions are completely defined but for some combinations of states and inputs, the output values may not be critical and thus left unspecified as shown in Table. 12.1(*b*). Such machines are also said to be incompletely specified.

When a state transition is unspecified, the future behaviour of the machine may become unpredictable. To avoid such a situation, it is assumed that when the machine is in any of its possible starting state, the input sequences are applied in such a way that no unspecified state is encountered except at the final step. It is important to note that all the outputs need not be specified for an input sequence, for a particular state; however, the next states should be specified except possibly for the last symbol of the sequence.

A machine with partially specified transitions can be described by another machine whose state transitions are completely specified. This is done by replacing all the dashes in the next state entries by 7 and adding a terminal state whose outputs are unspecified. For example, the specified behaviour of the machine shown in Table 12.1(*a*) can be modified as shown in Table 12.1(*c*) by following the above procedure.

When outputs are unspecified as shown in Table 12.1(*b*), they may be specified according to our convenience. If both dashes in Table 12.1(*b*) are replaced with '1's, then states *D* and *E* become equivalent since their next states and output are identical. Now, these two states can be combined by redirecting all the transitions, presently leading to *E* to *D* as shown in Table 12.1(*d*). However, if both dashes in Table 12.1(*b*) are replaced with '0's, then states *B* and *E* are equivalent and in addition, states *A*, *C* and *D* become equivalent. The blocks *BE* and *ACD* can be labeled as *X* and *Y* and the reduced table is as shown in Table 12.1(*e*).

12.5 PROBLEMS IN ASYNCHRONOUS CIRCUITS

The asynchronous sequential circuits have three disadvantages or problems, namely Cycles, Races and Hazards.

12.5.1 Cycles

If an input change induces a feedback transition through more than one unstable state, then such a situation (where a circuit goes through a unique sequence of unstable states) is called a *cycle*. To explain cycles, consider the PS/NS table shown in Tables 12.2(*a*) and (*b*).

Table 12.2(*a*) PS/NS table

Row	NS			
	x_1x_2 = 00	01	11	10
1	①			
2	1			
3	1			
4	1	②		

Table 12.2(*b*) PS/NS table

PS y_1y_2	NS (Y_1Y_2)			
	x_1x_2 = 00	01	11	10
1	0 0 0 0			
2	0 1			
3	1 1			
4	1 0	②		

In the table, x_1x_2 are input variables, y_1, y_2 are secondary state (Present State) variables and Y_1Y_2 are Next State variables. The circled and uncircled entries represent stable and unstable states respectively.

Assume that the circuit related to Table 12.2 is initially in stable state 2 (i.e. corresponding to input state $x_1x_2 = 01$ and secondary state $y_1y_2 = 10$). Now, when the input x_2 changes to 0 (i.e. $x_1x_2 = 00$ and $y_1y_2 = 10$), the next state (Y_1Y_2) becomes 11. Now, for the present state $y_1y_2 = 11$ and $x_1x_2 = 00$, the next state (Y_1Y_2) will be 01; for the present state $y_1y_2 = 01$ and $x_1x_2 = 00$, the next state (Y_1Y_2) will be 00; for the present state $y_1y_2 = 00$ and $x_1x_2 = 00$, the next state (Y_1,Y_2) will be 00 which is a stable state. Thus, the circuit goes through an unique sequence of unstable states because of an input change. Such a situation is called a *cycle.* When cycle exists in a state table of an asynchronous circuit, care must be taken to ensure that the circuit terminates in a stable state. Otherwise, the circuit goes from one unstable state to another until a new change in the input occurs.

12.5.2 Races

In the case of cycle, only one feedback variable is unstable at any time for a change in an input variable. When two or more feedback variables change value in response to a change in an input variable, then a RACE condition is said to exist in an asynchronous sequential circuit.

Races are of two types, namely (*i*) critical races and (*ii*) non-critical races. Critical races should be eliminated in a circuit whereas non-critical races may be tolerated.

Critical races Consider the excitation or PS/NS table shown in Table 12.3 to demonstrate the problem of critical races.

Table 12.3 PS/NS table

PS y_1y_2	**NS (Y_1 Y_2)** x_1x_2 = **00**	**01**	**11**	**10**
0 0	1 1 (1)	0 0 (5)	1 1 (9)	1 1 (13)
0 1	0 1 (2)	1 1 (6)	1 1 (10)	0 0 (14)
1 1	1 0 (3)	1 1 (7)	1 1 (11)	1 0 (15)
1 0	1 0 (4)	0 0 (8)	1 1 (12)	1 0 (16)

Assume that the circuit is in stable state 5 (i.e. $Y_1Y_2 = 00$ corresponding to $x_1x_2 = 01$ and $y_1y_2 = 00$) and input x_2 changes from 1 to 0. Now, the Next State variable Y_1Y_2 begins to switch from 00 to 11. But, due to unequal propagation delay, one of the next state variables becomes 1 and the other variable does not change from the value 0, and hence Y_1Y_2 will be either 01 or 10. If $Y_1Y_2 = 01$, then the circuit goes to stable state 2. If $Y_1Y_2 = 10$, then the circuit goes to stable state 4. Therefore, for an input change, i.e. from $x_1x_2 = 01$ to 00, depending on the relative switching speed of the next state variables Y_1 and Y_2, the final state will be either stable state 2 or 4. But, both the final states 2 and 4 are wrong for the given input change. This situation is called a *critical race,* which must be avoided in an asynchronous circuit.

It is important to note that for an input change, a particular circuit having critical race problem may always go to the same wrong state. Also, identical circuits with different propagation delays may go to different wrong states. Even if a gate is replaced in an asynchronous circuit, then it may lead to different circuit behaviour due to a small difference in propagation delay of the gate.

Non-critical races: Consider the Table 12.3 and assume that the circuit is in stable state 5 (i.e. $Y_1Y_2 = 00$). Now, if the input changes from $x_1x_2 = 01$ to 11, then the Next State variables (Y_1Y_2) of the circuit are suppose to switch from 00 to 11. Again, due to propagation delay, Y_1,Y_2 becomes either 01 or 10. Therefore, the circuit will either go to state (10) or state (12) and then finally to state 11 (eleven) which is a stable state. Here, irrespective of the Next State variable, i.e. Y_1 or Y_2 going to 1, the same correct final state 11 (eleven) is reached. Such a situation where the correct final stable state is reached after transition through unstable states is called a *Non-critical race.* Such non-critical races can be tolerated in an asynchronous circuit.

12.5.3 Hazards

Hazard is an unwanted transient i.e. spike or glitch that occurs due to unequal path or unequal propagation delays through a combinational circuit. There are two types of hazards, viz. (*i*) Static hazard and (*ii*) Dynamic hazard as shown in Fig. 12.2.

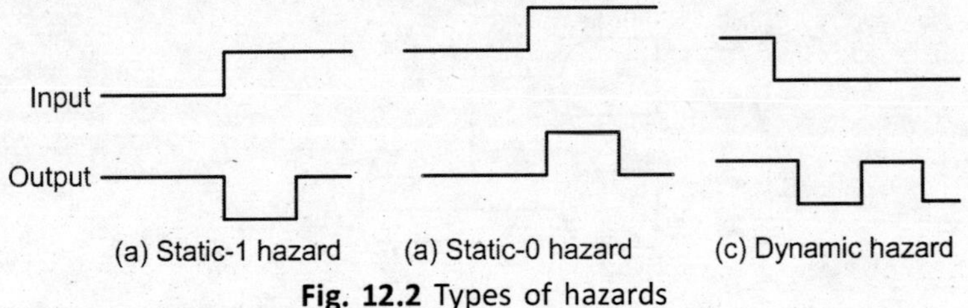

Fig. 12.2 Types of hazards

Similar to static and dynamic hazards caused by delay in combinational circuits, *essential hazards* occur in sequential circuits.

Static hazard Static hazard is a condition which results in a single momentary incorrect output due to change in a single input variable when the output is expected to remain in the same state. If the output momentarily goes to state '0' when the output is expected to remain in state '1' as per the steady state analysis, the hazard of this nature is known as *Static-1 hazard.* Similarly, if the output momentarily goes to state '1' when the output is expected to remain in state '0' as per the steady state analysis, it is known as *Static-0 hazard.*

To understand the static hazard, consider the K-map for the function $f(A, B, C) = S(0, 1, 2, 6)$ and the corresponding circuit shown in Fig. 12.3 (*a*) and (*b*).

When the input $ABC = 000$, the output, $f = 1$ due to HIGH at the output of upper AND gate. Now, when the input ABC changes to 010, the output f should remain in the '1' state due to HIGH at the output of lower AND gate. In other words, due to the change in the value of input variable B, switching of high output from upper to lower AND gate takes place and hence the output f is supposed to remain in HIGH state. But, due to unequal propagation delay, if the upper AND gate output changes to 0 shortly before the lower

AND gate output becomes '1', then, during this brief period, the output $f = 0$ momentarily. This situation is called *Static hazard.*

A static hazard can be removed by covering the adjacent cells with a redundant grouping that overlaps both groupings. In the K-map shown in Fig. 12.3(*a*), static hazard can be eliminated by covering adjacent cells corresponding to 000 and 010 as shown by dotted subcube. This leads to redundant grouping that overlaps both $\overline{A}\overline{B}$ and $\overline{D}\overline{C}$ groupings. The redundant term is $\overline{A}\overline{C}$ and the modified circuit is shown in Fig. 12.3(*c*). Now, when the input (*ABC*) changes from 000 to 010, the output will remain at '1' state (for both 010 and 000 inputs) because of high output at the newly added lower AND gate.

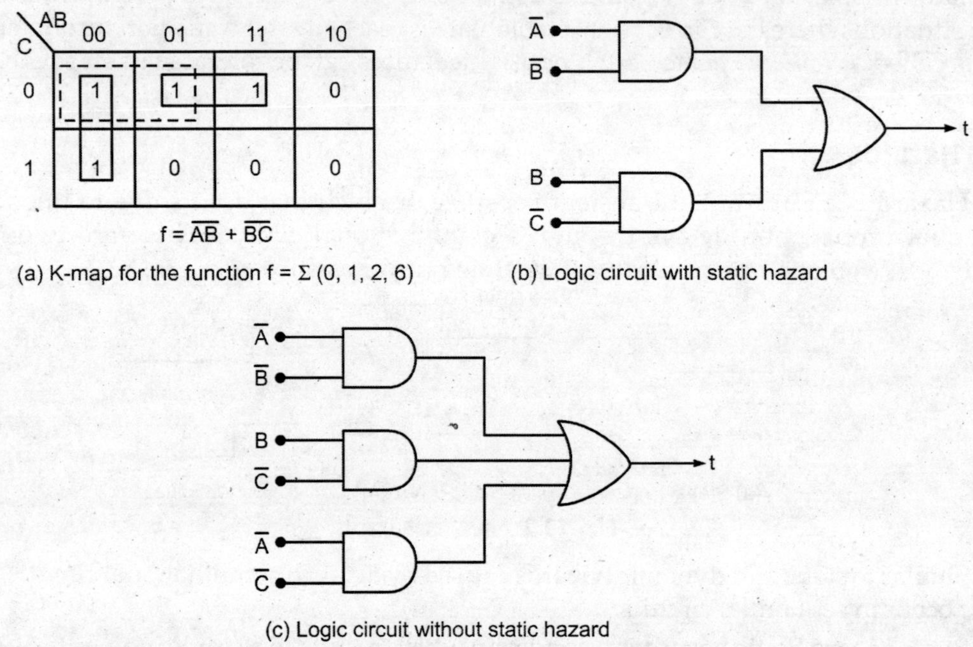

(a) K-map for the function f = Σ (0, 1, 2, 6)

(b) Logic circuit with static hazard

(c) Logic circuit without static hazard

Fig. 12.3

Dynamic hazard: When the output is supposed to change from 0 to 1 (or from 1 to 0), the circuit may go through three or more transients and produce more than one glitch. Such multiple glitch situations are known as *dynamic hazards*. Dynamic hazards can be eliminated by the same covering method, as explained earlier for static hazard elimination.

Essential hazard Essential hazard is a type of hazard that exists only in asynchronous sequential circuits with two or more feedbacks. Essential hazard occurs normally in toggling type circuits. It is an operational error generally caused by an excessive delay to a feedback variable in response to an input change, leading to a transition to an improper state. For example, an excessive delay through an inverter circuit in comparison to the delay associated with the feedback path may cause essential hazard. Such hazards cannot be eliminated by adding redundant gates as in static hazards.

On the other hand, the essential hazard can be eliminated by adjusting the amount of delay in the affected path. For this, each feedback loop must be designed with extra care to ensure that the delay in the feedback path is long enough compared to the delay of other signals that originate from the input terminals.

Hence, the essential hazard exists in an asynchronous sequential circuit due to a combination of delay and design specifications.

12.6 DESIGN OF HAZARD FREE SWITCHING CIRCUITS

Static, dynamic and essential hazards discussed in the previous sections can be eliminated by using different procedures discussed below.

12.6.1 Static Hazards Elimination

As discussed in the previous section, a transition between a pair of adjacent input combinations which correspond to identical outputs (i.e., both are 0's and 1's) contains a static hazard if it leads to the generation of momentary spurious output. Such hazards occur whenever there exists a pair of adjacent input combinations which produce the same output and there is no sub cube in *K*-map covering both combinations.

Therefore, to design a static hazard free switching circuit, all adjacent input combinations, having same output occur within some sub cube of the corresponding function. In Other words, every pair of adjacent 1 cells and every pair of adjacent 0 cells in the K-map of a switching function should be covered by at least one sub cube. The expression derived from such a collection of sub cubes is called hazard free switching function and it leads to the implementation of hazard free switching circuit.

Consider the switching function *f*(*A*, *B*, C) = $\sum_{in}(1,3,6,7)$ and the corresponding K-map in Fig. 12.4.

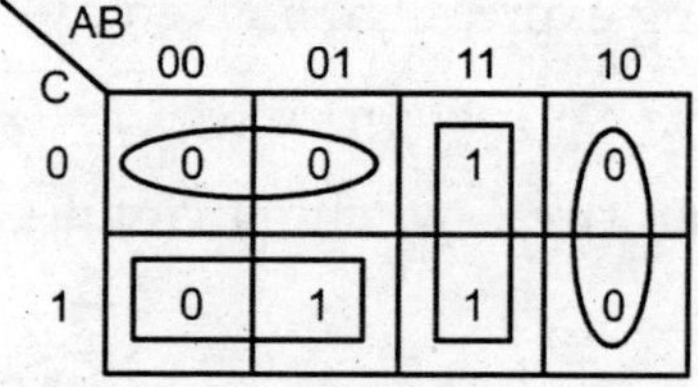

Fig. 12.4 K-map without hazard cover

From the K-map shown in Fig. 12.4, the simplified sum of product (SOP) and product of sum expression can be written as:

$$f = AB + \overline{A}C \text{ and } f = (\overline{A} + B)(A + C)$$

The corresponding switching circuits are shown in Fig. 12.5(a) and (b) respectively.

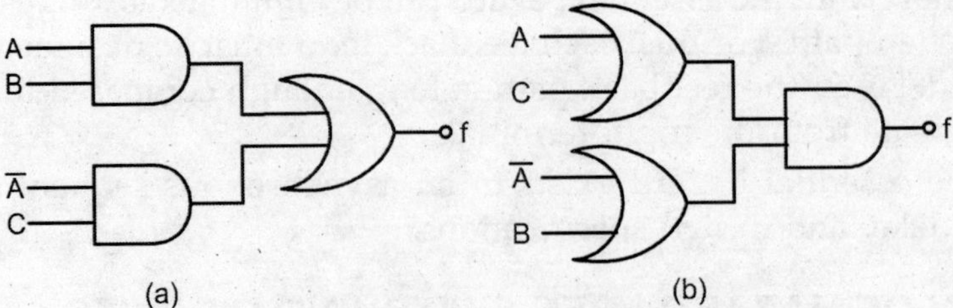

Fig. 12.5 Switching circuit with Hazards

In the circuit shown in Fig. 12.5(a), when the input (*ABC*) changes from 011 to 111 (or) 111 to 011, the output may momentarily go to '0' state due to the unequal propagation delay in the AND gates, rather than remaining constant at '1' as per the function value. Similarly, in the circuit shown in Fig. 12.5(b), when the input changes from 000 to 100 or 100 to 000, the output may momentarily go to '1' state due to the unequal propagation delay in the OR gates, rather than remaining constant at '0' as per the function value. These momentarily wrong outputs are also called glitches.

Now, in order to make the switching circuits shown in Fig. 12.5 hazard free, every pair of adjacent 1 cells and every pair of adjacent '0' cells in the *K*-map shown in Fig. 12.4 should be covered by at least one sub cube. This results in two additional dotted sub cubes as shown in Fig. 12.6.

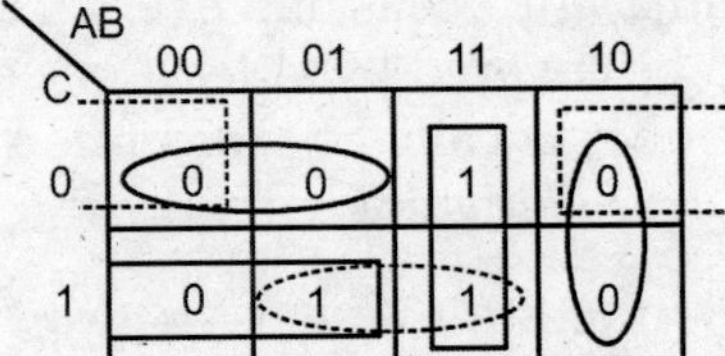

Fig. 12.6 K-map with hazard cover

The hazard free switching expression is given by

$$f = AB + \overline{A}C + BC; \quad f = (A+C)(\overline{A}+B)(B+C)$$

The corresponding hazard free switching circuits are shown in Fig. 12.7(a) and (b) respectively.

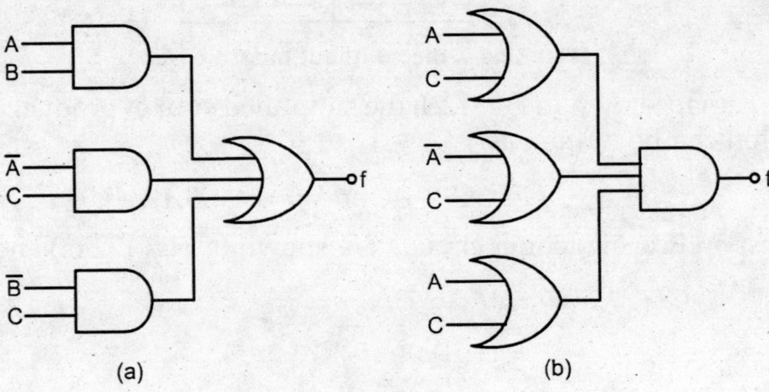

Fig. 12.7

12.6.2 Dynamic Hazards Elimination

Dynamic hazard can also be eliminated in a similar manner as that of static hazard elimination by covering every pair of 1 cells and every pair of '0' cells in the K-map by at least one sub cube.

12.6.3 Essential Hazards Elimination

Consider the flow table shown in Table 12.4 and the corresponding asynchronous sequential circuit shown in Fig. 12.8.

Table 12.4 Flow table

Present State	Next State (Y_1Y_2)	
y_1y_2	$x = 0$	$x = 1$
00	00	01
01	11	01
11	11	11

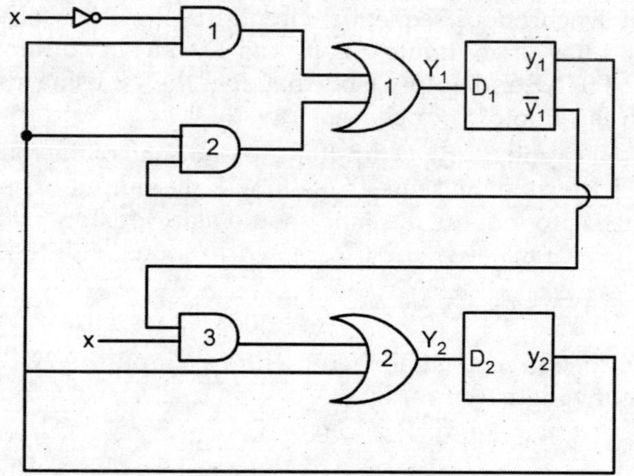

Fig. 12.8 Asynchronous sequential circuit

In the circuit shown in Fig. 12.8, y_1 depends on $\overline{x}$ while y_2 depends on x. Assume that the delay associated with the NOT gate is very large compared to the combined propagation delay of flip-flops and gates, and the present state (PS) of the circuit is 00 (i.e., y_1y_2 = 00). When the input (x) changes from 0 to 1, then Y_2 becomes 1 and thereby y_1 also becomes 1. This y_2 is fed back to the input of the first flip-flop through AND gate 1. If the change in input (x) is still not propagated through the NOT gate, then Y_1 becomes 1 and thereby y_1 also becomes 1. Therefore, the next state Y_1Y_2 is 11 instead of 01 as given in the flow table. This phenomenon, known as essential hazard, occurs due to the result of the input (x) change and y_2 changing faster than the propagation of the input through the NOT gate.

In the circuit shown in Fig. 12.8, the essential hazard can be eliminated by reducing the delay in the NOT gate or by increasing the delay in the feedback path of y_2 to AND gate 1.

■ REVIEW QUESTIONS ■

1. Differentiate synchronous and asynchronous sequential circuits.
2. Compare fundamental and pulse mode asynchronous sequential circuits.
3. Describe the design procedure for asynchronous sequential circuits.
4. Give the general model of asynchronous sequential circuits.
5. When are two states said to be equivalent states?
6. What are the advantages of merging process?
7. Describe cycles in asynchronous sequential circuits.
8. What do you mean by critical and non-critical races? How can they be avoided?
9. What are the different types of hazards in asynchronous circuits?
10. Differentiate static-0 and static-1 hazards.
11. Explain the method to eliminate static hazard in an asynchronous circuit with an example.
12. Explain the problems in asynchronous circuits.
13. What do you mean by hazard free asynchronous sequential circuits?
14. Define Races in asynchronous sequential circuits.
15. Design a clocked synchronous sequential circuit that checks a serial data line for even parity. The circuit should have two inputs, SYNC and DATA, in addition to CLK, and one output, ERROR. Devise a PS/NS and output table that does the job using just four states and design the logic circuit with flip-flop of your choice.
16. Design a sequential circuit, using *D* flip-flops to find each of the following sequences in a serial input signal (x). Each time the sequence is found, the output (Z) should be asserted. When a complete sequence is found, the machine should start looking for the sequence again without missing a bit. Changes in the signal (x) are synchronous with the negative going edge of the clock,

    (*a*) 000 (*b*) 010 (*c*) 110 (*d*) 0001 (*e*) 0011 (*f*) 1010
17. Design an asynchronous sequential circuit with two inputs x_1 and x_2 and with two outputs Z_1 and Z_2 for the following specifications,

    (*i*) When $x_1x_2 = 00$, the output $Z_1Z_2 = 00$

    (*ii*) When $x = 0$ and x_2 changes from 0 to 1, the output $Z_1Z_2 = 10$

    (*iii*) When $x_2 = 1$ and x_1 changes from 0 to 1, the output $Z_1Z_2 = 01$

    (*iv*) Otherwise, the output is constant.
18. Design an asynchronous sequential circuit with two inputs *X* and *Y* and one output *Z*. The output should be 1 if the number of 1 inputs on *X* and *Y* together since reset is a multiple of 4, and 0 otherwise. (Hint: Four states are sufficient). Realise the circuit with state assignments S0-S3 = 00, 01, 11, 10 and S0-S3 = 00, 01, 10, 11 and using *D* flip-flops. Compare the two solutions with respect to cost of the resulting sum-of-products excitation equations.
19. Design a sequential circuit with two inputs *A* and *B* and one output *Z*. The output *Z* is to be ASSERTED at the end of the input sequence $AB \rightarrow 00 \rightarrow 10 \rightarrow 11$ and remain ASSERTED until the end of the input sequence $AB \rightarrow 11 \rightarrow 01 \rightarrow 00$. Assume that only one input will change at a time.
20. Design a primitive state diagram and state table for an asynchronous circuit that is defined as a sample gate. This circuit, with two inputs control (*C*) and data (*D*) and one output (*Z*) operates as follows. The output is to be deasserted (low) when the *C* input is deasserted (low). When the

C input is asserted, the output (*Z*) moves to the level of the data (*D*) input and holds this level until *C* is deasserted. Assume initially that the two inputs never change simultaneously.

21. Design a primitive state diagram and state table for a circuit with two asynchronous input (*X* and *Y*) and one output *Z*. This circuit is to be designed so that if any change takes place on *X* and *Y*, *Z* is to change states. Assume initially that the two inputs never change simultaneously.
22. Design a JK Flip-Flop asynchronous sequential circuits with two inputs x_1 and x_2 and single output *Z*. The output *Z* = 1, if and only if the same input variable changes two or more times consecutively.
23. Design an asynchronous circuit that has two inputs x_1 and x_2 and one single output *Z*. The circuit is required to give an output whenever the input sequence 00, 10, 11 and 01 are received but only in that order.
24. Obtain a static hazard free asynchronous circuit for the following switching function
$$f = 2(0, 2, 4, 5, 8, 10, 14)$$
25. Describe incompletely specified state machines.
26. Distinguish completely and incompletely specified state machines.
27. How can essential hazards be eliminated?

❑❑❑

13 D/A and A/D Converters

13.1 INTRODUCTION

The data converters convert one form of data into another form. Real world processes produce analog signals which carry information pertaining to process variables such as voltage, current, charge, temperature and pressure. The rate of flow of such information may be very slow or very fast. It is difficult to store, manipulate, compare, calculate and retrieve such data with good accuracy using purely analog technology. But, computers can perform these operations quickly and efficiently using digital techniques. Therefore, it is necessary to convert the analog signals from various transducers into its equivalent digital data, which in turn act as the input for digital systems. Thus the requirement for converting analog signal into digital data emerged. The computers also need to communicate with people and physical processes through the use of analog signals, which necessitated the process of digital to analog conversion. This chapter discusses the most common D/A and A/D conversion techniques, D/A converter and A/D converter ICs, delta modulators and demodulators followed by sigma-delta converters and widely used ICs for D/A and A/D conversions.

13.2 ANALOG AND DIGITAL DATA CONVERSIONS

In the application of signal processing, the measurement and analysis of signals are very important to discover their characteristics. If the signal is unknown, the process of analysis begins with the acquisition of the signal. The most common technique of acquiring signals is by *sampling*. Sampling a signal is the process of acquiring its values only at discrete points in time.

The definitions related to the process of sampling and the subsequent analog and digital conversion processes are

(*i*) an *analog signal* is a signal that is defined over a continuous period of time, and in which the amplitude may assume a continuous range of values.

(*ii*) the term *quantization* refers to the process of representing a variable by a finite set of discrete values.

(*iii*) a *quantized variable* is the signal variable that can assume only finite distinct values.

(*iv*) a *discrete time signal* is the one that is defined at particular points of time only.

Therefore, the independent time variable is quantized. When the amplitude of a discrete-time signal is allowed to assume a continuous range of values, the function is called a *sampled-data* signal. A sampled data signal could result from sampling an analog signal at discrete points of time.

(*v*) a *digital signal* is a function, in which the time and amplitude are quantized. A digital signal is always represented by a sequence of *words*, where each word can contain a finite number of *bits* (binary digits).

A D/A converter (DAC) converts digital data into its equivalent analog data. The analog data is required to drive motors and other analog devices. An AD converter (ADC) converts analog data into its equivalent digital data i.e., binary data. The A/D converter and D/A converter are also called *data converters* and they are also available as monolithic integrated circuits.

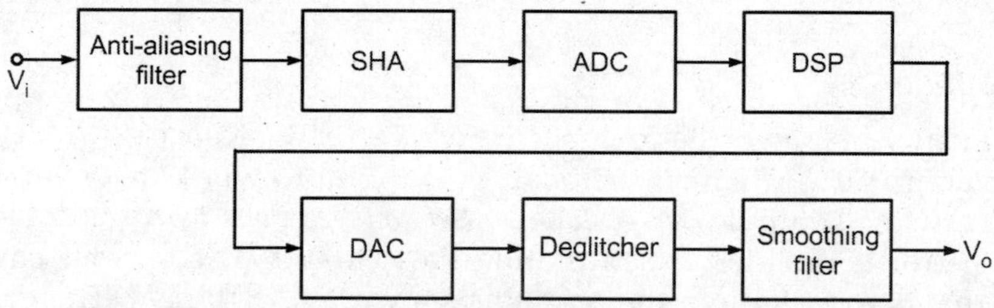

Fig. 13.1 Sampled data system using A/D and D/A converters

Figure 13.1 shows a typical application in which A/D and D/A conversions are employed. An analog input signal from a transducer is band-limited by *anti-aliasing* filter. The signal is then sampled at a frequency rate higher than twice the maximum frequency of the band-limited input signal. That is, when the A/D converter is operated at a rate of f_s samples/second, the highest frequency component of the input signal can be less than $f_s/2$. The A/D converters normally require the input to be held constant during the conversion process. Hence, a *Sample-and-Hold Amplifier* (SHA) is introduced in the loop as shown in Fig. 13.1 before A/D converter. The SHA freezes the band-limited signal just before the start of each conversion. The digital signal from A/D converter may be processed, transmitted and recorded in digital form by the digital signal processor (DSP) block. Then, the digital signal is converted into analog signal by D/A converter for use in analog form. The D/A converter is usually operated at the same frequency f_s as that of A/D converter. The output of D/A converter is commonly a staircase signal, which is passed through a smoothing filter to eliminate the quantization noise effects. A *deglitcher* may be introduced in the loop before the smoothing process, to remove any output glitches generated during input code variations.

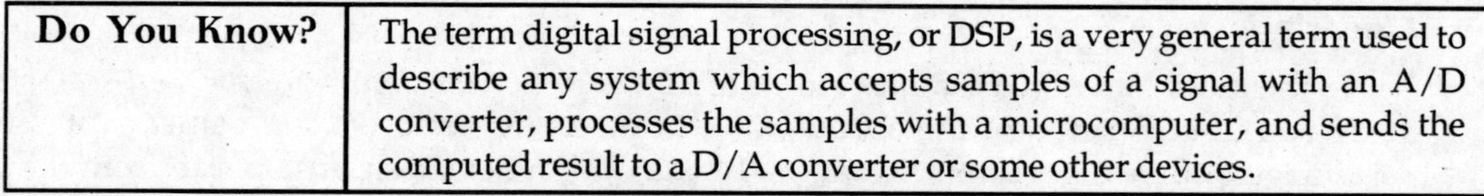

Do You Know?	The term digital signal processing, or DSP, is a very general term used to describe any system which accepts samples of a signal with an A/D converter, processes the samples with a microcomputer, and sends the computed result to a D/A converter or some other devices.

The schematic structure shown in Fig. 13.1 is prevalent either in full or in part in numerous applications such as digital signal processing, direct signal control, digital audio mixing, music and video synthesis, pulse-code modulation (PCM) communication, data acquisition and digital microprocessor based instrumentation.

13.3 SPECIFICATIONS OF D/A CONVERTER

The important specifications, namely; accuracy, offset voltage, monotonicity, resolution and settling time of D/A converter are discussed below.

Accuracy

The components in D/A converter circuits are prone to mismatches, drift, ageing, noise and other sources of errors. These factors lead to degradation in conversion performance.

Absolute accuracy defines the maximum deviation of the output from the ideal value and it is expressed in fractions of 1 LSB. The D/A converter manufacturers follow different ways of specifying accuracy. The D/A converter errors are classified as *static* and *dynamic error.*

Offset Voltage

The simplest kind of static errors are *offset error* and *gain error*. Ideally, the output of a D/A converter is 0 V when all the bits of binary input word are 0s. In practice, however, there is a very small output voltage called *offset voltage* or *offset error* as depicted in Fig. 13.2(*a*). The offset error is nullified by translating the actual A/D converter characteristics up or down so that it goes through the origin as shown in Fig. 13.2(*b*). The gain error shown in Fig. 13.2(*b*) is compensated by adjusting the scale factor *K*.

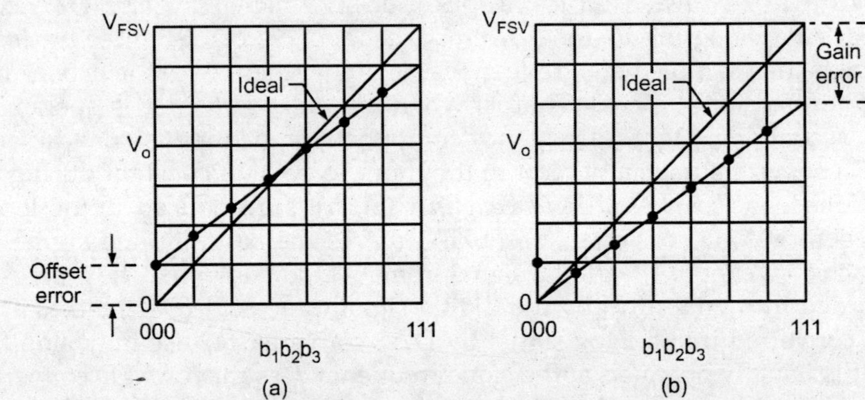

Fig. 13.2 (*a*) D/A converter offset and gain errors; (*b*) Nullifying the errors

Linearity

The most common dynamic errors are *fall-scale error* and *linearity error*. *Full-scale error* is the maximum deviation of the output value from its expected or *ideal* value, expressed in percentage of full-scale. *Linearity error* is the maximum deviation in step size from the ideal step size. More expensive D/A converters have full-scale and linearity errors as low as 0.001% of full-scale. General purpose D/A converters have accuracies in the range of 0.01 to 0.1%.

The linearity of a D/A converter is defined as the precision with which the digital input is converted into analog output. An ideal D/A converter produces equal increments in analog output for equal increments in digital input as shown in dotted line curve of the transfer characteristics of Fig. 13.3. However, in an actual D/A converter, the gain and offset errors due to resistors introduce non-linearity as shown by the solid line of the transfer characteristics of Fig. 13.3.

The linearity error measures the deviation of the output from the fitted straight line which passes through the measured output points. It is represented by $\in/\Delta$ as shown in Fig. 13.3. Commonly, the linearity of D/A converter is specified as less than ± 1/2 LSB meaning that $|\in| < (I/2)\Delta$.

Differential nonlinearity (DNL) error: For a D/A converter, the DNL error is the difference between the ideal and the measured output responses for successive D/A converter codes. An ideal D/A converter response would have analog output values exactly one code (1 LSB) apart (DNL = 0). A DNL specification of greater man or equal to 1 LSB guarantees monotonicity.

Integral nonlinearity (INL) error: For data converters, INL is the deviation of an actual transfer function from a straight line. After nullifying offset and gain errors, the straight line is either a best-fit straight line or a line drawn between the end points of the transfer function. The INL is often called *relative accuracy.*

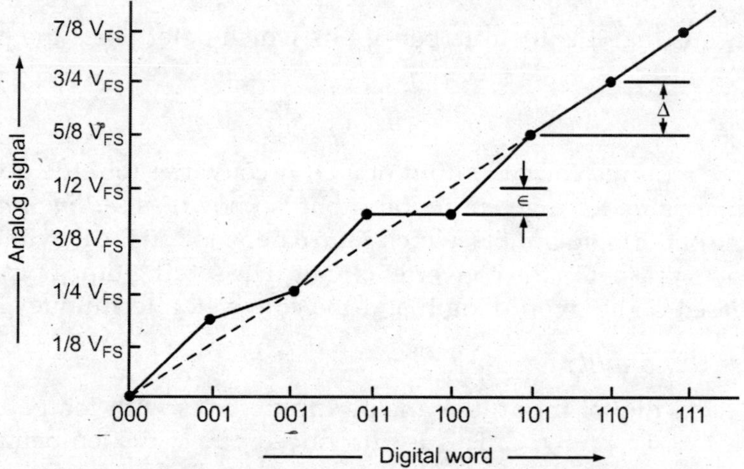

Fig. 13.3 Linearity error of a 3-bit D/A converter

Monotonicity

A D/A converter is monotonic if its output value increases as the binary inputs are incremented from one value to the next. That is, the staircase output can have no downward step as the binary input is incremented. Figure 13.3 shows the transfer curve for a non-monotonic D/A converter. The output decreases when the input word changes from 011 to 100.

The monotonic characteristic is important in control applications, without which, oscillations will result. If a D/A converter is identified to be monotonic, the error must be less than ± (1/2) LSB at each output level. Hence all the D/A converter ICs are designed to have linearity error of less than ± (1/2) LSB always.

Resolution (Step Size)

Resolution of D/A converter is defined as the smallest change that can occur in the analog output as a result of a change in the digital input. The resolution is always equal to the weight of the LSB and is also known as the step size, since it is the amount of V_o that will change when the digital input data goes from one step to the next.

Although resolution can be expressed as the amount of voltage or current per step, it is more useful to express it as a percentage of the full-scale output. The percentage resolution is given by

$$\text{Percentage resolution} = \frac{\text{step size}}{\text{full scale}} \times 100 \tag{13.1}$$

Percentage resolution can also be calculated as

$$\% \text{ resolution} = \frac{1}{\text{total number of steps}} \times 100 \tag{13.2}$$

For an n-bit digital input, the total number of steps is $(2^n - 1)$. Then,

$$\% \text{ resolution} = \frac{1}{(2^n - 1)} \times 100$$

This means that it is the number of bits which determines the percentage resolution of an A/D converter.

Settling Time

The time required for the output of a D/A converter to settle down to within ± (1/2) LSB of the final value for a given digital input is known as *settling time*. It depends on the switching time of the logic circuits, which inturn depends on the inevitable stray capacitances and inductances present in the converter circuit. The settling time normally ranges from 100 ns to 10 μs based on the word length and the conversion technique employed.

Temperature Sensitivity

For a fixed digital input, the analog input varies with temperature, normally from ± 50 ppm/°C to ± 1.5 ppm/°C. This is introduced due to the temperature sensitivity of the reference voltages, the resistors used in the converters, the op-amp and its offset voltage. Therefore, this factor determines the stability of D/A converter.

There are various types of D/A converters, namely, (*i*) weighted resistor type, (*ii*) R-2R ladder type, (*iii*) voltage mode R-2R ladder type, and (*iv*) inverted or current mode R-2R ladder type.

13.4 BASIC D/A CONVERSION TECHNIQUES

The D/A converter converts digital or binary data into its equivalent analog value. The input digital data for a D/A converter is an «-bit binary word *D*. The bit b_1 is called the most significant bit (MSB) and bit b_n the least significant bit (LSB). Then, the quantity D can be represented by $D = b_1 2^{-1} + b_2 2^{-2} + b_3 2^{-3} + ... + b_n 2^{-n}$. The D/A converter accepts the binary input D and produces an analog output, which is proportional to D using a reference voltage V_R. The converted analog value is either in voltage or current form. For a voltage output D/A converter, the conversion characteristic may be expressed as

$$V_o = KV_{FS} = KV_{FS}\left(b_1 2^{-1} + b_2 2^{-2} + b_3 2^{-3} + ... + b_n 2^{-n}\right) \quad (13.3)$$

where

V_o = output voltage

V_{FS} = full-scale range of voltage

K = scaling factor, usually unity

$b_1 ... b_n$ = n-bit binary fractional word with binary point located at the left

b_1 = most significant bit (MSB) of weight $V_{FS}/2$

b_n = least significant bit (LSB) of weight $V_{FS}/2^n$

The symbolic representation of an n-bit D/A converter is shown in Fig. 13.4.

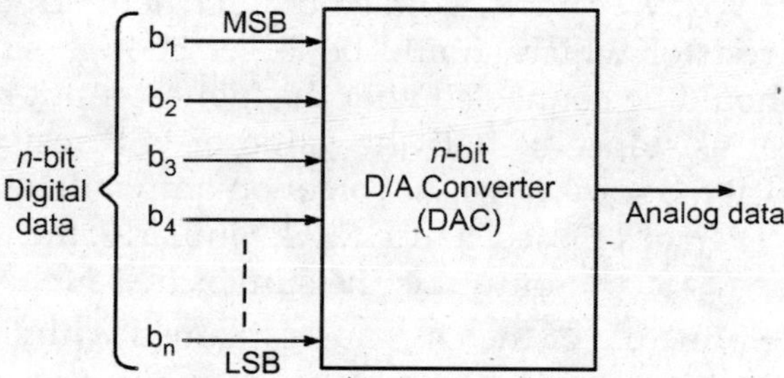

Fig. 13.4 *n*-bit D/A converter

13.4.1 Weighted Resistor Type D/A Converter

In the weighted resistor type D/A converter, each digital level is converted into an equivalent analog voltage or current. In a 4-bit D/A converter which accepts data from 0000 to 1111, there are 15 discrete levels of input, and hence it is convenient to divide the output analog signal into 15 levels. The LSB of the digital data causes a change in the analog output that is equal to 1/15th of the full-scale analog output voltage (V_R). Therefore, the weighted resistor network is designed in such a way that a 1 in LSB (2^0) position results in $V_R \times 1/15$ at the output. A 1 in the 2^1 bit position must cause a change in the analog output voltage that is equal to 2/15th of V_R (i.e., twice the size of the LSB). Similarly, a 1 in 2^2 and 2^3 bit positions must cause a change of $V_R \times 4/15$ V and $V_R \times 8/15$ V respectively as the analog output. It is important to note that the sum of the weights assigned to various bit positions of a 4-bit D/A converter must be equal to 1, i.e., $(1/15 + 2/15 + 4/15 + 8/15 = 15/15)\ V_R = V_{FS}$. In general, the weight assigned to the LSB is $1/(2^n-1)$, where n is the number of bits in the digital input.

Thus, the 4-bit weighted resistor network shown in Fig. 13.5 (*a*) performs the following D/A conversion:

1. The 2^0 bit is changed to 1/15th of V_R, 2^1 bit to 2/15th of V_R, 2^2 bit to 4/15th of V_R, and 2^3 bit to 8/15th of V_R.

2. These four voltages are added together to form the analog output voltage using an op-amp summer circuit.

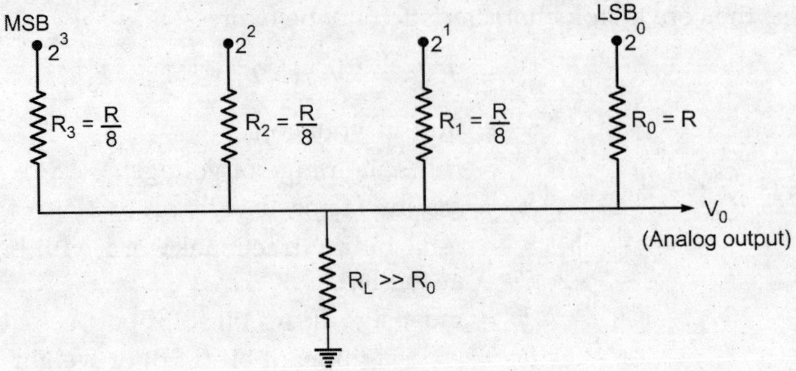

Fig. 13.5(*a*) Four-bit weighted resistor D/A converter

The resistor R_0, R_1, R_2 and R_3 voltage divider network connected with the op-amp and R_L is the load resistor which should be larger enough so as not to load the divider network. The LSB should be connected with the highest input resistance R_0 while the 2^1 bit is connected with a resistance of half the value of LSB resistor, i.e. $R_0/2$. Therefore, its current contribution at the summing junction of op-amp will be twice that of LSB. The 2^2 bit is connected with a 1/4th of LSB resistance $R_0/4$. Similarly, the MSB is connected with 1/8th of the LSB resistance $R_0/8$. The output is the sum of these four attenuated voltages.

The operating principle of the circuit is explained with the following illustration.

Illustration: The equivalent circuit of Fig. 13.5 (*a*), when applied with the digital data of 0001, is shown in Fig. 13.5 (*b*). The analog output voltage V_o can be calculated using Millman's theorem, which states that the voltage at any node in a resistive network is equal to the sum of the currents entering the node divided by the sum of the conductance connected at the node. Then the output voltage is expressed as

$$V_o = \frac{\frac{V_R}{R_1} + \frac{V_R}{R_2} + \frac{V_R}{R_3} + \frac{V_R}{R_4}}{\frac{1}{R_1} + \frac{1}{R_2} + \frac{1}{R_3} + \frac{1}{R_4}} \tag{13.4}$$

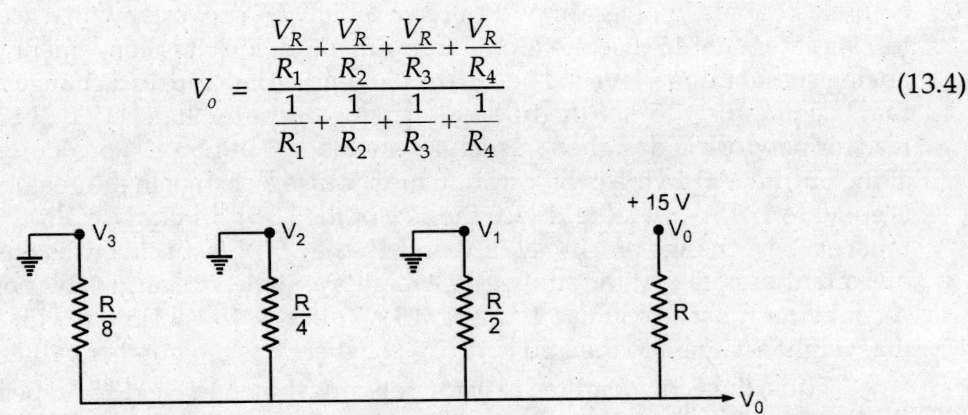

Fig. 13.5(*b*) Equivalent circuit of a 4-bit weighted resistor D/A converter for input 0001

For the weighted resistor network assuming $R_1 = R$, $R_2 = R/2$, $R_3 = R/4$ and $R_4 = R/8$, and applying the Millman's theorem to the circuit of Fig. 13.5 (*b*), we get

$$V_o = \frac{\dfrac{V_R}{R} + \dfrac{V_R}{R/2} + \dfrac{V_R}{R/4} + \dfrac{V_R}{R/8}}{\dfrac{1}{R} + \dfrac{1}{R/2} + \dfrac{1}{R/4} + \dfrac{1}{R/8}} \tag{13.5}$$

Figure 13.6 (*a*) shows the circuit of an n-bit D/A converter using op-amp as a summing amplifier. It employs a binary-weighted resistor network to generate the terms $b_i 2^{-1}$ where $i = 1, 2, \ldots n$. The circuit also uses *n*-electronic switches controlled by the binary input word $b_1, b_2, \ldots b_n$ and a reference voltage V_R. The switches are of single pole double throw (SPDT) type. If the binary input to a switch is 1, then the switch connects the resistance to the reference voltage $-V_R$. When the input bit to the switch is 0, it connects the resistor to ground.

Considering an ideal amp A, the output current I_o is given by

$$I_o = I_1 + I_2 + \ldots + I_n = \frac{V_R}{2^1 R} b_1 + \frac{V_R}{2^2 R} b_2 + \ldots\ldots + \frac{V_R}{2^n R} b_n$$

$$= \frac{V_R}{R}\left[b_1 2^{-1} + b_2 2^{-2} + \ldots\ldots + b_n 2^{-n}\right]$$

Then the output voltage $V_o = I_o R_f = \dfrac{V_R}{R}[b_1 2^{-1} + b_1 2^{-2} + \ldots\ldots + b_n 2^{-n}]R_f$ (13.6)

Using Eqn. (13.3) and Eqn. (13.6), it can be seen that if $R_f = R$, then $K = 1$ and $V_{FS} = V_R$.

The *n*-bit D/A converter circuit shown in Fig. 13.6 (*a*) uses a negative reference voltage, thus producing a positive staircase voltage. The analog output voltage waveform for 3-bit weighted resistor D/A converter is shown in the transfer characteristics of Fig. 13.6 (*b*) for an input binary word 000, 001, ... 111.

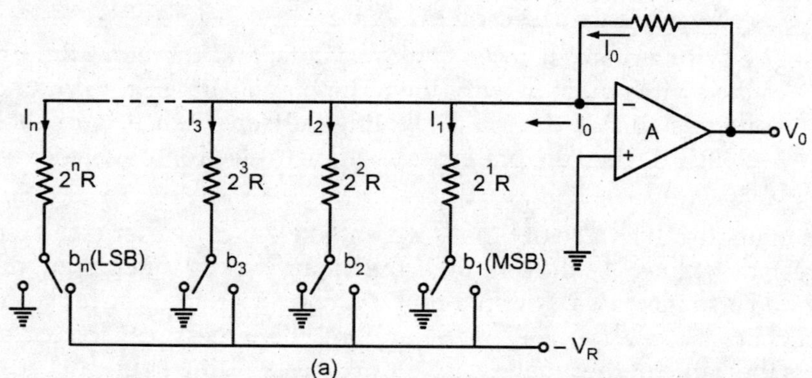

(a)

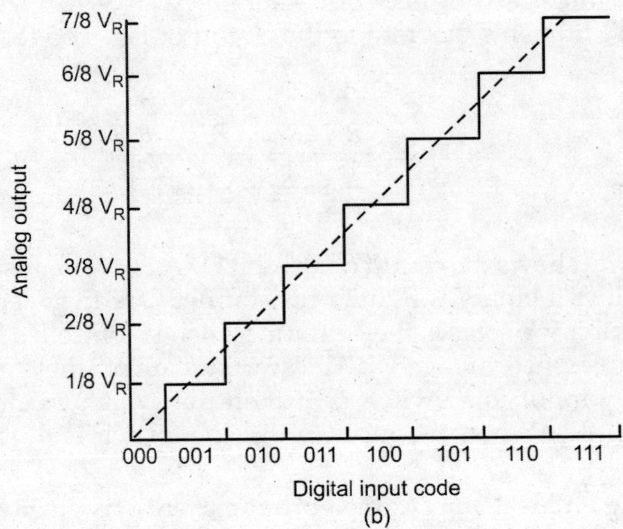

Fig. 13.6 Ann-bit weighted resistor D/A converter (*a*) Circuit diagram and (*b*) Transfer characteristic

It can be noted that

(*i*) The D/A converter output is the result of multiplying the analog signal V_R by the signal data. Therefore, if V_R is made variable, then the D/A converter is called a *multiplying D/A converter* [MDAC].

(*ii*) Higher the value of *n*, the finer the resolution of conversion is, and closer is the staircase to a continuous ramp waveform. DACs are available for word lengths ranging from 6 bits to 20 bits or more with 6, 8, 10, 12 and 14 bits being common.

(*iii*) The op-amp in Fig. 13.6 (*a*) can be connected in noninverting mode also.

(*iv*) The op-amp is operated as a *current-to-voltage* converter.

(*v*) Polarity of V_R is chosen depending on the switches to be operated.

(*vi*) The accuracy and stability of D/A converter are based on the accuracy of the resistors and their temperature dependence and the resistors have to handle varying currents based on bit values.

(*vii*) The switches are in series with resistors, and therefore the finite ON resistance of the switch must be very low. The bipolar transistor does not perform well as voltage switches due to the voltage offset when it is in saturation. Therefore MOSFET devices are preferred as efficient electronic switches which are discussed in section 13.5.

The main disadvantage of binary weighted D/A converter is the requirement of wide range of resistor values. As the length of the binary word is increased, the range of resistor values needed also increases. For an 8-bit D/A converter, the resistor values to be connected with the bits are $2^0R + 2^1R+ ... +2^7 R$. Therefore, the largest resistor corresponding to bit b_8 is 128 times the value of the smallest resistor corresponding to bit b_1. The fabrication of such large value of resistors of the order of MΩ is not practically possible in monolithic circuit fabrication. In addition, the voltage drop variations across such high value resistors due to the bias currents affect the accuracy. Therefore, the limitations in achieving and maintaining resistor ratios restrict the use of weighted resistor D/A converters to below 8-bit of word length.

The R-2R ladder type D/A converter is a better choice for practical applications and it overcomes such drawbacks.

Example 13.1. *A system uses a 12-bit word to represent the input signal. If the maximum peak-to-peak voltage at the output is set for 4 V, find the resolution of the system and the dynamic range.*

Solution: The 12-bit word can represent 2^{12} or 4096 levels, which are equally spaced across the 4 V range. Then, the step size = $\frac{4V}{4096}$ = 976 μV. Therefore the system can identify input changes as low as 976 μV.

The *dynamic range* gives the ratio of the largest value to the smallest value which can be converted. Therefore, the dynamic range $\frac{4V}{976\mu V}$ = 4096. The dynamic range in dB is given by 20 $\log_{10}$ 4096 = 72 dB.

Example 13.2. *An 8-bit D/A converter has an output of voltage range 0 to 2.55 V. Find the resolution of the system.*

Solution: The 8-bit D/A converter can identify 2^8 or 256 levels. Therefore, the output can have 256 different values starting from 00000000.

$$\text{Then the step size} = \frac{2.55\text{V}}{2^n - 1} = \frac{2.55}{255} = 10 \text{ mV}.$$

Therefore, the system can produce output changes as low as 10 mV.

Example 13.3. *A 4-bit R-2R ladder type D/A converter having resistor values of R = 10 kΩ and 2R = 20 kΩ, uses V_R of 10 V.*

Find (*a*) the resolution of the D/A converter, and

(*b*) I_o for a digital input of 1101.

Solution: Given n = 4, R = 10 kΩ and V_R = 10 V.

(*a*) $$\text{Resolution of 1 LSB} = \frac{1}{2^n} \times \frac{V_R}{R} = \frac{1}{2^4} \times \frac{1}{10 \times 10^3} = \frac{1}{16} \times 1 \text{ mA} = 62.5\ \mu\text{A}$$

(*b*) The output I_o for a digital input of 1101 is

I_o = 625 μA × 13 = 0.8125 mA (binary 1101 = decimal 13)

Example 13.4. *An 8-bit D/A converter has a resolution of 10 mV/bit. Find the analog output voltage for the inputs*

(a) 10001010, and (b) 00010000

Solution: The decimal equivalent value $D = b_8 2^7 + b_7 2^6 + b_6 2^5 + \ldots + b_1 2^0$

(*a*) For input =10001010,

$D = (1)2^7 + 0 + 0 + 0 + (1)2^3 + 0 + (1)2^1 + 0 = 128 + 8 + 2 = 138$

Therefore, $V_o = 138 \times 10$ mV/bit = 1.38 V

(*b*) For output = 00010000,

$D = 0 + 0 + 0 + (1)2^5 + 0 + 0 + 0 + 0 = 32$

Therefore, $V_o = 32 \times 10$ mV/bit = 0.32 V

13.4.2 R-2R Ladder D/A Converter

A wide range of resistor values is required in the design of binary weighted resistor D/A converter. In *R*-2*R* ladder D/A converter, resistors of only two values i.e., *R* and 2*R* are used. Hence, it is suited well for integrated circuit fabrication. The typical values of *R* used vary from 2.5 kΩ to 10 kΩ. The principle of operation of a ladder type network for 4-bit D/A conversion is shown in Fig. 13.7(*a*), with 4-bit binary input, $b_1\ b_2\ b_3\ b_4$, analog output V_o and one terminating resistor, 2*R*.

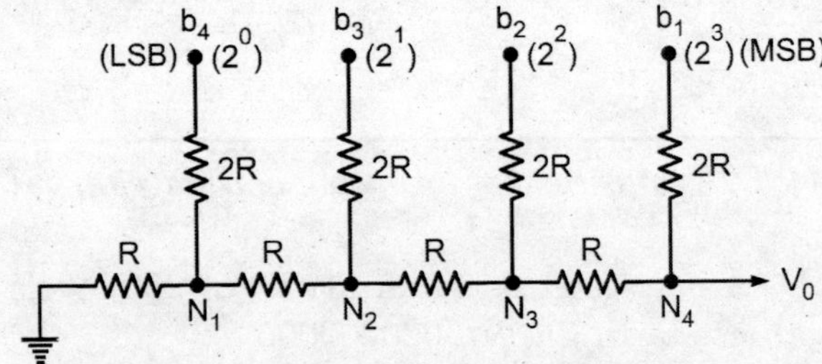

Fig. 13.7 (*a*) Four-bit R-2R ladder type D/A converter

In this ladder circuit, the output voltage is a weighted sum of digital inputs. For example, if the 4-bit binary input, $b_1\ b_2\ b_3\ b_4$, is 1000 i.e., if MSB is 1, while the other three inputs are 0, the circuit shown in Fig. 13.7(*a*) can be modified as shown in Fig. 13.7(*b*).

Here, the terminating resistor (2*R*) and the resistor connected to b_4 input (2*R*) are combined at node N_1 to form an equivalent resistor (*R*) as shown in the equivalent circuit of 1st stage in Fig. 13.7 (*c*).

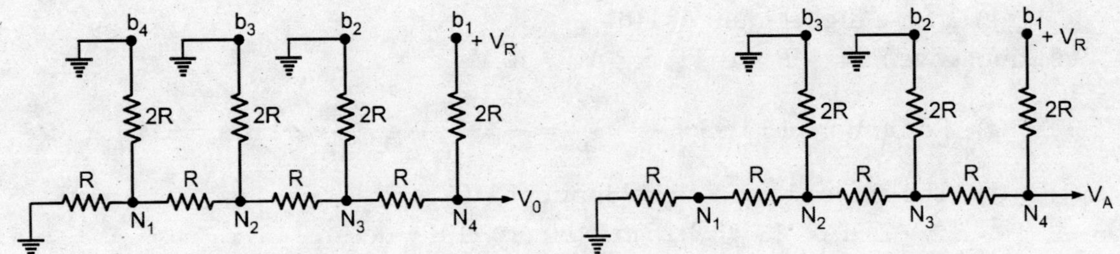

Fig. 13.7 (*b*) Equivalent circuit for binary input $b_1\ b_2\ b_3\ b_4$ = 1000
(*c*) Equivalent circuit of 1st stage

Then, at node N_2, the resistor connected with b_3 input (2*R*) can be combined with the resistor ($R + R = 2R$) to form the 2nd stage of equivalent circuit as shown in Fig. 13.7 (*d*).

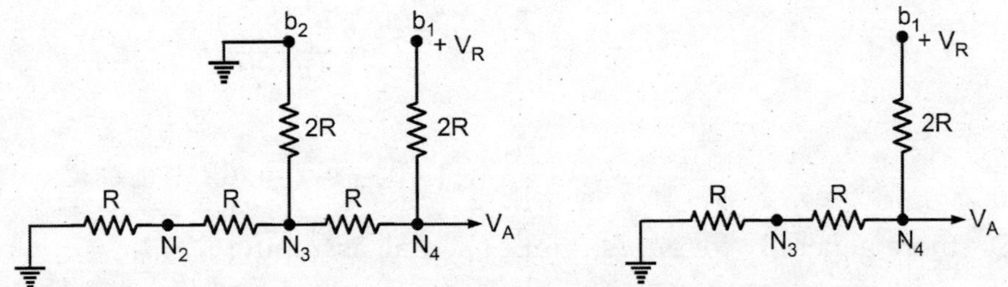

Fig. 13.7(*d*) Equivalent circuit of 2nd stage **Fig. 13.7 (*e*)** Equivalent circuit of 3rd stage

Similarly, at Node N_3, the equivalent resistor is R as shown in the equivalent circuit of stage 3 in Fig. 13.7(*e*). Then the analog output voltage V_o is given by

$$V_o = \frac{V_R \times 2R}{R + R + 2R} = \frac{V_R}{2} \tag{13.7}$$

Thus, for digital input $b_1\, b_2\, b_3\, b_4 = 1000$ i.e., when $MSB = 1$, the output is $V_R/2$. Similarly, it can be found that for digital input $b_1\, b_2\, b_3\, b_4 = 0100$, i.e., when second $MSB = 1$, the output is $V_R/4$; similarly, for $b_1\, b_2\, b_3\, b_4 = 0010$, the output is $V_R/8$ and for $b_1\, b_2\, b_3\, b_4 = 0001$, i.e., when LSB = 1, the output becomes $V_R/16$.

Since the resistive ladder is a linear network, the principle of superposition can be used to find the total analog output voltage for a particular digital input by adding the output voltages caused by the individual digital inputs. This can be represented for an n-bit D/A converter as follows:

$$V_o = \frac{V_R}{2^1} + \frac{V_R}{2^2} + \frac{V_R}{2^3} + \ldots + \frac{V_R}{2^n} \tag{13.8}$$

where n is the total number of bits at the input.

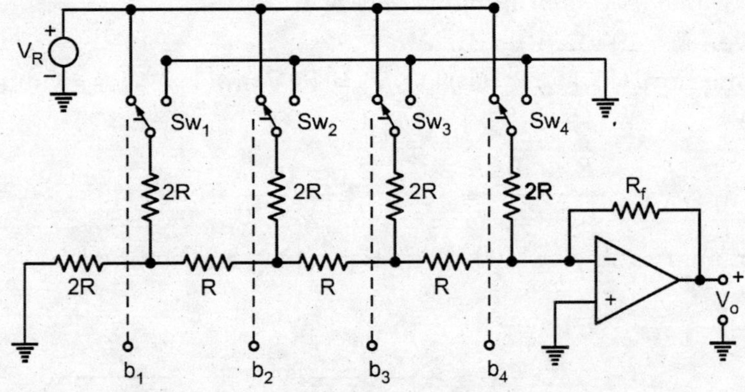

Fig. 13.8 Four-bit R/2R ladder D/A converter

Fig. 13.8 shows a practical circuit arrangement of a 4-bit D/A converter using an op-amp. The inverting input terminal of the op-amp acts as summing junction for the ladder inputs. Using Eqn. (13.8) the output voltage V_o is expressed by

$$V_o = V_R \frac{R_f}{R}\left(\frac{b_1}{2^1}+\frac{b_2}{2^2}+\frac{b_3}{2^3}+\frac{b_4}{2^4}\right) \quad (13.9)$$

$$= V_R \frac{R_f}{R\times 2^4}\left(b_1 2^3 + b_2 2^2 + b_3 2^1 + b_4 2^0\right)$$

Or, more generally for an n-bit input signal, assuming $R_f = R_1$

$$V_o = \frac{V_R}{2^n}\left(b_1 2^{n-1} + b_2 2^{n-2} + \ldots + b_n 2^0\right)$$

The resolution of the R/2R ladder type D/A converter with current output is given by

$$\text{Resolution} \quad I = \frac{1}{2^n}\times\frac{V_R}{R} \quad (13.10)$$

The resolution of the $R/2R$ ladder type D/A converter with voltage output is given by

$$\text{Resolution} \quad V = \frac{1}{2^n}\times\frac{V_R}{R}\times R_f \quad (13.11)$$

where R_f, is the feedback resistance of the op-amp.

Example 13.5. *Consider the R-2R 4-bit converter of Fig. 13.8 and assume the feedback resistance R_f of the op-amp is variable, the resistance $R = 10\ k\Omega$ and $V_R = 10\ V$. Determine the value of R_f, that should be connected to achieve the following output conditions.*

(a) The value of 1 LSB at the output is 0.5 V.

(b) An analog output of 6 V for a binary input of 1000.

(c) The full-scale output voltage of 12 V.

(d) The actual maximum output voltage of 10 V.

Solution: Given R = 10 kΩ, $V_R = 10$ V

(*a*) Using Eqn. (13.11), $R = 10$ kΩ, $V_R = 10$ V and $n = 4$ for a value of 1 LSB = 0.5 V, we have

$$\frac{R_f \times 10}{10^4 \times 2^4} = 0.5 \text{ or } R_f = \frac{10^4\times 2^4}{10}\times 0.5 = 8\text{k}\Omega$$

(*b*) For binary value of 1000, using Eqn. (13.9) and setting $b_1 = 1$ and $b_2 = b_3 = b_4 = 0$, we get

$$6 = \frac{R_f \times 10 \times 2^{-1}}{10^4} \text{ or } R_f = \frac{10^4}{10\times 2^{-1}}\times 6 = 12\text{k}\Omega$$

(*c*) For $V_{FS} = 12$ V, we get

$$\frac{R_f \times 10}{10^4} = 12 \text{ or } R_f = \frac{10^4}{10}\times 12 = 12\text{k}\Omega$$

(*d*) Let $b_1 = b_2 = b_3 = b_4 = 1$. Thus for getting a full scale voltage of 10 V.

$$\frac{R_f \times 10}{10^4}(2^{-1} + 2^{-2} + 2^{-3} + 2^{-4}) = 10$$

That is, $R_f = \dfrac{10^4}{10 \times 0.9375} \times 10 = 10.667\,\text{k}\Omega$

13.4.3 Voltage Mode R-2R Ladder

Figure 13.9 shows the alternate circuit arrangement of the *R*-2*R* ladder type called *voltage-mode R-2R ladder D/A converter*. The 2*R* resistors are switched between the two voltage levels V_L and V_H as determined by the bit values $b_1, b_2, \ldots b_n$. The output from ladder is obtained at the left most ladder node *S*, and buffered at the output of op-amp. The two voltages V_L and V_H can be any two voltage levels. As the input binary word changes from 0...0 (all 0 bits) to 1...1 (all 1 bits), the voltage of node *S* changes correspondingly in steps of $2^{-n}(V_H - V_L)$ from the minimum voltage of $V_o = V_L$ to the maximum of $V_o = V_H - 2^{-n}(V_H - V_L)$.

Advantages of *R* – 2*R* type D/A converters are

(*i*) more accurate selection and design of resistors *R* and 2*R* are possible and

(*ii*) the binary word length can be increased by adding required number of *R* – 2*R* sections.

13.4.4 Inverted or Current-Mode *R* – 2*R* Ladder D/A Converter

In weighted resistor and *R* – 2*R* ladder types of D/A converters the current flowing through the resistors changes as the input data changes. Power dissipation causes heating,

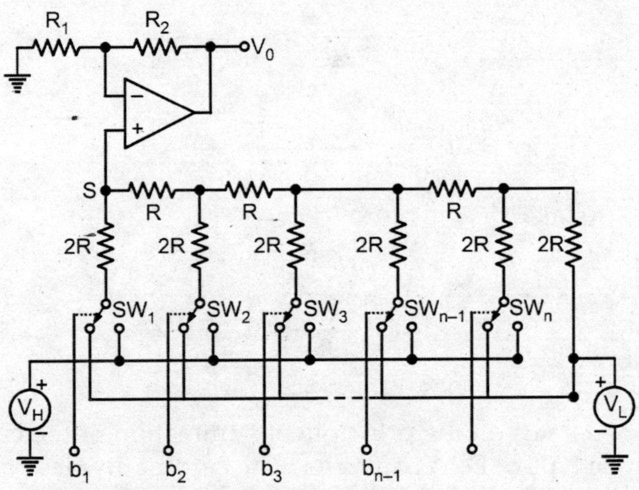

Fig. 13.9 Voltage-mode *R*-2*R* ladder D/A converter

and non-linearity of D/A conversion arises due to varying power dissipation values corresponding to bit patterns. This becomes a serious limitation as the word length increases. This is eliminated in the inverted *R* – 2*R* ladder type of D/A converter shown in Fig. 13.10.

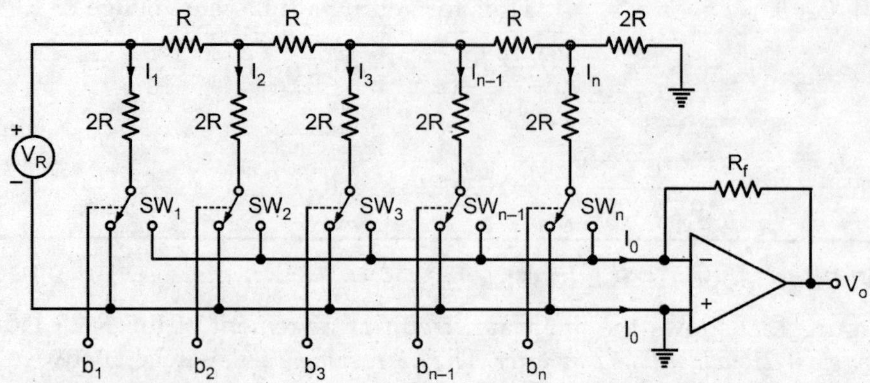

Fig. 13.10 Inverted or current-mode R-2R ladder D/A converter

The bit position of each of the subsequent MSBs and LSBs are interchanged. Each binary input is connected through the switch to either ground or to the inverting input terminal of op-amp, which is at virtual ground. Since both the positions of switch b_i are at ground potential, i.e. the actual or virtual ground, the current flow through any resistor is constant and it is independent of the input binary bit value.

These currents can be represented as

$$I_1 = \frac{V_R}{2R} \tag{13.12}$$

$$I_2 = \frac{(V_R/2)}{2R} = \frac{V_R}{4R} = \frac{I_1}{2} \tag{13.13}$$

$$I_3 = \frac{(V_R/4)}{2R} = \frac{V_R}{8R} = \frac{I_1}{4} \tag{13.14}$$

and

$$I_n = \frac{\left(V_R/2^{n-1}\right)}{2R} = \frac{I_1}{2^{n-1}} \tag{13.15}$$

The output voltage V_o is given by

$$V_o = -I_o \times R_f = -R_f\,(I_1 + I_2 + I_3 + \ldots + I_n)$$

$$= -\frac{V_R R_f}{R}(b_1 2^{-1} + b_2 2^{-2} + \ldots + b_n 2^{-n})$$

When $R_f = R$, $V_o = -V_R\,(b_1 2^{-1} + b_2 2^{-2} + \ldots + b_n 2^{-n})$ (13.16)

The circuit operates on the principle of summation of the currents. Hence it is called $R-2R$ current mode type of D/A converter. The current divides equally in successive nodes as indicated in Eqns. (13.12) to (13.15) and the current flow in individual arms of the network remains the same irrespective of the binary bit pattern. Therefore, currents are maintained constant in all the branches and the ladder node voltages also remain constant at V_R / 2^0, $V_R/1^1$, V_R /2^2... V_R /2^{n-1}. The op-amp is used as a current-to-voltage converter and the total current I_o is determined by the binary word.

The most important advantage of the current mode or inverted ladder type of D/A converter is that the stray capacitances do not affect the speed of response of the circuit due to the constant ladder node voltages. Hence, the speed performance is improved.

The advantage of this type of D/A converter is their capability of using any two voltage levels for the bit switching, neither of which need necessarily be *zero*.

Example 13.6. *The inverted R-2R ladder shown in Fig. 13.10 has* $R = R_f = 10$ *kW and* $V_R = 10$s *V. Calculate the total current delivered to the op-amp and the output voltage when the binary input is 1110.*

Solution: Using Eqns. (13.12) through (13.15), we get

$$I_1 = \frac{V_R}{2R} = \frac{10V}{2 \times 10 \times 10^3} = 0.5 \text{ mA}$$

$$I_2 = \frac{I_1}{2} = 0.25 \text{ mA and } I_3 = \frac{I_1}{4} = 0.125 \text{ mA}$$

Therefore, the current I_o is given by $I_o = I_1 + I_2 + I_3 = 0.5 + 0.25 + 0.125 = 0.875$ mA.

The output voltage $V_o = -0.875 \times 10^{-3} \times 10 \times 10^3 = -8.75$ V

13.5 MULTIPLYING D/A CONVERTERS (MDACS)

An A/D converter which uses a varying analog reference voltage instead of a fixed reference voltage V_R is called a *multiplying* D/A converter. It is known from Eqn. (13.9) that the output is the product of the digital word and the analog reference voltage V_R. This arrangement can be used as a *programmable attenuator*, when the binary word is made to represent a value less than *unity*. Then the output is a fraction of the input reference voltage V_R.

The simplified functional diagram of a multiplying D/A converter IC is shown in Fig. 13.11. The input digital word D is applied to $b_1 \ldots b_n$ terminals. The reference voltage V_R is applied to circuit. The output currents I_o and I'_o are functions of the digital word D. The value of the output is thus programmable from 0 V to $(1 - 2^{-n})\, V_R$.

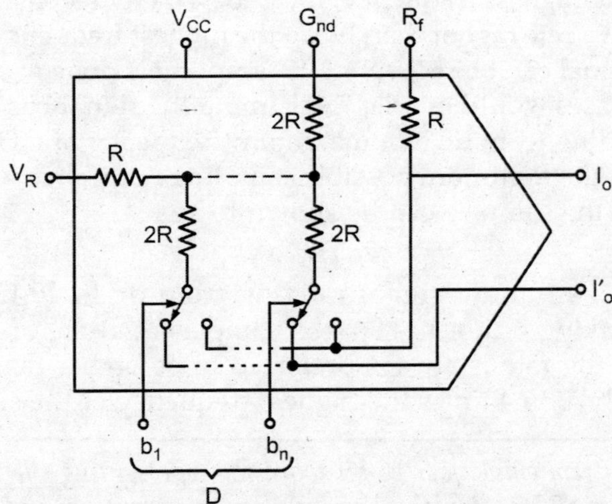

Fig. 13.11 Functional diagram of multiplying D/A converter

The reference voltage of a multiplying D/A converter can be varied over positive and negative values, including zero. This feature makes the multiplying D/A converters most suitable for digitally programmable applications such as programmable filters and oscillators.

13.6 SAMPLING PROCESS

Figure 13.12 shows the graphical representation of sampled input signal over time. The result of the sampling process is a series of the sampling instants and the amplitude of the signal at that instant of time. If the sampling is done at a constant rate, the resulting amplitude values may be used to reconstruct the signal. The accuracy of the conversion process will depend on two factors, namely, how frequently was the sampling done and the accuracy and resolution of the sample measurement.

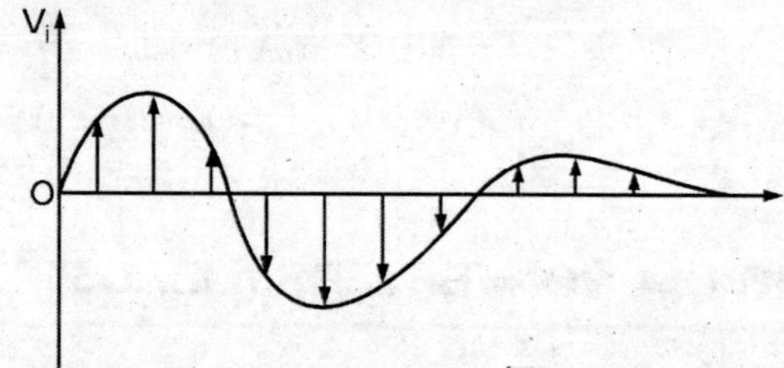

Fig. 13.12 Sampling the input signal over time

Shannon's sampling theorem

The minimum sampling rate used with any system bears great importance. Only a finite number of samples can be taken at any time, and a loss of information could result due to inadequate sampling frequency. The Shannon's sampling theorem establishes the theoretical basis for all the discrete sampling operations carried out on analog signals.

Shannon's sampling theorem states that if a signal is sampled for all time at a rate more than twice the highest frequency, it can be exactly reconstructed from the samples. If sampling is done at a rate higher than twice the highest frequency, the aliases do not overlap and the original signal can be recovered. If sampling is done at a slower rate than twice the frequency, then aliases would overlap. Assuming the signal has a spectrum with frequency components extending from dc to a maximum frequency of f_i Hz, the Shannon's sampling theorem states that the minimum possible sampling rate above which a bandpass signal can be recovered from the samples can be shown to be

$$f_s > 2f_i \text{ Hz}$$

For example, an audio signal ranging from dc to 20 kHz could theoretically be reconstructed by taking uniformly spaced samples at a rate of 40,000 samples/second. In practice, the sampling rate is always preferred to be higher than the theoretical minimum value, and normally, 3 to 4 times the highest frequency of the signal.

Example 13.7. *A system employs a 16-bit word for representing the input signal. If the maximum output voltage is set for 2 V, calculate the resolution of the system and its dynamic range.*

Solution: A 16-bit word represents 2^{16} or 65,536 levels. These levels are equally spaced across the 2 V range. Then, each step is given by

$$\text{Step size} = \frac{2\text{V}}{65536} = 30.52\,\mu\text{V}.$$

Therefore, the system can resolve voltage changes as low as 30.52 μV.

The *dynamic range* of a system represents the *ratio of the largest value obtainable to the smallest value*. Therefore,

$$\text{Dynamic range} = \frac{2\text{V}}{30.52\,\mu\text{V}} = 65536 = 20\log_{10} 65536 \cong 96 \text{ dB}.$$

Also the dynamic range may be calculated using the formula given by

Dynamic range = 6 dB × Number of bits = 6 dB × 16 bit = 96 dB.

Aliasing Error

When a signal f_i is sampled at a rate lower than $2f_i$, *aliasing error* occurs. The *aliasing error* is a phenomenon in which, the frequencies appear to be different from their true values and the signal can not be recovered correctly.

It is not always possible to exactly identify the frequency content of random data signals. Therefore, it is a common practice to pass the analog signals through a low-pass filter known as *prefilter* before sampling. The filter characteristics are designed such that it can reject all components equal to or greater than half the sampling rate. Without such a filter in the circuit, spurious input components outside the range of frequency of the signal could be shifted into the signal frequency range through the sampling and aliasing processes.

13.7 A/D CONVERTERS

An A/D converter does the inverse function of a D/A converter. It converts an analog signal into its equivalent n-bit binary coded digital output signal. The analog input is sampled at a frequency much higher than the maximum frequency component of the input signal. The digital output from an A/D converter can be in serial or parallel form.

The A/D converter accepts an analog input v_i and produces an output binary word $b_1, b_2 \ldots b_n$ of fractional value D such that

$$D = b_1 2^{-1} + b_2 2^{-2} + \ldots + b_n 2^{-n} \tag{13.17}$$

where b_1 is the MSB and b_n is the LSB. The symbolic representation of an n-bit A/D converter is shown in Fig. 13.13 (*a*). Two additional control pins START input and End of Conversion (EOC) output are provided with A/D converters. The START input initiates the conversion and the EOC announces when the conversion is complete. The output can be of parallel or serial form. Usually latches, control logic and buffers are provided to enable interfacing of the A/D converter to microprocessors or LCD/ LED displays directly.

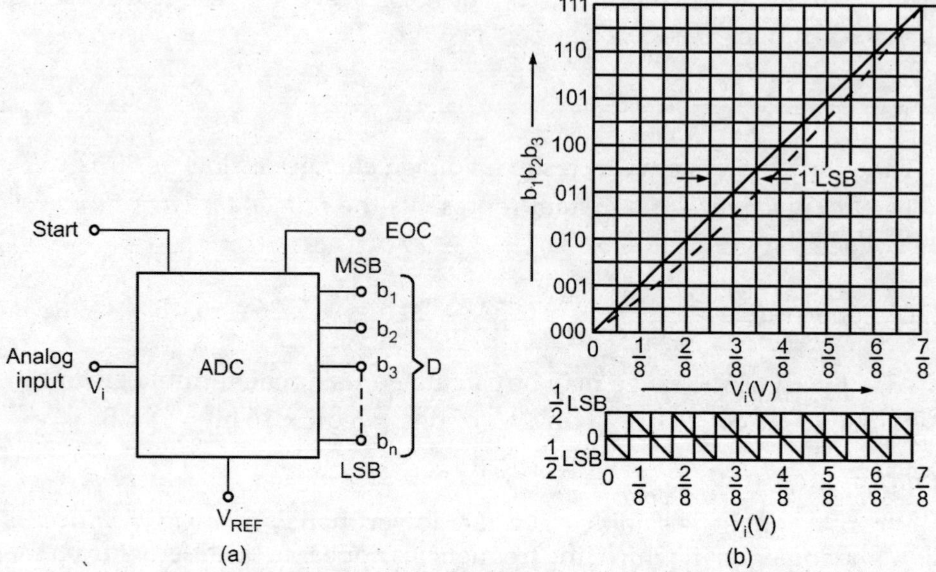

Fig. 13.13 A/D converter (*a*) Symbolic representation and (*b*) Ideal transfer characteristics and quantization noise for a 3-bit A/D converter

Figure 13.13 (*b*) shows the ideal characteristics of a 3-bit A/D converter with $V_{FS} = 1.0$ *V* where V_{FS} is the full-scale analog voltage. The A/D conversion process divides the analog input into 2^n intervals. These intervals are called *code ranges* and all the values of v_i falling within a code range are represented by the particular code. For instance, the code 110 corresponding to $V_i = 6/8$ V represents all inputs of value $6/8 + 1/16$ V. Hence the output can err by ± (1/2) LSB.

The general block diagram of an A/D converter is shown in Fig. 13.14. It consists of an antialiasing filter or prefilter, Sample-and-Hold amplifier, a quantizer and an encoder. The prefilter avoids the aliasing of high frequency signals. The Sample-and-Hold circuit holds the input analog signal into the A/D converter at a constant value during the conversion time. The quantizer segments the reference voltage signal into subranges. Typically, for an *n*-bit digital output code, there are 2^n subranges. The digital processor forms the encoder circuit which encodes the subrange into the corresponding digital bits. Therefore, the analog input signal is converted into an equivalent digital output code within the *conversion time.*

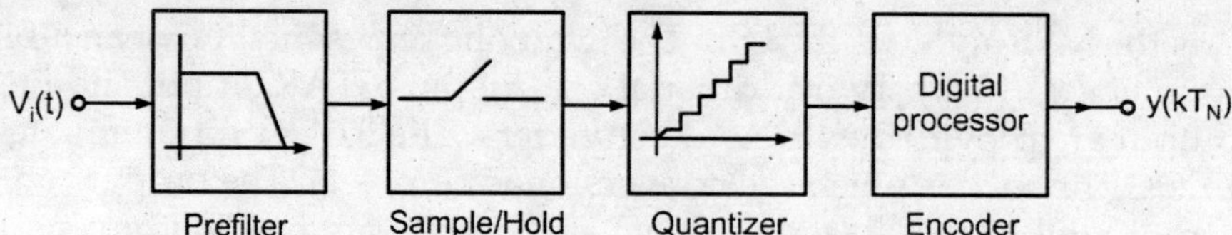

Fig. 13.14 General block diagram of an A/D converter

13.8 SPECIFICATIONS OF A/D CONVERTER

Some important specifications, namely, accuracy, differential linearity, conversion time, input voltage range and resolution of A/D converters are discussed below.

Resolution

The *resolution* refers to the finest minimum change in the signal which is accepted for conversion, and it is decided with respect to the *number of bits*.

It can be defined as *resolution* = $1/2^n$ where n is the number of digital output word bits. The ratio of the full-scale input voltage range V_{FS} to the resolution gives the minimum change of input voltage which can cause a change of 1 LSB at the output. This can be expressed as

$$\Delta v_i \text{ for 1 LSB} = \frac{V_{FS}}{2^n} \tag{13.18}$$

where V_{FS} is the full-scale input voltage range.

If the number of bits used to represent a signal is larger, then the resolution improves. For example, if an 8-bit word is used, a maximum of 256 distinct values are available. If a maximum analog signal amplitude of 1 V is used, then each step in the word represents $V/256 = 3.9$ mV. If 16 bits are used for the same 1 V range, then each step would produce $V/65536 = 15.26$ mV.

The digital output starts at 0 for an A/D converter. Therefore, the maximum full-scale input voltage which will cause the output to be all logic 1s is 1 LSB less than the full-scale voltage range.

$$v_{iFS} = V_{FS} - 1 \text{ LSB} \tag{13.19}$$

where v_{iFS} is the maximum input voltage which can cause all 1s at the output.

Example 13.8. *An 8-bit A/D converter accepts an input voltage signal of range 0 to 10 V.*

(a) What is the minimum value of the input voltage required to generate a change of 1 LSB?

(b) What input voltage will generate all 1s at the A/D converter output?

(c) What is the digital output for an input voltage of 4.8 V?

Solution:

(*a*) From Eqn. (13.18), 1 LSB = $\frac{10V}{2^8} = 39.1$ mV

(*b*) From Eqn. (13.19), $v_{iFS} = 10\text{ V} - 39.1\text{ mV} = 9.961$ V.

(*c*) The digital output for an applied input voltage of 4.8 V is given by

$$D = \frac{4.8\text{ V}}{39.1\text{mV}} = 122.76 \cong 123$$

Converting this to binary gives the digital output for the 8-bit A/D converter to be 01111011.

Quantization Error

A digital error in an A/D converter is based on the resolution of the digital system. In A/D conversion, a continuous analog voltage is represented by an equivalent set of digital numbers. When the digital numbers are converted back to analog voltage by a D/A converter, the output is a staircase waveform, which is a discontinuous signal composed of a number of discrete steps. The smallest digital step is due to the LSB and it can be made smaller only by increasing the number of bits in the digital representation. This error is called *quantization error*, or *digitizing error* and it is commonly the bit.

As shown in Fig. 13.13 (*b*) the digital output is 011 for all values of $\frac{3}{8}\text{V} \pm \frac{1}{2}$ LSB. Therefore there is an uncertainty about the exact value of v_i when the output is 011.

This uncertainty is called the *quantization error* and its value is $\pm\frac{1}{2}$ LSB.

Increasing the number of bits of A/D converter results in finer resolution and smaller quantization error.

Analog Error

Analog error in an A/D converter is mainly due to variations in the dc switching point of the comparator. The variations in switching are mainly due to offset, gain and linearity error of the operational amplifier used in the comparator. The other sources of analog error are the resistors in the A/D converter, the reference voltage source and the ripple and noise introduced by the circuit components

Linearity Error

This is an important measure of A/D converter performance. It is defined as a measure of the variation in voltage step size. This indicates the difference between the transitions for a minimum step of input voltage change. This is normally specified as a fraction of 1 LSB.

Differential Nonlinearity (DNL) Error

The analog input levels that trigger any two successive output codes should differ by 1 LSB (DNL = 0) for an A/D converter. Any deviation from 1 LSB value is defined as DNL error.

The counter type and continuous type A/D converters normally have better differential linearity than successive approximation type A/D converters.

Integral Nonlinearity (EVL) Error

Figure 13.15 shows an actual A/D converter characteristic with a missing code. The dotted curve represents the locus of the midpoints of the actual input step voltage ranges. This line is called the *code center line.* The maximum deviation of the code center line from the straight line passing through the end points of the ideal characteristics after nulling the offset and gain errors is called *integral nonlinearity error* (INL).

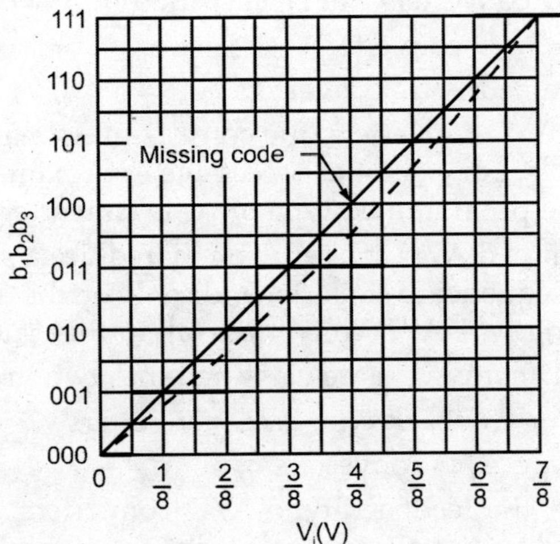

Fig. 13.15 A/D converter characteristic with a missing code

Dither

The performance of A/D converters can be improved using *dither* *<http://en.wikipedia.org/wiki/Dither>*. This is a very small amount of random noise (white noise <http: / / en.wikipedia.org / wiki / White_noise>) which is added to the input before A/D conversion. Its amplitude is set to half of the LSB value. Its effect is to cause the state of the LSB to randomly oscillate between 0 and 1 in the presence of very low levels of input, rather than sticking at a fixed value. Instead of the signal simply getting cut-off altogether at this low level (which is only being quantized to a resolution of 1 bit), it extends the effective range of signals that the A/D converter can convert, at the expense of a slight increase in noise. Thus, the quantization error is diffused across a series of noise values which is far less objectionable than a hard cut-off. The result is an accurate representation of the signal over time. A suitable filter at the output of the system can thus recover this small signal variation.

Conversion Time

The time required for an A/D converter to convert an analog input value into its equivalent digital data is called the *conversion time.*

Input Voltage Range

It is the range of voltage that an A/D converter can accept as its input without causing any overflow in the digital output.

13.9 CLASSIFICATION OF A/D CONVERTERS

The A/D converters (ADC) can be classified based on their operational features as follows.

Type. I: The A/D converters can be classified into two groups as

(*a*) Programmed A/D converters

(*b*) Non-programmed A/D converters

In programmed A/D converters, the conversion is made in a fixed number of steps, with equal time intervals. For example, successive approximation type of A/D converter is a typical example of the programmed type of A/D converter.

The non-programmed A/D converters may require a sequence of steps initially, and the time interval of the sequence of steps depends only on the response time of the conversion circuitry. The integrating type A/D converters fall in this category.

Type II: The A/D converters are classified into two groups as

(*a*) Closed-loop or feedback type A/D converters

(*b*) Open-loop type A/D converters

In closed-loop or the feedback type A/D converters, the analog voltage generated internally as a function of digital input is feedback to one input of the comparator. This voltage is compared with the analog voltage under conversion. When the input voltage and the feedback voltages are equal, the conversion is said to be complete. All D/A converter based A/D converters belong to this category.

In open-loop converters, a direct comparison is made between the analog input voltage and a set of reference analog voltages. The result of the comparison forms a digital word at the output. The flash type A/D converters are typical examples of open-loop converters.

Type III: The A/D converters are classified into two groups as

(*a*) Capacitor-charging type A/D converters

(*b*) Discrete voltage comparison type A/D converters

The capacitor charge-balancing type of A/D converter operates on the principle of charging the capacitor at a rate proportional to the input voltage, while simultaneously pulling out discrete charge packets out of the capacitor at a rate such that the net charge flow is always zero. The capacitor balancing integrating type of A/D converters belong to this category.

Discrete voltage comparison type employs the principle of generation of discrete voltages whose levels are equivalent to digital words. The comparison for these discrete voltage levels is then made with the analog input voltage to determine the equivalent digital word. A/D converters based on weighted capacitor of D/A converters fall under this category.

Type IV: The A/D converters are classified into two groups based on their conversion techniques as

(*a*) Direct type A/D converters

(*b*) Integrating type A/D converters.

The direct type A/D converters compare a given analog signal with an internally generated equivalent analog signal. Flash (comparator) type A/D converter, Counter type A/D converter, Tracking or Servo operated A/D converter and Successive approximation A/D converter are direct type of A/D converters.

The integrating type of A/D converters performs the A/D conversion in an indirect manner. It is a special class of converter which uses either a reference voltage, or which integrates the signal during the conversion process. Therefore, they do not require S/H circuit at the input. The process of integrating the signals also improves the signal-to-noise ratio for certain type of analog signals. The integrating A/D converters are suitable only for very low frequency signals. For example, many of the digital voltmeters employ an integrating A/D converter in the circuit. The charge balancing type arid dual slope A/D converters fall in this category.

13.9.1 Simultaneous Type (Flash type) A/D Converter

The simultaneous type A/D converter is based on comparing an unknown analog input voltage with a set of reference voltages. To convert an analog signal into a digital signal of n output bits (2^n –1) number of comparators are required. For example, a 2-bit A/D converter requires 3 or (2^2 – 1) comparators, while a 3-bit converter needs 7 or (2^3 – 1) comparators. The block diagram of a 2-bit simultaneous type A/D converter is shown in Fig. 13.16.

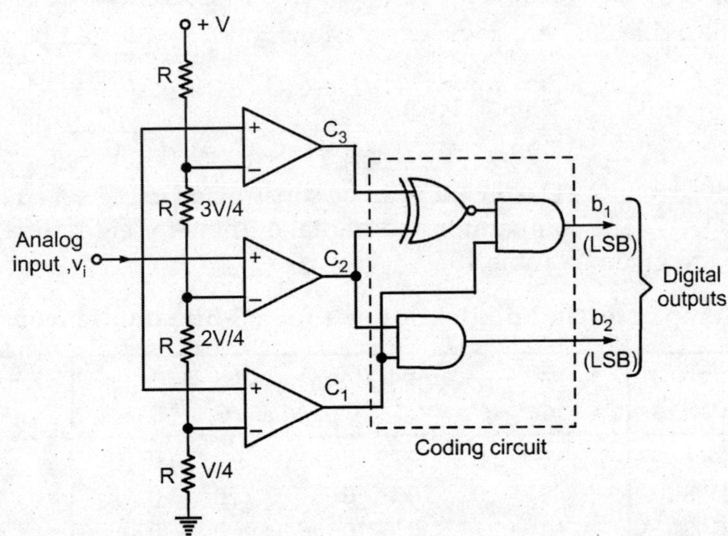

Fig. 13.16 Block diagram of a 2-bit simultaneous type A/D converter

As shown in Fig. 13.16, the three op-amps are used as comparators. The noninverting inputs of all the three comparators are connected to the analog input voltage. The inverting input terminal of the op-amps are connected to a set of reference voltages $V/4$, $2V/4$ and $3V/4$ respectively, which are obtained using a resistive divider network and power supply $+V$.

The output of a comparator is in *positive* saturation state when the voltage at the noninverting input terminal is more than the voltage at the inverting terminal and it is in *negative* saturation state otherwise. When the analog input voltage is less than $V/4$, the voltage at the noninverting terminals of the three comparators is less than their respective inverting input voltages, and hence, the comparator outputs are $C_1C_2C_3$ = 000.

When the analog input is between $V/4$ and $V/2$, the comparator outputs are $C_1C_2C_3$ = 100. Table 13.1 shows the comparator outputs for different ranges of analog voltage and their corresponding digital outputs

Table 13.1: Comparator and digital outputs for a 2-bit simultaneous type A/D converter

Analog Input Voltage (V_i)	Comparator Outputs C_1	C_2	C_3	Digital Outputs b_2	b_1
$0 \le V_i \le V/4$	0	0	0	0	0
$V/4 \le V_i \le V/2$	1	0	0	0	1
$V/2 \le V_i \le 3V/4$	1	1	0	1	0
$3V/4 \le V_i \le V$	1	1	1	1	1

Since there are four ranges of analog input voltages, this can he coded using a 2 bit digital output (b_2, b_1) as shown in Table 13.1. The coding circuit for encoding the three comparator outputs into two digital outputs is shown inside the dotted square of Fig. 13.16 using the simplified expressions for b_1 and b_2 as discussed below.

From the Table 13.1, the logic expressions for b_2 and b_1 can be written as

$$b_2 = C_1C_2\overline{C_3} + C_1C_2C_3 = C_1C_2(\overline{C_3} + C_3) = C_1C_2 \tag{13.20}$$

$$b_1 = C_1\overline{C_2C_3} + C_1C_2C_3 = C_1(\overline{C_2 \oplus C_3}) \tag{13.21}$$

Similarly, a 3-bit A/D converter can be constructed using seven ($2^3 - 1$) comparators as shown in Fig. 13.17. The comparator and digital outputs for eight different ranges of analog input voltage are given in Table 13.2.

Table 13.2: Comparator and digital outputs for a 3-bit simultaneous type A/D converter

Analog input voltage (V_i)	Comparator Outputs C_1	C_2	C_3	C_4	C_5	C_6	C_7	Digital Outputs b_3	b_2	b_1
$0 \le V_i \le V/8$	0	0	0	0	0	0	0	0	0	0
$V/8 \le V_i \le 2V/8$	1	0	0	0	0	0	0	0	0	1
$V/8 \le V_i \le 3V/8$	1	1	0	0	0	0	0	0	1	0
$V/8 \le V_i \le 4V/8$	1	1	1	0	0	0	0	0	1	1
$V/8 \le V_i \le 5V/8$	1	1	1	1	0	0	0	1	0	0
$V/8 \le V_i \le 6V/8$	1	1	1	1	1	0	0	1	0	1
$V/8 \le V_i \le 7V/8$	1	1	1	1	1	1	0	1	1	0
$V/8 \le V_i \le V$	1	1	1	1	1	1	1	1	1	1

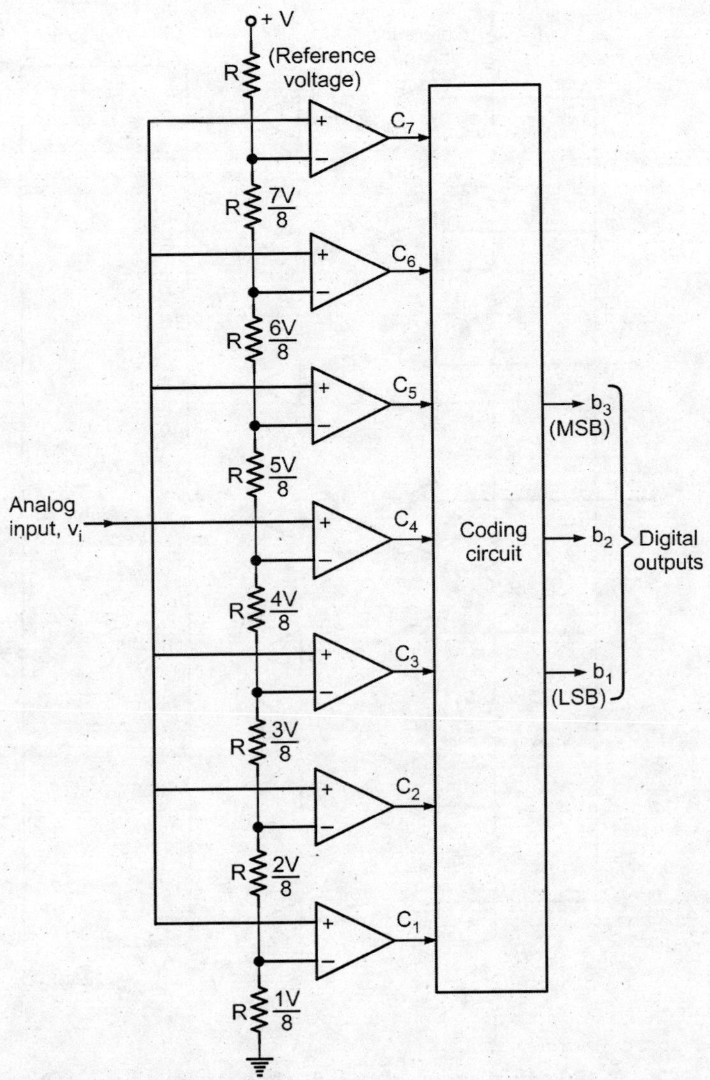

Fig. 13.17 Block diagram of a 3-bit simultaneous type A/D converter

From the Table 13.2, it is clear that the logic expressions for b_3, b_2 and b_1 are complex due to their dependence on seven input variables (C_1 C_2...C_7). Hence, the coding circuit is implemented using a priority encoder. The IC 74148 is an 8-to-3 priority encoder with active LOW inputs and outputs. Since the comparator outputs are active HIGH, they are connected to the inputs of encoder through inverters and the outputs of encoder are inverted once again to get active HIGH digital outputs b_3, b_2 and b_1 as shown in Fig. 13.18.

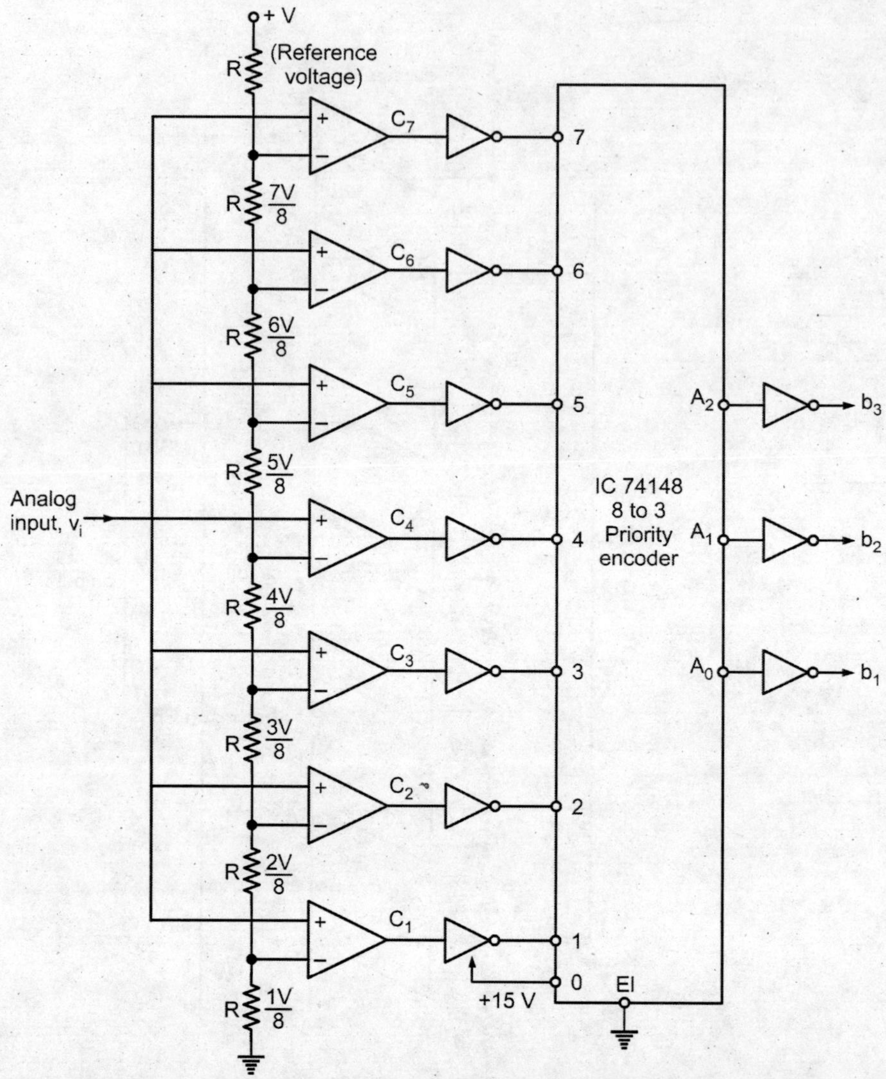

Fig. 13.18 Logic diagram of a 3-bit simultaneous type A/D converter

Advantages

(*i*) Simultaneous type A/D converter is the fastest because A/D conversion is performed simultaneously through a set of comparators. Hence, it is also called *flash type* A/D converter. Typical conversion time is 100 ns or less.

(*ii*) The construction is simple and easier to design.

Disadvantages

The simultaneous type A/D converter is not suitable for A/D conversion with more than 3 or 4 digital output bits. Then $(2^n - 1)$ comparators are required for an *n*-bit A/D converter and the number of comparators required doubles for each added bit.

13.9.2 Counter Type A/D Converter

The counter type A/D converter is constructed using only one comparator with a variable reference voltage. The variable reference voltage can be obtained by a sequence or binary counter and a D/A converter. The block diagram for an *n*-bit counter type A/D converter is shown in Fig. 13.19 (*a*).

The operation of the counter type A/D converter is as follows. The *n*-bit binary counter is initially set to 0 by the *Reset* switch which is normally active LOW. Therefore, the digital output is zero and the analog equivalent V_r is also 0. When Reset signal is released (HIGH), the clock pulses gated through the AND gate are counted by the binary counter. The D/A converter converts the digital output to an analog voltage and connects it as the inverting input to the comparator. The output of the comparator enables the AND gate to pass the clock. The number of counted pulses increases with time and the analog output V_r from DAC is a rising staircase waveform as shown in Fig. 13.19(*b*).

The counting will continue until the reference voltage V_r equals and just rises more than V_i. Then the comparator output becomes LOW and this disables the AND gate from passing the clock. The counting stops at the instance $V_i < V_r$ and at that instant the digital output of the comparator represents the analog input voltage V_i. Then the clock is inhibited, the counter stops its progress and the conversion is said to be complete. The numbers stored in the *n*-bit counter is the equivalent *n*-bit digital data for the given analog input voltage.

In this A/D converter, the counter advances by one count for every clock pulse, and therefore, the clock speed decides the conversion speed. For example, if a 100 kHz clock is used in an 8-bit A/D converter, the counter advances for every step and it will take 2.56 ms (i.e., $2^8 \times \frac{1}{100\,\text{kHz}} = 2.56$ ms) to reach the full-scale digital output (i.e., $2^8 \times 10\ \mu s = 256$ ms). Normally, the time required to reach one half of the full-scale voltage is called *average conversion* time. Hence, the average conversion time of the above A/D converter is 1.28 ms.

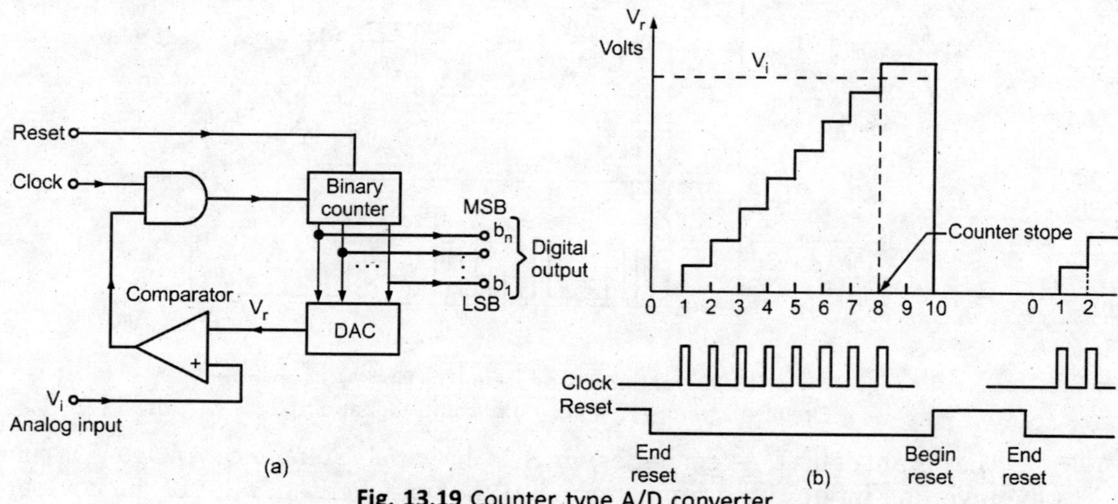

Fig. 13.19 Counter type A/D converter
(*a*) Block diagram (*b*) Its output staircase waveform

Advantages

(*i*) The counter type A/D converter is very simple and needs less hardware compared to the simultaneous type A/D converter.

(*ii*) This is suitable for digitizing applications with high resolution .

Disadvantages

In counter type A/D converter, the conversion time is very long, variable and proportional to the amplitude of the analog input voltage. Since the counter always counts from 0 through a normal sequence, a maximum of 2^n counts are required to convert a full-scale analog input voltage. Hence, for an *n*-bit A/D converter, the average conversion time is $2^n/2^{n-1}$ times the clock period, which can be very long for large value of *n*.

13.9.3 Continuous Type (Servo Tracking) A/D Converter

The main drawback of very long conversion time of the counter type A/D converter can be eliminated by counting from the previously counted value, instead of resetting the counter for each conversion. This requires an UP/DOWN counter mechanism and additional control circuitry as shown in the continuous or servo tracking A/D converter of Fig. 13.20(*a*).

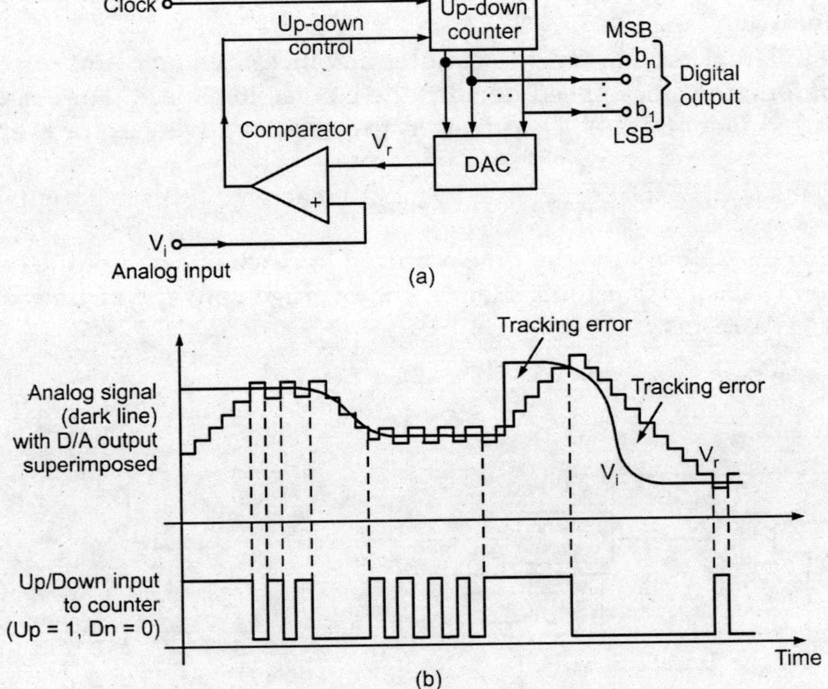

Fig. 13.20 Continuous or Servo tracking A/D converter
(*a*) Circuit diagram; (*b*) Analog input and digital output waveform

The output of the D/A converter which is the variable reference voltage V_r is connected to the inverting input of the comparator and the noninverting input of the comparator is connected to the unknown analog input voltage V_i. When the analog input voltage V_i is greater than the variable reference voltage V_r, the output of the comparator is in HIGH state. The counter is then made to count UP, and the D/A converter output increases. The new *n*-bit digital data is converted and compared with the unknown analog voltage. The process of counting-up continues until V_r is less than V_i. When V_r becomes equal and just more than V_i, the comparator output becomes LOW, and the counter starts counting DOWN. The process continues with the digital output moving UP and DOWN about the correct final digital value. The converted digital data is available in the *n*-bit counter.

Figure 13.20 (*b*) shows the waveforms of (*i*) the analog input signal, (*ii*) the UP/DOWN input signal to the counter and (*iii*) the D/A converter output superimposed on (*i*) and (*ii*) to show the response characteristics of the converter.

In practice, the analog input voltage is within 1 LSB of variable reference voltage leading to oscillation between two adjacent digital values. This can be eliminated by adjusting the comparator in such a way that the comparator output (UP) line will not reach HIGH unless the analog input voltage is higher than the variable reference voltage by 1/2 LSB.

Advantage

The continuous type A/D converter is faster than the counter type A/D converter as the conversion starts from the previous counted value instead of resetting the counter every time.

Disadvantages

(*i*) Additional logic is required to control the circuit for performing UP/DOWN counting operations.

(*ii*) Conversion time is variable and it also depends on the last converted value.

(*iii*) The tracking continues efficiently as long as the analog input changes slowly. When the analog input changes rapidly, the tracking can not be achieved in tune with the change in analog input. This is called as *the tracking error* as shown in Fig. 13.20(*b*).

13.9.4 Successive Approximation Type A/D Converter

The conversion time is maintained constant in successive approximation type A/D converter, and it is proportional to the number of bits in the digital output, unlike the counter and continuous type A/D converters.

The basic principle of this A/D converter is that the unknown analog input voltage is approximated against an *n*-bit digital value by trying one bit at a time, beginning with the MSB. The principle of successive approximation process for a 4-bit conversion is shown in Fig. 13.21.

This type of A/D converter operates by successively dividing the voltage range by half, as explained in the following steps.

(*i*) The MSB is initially set to 1 with the remaining three bits set as 0. The digital equivalent is compared with the unknown analog input voltage.

(*ii*) If the analog input voltage is higher than the digital equivalent, the MSB is retained as 1 and the second MSB is set to 1. Otherwise, the MSB is reset to 0 and the second MSB is set to 1.

(*iii*) Comparison is made as given in step 1 to decide whether to retain or reset the second MSB. The third MSB is set to 1 and the operation is repeated down to LSB and by this time, the converted digital value is available in the SAR.

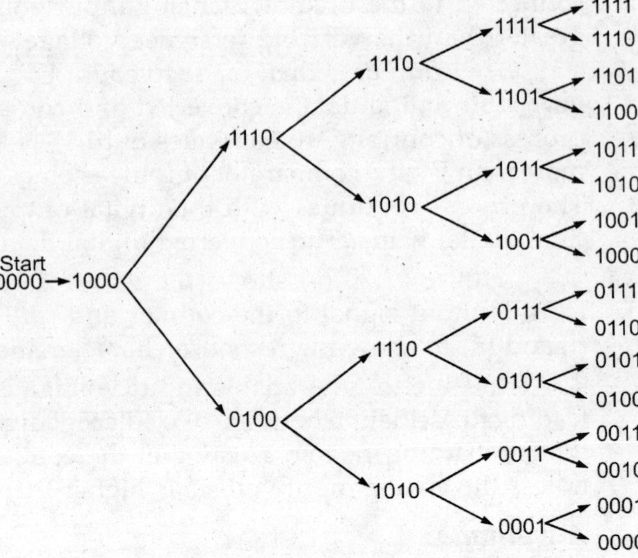

Fig. 13.21 Successive approximation principle for 4-bit digital output

From Fig. 13.21, it can be seen that the conversion time is constant (i.e., four cycles for 4-bit A/D converter) for various digital outputs. This method uses a very efficient search strategy to complete an *n*-bit conversion in just *n*-clock periods. Therefore, for an 8-bit successive approximation type A/D converter, the conversion requires only 8 cycles, irrespective of the amplitude of analog input voltage.

The functional block diagram of successive approximation type A/D converter is shown in Fig. 13.22. The circuit employs a *successive approximation register* (SAR) which finds the required value of each successive bit by *trial and error* method. The output of the SAR is fed to an *n*-bit D/A converter. The analog output equivalent of the D/A converter is applied to the noninverting input of the comparator, while the other input of the comparator is connected with the unknown analog input voltage V_i under conversion. The comparator output is used to activate the successive approximation logic of SAR.

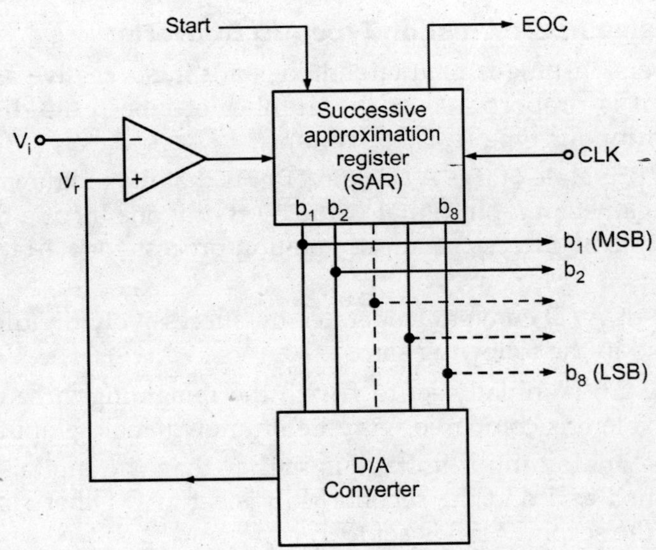

Fig. 13.22 Functional block diagram of a successive approximation type A/D convener

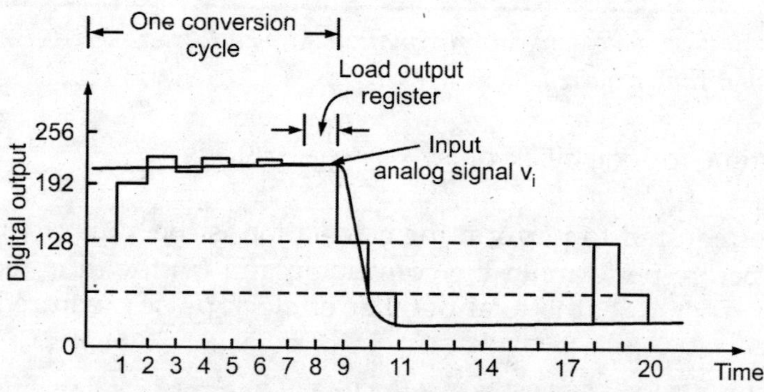

Fig. 13.23 Output response for an analog input

When the START command is applied, the SAR sets the MSB (b_n) of the digital signal, while the other bits are made zero, so that the trial code becomes 1 followed by zeros. For example, for an 8-bit A/D converter the trial code is 10000000. The output of the SAR is converted into analog equivalent V_r and gets compared with the input signal V_i . If V_i is greater than that of the D/A converter output, then the trial code 10000000 is less than the correct digital value. The MSB is retained as 1 and the lower significant bit is made 1 and the testing is repeated. If the analog input V_i is now less than the D/A converter output, then the value 11000000 is greater than the exact digital equivalent. Therefore, the comparator resets the second MSB to 0 and proceeds to the next most significant bit. This process is repeated for all the remaining lower bits in sequence until all the bit positions are tested. The EOC signal is sent out when all the bits are scanned and the value of D/A converter output just crosses V_i.

Table 13.3 shows the flow of conversion sequence and Fig. 13.23 shows the output response with the associated waveforms. It can be observed that the D/A converter output voltage gets successively closer to the analog input voltage V_i . For an 8-bit A/D converter, it requires 8 pulses to compute the output irrespective of the value of the analog input.

Table 13.3: Successive approximation conversion sequence

Correct digital representation	Successive approximation register (SAR) output V_r ad different stages in the conversion	Comparator output
11010100	10000000	1 (initial output)
	11000000	1
	11100000	0
	11010000	1
	11011000	0
	11010100	1
	11010110	0
	11010101	0
	11010100	

Example 13.9. *An 8-bit successive approximation A/D converter is driven by a 2 MHz clock signal. Find the conversion time required.*

Solution: The time for one clock pulse = $\frac{1}{2\,\text{MHz}} = 0.5\mu\text{s}$.

The time required to perform the calculation is the sum of (*i*) the time required for resetting SAR before performing the conversion and (*ii*) the time required for performing the conversion. Therefore, the total number of clock pulses required for the conversion is given by

$$(8 + 1 = 9) \text{ clock cycles } = 9 \times 0.5\ \mu\text{s} = 4.5\ \mu\text{s}$$

A comparison between an 8-bit continuous type A/D converter and an 8-bit successive approximation A/D converter is shown hi Fig. 13.24. Generally the successive approximation technique is more versatile and superior. Only *n* number of comparisons are needed for an A/D conversion process for an *n*-bit digital output.

Successive approximation ICs are available as monolithic circuits. The AD7582 from Analog Devices Corporation provides a 28-pin DIP CMOS package for 12-bit A/D conversion using successive approximation technique.

13.9.5 Single Slope Type A/D Converter

If short conversion time is not important, one can consider single or dual slope type A/D converter. These converter techniques are based on comparing the unknown analog input voltage with a reference voltage that begins at 0 V and increases linearly with time. The time required for the reference voltage to reach the value of unknown analog input voltage is proportional to the amplitude of unknown analog input voltage. This time period can be measured using a digital counter.

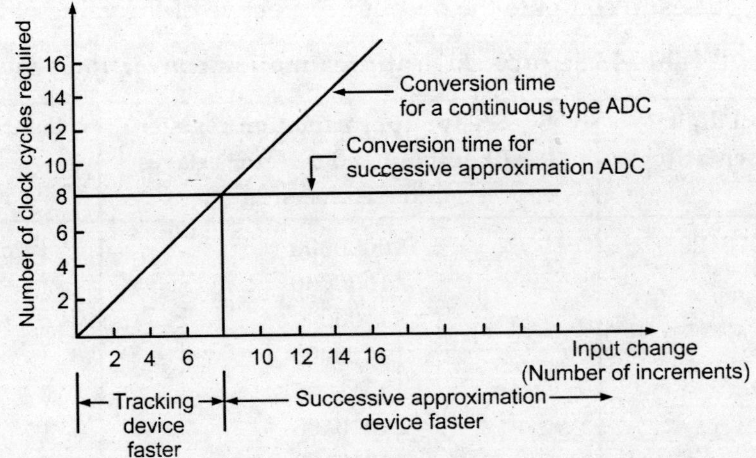

Fig. 13.24 Speed comparison of successive approximation and tracking A/D converters

The block diagram of single slope type A/D converter is shown in Fig. 13.25.

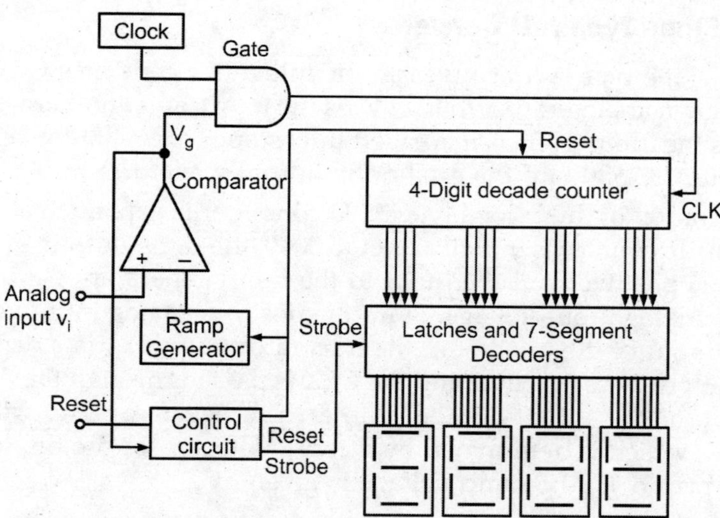

Fig. 13.25 Block diagram of a single slope type A/D converter

The main circuit of this converter is a ramp generator, which, on receiving a RESET from the control circuit increases linearly with time from 0 V to a maximum voltage V_m. For example, if V_m =10 V and it takes 1 ms to move from 0 V to 10 V, then the slope is 10 V/ms. Such a ramp generator can be either an op-amp based integrator circuit or D/A converter driven by a sequence binary counter, whose output waveform is a staircase increasing linearly.

The operation of this converter is explained as follows. Assume that a positive analog input voltage V_i is applied at the noninverting input of the comparator. Now, when a RESET signal is applied to the control logic, the 4-digit decade counter resets to 0 and the ramp voltage begins to increase. Since V_i is positive, the comparator output is in HIGH state. This allows the CLK pulse to pass to the input of the 4-digit counter through the AND gate and the counter is incremented. This process continues until the analog input voltage is greater than the ramp generator voltage. When the ramp generator voltage is equal to the analog input voltage, the comparator output becomes negatively saturated or logic 0 and the clock is prevented from passing through the gate, ceasing the counter operation. Then, the control circuit generates a STROBE signal, which latches the counter value in the 4-digit latch, which is displayed on 7-segment displays. The displayed value is then equivalent to the amplitude of analog input voltage.

For example, if CLK value is 1 MHz and the slope of the ramp generator is 1 V/ms, the 4-digit decade counter reaches its full-scale value i.e., 9999 in 9999 μs (i.e., 9.999 ms). It means when the time is 9.999 ms, the ramp generator voltage reaches 9.999 V. So, this single slope type A/D converter can display any analog input value from 0 V to 9.999 V If V_i = 5.62 V, the counter requires 5620 clock pulses to advance from 0000 to 5620, the ramp voltage rising to 5.62 V. Therefore, at the end of conversion, the display will be 5620. Now, by activating the decimal point of most significant seven segment display, it will directly read as 5.620 V.

The single slope converter has a disadvantage due to the component value errors and the clock errors. The integrated output voltage is a function of the product of R and C. Therefore, changes in the value of capacitance and resistance due to temperature affect the integrated output and introduce errors. The drift in clock frequency also causes errors.

13.9.6 Dual Slope Type A/D Converter

In dual slope type A/D converter, the integrator generates two different ramps, one with the unknown analog input voltage V_i as the input, and another with a known reference voltage $-V_R$ as the input. Hence, it is called dual slope type A/D converter. Its logic diagram is shown in Fig. 13.26(a) and the dual ramp output waveform in Fig. 13.26(b).

The operation of dual slope type A/D converter is explained as follows. Assume that the 4-digit decade counter is initially reset to 0000, the ramp output V_S is reset to 0 V, analog input voltage is positive, and the input to the ramp generator or integrator is switched to the unknown analog input voltage. Since the positive analog input voltage is connected to the inverting input of the integrator, the integrator output V_S is a negative ramp while the comparator output V_g is positive, and the CLK is passed through the AND gate. This results in counting-up of the 4-digit decade counter. The negative ramp will proceed for a fixed time period T_1, which is determined by a count detector for the time period T_1. At the end of fixed time period T_1, the ramp voltage is given by

$$V_S = \frac{V_i}{RC} \times T_1 \tag{13.22}$$

where RC is the time constant of the ramp generator circuit.

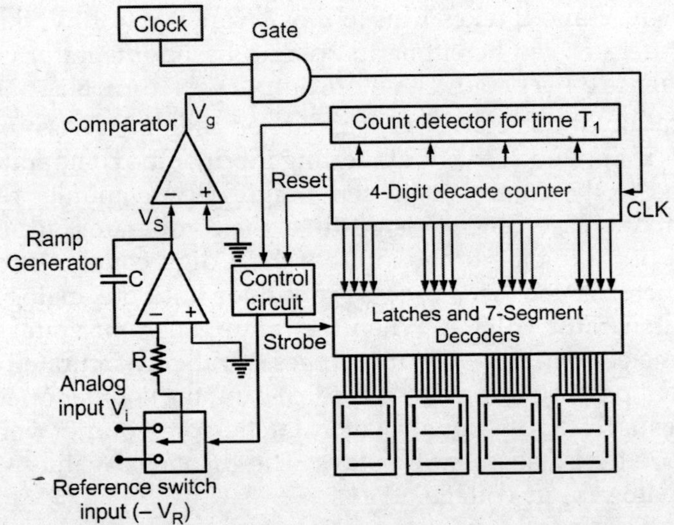

Fig. 13.26(a) Logic diagram of dual slope type A/D converter

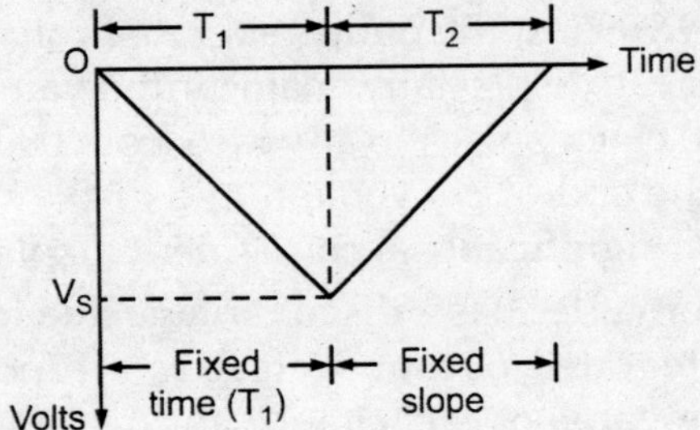

Fig. 13.26(b) Dual ramp output waveform

When the counter reaches the fixed count at time period T_1, the count detector gives a signal to the control circuit which in turn resets the counter to 0 and switches the integrator input to a negative reference voltage ($-V_R$). Now, the ramp generator begins at $-V_S$ and increases upward until it reaches 0 V. During this time, the counter gets advanced. When V_S reaches 0 V, the comparator output will become 0 and the CLK is inhibited from passing through the AND gate. Now, the conversion cycle is said to be completed and the positive ramp voltage is given by

$$VS = -\left(\frac{-V_R}{RC} \times T_2\right) \tag{13.23}$$

where V_R and RC are constants and the time period T_2 is variable.

Since the ramp generator voltage starts at 0 V, decreasing down to $-V_S$ and then increasing up to 0 V, the amplitude of negative and positive ramp voltages can be equated as follows:

$$-\frac{V_i}{RC} \times T_1 = \frac{V_R}{RC} \times T_2 \tag{13.24}$$

$$-V_i = V_R \times \frac{T_2}{T_1}$$

From the above equation, it is clear that the unknown analog input voltage is proportional to the time period T_2, because V_R is a known reference voltage and T_1 is the predetermined time period. Also, the contents of the 4-digit decade counters at the end of conversion reflect the variable time period T_2.

For example, consider the frequency of CLK is 1 MHz, the reference voltage is –1.0 V, the fixed time period T_1 is 1 ms and the *RC* time constant is set at *RC* = 1 ms. Assuming the unknown analog input voltage amplitude as V_i = 5 V, during the fixed time period T_1 the integrator output V_S will go down to

$$V_s = \frac{V_i}{RC} \times T_1 = \frac{-5}{1\,\text{ms}} \times 1\,\text{ms} = -5\text{ V}$$

Then, during the time period T_2, the V_S will integrate all the way back to 0 V.

That is, $$T_2 = \frac{V_s}{V_R} \times RC = \frac{5}{1} \times 1\,\text{ms} = 5\text{ ms} = 5000\ \mu\text{s}$$

Hence, the 4-digit counter value is 5000, and by activating the decimal point of MSD seven segment displays, the display can directly read as 5 V.

13.9.7 Analog-to-Digital Converter using Voltage-to-Time Conversion

An analog signal can be converted into digital signal by counting the pulses from a variable frequency source whose frequency is dependent on the analog input signal value. The counting is done for a fixed period of time. Alternatively, the pulses from a fixed frequency source can be counted for a variable period of time, and the time period is then dependent on the analog signal under conversion.

Fig. 13.27 (*a*) shows such an A/D converter. It employs an integrator, a Sample-and-Hold (S/H) circuit, a voltage comparator and a high-speed counter. A negative reference voltage V_R is applied to the integrator, which integrates the voltage V_R and provides a positive polarity output. The analog signal input under conversion $V_i(t)$ is sampled at a rate fixed by the control voltage V_c, and the sampled signal at any instant V_i is applied as input to the noninverting terminal of comparator. The integrator output V_s is connected to the inverting input of comparator.

When the integrated voltage V_s is less than the analog voltage sample V_i as shown in Fig. 13.27(*b*), the comparator output is at *positive* saturation, or a logic 1.

A fixed frequency clock V_{CL} is applied to the high-speed counter through the AND gate G. The AND gate is enabled for the duration from $t = 0$ when $V_i = 0$ to the time $t = T$ when $V_s = V_i$.

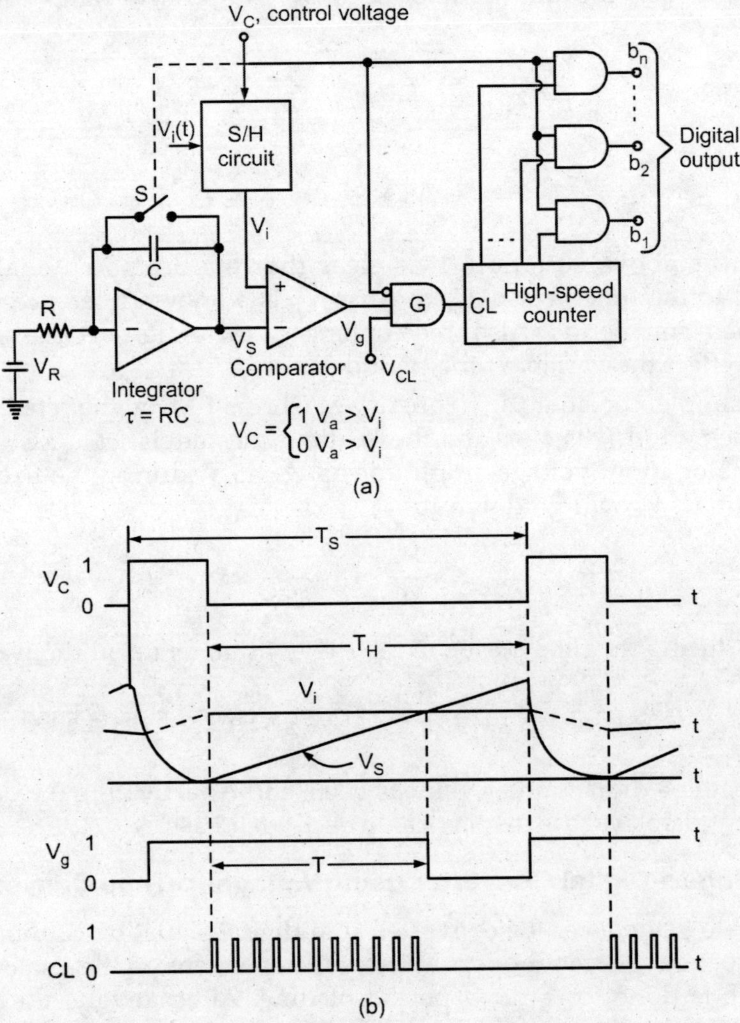

Fig. 13.27 An A/D converter using a voltage-to-time converter (*a*) Functional block diagram; (*b*) Input and output waveform

We know that $Vs = \frac{V_R t}{\tau}$ where the time constant is given by $t = RC$ for the integrator.

Therefore, at time $t = T$ when $V_s = V_i$, we can infer

$$T = \frac{\tau V_i}{V_R} \quad (13.26)$$

Assuming F_C is the clock frequency, the count output N obtained during the time interval 0 to T is given by

$$N = F_C T = \frac{\tau F_C V_i}{V_R} \quad (13.27)$$

Hence, the count value N is proportional to V_i.

Figure 13.27(*b*) shows the waveforms of the sampling control voltage V_c, output of comparator V_g and the clock output CL. The operation of the circuit can be analyzed as follows.

The Sample-and-Hold circuit samples the positive input voltage $V_i(t)$ for every T_A. Then the sampled voltage V_i is held for a time duration indicated by T_H in Fig. 13.27(b). During the period T_H, the switch S is held *open,* and the integrator operates with its output following a ramp voltage waveform V_s. When $V_s < V_i$, the comparator output V_g is at logic 1, and the gate G is enabled, with V_c in 0 state. This continues for a time interval T and during this time, the clock pulses are passed by the gate G to the high-- speed counter. Thus the digital output of the counter is directly proportional to T. During the time interval T_A, the gate is disabled, and the digital output is read from the counter. The switch S is closed during T_A, and the capacitor discharges, resetting V_s to 0 V thus starting a new conversion.

■ REVIEW QUESTIONS ■

1. What do you mean by data converters?
2. Define D/A conversion.
3. Distinguish between the full-scale error and the linearity error of D/A converter.
4. Describe the various specifications of D/A converter.
5. What do you understand by offset error in a D/A converter?
6. Define (*i*) Monotonicity and (*ii*) Settling time of a D/A converter.
7. What is meant by the resolution of a D/A converter?
8. List the essential parts of a D/A converter.
9. What is the principle of multiplying D/A converters? Explain.
10. Explain the 4-bit weighted resistor type D/A converter in detail.
11. What are the limitations of weighted resistor type D/A converter?
12. How many resistors are required for an 8-bit weighted resistor D/A converter? What are those resistance values, assuming the smallest resistance is R?
13. What are the advantages of R – 2R ladder type D/A converter over weighted resistor type?

14. Explain a 4-bit R – 2R ladder type D/A converter in detail.
15. What is the advantage of inverted R – 2R ladder network D/A converter over R-2R ladder D/A converter?
16. Differentiate between current-mode and voltage-mode R – 2R ladder D/A converters. Explain.
17. What is the need for A/D converter?
18. What are the different types of A/D converters?
19. How do you classify the A/D converters based on their operational features?
20. Describe various specifications of an A/D converter.
21. What do you mean by quantization error in an A/D converter?
22. Define aliasing error.
23. What are the sources of analog error in an A/D converter?
24. What is meant by differential linearity of an A/D converter?
25. Define the conversion time of an A/D converter.
26. What is the resolution of an A/D converter?
27. What is the minimum quantization error that can be achieved in an A/D converter?
28. Explain a typical simultaneous type A/D converter in detail.
29. Design a 3-bit simultaneous type A/D converter.
30. What are the limitations of simultaneous type A/D converter?
31. What are servo-tracking A/D converters? Why are they called so? How is it better than counter type A/D converter?
32. Which type of A/D converter is faster? Why?
33. What is Flash type A/D converter? Why is it called so?
34. With a neat block diagram, explain the counter type A/D converter in detail.
35. What are the advantages of counter type A/D converter?
36. Explain continuous type A/D converter with a neat block diagram.
37. What are the advantages and disadvantages of continuous type A/D converter?
38. What are the advantages of continuous type A/D converter over counter type A/D converter?
39. Describe the principle of successive approximation type A/D converter.
40. With a neat block diagram, explain successive approximation type A/D converter in detail.
41. Explain the construction and working of single slope type A/D converter in detail.
42. Describe the operation of dual slope A/D converter with necessary diagrams.
43. Distinguish between single slope and dual slope types of A/D converters.

❑❑❑

Clock Generators

14.1 INTRODUCTION

The synchronous sequential circuits depend on frequencies of clock waveforms for their operation. The clock waveforms are square waves with sharp, leading and trailing edges. The square waves are generated by using square-wave relaxation oscillators called astable multivibrators. Here the active devices such as BJTs, FETs etc. are operated in their relaxed modes of saturation and cut-off.

Multivibrators are two stage switching circuits in which the output of the first stages is fed to the input of the second stage and vice versa. The outputs of two stages are complementary. Multivibrators are of three types namely (*i*) Astable multivibrator (*ii*) Monostable multivibrator and (*iii*) Bistable multivibrator.

14.2 ASTABLE MULTIVIBRATOR

The astable multivibrator or free run oscillator generates square wave without any external triggering pulse. It has no stable states, i.e., it has two quasi stable states. It switches back and forth from one stable to the other, remaining in each state for a time depending upon the discharging of a capacitive circuit.

14.2.1 Astable Multivibrator Using BJT

Fig. 14.1(*a*) shows a basic symmetrical Bipolar Junction Transistor (BJT) astable multivibrator in which components in one half of the circuit are identical to the components in the other half. The square wave output can be taken from collector point of Q_1 or Q_2. The waveforms at base and collector of transistors Q_1 and Q_2 are shown in Fig. 14.1(*b*).

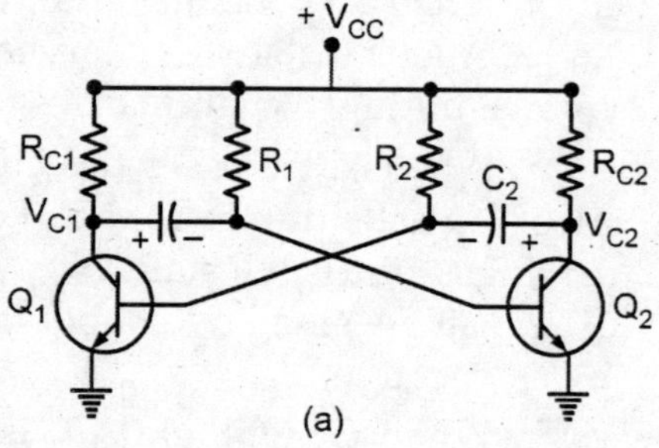

(a)

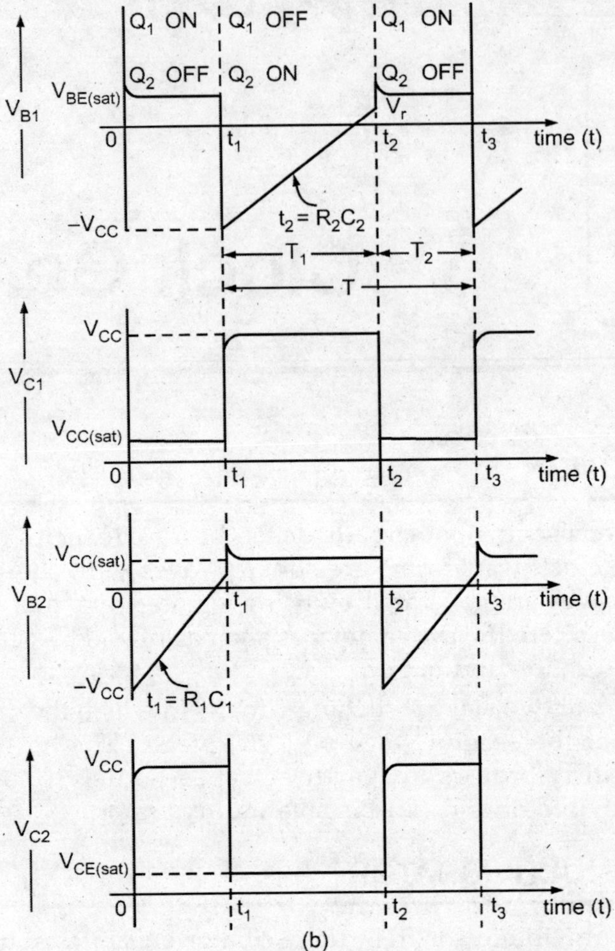

Fig. 14.1 Astable multivibrator (*a*) circuit diagram and (*b*) waveforms at base and collector of Q_1 and Q_2

When the supply voltage $+V_{CC}$ is applied, one transistor will conduct more than the other due to some circuit imbalance. Initially let us assume that Q_1 is conducting and Q_2 is cutoff. Then V_{C1}, the output of Q_1 is equal to $V_{CE(sat)}$ i.e. approximately zero volt and V_{C2} = + V_{CC}. At this instant, C_1 charges exponentially with a time constant R_1C_1 towards the supply voltage through R_1 and correspondingly V_{B2} also increases exponentially towards V_{CC}.

When V_{B2} crosses the cut-in voltage, Q_2 starts conducting and V_{C2} falls to $V_{CE(sat)}$. Also, V_{B1} falls due to capacitive coupling between collector of Q_2 and base of Q_1, thereby driving Q_1 into OFF state. Now, the rise in voltage V_{C1} is coupled through C_1 to the base of Q_2, causing a small overshoot in voltage V_{B2}. Thus Q_1 is OFF and Q_2 is ON. At this instant, V_{B1} is negative, $V_{C1} = V_{CC}$, $V_{B2} = V_{BE(sat)}$ and $V_{C2} = V_{CE(sat)}$.

When Q_1 OFF and Q_2 is ON, the voltage V_B, increases exponentially with a time constant R_2C_2 towards V_{CC}. Therefore, Q_1 is driven into saturation and Q_2 is cutoff. Now, the voltage levels are : $V_{B1} = V_{BE(sat)}$, $V_{C1} = V_{CE(sat)}$, V_{B2} is negative and $V_{C2} = V_{CC}$.

It is clear that when Q_2 is ON, the falling voltage V_{C2} permits the discharging of the capacitor C_2 which drives Q_1 into cutoff. The rising voltage of V_{C1} feeds back to the base of Q_2 tending to turn it ON. This process is said to be regenerative.

The ON time for Q_1 is $T_1 = R_1C_1 \ln 2 = 0.693\, R_1C_1$

Similarly, the ON time for Q_2 is $T_2 = R_2C_2 \ln 2 = 0.693\, R_2C_2$

The total period of the wave is $T = T_1 + T_2 = 0.693\,(R_1C_1 + R_2C_2)$

If $R_1 = R_2 = R$ and $C_1 = C_2 = C$, we have a symmetrical multivibrator, with outputs at the two collectors having the same waveforms but out of phase with each other.

Therefore $T = 1.386\, RC$ and $f = \dfrac{1}{T} = \dfrac{1}{1.386\, RC}$.

To ensure oscillations, the value of resistors should satisfy the following conditions.

$R_1 \le h_{FE(min)}R_{C1}$ and $R_2 \le h_{FE(min)}\, R_{C2}$

where $h_{FE(min)}$ is the minimum value of d.c. current gain of transistor Q_1 and Q_2.

Applications

1. The astable multivibrator is used as square wave generator, voltage to frequency converter and in pulse synchronisation, as clock for binary logic signals and so on.
2. Since it produces square waves, it is a source of production of harmonic frequencies of higher order.
3. It is used in the construction of digital voltmeter and SMPS.
4. It can be operated as an oscillator over a wide range of audio and radio frequencies.

Example 14.1. *If an astable multivibrator has $C_1 = C_2 = 1000$ pF and $R_1 = R_2 = 20$ kΩ, calculate the frequency of oscillation.*

Solution: The frequency of a symmetrical astable multivibrator is

$$f = \frac{1}{1.386RC} = \frac{1}{1.386 \times 20 \times 10^3 \times 1000 \times 10^{-12}} = 36.25 \text{ kHz}$$

Example 14.2. *Determine the period and frequency of oscillation for an astable multivibrator with* component values $R_1 = 2$ *kΩ, $R_2 = 10$ kΩ, $C_1 = 0.01$ μF and $C_2 = 0.05$ μF.*

Solution: The period of oscillation for an asymmetrical astable multivibrator is

$$T = 0.693(R_1C_1 + R_2C_2)$$

$$= 0.693(2 \times 10^3 \times 0.10 \times 10^{-6} + 10 \times 10^3 \times 0.05 \times 10^{-6})$$

$$= 360.36 \text{ μs}$$

Therefore, the frequency of oscillation, $f = \dfrac{1}{T} = \dfrac{1}{360.36 \times 10^{-6}} = 2.775 \text{ kHz}$

Example 14.3. *Determine the value of capacitors to be used in an astable multivibrator to provide a train of pulses of 2 μs wide at a reception rate of 100 kHz, if $R_1 = R_2 = 20$ kΩ.*

Solution: The period of oscillation,

$$T = \frac{1}{f} = \frac{1}{100 \times 10^3} = 10\ \mu s$$

Given $T_1 = 2\ \mu s$

Hence, $T_2 = T - T_1 = 8\ \mu s$

Since $T_1 = 0.693\ R_1C_1$,

$$C_1 = \frac{T_1}{0.693 R_1} = \frac{2 \times 10^{-6}}{0.693 \times 20 \times 10^3} = 145\ \text{pF}$$

Since $T_2 = 0.693 R_2C_2$,

$$C_2 = \frac{T_2}{0.693 R_2} = \frac{8 \times 10^{-6}}{0.693 \times 20 \times 10^3} = 580\ \text{pF}$$

14.2.2 General Description of the IC 555 Timer

The 555 integrated circuit timer was first introduced by Signetics Corporation as Type SE555/NE555. It is available in 8-pin circular style TO-99 Can, 8-pin mini-DIP and 14-pin DIP as shown in Fig. 14.2. The 555 IC is widely popular and various manufacturers provide the IC. The IC 556 contains two 55 timers in a 14-pin DIP package, and Exar's XR-2240 contains a 555 timer with a programmable binary counter in a single 16-pin package.

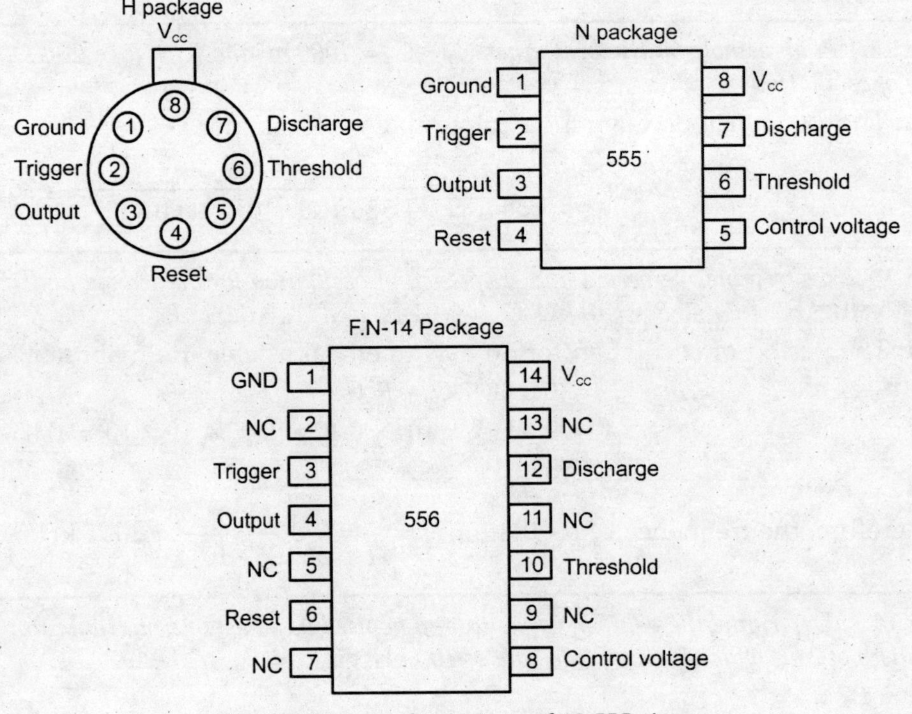

Fig. 14.2 Pin configurations of IC 555 timer

The 555 timer can be operated with a dc supply voltage ranging from +5 V to +18 V. This feature makes the IC compatible to TTL/CMOS logic circuits and op-amp based circuits. The IC 555 timer is very versatile and its applications include oscillator, pulse generator, square the ramp wave generator, one-shot multivbrator, safety alarm and timer circuits, traffic light controllers etc. The 555 timer can provide time delay, ranging from microseconds to hours.

Figure 14.3 shows the functional block diagram of 555 IC timer. The positive dc power supply terminal is connected to pin 8 (V_{CC}) and negative terminal is connected to pin 1 (GND). The ground pin acts as a common ground for all voltage references while using the IC. The output (pin 3) can assume a HIGH level (typically 0.5 V less than V_{CC}) or a LOW level (approximately 0.1 V).

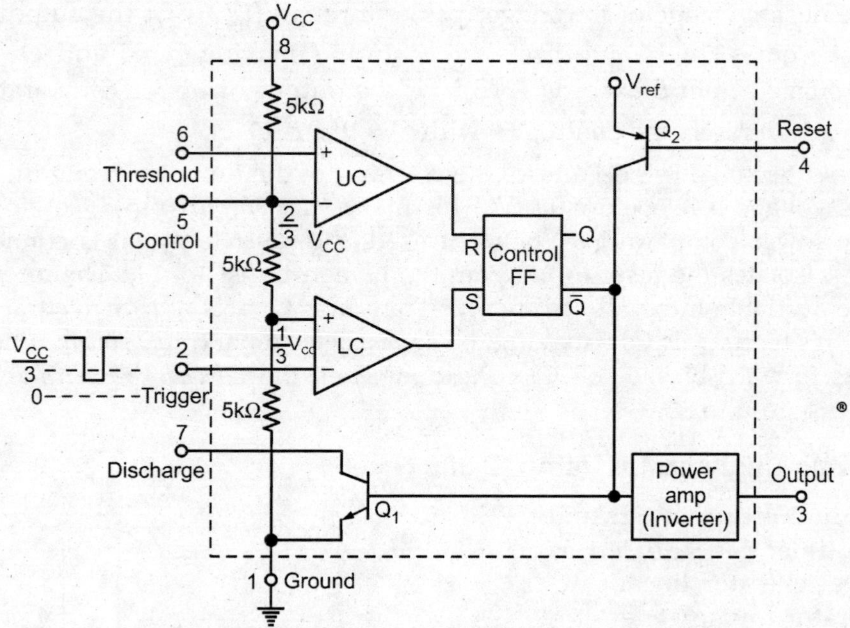

Fig. 14.3 Functional block diagram of IC 555 timer

Two comparators, namely, upper comparator (UC) and lower comparator (LC) are used in the circuit. There 5 kΩ internal resistors provide a potential divider arrangement. It provides a voltage of $(2/3)V_{CC}$ to the (–) terminal of the upper comparator and $(1/3)V_{CC}$ to the (+) input terminal of the lower comparator. A *control* voltage input terminal (pin 5) accepts a modulation control input voltage applied externally. Pin 5 is connected to ground through a bypassing capacitor of 0.1 μF. It bypass the noise or ripple from the supply. The (+) input terminal of the UC is called the *threshold terminal* (pin 6) and the (–) input terminal of the LC is the *trigger terminal* (pin 2). The operation of the IC can be summarized as shown in Table 14.1.

Table 14.1 States of Operation of IC 555

Sl. No.	*Trigger (pin 2)*	*Threshold (pin 6)*	*Output state (pin 3)*	*Discharge state (pin 7)*
1.	Below $(1/3)V_{CC}$	Below $(2/3)V_{CC}$	High	Open
2.	Below $(1/3)V_{CC}$	Below $(2/3)V_{CC}$	Last state remains	Last state remains
3.	Above $(1/3)V_{CC}$	Below $(2/3)V_{CC}$	Last state remains	Last state remains
4.	Above $(1/3)V_{CC}$	Below $(2/3)V_{CC}$	Low	Ground

The standby (state) state makes the output $\overline{Q}$ of flip-flop (*FF*) HIGH. This makes the output of inverting power amplifier LOW. When a negative going trigger pulse is applied to pin 2, as the negative edge of the trigger passes through $(1/3)V_{CC}$, the output of the lower comparator becomes HIGH and it sets the control *FF* making $Q = 1$ and $\overline{Q} = 0$. When the threshold voltage at pin 6 exceeds $(2/3)V_{CC}$, the output of upper comparator goes goes HIGH. This action resets the control *FF* with $Q = 0$ and $\overline{Q} = 1$.

The *reset* terminal (pin 4) allows the resetting of the timer by grounding the pin 4 or reducing its voltage level below 0.4 V. This makes the *output* (pin 3) low overriding the operation of lower comparator. When not used, the *Reset* terminal is connected to V_{CC}. Transistor Q_2 isolates the reset input from the *FF* and transistor Q_1. The reference voltage V_{ref} is made available internally from V_{CC}. Transistor Q_1 acts a *discharge* transistor. When *output* (pin 3) is high, Q_1 is OFF making the *discharge* terminal (pin 7) *open*. When the output is low, Q_1 is forward-biased to ON condition. Then the *Discharge* terminal appears as a short-circuit to ground.

14.2.3 Astable Multivibrator Using IC 555 Timer

The monolithic integrated circuit 555 timer can be used as square-wave generator, linear saw-tooth generator, pulse generator, time delay degenerator etc. The 555 timer circuit to generate square waveform (in astable mode) and triangular waveform is shown in Fig. 14.4. The capacitor C_i charges through the resistances R_1 and R_2 and discharges through the resistance R_2 only. The duty cycle can be controlled by the ratio of resistances R_1 and R_2. Here the capacitor charges and discharges between $(1/3)V_{CC}$ and $(2/3)V_{CC}$. The charging and discharging times are independent of supply voltage.

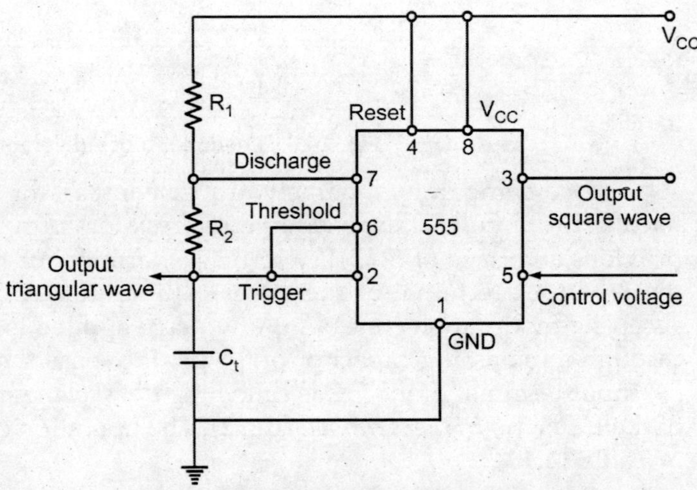

Fig. 14.4 Square and triangular waveforms generator using 555 Timer

The charging time of the capacitor is $t_1 = 0.693\,(R_1 + R_2)C_t$

The discharging time of the capacitor is $t_2 = 0.693\,R_2C_t$

Therefore, the time period of the waveform is given by

$$T_1 = t_1 + t_2 = 0.693\,(R_1 + 2R_2)C_1$$

The frequency of the oscillation is given by

$$f = 1/T = 1.44/C_1\,(R_t + 2R_2)$$

where R_1 and R_2 are in ohms and C_t is in farads.

The 555 timer can also be used for construction of Monostable multivibrator and Schmitt trigger.

Example 14.4. *Determine the frequency of oscillation if the duty cycle D = 20% and the ON period* T_1 = *1 ms.*

Solution:

The duty cycle is $$D = \frac{T_1}{T_1 + T_2} \times 100\%$$

Therefore, $$\frac{20}{100} = \frac{1 \times 10^{-3}}{T_1 + T_2}$$

Here the total period $T = T_1 + T_2 = 5$ ms

Therefore, the frequency of oscillation, $f = \frac{1}{T} = 200$ Hz

---

Example 14.5. *Design an astable multivibrator using 555 timer for a frequency of 1 kHz and a duty cycle of 70%.*

Solution:

The ON period, $T_{ON} = 0.693\,(R_1 + R_2)\,C_t$

Similarly, the OFF period $T_{OFF} = 0.693R_2C_1$

The total period T is given by $T = T_{ON} + T_{OFF} = 0.693\,(R_1 + 2R_2)C_t$

Therefore, the duty cycle D is given by

$$D = \frac{T_{ON}}{T_{ON} + T_{OFF}} = \frac{T_{ON}}{T} = \frac{0.693\,(R_1 + R_2)C_t}{0.693(R_1 + 2R_2)C_t} = \frac{R_1 + R_2}{R_1 + 2R_2}$$

Given $D = 0.7$, we have

$$D = \frac{R_1 + R_2}{R_1 + 2R_2} = \frac{7}{10}$$

Therefore,

$$R_1 = \frac{4}{3}R_2$$

The period of oscillation, $T = 0.693 \times \frac{10}{3} R_2 \times 10^{-7}$

Substituting the value of T and solving, we get

$$R_2 = \frac{1 \times 10^4}{0.693 \times \frac{10}{3}} = 470\ \Omega$$

Therefore, $$R_2 = \frac{4}{3} \times 470 = 680\ \Omega$$

14.2.4 Astable Multivibrator Using NAND / NOT Gates

The principle of the astable multivibrator using NAND gates and NOT gates is incorporated in the circuit shown in Figs. 14.5(*a*) and 14.5(*b*) respectively, which show two forms of astable multivibrator using inverter gates.

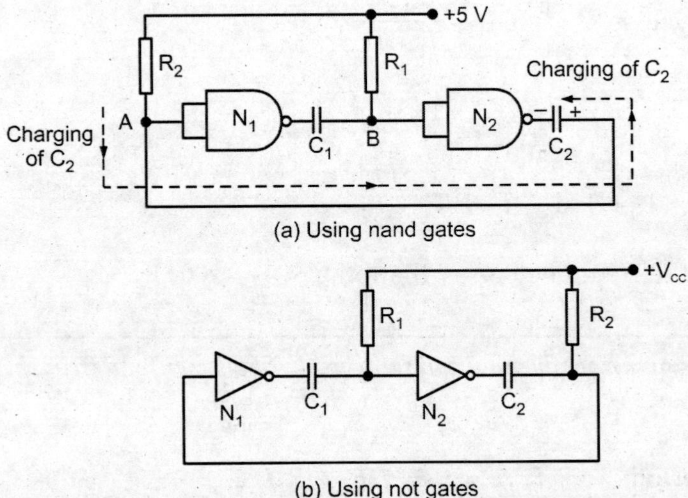

Fig.14.5 Astable multivibrator using inverter gates

Let us assume that initially the gate N_1 is OFF and N_2 is ON. As N_2 is ON, its output voltage is at V_{CES} (= 0.2 V). Therefore, C_2 will get charged to V_{CC} through R_2, as shown in Fig. 14.5(*a*), and the voltage at point A (input of gate N_1) increases. When this voltage reaches the cut-in voltage of gate N1 (2.5 V for TTL NANDs), it turns ON and its output suddenly drops to V_{CES}. As C_1 acts momentarily as a short circuit, this drop in V_{CC} to V_{CES} is conveyed to point B, and N_2 turns OFF immediately. Thus, at this state, N_1 is ON and N_2 is OFF.

14.2.5 Astable Multivibrator Using NAND/NOT Gates with R Returned to Ground

The astable multivibrator circuit using NAND gates shown in Fig. 14.5(*a*) is modified by returning R to ground instead of V_{CC}. Here, the waveform is inverted in comparison with that shown in Fig. 14.6. This is because the capacitors now get charged and discharged in the reverse order through the OFF and ON gates respectively as shown in Fig. 14.7.

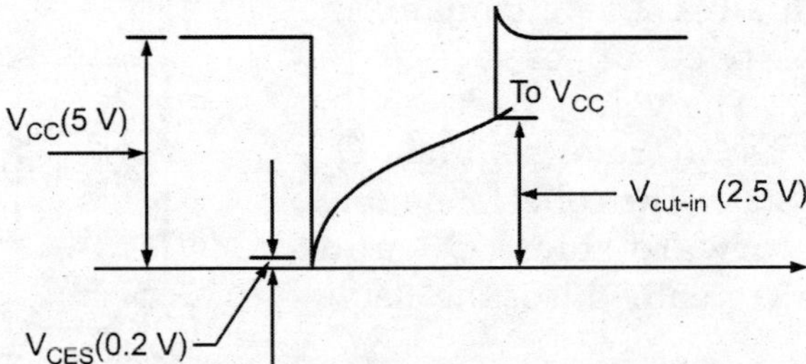

Fig. 14.6 Waveform for calculating frequency

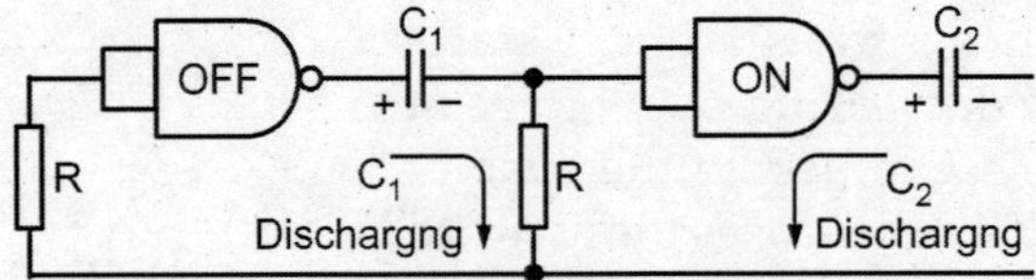

Fig. 14.7 Astable multivibrator using NAND gates with *R* returned to ground

Example 14.6. *Design an astable multivibrator using 7400 NAND gates to produce 1 kHz square wave.*

Solution:

The period of oscillation, $T = \dfrac{1}{f} = 1.386\,\text{RC}$

Substituting $f = 10^3$, we get

$10^3 = 1.386$ RC

Let $C = 0.1$ μF. Therefore, $R = \dfrac{10^{-3}}{1.386 \times 0.1 \times 10^{-6}} = 6.8\,\text{k}\Omega$

14.3 MONOSTABLE MULTIVIBRATOR

Monostable multivibrator has one stable state and one quasi stable state. It is also known as monoshot or one-shot multivibrator or univibrator. It remains in its stable state until an input pulse triggers it into its quasi stable state for a time duration determined by discharging an RC circuit and the circuit returns to its original stable state automatically

It remains there until the next trigger pulse is applied. Thus a monostable multivibrator cannot generate square waves of its own like an astable multivibrator. Only external trigger pulses will cause it to generate the rectangular waves.

14.3.1 Monostable Multivibrator Using BJT

Fig. 14.8(*a*) shows the circuit of a transistor monostable multivibrator and the output waveforms are shown in Fig. 14.8(*b*). It consists of two identical transistors Q_1 and Q_2 with equal collector resistances of R_{C1} and R_{C2} The output of Q_2 is coupled to the input at the base of Q_1 through a resistive attenuator in which C_1 is a small speed up capacitor to speed up the transition. The values of R_2 and $-V_{BB}$ are chosen, so as to reverse bias Q_1 and keep it in the OFF state. The collector supply $+V_{CC}$ and R will forward bias Q_2 and keep it in the ON state. Actually this is the stable state for the circuit.

When a positive trigger pulse of short duration and sufficient magnitude is applied to the base of Q_1 through C_2, transistor Q_1 starts conducting and thereby decreasing the voltage at its collector V_{C1} which is coupled to the base of Q_2 through capacitor C. This decreases the forward bias on Q_2 and its collector current decreases. The increasing positive potential on the collector of Q_2 is applied to the base of Q_1 through R_1. This further increases the base potential of Q_1 and Q_1 is quickly driven to saturation and Q_2 to cut off.

The capacitor C is charged to approximately $+V_{CC}$ through the path V_{CC}, R and Q_1. As the capacitor C discharges, the base of Q_2 is forward biased and collector current starts to flow into Q_2. Thus Q_2 is quickly driven to saturation and Q_1 is cut off. This is the stable state for the circuit and remains in this condition until another trigger pulse causes the circuit to switch over the states.

The duration of the output pulse of the monostable multivibrator is given by $T = 0.693RC$

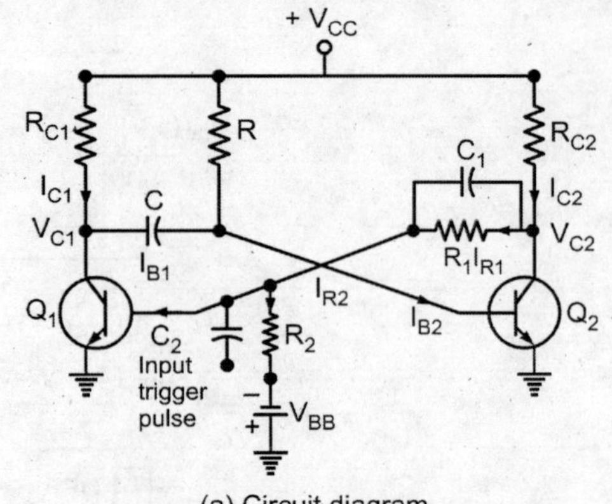

(a) Circuit diagram

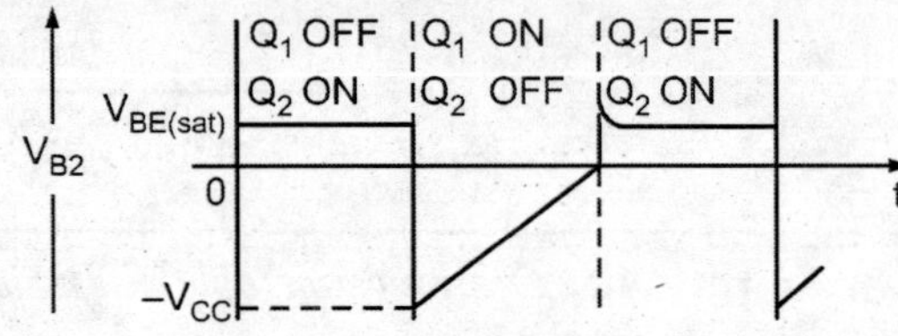

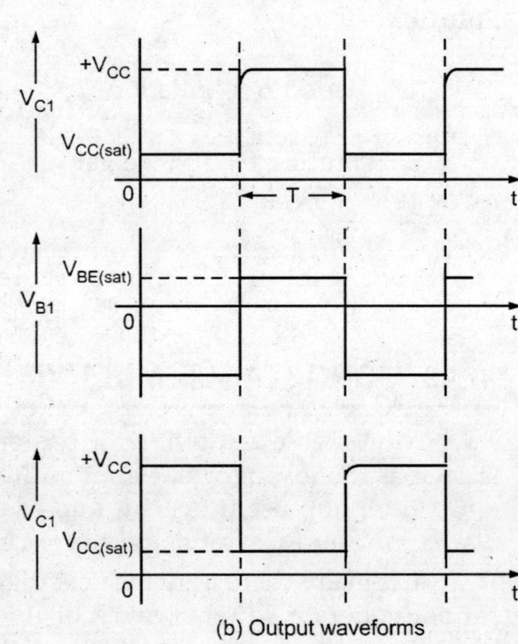

(b) Output waveforms

Fig. 14.8 Monostable multivibrator

Applications

1. The monostable multivibrator is used to function as an adjustable pulse width generator.
2. It is used to generate uniform width pulses from a variable width input pulse train.
3. It is used to generate clean and sharp pulses from the distorted pulses.
4. It is used as a time delay unit since it produces a transition at a fixed time after the trigger signal.

Example 14.7. *Calculate the component values of a monostable multivibrator developing an output pulse of 140 μs duration. Assume* $h_{FEmin} = 20$, $I_{C(sat)} = 6\ mA$, $V_{CC} = 6\ V$, $V_{BB} = -1.5\ V$.

Solution:

At stable state, Q_2 is ON and Q_1 is OFF.

$$R_{C_2} = R_{C_1} = \frac{V_{CC} - V_{CE(sat)}}{I_{C(sat)}} = \frac{6 - 0.3}{6 \times 10^{-3}} = 950\ \Omega$$

$$R_{B2(sat)} = \frac{I_{C(sat)}}{h_{FE(min)}} = \frac{6 \times 10^{-3}}{20} = 0.3\ \text{mA}$$

Also, $$I_{B1(sat)} = 0.3\ \text{mA}$$

$$R = \frac{V_{CC} - V_{BE(sat)}}{I_{B2(sat)}} = \frac{6 - 0.7}{0.3 \times 10^{-3}} = 17.67\ \text{k}\Omega$$

[$\because V_{BE(sat)} = 0.7$ V for Si Transistor]

At quasi-stable state, Q_1 is ON and Q_2 is OFF

$$T = 0.693\ RC$$

Therefore, $$C = \frac{T}{0.693R} = \frac{140 \times 10^{-6}}{0.693 \times 17.67 \times 10^{3}} = 0.114\ \mu\text{F}$$

Assume $$I_{B1(sat)} = I_{R2}$$

Therefore,

$$I_{R1} = I_{B1(sat)} + I_{R2} = 0.3\ \text{mA} + 0.3\ \text{mA} = 0.6\ \text{mA}$$

$$V_{CC} = V_{BE(sat)} + I_{R1}(R_{C2} + R_1)$$

$$6 = 0.7 + 0.6 \times 10^{-3}\ (950 + R_1)$$

Therefore

$$R_1 = \frac{V_{CC} - V_{BE(sat)}}{I_{R1}} - R_{C2}$$

$$= \frac{6 - 0.7}{0.6 \times 10^{-5}} - 950 = 7.883\ \text{k}\Omega$$

$$R_2 = \frac{V_{BE(sat)} - (-V_{BB})}{I_{R2}}$$

$$= \frac{0.7 + 1.5}{0.3 \times 10^{-3}} = 7.33\ \text{k}\Omega$$

The speed up capacitor C_1 is chosen such that $R_1C_1 = 1\ \mu$s and hence,

$$C_1 = \frac{10^{-6}}{7.883 \times 10^{3}} = 126.9\ \text{pF}$$

14.3.2 Monostable Multivibrator Using IC 555 Timer

The monostable multivibrator using 555 timer is shown in Fig. 14.9. In the standby state, referring to Fig. 14.10(*a*), the control flip-flop holds Q_1 ON, thus clamping the external timing capacitor *C* to ground. The output (pin 3) during this time is at ground potential, or LOW. The three 5 kΩ internal resistors act as voltage dividers providing bias voltages of $(2/3)V_{CC}$ and $(1/3)V_{CC}$ respectively. Since these two voltages fix the necessary comparator threshold voltages, they also aid in determining the timing interval.

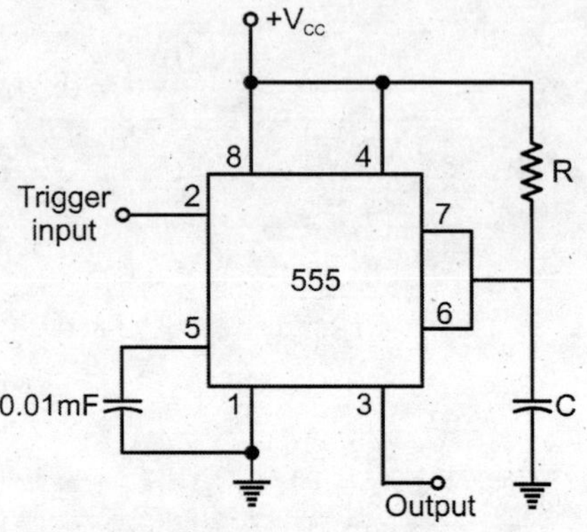

Fig. 14.9 Monostable multivibrator using IC 555 timer

Since the "lower" comparator is biased at $(1/3)V_{CC}$, it remains in the standby state so long as the trigger input (pin 2) is held above $(1/3)V_{CC}$. When triggered only by a negative going pulse, the lower comparator sets the internal flip-flop which releases the short circuit across the timing capacitor, thus turning Q_1 OFF, and the output goes HIGH (approximately equal to V_{CC}). Since the timing capacitor is now undamped, the voltage across it now rises exponentially through *R* towards V_{CC}, with a time constant RC. After a period of time, the capacitor voltage will equal $(2/3)V_{CC}$, and the "upper" comparator resets the internal flip-flop, which in turn discharges the capacitor rapidly to ground potential, turning Q_1 ON. As a consequence, the output now returns to the standby state or ground.

Since the "lower" comparator is biased at $(1/3)V_{CC}$, it remains in the standby state so long as the trigger input (pin 2) is held above $(1/3)V_{CC}$. When triggered only by a negative going pulse, the lower comparator sets the internal flip-flop which releases the short circuit across the timing capacitor, thus turning Q_1 OFF, and the output goes HIGH (approximately equal to V_{CC}). Since the timing capacitor is now undamped, the voltage across it now rises exponentially through *R* towards V_{CC}, with a time constant RC. After a period of time, the capacitor voltage will equal $(2/3)V_{CC}$, and the "upper" comparator resets the internal flip-flop, which in turn discharges the capacitor rapidly to ground potential, turning Q_1 ON. As a consequence, the output now returns to the standby state or ground.

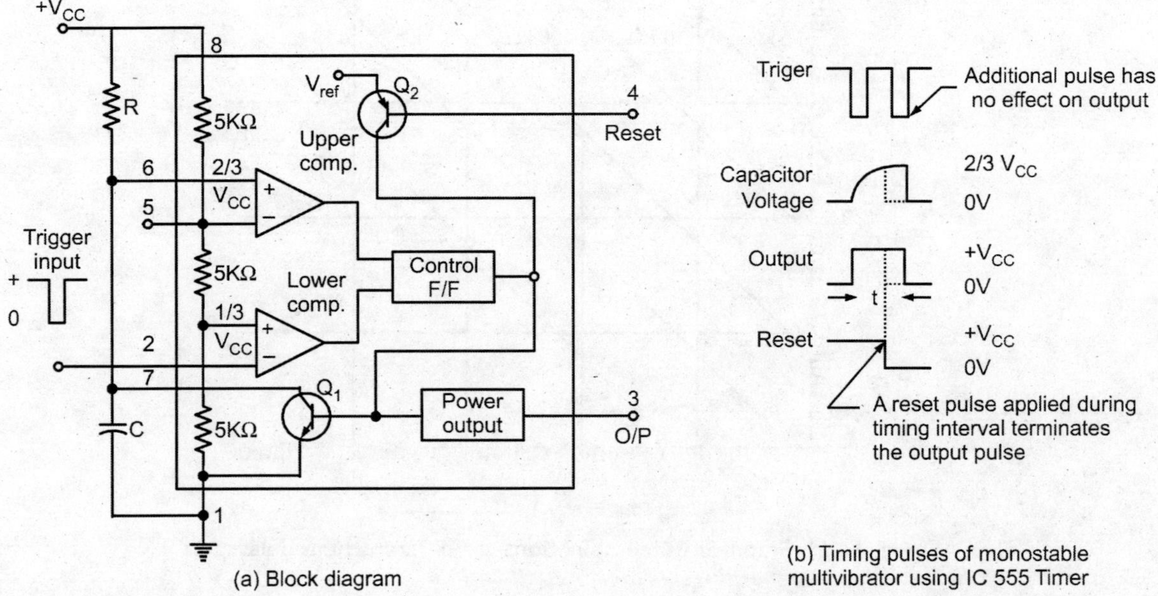

(a) Block diagram

(b) Timing pulses of monostable multivibrator using IC 555 Timer

Fig. 14.10

The 555 monostable timing sequence is shown in Fig. 14.10(*b*). The circuit triggers only on a negative going pulse when the level is less than $(1/3)V_{CC}$. Once triggered, the output will remain HIGH until the set time has elapsed, even if it is triggered again during this interval. Since the external capacitor voltage changes exponentially from 0 to $(2/3)V_{CC}$,

$$\Delta V = V_{CC}(1 - e^{-t/RC})$$

$$\frac{2}{3}V_{CC} = V_{CC}(1 - e^{-t/RC}) \qquad \text{...(14.1)}$$

or

$$t = -RC \ln\left(\frac{1}{3}\right)$$

When the output is HIGH, the time interval becomes

$$t = 1.1 \text{ RC (seconds)} \qquad (14.2)$$

Fig. 14.11 shows a graph of the various combinations of *R* and *C* necessary to produce a given time delay. Since the charging rate and comparator thresholds are both directly proportional to the supply voltage, the timing interval given by Eqn. 14.2 is independent of the supply voltage.

For proper monostable operation with the 555 timer, the negative-going trigger pulse width should be kept shorter compared to the desired output pulse width. The values for external timing resistor and capacitor can either be determined from Eqn. 14.2 or from the graph shown in Fig. 14.11.

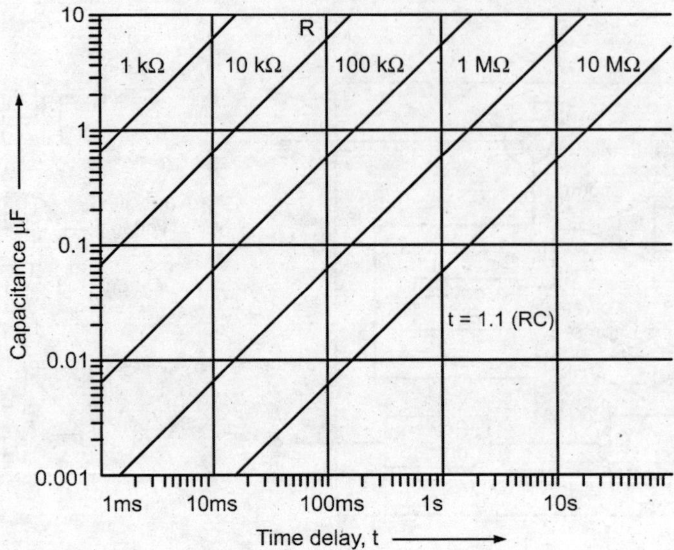

Fig. 14.11 Graph of RC combinations for different time delays

14.3.3 Monostable Multivibrator Using IC 74121

The block diagram of monostable multivibrator using IC 74121 is shown in Fig. 14.12. It shows two active-low inputs $\overline{A}_1$ and $\overline{A}_2$ and an active-high input B. There is an internal resistor of 2 kΩ value. This circuit has one permanent output (Q) state (logic 0). This permanent state can be temporarily changed to logic 1 state by the application of an external trigger pulse. The output will remain in this high state till an externally added capacitor helps to return the output back to the low state at the end of a predetermined interval.

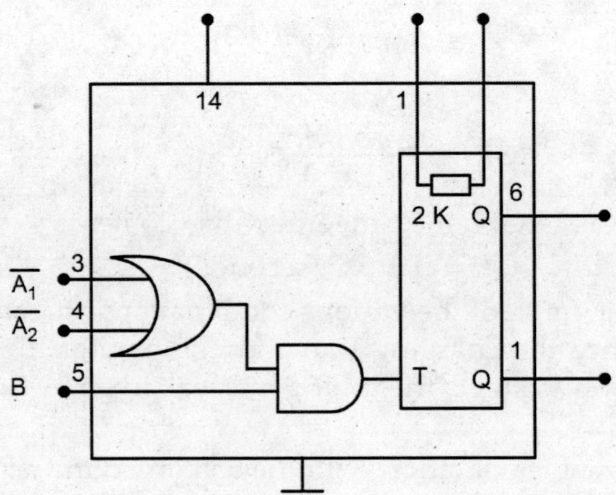

Fig. 14.12 Block diagram of IC 74121 monoshot

The circuit of a monostable multivibrator using 74121 is shown in Fig. 14.13. Initially, the inputs A_1 and A_2 are kept at low state (logic 0). When a positive trigger pulse is applied to the B input, the circuit output will jump to logic 1 from logic 0. The external capacitor C_E which would have already charged to V_{CC} through R_E, will now discharge and at the end of the discharge interval, the output Q returns to its logic 0 (permanent) state.

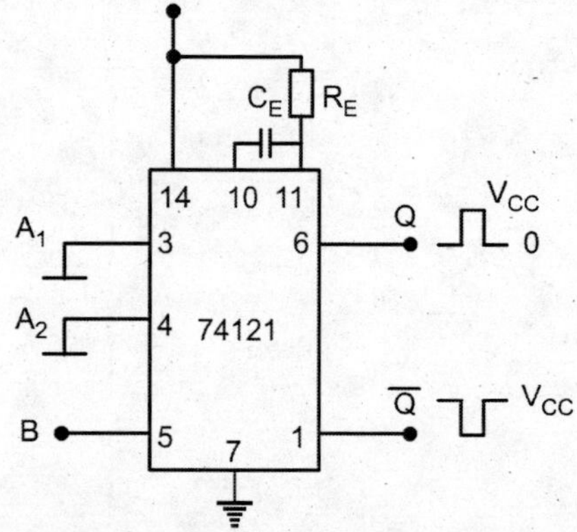

Fig. 14.13 IC 74121 as a monoshot circuit

14.4 BISTABLE MULTIVIBRATOR USING BJT

The bistable multivibrator is also referred to as flip-flop, Eccles-Jordan circuit, trigger circuit or binary. It has two stable states. A trigger pulse applied to the circuit will cause it to switch from one state to the other. Another trigger pulse is then required to switch the circuit back to its original state.

Fig. 14.14(*a*) shows the circuit of a bistable multivibrator using two NPN transistors. In this circuit the output (collector point) of a transistor Q_2 is coupled to the base of transistor Q_1 through a resistor R_1. Similarly the output of Q_1 is coupled to the base of Q_2 through resistor R_1. The main purpose of capacitors C_1 and C_2 is to improve the switching characteristics of the circuit by passing the high frequency components of these square wave pulses. This allows fast rise and fall times, so that these square waves will not be distorted. C_1 and C_2 are thus called "Commutating capacitors."

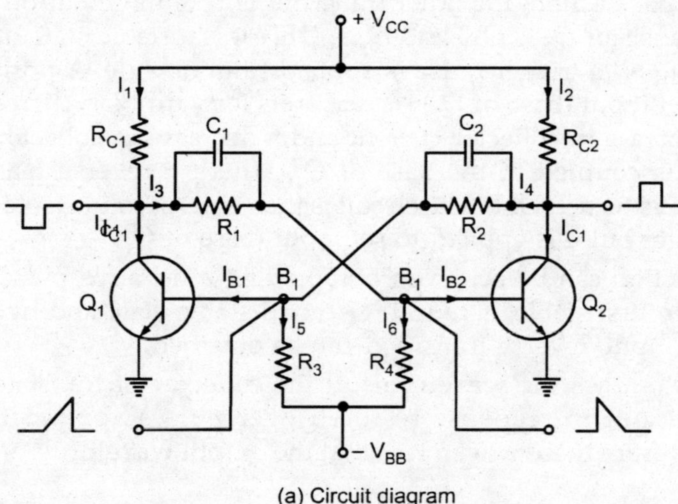

(a) Circuit diagram

Fig. 14.14(*a*) Bistable multivibrator

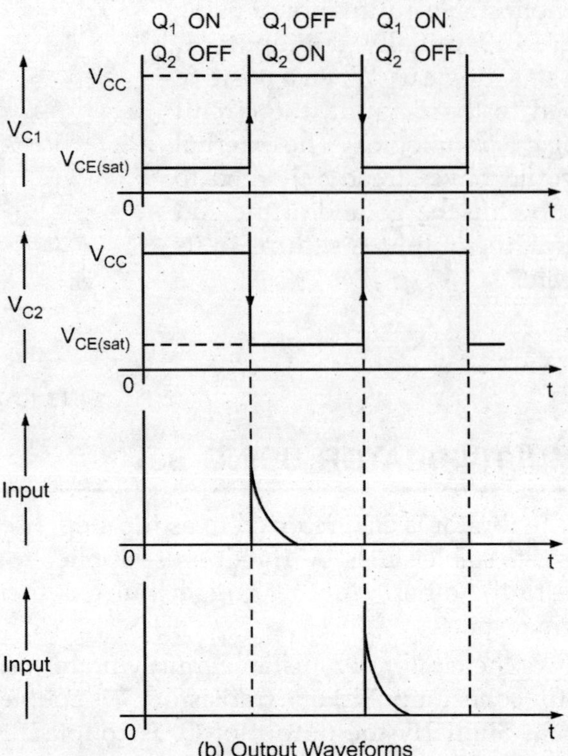

(b) Output Waveforms

Fig. 14.14(*a*) Bistable multivibrator

When the circuit is first switched on, one of the transistors will start conducting more than the other. This transistor is thus driven into saturation (i.e., ON). Then because of the regenerative feed back action, the other transistor is taken into cut off (i.e., OFF) state. Let us assume that transistor Q_1 is ON and Q_2 is OFF. It is a stable state of the circuit and will remain in this state till a trigger pulse is applied from outside. A positive triggering pulse applied to the reset input (base of Q_2) increases its forward bias, thereby turning transistor Q_2 ON and an increase in collector current and a decrease in collector voltage. The fall in collector voltage is coupled to the base of Q_1, where it reverse biases the base-emitter circuit and Q_1 is thus turned OFF. The circuit is then in its second stable state and remains so till a positive trigger pulse is applied to set input (base of Q_1).

A similar action can be achieved by applying a negative pulse at the set input for transition from the first stable state to the second stable state and by applying a negative pulse at the reset input, reverse transition can be obtained.

Fig. 14.14(*b*) shows the waveforms at the collector of transistor Q_1 and Q_2 of the bistable multivibrator in response to the trigger pulses applied to the set and reset input. It is evident from these waveforms that the output waveforms are the complement of each other.

Applications

1. The bistable multivibrator is used as memory element in shift registers, counters and so on.
2. It is used to generate square waves of symmetrical shape by sending regular triggering pulse to the input. By adjusting the frequency of the input trigger pulse, the width of the square wave can be altered.
3. It can also be used as a frequency divider (as a divide by two counter)

Example 14.9. *Calculate the stable state currents and voltages for the bistable multivibrator having $V_{CC} = 12$ V, $R_{C_1} = R_{C_2} = 2\ k\Omega$, $R_1 = R_2 = 15\ k\Omega$, $R_3 = R_4 = 100\ k\Omega$. Assume that a transistor having a minimum h_{fe} of 20.*

Solution: Referring to Fig. 14.14,

$$V_{B1} = V_{BB}\frac{R_2}{R_2+R_3} = \frac{-12\times 15\times 10^3}{115\times 10^3} = -1.56\text{ V}$$

Since V_{B1} is less than $V_{BE(\text{cut-off})}$ i.e., 0.7 V for silicon transistor, Q_1 is OFF.

Therefore, $I_{B1} = 0$ and $I_{C1} = 0$

$$I_2 = I_4 + I_{C2}$$

$$I_{C2} = I_2 - I_4$$

$$= \left[\frac{V_{CC} - V_{C2(\text{sat})}}{R_{C2}}\right] - \left[\frac{V_{C2(\text{sat})} - (V_{BB})}{R_2 + R_3}\right]$$

$$= \left[\frac{12 - 0.3}{2.2\times 10^3}\right] - \left[\frac{0.3 + 12}{115\times 10^3}\right] \quad \text{(Since } Q_2 \text{ is ON, } V_{C2(\text{sat})} = 0.3\text{ V)}$$

$$= 5.42\text{ mA}$$

$$I_{B2} = \left[\frac{I_{C2}}{h_{fe(\min)}}\right] = \left[\frac{5.35\times 10^{-3}}{20}\right] = 0.27\text{ mA}$$

$$I_1 = I_3 + I_{C1}$$

$$= I_3,\ \text{as } I_{C1} = 0$$

$$I_3 = I_{B2} + I_6$$

$$I_6 = \left[\frac{V_{B2} - (-V_{BB})}{R_4}\right]$$

$$V_{B2} = V_{BE2(\text{on})} = 0.7\text{ V}$$

Therefore,

$$I_6 = \left[\frac{0.7 + 12}{100\times 10^3}\right] = 1.127\text{ mA}$$

$$I_3 = I_{B2} + I_6 = 0.27 + 0.127 = 0.397\text{ mA}$$

$$V_{C1} = V_{CC} - I_1 \times R_{C1}$$

$$= 12 - (0.397\times 10^{-3}\times 2.2\times 10^3) = 11.1\text{ V}$$

14.5 SCHMITT TRIGGER

Schmitt trigger is a wave shaping circuit, used for generation of a square wave from a sine wave input. It is a bistable circuit in which two transistor switches are connected regeneratively.

14.5.1 Schmitt Trigger Using BJT

Fig. 14.15 shows the circuit of a Schmitt trigger with the input and output waveforms. It consists of two identical transistors Q_1 and Q_2 coupled through an emitter resistor R_E. The resistors R_1 and R_2 form a voltage divider across V_{C1} and ground. This provides a small forward bias to the base-emitter junction of the transistor Q_2. When the supply is switched ON, with no input signal, transistor Q_2 starts conducting. The rise in current (I_E) of Q_2 causes a voltage drop across R_E i.e., $V_{RE} = I_E R_E$. This voltage provides a reverse bias across the emitter-base junction of Q_1 and it is driven into cut off state. Since Q_1 is in the OFF state, the voltage at its collector rises to V_{CC}. Since the collector of Q_1 is coupled to the base of Q_2 through the resistor R_1, the forward bias for the transistor Q_2 is increased. Thus Q_2 is driven into saturation. At this instant the collector voltage levels are $V_{C1} = V_{CC}$ and $V_{C2} = V_{CE(sat)}$.

Consider an a.c. signal of sinusoidal or triangular variation applied to the base of Q_1. When the voltage increases above zero, nothing will happen till it crosses the Upper Trigger Level (UTL). As the input voltage increases above UTL i.e., $V_{in} > V_{RE} + V_{BE1}$, Q_1 conducts. The point at which Q_1 starts conducting is known as Upper Trigger Point (UTP). As the transistor Q_1 conducts, its collector voltage falls below V_{CC}. Since the collector of Q_1 is coupled to base of Q_2, the forward bias to Q_2 is reduced. This in turn reduces the current of transistor Q_2 and hence the voltage drop across R_E. As a result, the reverse bias of transistor Q_1 is reduced and it conducts more which drives Q_2 to nearer to cut off. This process continues till Q_1 is driven into saturation and Q_2 is cut off. At this instant, the collector voltage levels are $V_{C1} = V_{CE(sat)}$ and $V_{C2} = V_{CC}$.

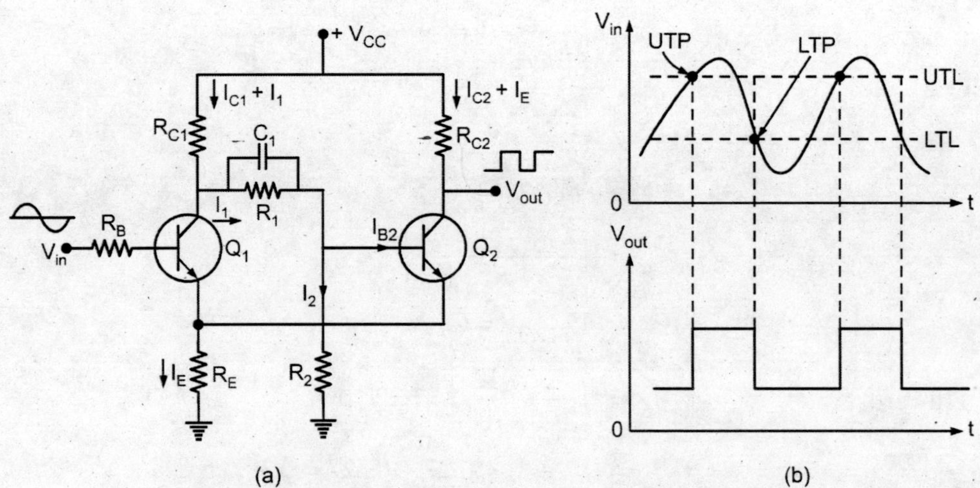

Fig. 14.15(*a*) Circuit Diagram of Schmitt trigger using BJT
(*b*) Input and Output waveforms of Schmitt trigger using BJT

The transistor Q_1 will continue to conduct till the input voltage crosses the Lower Trigger Level (LTL). When the input voltage becomes equal to LTL, the emitter-base junction of Q_1 becomes reverse biased i.e., $V_{in} < V_{RE} + V_{BE1}$. Hence its collector voltage starts rising towards V_{CC}. This forward biases Q_2 and it starts conducting. The point at which Q_2 starts conducting is called Lower Trigger Point (LTP). Then Q_2 is very quickly driven into saturation and Q_1 is cut-off. At this instant the collector voltage levels are $V_{C1} = V_{CC}$ and $V_{C2} = V_{CE(sat)}$ No change in state will occur during the negative half cycle of the input voltage.

The difference between UTP and LTP is known as Hysteresis voltage (V_H) as shown in Fig. 14.16. V_H is also known as "Dead Zone" of the Schmitt Trigger. The lagging of the lower threshold voltage from the upper threshold voltage is known as the hysteresis.

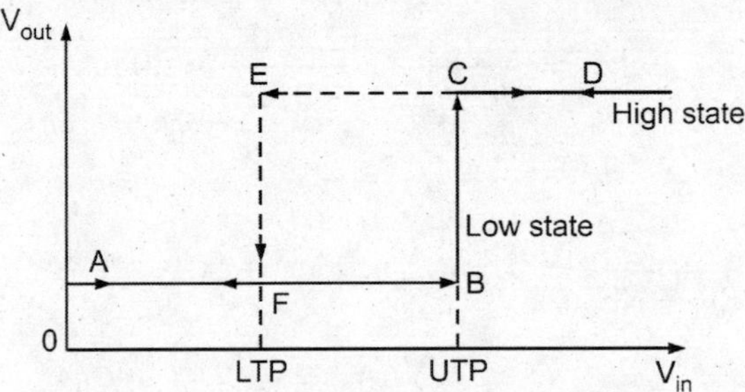

Fig. 14.16 Hysteresis of Schmitt trigger

Applications

1. Schmitt trigger is used for wave shaping circuits.
2. It can be used for generation of a rectangular waveforms with sharp edges from a sine wave or any other waveform.
3. It can be used as a voltage comparator.
4. The hysteresis in Schmitt trigger is valuable when conditioning noisy signals for using digital circuits. The noise does not cause false triggering and so the output will be free from noise.
5. The hysteresis can be eliminated by keeping $R_{C1} = R_{C2}$ in the circuit or by introducing another resistor between the two emitters. Reducing the hysteresis increases the rise and fall times of the output. This makes the triggering more sensitive to small noise fluctuations present in the input signal.

Example 14.10. *Design a Schmitt trigger circuit to have $V_{CC} = 12$ V, UTP = 5 V, LTP = 3 V and $I_C = 2$ mA, using two silicon NPN transistors with $h_{FE(min)} = 100$ and $I_2 = 0.1I_{C2}$.*

Solution: Referring to Fig. 14.15,

$$\text{UTP} = V_{B2} = 5\text{ V}$$

Voltage across R_E is $V_E = V_{B2} - V_{BE} = 5 - 0.7 = 4.3$ V

$$I_E = I_C = 2\text{ mA}$$

$$R_E = \frac{V_E}{I_E} = \frac{4.3}{2 \times 10^{-3}} = 2.15\text{ k}\Omega$$

Taking Q_2 Saturated, $V_{CE(sat)} = 0.2$ V typically

$$I_C R_{C2} = V_{CC} - V_E - V_{CE(sat)} = 12 - 4.3 - 0.2 = 7.5 \text{ V}$$

$$R_{C2} = \left[\frac{7.5}{2 \times 10^{-3}}\right] = 3.75 \text{ k}\Omega$$

$$I_2 = 0.1 I_{C2} = 0.1 \times 2 \times 10^{-3} = 0.2 \text{ mA}$$

$$R_2 = \left[\frac{V_{B2}}{I_2} = \frac{5}{0.2 \times 10^{-3}} = 25 \text{ k}\Omega\right]$$

$$I_{B2} = \frac{I_{C2}}{h_{FE(\min)}} = \frac{2 \times 10^{-3}}{100} = 20 \text{ }\mu\text{A}$$

$$I_1 = I_2 + I_{B2}$$

$$\frac{V_{CC} - V_{B2}}{R_{C1} + R_1} = I_1 = 0.2 \times 10^{-3} + 20 \times 10^{-6}$$

$$\frac{12 - 5}{R_{C1} + R_1} = 0.22 \times 10^{-3}$$

$$R_{C1} + R_1 = \frac{7}{0.22 \times 10^{-3}} = 31.8 \text{ k}\Omega$$

When Q_1 is ON, $V_1 = \text{LTP} = V_{B2} = 3$ V

$$I_1 = \frac{V_{B2}}{R_2} = \frac{3}{25 \times 10^{-3}} = 0.12 \text{ mA}$$

$$I_{C1} = I_E = \frac{V_{B1} - V_{BE}}{R_E} = \frac{3 - 0.7}{2.15 \times 10^3} = 1.07 \text{ mA}$$

$$V_{CC} = R_{C1}(I_{C1} + I_1) + I_1(R_1 + R_2)$$

$$I_2 = R_{C1}(I_{C1} + I_1) + I_1(31.8 \times 10^3 - R_{C1} + R_2)$$

$$= R_{C1} I_{C1} + I_1(31.8 \times 10^3 + 25 \times 10^3)$$

$$R_{C1} = \frac{12 - 0.12 \times 10^{-3} \times 56.8 \times 10^3}{1.07 \times 10^{-3}} = 4.84 \text{ k}\Omega$$

$$R_1 = (31.8 - 4.84) \times 10^3 = 26.96 \text{ kW}$$

$$R_B < h_{FE} R_E$$

$$R_B = \frac{h_{FE} R_E}{10} = \frac{100 \times 2.15 \times 10^3}{10} = 21.5 \text{ k}\Omega$$

14.5.2 Schmitt Trigger Using IC 555 Timer

The IC 555 timer can be used to function as a variable-threshold Schmitt trigger. Since the internal circuitry has a high output impedance and latching capability, the threshold voltage can be adjusted over a wide range with simultaneously open-collector/totempole outputs.

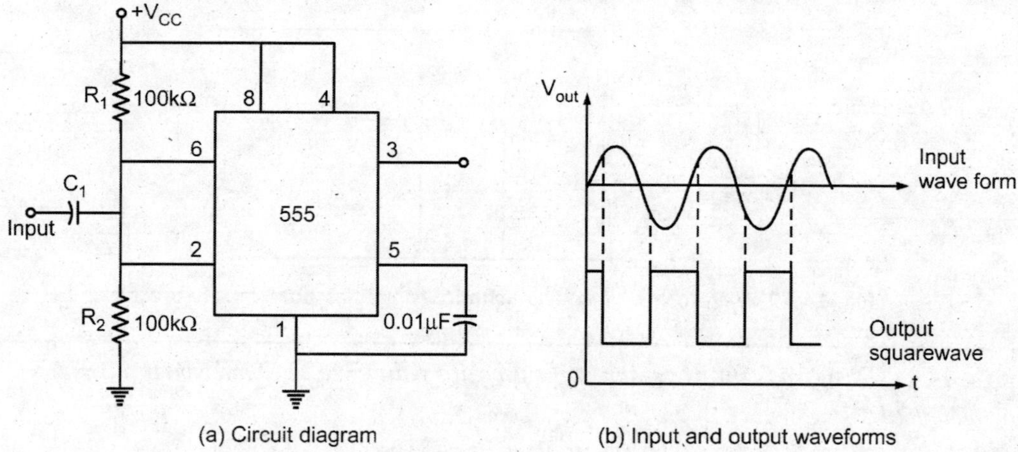

Fig. 14.17 Schmitt trigger using IC 555 Timer

Another Schmitt trigger circuit is shown in Fig. 14.17(*a*) where the two internal comparator inputs (pins 2 and 6) are connected together and externally biased at $(1/2)V_{CC}$ through R_1 and R_2. Since the upper comparator at pin 6 will trip at $(2/3)V_{CC}$, and the lower comparator at $(1/3)\ V_{CC}$, the bias provided by R_1 and R_2 is centered within these two thresholds.

A sine-wave input of sufficient amplitude to exceed the reference levels causes the internal flip-flop to alternatively set and reset, generating a square wave output. As long as R_1 equals R_2, the 555 timer will automatically be biased for any supply voltage in the range of 5 to 16 V. From the curve shown in Fig. 14.17(*b*), it is observed that there is a 180° phase shift.

Unlike a conventional multivibrator type of square wave generator that divides the input frequency by 2, the main advantage of Schmitt trigger is that it simply converts the sine-wave signal into square wave signal without division.

14.5.3 Schmitt Trigger Using IC 74132

Fig. 14.18 shows IC 74132 Quad two-input NAND Schmitt trigger to work as a free running astable multivibrator for the generation of square waves.

Assuming that the output is initially at logic 1, the capacitor C starts charging towards V_{CC} through R as shown in Fig. 14.18. When the capacitor voltage V_C is greater than UTP, the Schmitt trigger turns ON and its output goes LOW. Now, C discharges through R, and when V_C is less than LTP of the Schmitt trigger, it triggers back to its high state. The process repeats to produce a square wave. The waveforms are shown in Fig. 14.19.

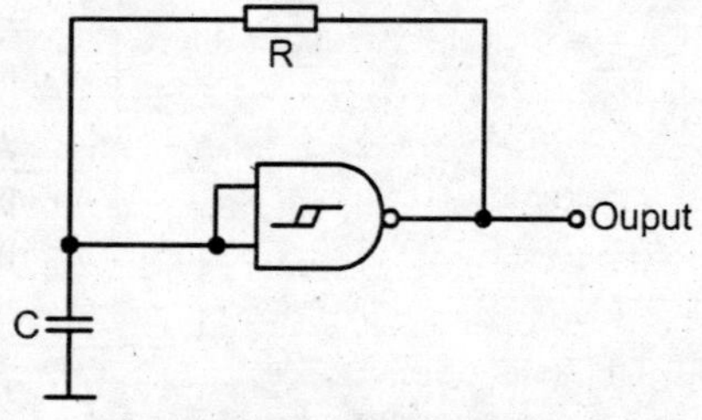

Fig. 14.18 Schmitt trigger using NAND gates

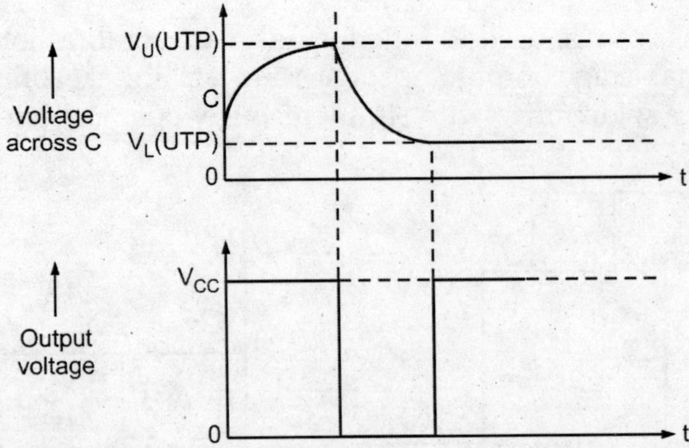

Fig. 14.19 Waveforms of NAND Schmitt trigger astable multivibrator

Example 14.11. *Design a Schmitt astable multivibrator using 74132 Quad two input NAND gate for a frequency of 1 kHz.*

Solution:

$$\text{Half period of oscillation, } \frac{T}{2} = RC \ln \frac{V_U - V_{CC}}{V_L - V_{CC}}$$

Substituting $V_U = 4$ V, $V_L = 2$ V and $V_{CC} = 5$ V, we get

$$\frac{T}{2} = 0.693 \text{ RC}$$

Therefore, the period of oscillation, T = 1.386 RC

Let C = 0.1 μF. Hence R = 6.8 kΩ

14.6 CRYSTAL CLOCK GENERATORS

The crystal-based astable multivibrators are used for every precise clock application. Crystal oscillators are used as the clocks in microprocessors and microcomputers. Fig. 14.20 shows a clock generator using NAND gates in which crystal is the controlling element. This is a modification of the astable multivibrator using NAND gates.

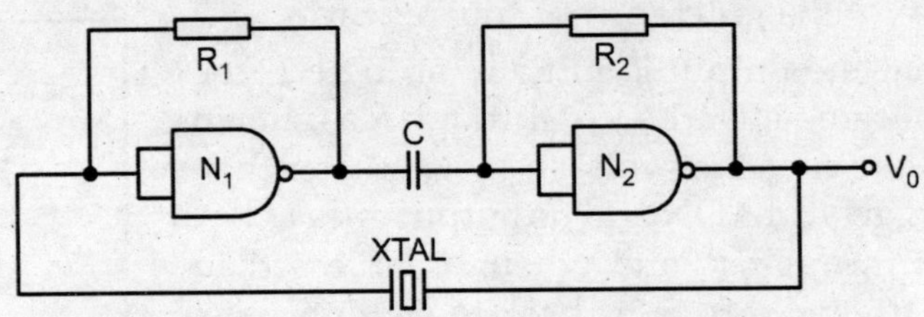

Fig. 14.20 Crystal clock generator using NAND gates

In Fig. 14.20, two NAND gates are connected through *RC* coupling and they act as high gain *RC*-coupled amplifiers as explained in *section 14.2.* The feedback is provided through the crystal which acts as an RLC circuit. Under resonance condition, the capacitive and inductive reactances of the crystal cancel each other so that the crystal behaves as a mere resistance (= 1 kΩ). Hence, as explained in *section 14.2,* this circuit produces square waveforms at the resonant frequency of the crystal. This frequency of oscillation is highly stable.

■ REVIEW QUESTIONS ■

1. What is a multivibrator? On what basis are multivibrators classified?
2. With a neat sketch, explain the working of an astable multivibrator. On what factors does the frequency of the output waves depend?
3. Draw the switching waveform for the astable multivibrator.
4. List the applications of astable multivibrator.
5. In a transistorised astable multivibrator, $R_{C1} = R_{C2} = 2$ kΩ, $R_1 = R_2 = 50$ kΩ and $C_1 = C_2 = 0.01$ μF. Find the frequency of the square wave generated.
6. Calculate the value of resistors R_1 and R_2 in an astable multivibrator to provide a symmetrical oscillation of 100 kHz, if $C_1 = C_2 = 330$ pF.
7. Design an astable multivibrator circuit that gives out a 12 V output pulse waveform where $t_p = 10$ μs and pulse repetition frequency (PRF) = 20 kHz. The two NPN transistors are used with $I_{C(max)} = 20$ mA, $\beta = 50$ and $V_{BE(ON)} = V_{CE(ON)} = 0$ V.
8. Draw the internal structure of IC 555 Timer.
9. Derive the frequency of oscillation of an astable multivibrator using IC 555 Timer.
10. Describe the theory behind astable multivibrator using (*a*) NOT gates and (*b*) NAND gates.
11. Design an astable multivibrator using NAND gates for a frequency of 4 kHz.
12. What is a monostable multivibrator? Explain its working with the help of waveforms.
13. Which multivibrator would function as a time delay unit? Why?
14. Explain, with the circuit diagram, the operation of monostable multivibrator using transistors. Sketch the input-output waveforms.
15. Mention the applications of one shot multivibrator.
16. Derive the expression for the period of a pulse generated when 555 Timer is used as a monostable multivibrator.
17. Explain the operation of IC 74121 monoshot.
18. With a neat sketch, explain the operation of a bistable multivibrator.
19. List the applications of bistable multivibrator.
20. Calculate the component values of a symmetrical binary BJT circuit with the following values: $h_{fe(min)} = 20$, $V_{CC} = 12$ V, $V_{BB} = 12$ V. Assume $I_{C(sat)} = 5.35$ mA.
21. Which multivibrator would be useful for each of the following purpose?
    (*a*) harmonic generation of square waves
    (*b*) time-delay unit
    (*c*) memory device
    (*d*) frequency division

(*e*) adjustable pulse width generator

(*f*) reference clock to synchronize timings hi digital systems.

22. How a sine wave is converted into a square wave?
23. How a Schmitt trigger is different from a multivibrator?
24. Explain "Lower" and "Upper" Threshold voltages in a Schmitt trigger.
25. What is meant by Hysteresis voltage in a Schmitt trigger?
26. Under what condition would a Schmitt trigger circuit operate as an amplifier?
27. Giving circuit details, explain the operation of a Schmitt trigger using transistors. Show how Schmitt trigger can be used for wave-shaping purposes.
28. List the applications of Schmitt trigger.
29. Draw any one of a Schmitt trigger circuit and explain with waveforms.
30. A Schmitt trigger circuit has the following components. The supply is V_{CC} = 15 V. Analyse the circuit to determine the UTP and LTP.
31. Design a Schmitt trigger circuit with the following characteristics V_{CC} = 15 V, UTP = 5 V, I_{C2} = 5 mA and LTP = 3 V. Two silicon transistors with $h_{fe(min)}$ = 20 and one 15 V d.c. source are given. Assume all junction voltages are equal to zero. I_2 = 10% of I_{C2}; I_{CBO} = 0. Determine R_1, R_2, R_E, R_{C1}, R_{C2} and R_1.
32. How are symmetrical output waveforms produced in multivibrators?
33. Describe the theory behind Schmitt trigger using IC74132.
34. Explain the basic principle of generating square waves in clock generator using NAND gates and a quartz crystal.

Applications of Digital Circuits

15.1 INTRODUCTION

Digital ICs are indispensable in the realization of electronic systems used in communications, computers, controls and instrumentation. For the design of electronic circuits, digital ICs are mainly used because they have many important characteristics such as logic flexibility, high operating speed, availability of complex function, noise immunity, low cost, compactness and wide operating temperature range. In this chapter, the design of various digital systems such as frequency counter, time meter, bar graph display system, multiplexed display system, dot matrix display system, digital voltmeter and digital multimeter using digital ICs have been discussed.

15.2 FREQUENCY COUNTER

Frequency counter (Frequency meter) is a digital instrument that measures the frequency of any periodic waveform. Frequency counter has high sensitivity, resolution, accuracy and stability. It can perform a wide range of functions, such as frequency measurement, period measurement, ratio of frequencies measurement, scaling, time interval measurement etc.

The block diagram of a basic frequency counter is shown in Fig. 15.1. Frequency measurement is performed by totalling the number of input cycles for a known period of time so that the resulting total count is proportional to the unknown input frequency. The time reference (gate ENABLE signal having a known period '*t*')

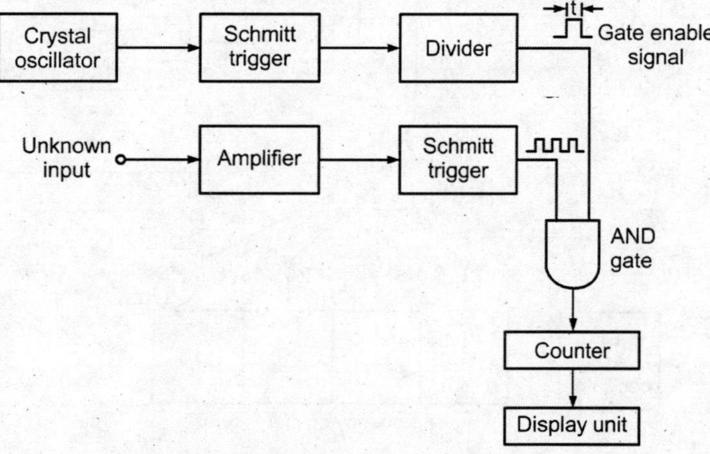

Fig. 15.1 Block diagram of frequency counter

is obtained from a precise and highly stable quartz crystal oscillator, a wave shaper (Schmitt trigger) and a set of dividers.

The input signal is applied to a Schmitt trigger circuit so that it will provide uniform pulses. This pulse train is given to the counter through an AND gate and this gate is controlled by the time reference. The number of pulses totalled in the counter for the selected time period '*t*' represents the input signal frequency. Finally, it is displayed on the numeric display device at the output of the counter. The frequency is displayed for a finite time period until a new sample is taken. The display time of the frequency measurement is determined by the sample rate control. The resetting of the counter and the next measurement cycle are also initiated by using the sample rate control.

For example, if the time reference is one second and the input signal is a 800 Hz square wave, at the end of one second, the counter would have counted up to 800 which is exactly the frequency of the input signal.

15.2.1 5-digit Frequency Counter

The logic diagram of 5-digit frequency counter is shown in Fig. 15.2. Here, a 1 kHz oscillator signal is divided into 100 Hz, 10 Hz, 1 Hz and 0.1 Hz time base signals using a 4-stage frequency divider. Any of these time base signals is selected with the help of rotary selector switch and connected to the clock input of a *JK* flip-flop. The *JK* flip-flop divides the frequency of the time base signal by a factor 2 so that the ON time of ENABLE signal is equal to period of the respective time base signal. For example, if the selected time base signal is 1 Hz, then the frequency of the ENABLE signal is 0.5 Hz i.e., its period is 2 s and therefore its ON time is 1 s.

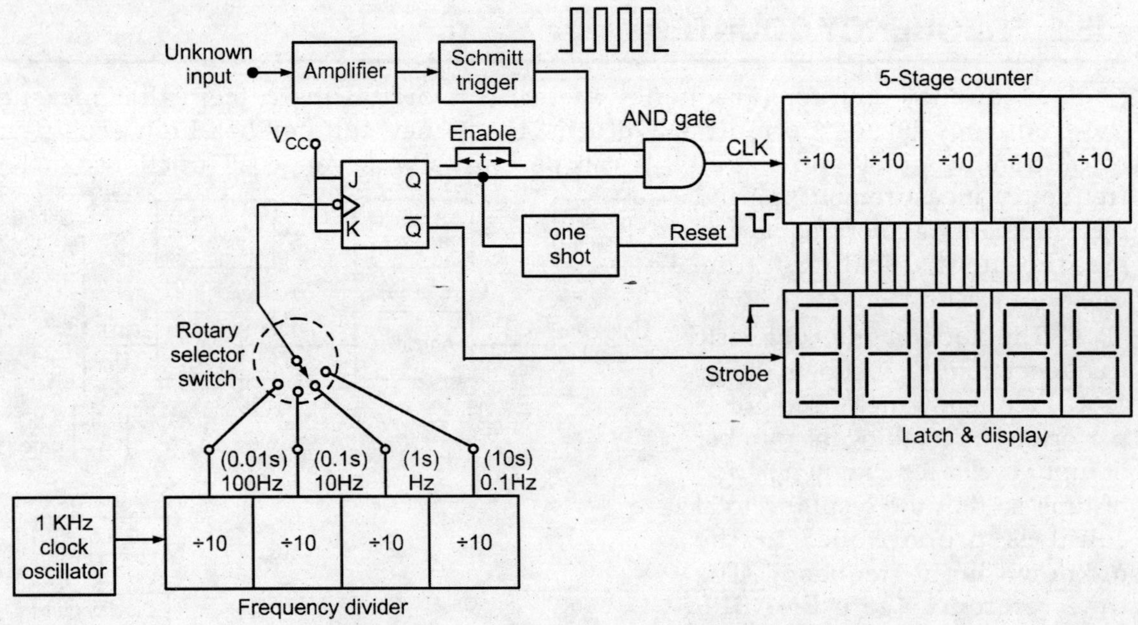

Fig. 15.2 Logic diagram of 5-digit frequency counter

When the ENABLE signal is HIGH, the unknown input signal is passed through the AND gate and acts as CLOCK input of the 5-stage counter and thereby the counter advances its value. When the ENABLE signal becomes LOW, he counter ceases its operation. Simultaneously, a low to high transition at $\bar{Q}$ strobes or latches the counter contents in the displays latches. Also, the high to low transition at Q output triggers the one-shot mono stable multivibrator and its output resets the content of the 5-stage counter to 00000 and it gets ready for another cycle of frequency measurement. Now, the value displayed in the 5-digit 7-segment display is proportional to the frequency of the unknown input signal.

Consider the frequency measurement of unknown input signal with different known time base signals. Let us assume that the frequency of unknown input is 1000 Hz (i.e., period = 1 ms). If the time base signal is 100 Hz (i.e., t = 0.01 s = 10 ms), then 10 number of 1 ms pulses of known input are passed through the AND gate and the counter value is 10. Now, in order to get the actual frequency value of 1000 Hz, a multiplication factor of 100 (× 100) has to be used. If 10 Hz time base signal (i.e., t = 0.1 s = 100 ms) is selected, then 100 pulses of known input are passed through the AND gate and the counter value is 100. Now, a multiplication factor of 10 (× 10) has to be used to get the actual frequency value. If 1 Hz time base signal (i.e. 1 = 1 s =1000 ms) is selected, then 1000 pulses of known input are passed and the counter value will be 1000 which is exactly equal to the actual frequency value. The required multiplication factor in this case is 1 (× 1). If 0.1 Hz time base signal is selected (i.e., t = 10 s =10,000 ms), then 10,000 pulses will be passed and the displayed value is 10,000. Now, to get the actual frequency value of 1000 Hz, either a multiplication factor of 0.1 or a division factor of 10 (÷ 10) has to be used or a decimal point should be displayed between Ten and Unit digits of display as 1000.0.

15.3 TIME METER

With some modifications, the frequency counter can be converted into a Time meter (Time counter). Time meter is an instrument to measure time. The block diagram of a time meter is shown in Fig. 15.3. It shows the construction of an instrument which is used to measure the period of any periodic waveform.

The unknown signal is passed through an amplifier to produce a periodic waveform that is compatible with TTL circuits and is then applied to a JK flip-flop. The output of this flip-flop is used as the Enable gate signal since it is high for a time 't' that is exactly equal to the time period of the unknown input signal.

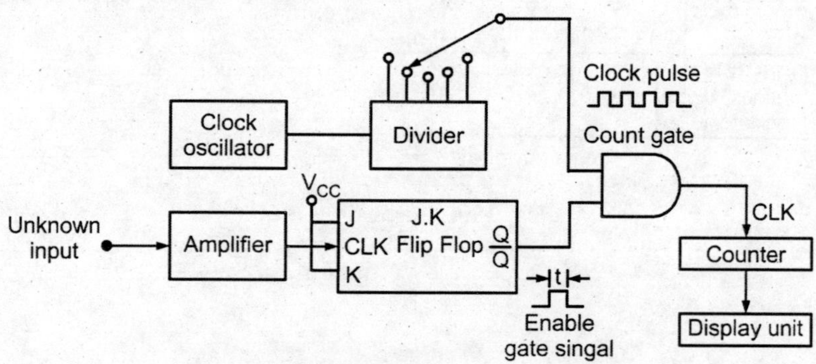

Fig. 15.3 Block diagram of time meter

The oscillator and divider provide a series of pulses that are passed through the count gate and serve as the clock for the counter. The contents of the counter and display unit will then be proportional to the time period of the unknown input signal.

For example, let the unknown input signal be at a frequency of 10 kHz and the clock pulses from the divider are 0.1 μs in width at the rate of one pulse per μs. The JK flip-flop produces an output pulse train with ON time '*t*' which is equal to $1/f$. Therefore, $t = \frac{1}{10 \times 10^3} = 100\ \mu s$. Hence the counter and display will read 100. This means 100 μs, since 100 of these 0.1 μs pulses will pass through the count gate during 100 us that ENABLE gate signal is high.

15.3.1 4-digit Time Meter

The logic diagram of 4-digit Time/Period Meter is shown in Fig. 15.4. Here, a 1 MHz clock oscillator signal (i.e., 1 μs signal) is divided into 100 kHz (10 μs), 10 kHz (100 μs) and 1 kHz (1 ms) time base signals using a 4-stage frequency divider. Any of the time base signals can be connected to the clock input of AND gate using a rotary selector switch. The unknown input signal is divided in frequency by a factor 2 using a JK flip-flop so that the ON time of the ENABLE signal is equal to the period *T* of the unknown input signal.

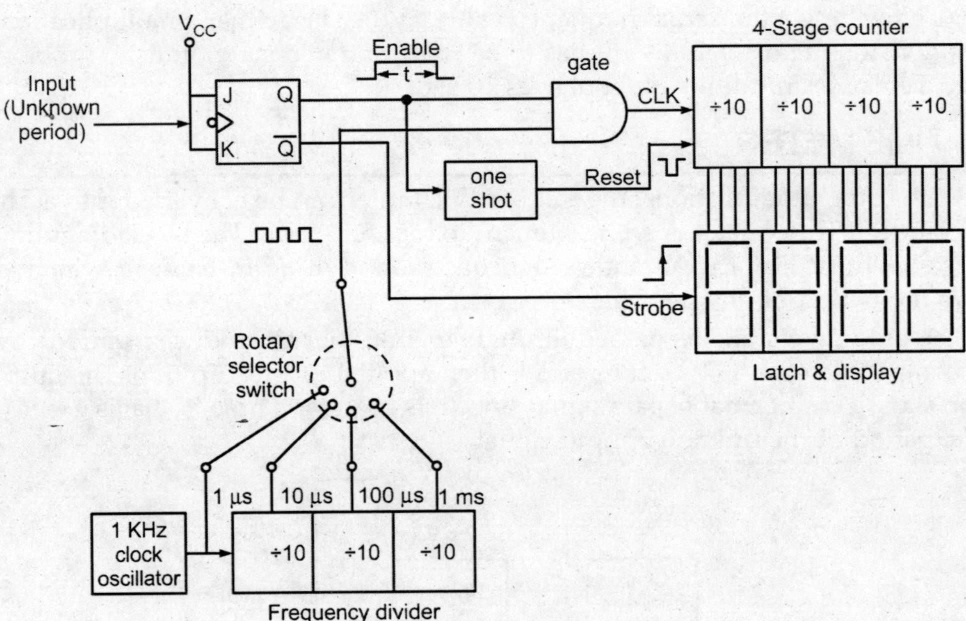

Fig. 15.4 4-digit time meter

When the ENABLE signal is HIGH, the clock pulses are passed through the AND gate and acts as clock input for 4-stage counter and the counter advances its value. When the ENABLE signal becomes LOW, the counter ceases its operation. Simultaneously, the LOW to HIGH transition at $\bar{Q}$ strobes the counter value into the latches. Also, the HIGH to LOW transition at ENABLE (i.e., Q) signal triggers the one-shot and it resets the counter value to 0000, so that it gets ready for next cycle of period measurement. Now, the value displayed in the 4-digit 7-segment display is proportional to the period of unknown input signal.

The above 4-digit Time/Period measurement unit has a full-scale value 9999. When the rotary selector switch is connected with 1 μs time base signal, the full-scale value is 9999 × 10 μs; time base signal the full-scale value is 9999 × 10 μs; for the 0.1 ms time base signal, the full-scale value is 9999 × 0.1 ms = 999.9 ms and for 1 ms time base signal, the full-scale value is 9999 × 1 ms.

15.4 BAR GRAPH DISPLAY SYSTEM

Bar graph displays are used in electronic instruments and appliances as indicators of voltage, current or any other parameter being measured/indicated. In bar graph display system, a closely packed array of LEDs which are driven independently to emit light is used. The length of the illuminated array corresponds to the strength of the input parameter being measured.

The block diagram of a typical bar graph display system to drive a set of 10 LEDs is shown in Fig. 15.5. No LED is illuminated when the input voltage V_{in} is less than 0.4 V ($V_{in} < 0.4$ V). LED D_1 is illuminated when the input voltage is greater than 0.4 V and less than 0.8 V (i.e., 0.4 V < V_{in} < 0.8 V); Two LEDs D_1 and D_2 are illuminated when 0.8 V < V_{in} < 1.2 V etc., and all LEDs are turned ON when the input voltage is greater than 4 V (i.e., V_{in} > 4 V).

As shown in Fig. 15.5, a resistive voltage divider with 5 V as the reference voltage is connected using ten 1 kΩ resistors and one 2.5 kΩ resistor so that the voltage drop across each 1 kΩ resistor is 0.4 V. Here, the op-amps are connected as comparators and based upon the comparator inputs, the LEDs will be driven. For example, if the input voltage (V_{in}) is less than 0.4 V, all the comparator outputs are in LOW state and thereby no LEDs will be turned ON. When the input voltage V_{in} is greater than 0.4 V and less than 0.8 V, then output of the comparator 1 (C_1) is in HIGH state while the outputs of the other comparators are in LOW state. This results in the illumination of LED D_1. Similarly, for 0.8 V < V_{in} < 1.2 V, the output of C_1 and C_2 are in HIGH state while the other comparator outputs are in LOW state. This leads to the illumination of two LEDs D_1 and D_2. Similarly, when V_{in} > 4 V, all comparator outputs are in HIGH state and thereby all LEDs are turned ON. Thus, the length of illuminated array will vary in proportion to the strength of input parameter.

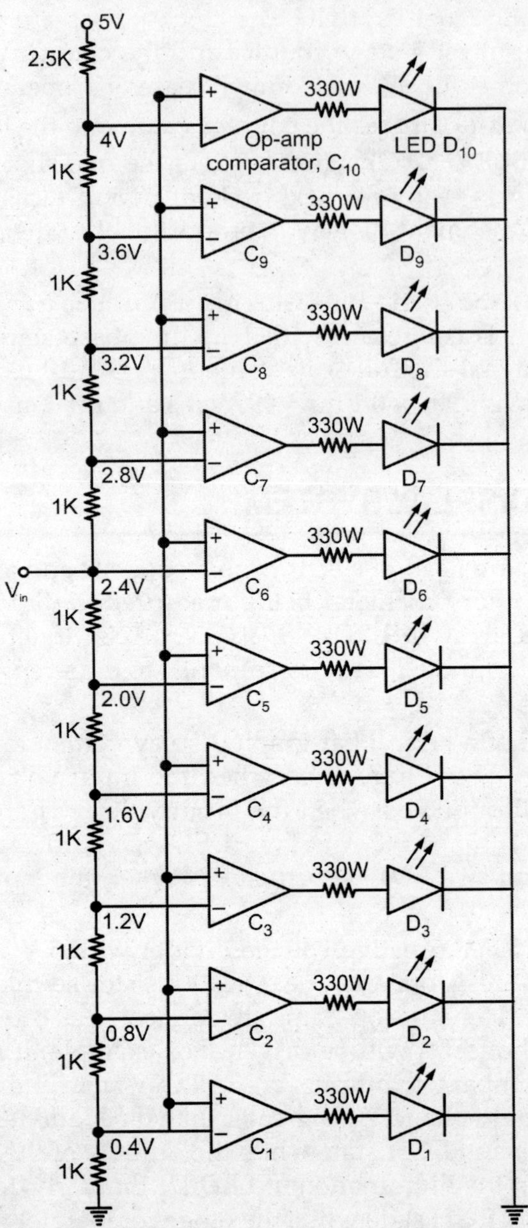

Fig. 15.5 Block diagram of a typical bar graph display system

15.5 MULTIPLEXED DISPLAY SYSTEM

Digital instruments such as frequency counter, function generator and digital voltmeter invariably have 7-segment displays to display their corresponding output values. Normally, each 7-segment display requires a separate BCD to seven segment decoder/driver in non-multiplexed display system. (Fig. 15.6(a) and Fig. 15.7(a)).

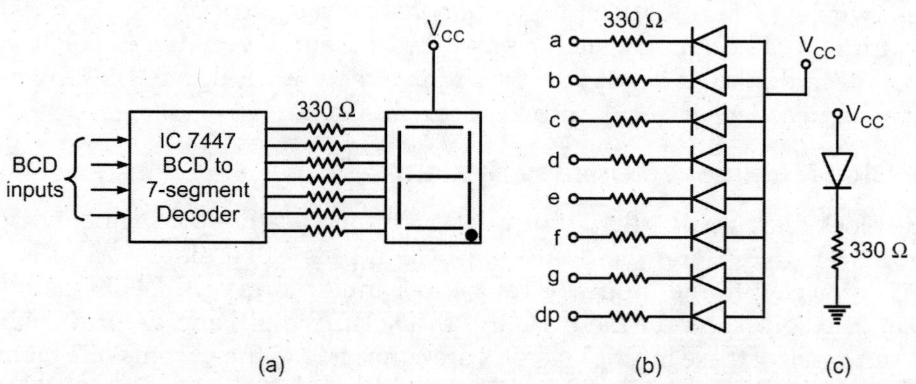

Fig. 15.6 (*a*) Single digit display system (Common anode), (*b*) Equivalent circuit of 7-segment display and (*c*) Equivalent circuit of an illuminated segment

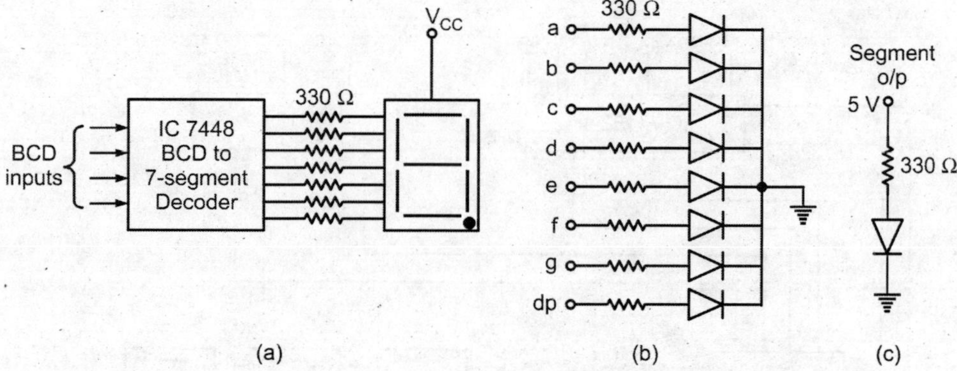

Fig. 15.7 (*a*) Single digit display system (Common cathode), (*b*) Equivalent circuit of 7-segment display and (*c*) Equivalent circuit of an illuminated segment

As shown in Fig. 15.6(*b*), in common anode displays, all the anodes of LEDs are connected together to a power supply (V_{CC}). Therefore, to illuminate an LED segment, the 7-segment decoder for common anode display (IC 7447) will output a LOW state (i.e., 0 V = GND) at its segment output. Similarly, in the case of common cathode displays, as shown in Fig. 15.7(*b*), all cathodes of LEDs are connected together to ground (GND). Therefore, to illuminate a segment in common cathode display, the 7-segment decoder will output a HIGH state (i.e., 5 V = V_{CC}) at its segment output. Based on these discussions, the equivalent circuit of an illuminated segment for common anode and common cathode displays are shown in Fig. 15.6(*c*) and 15.7(*c*) respectively. If V_{CC} is 5 V and there is a 0.8 V drop across an illuminated segment, then a current of 12.6 mA [(5 V-0.8 V)/330 Ω = 12.6 mA] is required to illuminate a segment. The maximum current requirement occurs when the number 8 with decimal point is displayed. Under this condition, a current of 101 mA (8 × 12.6 mA) is required. The 7-segment decoder (IC 7447/7448) requires a current of about 64 mA. Therefore, a total current of 165 mA is required for a single digit display system.

If V number of 7-segment displays are connected in a digital system in a non-multiplexed manner, then it consumes a total current of n × 165 mA. For example, a digital instrument with six 7-segment displays will consume a current of 6 × 165 = 990 mA ≈ 1 A. The consumption of current by such a small instrument is very large. This can be greatly reduced in the multiplexed display system. Moreover, with additional control circuits, only one 7-segment decoder/driver is sufficient for all displays.

15.5.1 4-Digit Multiplexed Display System

The block diagram of 4-digit multiplexed common cathode 7-segment display system is shown in Fig. 15.8. Here, the dual 4-bit BCD inputs to be displayed in a multiplexed manner is connected to the inputs of four 4 to 1 multiplexers (IC 74153) as: 1A 1B 1C 1D (MSD); 2A 2B 2C 2D (second MSD); 3A 3B 3C 3D (Third digit) and 4A 4B 4C 4D (LSD). The select lines *A* and *B* of these multiplexer ICs are connected to the outputs of 2-bit multiplexing counter, constructed using JK flip-flops. The output of multiplexing counter also acts as the input for a 2 to 4 decoder IC 74155 whose outputs are digit select inputs (DS_0-DS_3) of the four 7-segment displays.

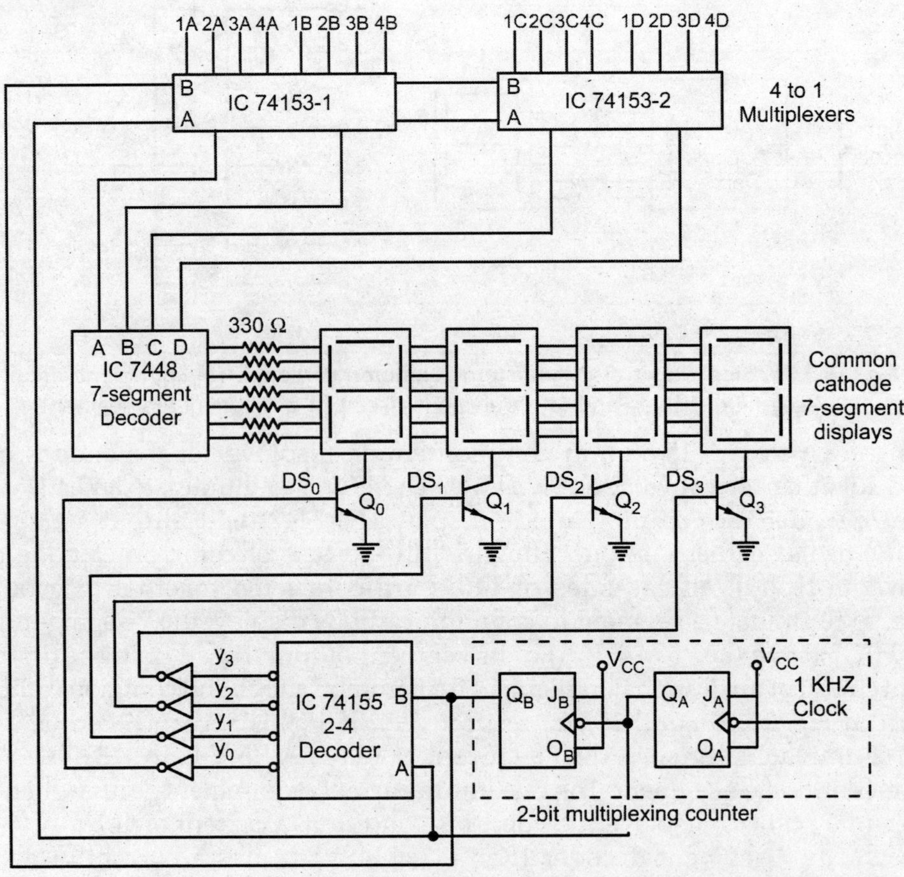

Fig. 15.8 4-digit multiplexed common cathode 7-segment display system

Operation

The multiplexing counter counts from 00-01-10-11 in a cyclic manner. When it is 00 (i.e., BA = 00), the first input in each of the dual 4 to 1 multiplexers (i.e, 1A, 1B, 1C and 1D) is selected and available at their respective outputs and at the inputs of *BCD* to 7-segment decoder IC 7448. Therefore, when. *BA* = 00, the 7-segment code of the most significant *BCD* digit inputs are available at the 7-segment display inputs. Also, when *BA* = 00, the Y_0 output of the 2 to 4 decoder IC 74155 is activated and thereby the digit select transistor Q_0 is in ON state while the other transistors Q_1, Q_2 and Q_3 are in the OFF condition. This results in the display of most significant decimal digit in the first display while other displays are deactivated. This display continues until *BA* = 00 and this depends on the clock frequency of multiplexing counter. Since it is 1 kHz, *BA* = 00 for a time period of exactly 1 ms.

When *BA* becomes 01, the second input in each of the dual 4 to 1 multiplexers (i.e, 2*A*, 2*B*, 2*C* and 2*D*) are selected and available at the inputs of IC 7448 and thereby 7-segment code of second MSB is available at the inputs of 7-segment displays. Simultaneously, when *BA* = 01, Y_1 of 2 to 4 decoder is activated and second display is selected using a digit select transistor Q_1 while other displays are deactivated. Therefore, second decimal digit is displayed in the second display for a period of next 1 ms (i.e., 1 ms to 2 ms). Similarly, when *BA* = 10, the third decimal digit is displayed in the third 7-segment display for the next 1 ms (i.e., 2 ms to 3 ms). When *BA* = 11, the LSD is displayed in the fourth 7-segment display for the next 1 ms (i.e., 3 ms to 4 ms). Again the multiplexing counter content becomes 00 and hence, MSD will be displayed in the first 7-segment display from fourth ms to fifth ms. Thus, the cycle continues for every 4 ms.

Similarly, multiplexed display system with more than 4 decimal digits can be constructed by using more bit multiplexing counters and replacing 4 to 1 multiplexer with a mutliplexer that has more inputs. Also, with slight modification in the circuit, the common anode 7-segment display can be used.

15.5.2 Advantages of Multiplexed Display System

1. Only one 7-segment decoder/driver is required irrespective of the number of displays.
2. Since only one 7-segment display is activated at any time, the current requirement is always for a single display only and independent of number of displays.

15.5.3 IC 74925 - 4-Digit Counter with Multiplexed 7-segment Display Decoder/Driver

The IC 74925 is a LSI display having 4-digit counter with multiplexed 7-segment display decoder/driver in a single package. The internal logic diagram of IC 74925 is shown in Fig. 15.9.

As shown in Fig. 15.9, a 4-digit counter is available inside the chip. A positive pulse on the RESET input will reset the 4-digit counter and for every negative clock transition, the counter advances its value. A negative pulse on the ENABLE latch will latch the counter values into the four 4-bit latches. The four numbers stored in the latches are multiplexed and decoded as 7-segment outputs at *a, b, c, d, e, f* and *g*. Hence, only four 7-segment displays and 4-digit selectors i.e., four NPN transistors are required as external components to construct 4-digit counter with multiplexed display, as shown in Fig. 15.10.

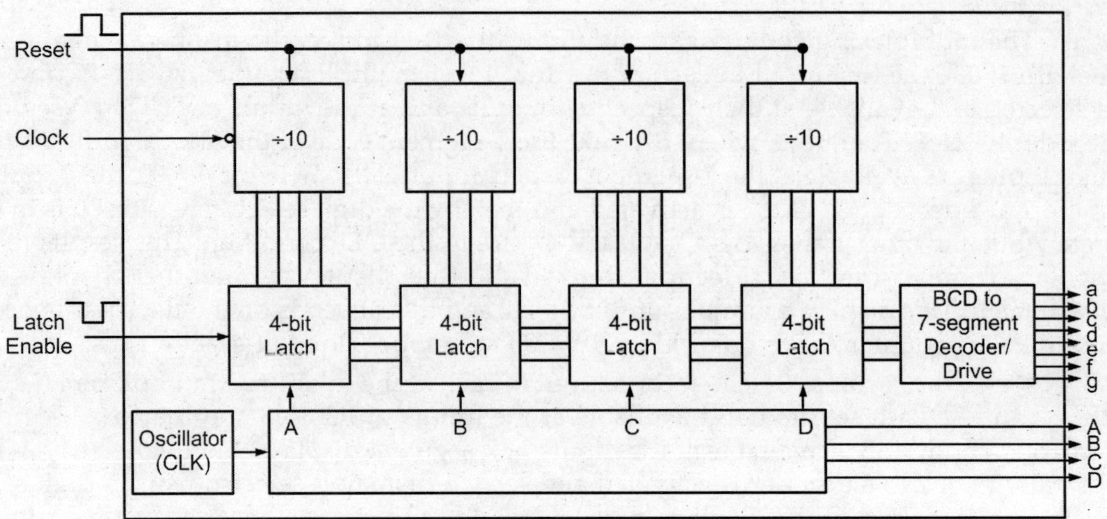

Fig. 15.9 Internal logic diagram of IC 74925

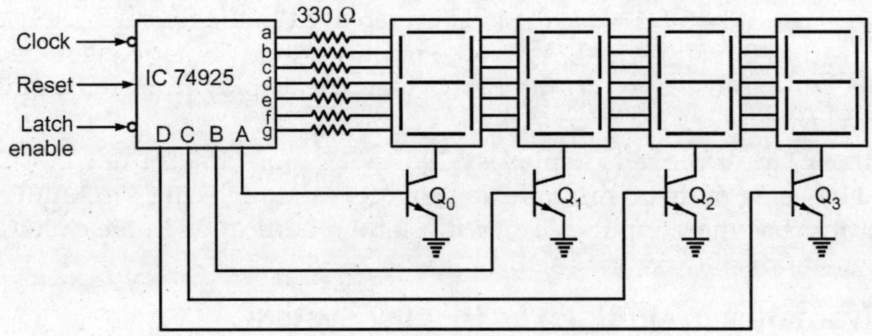

Fig. 15.10 4-digit converter with multiplexed display

15.6 DOT MATRIX DISPLAY SYSTEM

Dot matrix display has a superior character font and useful for alphanumeric displays. In the dot matrix displays, LEDs are organized in rows and columns. To minimize the hardware requirement of dot matrix display system, dynamic or strobed display is used. This can be done in two ways: (*i*) horizontal strobing (one column at a time) and (*ii*) vertical strobing (one row at a time). Either of them can be used for refreshing dot matrix display.

The block diagram of a 2-digit 5 × 7 dot matrix display system using horizontal strobing is shown in Fig. 15.11. Here, the two digit alphabets or numerals to be displayed in the dot matrix display is given at the inputs of IC 74157 which is actually a quad 2 to 1 multiplexer arranged as a 4-bit or nibble multiplexer. Based on the select signal obtained from the output of a *JK* flip-flop (i.e., a 1-bit counter), either 4-bits 1A 2A 3A 4A or 1B 2B 3B 4B will be available at the outputs of the multiplexer and act as address inputs of a PROM/PLA. The 4-bit alphanumeric character, corresponding to each digit, to be displayed must be converted into the dot selection code (i.e., dot matrix display data) to activate LED dots, one column at a time in five steps for horizontal strobing. Since 5 × 7 dot matrix LED display with 5 columns and 7 rows is used, the code converter should have 7 output lines and it has to provide current sinking path to the chosen LEDs. PROM/PLA is used to convert alphanumeric character into dot matrix display code. To display the data from PROM/PLA, one column at a time has to be selected using 10 AND gates as shown in the Fig. 15.11. The outputs of the *JK* flip-flop along with a 5-bit ring counter will have the inputs for the column selector gates. For example, when $Q = 0$ and ring counter output is 00001 (i.e., LSB = 1), then column 1 in digit 1 is selected. Similarly, for $Q = 0$ and for different ring counter outputs, other columns are selected. When $Q = 1$ and for different ring counter outputs, the columns of second digit will be selected. Thus, the dot matrix display data from PROM is displayed in one column at a time in digit 1 or digit 2.

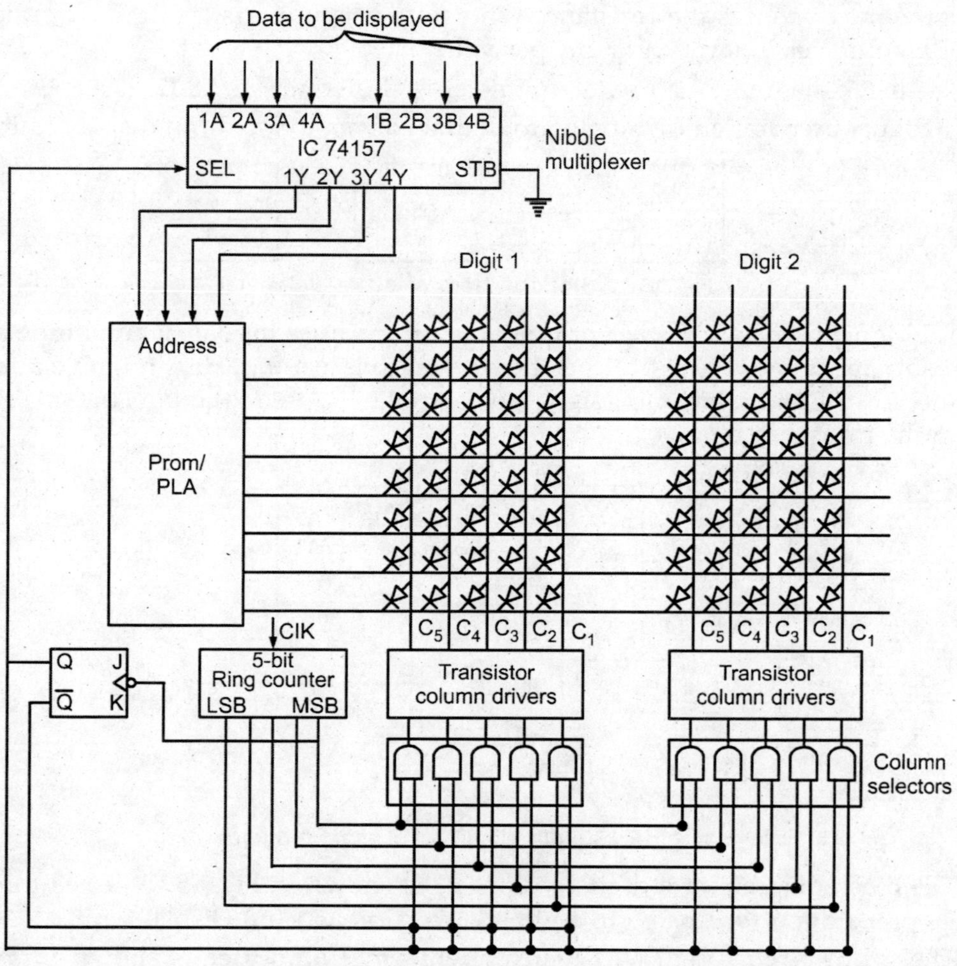

Fig. 15.11 Block diagram of 2-digit 5 × 7 dot matrix display system

15.7 DIGITAL VOLTMETER

Digital voltmeter (DVM) is used to measure voltage between two points. It displays dc or ac voltage as discrete numerals. It is a useful laboratory instrument for several applications. It is also a useful building block of digital instrumentation systems. The utility of a DVM can be easily extended for multiple functions, such as a Digital Multimeter (DMM), by the addition of simple auxiliary hardware. The DVM is often used in data processing system.

Do You Know?	DVM is a versatile and accurate instrument that is employed in many laboratory measurement applications.

An ideal voltmeter has an infinitely high input resistance so that it does not draw any current from the circuit. Consider a meter which has a low input resistance of 1000 ohms. It cannot give an accurate value of the voltage across a resistance of the same magnitude because the meter shunts the resistance. Therefore it is important to measure the loading effect of a voltmeter in terms of ohms per volt.

The block diagram of a digital voltmeter is shown in Fig. 15.12. It has three stages: (*i*) Signal preparation (*ii*) Analog to digital conversion and (*iii*) display unit.

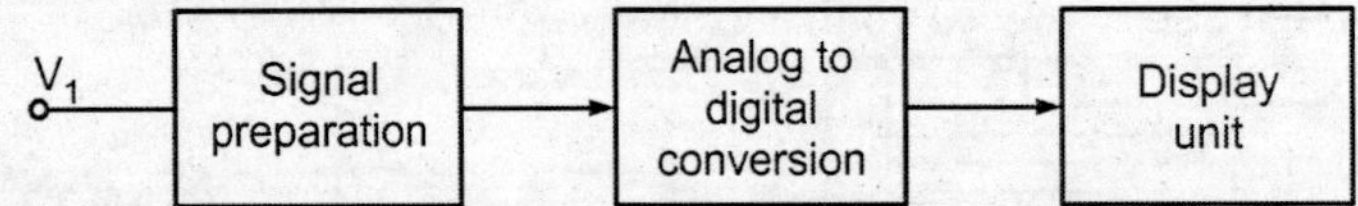

Fig. 15.12 Block diagram of digital voltmeter

The signal preparation stage or input circuit modifies the signal amplitude according to the requirement and it also protects the source from loading. Here resistive attenuator is used to decrease the large incoming signal and amplifier is, used to amplify the small incoming signal to the measurable range.

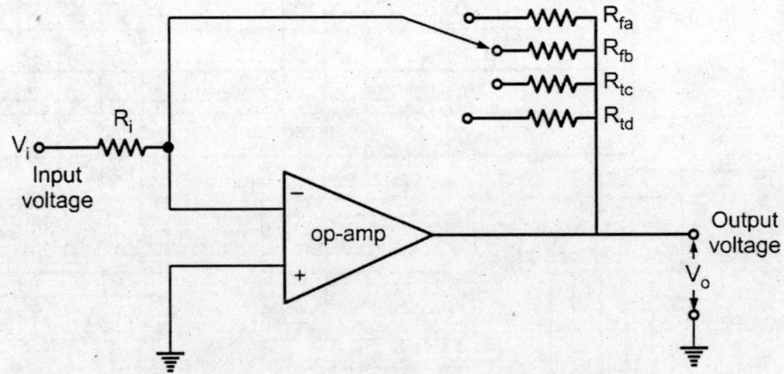

Fig. 15.13 Input circuit for digital voltmeter

The block diagram of a digital voltmeter is shown in Fig. 15.12. It has three stages: (*i*) Signal preparation (*ii*) Analog to digital conversion and (*iii*) Display unit.

The input circuit for DVM using operational amplifier is shown in Fig. 15.13. Operational amplifier is a multistage integrated circuit. Amplification is controlled with negative feedback since amplifier gain '*A*' is proportional to the ratio of the feed back and input resistors.

$$A = \frac{R_f}{R_i}$$

In the next two stages, the analog input signals are typically converted into digital signals in the form of binary or binary coded decimal (BCD) data and suitably displayed in the display unit. The block diagram of a dual slope A/D converter with display unit is shown in Fig. 15.14.

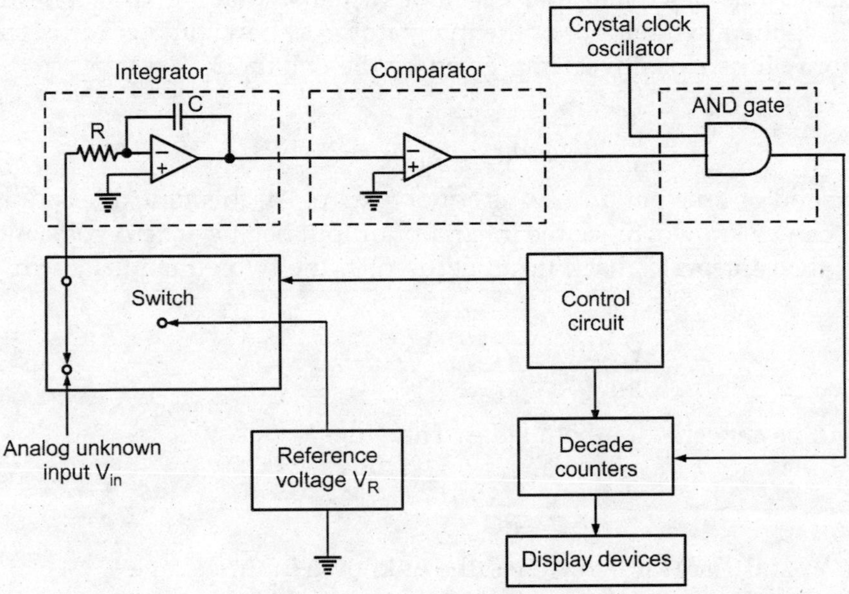

Fig. 15.14 Block diagram of dual slope A/D converter with display

The unknown voltage V_{in} is given to the input of the integrator through the selection switch for a known time period T. The output from the integrator is given by the equation

$$V_C = \frac{V_{in}}{RC} T$$

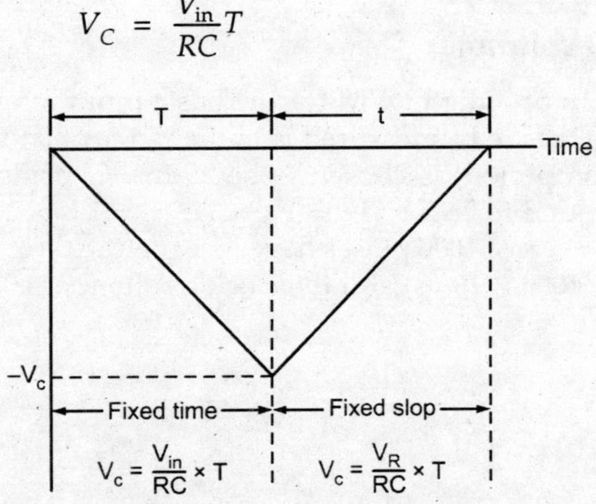

Fig. 15.15 Output of integrator

Here R and C are the resistance and capacitance values in the integrator. As the input voltage V_{in} is positive, the integrator output will be the negative ramp as shown in Fig. 15.15. The output from the integrator is compared with zero volts reference in the comparator. The output from the comparator will be a positive voltage. For the entire time period T, the AND gate is opened and during this period, the pulses from the crystal clock oscillator is counted in the counter.

At the end of the known time period T, the counters are reset by the control circuit and at the same time it activates the selection switch so that the known negative reference voltage V_R is applied to the input of the integrator. As shown in Fig. 15.15, the output from the integrator will be a positive ramp given by the equation

$$V_C = \frac{V_R}{RC}t$$

At the end of time period t, V_C is equal to zero. All this time, the counter is counting and hence t can be known. Since the integrator output begins at zero volts, integrates down to $-V_C$ and then integrates back up to zero volts, the two equations given for V_C can be equated as

$$\frac{V_{in}}{RC}T = \frac{V_R}{RC}t$$

RC will be cancelled on both sides. Therefore

$$V_{in} = \frac{V_R}{RC}t$$

Since V_R and T are known values, the unknown input voltage V_{in} which is the content of the counter is proportional to the variable time period t.

A complete DVM on a single chip (IC) is available. They include A/D conversion circuitry and the necessary timing, counting and display circuitry. Examples are the low power 3 1/2-digit ICL7136 and the 4 1/2-digit ICL7129; both use LCD7-segment displays and run from a single 9-volt battery.

15.7.1 3-digit Digital Voltmeter

The block diagram of 3 digit DVM for the basic input range of V_{in} = IV is shown in Fig. 15.16. The input voltage to be measured is in the range of 0 to IV. The first time interval (i.e., the first integration period) is chosen to be 20 ms. Consider t_{max} = 10 ms. From the above equation, V_R is found to be 2 V ($V_R = V_{in} \times T/t_{max}$ = 1 V × 20 ms/10 ms = 2 V). For IV input, during this t_{max} = 10 ms, 1000 pulses have to be counted for a 3-digit display. It means that t_{clk} = 10 μs (10 ms/1000 = 10 μs) and the clock frequency is 0.1 MHz.

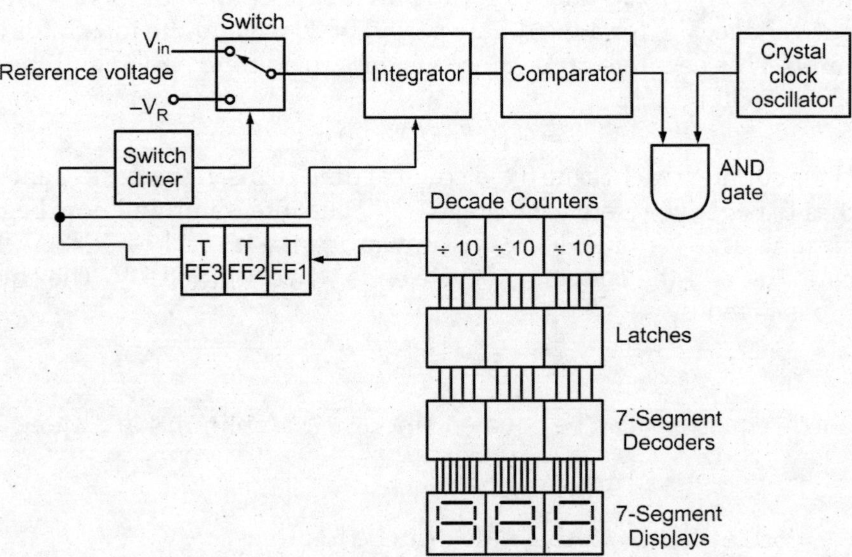

Fig. 15.16 Block diagram of 3-digit digital voltmeter

Therefore, a maximum of 30 ms (i.e., $T + t_{max}$) is necessary for each measurement cycle. For the above parameters *i.e.*, f_{clk} = 0.1 MHz; T = 20 ms; t_{max} = 10 ms, the waveforms at the outputs of clock oscillator, Integrator, Comparator, Gate and Trigger flip-flop outputs are shown in Fig. 15.17.

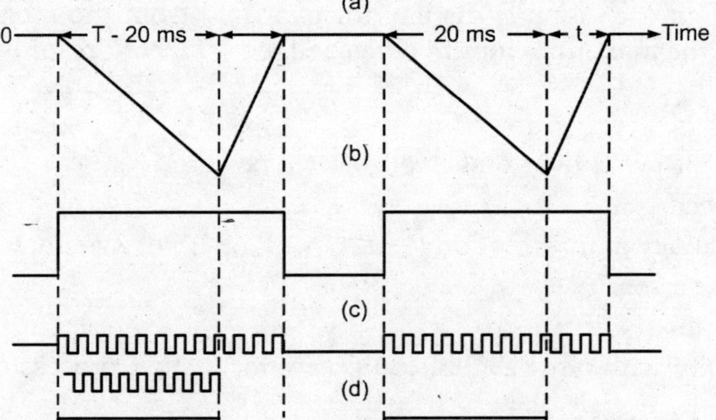

Fig. 15.17 Waveforms of 3-digit digital voltmeter output of (*a*) Crystal clock, (*b*) Integrator, (*c*) Comparator, (*d*) AND gate and (*e*) Trigger flip-flop

As shown in Fig. 15.16 and 15.17, the switch is connected first with an unknown positive input voltage, V_{in}, for a fixed time period of 20 ms and the integrator output will be a negative ramp. This negative ramp proceeds for a fixed time period and the voltage at the end of this 20 ms will be

$$V_{Int} = -(V_{in}/RC) \times 20 \text{ ms}$$

When the counter reaches the fixed count at time 20 ms, it is reset to 0 and the integrator input switches to negative reference voltage $-V_R$. The integrator will now begin to generate a ramp beginning at $-V_{Int}$ and increases upward until it reaches 0 V. All this time, the counter is counting and the conversion cycle ends when $V_{Int} = 0.0$ V. The equation for this positive ramp is given by

$$V_{Int} = \frac{V_R}{RC}t$$

The above equations can be equated since their amplitudes are equal.

$$\frac{V_{in}}{RC} \times 20 \text{ ms} = \frac{V_R}{RC}t$$

For $V_{in} = 650$ mV, the variable time t can be found as

$$t = \frac{V_{in}}{V_R} \times 20 \text{ ms} = \frac{0.65}{2} \times 20 \text{ ms} = 6.5 \text{ ms}$$

Also, here $t_{ctk} = 10$ μs. During the above variable time of 6.5 ms (i.e., 6500 μs), 650 (6500 μs/10 μs) clock pulses are passed through the AND gate and the counter value of 650 is displayed in the 3-digit display. The unit is mV and the full-scale output voltage that can be measured is + 999 mV. If the input voltage exceeds this value, over range indication can be provided. The above 3-digit DVM can be extended as $3\frac{1}{2}$ digit DVM by providing additional 1/2 digit (i.e., 0 or 1) display for overflow from the most significant decade counter. For this, the maximum input voltage will be + 1.999 V, resulting in $t_{max} = 19.99$ ms.

15.7.2 Merits of DVM

The merits of DVM over other voltmeter types are:

- Greater speed
- Higher accuracy and resolution (1 part in 10^3 or 10^4 or even better)
- No parallax error
- Reduced human error
- Compatibility with other digital equipment for further processing and recording.

15.7.3 Demerits of DVM

- While measuring under 1 mV for a.c. and under 100 μV for d.c. the noise and drift will cause the last digit or last but one digit to fluctuate a lot and hence it is difficult to estimate average reading under these conditions.
- Adjustment of the reading to a certain value, even to a minimum or maximum, is not convenient with a digital instrument.

15.8 DIGITAL MULTIMETER (DMM)

Multimeter is an instrument used to measure voltage, current and resistance. There are two types of multimeters, viz. analog and digital. Digital multimeter is commonly used in laboratory and workshop because of its high input resistance, greater accuracy, better resolution and easy readability. The DMM contains instruments for the measurements of voltage, current and resistance. The block diagram of a digital multimeter is shown in Fig. 15.18.

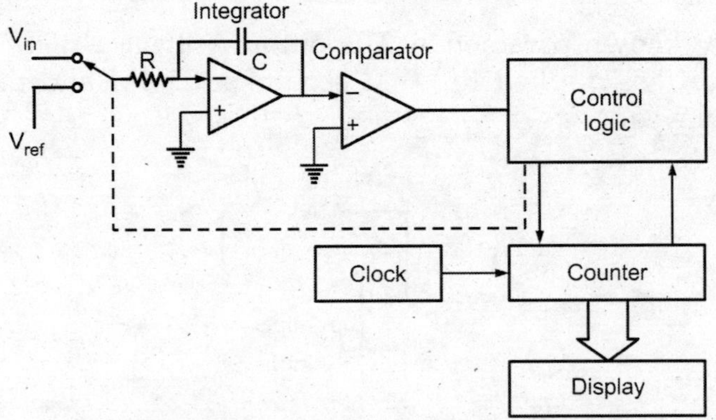

Fig. 15.18 Block diagram of a digital multimeter

Note: *All quantities other than dc voltage are first converted into an equivalent dc voltage by some device.*

15.8.1 Measurement of Voltage

The digital voltmeter (DVM) for measurement of voltage is discussed in section 15.7. The same principle is used in DMM for the measurement of voltage.

15.8.2 Measurement of Current

A series of current sensing resistors are used to measure either d.c. or a.c. current. The current to be measured is passed through one of the sensing resistors and the DMM digitizes the voltage developed across the resistor.

For example, referring to Fig. 15.19, the output voltage of a current to voltage converter is given by

$$V_O = -I_S R_f$$

where R_f is the known resistance. The output voltage V_0 which is proportional to the unknown source current I_S is applied to DVM section of DMM and the value of current I_S is displayed.

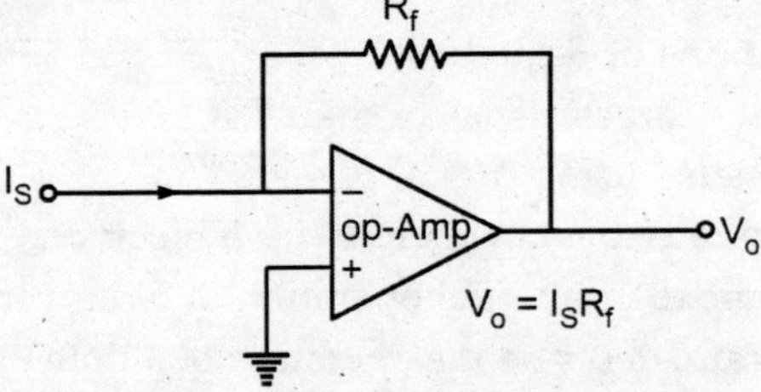

Fig. 15.19 Current to voltage converter

15.8.3 Measurement of Resistance

The DMM measures the resistance by applying a known current from an internal current source to the unknown resistance and then digitizing the resulting voltage developed.

For example, in Fig. 15.20, the output voltage of a scale changer is given by

$$V_O = \frac{-R_f}{R_i} V_i$$

where V_i and R_i are known parameters. The output voltage which is proportional to the unknown resistance R_f is applied to DVM section of DMM and the value of unknown resistance R_f is displayed.

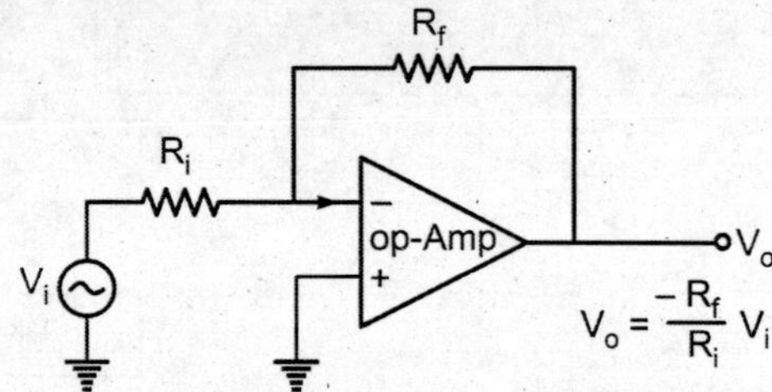

Fig. 15.20 Scale changer

Most of the DMMs are similar in terms of voltage, current and resistance measurements. They differ only in terms of accuracy, selection of ranges and band width. There are some DMMs having built-in capacitance measuring circuitry also. Most DMMs have protection from input overload by using circuit breakers, fuses, autoranging and diode clipper circuit. The display used can be either Liquid Crystal Display (LCD) or Light Emitting Diode (LED) display.

Applications

A DMM is typically used for measurement of voltage, current and resistance. It is also used to test whether the diode, transistor or SCR is good or faulty and to check circuit continuity.

For example, to check a diode, the resistance is measured in one direction and then in other direction. In the forward-biased direction, a low resistance is indicated and in the reverse-biased direction, a high resistance is indicated.

■ REVIEW QUESTIONS ■

1. Mention a few applications of digital circuits.
2. What are the important characteristics of digital ICs?
3. Name a few digital systems using digital ICs.
4. Explain the operation of a frequency meter with a block diagram.
5. Explain, with a neat logic diagram, the operation of 5-digit frequency counter.
6. With a neat block diagram, describe the operation of a time meter.

7. Explain, with a neat logic diagram, the operation of a 4-digit time meter.
8. Draw the block diagram of a typical bar graph display system and explain its operation.
9. Explain clearly the function of single digit multiplexed display system.
10. With necessary diagrams, explain clearly the function of four digit multiplexed display system.
11. What are the advantages of multiplexed display system over non-multiplexed display system?
12. Draw the internal logic diagram of IC74925 and explain its operation.
13. Explain the function of dot matrix display system with necessary diagrams.
14. Explain the basic principle of a digital voltmeter.
15. State the merits and demerits of a digital voltmeter over an analog voltmeter.
16. Explain the operation of a basic digital multimeter.

HDL for Digital Circuits

16.1 INTRODUCTION TO HARDWARE DESCRIPTION LANGUAGE (HDL)

VHDL is an acronym which stands for *VHSIC Hardware Description Language* and VHSIC stands for *Very High Speed Integrated Circuits.* The acronym itself captures the entire theme of the language and it describes the hardware in the same manner as does the schematic.

Do You Know?	A hardware description language is a language that describes the hardware of a digital system in its texual form. It can be used to represent logic diagrams, Boolean expressions and other more complex digital circuits. There are two applications of HDL processing: 1. Simulation and 2. Synthesis

VHDL is being used for documentation, verification, and synthesis of large digital designs. This key feature of the VHDL saves a lot of design effort, since the same VHDL code can theoretically achieve all three of these goals. In addition to being used for each of these purposes, VHDL can be used in three different approaches for describing the hardware. These three different approaches are the *dataflow, structural* and *behavioral* methods of hardware description. A mixture of these three methods can also be employed to arrive at an efficient design.

16.2 DATA FLOW DESCRIPTION

The first of the three approaches, namely, the *dataflow* description for design with VHDL is discussed below.

Building Blocks: To make the designs more modular and effective, a design is typically decomposed into several blocks. These blocks are then connected together to form a complete design. Using the schematic capture approach to design, this might be done with a block diagram editor. Every portion of a VHDL design is considered a block. A VHDL design may be completely described in a single block, or it may be decomposed in several blocks. Each block in VHDL is analogous to an off-the-shelf part and is called an *entity.* The entity describes the interface to that block and a separate part associated with the entity describes how that block operates. The interface description is similar to a pin description in a data book, specifying the inputs and outputs to the block. The description of the operation of the part is synonymous to a schematic for the block.

An example of an entity declaration in VHDL is given below.

```
entity andgate is                    - Declaration of new entity.
  port (A, B:in std_logic;           - Set of i/p and o/p pins declaration.
    Y:out std_logic);
end andgate;                         - End of entity.
```

The first line indicates a declaration of a new entity, whose name is *andgate*. The last line marks the end of the declaration. The lines in between, called *import* clause, describe the interface to the design. The port clause contains a list of interface declarations. Each interface declaration defines one or more signals that are inputs or outputs to the design.

Each interface declaration contains a list of names, a mode and a type. In the first interface declaration shown in the example above, two input signals, namely, *A* and *B* are defined. The list to the left of the colon contains the names of the signals, and to the right of the colon is the mode and type of the signals. The mode specifies whether this is an input (*in*), output (*out*), or both (*inout*). The type specifies the kind of values the signal can carry. The signals *A* and *B* are of mode *in* (inputs) and type *std_logic* (standard logic). The signals *Y* is declared as mode *out* (outputs) and of the type *std_logic*. Notice the particular use of the semicolon in the port clause. Each interface declaration is followed by a semicolon, except the last one, and the entire port clause has a semicolon at the end. Comments can be added at the end of a VHDL statement or as an individual line by itself preceded by the '–' symbol, as shown in the above example.

The second part of the description of the *andgate* design is a description of how the design operates. This is defined by the *architecture* declaration. The following is an example of an architecture declaration for the *andgate* entity.

```
architecture arch_and of andgate is      - Declaration of new architecture.
begin
Y<=A and B;                              - Logical AND operation
end;                                     - End of architecture.
```

The first line of the declaration indicates that this is the definition of a new architecture called *arch_and* and it belongs to the entity named *andgate*. So this architecture describes the operation of the *andgate* entity. The lines in between the *begin* and the *end* describe the operation of the AND gate.

In the above example, the architecture part describes the internal operation of the AND gate using the *dataflow* approach. The data flow approach indicates how the data flows from the inputs to the outputs. In VHDL, this is accomplished by the signal assignment statement. It is important to note that the signal assignment operator in VHDL specifies a relationship between signals, not a transfer of data as in programming languages. The example architecture cited above consists of only one signal assignment statement.

A signal assignment statement describes how the data flows from the signals on the right side of the <= operator to the signal on the left side. The signal assignment in the example means that the data coming from signals *A* and *B* flow through an AND gate to determine the value of the signal *Y*. The *and* represents a built-in component called an operator, because it operates on some data to produce new data. The right side of the <= operator is called an expression. The value of the expression is determined by evaluating the expression. Evaluation of the expression is performed by substituting the values of the signals in the expression and computing the result of each operator in the expression.

The following is another example of an entity declaration in VHDL for SR latch. The schematic diagram of the SR latch is shown in Fig. 16.1.

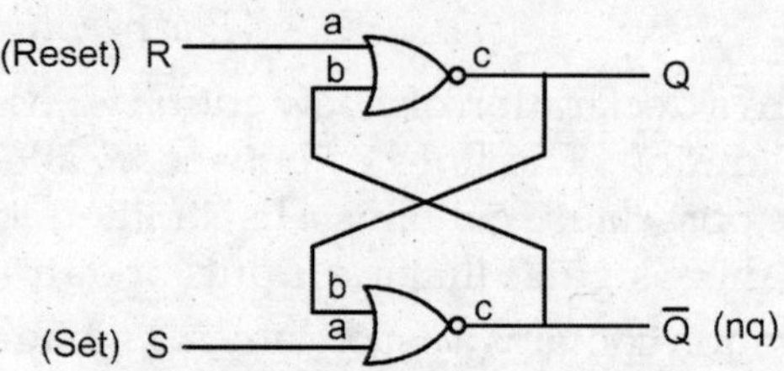

Fig. 16.1 Schematic of SR latch

```
entity latch is
  port (s,r: in bit;
    q,nq: outbit);
end latch;
```

The first line indicates a definition of a new entity, whose name is latch. The last line marks the end of the definition. The lines in between, called the port clause, contain a list of interface declarations. Each interface declaration defines one or more signals that are inputs or outputs to the design and contains a list of names, a mode and a type.

In the first interface declaration of the example, two input signals are defined, *s* and *r*. The signals *s* and *r* are of mode *in* (inputs) and type *bit* (binary). Next, the signals *q* and *nq* are defined to be of the mode *out* (outputs) and of the type *bit*.

The signals in the above example are defined to be of the type bit. The type bit is a predefined type that can have two values represented by '0' and '1'. This type is used to represent two level logic signals.

The second part of the description of the latch design is the architecture declaration, which describes how the design operates. The architecture declaration through data flow approach for the latch entity is given as follows.

```
architecture data flow of latch is
begin
  q<=rnornq;
  nq<=s nor q;
end dataflow;
```

The first line of the declaration indicates that this is the definition of a new architecture called *data flow* and it belongs to the entity named latch. Therefore, this architecture describes the operation of the latch entity. The lines in between the *begin* and the *end* describe the operation of the latch.

The architecture part describes the internal operation of the design. The data flow approach indicates how data flows from the input to the output. In VHDL this is accomplished with the signal assignment statement. This example **architecture** consists of two signal assignment statements. The first signal assignment in the example represents the data coming from signals *r* and *nq* flow through a NOR gate to determine the value of the signal *q*. The *nor* represents a built-in component called an operator, because it operates on a data to produce a new data. The second signal assignment, similar to the first, indicates that the signal *nq* is produced from data *s* and *q* flowing through, or processed by the *nor* operator.

The next section explains how to specify the latch operation using the structural approach.

16.3 STRUCTURAL DESCRIPTION

The structural description is the second type of the three paradigms for describing hardware design with VHDL. The following sections discuss this approach of the VHDL design.

Connecting Blocks: The basic building blocks of the design are defined using *entities* and their associated architectures. Once this has been done, the basic building blocks can be combined together to form bigger designs. This section describes how to combine and link these blocks together in a structural manner.

The structural description of a design is simply a textual description of the schematic. A list of components and their interconnections in a language is sometimes called a *netlist*. The structural description of a design in VHDL is one of such means of specifying netlists.

Here, the operation of the latch entity used in the previous section is identified by connecting some previously defined entities. The entity declaration for the latch was:

```
entity latch is
  port (s, r: in bit;
q,nq: out bit);
end latch;
```

The architecture declaration of the structural description is different from that of data flow description. For the latch example, assume that an entity named *nor gate* has been defined that will be used in the design. The same connections that occurred in the schematic shown in Fig. 16.1 can be rewritten using VHDL with the following architecture declaration:

```
architecture structure of latch is
  component nor_gate
    port (a,b: in bit;
    c: out bit);
  end component;
begin
a1: nor_gate
  port map (r,nq,q);
a2: nor_gate
  port map (s,q,nq);
end structure;
```

The lines between the first and the keyword *begin* identify the component declaration. It describes the interface of the entity *nor gate* that is chosen as a component in (or part of) this design. Between the *begin* and the *end* keywords, the first two lines and second two lines define two component instances.

There is an important distinction between an entity, a component, and a component instance in VHDL. The entity describes a design interface, the component describes the interface of an entity that will be used as an instance (or a sub-block), and the component instance is a distinct copy of the component that has been connected to other parts and signals.

In this example, the component *nor gate* has two inputs (*a* and *b*) and an output (*c*). There are two instances of the *nor gate* component in this architecture corresponding to the two *nor* symbols in the schematic. The first instance represents the top *nor gate* in the schematic. The first line of the component instantiation statement gives this instance a name, *a*1, and specifies that it is an instance of the component *nor gate*. The second line describes how the component is connected to the reset of the design using the port map clause. The port map clause specifies the signals of the design to connect to the interface of the component in the same order as they are listed in the component declaration. The interface is specified in order as *a*, *b*, followed by *c*. Hence, this instance connects *r* to *a*, *nq* to *b*, and *q* to *c*. This corresponds to the way the top gate in the schematic is connected. The second instance, named *a*2, connects *s* to *a*, *q* to *b*, and *nq* to *c* of another instance of the same *nor_gate* component in the same manner as shown in the schematic.

16.4 SIMULATORS

A simulator that uses the dataflow description to model the design is discussed below. The VHDL standard not only describes how the designs are specified, but also how they should be interpreted. The scheme used to model a VHDL design is called *discrete event time simulation*. When the value of a signal changes, an event has occurred on that signal. If data flows from signal *A* to signal *B*, and an event has occurred on signal *A* (i.e., *A*'s value changes), then the possible new value of *B* has to be determined. This is the foundation of the discrete event time simulation. The values of signals are updated only when certain events occur at discrete instances of time.

Since one event causes another, simulation proceeds in sequence. The simulator maintains a list of events that needs to be processed. In each round, all the events in the list are processed. Any new event that is produced is placed in a separate list (and are said to be scheduled) for processing at a later round. Each signal assignment is evaluated once, when simulation begins to determine the initial value of each signal.

The event time simulation proceeds for the previous example of SR latch is discussed as follows. The internal operation of the latch was essentially captured using the following two statements.

```
q<=r nor nq;
nq<=s nor q;
```

Since data flows from *r* and *nq* to *q*, that means *q* depends on *r* and *nq*. In general, given any signal assignment statement, the signal on the left side of the <= operator depends on all the signals appearing on the right side. If a signal depends on another signal that had an event occurred-on, then the expression in the signal assignment is re-evaluated. If the result of the evaluation is different from the current value of the signal, an event will be scheduled (added to the list of events to be processed) to get the signal updated with the new value. Thus, if an event occurs on *r* or *nq*, then the *nor* operator is evaluated, and if the result is different from the current value of *q*, then an event will be scheduled to update *q*.

Suppose, at a particular moment during a simulation of the SR latch example, the values of the signals are *s* = ‘0’, *r* = ‘0’, *q* = ‘1’, and *nq* = ‘0’. If the value of the signal *r* changes (due to some event external to the design) to the value ‘1’, since *q* depends on *r*, it is required to reevaluate the expression *r nor nq*, which now evaluates to ‘0’. Since the value of *q* must be changed to ‘0’, a new event will be scheduled on the signal *q*. During the next round, the event scheduled for *q* is processed and value of *q* is updated to be ‘0’. Also, since *nq* depends on *q*, the expression *s nor q* must be reevaluated. The result of the expression is ‘1’. Hence, an event is scheduled to update the value of *nq*. During the next round, when the event on *nq* is processed, the expression for *q* will be evaluated again, since it depends on *nq*. However, the result of the expression will be ‘0’ and no new event will be scheduled, because *q* is already ‘0’. Since, no new events were scheduled, there will be no more events which will occur internally to the latch.

Now, suppose an external event causes *r* to return to the value ‘0’. Since *q* depends on *r*, *r nor nq* is evaluated again. The result of this expression is ‘0’ and *q* is already ‘0’, so no events are scheduled. This correctly models the *SR latch* as expected. When the signal *r* became active (‘1’) the output of the latch was reset, and when *r* became inactive (‘0’) the output remained unchanged.

The simulation rounds described in these last two paragraphs can be summarized as follows:

start : r=‘0’,r=‘0’,q=‘1’,nq=‘0’

round 1 : r=‘1’,s=‘0’,q=‘1’,nq=‘0’, The value ‘0’ is scheduled on q.

round 2 : r=‘1’,s=‘0’,q=‘0’,nq=‘0’, The value ‘1’ is scheduled on nq.

round 3 : r=‘1’,s=‘0’,q=‘0’,nq=‘1’, No new events are scheduled.

round 4 : r=‘0’,s=‘0’,q=‘0’,nq=‘1’, No new events are scheduled.

The Delay Model: The example of the last section shows how a functional simulation proceeds. It is called a *functional simulation,* since it models only the design functions without involving any timing considerations. This is in contrast to a timing simulation, which models the internal delays that are present in real circuits. This section explains how VHDL can be used to model time delays to obtain a timing simulation.

Two models of delay are used in VHDL, such as, (*i*) inertial delay model and (*ii*) transport delay model. The inertial delay model is specified by adding an *after* clause to the signal assignment statement. For example, suppose that a change on the input of a *nor* gate would cause the output to change after a delay of 1*ns*. To model this delay in the SR latch example, the two signal assignments can be replaced with the following two statements.

```
q<=r nor nq after 1 ns;
nq<s nor q after 1 ns;
```

Now during simulation, say, the signal *r* changes and it will cause the signal *q* to change. Rather than scheduling the event on *q* to occur during the next round, it is scheduled to occur at after 1 *ns* from the current time. Thus, the simulator must maintain a current time value. When no more events exist waiting to be processed at the current time value, the time is updated to the time of the next earliest event and all events scheduled for that time will be processed. A timing diagram for this modified SR latch produced by a simulator might be as shown in Fig. 16.2.

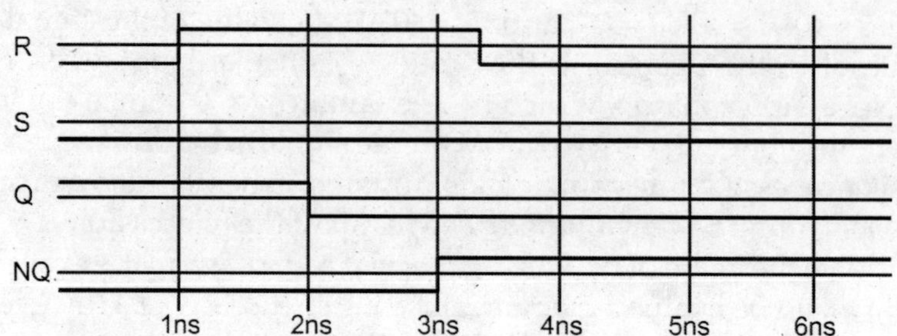

Fig. 16.2 Timing diagram of SR latch using Inertial delay model

Notice that the change did not occur in *q* until 1 *ns* after the change in *r*. Likewise, the change in *nq* did not occur until 1 *ns* after the change in *q*. Thus, the *"after 1ns"* models an internal delay of the NOR gate.

Typically, when a component has some internal delay and an input changes for a time less than this delay, then no change in the output will occur. This is also the case for the inertial delay model. The timing diagram shown in Fig. 16.3 would be produced using the inertial delay model, if the '1' pulse on the signal *r* was shortened (to anything less than 1 *ns*) from the previous example.

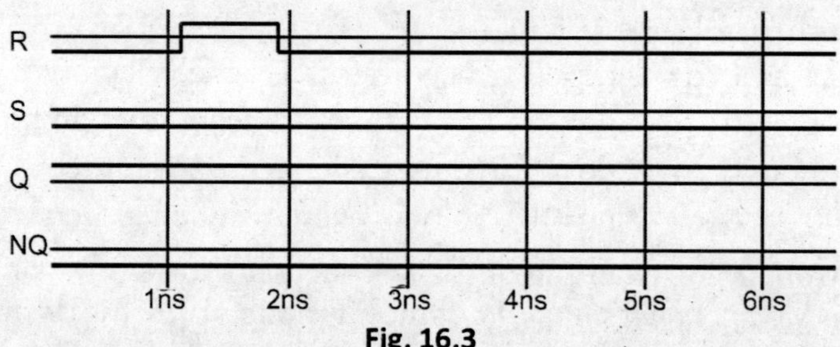

Fig. 16.3

The value of *q* never changed because the change in *r* did not last long enough.

Although most often the inertial delay is desired, sometimes all changes on the input should have an effect on the output. For example, a bus experiences a time delay, but will not "absorb" short pulses as with the inertial delay model. As a result, VHDL provides the transport delay model. The *transport delay model* just delays the change in the output by the time specified in the *after* clause. The transport delay model can be chosen instead of the inertial delay model by adding the keyword *transport* to the signal assignment statement.

The SR latch example could be modified to use the transport delay model by replacing the signal assignments with the following two statements.

```
q<=transport r nor nq after 1ns;
nq<=transport s nor q after 1ns;
```

If the transport delay model were used, the result of the same simulation shown in the Fig. 16.3 would result in the timing diagram, shown in Fig. 16.4.

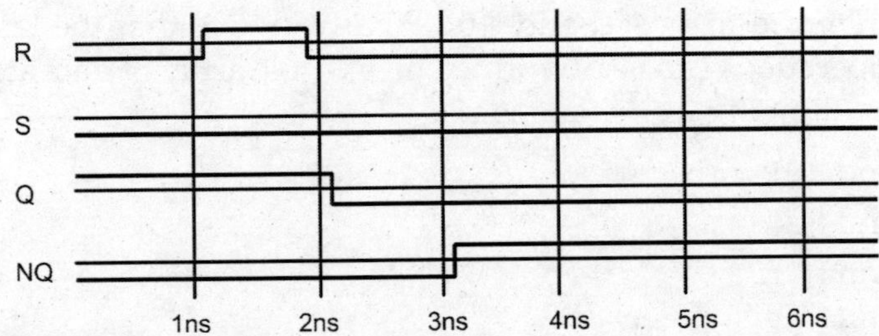

Fig. 16.4 Timing diagram of SR Latch using transport delay model

16.5 OTHER TYPES

Most of the signals in the examples so far have been of the type *bit*. VHDL provides a mechanism for defining new types which represent a collection of several data items of the same type. These kinds of types are called arrays.

Bit_vector: In VHDL, it is often required to use several bit signals together to represent a binary number in a design. There is a predefined array type called *bit_vector* which represents a collection of bits. The following example demonstrates how the *bit_vector* type can be used to define a 1-to-4-line demultiplexer.

```
entity demux is
  port (e: in bit_vector (3 downto 0);        - enables for each output
s: inbit_vector(l downto 0);                  - select signals
d: out bit_vector (3 downto 0));              - four output signals
end demux;
architecture rtl of demux is
  signal t: bit_vector (3 downto 0);          - an internal signal
begin
  t(4)<=s(1) and s(0);
  t(3)<=s(1) and not s(0);
  t(2)<=not s(1) and s(0);
  t(1)<=not s(1) and not s(0);
  d<=e and t;
end rtl;
```

As discussed earlier, comments can be added at the end of a VHDL statement or on a line by itself preceded by the — symbol. First notice how the *bit_vector* is used in the definition of a signal. The definition of *s* indicates that *s* is a *bit_vector* and the (1 *downto* 0) part specifies that the signal *s* contains two bits numbered 1 downto 0. Similarly, *d* and *e* are arrays of 4-bits numbered from 3 downto 0. Second, notice that signals, such as *t* can be declared within an architecture that is not visible from outside this entity. These internal signals are created with signal declarations as the one shown in the example. They contain the key word *signal* followed by a list of names of the signals to create, followed by the type of the signals. Third, the architecture refers to the individual bits in *t* and *s* by number. The two bits in s are number 1 and 0, so they are referred to as *s*(1) and *s*(0). Finally, notice that the last signal assignment demonstrates that operations can be performed on a whole array of data at once. This statement is equivalent to the following four statements.

```
d(3)<=e(3) and t(3);
d(2)<=e(2) and t(2);
d(1)<=e(1) and t(1);
d(0)<=e(0) and t(0);
```

Each data item in an array is called an *element*. The number of elements in a signal of an array type is indicated by the range that follows the type name. The elements are numbered according to the range, and each element is referred to as an individual by number. Operations can be performed on an array as a whole (applying to every element in the array), or they can be performed using individual elements of the array, independent of the others.

When operations are performed on whole vectors, the vectors must have the same number of elements. If they do not, the simulator will report an error and stop the simulation. In an operation between vectors, elements are matched as they are numbered from left to right. Thus, if a variable v1 has elements 0 to 1 and variable v2 has elements 1 downto 0. Then

```
v1:=v2;
```

would assign v2(1) to v1(0) and v2(0) to v1(1).

Time: Another predefined type is *time.* This type is used to represent values of time. Already constant values were used for this type in the after clause. Time is an example of a physical type. All values of a physical type have two parts, a number and a unit name. The type *time* includes the following predefined unit names *sec* (seconds), *ms* (milliseconds), ms (microseconds), *ns* (nanoseconds), *ps* (picoseconds), and *fs* (femtoseconds).

There are several other types predefined in VHDL including types for integers and real numbers. These are mentioned later in the sections related to behavioral descriptions.

16.6 OTHER OPERATORS

The previous section presented a few types that are available in VHDL. There are also several other built-in operators that can be used with those types as described below.

Logical operators: The logical operators NOT, AND, OR, NAND, NOR, and XOR can be used with any bit type or bit_vectors. When used as operators on bits, they have their usual meaning. When used with bit_vectors, the bit_vectors must have the same number of elements, and the operation is performed bitwise. For example, "00101001" xor "11100101" results in "11001100".

It is important to note that '0' and '1' represent constant bit values, while constant bit_vectors can be written in VHDL as a list of bit values in double quotes. For example, if d is a bit_vector (1 to 4) the following statement gives d the permanent values *d(1)='1', d(2)='1', d(3)='0', and d(4)='0'.*

d<="1100";

Hexadecimal can also be used as a shortcut as in the following example.

d<=X"C";

Since *C* is the hexadecimal number 12, which in binary is 1100, this statement is equivalent to the one preceding it. The *X* in the front indicates that the number is in hexadecimal instead of the normal binary.

Arithmetic operators: The typical *arithmetic operators* are available for integers, such as +, –, * (multiplication), and / (division). Although these operations are not built-in for bit_vectors, they are often provided in libraries that come with VHDL software. They are used with bit_vectors by interpreting them as a binary representation of integers, which may be added, subtracted, multiplied or divided.

Relational operators: Also predefined are the normal relational operators. They are =, /=, <, <=, > and >= and have their usual meanings (/= denotes the not equal operator). The result of all these operators is a Boolean value (TRUE or FALSE). The arguments to the = and /= operators may be of any type. The arguments of the <, <=, > and >= operators may be any scalar type (integer, real, and physical types) or the bit_vector type. If the arguments are bit_vectors, then the arguments must be the same length and the result is TRUE only if the relation is true for each corresponding element of the array arguments.

Concatenation operator: The & operator is a built-in VHDL operator that performs the concatenation of bit_vectors. For example, with the following declarations:

signal a: bit_vector (1 to 4);

signal b: bit_vector (1 to 8);

The following statement would connect *a* to the right half of *b* and make the left half of *b* constant '0'.

b<="0000" & a;

The & appends the *a* to the end of the "0000" to form a result that contains 8-bits.

16.7 BEHAVIOURAL DESCRIPTION

The behavioral approach to modeling hardware components is different from the other two methods in that, it does not necessarily in any way reflect how the design is implemented. It is basically the *black box approach* to modeling. It accurately models what happens on the inputs and the way it is guided towards outputs of the black box. It does not deal with what is inside the box or how it works. The behavioral description is usually used in two ways in VHDL. First, it can be used to model complex components that would be tedious to model using the other methods. This might be the case, for example, if it is required to simulate the operation of the custom design connected to a commercial part like a microprocessor. In this case, the microprocessor is complex and its internal operation is irrelevant (only the external behavior is important) so it would probably be modeled using the behavioral style. Second, the behavioral capabilities of VHDL can be more powerful and is more convenient for some designs. In this case, the behavioral description will likely imply some structure of the implementation.

The Process Statement: Behavioral descriptions are supported with the process statement. The process statement can appear in the body of an architecture declaration just as the signal assignment statement does. The contents of the process statement can include sequential statements like those found in software programming languages. These statements are used to compute the outputs of the process from its inputs. Sequential statements are often more powerful, but sometimes have no direct correspondence to a hardware implementation. The process statement can also contain signal assignments in order to specify the outputs of the process.

The following example of the process statement is trivial and would not normally be done in a process statement. However, it allows to examine the process statement without learning any sequential statements.

```
compute_xor: process (b,c)
begin
 a<=b xor c;
end process;
```

The first part 'compute_xor:' is used to name the process. This part is optional. Next is the keyword *process* that starts the definition of a process. Following the keyword is a list of signals in parenthesis, called the *sensitivity list*. Since the contents of the process statement may not have indicated any structural characteristics, there is no way to know when the process should be reevaluated to update the outputs. The signal sensitivity list is used to identify the signals that should cause the process to be reevaluated. Whenever any event, occurs on one of the signals in the sensitivity list, the process is reevaluated. A process is evaluated by performing each statement that it contains. These statements (the body of the process) appear between the *begin* and the *end* keywords.

This example process contains one statement, namely, the signal assignment. Unlike signal assignments that appear outside the process statement, this signal assignment is only evaluated when events occur on the signals in the sensitivity list of the process, regardless of which signals appear on the right hand side of the <= operators. This means that it is crucial to make sure that the proper signals are in the sensitivity list. The statements in the body of the process are performed (or executed) in order, from the first to the last. When the last statement has been executed, the process is finished and is said to be suspended. When an event occurs on a signal in the sensitivity list, the process is said to be resumed and the statements will be executed from top to bottom again. Each process is executed once at the beginning of a simulation to determine the initial values of its outputs.

Using Variables: There are two major kinds of objects used to hold data. The first kind, used mostly in structural and data flow descriptions, is the signal. The second, which can only be used in processes, is called a *variable.* A variable behaves in a manner that would be expected in a software programming language, which is much different than the behavior of a signal.

Although variables represent data like the signal, they do not have, or cause events and they are modified differently. Variables are modified with the variable assignment. For example,

```
a:=b;
```

assigns the value of *b* to *a*. The value is simply copied to *a* immediately. Since variables may only be used in processes, the variable assignment statement may only appear in a process. The assignment is performed when the process is executed, as explained in the previous section.

The following example shows how a variable is used in a process.

```
count: process (x)
  variable cnt: integer := – 1;
begin
  cnt:=cnt+1;
end process;
```

Variable declarations appear before the *begin* keyword of a process statement as shown in the example. The variable declaration is the same as the signal declaration except that the key word *variable* is used instead of *signal*. The declaration in this example includes an optional part, which specifies the initial value of the variable, when a simulation begins. The initialization part is included by adding the := and some constant expression after the type part of the declaration. This initialization part may also be included in signal declarations. The variable *cnt* is declared to be of the type integer. The integer type represents negative and positive integer values.

The process in the example contains one statement, the assignment statement. This assignment computes the value of *cnt* plus one and immediately stores that new value in the variable *cnt*. Thus, *cnt* will be incremented by one each time this process is executed. Remember, from the last section, that a process is executed once at the beginning of simulation and then each time an event occurs on any signal in its sensitivity list. Since the value is initialized to –1, and the process is executed once before beginning simulation, the value of *cnt* will be 0 when simulation begins. Once simulation begins, *cnt* will be incremented by one each time the signal *x* changes, since *x* is in the sensitivity list. If *x* is a bit signal, then this process will count the number of rising and falling edges that occur on the signal *x*.

Sequential Statements: There are several statements that may only be used in the body of a process. These statements are called sequential statements, because they are executed sequentially. That is, one after the other, as they appear in the design from the top of the process body to the bottom.

The following example illustrates the *if* statement and a common use of the VHDL attribute.

```
count: process (x)
  variable cnt: integer :=0;
begin
  if(x='i'andx'last_value='0')then
    cnt:=cnt+l;
  end if;
end process;
```

This *if* statement has two main parts, the *condition* and the *statement body*. A condition is any *Boolean* expression (an expression that evaluates TRUE or FALSE, such as expressions using relational operators). The condition in the example uses the attribute last_value, which is used to determine the last value that the signal had. Attributes can be used to obtain a lot of auxiliary information about signals. The value of an attribute for a particular signal is obtained by specifying the name of the signal, followed by *a'* (called a tick) and the name of the attribute desired. Thus, the condition in the example is true only if the current value of *x* is '1' and its previous value was '0'. Since this statement will only be executed when an event has occurred on *x* (i.e., *x* has just changed), this condition will be true when a rising edge occurs on *x*. This is because *x* just changed, it was a '0' and now it is a '1'. The statement body of the *if* statement is just a list of sequential statements that appear between the key words *then* and *end if.*

The execution of the *if* statement begins by evaluating the condition. If the condition evaluates to the value TRUE, then the statements in the statement body will be executed. Otherwise, the execution will continue after the *end if and* the statement body of the *if* statement is skipped. Thus, the assignment statement in this example is executed every time there is a rising edge on the signal *x*, counting the number of rising edges.

An example of another common form of the *if* statement is·

```
if(inc='1') then
  cnt:=cnt+1;
else
cnt:=cnt – 1;
end if;
```

This form has two statement bodies. If the condition is TRUE, the first list of statements is executed (between the *then* and the *else*) and the second list of statements (between the *else and* the *end if*) is not. Otherwise, the second statement list is executed and the First is not. Thus, this example will increment *cnt*, if *inc* is '1' and decrement it otherwise.

The last statement that will be discussed is the *loop* statement. One form of the loop statement shall be explained, often called as *for* statement. The *for* statement is used to execute a list of statements several times. The following example uses a *loop* statement to compute the even parity of a bit vector.

```
signal x : bit_vector (7 downto 0);
...
process (x)
variable p: bit;
begin
  p:='0'
  for i in 7 downto 0 loop
  p:=p xor x(i);
  end loop;
end process;
```

The signal x is an 8-bit signal representing a *byte*. The variable p is used to compute the parity of this byte. The first part of the *for* loop *i* in 7 *downto* 0 is called the parameter specification. It specifies how many times the loop body will be executed and creates a temporary variable. It begins with the name of the temporary variable that will be created, in this case it is *i*. This is followed by the key word *in* and then a range of values as seen before. The body of the loop is executed once for every value in the range specified. The value of the temporary variable is assigned one of the values in the range each time the loop body is executed. In this example, the assignment will be executed first with $i = 7$ then again with $i = 6$, and again with $i = 5$, and so on downto 0. This loop statement behaves in the same manner as the following statements.

```
P:='0';
p:=p xor x(7);
p:=p xor x(6);
p:=p xor x(5);
p:=p xor x(4);
p:=p xor x(3);
p:=p xor x(2);
p:=p xor x(1);
p:=p xor x(0);
```

It is to be observed that how the temporary variable *i* was used in the statement body of the loop to operate on different elements of the vector x each time the body of the loop is executed. This is a very common use of the loop statement. Although this loop contains only one statement, there may be many statements in the loop body.

Signals and Processes: The issue of concern is to avoid confusion about the difference between how a signal assignment and a variable assignment would behave in the process statement. A signal assignment merely schedules an event to occur on a signal and does not have an immediate effect. When a process is resumed, it executes from top to bottom and no events are processed until after the process is complete. This means that if an event is scheduled on a signal during the execution of a process, the event can be processed after the process has been completed at the earliest. In the following example, this behavior is illustrated. In the following process two events are scheduled on signals x and z.

```
...
signal x,yz: bit;
...
process (y)
begin
 x<=y;
 z<=notx;
end process;
```

If the signal y changes, then an event will be scheduled on x to make it the same as y. Also, an event is scheduled on z to make it the opposite of x. The value of z will not be the opposite of y, because when the second statement is executed, the event on x has not been processed yet, and the event scheduled on *z* will be the opposite or *not* of the value of *x* before the process begins.

This is pointed out because this is not necessarily the intuitive behavior and because variables operate differently. For example, in

```
process (y)
variable x,z : bit;
begin
  x:=y;
  z:=not x;
end process;
```

The value of the variable *z* would be the opposite of the value of *y* because the value of the variable *x* is changed immediately.

Program Output: In most programming languages, there is a mechanism for printing text on the monitor and getting the input from the user through the keyboard. Even though the simulator allows monitoring the value of signals and variables of the design, it will be encouraging to be able to output certain information during simulation. It is not provided as a language feature in VHDL, but rather as a standard library that comes with every VHDL language system. In VHDL, common code can be put hi a separate file to be used by many designs. This common code is called a library. In order to use the library that provides input and output capabilities, the following statement has to be included

```
use textio.all;
```

immediately before every architecture that uses input and output. The name of the library is *textio* and this statement indicates that everything or all of the *textio* library can be utilized. Note that although it is not part of the language, the library is standard and will be the same regardless of the VHDL tools.

The text is input and output using *textio* via a variable of the type line. Since variables are used for textio, input and output is done in processes. The procedure for outputting information is to first place it in text form into the variable of type line and then to request that the line be output. This is shown in the following example.

```
architecture behaviour of check is
begin
  process (x)
    variable s : line;
    variable cnt : integer:=0;
begin
if (x='1' and x'last_value='0') then
  cnt:=cnt+1;
if (cnt>MAX_COUNT) then
  write(s, "Counter overflow -");
  write(s,cnt);
```

```
          writeline(output,s);
        end if;
      end if;
    end process;
  end behaviour;
```

The *write* function is used to append text information at the end of a line variable which is empty when the simulator is initialized. The function takes two arguments, the first is the name of the line to append to, and the second is the information to be appended. In the example, s is set to "Counter overflow –", and then the current value of *cnt* is converted to text and added to the end of that. The writeline function outputs the current value of a line to the monitor, and empties the line for re-use. The first argument of the writeline function just indicates that the text should be output to the screen. If MAX_COUNT were a constant equal to 15 and more than 15 rising edges occur on the signal *x*, then the message

Counter overflow – 16

would be printed on the screen.

The write statement can also be used to append constant values and the value of variables and signals of the type's bit, bit_vector, time, integer, and real. Keyboard input is more complex than output.

16.8 VHDL LIBRARIES

The libraries can be declared in VHDL using two lines of code, one containing the *name* of the library and the other line containing a *use* clause as follows.

library library_name;

uselibrary_name.package_name.all;

At least three packages are usually needed in a design from three different libraries. They are (*i*) *ieee.std_logic_1164* from the *ieee* library, (*ii*) *standard* from the *std* library and (*iii*) *work* from *work* library. The library declarations for the above three ii different packages are given as follows.

```
library ieee;
use ieee.std_logic_1164.all;
library std;
use std.standard.all;
library work;
use work.all;
```

The libraries *std* and work shown above are made visible by default, so there is no need to declare them; only the *ieee* library must be explicitly written. However, the latter is only necessary, when the *std_logic* ("or *std_ulogic*) data type is employed in the design.

The purpose of three packages/libraries mentioned above is the following: the *std_logic_1164* package of the *ieee* library specifies a multi-level logic system; *std* is a resource library (data types, text i/o, etc.) for the VHDL design environment; and the *work* library is where the user designed programs are saved (the .vhd file, plus all files created by the compiler, simulator, etc.).

The *ieee* library contains several packages, including the following: *std_logic_1164:* Specifies the *std_logic* (8 levels) and *std_ulogic* (9 levels) multivalued logic systems. The 8 levels of *std_logic* are: 'X', —Forcing Unknown, '0', —Forcing 0, '1', —Forcing 1, 'Z', —High Impedance, 'W', —Weak Unknown, 'L', — Weak 0, 'H', — Weak 1 and '-' — Don't care. The 9 levels of *std_ulogic* are: 'U', — Uninitialized plus the 8 levels of *std_logic*.

std_logic_arith: specifies the SIGNED and UNSIGNED data types and related arithmetic and comparison operations. It also contains several data conversion functions, which allow one type to be converted into another: *conv_integer (p), conv_unsigned (p,b), conv_signed (p,b), conv_std_logic_vector (p,b)*.

std_logic_signed: contains functions that allow operations with *std_logic_vector* data to be performed as if the data were of type SIGNED.

std_logic_unsigned: contains functions that allow operations with *std_logic_vector* data to be performed as if the data were of type UNSIGNED.

16.9 VHDL PROGRAMS FOR COMBINATIONAL CIRCUITS

16.9.1 VHDL Programs for Logic Gates

The VHDL program for logic gates are written following the *dataflow approach.*

AND gate: The VHDL program for the *AND gate* shown in Fig. 3.3(*c*) can be written as follows:

```
library ieee;                          - std_logic_1164 is defined by IEEE.
use ieee.std_logic_1164.all;           - Package that defines Data types :
                                         std_logic, std_ulogic_std_logic_vector
                                         & std_ulogic_vector.
entity andgate is                      - Declaration of entity.
port (A,B:in std_logic;                - Set of i/p and o/p pins declaration.
    Y:out std_logic);
end andgate;                           - End of entity.
architecture arch_and of andgate is
begin
Y<=A and B;                            - Logical AND operation
end;                                   - End of architecture.
```

OR gate: The VHDL program for the *OR gate* shown in Fig. 3.1(*c*) can be written as follows:

```
library ieee;
use ieee.std_logic_1164.all;
entity orgate is
```

```
port(A,B:in std_logic;
  Y:out std_logic);
end orgate;
architecture arch_or of orgate is       - Set of interconnected components and
                                          assignment statements
begin
Y<=A or B;                              - Logical OR operation
end;
```

NAND gate: The VHDL program for the *NAND gate* shown in Fig. 3.5(*a*) can be written as follows:

```
library ieee;
use ieee.std_logic_1164.all;
entity nandgate is;
port (A, B: in std_logic;
  Y:out std_logic);
end nandgate;
architecture arch_nand of nandgate is
begin
Y<=A nand B;                            - Logical NAND operation
end;
```

NOR gate: The VHDL program for the *NOR gate* shown in Fig. 3.6(*a*) can be written as follows:

```
library ieee;
use ieee.std_logic_1164.all;
entity norgate is
port (A, B: in std_logic;
  Y: out std_logic); end norgate;
architecture arch_nor of norgate is
begin
Y<=A nor B;                             - Logical NOR operation
end;
```

NOT gate: The VHDL program for the *NOT gate* shown in Fig. 3.4(*b*) can be written as follows:

```
library ieee;
use ieee.std_logic_1164.all;
entity notgate is
port (A: in std_logic;
  Y: out std_logic);
```

```
end notgate;
architecture arch_not of notgate is
begin
Y<=not A;                              - Logical NOT operation
end;
```

XOR gate: The VHDL program for the *XOR gate* shown in Fig. 3.14 can be written as follows:

```
library ieee;
use ieee.std_logic_1164.all;
entity xorgate is
port (A, Control input: in std_logic;
  Y: out std_logic);
end xorgate;
architecture arch_xor of xorgate is
begin
Y<= A xor Control input;               - Logical X-OR operation
end;
```

XNOR gate: The VHDL program for the *XNOR gate* shown in Fig. 3.15 can be written as follows. Here, note that a signal *c* is declared for the purpose of the intermediate X-OR operation.

```
library ieee;
use ieee.std_logic_1164.all;
entity xnorgate is
port (A, B : in std_logic;
  Y: out std_logic);
end xnorgate;
architecture arch_xnor of xnorgate is
signal c: std_logic;
begin
c<=a xor b;                            - Logical X-OR operation
Y<= not c;                             - Logical NOT operation
end;
```

16.9.2 Adders and Subtracters

Half Adder: The VHDL program for the *Half Adder* shown in Fig. 5.1(*b*) can be written as follows. This program follows *dataflow approach.*

```
library ieee;
use ieee.std_logic_1164.all;
entity halfadder is
```

```
port(
A, B: in std_logic;
Sum, Carry: out std_logic);
end halfadder;
architecture halfadder_arch of halfadder is
begin
Sum<=A xor B;                    - Logical X-OR operation
Carry<=(A and B);                - Logical AND operation
end halfadder_arch;
```

Half Subtractor: The VHDL program for the *Half Subtracter* shown in Fig. 5.6(*b*) can be written as follows. This program follows *dataflow approach.*

```
library ieee;
use ieee.std_logic_1164.all;
entity halfsubtractor is
port(X, Y:in std_logic;
Difference, Barrow:out std_logic);
end halfsubtractor;
architecture halfsubtractor_arch of halfsubtractor is
begin
Difference <=X xor Y;            - Logical X-OR operation
Barrow<= (not X) and Y;          - Logical AND operation of one
end;                             - i/p with not of other i/p
```

Full Adder: The VHDL program for the *Full Adder* shown in Fig. 5.3 can be written as follows. This program follows *dataflow approach.*

```
library ieee;
use ieee.std_logic_1164.all;
entity fulladder is
port (A, B, Cin: in std_logic;
Sum, Carry: out std_logic);
end fulladder;
architecture adder_arch of fulladder is
begin
Sum <= A xor B xor Cin;
Carry <= (A and B) or (B and Cin) or (Cin and A);
end;
```

Full Subtracter: The VHDL program for the *Full Subtracter* shown in Fig. 5.9 can be written as follows. This program follows *dataflow approach.*

```
library ieee;
use ieee.std_logic_1164.all;
entity fullsub is
port(X,Y, Bin:instd_logic;
Diff, Bout:out std_logic);
end fullsub;
architecture fullsub_arch of fullsub is
begin
Diff<=X xor Y xor Bin;
Bout<=(Y and Bin)or((not X) and Bin)or((not X) and Y);
end fullsub_arch;
```

Four Bit Parallel Binary Adder: The VHDL program for the *Four Bit Parallel Binary Adder* shown in Fig. 5.10 can be written as follows. This program follows *structural approach.*

```
library ieee;
use ieee.std_logic_1164.all;
entity adder_4bit is
port (A, B:in std_logic_vector(0 to 3);
S:out std_logic_vector(0 to 3);
Cout:out std_logic);
end adder_4bit;
architecture adder_4bit_arch of adder_4bit is
component fulladder is                          - Full Adder program is used as
                                                - a component.
port(A, B, C:mstd_logic;
  Sum, Carry:outstd_logic);
end component;
signal Cin,C1,C2,C3,C4:std_logic;               - Holds current value and a set
begin                                           - of possible future values.
Cin<='0';
a1:fulladder port map (A(0),B(0),Cin,S(0),C 1); - Calling of full adder program
a2:fulladder port map (A(1),B(1),C1,S(1),C2);   - first three bits i/p; – last 2
                                                  bit o/p
a3:fulladder port map(A(2),B(2),C2,S(2),C3);    - Carry map is passed as i/p
a4:fulladder port map(A(3),B(3),C3,S(3),C4);    - to the next map
Cout<=C4;
end;
```

16.9.3 Multiplexer & Demultiplexer

4-to-1 Multiplexer: The VHDL program for the *4-to-1 Multiplexer* shown in Fig. 6.2 can be written as follows. This program follows *dataflow approach.*

```
library ieee;
use ieee.std_logic_1164.all;
entity mux is
port(S1,S0,D0,D1,D2,D3:in std_logic;
Y:out std_logic);
end mux;
architecture arch_mux of mux is
begin
Y<=((not S1)and(not S0)and D0)or          - Any one of D0, D1, D2 or D3
((not S1) and S0 and D1)or                - is taken as output, according to
(S1 and(not S0)and D2)or                  - values of S1 and S2.
(S1 and S0 and D3);
endarch_mux;
```

1-to-4 Demultiplexer: The VHDL program for the *1-to-4 Demultiplexer* shown in Fig. 6.7 can be written as follows. This program follows *dataflow approach.*

```
library ieee;
use ieee.std_logic_1164.all;
entity demux is
port(D,S1,S0:in std_logic;
  Y0,Y1,Y2,Y3:outstd_logic);
end demux;
architecture demux_arch of demux is
begin
  Y0<=D and (not S1) and (not S0);        - The i/p is given to any one of
  Y1<=D and (not S1) and (S0);            - the o/p, which is selected by
  Y2<=D and (S1) and (not S0);            - the values of S0 and S1
  Y3<=D and (S1) and (S0);
end;
```

16.9.4 Encoder and Decoder

Octal-to-Binary Encoder: The VHDL program for the *Octal-to-Binary Encoder* shown in Fig. 6.26 can be written as follows. This program follows *dataflow approach.*

```
library ieee;
use ieee.std_logic 1164.all;
entity octal_binary encoder is
```

```
port(D:in std_logic_vector(0 to 7);
 Y:out std_logic_vector(0 to 2));
end octal_binary encoder;
architecture encoder_arch octal_binary encoder is
begin
 Y(0)<=D(1) or D(3) or D(5) or D(7);
 Y(1)<=D(2) or D(3) or D(6) or D(7);
 Y(2)<=D(4) or D(5) or D(6) or D(7);
end encoder_arch;
```

3-to-8 Decoder: The VHDL program for the *3-to-8 Decoder* shown in Fig. 6.12 can be written as follows. This program follows *dataflow approach.*

```
library ieee;
use ieee.std_logic_1164.all;
entity decoder is
port(A,B,C:instd_logic;
 D: out std_logic_vector(0 to 7));
end decoder;
architecture decoder_arch of decoder is
begin
 D(0)<=(not A) and (not B) and (not C);
 D(1)<=(not A) and (not B) and C;
 D(2)<=(not A) and B and(not C);
 D(3)<=(not A) and B and C;
 D(4)<=A and (not B) and (not C);
 D(5)<=A and (not B) and C;
 D(6)<=A and B and (not C);
 D(7)<=A and B and C;
end;
```

16.9.5 4-Bit Parity Checker

The VHDL program for the *4-bit Parity Checker* shown in Fig. 6.30(*a*) can be written as follows. This program follows *dataflow approach.* Here, the output is 1 when the number of 1's in the inputs *W, X, Y* and *Z* is odd and 0 when the number of 1's in the inputs is even.

```
library ieee;
use ieee.std_logic_1164.all;
entity parity_even is
port(W,X,Y,Z:in std_logic;
Output:out std_logic);
```

```
end parity_even;
architecture arch_parity_even of parity_even is
begin
  output<=((W xor X) xor Y) xor Z;
end;
```

16.9.6 4-bit Magnitude Comparator

The VHDL program for the *4-bit Magnitude Comparator* shown in Fig. 6.48 can be written as follows. This program follows *dataflow approach.*

```
library ieee;
use ieee.std_logic_1164.all;
entity comparator is
port(A3,A2,A1,A0,B3,B2,B1,B0:in std_logic;
  E,GT,LT:out std_logic);
end comparator;
architecture arch_comparator of comparator is
signal E3,E2,E1,E0,GT3,GT2,GT1,GT0,LT3,LT2,LT1,LT0:std_logic;
begin
  E3<=A3 xnor B3;
  E2<=A2 xnor B2;
  E1<=A1 xnor B1;
  E0=A0 xnor B0;
  E<= E3 and E2 and E1 and E0;
  GT3<=A3 and (not B3);
  GT2<=E3 and A2 and (not B2);
  GT1<= E3 and E2 and A1 and (not B1);
  GT0<=E3 and E2 and E1 and A0 and (not B0);
  GT<=GT3 or GT2 or GT1 or GTO;
  LT3<=(not A3) and B3;
  LT2<=E3 and (not A2) and B2;
  LT1<= E3 and E2 and (not A1) and B1;
  LT0<=E3 and E2 and E1 and (not A0) arid B0;
  LT<=LT3 or LT2 or LT1 or LT0;
```

16.9.7 Code Converters

4-bit Binary-to-Gray Code Converter: The VHDL program for the *4-bit Binary-to-Gray Code Converter*, shown in Fig. 6.41 can be written as follows. This program follows data flow *approach.*

```
library ieee;
```

```
use ieee.std_logic_1164.all;
entity Binary-to-gray is
port(B0,B1,B2,B3:in std_logic;
G:out std_logic_vector(0 to 3));
end Binary-to-gray;
architecture arch_Binary-to-gray of Binary-to-gray is
begin
  G(3)<=B3;
  G(2)<=B3 xor B2;
  G(1)<=B2 xor B1;
  G(0)<=B1 xor B0;
end;
```

4-bit Gray Code-to-Binary Converter: The VHDL program for the *4-bit Gray code-to-Binary Converter* shown in Fig. 6.44 can be written as follows. This program *follows dataflow approach.*

```
library ieee;
use ieee.std_logic_1164.all;
entity Gray-to-Binary is
port(G3,G2,G1,G0:in std_logic;
  B:out std_logic_vector(0 to 3));
end Gray-to-Binary;
architecture arch_Gray-to-Binary of Gray-to-Binary is
begin
  B(3)<=G3;
  B(2)<=G3 xor G2;
  B(1)<=B(2) xor G1;
  B(0)<B(1) xor G0;
end;
```

16.10 HDL FOR SEQUENTIAL LOGIC CIRCUITS

16.10.1 Realization of Flip-Flops

SR-Flip-Flop: The VHDL program for the *SR-Flip-Flop* shown in Fig. 7.4(*b*) can be written as follows. This program follows *dataflow approach.*

```
library ieee;
use ieee.std_logic_1164.all;
entity srff1 is                                    - Declaration of entity.
port(S,R:in std_logic;Q, NQ:inout std_logic);  - Set of i/p & o/p in declaration
```

```
end srff1;                                    - End of entity declaration
architecture srff arch of srff1 is
begin
  Q<=R nor NQ;
  NQ<=S nor Q;
end ;
```

Clocked SR-Flip-Flop: The VHDL program for the *Clocked SR-Flip-Flop* shown in Fig. 7.10 can be written as follows. This program follows *structural approach.*

```
library ieee;
use ieee.std_logic_1164.all;
entity clksr is
port (S,R,CLK: in std_logic;
  M,N: inout std_logic;
  Q, NQ:inout std_logic);
end clksr;
architecture clksr_arch of clksr is
component srffl is
port(S,R : in std_logic;                      - SR FF program is used as component
  Q,NQ:inout std_logic);
end component;
begin
  M <= S and Clk;                             - AND logic operation S with CLK
  N <= R and CLK;                             - AND logic operation R with CLK
  a1: srff1 port map (M, N, Q, NQ);           - SR FF component is used at
end;                                          - instance a1
```

D-Flip-Flop: The VHDL program for the *D-Flip-Flop* can be written as follows. This program follows *behavioral approach* as per truth table given in Table 7.6.

```
library ieee;
use ieee.std_logic_1164.all;
entity dfff1 is
port(D,CLK,reset:in std_logic;
  Q:out std_logic);
enddfffl;
architecture areh_dfiipflop of dfff1 is
begin
process(CLK)
begin
  if (CLK'event and CLK='1') then             - CLK= 1:
```

```
    ifreset='0'then                          - Reset = 0    q = 0
        Q<-'0';                              - Reset = 1    q = D
  else
        Q<=D;
      end if;
    end if;
  end process;
end;
```

JK-Flip-Flop: The VHDL program for the *JK-Flip-Flop* can be written as follows. This program follows *behavioral approach* as per truth table given in Table 7.9.

```
library ieee;
use ieee.std_logic_1164.all;
entity jkff1 is
port (J,K,CLK: in std_logic;
  Q, NQ: inout std_logic);
end jkff1;
architecture jkff_arch of jkff1 is
begin
process(CLK,J,K)
begin
if (CLK='1' and CLK'event)then
if (J='0' and K='0') then
  Q<=Q;
  NQ<=NQ;
elsif (J='1' and K='0')then
  Q<='1';
  NQ<='0';
elsif(J='0' and K='1')then
  Q<='0';
  NQ<='1';
elsif(J='1' and K='1')then
  Q<=not Q;
  NQ<=not NQ;
end if;
end if;
end process;
end;
```

T-Flip-Flop: The VHDL program for the *T-Flip-Flop* can be written as follows. This program follows *behavioral approach* as per truth table given in Table 7.12.

```
library ieee;
use ieee.std_logic_1164.all;
entity tff1 is
port(T,CLK:instd_logic;
  Q, NQ:inout std_logic);
  end tff1;
architecture arch_tff of tff1 is
begin
process (CLK,T)
begin
if (CLK='1' and CLK'event) then
  if(T='1')then
    Q<=not Q;
    NQ<= not (Q) after 0.5 ns;
  else
    Q <=Q;
    NQ <= not (Q) after 0.5 ns;
end if;
end if;
end process;
end;
```

16.10.2 Realization of Shift Registers

4-bit Serial in Serial out Shift Register: The VHDL program for the *4-bit Serial in Serial out Shift Register* shown in Fig. 9.4(*a*) can be written as follows. This program follows *structural approach*. When clock pulse is applied, output of one flip-flop is given as input of next flip-flop. The serial output is taken from the last flip-flop.

```
library ieee;
use ieee.std_logic_1164.all;
entity siso is
port(D:instd_logic;
reset,CLK:in std_logic;
  Q:out std_logic);
end siso;
architecture arch_siso of siso is
signal QA,QB,QC,QD:std_logic;
component dff1 is
port(D,CLK,reset:in std_logic;            - D Flip-Flop program is used as
  Q:out std_logic);                       - Component
```

```
end component;
begin
a1:dff port map (D,CLK,reset,QA);
a2:dff port map (QA,CLK,reset,QB);
a3:dff port map (QB,CLK,reset,QC);
a4:dff port map (QC,CLK,reset,QD);
end;
```

4-bit Serial in Parallel out Shift Register: The VHDL program for the *4-bit Serial in Parallel out Shift Register* can be written as follows. This program follows *behavioral approach*. When clock pulse is applied, output of one flip-flop is given as input of the next one and the output is obtained from all the flip-flops.

```
library ieee;
use ieee.std_logic_1164.all;
use ieee.std_logic_unsigned.all;
entity shiftreg is
port(CLK,reset,d,enable:instd_logic;
shiftedop: out std_logic_vector(3 downto 0));
end shiftreg;
architecture arch_shiftreg of shiftreg is
signal a:std_logic_vector(3 downto 0);
begin
process(CLK,reset)
begin
   if reset='0' then
     a<=(others=>'0');                    - Signal a is cleared, i.e., a = 0000
   elsifCLK'event and CLK='1' then
     if enable='1' then
       a<=shl(a,"1");                     - Shift left a by one bit
     a(0)<=d;
   end if;
 end if;
end process;
shiftedop<=a;
end;
```

4-bit Parallel in Serial out Shift Register: The VHDL program for the *4-bit Parallel in Serial out Shift Register* using *behavioral approach* can be written as follows.

```
libraryieee;
use ieee.std_logic_1164.all;
```

```
entity dpigo is
port(CLK,load: in std_logic;
d: in std_logic_vector (3 downto 0);
dout: out std_logic);
end dpiso;
architecture arch_dpiso of dpiso is
signal reg: std_logic_vector (3 downto 0);
begin
process (CLK)
begin
if (CLK'event and CLK='1') then
if (load ='1') then reg <= d;
else reg <= reg (2 downto 0) & '0';
end if;
end if;
end process;
dout <= reg (3);
end;
```

4-bit Parallel in Parallel out Shift Register: The VHDL program for the *4-bit Parallel in Parallel out Shift Register* shown in Fig. 9.13 can be written as follows. This program follows *structural approach.*

```
library ieee;
use ieee.std_logic_1164.all;
entity pipo is
port(D:in std_logic_vector(0 to 3);
reset,CLK:instd_logic;
  Q:outstd_logic_vector(0to3));
end pipo;
architecture arch_pipo of pipo is
component dff1 is
port(D,CLK,reset:in std_logic;
  Q:out std_logic);
end component;
begin
a1:dff1 port map (D(0),CLK,reset,Q(0));    - Delay FF component is used at
a2:dff1 port map (D(1),CLK,reset,Q(1));    - instances a1, a2, a3 and a4
a3:dff1 port map (D(2),CLK,reset,Q(2));
a4:dff1 port map (D(3),CLK,reset,Q(3));
```

```
end;
```

16.10.3 Counters

4-bit Asynchronous / Ripple Counter: The VHDL program for the *4-bit Asynchronous / Ripple Counter* shown in Fig. 8.1 can be written as follows. This program follows *structural approach.*

```
library ieee;
use ieee.std_logic_1164.all;
entity ripple_counter is
port(Vcc,CLK,reset:in std_logic;
Q, NQ:inout std_logic_vector(0 to 3)); end ripple_counter;
architecture arch_ripple_counter of ripple_counter is
component jkff1 is
port (J,K,CLK,reset: in std_logic;
  Q, NQ: inout std_logic);
end component;
begin
  a:jkff1 port map (Vcc,Vcc,CLK, reset,Q(0), NQ (0));
  b:jkff1 port map (Vcc,Vcc,Q(0),reset,Q(1), NQ (1));
  c:jkff1 portmap (Vcc,Vcc,Q(1),reset,Q(2), NQ(2));
  d:jkff1 portmap (Vcc,Vcc,Q(2),reset,Q(3),NQ(3));
end;
```

The external clock pulse is applied to the first flip-flop only and, '1' signal (Vcc) is given to JK inputs of all the Flip-Flops. The output of one flip-flop is connected to clock input of the next flip-flop.

4-bit Synchronous Binary Counter: The VHDL program for the *4-bit Synchronous Binary Counter* shown in Fig. 8.11 can be written as follows. This program follows *structural approach.*

```
library ieee;
use ieee.std_logic_1164.all;
entity binarycounter is
port (Vcc,CLK:in std_logic;
  Q,NQ:inout std_logic_vector(0 to 3));
end binarycounter;
architecture arch_binarycounter of binarycounter is
signal X1,Y1:std_logic;
component jkff1 is
port(J,K,CLK:in std_logic;
Q,NQ: inout std_logic);
end component;
```

```
component andgate is
port(A,B:in std_logic;
  Y:out std_logic);
end component;
component andgates is
port(A,B,C:in std_logic;
  Y:out std_logic);
end component;
begin
  a:jkff1port map (Vcc,Vcc,CLK,Q(0),NQ(0));
  b:jkff1 port map (Q(0),Q(0),CLK,Q(1),NQ(1));
  c:andgate port map (Q(1),Q(0),X1);
  d:jkff1 port map (X1,X1,CLK,Q(2),NQ(2));
  f:andgates port map (Q(0),Q(1),Q(2),Y1);
  g:jkff1 port map (Y1,Y1,CLK,Q(3),NQ(3));
end;
```

3-bit Synchronous Up/Down Counter: The VHDL program for the *3-bit Synchronous Up/Down Counter* shown in Fig. 8.34 can be written as follows. This program follows structural approach.

```
library ieee;
use ieee.std_logic_1164.all;
entity updowncounter is
port(Vcc,Up,CLK,reset:instd_logic;
  Q, NQ:inout std_logic_vector(0 to 2));
end updowncounter;
architecture arch_updowncounter of updowncounter is
signal X1,Y1,Z1,A1,B1,C1,D1:std_logic;
component jkff1 is
port (J,K,CLK: in std_logic;
  Q, NQ: inout std_logic);
end component;
component andgate is
port(A,B:in std_logic;
  Y:out std_logic);
end component;
component notgate is
port(A:in std_logic;
  Y:out std_logic);
```

```
end component;
component orgate is
port(A,B:in std_logic;
  Y:out std_logic);
end component;
begin
  ff1:jkff1 port map(Vcc,Vcc,CLK,Q(0),NQ(0));
  ag1:andgate port map(Q(0),Up,X1);
  ng1:notgate port map (Up,Y1);        -  Up control i/p is inverted to get
  ag2:andgate port map(NQ(0),Y1,Z1);   -  Down (active low)
  og1 :orgate port map(X1,Z1,A1);
  ff2:jkff1 port map(A1,A1,CLK,Q(1),NQ(1));
  ag3:andgate port map(Q(1),X1,B1);
  ag4:andgate port map(Z1,NQ(1),C1);
  og2:orgate port map(B1,C1,D1);
  ff3:jkff1 port map(D1,D1,CLK, Q(2),NQ(2));
end;
```

When Up = 1, count starts from 0 to $2^n - 1$, where n is number of Flip-Flops. When Up = 0 (i.e., down mode), count starts from $2^n - 1$ to 0.

4-bit Ring Counter: The VHDL program for the *4-bit Ring Counter* shown in Fig. 9.19 can be written using *behavioral approach* following Table 9.7 as follows.

```
library ieee;
use ieee.std_logic_1164.all;
entity ringcount is
port (CLK,CLR: in std_logic; Q: inout std_logic_vector (0 to 3));
end ringcount;
architecture ringcountarch of ringcount is
begin
process (CLK,CLR)
begin
if(CLR='0') then
  Q<="1000";
elsif(CLR='1') then
 if (CLK = '1') and CLK'event then
            Q(0)<=Q(3);
            for i in 0 to 2 loop
            Q(i + 1)<=Q(i);
            end loop;
```

```
  end if;
 end if;
 end process; end ringcountarch;
```

4-bit Johnson Counter: The following is D-Flip-Flop program that follows *behavioral approach* and is used as component in Johnson counter.

```
library ieee;
use ieee.std_logic_1164.all;
entity dff2 is
port(D,CLK,reset:in std_logic;
  Q,NQ:inout std_logic);
enddff2;
architecture arch_dflipflop of dff2 is
begin
process (clk)
begin
if clk'event and clk='1' then—CLK = 1:
if reset='0' then                              - Reset = 0 q = 0
     Q<='0';                                    - Reset = 1 q = D
       NQ<='1';
     else
       Q<=D;
  NQ<=not (Q);
     end if;
   end if;
  end process;
end;
```

The VHDL Program for the *4-bit Johnson Counter* shown in Fig. 9.25 can be written as follows that uses the above D-Flip-Flop program as component.

```
entity Johnson is
port  (CLK,reset:in std_logic;
      Q,NQ: inout std_logic_vector(0 to 3));
end Johnson;
architecture arch_johnson of Johnson is
component dff2 is
port  (D,CLK,reset:in std_logic;
      Q, NQ :inout std_logic);
end component;
begin
```

```
        a:dff2 port map (NQ(3),CLK,reset,Q(0),NQ(0));
        b:dff2 port map (Q(0), CLK,reset,Q(1),NQ(1));
        c:dff2 port map (Q(1), CLK,reset,Q(2),NQ(2));
        d:dff2 port map (Q(2), CLK,reset,Q(3),NQ(3));
end;
```

16.11 VHDL PROGRAM FOR ARITHMETIC LOGIC UNIT

An Arithmetic logic unit is a logic circuit that performs various Boolean and arithmetic operations on n-bit operands. In this design, ALU has two 4-bit data inputs, A and B, 3-bit select inputs, S and four-bit output F. The output is defined by various arithmetic and Boolean operations on the inputs A and B as shown in the Table 16.1.

Table 16.1 4-bit ALU operation

Operation	Inputs (S3S2S1S0)	Outputs (F)
Clear	0000	000
ADD	0001	A + B
SUB	0010	A – B
MUL	0011	A B
DIV	0100	A/B
AND	0101	A AND B
OR	0110	A OR B
NOT	0111	NOT A
NAND	1000	A NAND B
NOR	1001	A NOR B
XOR	1010	A XOR B
XNOR	1011	A XNOR B
Preset	1100	1111

```
library ieee;
use ieee.std_logic_1164.all;
use ieee.std_logic_unsigned.all
entity alu is
port(A,B : in std_logic_vector(3 downto 0);
   S : in std_logic_vector(3 downto 0);
   F : out std_logic_vector(3 downto 0));
end alu;
architecture arch_alu of alu is begin
process(S,A,B)
begin
```

```
if S = "0000" then
   F<="0000";
elsif S = "0001" then
   F<=A + B;
elsif S = "0010" then
   F<=A – B;
elsif S = "0011" then
   F<=A   B;
elsif S = "0100" then
   F<=A/B;
elsif S = "0101" then
   F<=A and B;
elsif S = "0110" then
   F<=A or B;
elsif S = "0111" then
   F<=not A;
elsif S = "1000" then
   F <= A nand B;
elsif S = "1001" then
   F<=A nor B;
elsif S = "1010" then
   F<=A xor B;
elsif S = "1011" then
F<=A xnor B;
elsif S = "1100" then
end if;
end process;
end;
```

■ REVIEW QUESTIONS ■

1. What is an entity in VHDL?
2. What is the significance of architecture declaration in VHDL?
3. Describe dataflow approach for hardware description with an example.
4. How is the signal assignment done?
5. What is the significance of the signal assignment?
6. What is structural description?
7. Distinguish among entity, component and component instance.
8. What do you understand by *event time simulation*?

9. Distinguish between functional and timing simulations?
10. What is inertial delay model?
11. Explain the transport delay model with an example.
12. What is the significance of the transport delay model?
13. What are the different *'types'* used in VHDL?
14. What is meant by the types 'bit' and 'bit Vector'?
15. What is the 'time' type?
16. Discuss the logical, arithmetic and relational operators used in VHDL.
17. What is the use of Concatenation operator?
18. Describe the behavioral approach for hardware description.
19. How can the behavioral approach also be called as black box approach?
20. Write a short note on process statement.
21. How are the variables declared in VHDL?
22. What are sequential statements?
23. Write a short note on *'loop'* and *'for'* statements.
24. Distinguish between signal and variable assignment in a process.
25. What is the use of *'write'* function?
26. What are the needs of libraries in VHDL?
27. List three packages needed in a design.
28. Distinguish between std_logic and std_ulogic.
29. What are the levels supported by std_ulogic?
30. Develop a VHDL program for a 4-bit parallel binary subtracter using structural approach.
31. Implement BCD to Excess - 3 code converter, shown in Fig. 3.29 using VHDL.
32. Write a VHDL program for 8-to-1 mux and use it as a component for the realization of 64-to-1 mux.
33. Implement a BCD Adder shown in Fig. 5.22 using VHDL.
34. Develop a VHDL program for BCD Adder shown in Fig. 5.23 using 4-bit Binary Adder as a component.
35. Realize 1-to-8 Demux using VHDL.
36. Develop a VHDL code for BCD-to-decimal decoder.
37. Develop a VHDL code for BCD-to-7 Segment decoder.
38. Implement 8-to-3 priority encoder according to Truth Table of Table 6.16 following the behavioral approach.
39. Realize a Binary-to-BCD converter using VHDL.
40. Realize the SR, D and T-flip-flops using JK-flip-flop using VHDL.
41. Implement a mod-10 synchronous counter shown in Fig. 8.25 using VHDL.
42. Realize a 4-bit bi-directional Shift register shown in Fig. 9.16 using VHDL.
43. Implement a 5-bit sequence Generator shown in Fig. 9.23 using VHDL.
44. Implement a mod-10 shift counter using VHDL.

❑❑❑

APPENDIX

Appendix A
Conversion Table

DECIMAL—HEXADECIMAL—BINARY

Dec	*Hex*	*Bin*	*Dec*	*Hex*	*Bin*	*Dec*	*Hex*	*Bin*
0	0	00000000	36	24	00100100	72	48	01001000
1	1	00000001	37	25	00100101	73	49	01001001
2	2	00000010	38	26	00100110	74	4a	01001010
3	3	00000011	39	27	00100111	75	4b	01001011
4	4	00000100	40	28	00101000	76	4c	01001100
5	5	00000101	41	29	00101001	77	4d	01001101
6	6	00000110	42	2a	00101010	78	4e	01001110
7	7	00000111	43	2b	00101011	79	4f	01001111
8	8	00001000	44	2c	00101100	80	50	01010000
9	9	00001001	45	2d	00101101	81	51	01010001
10	a	00001010	46	2e	00101110	82	52	01010010
11	b	00001011	47	2f	00101111	83	53	01010011
12	c	00001100	48	30	00110000	84	54	01010100
13	d	00001101	49	31	00110001	85	55	01010101
14	e	00001110	50	32	00110010	86	56	01010110
15	f	00001111	51	33	00110011	87	57	01010111
16	10	00010000	52	34	00110100	88	58	01011000
17	11	00010001	53	35	00110101	89	59	01011001
18	12	00010010	54	36	00110110	90	5a	01011010
19	13	00010011	55	37	00110111	91	5b	01011011
20	14	00010100	56	38	00111000	92	5c	01011100
21	15	00010101	57	39	00111001	93	5d	01011101
22	16	00010110	58	3a	00111010	94	5e	01011110
23	17	00010111	59	3b	00111011	95	5f	01011111
24	18	00011000	60	3c	00111100	96	60	01100000
25	19	00011001	61	3d	00111101	97	61	01100001
26	1a	00011010	62	3e	00111110	98	62	01100010
27	1b	00011011	63	3f	00111111	99	63	01100011
28	1c	00011100	64	40	01000000	100	64	01100100
29	1d	00011101	65	41	01000001	101	65	01100101
30	1e	00011110	66	42	01000010	102	66	01100110
31	1f	00011111	67	43	01000011	103	67	01100111
32	20	00100000	68	44	01000100	104	68	01101000
33	21	00100001	69	45	01000101	105	69	01101001
34	22	00100010	70	46	01000110	106	6a	01101010
35	23	00100011	71	47	01000111	107	6b	01101011

Dec	*Hex*	*Bin*	*Dec*	*Hex*	*Bin*	*Dec*	*Hex*	*Bin*
108	6c	01101100	149	95	10010101	190	be	10111110
109	6d	01101101	150	96	10010110	191	bf	10111111
110	6e	01101110	151	97	10010111	192	c0	11000000
111	6f	01101111	152	98	10011000	193	c1	11000001
112	70	01110000	153	99	10011001	194	c2	11000010
113	71	01110001	154	9a	10011010	195	c3	11000011
114	72	01110010	155	9b	10011011	196	c4	11000100
115	73	01110011	156	9c	10011100	197	c5	11000101
116	74	01110100	157	9d	10011101	198	c6	11000110
117	75	01110101	158	9e	10011110	199	c7	11000111
118	76	01110110	159	9f	10011111	200	c8	11001000
119	77	01110111	160	a0	10100000	209	c9	11001001
120	78	01111000	161	a1	10100001	202	ca	11001010
121	79	01111001	162	a2	10100010	203	cb	11001011
122	7a	01111010	163	a3	10100011	204	cc	11001100
123	7b	01111011	164	a4	10100100	205	cd	11001101
124	7c	01111100	165	a5	10100101	206	ce	11001110
125	7d	01111101	166	a6	10100110	207	cf	11001111
126	7e	01111110	167	a7	10100111	208	d0	11010000
127	7f	01111111	168	a8	10101000	209	d1	11010001
128	80	10000000	169	a9	10101001	210	d2	11010010
129	81	10000001	170	aa	10101010	211	d3	11010011
130	82	10000010	171	ab	10101011	212	d4	11010100
131	83	10000011	172	ac	10101100	213	d5	11010101
132	84	10000100	173	ad	10101101	214	d6	11010110
133	85	10000101	174	ae	10101110	215	d7	11010111
134	86	10000110	175	af	10101111	216	d8	11011000
135	87	10000111	176	b0	10110000	217	d9	11011001
136	88	10001000	177	b1	10110001	218	da	11011010
137	89	10001001	178	b2	10110010	219	db	11011011
138	8a	10001010	179	b3	10110011	220	dc	11011100
139	8b	10001011	180	b4	10110100	221	dd	11011101
140	8c	10001100	181	b5	10110101	222	de	11011110
141	8d	10001101	182	b6	10110110	223	df	11011111
142	8e	10001110	183	b7	10110111	224	e0	11100000
143	8f	10001111	184	b8	10111000	225	e1	11100001
144	90	10010000	185	b9	10111001	226	e2	11100010
145	91	10010001	186	ba	10111010	227	e3	11100011
146	92	10010010	187	bb	10111011	228	e4	11100100
147	93	10010011	188	bc	10111100	229	e5	11100101
148	94	10010100	189	bd	10111101	230	e6	11100110

Dec	*Hex*	*Bin*	*Dec*	*Hex*	*Bin*	*Dec*	*Hex*	*Bin*
231	e7	11100111	240	f0	11110000	249	f9	11111001
232	e8	11101000	241	f1	11110001	250	fa	11111010
233	e9	11101001	242	f2	11110010	251	fb	11111011
234	ea	11101010	243	f3	11110011	252	fc	11111100
235	eb	11101011	244	f4	11110100	253	fd	11111101
236	ec	11101100	245	f5	11110101	254	fe	11111110
237	ed	11101101	246	f6	11110110	255	ff	11111111
238	ee	11101110	247	f7	11110111			
239	ef	11101111	248	f8	11111000			

Appendix B
Summary of Laws of Boolean Algebra

Indempotent

$A + A = A$ $\quad$ $A \ A = A$

Associative

$(A + B) + C = A + (B + C)$ $\quad$ $(A \ B) \ C = A \ (B \ C)$

Commutative

$A + B = B + A$ $\quad$ $A \ B = B \ A$

Distributive

$A + (B \ C) = (A + B) \ (A + C)$ $\quad$ $A \cdot A = A$

Identify

$A + 0 = A$ $\quad$ $A + 1 = 1$

$A \ 0 = 0$ $\quad$ $A \ 1 = A$

Complement

$A + \overline{A} = 1$ $\quad$ $\overline{\overline{A}} = A$

$A \cdot \overline{A} = 0$ $\quad$ $\overline{1} = 0$

DeMorgan's

$\overline{A + B} = \overline{A} \cdot \overline{B}$ $\quad$ $\overline{A \cdot B} = \overline{A} + \overline{B}$

Duality

Interchanging union and intersection operators, as well as Universal and Null sets. The resulting equation is equivalent to the original.

Appendix C
Logic Gates: Implementation and Truth Tables

Gate symbol	*Gate implementation*	*Truth table*
INVERTER  7404	7400 7402	A X 0 1 1 0
BUFFER 7407	7404 7404 7408	A X 0 0 1 1
AND 7408	7400 7407 7400 7400	A B X 0 0 0 0 1 0 1 0 0 1 1 1
NAND 7400	7400 7400	A B X 0 0 0 0 1 1 1 0 1 1 1 1
OR 7432	7402 7404 7402 7402	A B X 0 0 0 0 1 1 1 0 1 1 1 1

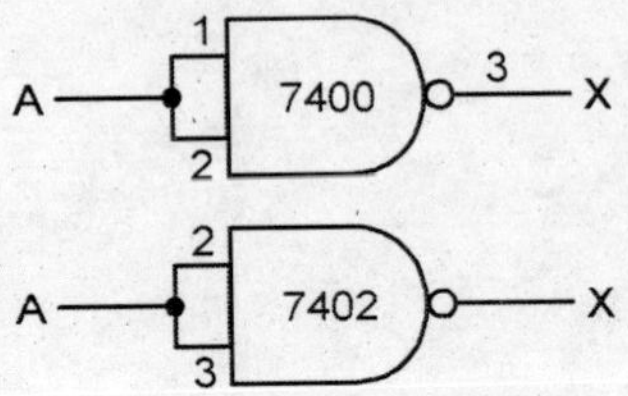

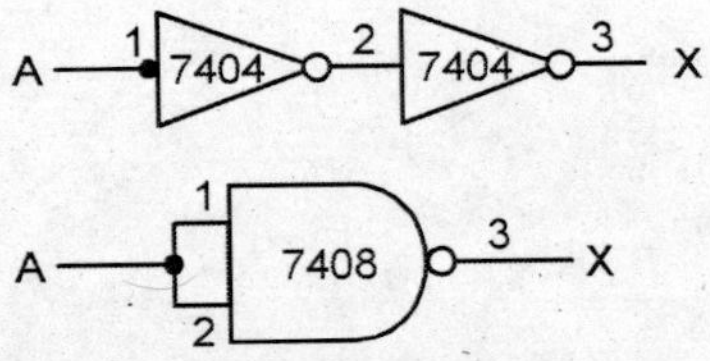

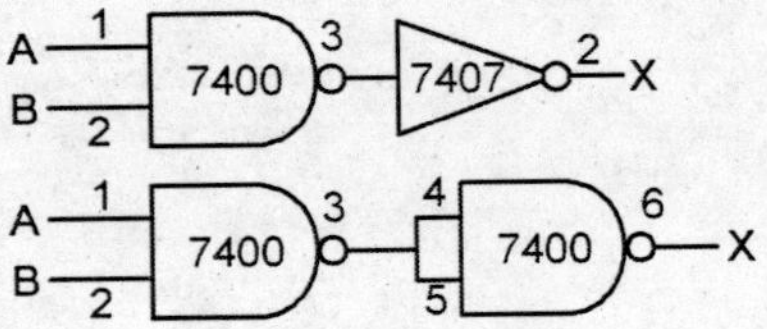

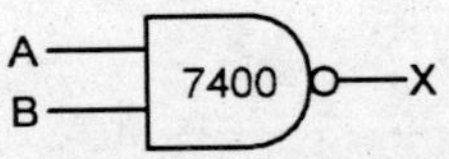

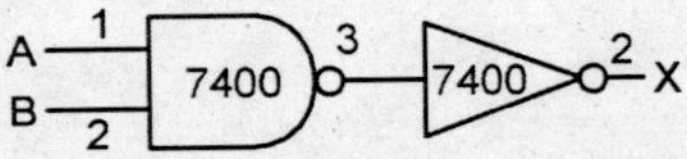

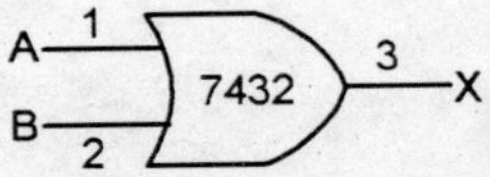

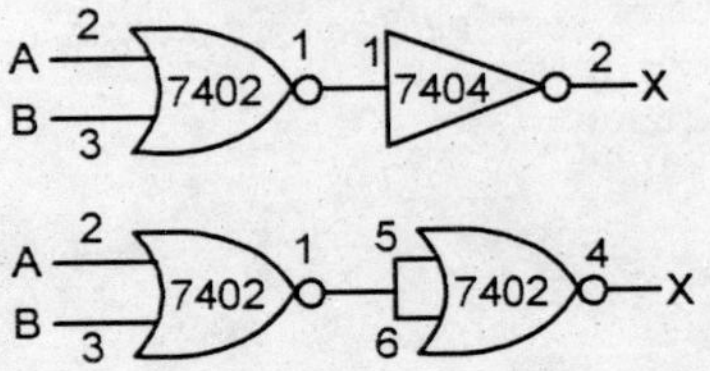

Appendix D
74XX Series TTL

7400 Quad 2-input NAND gate
7401 Quad 2-input NAND gate (Open Collector)
7402 Quad 2-input NOR gate
7403 Quad 2-input NAND gate (Open Collector)
7404 Hex Inverter
7405 Hex Inverter (Open Collector)
7406 Hex. Inverter buffer/driver (Open Collector)
7407 Hex buffer/driver (Open Collector)
7408 Quad 2-input AND gate
7409 Quad 2-input AND gate (Open Collector)
7410 Triple 3-input NAND gate
7411 Triple 3-input AND gate
7412 Triple 3-input NAND gate (Open Collector)
7413 Dual 4-input NAND Schmitt Trigger
7414 Hex Inverter Schmitt Trigger
7415 Triple 3-input AND gate (Open Collector)
7416 Hex Inverter buffer/driver (Open Collector)
7417 Hex buffer/driver (Open Collector)
7418 Dual 4-input NAND Schmitt Trigger
7419 Hex Inverter Schmitt Trigger
7420 Dual 4-input NAND gate
7421 Dual 4-input AND gate
7422 Dual 4-input NAND gate (Open Collector)
7423 Expandable dual 4-input NOR gate with strobe
7424 Quad 2-input NAND Schmitt Trigger
7425 Dual 4-input NOR gate with strobe
7426 Quad 2-input TTL-MOS interface NAND gate (Open Collector)
7427 Triple 3-input NOR gate
7428 Quad 2-input NOR buffer
7430 8-input NAND gate
7431 Delay Element
7432 Quad 2-input OR gate
7433 Quad 2-input NOR buffer (Open Collector)
7437 Quad 2-input NAND buffer
7438 Quad 2-input NAND buffer (Open Collector)
7439 Quad 2-input NAND buffer (Open Collector)
7440 Dual 4-input NAND buffer
7441 BCD to Decimal decoder-Nixie Driver
7442 BCD to Decimal decoder
7443 Excess 3 to decimal decoder
7444 Gray to decimal decoder
7445 BCD to Decimal decoder/Driver (Open Collector)
7446 BCD to Seven segment decoder/ Driver (Active Low-open collector output)
7447 BCD to Seven segment decoder/ Driver (Active Low-open collector output)
7448 BCD to Seven segment decoder/ Driver (Active High output)

7449 BCD to Seven segment decoder/ Driver (Active High-open collector output)

7450 Expandable dual 2 input 2 wide AND-OR-INVERT gate

7451 Dual 2 input 2 wide AND-OR-INVERT gate

7452 Expandable 4 wide AND-OR gate

7453 Expandable 2 input 4 wide AND-OR-INVERT gate

7454 4 Wide AND-OR-INVERT gate

7455 Expandable 4 input 2 wide AND-OR-INVERT gate

7456 Frequency divider (50 to 1)

7457 Frequency divider (60 to 1)

7459 Dual 2-3 input 2 wide AND-OR-INVERT gate

7460 Dual 4 input Expander

7461 Triple 3 input Expander

7462 2-3-3-2 input 4 wide AND-OR Expander

7463 Hex current sensing Interface gate

7464 4-2-3-2 input 4 wide AND-OR-INVERT gate

7465 4-2-3-2 input 4 wide AND-OR-INVERT gate (Open Collector)

7468 Dual 4 bit decade or binary counter

7469 Dual 4 bit decade or binary counter

7470 Positive Edge Triggered JK Flip-Flop

7472 JK master-slave Flip-Flop

7473 Dual JK Flip-Flop

7474 Dual D Flip-Flop (Positive Edge Triggered)

7475 Quad Bistable Latche

7476 Dual JK Flip-Flop

7478 Dual JK Flip-Flop

7480 Gates full adder

7482 2-bit binary full adder

7483 4-bit binary full adder

7485 4-bit magnitude comparator

7486 Quad 2-input EXCLUSIVE-OR gate

7489 64 bit random-access read-write memory

7490 Decade counter

7491 8-bit Serial-in Serial-out Shift register

7492 Divide by 12 Counter

7493 4-bit binary ripple counter

7494 Dual 4-bit Serial-in Parallel-in Serial-out Shift register

7495 4-bit shift register

7496 5-bit Shift register

7497 6-bit synchronous binary rate multipliers

7498 4-bit Data selectors/storage registers

7499 4-bit PISO/SIPO Right shift Left shift register

74100 8-bit bistable latche

74101 Negative EdgeTriggered JK Flip-Flop with Preset

74102 Negative EdgeTriggered JK Flip-Flop with Preset and clear

74103 Dual Negative EdgeTriggered JK Flip-Flop with clear

74104 JK master-slave Flip-Flop

74105 JK master-slave Flip-Flop 74107 Dual JK Flip-Flop with clear

74109 Dual JK positive-edge-triggered Flip-Flop with preset and clear

74110 JK master slave Flip-Flop with data lock-out 74116 Dual 4-bit latches with clear

74120 Dual pulse synchronizers/driver

74121 Monostable multivibrator

74122 Retriggerable Monostable Multivibrator with clear

74123 Dual retriggerable Monostable multivibrator

74124 Dual voltage controlled oscillators

74125 Three-state quad bus buffer

74126 Three-state quad bus buffer

74128 Quad 2-input NOR Buffer
74132 Quad 2-input NAND Schmitt trigger
74133 13-input NAND gate
74134 Tristate 12-input NAND gate
74135 Quad Ex-OR/NOR gate
74136 Quad 2-input EXCLUSIVE-OR gate (Open Collector)
74137 3 to 8 decoder/demultiplexer with Address Latches
74138 3 to 8 decoder/demultiplexer
74139 Dual 2 to 4 decoder/demultiplexer
74140 Dual 4-input NAND 50 ohm line driver
74141 BCD-to-decimal decoder/driver
74142 BCD counter-latch-driver/Decoder
74143 4-bit counter/latch, 7-seg LED driver (Active Low output)
74144 4-bit counter/latch, 7-seg LED driver (Open Collector)
74145 BCD-to-decimal decoder/driver
74147 10 to 4 line priority encoder
74148 8 to 3 line Priority encoder
74150 16-line-to-1-line multiplexer
74151 8 to 1 multiplexer
74152 8 to 1 selector-multiplexer (with complemented output)
74153 Dual 4 to 1 multiplexer
74154 4-line-to-16-line decoder/ demultiplexer
74155 Dual 2 to 4 demultiplexer/decoder
74156 Dual 2 to 4 demultiplexer/decoder
74157 Quad 2 to 1 data selector/multiplexer
74158 Quad 2 to 1 data selector/multi plexer (Inverted output)
74159 4 line to 16 line decoder/ emultiplexers (Open collector)
74160 Decade synchronous counter with direct clear
74161 Synchronous 4-bit Binary counter with direct clear
74162 Synchronous decade counter
74163 Synchronous 4-bit binarycounter
74164 8-bit serial in-parallel-out shift register
74165 Parallel-load 8-bit serial shift register (8 bit PISO)
74166 Parallel load 8-bit PISO shift register
74167 Synchronous decade rate multipliers
74168 Presettable synchronous Up/down decade counter
74169 Presettable synchronous Up/down binary counter
74170 4 to 4 register file (Open Collector)
74173 4-bit three state D type register
74174 Hex D flip-flop with clear
74175 Quad D flip-flop with clear
74176 35 MHz presettable decade and Binary counter
74177 35 MHz presettable binary counter
74179 4-bit parallel-access shift register
74180 9-bit odd-even parity generator/ checker
74181 4-Bit Arithmetic-logic Unit
74182 Look-ahead carry generator
74184 BCD-to-Binary Converter
74185 Binary-to-BCD converter
74189 Three state 64-bit random-access-memory
74190 Synchronous up-down decade counter
74191 Synchronous binary up-down counter
74192 Synchronous Binary up-down counter
74193 Binary up-down counter
74194 4-bit Bidirectional shift register
74195 4-bit parallel-access shift register
74196 50 MHz Presettable decade or binary counter
74197 Presettable decade or binary counter
74198 8-bit shift register

74199 8-bit shift register
74221 Dual one-shot Schmitt trigger
74251 Three-state 8-to-l multiplexer
74259 8-bit addressable latch
74276 Quad JK flip-flop
74279 Quad $\overline{S} - \overline{R}$ Latches
74283 4-bit binary full adder with fast carry
74284 4-bit by 4-bit parallel Binary multiplier
74285 4-bit by 4-bit parallel Binary multiplier
74365 Three-state Hex buffer/Driver
74366 Three-state Hex inverter buffer
74367 Three-state Hex buffer/Driver
74368 Three-state Hex inverter buffer
74390 Individual clocks with flip-flops provide Dual ÷ 2 and ÷ 5 counter
74393 Dual 4-bit binary counter with Individual clock

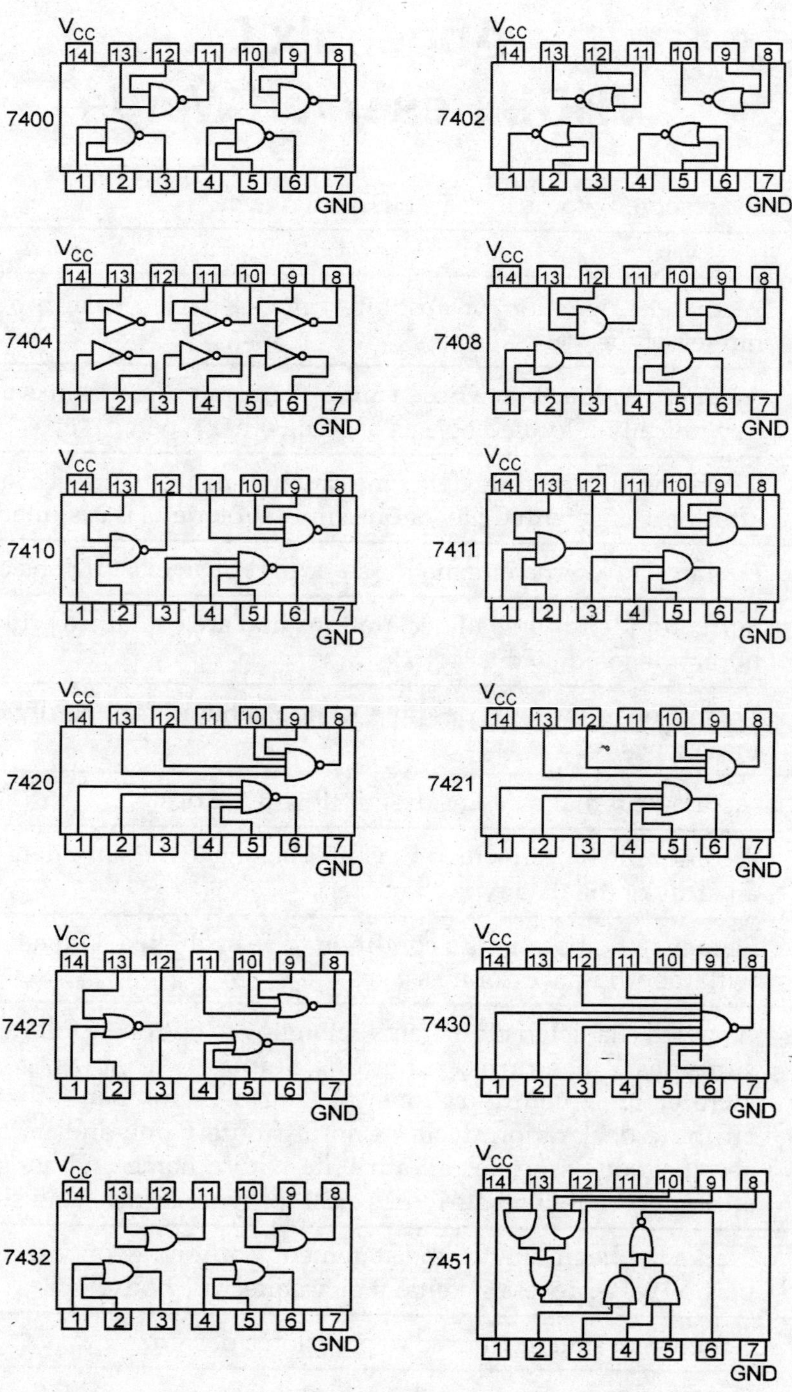
7400
7402
7404
7408
7410
7411
7420
7421
7427
7430
7432
7451
VCC
GND

Appendix E
VHDL Reserved Words

Following are the reserved words in the VHDL language.

Reserved Word	*Purpose*
`abs`	Arithmetic operator for absolute value. Unary operator, predefined for any numeric type.
`access`	A variety of data type whose values are pointers to (or links to, or addresses of) dynamically-allocated objects of some other type.
`after`	Clause used to include delay information in a signal assignment. If there is no after clause, default delay of one simulation delta is assumed.
`alias`	Declares an alternate name for all or part of an existing object.
`all`	Suffix for identifying all declarations that are contained within the package or library denoted by the prefix.
`and`	Logical operator for types bit and Boolean and for one-dimensional arrays of these types.
`architecture`	Statement that contains description of the design.
`array`	A composite type in which all values belong to the same data type (e.g., string is an array of the data type character).
`assert`	Statement that presents a condition to be evaluated. Often used in conjunction with reporting of error messages.
`attribute`	A named characteristic of items belonging to one of the following classes: Types, subtypes Procedures, functions Signals, variables, constants Entities, architectures, configurations, packages Components Statement labels An attribute declaration declares an attribute name and its type. An attribute specification associates an attribute with a name and assigns a value to the attribute. Predefined attributes exist for types, arrays, and signals.
`begin`	Marks the beginning of the statement portion (as opposed to the declarative portion) of a process statement or architecture body.
`block`	Concurrent statement used to partition a design.
`body`	Conjoined with package, A package body stores the definitions of functions, procedures and the complete constant declarations for any deferred constants

	that appear in a corresponding package declaration. The name of the package body is the same as that the package declaration to which it refers.
`buffer`	A mode that enables a port to be read and updated within the entity model. A buffer port cannot have more than one source, and can be connected only to another buffer port or to a signal with no more than one source.
`bus`	A kind of signal that represents a hardware bus. When all drivers to the signal become disconnected, the signal's value is determined by calling the resolution function with all the drivers off. Any previous value is lost. Bus signals may be either a port or locally declared signal.
`case`	A form of conditional control that selects statements for execution based on the value of a given expression.
`component`	Declaration made in a top level entity to instantiate lower-level entities.
`configuration`	Associates particular component instances with specific design entities, and associates entity declarations with specific architectures.
`constant`	A class of data object. Constants can hold a single value of a given type. If the value is not specified, the constant is a deferred constant, and can appear inside a package declaration only.
`disconnect`	Specifies the disconnect time for a guarded signal.
`down to`	Specifies direction in a range.
`else`	Optional clause in an if statement. An else clause specifies alternative statements when the if clause and any elsif clauses evaluate false.
`elsif`	Clause in an if statement that poses an alternative condition when the if clause evaluates to false.
`end`	Marks the end of a statement, subprogram, or declaration of a library unit.
`entity`	Specifies input and output definitions of the design.
`exit`	Causes execution to jump out of the innermost loop or the loop whose label is specified.
`file`	A category of data type. File types provide a way for a VHDL design to communicate with the host environment. File type is declared with a file type definition, while files are declared with a file declaration.
`for`	Used to iterate a predetermined number of replications in replicated logic, such as generate and loop statements. Also used in specifying blocks, components, and configurations, and in specifying time expression in a time-out clause.
`function`	A subprogram used for computing a single value. Functions are always

	terminated by a return statement, which returns a value. Functions are specified with a subprogram specification.
`generate`	Replicates one or more concurrent statement. Can be in for or if format.
`generic`	Passes environment information to subcomponents; can be declared in the same constructs in which ports can be declared. Generics are of the object class constant. The declaration of a generic may also include a default value, which will be used if an actual value is missing in the generic map.
`guarded`	Option for a concurrent signal assignment. The guarded option specifies that the signal assignment statement will execute only when the guard condition of the block statement that contains the assignment is true.
`if`	Conditional logic statement. Presents a condition to be evaluated as true or false.
`impure`	Option for a function in a subprogram specification. Use of this reserved word extends the scope of variables and signals declared outside of the function to be available to that function, resulting in the possibility that the function may return different values when called multiple times with the same actual parameter values.
`in`	Port mode that allows the port to be read only. If no mode is specified, in is assumed.
`inertial`	An option for specifying delay mechanism in a signal assignment statement. Inertial delay is characteristic of switching circuits: a pulse whose duration is shorter than the switching time of the circuit will not be transmitted or in the case that a pulse rejection limit is specified, a pulse whose duration is shorter than that limit will not be transmitted.
`in out`	Port mode that allows a bidirectional port to be read and updated within the entity model.
`is`	Reserved word that equates the identity portion to the definition portion of a declaration.
`label`	An entity class, to be stated during attribute specification of user-defined attributes.
`library`	A context clause that makes visible the logical names of design libraries that can be referenced within a design unit. The following library clause is implied for every design unit: `library std, work`
`linkage`	A port mode similar to in out used to connect VHDL ports to non-VHDL ports.
`literal`	An entity class, to be stated during attribute specification of user-defined attributes.

`loop`	Statement used to iterate through a set of sequential statements.
`map`	With port or generic, associates port names within a block (local) to names outside a block (external). A port of mode may be left unconnected either by omitting it from the port map, or by connecting it to the reserved word open. In either case, the corresponding port declaration must include a default value.
`mod`	Arithmetic operator for modulus. Modulus is predefined for any integer type; the operands and the result are of the, same type. The result of a mod operator has the sign of the second operand and is defined (for some integer n) as: `a mod b = a - b*n`
`nand`	Logical operator for types bit and Boolean and for one-dimensional arrays of these types. Complement of and.
`new`	An allocator that enables objects of a specific type to be created dynamically. These dynamically-created objects are accessed by access types.
`next`	Statement that causes the current iteration of the specified loop to be prematurely terminated, resuming execution with the next iteration of the loop.
`nor`	Logical operator for types bit and Boolean and for one-dimensional arrays of these types. Complement of or.
`not`	Unary logical operator for types bit and Boolean.
`null`	Sequential statement that causes no action to take place; execution continues with the next statement
`of`	Reserved word used to link an identifier to its entity name, and used when specifying type mark in a file type definition.
`on`	Used to introduce the sensitivity list in the sensitivity clause of a wait statement.
`open`	An entity aspect, used as a binding indication to indicate that binding is not yet specified and that it is to be deferred.
`or`	Logical operator for types bit and Boolean and for one-dimensional arrays of these types.
`others`	When used as the last branch of case statement, used to cover all values not specified by when statements. Can also be used as part of the right-hand side of a signal or variable assignment statement for array types. This assigns values to array elements not otherwise assigned.
`out`	Port mode that enables the port to be updated only. It cannot be read.
`package`	Optional library unit for making shared definitions (usually type definitions). You must issue a use-statement to make the package available to other parts of the design.

`port`	Signals through which an entity communicates with the other models in its external environment.
`postponed`	Option for a concurrent signal assignment or process statement.
`procedure`	Subprogram used to partition large behavioral descriptions. Procedures can return zero or more values.
`process`	A process represents a level of hierarchy in a design. The statements contained in the `process_statement_partrun sequentially` (from top to bottom) rather than concurrently. If the process includes the optional `sensitivity_list`, the `process_statement_part` is executed only-when there is an event on one or more of the signals listed in the `sensitivity_list`. For simulation, the `process_statement_part` of all processes is executed once when the simulation initializes. Processes with sensitivity lists will not execute again until there is an event on one of the signals in the `sensitivity_list`. Processes without a sensitivity list will continue to re-execute their `process_statement_part` for the remainder of the simulation. This implies that the `process_statement_part` should include at least one wait statement. Otherwise, the simulation time will not advance and the simulator will appear to be frozen. For synthesis, processes may be used to infer either sequential (clocked) or combinational logic. Processes intended to infer combinational logic should include in the `sensitivity_list` all signals that affect the behavior of the `process_statement_part`. This includes not only signals appearing on the right-hand side of signal or variable assignment statements, but also signals or variables appearing as part of conditional statements such as if or case. Processes that infer synchronous logic should include the clock signal and any asynchronous controls (asynchronous resets or presets) in the sensitivity list. In general, processes may or may not be synthesizable, depending on the details of how they are written. See the IEEE VHDL User Manual for details.
`pure`	Option for a function in a subprogram specification. A pure function will disallow the use of any signals or variables declared outside of the function. All functions are pure unless specified as impure.
`range`	Parameter used when specifying subtypes in an array type declaration.
`record`	A composite data type in which the collection of values may belong to the same or different types.
`register`	A kind of signal which models a latch. If all drivers to such a signal are disconnected, the signal retains its old value.

`reject`	An option for specifying delay mechanism in a signal assignment statement. Every inertially delayed signal assignment has a pulse rejection limit. If the delay mechanism specifies inertial delay, and if the reserved word reject followed by a time expression is present, then the time expression specifies the pulse rejection limit. In all other cases, the pulse rejection limit is specified by the time expression associated with the first waveform element. Not supported for synthesis.
`rem`	Arithmetic operator for remainder. Remainder is predefined for any integer type; the operands and the result are of the same type. The result of a rem operator has the sign of the first operand and is defined as: `a rem b = a - (a/b)*b`
`report`	Statement for generating report messages. Not supported for synthesis.
`return`	Statement that causes a subprogram to terminate, returning control back to the calling object. All functions must have a return statement, and the value of the expression in the return statement is returned to the calling program. For procedures, objects of mode out and inout return their values to the calling program.
`rol`	Shift operator: rotate left. Shift operators are defined for any one-dimensional array type whose element type is either bit or Boolean. The arguments of rol are the array that will be rotated and the amount by which it will be rotated.
`ror`	Shift operator: rotate right. Shift operators are defined for any one-dimensional array type whose element type is either bit or Boolean. The arguments of rol are the array that will be rotated and the amount by which it will be rotated.
`select`	Expression whose value determines different values for a target signal in a selected signal assignment statement.
`severity`	A predefined type in the language with values note, warning, error, and failure.
`shared`	An type of variable that can be declared only in entities, architectures, and generates. A shared variable can be accessed by all three of the subprograms/processes local to the declarative region.
`signal`	Represents a wire or a placeholder for a value. Signals are assigned in signal assignment statements, and declared in signal declarations. Note that signal assignments always occur with some amount of delay. In the absence of the optional `delay_mechanism`, signal assignments will occur one delta delay after the signal assignment statement is executed. This fact has major implications when a signal assignment is executed as part of a block of sequential statements within a process.
`sla`	Shift operator: shift left arithmetic. Shift operators are defined for any one-dimensional array type whose element type is either bit or Boolean. The

	arguments of sla are the array that will be shifted and the amount by which it will be shifted. This shift operator will fill with the leftmost bit.
`sll`	Shift operator: shift left logical; Shift operators are defined for any one-dimensional array type whose element type is either bit or Boolean. The arguments of sll are the array that will be shifted and the amount by which it will be shifted. This shift-operator will fill with zeros.
`sra`	Shift operator: shift right arithmetic. Shift operators are defined for any one-dimensional array type whose element type is either bit or, Boolean. The arguments of sra are the array that will be shifted and the amount by which it will be shifted. This shift operator will fill with the rightmost bit.
`srl`	Shift operator: shift right logical. Shift operators are defined for any one-dimensional array type whose element type is either bit or Boolean. The arguments of sri are the array that will be shifted and the amount by which it will be shifted. This shift operator will fill with zeros.
`subtype`	A declaration that defines a base type and a constraint. The constraint specifies a subset of values for the base type. An object is said to belong to a subtype if it is of the base type and if it satisfies the constraint.
`then`	Introduces statements to execute when the preceding if or elsif statement evaluates true.
`to`	Specifies direction in a range.
`transport`	An option for specifying delay mechanism in a signal assignment statement. Transport delay is characteristic of hardware devices, such as transmission lines, that exhibit nearly infinite frequency response: any pulse is transmitted, no matter how short its duration. Not supported for synthesis.

`type`	Data type. Each data type has a set of values and a set of operations associated with it. User-defined types are created with type declarations. Predefined types can be divided into several categories: scalar, composite, access, and file. In addition, there are non-predefined types established by the IEEE standard 1164.
	Each of these types is listed below.
	• **Scalar types**
	  • Enumerated
	  • Character (literals: 128 characters of the ASCII character set)
	  • Bit Boolean (literals: true, false)
	  • Severity_level (literals: note, warning, error, failure)
	  • Numeric
	  • Integer
	  • Physical
	  • Floating_point

- **Composite Types**
  - Array
  - String (1-dimensional array of type character)
  - Bit_vector (1-dimensional array of type bit)
  - Record
  - Access Types (see access)
  - File Types (see. file)
- **Non-predefined Types**

  In-addition to the predefined types, IEEE standard 1164 adds the following types that are commonly used for modeling digital logic:
  - Std_ulogic (an enumerated type with the values 'U', 'X', '0', '1', 'Z', 'W', 'L', 'H', '–')
  - Std_logic (same as std_ulogic except that this is a resolved type)
  - Std_ulogic_vector (an array of std_ulogic)
  - Std_logic_vector (an array of std_logic)
- **Predefined-Types**

  Types defined by the IEEE standard 1076.3 are:
  - Unsigned (an array of std_logic)
  - Signed (an array of std_logic)
  - Overloaded arithmetic and conversion operators for types unsigned and signed are defined in the package numeric_std.

More information on selecting a data type can be found in the Design Considerations section of the IEEE VHDL Reference Manual.

unaffected	Concurrent statement that causes no action to take place; execution continues with the next statement.
units	An entity class, to be stated during attribute specification of user-defined attributes. Also used in physical type definition statement
units	Part of the condition clause of a wait statement.
use	Clause that makes the contents of a package visible from inside an entity or an architecture.
variable	Declared inside a process statement with a variable declaration; assigned with a variable assignment statement. Variables are created at the time of elaboration and retain their values throughout the entire simulation run. Note that variable assignments occur without delay (unlike signal assignments). This has major implications when variable assignments are used as part of a block of sequential statements within a process. See the VHDL User Manual for details.
wait	Suspends evaluation of a process. There are three basic forms of a wait statement:

	`wait on sensitivity_list;`
	`wait until boolean_expression;`
	`wait for time_expression;`
	These forms can be combined:
	`wait on sensitivity_list until`
	`boolean_expression for time_expression;`

when	Used to present choices for conditional logic in a case statement.
while	Used to iterate replications in replicated a loop statement. Also used for conditional logic in selected waveforms.
with	Introduces the select expression of a selected signal assignment statement.
xnor	Logical operator for types bit and Boolean and for one-dimensional arrays of these types. Logical exclusive or.
xor	Logical operator for types bit and Boolean and for one-dimensional arrays of these types. Logical exclusive or.

Appendix F
VHDL Defined Symbols

Symbol	Meaning
+	Addition, or positive number
–	Subtraction, or negative number
/	Division
=	Equality
<	Less than
>	Greater than
&	Concatenator
|	Vertical bar
;	Terminator
#	Enclosing based literals
(	Left parenthesis
)	Right parenthesis
.	Dot notation
:	Separates data object from type
"	Double quote
'	Single quote or tick mark
=>	Exponentiation
=>	Arrow meaning "then"
=>	Arrow meaning "gets"
:=	Variable assignment
/=	Inequality
>=	Greater than or equal to
<=	Less than or equal to
<=	Signal assignment
0	Box
—	Comment

VHDL Concurrent operations

Keyword	Example	Meaning
not	not a	&&
and	a and b	a . b
or	a or b	a + b
xor	a xor b	$a \oplus b$
nand	a nand b	$\overline{a . b}$
nor	a nor b	$\overline{a + b}$
xnor	a xnor b	$\overline{a \oplus b}$

LAB WORK
&
VIVA QUESTIONS

LAB WORK

EXPERIMENT 1: To study and verity truth tables of OR, AND, NOT, NAND, NOR, EX-OR gates.

To study and verity truth tables of OR, AND, NOT, NAND, NOR, EX-OR gates.

Objectives:

1. To study basic gates.
2. To study universal gates.
3. To verity truth tables of basic and universal gates.

Theory: OR AND, NOT gates are known as basic gates. NAND, NOR gates are known as universal gates. By using either NAND gates only any logical expression can be realized.

OR gate has two or more inputs but only one output. If all inputs are low, the output is low. If any of the inputs is high, the output is high. OR gate is written as

$$Y = A + B = A \text{ } OR \text{ } B$$

AND gate has two or more inputs but only one output. All inputs must be get a high output thus,

$$Y = AB = A \cdot B = A \text{ AND } B$$

NOT gate has one input and one output. The output is always opposite of input. Thus

$$Y = \overline{A} = \text{NOT } A$$

If $A = 0$, $Y = 1$. If $A = 1$, $Y = 0$

NOR gate is combination of OR gate and NOT gate. It has two or more inputs but only one output. The boolean expression for 2 input NOR gate is

$$Y = \overline{A + B}$$

NAND gate is a combination of AND gate and NOT gate. It has two or more inputs but only one output. The boolean expression for 2 input NAND gate is

$$Y = \overline{A \cdot B}$$

Exclusive OR gate (Written as XOR) recognises works which have an odd number of ones. The upper AND gate gives an output $\bar{A}B$ and the lower AND gate gives an output $A\bar{B}$. Thus

$$Y = \bar{A}B + A\bar{B}\ A \text{ XOR } B = A \oplus B$$

Circuit Diagram: Fig 1.2 shows the circuit diagram and truth table of OR, AND, NOT, NAND, NOT and XOR gates.

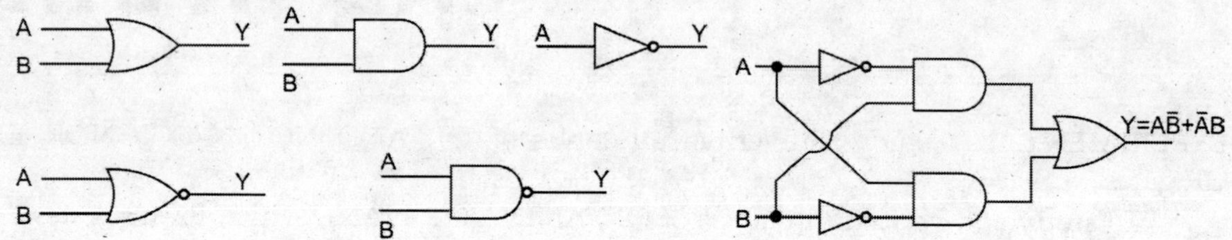

Fig. 1.1 (a) Two input OR gate (b) Two input AND gate (c) NOT gate (d) Two input NOR gate (e) Two input NAND gate (f) XOR gate

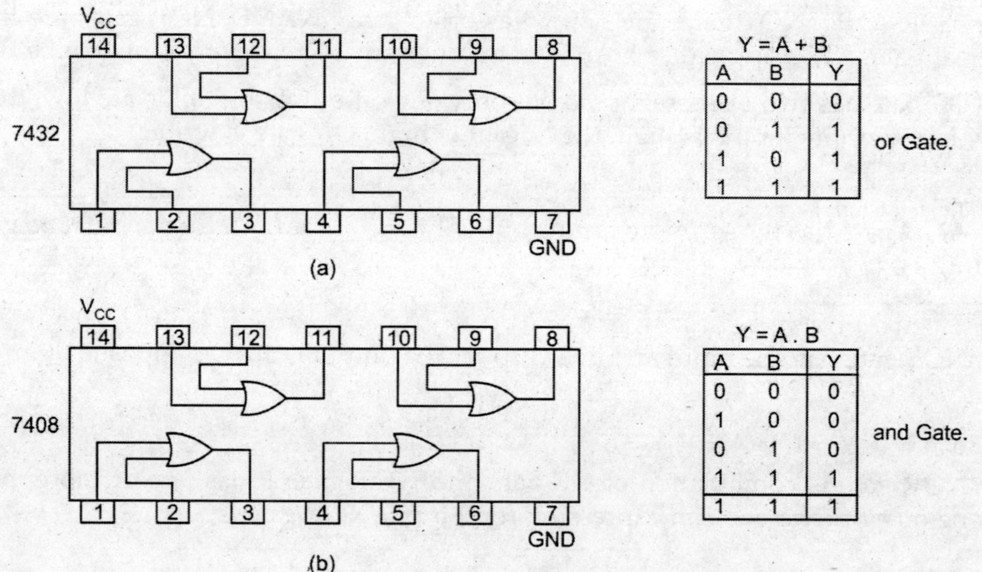

Y = A + B

A	B	Y
0	0	0
0	1	1
1	0	1
1	1	1

Y = A . B

A	B	Y
0	0	0
1	0	0
0	1	0
1	1	1
1	1	1

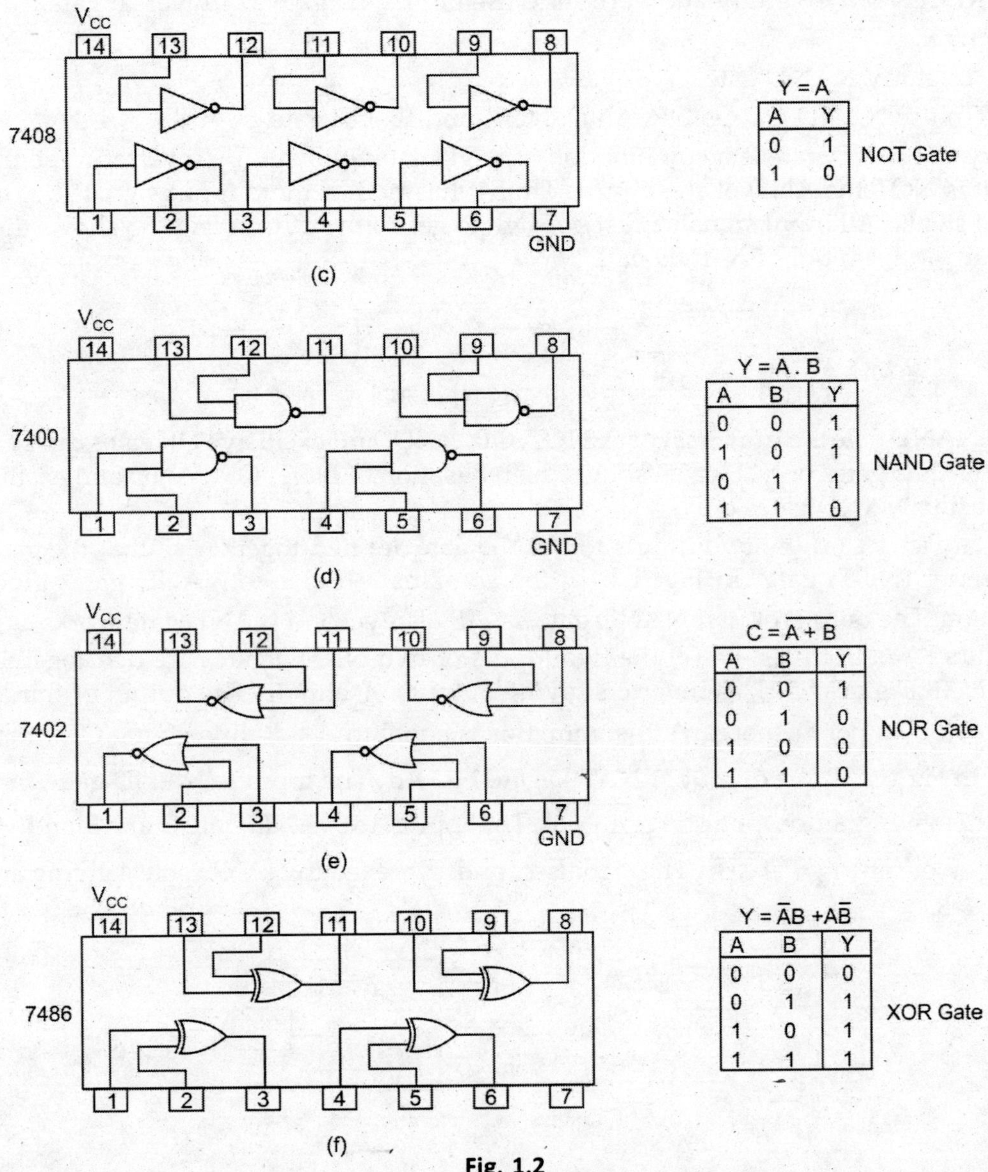

Fig. 1.2

Apparatus: IC 7432 (OR gate), 7404 (NOT gate), 7402(NOR gate), 7400(NAND gate), IC 7486(quad XOR gate) logic trainer board, binary switches.

Procedure: Insert IC 7432 on the IC base of the experimental trainer board. Give + 5 V input to pin 14 and ground pin 7. Give inputs *A* and *B* to pins 1 and 2 using binary switches. Connect the output pin 3 to LED. Switch on the trainer. Observe LED output. Try with different combinations of input *A*, *B*. Prepare truth table. Repeat for other gates.

EXPERIMENT 2: To study and verify the use of NAND gate as universal gate.

Objective:

1. To study NAND gate
2. To use NAND gate as OR, AND, NOT, and Ex-OR gate.

Theory: A NAND gate is a combination of AND gate followed by inverter). Therefore, the output is NOT the AND of the inputs. Thus it has two or more input signals but only one output signal. All input signals must be High to get a Low output. Fig. 1.3 shows the logical meaning and symbol of NAND gate.

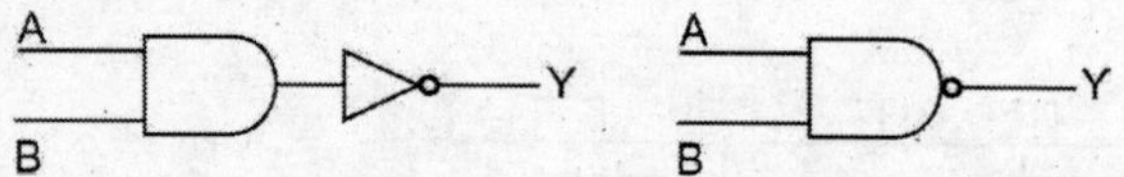

Fig. 1.3 Two input NAND gate (a) Logical meaning (b) symbol

NAND gate is a universal gate. NOT, OR, AND, and exclusive OR gates can be realized using NAND gates only. Fig. 1.4 shows the realisation of NOT, OR., AND and exclusive OR gates using NAND gates only.

In Fig. 1.4 (a) the two inputs to NAND gate are tied together so that the gate works as inverter (NOT) gate In Fig. 1.4 (b) the combination of two NAND gates gives AND operation. The output of first NAND gate. is $\overline{AB}$. The second NAND gate works as inverter and thus $Y = AB$. In Fig 1.4 (c) the two inputs of two NAND gates are tied together. Thus these NAND gates work as inverters giving outputs $\bar{A}$ and $\bar{B}$. The output of third NAND gate is $\overline{\bar{A}\bar{B}}$. As per Demorgan's theorem, this is equal to $A + B$ thus giving OR function.

In Fig 1.4 (d) the output of NAND gate 1 is $\overline{AB}$. The input to NAND gate 2 are A and $\overline{AB}$ thus giving an output of Y_1 $\overline{A \cdot \overline{AB}}$. The inputs to NAND gate 3 are B and $\overline{AB}$ thus giving an output $Y_2 = \overline{B \cdot \overline{AB}}$. The signals Y_1 and Y_2 are fed to NAND gate 4 giving an output $Y = A \oplus B$.

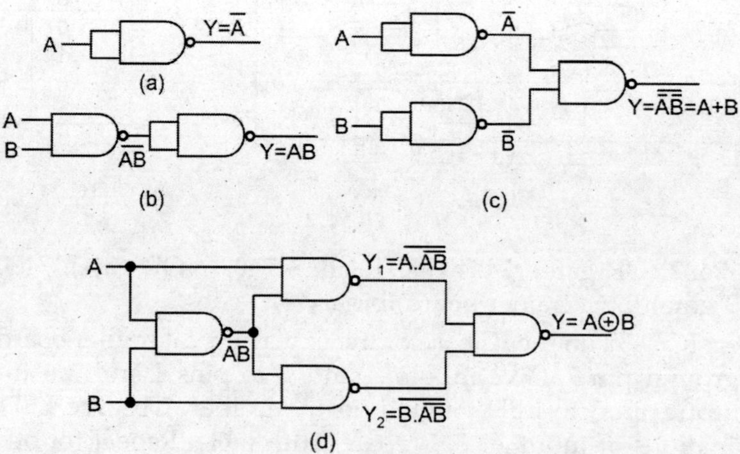

Fig. 1.4: Realisation of (a) NOT (b) AND
(c) OR (d) Ex-OR gate using NAND gates only

Circuit Diagram: Fig 1.4 shows the circuit diagram using NAND gates as NOT, AND, OR and XOR gates.

Apparatus: IC 7400 (quad two input NAND gate) logic trainer board, binary switches.

Procedure: Insert IC 7400 On the IC base of logic trainer board Give + 5 V supply to pin 14 and ground pin 7.

(a) Join pins 1 and 2 together and give input to this common terminal using binary switch. Connect the output pin 3 to LED. Switch on the trainer. Observe the LED output.

(b) Give inputs *A* and *B* to pins 1 and 2 using binary switches. Join pins 4 and 5 together and connect pin 3 to this common terminal. Connect pin 6 to LED. Switch on the trainer and observe LED output.

(c) Join pins 1 and 2 connect input *A* to this terminal using binary switch, Join pins 4 and 5 together and connect input *B* to this terminal using binary switch. Connect pins 3 and 6 to pins 9 and 10 respectively and pin 8 to LED. Switch on the trainer and observe LED output.

(d) Connect inputs *A* and *B* to pins 1 and 2 using binary switches. Also connect input *A* to pin 4 and input *B* to pin 3 to pins 5 and 10. Connect pins 6 and 8 to pins 12 and 13 repectively and connect pin 11 to LED. Switch on the trainer and observe LED output for various combinational of *A* and *B*.

EXPERIMENT 3: Study and use of NOR gate as universal gate.

Objective:

1. To study NOR gate
2. To use NOR gate as NOR, OR, AND and Ex-OR gate.

Theory: NOT gate is a combination or OR gate and NOT gate. It has two or more input signals but only one output signal. Fig 1.5 shows the logical meaning and symbol for two input NOR gate.

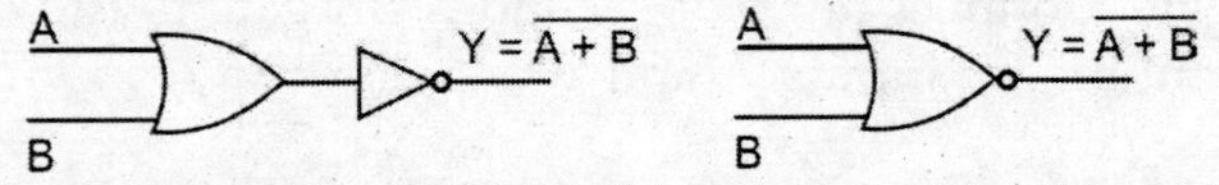

Fig. 1.5: Two point NOR gate (a) logical meaning (b) symbol

NOR gate is a universal gate. NOT, OR, AND and exclusive OR gates can be realized using NOR gates only.

In Fig. 1.6 (a) the two inputs to a NOR gate are tied together so that the gate works as inverter (NOT) gate. In Fig. 1.6 (b) the output of NOR gate is negated by the second gate to give OR function. In Fig. 1.6 (c) both the input A and B are negated and then fed to third NOR gate. As per Demorgan's theoram the output $Y = \overline{\overline{A} + \overline{B}} = AB$. In Fig. 1.6 (d) NOR gate 1 gives the output $\overline{A+B}$. This output is fed to gates 2 and 3 along with signals *A* and *B* respectively. The output of gate 2 is $\overline{A + \overline{A+B}}$ and that of gate 3 is $\overline{B + \overline{A+B}}$. The outputs of gates 2 and 3 are fed as inputs to gate 4. The output of gate 4 is $(\overline{A+\overline{A+B}})(\overline{B+\overline{A+B}})$. This is negated by gate 5 to give the final output (after using logic) simplification) as $A \oplus B$.

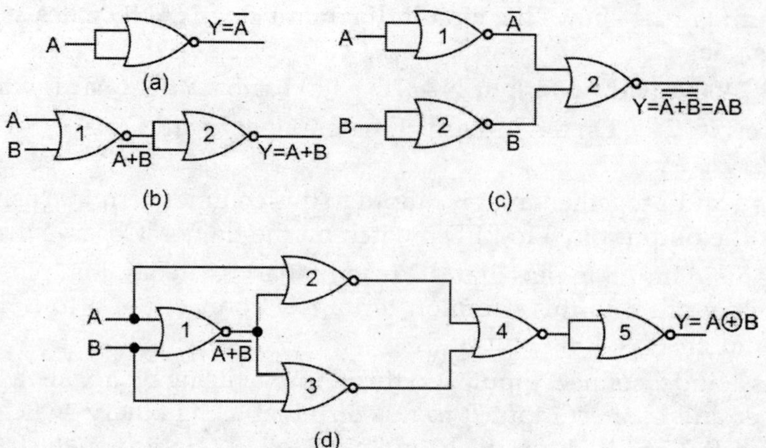

Fig. 1.6: Realization of (a) NOT (b) OR (c) AND (d) Ex-OR gate using NOR gates only

Circuit Diagram: Fig. 1.6 [(a) to (d)] shows the circuit diagrams using NOR gate to realize NOT, OR, AND and exclusive OR gate.

Apparatus: IV 7402 (quad 2 input NOR gates) 2 number.

Procedure: Insert IC 7402 on IC bases of logic trainer board. Give + 5 V supply to pins 14 and ground pins 7.

(a) Join pins 2 and 3 together and give input to this common terminal using binary switch. Connect output pin 1 to LED. Switch on the trainer. Observe the LED output for inputs 1 and 0.

(b) Give inputs *A* and *B* to pins 2 and 3. Join pins 5 and 6 and connect pin 1 to this common connection of pins 5 and 6. Observe LED output for different inputs *A* and *B*.

(c) Give input A to pins 2 and 5 and input B to pins 3 and 9. Connect pin 1 pins 6 and 8. Connect pin 4 to pin 11 and pin 10 to pin 12. Connect pin 13 to pin 13 to pins 2 and 3 of second NOT gate. Connect pin 1 of second NOR gate to LED. Observe the output of LED for different binary inputs *A* and *B*

EXPERIMENT 4: Construction of half adder circuit using Ex-OR and NAND gates and verification of its operation.

Objective:

1. To study the working of half circuit.
2. To construct half adder circuit using Ex-OR and NAND gates and verify its operation.

Theory: Fig. 1.7 shows the half adder circuit. It can add two bits at one time. It uses one AND gate and one exclusive OR gate. The two outputs are SUM and CARRY. When A or B 0(and the other is 1) it is recognised by Ex-OR gate* and the output is 1. When both A and B are 0. Ex-OR gate gives the output as 0. When both A and B are 1, the AND gate gives and output of 1 which is carried to the next higher column. The boolean expressions for the output are

*An Ex-OR gate recognises odd number of 1's in the input.

$$\text{SUM} = A \oplus B$$
$$\text{CARRY} = AB$$

Table 1.1 gives truth table for half adder.

Table 1.1 Truth table for half adder

A	*B*	Carry	Sum
0	0	0	0
0	1	0	1
1	0	0	1
1	1	1	0

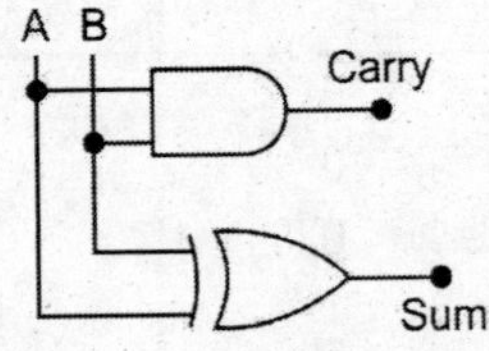

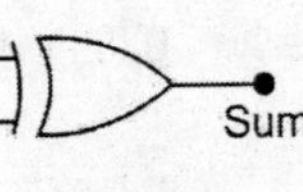

Fig. 1.7

A B

Carry

Sum

Fig. 1.8 Half adder circuit using Ex-OR and NAND gates

Circuit Diagram: Fig. 1.8 shows the circuit diagram for a half adder using Ex-OR and NAND gates. NAND gate uses less power than AND gate and is preferred over AND gate in most applications.

Apparatus: IC 7486 (quad ex-OR), IC 7400(NAND gate), logic trainer board, binary switches.

Procedure: Connect the ex-OR and NAND gates as shown. The two NAND gates give an AND operation as discussed in experiment 2 Connect LEDs to the SUM and CARRY outputs. Give binary inputs *A* and *B* using binary switches and observe LED output. Prepare truth table.

EXPERIMENT 5: Construction of full adder circuit using Ex-OR and NAND gates and verification of its operation.

Objectives:

1. To study the working of full adder circuit.
2. To construct full adder circuit using ex-OR and NAND gates and verify its operation.

Theory: Fig. 1.9 shows a logic circuit known as full adder. It used two half adders and one OR gate and can add 3 bits at a time. The outputs are SUM and CARRY. When there is odd number of 1 s in the input it is recognised by ex-OR gate and the output is 1. Thus

$$\text{SUM} = A \text{ XOR } B \text{ XOR } B$$
$$\text{CARRY} = AB \text{ or AC or BC}$$

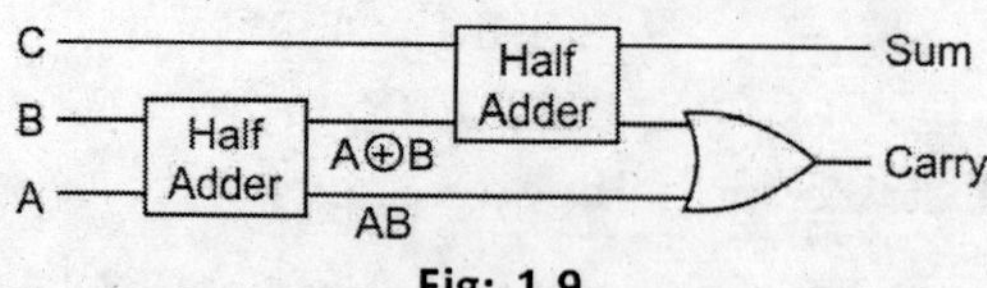

Fig: 1.9

When we have odd number of 1 s in the input, SUM is 1. When two or more inputs are 1, CARRY is 1. The booolean expressions are

$$SUM = A \oplus B \oplus C$$

$$CARRY = AB + AC + BC$$

Table 1.2 shows the truth table

Table 1.2 Truth table for full adder

A	*B*	*C*	CARRY	SUM
0	0	0	0	0
0	0	1	0	1
0	1	0	0	1
0	1	1	1	0
1	0	0	0	1
1	0	1	1	0
1	1	0	1	0
1	1	1	1	1

Circuit Diagram: Fig 1.10 shows the circuit diagram of a full adder using ex-OR and NAND gates. The NAND gates are connected to give AND and OR functions.

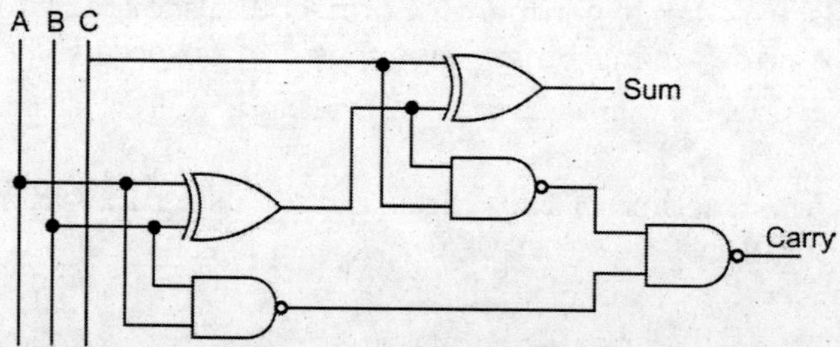

Fig: 1.10 Full adder using Ex-OR and NAND gates

Apparatus: IC 7400 (quad 2 input NAND gates), IC 7486 (quad Ex-OR gate), logic trainer board, binary switches.

Procedure: Connect the circuit as shown in Fig. 1.10. Give binary inputs A, B, C using binary switches and observe outputs of LEDs. Prepare truth table.

EXPERIMENT 6: Construction of 4 bit IC adder and verification of its operation.

Objectives:

1. To study the working of 4 bit IC adder.
2. To verify the operation of 4 bit IC adder.

Theory: Fig. 1.11. shows IC 7483 adder (4 bit). It has inputs A_0, A_1, A_2, A_3 for the first 4 bit number and inputs B_0, B_1, B_2, B_3 for the second 4 bit number. For adding two 4 bit numbers C_0 is kept low. S_0, S_1, S_2, S_3 are the SUM outputs and C_4 is the CARRY output. The adder can add any two bit numbers upto 1111 and 1111(decimal 15) to give an output 11110 (decimal 30). IC 7483 can be cascaded by connecting C_4 output of the first IC to C_0 CARRY input ot the next IC thus forming 8 bit adder. Further cascading can give 12 bit, 16 bit adders and so on.

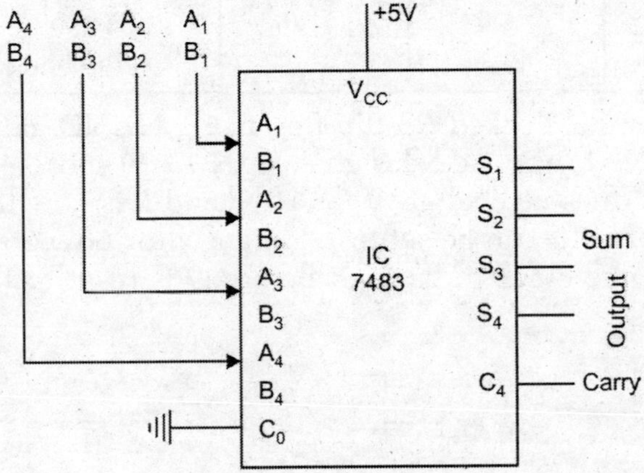

Fig. 1.11 4-bit adder

Circuit Diagram: Fig. 1.11. shows the circuit diagram.

Apparatus: IC 7483, logic trainer board, binary switches.

Procedure: Insert IC 7483 on the logic trainer board. Give + 5 V input to the supply pin and ground C_0 pin. Connect LEDs to S_0, S_1, S_2, S_3, S_4, and C_4. Give inputs A_4 A_3 A_2 A_1 and B_4 B_3 B_2 B_1 to IC using binary switches. Vary these inputs from 0000 to 1111 and observe LED outputs at S_0, S_1, S_2, S_3, S_4 and C_4. Prepare truth table.

EXPERIMENT 7: Construction and verification of operation of SR, clocked SR and D flip flops.

Objectives:

1. To study the working of SR, clocked SR and D flip flops.
2. To construct SR, clocked SR and D flip-flops using NAND gates and verifity the operations.

Theory: Fig. 1.12 shows the circuit of an SR flip-flop using NAND gates. It has two inputs *R* and *S* (called reset and set). The two outputs are *Q* and $\overline{Q}$ (i.e., NOT *Q*). The two inputs are complementary. If $Q = 0$, $\bar{Q} = 1$ and if $Q = 1$ $\bar{Q} = 0$. Table 1.3 shows the truth table. A high *R* and High *S* results in inactive state (i.e., no change). A Low *R* and High *S* results in set state while High *R* and Low *S* results in reset state. If Both *R* and *S* are Low the output is indeterminate. This is called race condition.

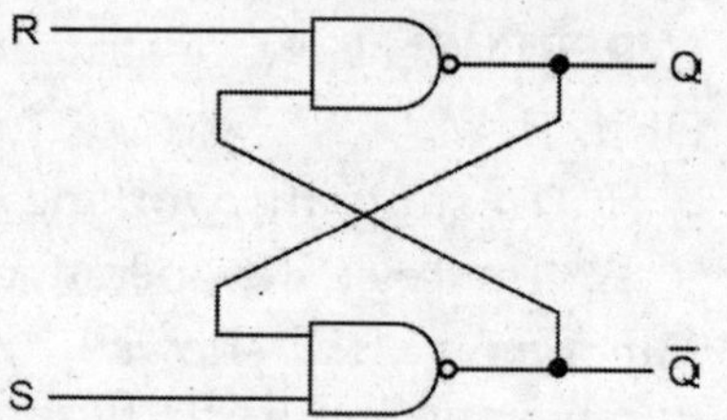

Fig. 14.12

Table 1.3 NAND R-S flip-flop

R	S	Q	Comment
0	1	1	Set
1	0	0	Reset
1	1		No change

Fig. 1.13 (a) shows a clocked RS flip flop using NAND gates. This circuit uses two NAND gates (N_3 and N_4) to apply clock signal. Fig. 1.13 (b) shows its symbol. When clock is Low, the output *Q* indicates no change. If *S* = High and *R* is Low the flip flop must wait till CLK becomes High before *Q* can be set to 1. Similarly for Low *S* and High *R* the flip flop must wait for clock to be clock to be high before *Q* is reset to Low (i.e., 0). Fig.1.13 (b) shows its symbol.

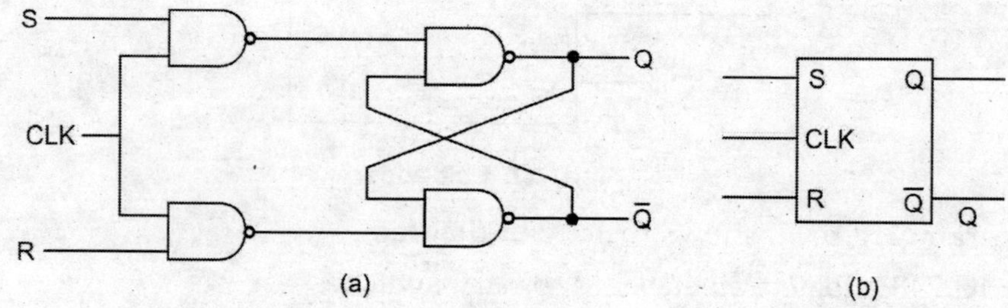

Fig. 1.13 Clocked RS flip flop using NAND gates (a) circuit (b) symbol

Fig. 1.14 shows edge triggered *D* flip flop using NAND gates. *A D* flip flop is an improvement on RS Flip flop to avoid race condition. It can be level clocked or edge triggered. The edge triggered one causes a change in the output at a unique instant. The clock provides square wave singal. The RC circuit converts the signal into spikes so that the triggering occurs at the positive spike. Table 1.4 shows the truth table.

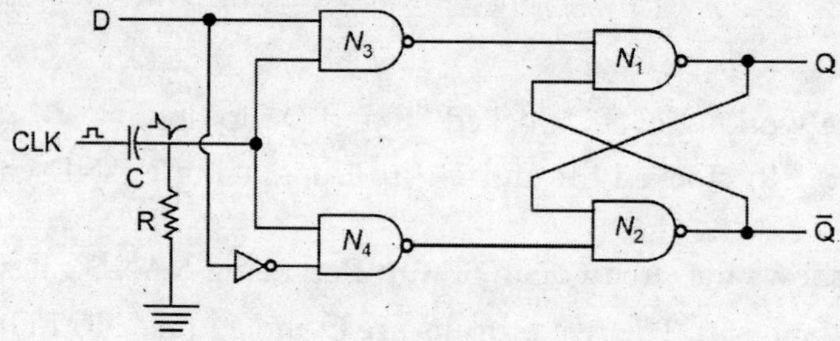

Fig. 1.14 Edge trigger D flip flop

Table 1.4 Truth table for edge triggered *D* flip-flop

CLK	*D*	*D*
0	×	No change
1	×	No change
↓	×	No change
↑	0	0
↑	1	1

Circuit Diagram: Fig. 1.12, 1.13, 1.14 show the circuit diagram of RS, clocked RS and edge triggered *D* flip flop using NAND gates.

Apparatus: IC 7400 (quad two input NAND gate), logic trainer board, binary switches resistor capacitor.

Procedure: Insert IC 7400 on IC base of logic trainer board. Give + 5 supply to pin 14 and ground pin 7.

(a) From RS flip flop (Fig. 1.12) connect LED to output Q. Give binary inputs. R and S using binary switches and observe LED output.

(b) From clocked RS flip flop (Fig 1.13).

(c) From D flip flop. Using suitable values of R and C form a circuit for edge triggering (Fig. 1.14).

EXPERIMENT 8: Display devices and their decoders/drives.

Objectives:

1. To study BCD to seven segment decoder.
2. To study seven segmentdisplay.
3. To verfity the working of BCD to seven segment decoder/driver and seven segment display.

Theory: An LED (light emitting diode) is a display device.

A group of seven LEDs is formed to display the numbers 0 to 9. Each LED is called a segment because it forms a part of the character being displayed.

A decoder converts binary or BCD words into alphanumeric characters. A 4 bit BCD decorder has a BCD input (4 lines) and 10 output lines for the decimal output.

BCD decoder/seven segment driver is a combination of BCD-decimal decoder and a seven segment display driver. The input is BCD. The outputs drive 7 segment display unit formed by LEDs. The decimal output is displayed. The seven segment display can be connected in common anode or common cathode configuration.

Circuit Diagram: Fig. 1.15 show the circuit of BCD decoder/seven segment driver IC 7447 A is a BCD to seven segment decoder and gives active low output. The outputs *a, b, c, d, e, f, g* are connected to the terminals *a, b, c, d, e, f, g* of the seven segment display. Since the decoder in an active low device, common anode connection of seven segment display is used.

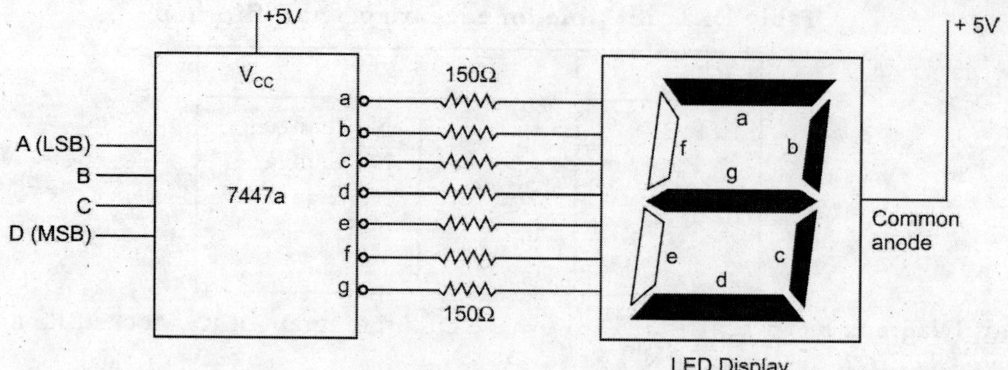

Fig. 1.15 BCD-seven segment decoder/driver

Apparatus: IC 7447A, LED display unit, 150 Ω resistances (7 number), logic trainer board, binary switches.

Procedure: Insert IC 7447 A on the logic trainer board and connect output terminals of IC 7447 A to corresponding terminals of 7 segment display through 150 Ω resistance as shown in Fig. 1.15. Give + 5V supply to IC 7447 and LED display. Give BCD inputs 0000 to 1001 to IC 7447 A and observe the display in each case.

EXPERIMENT 9: Study decimal to binary encoder and verify its truth table.

Objectives:

1. To study decimal to binary encoder.
2. To study truth table of decimal to binary encoder.

Theory: An encoder is a device which converts alphanumeric characters to binary code. An encoder may be decimal to binary, decimal to BCD, hexadecimal to binary, octal to BCD etc. When encoding from decimal to binary or BCD etc., it is possible that more than one key is pressed simultaneously. A priority encoder gives priority to the highest order data line and encodes it. IC 74147 is a decimal to BCD priority encoder. It has 10 input lines (for decimal numbers 0 to 9) and 4 output lines (for BCD output).

Circuit Diagram: Fig. 1.16 shows the block diagram of decimal to BCD priority encoder. Connect LEDs at the 4 output terminals.

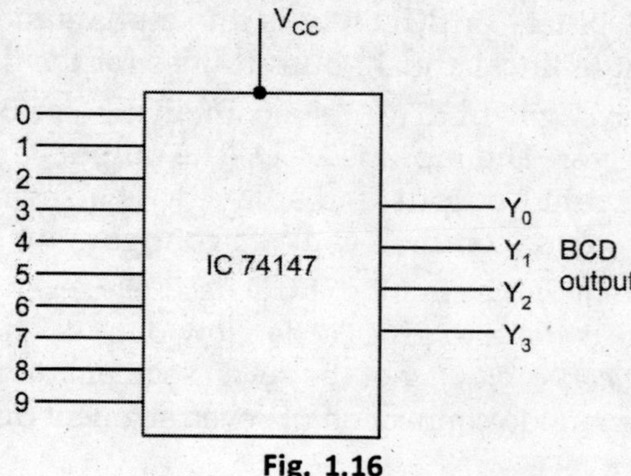

Fig. 1.16

Apparatus: IC 74147, key board, + 5 V regulated power supply. LEDs (4 number), logic trainer board.

Procedure: Connect key board at input of IC 74147. Insert IC on the logic trainer board. Connect LEDs the output terminals of IC. Give various inputs 0 to 9 and observe LED output. Prepare a truth table. Verify that the IC is giving priority to higher order number by pressing two keys together and observing LED output.

EXPERIMENT 10: To study the working of multiplexer and its operation as a logic function generator.

Objectives:

1. To study a multiplexer .
2. To use a multiplexer as a logic function generator.

Theory:

A multiplexer is a device having many inputs and one output. By applying control signals any one of the inputs can be steered to the output it is also known as data selector it is basically a single pole *N* throw selector switch. The number of select inputs is m where $2^m = N$.

Digital multiplexers are available as IC. IC 74151 A is an 8 input Mux with 3 select inputs.

Multiplexer is a very convenient logic circuit used in combination logic circuit. The procedure to use it to generate a given min term logic expression is as under.

1. The select lines are used as data input lines.
2. Input lines corresponding to given min terms are connected to logic 1 level.
3. The remaining input lines re connected to logic (i.e., ground).

Circuit Diagram: Fig. 1.17 show the block diagram of IC 74151 A connected to generate the logic function $Y = \overline{C}\overline{B}A + \overline{C}BA + C\overline{B}A + CB\overline{A}$. This function is terms of min terms is

$$Y = \Sigma m\ (1, 3, 5, 6)$$

Connect the input lines 1, 3, 5, 6 to + V_{CC} and the remaining input lines 0, 2, 4, 7 to ground. Connect LED at output terminal.

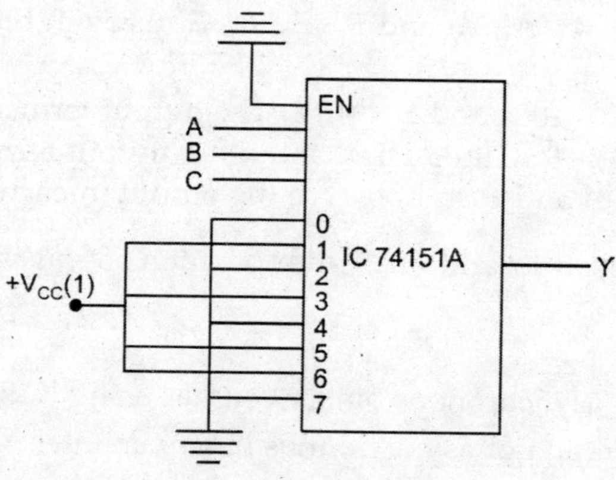

Fig. 1.17

Apparatus: Logic trainer board , IC 74151 A, binary switches, LED.

Procedure: Insert IC on the logic trainer board. Connect the IC as shown in Fig. 1.17. Give binary inputs *A, B, C* through binary switches. Observe the LED output. Select some more logic Function with three variables. Reconnect the terminals of IC for each logic function and verify the operation.

EXPERIMENT 11: To study the use of IC comparator and to verify ita operation.

Objectives:

1. To study the working of IC 7485(Magnitude comparator).
2. To verify the operation of IC 7485.

Theory: The function of a mangnitude comparator is to compare the magnitudes of two binary numbers and indicate which one is the bigger of the two. An IC magnitude comparator had inputs as binary numbers *A* and *B*. Its output indicates if $A > B$ or $A = B$ or $B > A$. IC 7485, is a 4 bit comparator.

Circuit Diagram: Fig. 1.18 shows the symbol of IC 7485. Connect the terminals A_0, A_1, A_2, A_3 to binary input *A* and B_0, B_1, B_2, B_3 to binary input *B*. Connect LEDs at the output.

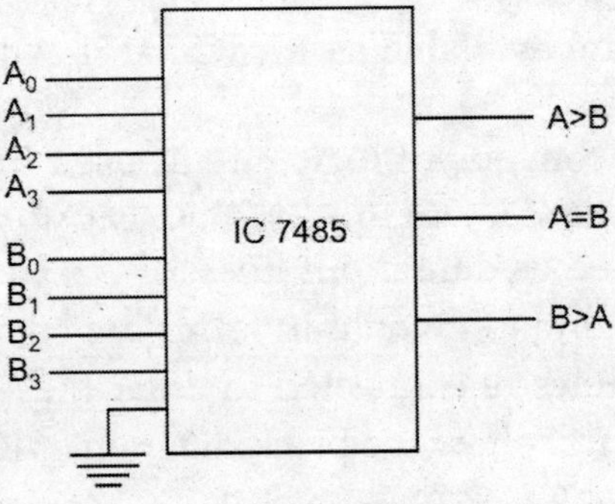

Fig. 1.18

Apparatus: IC 7485, logic trainer board, binary switches, regulated power supply, LEDs.

Procedure: Insert IC 74HC85 on the logic trainer board. Connect cascading inputs A > B and A < B together to the ground. Connect cascading input A = B to + 5 V regulatid power supply. Give inputs A_3, A_2, A_1, A_0 and B_3, B_2, B_1, B_0 through binary switches and observe LEDs at the output.

Give inputs say *A* = 1001 and *B* = 0101. The output terminal A > B will be High and other output terminal $A > B$ will be High and other output terminals will be Low. Repeat with different values of *A* and *B* and observe the output in each case.

EXPERIMENT 12: To construct and verify the operation of 4 bit asynchronous binary counter using IC 7473 (JK flip flop).

Objectives:

1. To construct 4 bit asynchronous binary counter using IC 7473 (JK flip flop)
2. To verify the operation of asynchronous binary counter.

Theory: The function of a counter is to count the number of clock pulses. A counter consists of flip flops. An asynchronous binary counter is also ripple counter.

Fig. 1.19 shows the circuit of a 4 bit ripple counter consisting of 4 edge triggered JK flip flop. As indicated by small circles at the CLK input of flip flop, the triggering occurs when CLK input gets a negative edge. Q_0 is the least significant bit (LSB) and Q_3 is the most significant bit (MSB). The flip flops are connected in series. The Q_0 output is connected to CLK terminal of second flip flop. The Q_1 output is connected to CLK terminal of third flip flop and so on. By adding more flip flop, a counter of any length can be built. It is known as a ripple counter because the carry moves through the flip flops like a ripple on water. Initially, CLR is made low and all flip flops reset giving and output Q = 0000. When CLR becomes high, the counter is ready to start, As LSB receives its clock pulse, its output changes form 0 to 1 and the total output Q = 0001. When second clock pulse arrives, Q_0 reset and carries (i.e., CLK pulse changes Q_0 goes from 1 to 0 and, second flip flop will receive CLK input). Now the output is Q = 0010. The third CLK pulse changes Q_0 to 1 giving a total output Q = 0011. The forth CLK pulse causes Q_0 to reset and carry and Q_1 also resets and carries giving a total output Q = 0100 and the process goes on. The action is shown in Table 1.5. The number of output states of a counter are known as modulus (ormod). A ripple counter with 4 flip flop can count from 0 to 15 and is, therefore, known as mod-16 counter while one with 6 flip flops can count from 0 to 63 and is a mod-64 counter and so on.

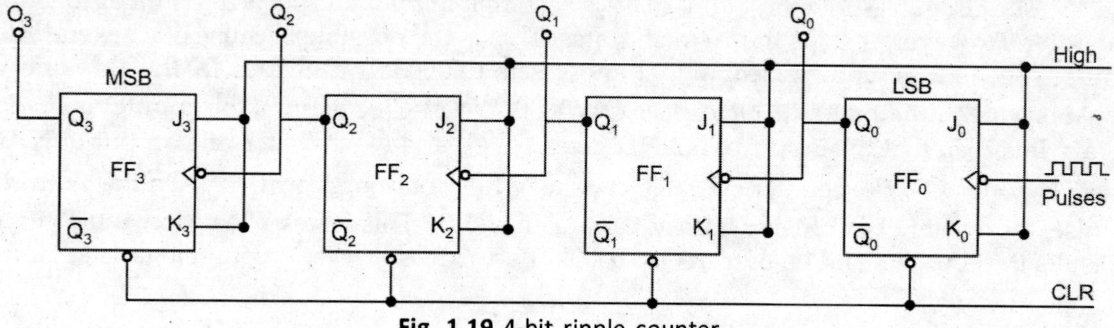

Fig. 1.19 4-bit ripple counter

Table 1.5 4-bit Ripple Counter

Count	Q_3	Q_2	Q_1	Q_0	Count	Q_3	Q_2	Q_1	Q_0
0	0	0	0	0	8	1	0	0	0
1	0	0	0	1	9	1	0	0	1
2	0	0	1	0	10	1	0	1	0
3	0	0	1	1	11	1	0	1	1
4	0	1	0	0	12	1	1	0	0
5	0	1	0	1	13	1	1	0	1
6	0	1	1	0	14	1	1	1	0
7	0	1	1	1	15	1	1	1	1

Ripple counters are simple to fabricate but have the problem that the carry has to propagate through a number of flip flops. The delay times of all the flip flops are added. Therefore, they are very slow for some applications. Another problem is that unwanted pulses occur at the output of gates.

Circuit Diagram: Fig. 1.19 shows the circuit diagram. Connect LEDs at output Q_0, Q_1, Q_2, Q_3.

Apparatus: Logic trainer board, IC 7473 (4 number),4 number LEDs clock generator.

Procedure: Insert IC on IC base of logic trainer board and construct the circuit as shown in Fig. 1.19. Clear the output to 0000. Apply clock pulses using clock generator. Observe the LEDs and verify Table 1.5.

EXPERIMENT 13: To construct and verify the operation of Decade counter.

Objectives:

1. To study decade counter.
2. To verify the operation of decade counter.

Theory: The decade counter is also known as mode 10 counter or BCD counter. It counts from 0 to 9. Thus, it requires 10 pulses for resetting. It is used in frequency counter, digital voltmeters, wristwatches, etc.

The circuit is shown in Fig. 1.20. It is seen that in many respects it is similar to a ripple counter. However, it skips the 10 to 15 states. This is the possible because on the tenth clock pulse it generates its own clear signal and resets to 0000. It uses four JK flip flop and one NAND gate. Initially, low $\overline{CLR}$ causes $Q = 0000$. When $\overline{CLR}$ is high, the counter is ready to start. It is seen that the inputs of NAND gates are Q_1 and Q_3 so that Y, the output of NAND gate is $Y = \overline{Q_1 Q_3}$. The counter operates as usual while counting from 0 to 9. On the tenth clock pulse, the output $Q = 1010$, i.e., both Q_1 and Q_3 are high. This forces Y to be low which in turn results the counters and its output $Q = 0000$. Then Y goes high and counter starts again.

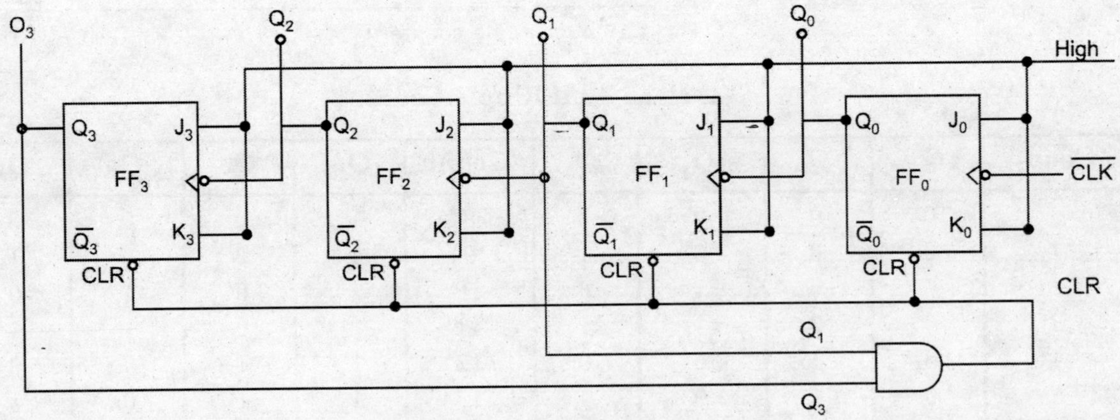

Fig: 1.20

Circuit Diagram: Fig. 1.20 shows the circuit diagram. Connect LEDs at output Q_0, Q_1, Q_2, and Q_3. Use JK flip flops 7473.

Apparatus: JK flip flops 7473 (4 number), logic trainer board, LEDs, clock generator, NAND gate.

Procedure: Insert flip flops in IC bases of logic trainer board and connect as in Fig. 1.20. Give + 5 V supply. Apply clock pulses and verify the operation by observing the LED output.

EXPERIMENT 14: To construct and verify the operation of synchronous counter.

Objectives:

1. To study synchronous counter.
2. To construct synchronous counter.
3. To verify the operation of synchronous counter.

Theory: In a synchronous counter, all the flip flops are clocked together. Fig. 1.21 shows a synchronous counter having positive edge triggered JK flip flop. Since all the flip flops are clocked together the delay time is less. The flip flop corresponding to least significant bit (LSB) has its input JK fed form voltage + V_{CC}. Therefore, it responds to each positive clock. edge. However, the other toggles on positive clock edge only if Q_0 is 1. The Q_2 flip flop toggles on positive clock edge only when Q_0 and Q_1 are (due to presence of And circuit) and so on.

Thus, each flip flop toggles on the next positive clock edge if all lower bits are 1.

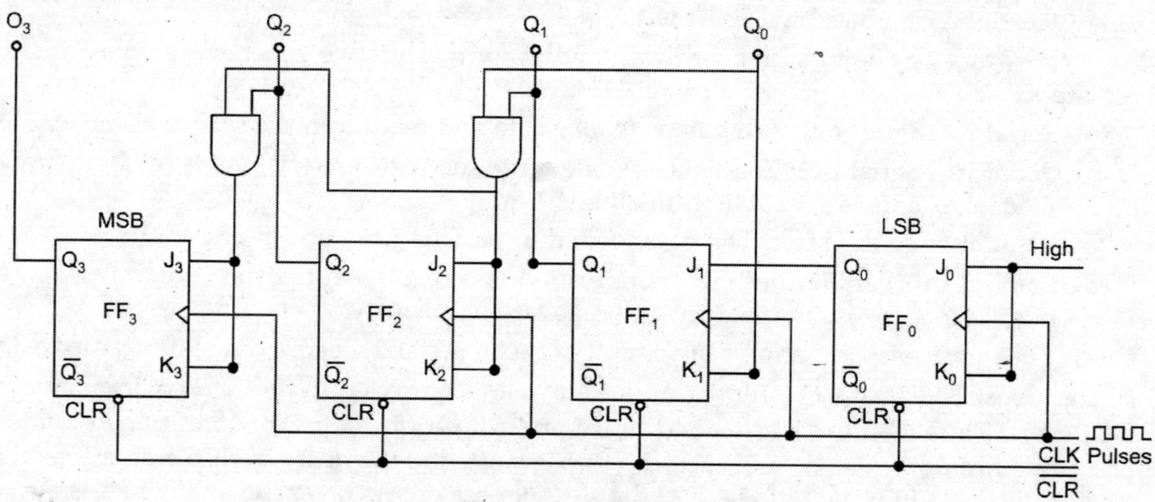

Fig. 1.21 Synchronous counter

A low $\overline{CLR}$ signal resets the counter to 0000. When $\overline{CLR}$ goes high, the counter is ready to start. The first positive clock edge sets Q_0 to 1 so that Q = 0001. At second positive clock edge Q_1 and Q_0 toggle and Q = 0010. The third positive clock edge increases the count by 1 so that Q = 0011.

The successive Q outputs are 0100, 0101 and so on upto 1111(i.e., decimal 15). The next positive clock edge resets the counters to 0000 and the cycle is repeated. More flip flops can be added to increases the count.

Circuit Diagram: Fig. 1.21 shows the circuit diagram. Connect LEDs at output Q_0, Q_1, Q_2 and Q_3.

Apparatus: JK flip flops 7473 (4 number), AND gates (IC 7408), LEDs, clock generator.

Procedure: Insert the ICs in the IC bases of logic trainer board and construct the circuit as in Fig. 1.21. Apply clock pulses and verify the operation of counter by observing the LED output.

EXPERIMENT 15: Construction of 4 bit serial in serial out shift register using JK flip flops.

Objectives:

1. To study the function of shift register.
2. To study classification of shift registers.
3. To construct 4 bit serial in-serial out shift register using JK flip flops.
4. To verify the operation of 4 bit serial in -serial out shift register.

Theory: A register is a device capable of storing a bit. We know that a flip flop has a memory characteristic. Thus flip flops can be connected together to form a register. A shift register consists of a group of flip flops connected so that each flip flop transfers ins bit of data to the next flip flop of the register when a clock pulse arrives. Data may have ot be shifted left of shifted right, Thus we have shift left and shift right registers.

The data can be in serial form of parallel form. Thus we can have 4 types of shift registers:

(a) **Serial in - Serial out:** The data is loaded into and read from the shift register serially.

(b) **Serial in - Serial out:** The data is loaded into the register serially but read in parallel mode, i.e., data is available from all bits simultaneously.

(c) **Parallel in - Serial out:** The data loaded in parallel from and read serially.

(d) **Parallel in - Parallel out:** The data loaded and read from register in parallel, i.e., all bits are loaded simultaneously and read simultaneously.

Circuit Diagram: Fig. 1.22 shows the circuit of a 4 bit serial in - serial out shift register using JK flip flops. The Q and $\bar{Q}$ outputs of one stage are connected to the J and K inputs of the next stage. Thus the inputs to second, third and fourth flip flops are conditioned by their preceding flip flops. The data in each flip flop is shifted to the next flip flop on the arrival of a clock pulse. As indicated in Fig. 1.22 the shift of data occurs on each negative transition of the clock. It is seen that the circuit in Fig. 1.22 is a shift right register. If the contents of a shift register are to be considered as a binary number LSB* Must be on the left and MSB** on the right. Since it is a 4 bit register, 4 clock pulses are required to shift the data through this register.

* Least significant bit.

** Most significant bit.

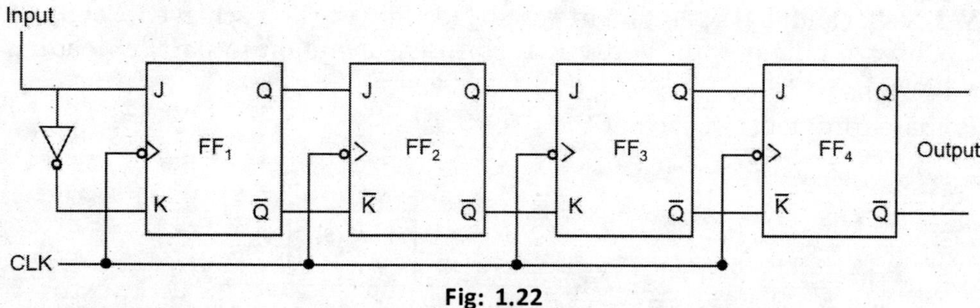

Fig: 1.22

Apparatus: JK flip flops 7473 (4 number), clock generator, logic trainer board binary switches, LED.

Procedure: Insert the JK flip flop on IC bases of logic trainer board. Content clock generator. As shown. Connect the Q the $\bar{Q}$ outputs from one flip flop to J and K inputs to the next flip-flop. Connect LED to the output of FF_4.

EXPERIMENT 16: Verification of truth table of IC Universal shift register.

Objectives:

1. To study IC universal shift register.
2. To study truth table of IC universal shift registers.

Theory: A universal shift register (also called bidirectional shift register) can shift data in both directions, i.e., to left as well as to right. The logic gates are arranged in such a way that data bits can be transferred from one stage to the next in either direction depending on control line input.

A universal shift registers can be constructed from flip flops. It is also available in IC form. IC 74194 is a 4 bit universal shift register. Fig. 1.23. Shows its symbol and pin connections. A, B, C, D are inputs. Q_A, Q_B, Q_C, Q_D are output. Parallel loading is achieved by applying 4 bits of data simultaneously to inputs A, B, C, D and High to S_0, S_1 inputs. At positive edge of clock, parallel data is loaded into register.

When S_0 is high and S_1 is Low, shift right operation occurs at positive edge of clock. In this mode serial data can be entered at SR SER(shift-right serial input). When S_0 is Low and S_1 is High shift left operation occurs at positive edge of clock. In this mode serial data can be entered at shift left serial input (SL SER).

Apparatus: IC 74194, logic trainer board, binary switches LEDs clock generator.

Procedure: Insert IC 74194 in the IC base of logic trainer board. Connect LEDs to outputs Q_A, Q_B, Q_C and Q_D.

(a) Give input at SR SER (shift right serial input). $S_0 = 1$ and $S_1 = 0$. Observe LEDs output with the application of each clock pulse.

(b) Give input at SL SER (shift left serial input). Set $S_0 = 0$ and $S_1 = 1$. Observe LEDs output with the application of each clock pulse.

(c) Give input A, B, C, D Set SL SER = 0 and SL SER = 0. Set $S_0 = 1$ and $S_1 = 1$. Observe LEDs output with the application of each clock pulse. Change $S_1 = 0$. Observe LEDs. output.

Verify that shift left operation in parallel loading is being achieved. Set $S_0 = 0$ and $S_1 = 1$. Observe LED output. Verity that shift left operation in parallel loading is being achieved.

Prepare truth table in each of the steps *a, b, c*.

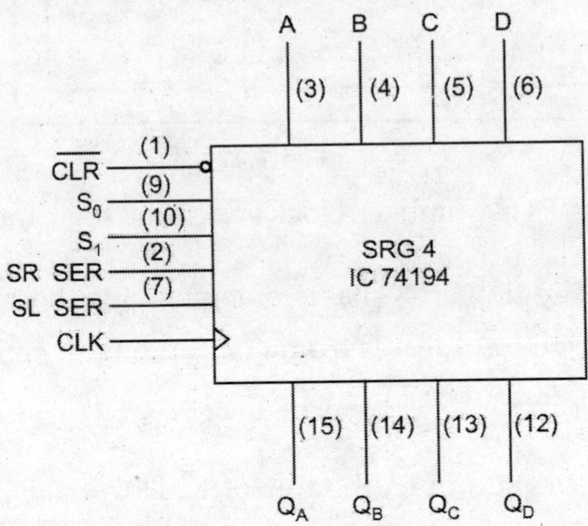

Fig. 1.23 4-bit IC universal shift register.

EXPERIMENT 17: To construct and test D/A converter using IC.

Objectives:

1. To study D/A converter .
2. To study DAC 80 D/A converter.
3. To test DAC 80 D/A converter.

Theory: Digital signals and digital transmission have many advantages. However, most of the information used by us is analog, If an analog signal is to be transmitted through digital means we need and analog to digital (A/D) converter at sending end digital to analog (D/A) converter at the receiving end.

The most commonly used method of digital to analog conversion is that using a binary ladder network. This network requires only two resistance values. Fig. 1.24 shows the network. From node 1 the resistance to digital source is 2R and resistance to ground is also 2R. It can be verified that resistance to source and ground from each of the reaming nodes. (i.e, 2, 3, 4) is also 2R. A digital input 0001 means that *D* is connected to *V* and *A, B, C* are grounded. The output voltage V_0 is V_{16}. A digital input 0010 means that *C* is connected to + V and *A, B, D* are grounded giving an output of 2V/16. As the varies from 0000 to 1111 the output varies from to 0 to *V* in steos of V_{16}.

D/A converts are available as ICs. DAC 0808 is an 8 bit device having a reference current source, resister network. and 8 transistor switches to contect the input binary signals to the internal circuit. Fig. 1.25 shows the pin diagram of DAC 0808.

Pin 1 is not used (NC means nq connection). Pin 2 is ground. Pin 3 (V_{EE}) is – 15 V. Pin 4 is the ground return for the current output of the ladder. This pin is usually connected to op-amp. Pins 5 to 12 are for 8 bit binary input. Pin 13 (V_{CC}) is for + 5 V. Pin 14 is connected to a positive supply through a resistance. Pin 15 is grounded through resistance. A 15 pF capacitor is connected between pins 16 and 3 for frequency compensation.

Circuit Diagram: Fig. 1.26 shows the circuit diagram.

Apparatus: DAC 0808, 5 V supply, 15 kΩ resistance (2 number, 1Ω resistance, 15 pF capacitor, binary switches, CRO, op-amp 741, binary switches, logic trainer board.

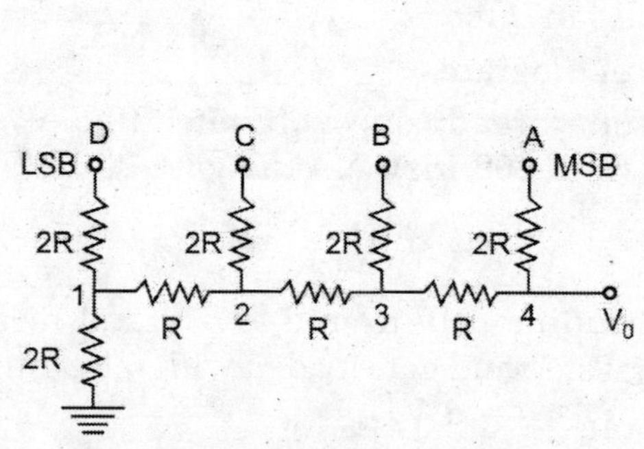

Fig. 1.24 Binary ladder network for D/A conversion

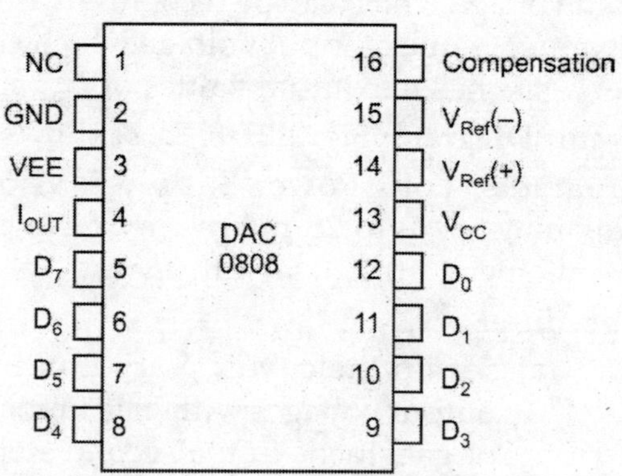

Fig. 1.25 Pin diagram of DAC 0808

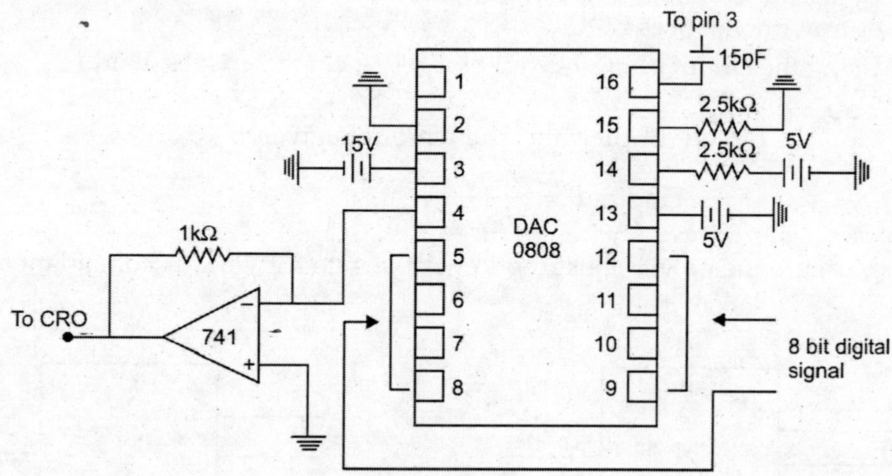

Fig. 1.26

Procedure: Insert IC 0808 and op-amp 741 on IC bases and set up the circuit as shown in Fig. 1.26. Give 8 bit digital signal input ot pins 5 to 12. Set the digital input to 0000 to 0000 and observe the output on CRO screen. Change digital input and observe analog voltage on screen in every case.

EXPERIMENT 18: To study static characteristics of 2 input NAND gate.

Objectives:

1. To measure and plot voltage transfer characteristics of 2 input NAND gate.
2. To find $I_{OH(max)}$ and $I_{OL(MAX)}$ of 2 input NAND gate.
3. To find output impedances Z_{OH} and output impedances Z_{OL} of two input NAND gate.

Theory: Digital ICs operate preperly if the input signal voltage is within certain limits. As discussed in chapter 5 every digital IC has a minimum value of V_{IH} (i.e., minimum value of input voltage which can be recognised as logic 1 by the gate) and maximum value of V_{IL} (i.e., maximum value of input voltage which can be recognised as logic 0 by the gate). In addition every gate has maximum values of I_{OH} and I_{OL} and fan out.

Circuit Diagram: Fig. 1.27 (*a, b, c*) shows the circuit diagram.

Apparatus: Four Power Supply 5 V, 500 mA, one bread, one voltmeter (0-5 V), one milliammeter (0 to 25 mA), One micro-ammeter (0 to 500 μA), NAND gate IC 7400, one potentiometer 10 k, 0.5 watt.

Procedure:

(a) Set up the circuit. Vary input voltage continuously from 0 to 5 V and measure output voltages with gate unloaded in first step and gate loaded with rated number of gate loads in the second step. Find $V_{OH(min)}$ and $V_{OL(max)}$.

(b) Using the circuit of Fig 1.27 (a) plot $I_{OH} - V_{OH}$ characteristics. Find $I_{OH(max)}$ and output impedances.

(c) Using the circuit of Fig 1.27 (b) plot $I_{OL} - V_{OL}$ characteristics. Find $I_{OL(max)}$ and output impedances ZOL.

(d) Using the circuit of Fig 1.27 (c) plot $I_I - V_I$ characteristics. Find $I_{I(max)}$ at which $V_I - V_{OL(max)}$.

(e) Find the fan out of the gate. The fan out is given by

$$\text{Find out} = \frac{I_{OL(max)}}{I_{IL(max)}}$$

(f) Measure the power consumption of the gate for various combination of input.

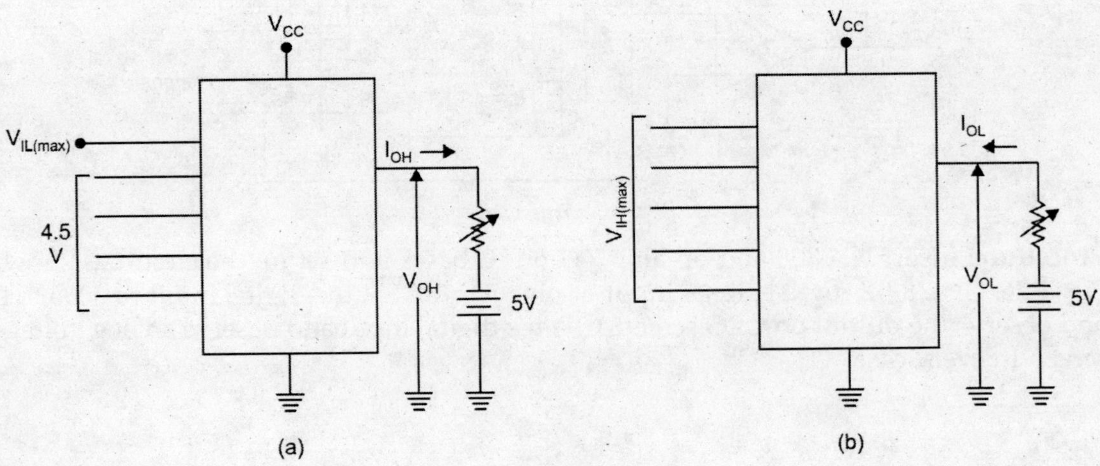

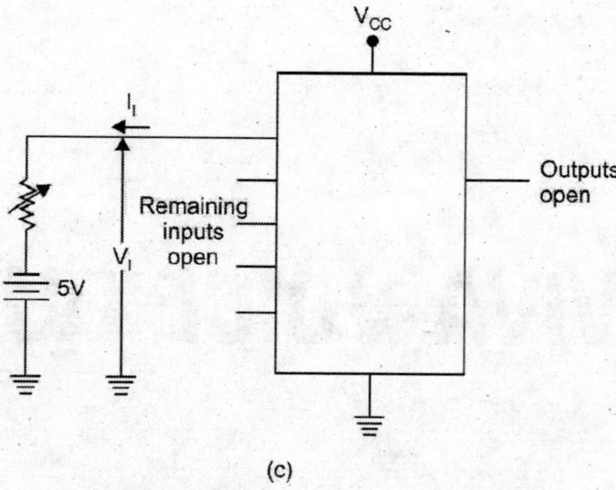

(c)

Fig. 1.27 Static characteristics of IC gates.

❑❑❑

VIVA-VOICE QUESTIONS

qqq

VIVA-VOCE QUESTIONS

1. Differentiate between analog and digital signals. Give examples.
2. How is an analog signal converted to digital signal?
3. What are the advantages of digital signals?
4. What is a pulse waveform? How are rise time and fall time measures? What is pulse duration?
5. Differentiate between positive and negative logic.
6. Name some digital trouble shooting instruments.
7. Give some examples of two state devices.
8. What is binary number system? How can you convert (a) binary to decimal (b) decimal to binary.
9. Tell the rules of binary addition and subtraction.
10. What are signed binary numbers?
11. What are 1's complement and 2's? How are they obtained? What is the advantages of 2's complement arithmetic?
12. What are binary fractions? Tell the weights of digits after binary point?
13. What are floating point numbers? How is this representation used? What is excess-32 format?
14. What is hexadecimal number system? What digits are used in this system?
15. What is octal number system? What digits are used in this system?
16. What is meant by BCD? How is it different from binary?
17. What is gray code? What is its special feature? Where is it used?
18. What is excess 3 code? What is its special feature?\
19. What is alphanumeric code? Why is such a code necessary? Name the commonly used alphanumeric codes.
20. How is error detection and correction done in digital signals? What is parity? Explain hamming;s code.
21. What is OR gate? Draw its symbol and write truth table for 3 input OR gate.
22. What is AND function? Draw symbol of 2 input AND gate and write its truth table.
23. What is NOT gate? Draw its symbol.
24. Draw symbol of 2 input NOT gate and write its truth table.

25. Draw symbol of 2 input NAND gate and write its truth table.
26. What exclusive OR gate? Draw its symbol. Write the truth table of 2 input exclusive OR gate.
27. Why are NOR and NAND gates called universal gates? How can OR, AND, NOT functions be obtained from NOR and NAND gates. Draw circuits for these equations.
28. What is meant by POS and SOP forms of logic equations?
29. What are min terms and max terms?
30. What is Karnaugh map? Why is it used?
31. What is Demorgan's theoram?
32. Differentiate between LSI, MSI, LSI, and VlSI.
33. Name the logic families being used nowadays.
34. Give the various specifications of digital ICs.
35. Name the various TTL sub families.
36. What is interfacing? What consideration must be kept in view while interfacing different IC families?
37. What is meant by (a) totem pole output (b) wired AND connection?
38. What is the tristate logic?
39. What is the difference between half adder and full adder? Give their applications.
40. Show circuit of half adder and explain its working.
41. Show circuit of full adder and explain its working.
42. What 1's complement? What 2's complement? What are their uses.
43. How does 2's complement adder/subtracter work? Explain.
44. What is magnitude comparator? What is its utility?
45. How is multiplication done in computers ? Explain.
46. What is the difference between LED and LCD?
47. What is VF display? Explain its principle.
48. What is 7 segment display? Where is it used.
49. Differentiate between common anode and common cathode connection of 7 segment display.
50. Discuss the principle of encoder.
51. What is meant by decoder/driver?
52. What is code converter?
53. How does a docoder work? Explain.
54. Differentiate between a multiplexer and demultiplexer? Why are the needed?
55. Why are select lines nedded in multiplexers and demultiplexers?
56. Explain the working of multiplexer.
57. On what principle does a demultiplexer work?
58. How multiplexer can be used as a logic function generator? Explain.
59. What is multiplexer tree? Why is it nedded?
60. What is a flip flop? What is it used?
61. Name some commonly used flip flops.
62. Differentiate between edge triggering and level clocking.
63. What is the difference between positive edge triggering and negative edge triggering?
64. Explain the principle of JK flip flop.

65. Draw the circuit of JK master slave flip flop and explain its working.
66. What is meant by race condition in flip flops?
67. Differentiate between synchronous and asynchronous counters,
68. Draw a circuit of ripple counter and explain its working.
69. What is meant by mode of counter?
70. What is a Mod-10 counter? Draw its circuit?
71. Explain the difference between up counters and down counters.
72. What is meant by presettable counter and programmable modulus?
73. Explain the words 'Register.' 'Shift manager'.
74. What is the difference between shift left and shift right registers?
75. What is a universal shift register? Draw its circuit.
76. Explain the difference between serial loading and parallel loading.
77. Tell the difference between RAM and ROM.
78. What is meant by PROM, EPROM, EEPROM?
79. Explain the terms Bubble memory, CCD, static RAM, Dynamic RAM.
80. What is the function of D/A and A/D converts?
81. Draw the circuit of ladder type D/A converter and explain its working.
82. What is the necessity of using op-amp along with D/A converter?
83. What are the various A/D converts?
84. Draw the circuit of A/D parallel converter and explain its working.
85. What is an analog comparator? What is its principle?
86. Explain the working of stair type A/D converter.
87. What is successive approximation A/D converter?
88. Explain the difference between single slope and dual slope A/D converts.
89. List the performance characteristics of A/D converts.
90. Which type of D/A converter is available in IC form? Name some ICs for D/A conversion.
91. What is ALU ? Tell some of the operations carried out by ALU.
92. What is meant by logic decisions? What is their necessity?
93. Tell the main features of IC 74181.
94. How many words can IC 74181 accept? What is their nit length?
95. What is the purpose of select lines and mode line in ALU 74181?
96. Explain the terms serial input serial output.
97. Draw a simple block diagram of microprocessor 8085 A.
98. What is the meaning of term multiplexing?
99. Tell the main features of microprocessor or 8085 A.
100. What is the purpose of using a crystal in 8085 A.